Numerical Analysis Using R

This book presents the latest numerical solutions to initial value problems and boundary value problems described by ODEs and PDEs. The author offers practical methods that can be adapted to solve wide ranges of problems and illustrates them in the increasingly popular open source computer language R, allowing integration with more statistically based methods.

The book begins with standard techniques, followed by an overview of "high-resolution" flux limiters and WENO to solve problems with solutions exhibiting high-gradient phenomena. Meshless methods using radial basis functions are then discussed in the context of scattered data interpolation and the solution of PDEs on irregular grids. Three detailed case studies demonstrate how numerical methods can be used to tackle very different complex problems.

With its focus on practical solutions to real-world problems, this book is useful to students and practitioners in all areas of science and engineering, especially those using R. R Code is available for download from the book's home page.

Graham W. Griffiths is a visiting professor in the School of Engineering and Mathematical Sciences, City University London. His primary interests are in numerical methods and climate modeling, on which he has previously published four books. Griffiths was a founder of Special Analysis and Simulation Technology Ltd. and later became vice president of operations and technology with AspenTech. He is a Chartered Engineer and a Fellow of the Institute of Measurement and Control and was granted Freedom of the City of London in 1995.

NUMERICAL ANALYSIS USING R

Solutions to ODEs and PDEs

Graham W. Griffiths
City University, United Kingdom

CAMBRIDGE
UNIVERSITY PRESS

32 Avenue of the Americas, New York, NY 10013

Cambridge University Press is part of the University of Cambridge.

It furthers the University's mission by disseminating knowledge in the pursuit of education, learning, and research at the highest international levels of excellence.

www.cambridge.org
Information on this title: www.cambridge.org/9781107115613

© Graham W. Griffiths 2016

This publication is in copyright. Subject to statutory exception
and to the provisions of relevant collective licensing agreements,
no reproduction of any part may take place without the written
permission of Cambridge University Press.

First published 2016

Printed in the United States of America by Sheridan Books, Inc.

A catalog record for this publication is available from the British Library.

Library of Congress Cataloging in Publication Data
Names: Griffiths, Graham W.
Title: Numerical analysis using R : solutions to ODEs and PDEs / Graham W. Griffiths, City University, United Kingdom.
Description: New York, NY : Cambridge University Press, 2016. | Includes bibliographical references and index.
Identifiers: LCCN 2015046150 | ISBN 9781107115613 (hardback : alk. paper)
Subjects: LCSH: Initial value problems – Data processing. | Boundary value problems – Data processing. | Differential equations – Data processing. | Differential equations, Partial – Data processing. | Numerical analysis. | R (Computer program language)
Classification: LCC QA378 .G76 2016 | DDC 518.0285/5133–dc23
LC record available at http://lccn.loc.gov/2015046150

ISBN 978-1-107-11561-3 Hardback

Cambridge University Press has no responsibility for the persistence or accuracy of URLs for external or third-party Internet Web sites referred to in this publication and does not guarantee that any content on such Web sites is, or will remain, accurate or appropriate.

To the memory of my dear son, Paul W. Griffiths (1977–2015).

Contents

Preface		*page* xv
1	**ODE Integration Methods**	1
	1.1 Introduction	1
	1.2 Euler Methods	11
	1.2.1 Forward Euler	11
	1.2.2 Backward Euler	12
	1.3 Runge–Kutta Methods	12
	1.3.1 RK Coefficients	15
	1.3.2 Variable Step Size Methods	19
	1.3.3 SHK: Sommeijer, Van Der Houwen, and Kok Method	36
	1.4 Linear Multistep Methods (LMMs)	37
	1.4.1 General	37
	1.4.2 Backward Differentiation Formulas (BDFs)	38
	1.4.3 Numerical Differentiation Formulas (NDFs)	44
	1.4.4 Convergence	46
	1.4.5 Adams Methods	60
	1.5 Truncation Error and Order of Integration	61
	1.5.1 LMM Truncation Error	62
	1.5.2 Verification of Integration Order	66
	1.6 Stiffness	69
	1.7 How to Choose a Numerical Integrator	69
	1.A Installation of the R Package `Ryacas`	70
	1.B Installation of the R Package `rSymPy`	71
	References	72
2	**Stability Analysis of ODE Integrators**	74
	2.1 General	74
	2.1.1 Dahlquist Barrier Theorems	75
	2.2 Dahlquist Test Problem	75
	2.3 Euler Methods	76

		2.3.1	Forward Euler	76
		2.3.2	Backward Euler	76
	2.4	Runge–Kutta Methods		76
		2.4.1	RK-1: First-Order Runge–Kutta	76
		2.4.2	RK-2: Second-Order Runge–Kutta	79
		2.4.3	RK-4: Fourth-Order Runge–Kutta	80
		2.4.4	RKF-54: Fehlberg Runge–Kutta	83
		2.4.5	SHK: Sommeijer, van der Houwen, and Kok	85
	2.5	Linear Multistep Methods (LMMs)		87
		2.5.1	General	87
		2.5.2	Backward Differentiation Formulas (BDFs)	89
		2.5.3	Numerical Differentiation Formulas (NDFs)	95
		2.5.4	Adams Methods	97
	References			101
3	**Numerical Solution of PDEs**			102
	3.1	Some PDE Basics		102
	3.2	Initial and Boundary Conditions		103
	3.3	Types of PDE Solutions		105
	3.4	PDE Subscript Notation		105
	3.5	A General PDE System		106
	3.6	Classification of PDEs		107
	3.7	Discretization		109
		3.7.1	General Finite Difference Terminology	109
		3.7.2	The Mesh	111
		3.7.3	Nonuniform Grid Spacing	112
		3.7.4	The Courant–Friedrichs–Lewy Number	112
		3.7.5	The Stencil	112
		3.7.6	Upwinding	113
	3.8	Method of Lines (MOL)		114
		3.8.1	Introduction	114
		3.8.2	Finite Difference Matrices	115
		3.8.3	MOL 1D: Cartesian Coordinates	123
		3.8.4	MOL 2D: Cartesian Coordinates	141
		3.8.5	MOL 2D: Polar Coordinates	175
	3.9	Fully Discrete Methods		194
		3.9.1	Introduction	194
		3.9.2	Overview of Some Common Schemes	194
		3.9.3	Results from Simulating a Hyperbolic Equation	197
	3.10	Finite Volume Method		207
		3.10.1	General	207
		3.10.2	Application to a 1D Conservative System	208
		3.10.3	Application to a General Conservation Law	210
	3.11	Interpretation of Results		210
		3.11.1	Verification	210
		3.11.2	Validation	211
		3.11.3	Truncation Error	211

	3.A	Appendix: Derivative Matrix Coefficients	211
		3.A.1 First Derivative Schemes	211
		3.A.2 Second Derivative Schemes	213
		3.A.3 Third Derivative Schemes	215
		3.A.4 Fourth Derivative Schemes	216
	3.B	Appendix: Derivative Matrix Library	217
		3.B.1 Example	220
	References		222

4 PDE Stability Analysis ... 225
- 4.1 Introduction — 225
- 4.2 The Well-Posed PDE Problem — 226
- 4.3 Matrix Stability Method — 231
  - 4.3.1 Semi-Discrete Systems — 231
- 4.4 Von Neumann Stability Method — 242
  - 4.4.1 General — 242
  - 4.4.2 Fully Discrete Systems — 243
  - 4.4.3 Semi-Discrete Systems — 253
- 4.5 Unstructured Grids — 260
- 4.A Fourier Transforms — 261
- References — 262

5 Dissipation and Dispersion ... 264
- 5.1 Introduction — 264
- 5.2 Dispersion Relation — 264
- 5.3 Amplification Factor — 265
- 5.4 Dissipation — 266
- 5.5 Dispersion — 267
- 5.6 Dissipation and Dispersion Errors — 269
  - 5.6.1 The 1D Advection Equation, Semi-Discrete Upwind — 269
  - 5.6.2 The 1D Advection Equation, Semi-Discrete Second-Order Upwind — 270
  - 5.6.3 The 1D Advection Equation, Fully Discrete Upwind — 275
  - 5.6.4 The 1D Advection Equation, Fully Discrete Lax–Friedrichs (LxF) — 276
- 5.7 Group and Phase Velocities — 277
  - 5.7.1 Exact Relationships for the Basic PDE — 278
  - 5.7.2 Semi-Discrete, First-Order Upwind Discretization — 278
  - 5.7.3 Semi-Discrete Leapfrog Discretization — 279
  - 5.7.4 Fully Discrete Leapfrog Discretization — 280
- 5.8 Modified PDEs — 282
- References — 284

6 High-Resolution Schemes ... 285
- 6.1 Introduction — 285
- 6.2 The Riemann Problem — 285
- 6.3 Total Variation Diminishing (TVD) Methods — 286
  - 6.3.1 TVD Numerical Integration — 287

6.4	Godunov Method	288
	6.4.1 Godunov's Theorem	290
6.5	Flux Limiter Method	292
	6.5.1 How Limiters Work	293
	6.5.2 Limiter Functions	294
6.6	Monotone Upstream-Centered Schemes for Conservation Laws (MUSCL)	298
	6.6.1 Linear Reconstruction	298
	6.6.2 Kurganov and Tadmor Central Scheme	303
	6.6.3 Piecewise Parabolic Reconstruction	312
	6.6.4 Solutions to the Euler Equations	315
6.7	Weighted Essentially Nonoscillatory (WENO) Method	324
	6.7.1 Polynomial Reconstruction: Finite Volume Approach	325
	6.7.2 Polynomial Coefficients	327
	6.7.3 Polynomial Reconstruction: Finite Difference Reconstruction	331
	6.7.4 WENO Reconstruction	331
	6.7.5 Alternative Calculation for Substencil Coefficients	332
	6.7.6 Weights	335
	6.7.7 Smoothness Indicators	336
	6.7.8 Calculation of Smoothness Indicator Coefficients	336
	6.7.9 Flux Splitting	337
	6.7.10 Implementation of a WENO Finite Volume Scheme	337
	6.7.11 Scalar Problems	342
	6.7.12 Euler Equation Problems	343
	6.7.13 2D Examples	347
6.8	Further Reading	351
6.A	Eigenvalues of Euler Equations	351
6.B	R Code for Simulating 1D Scalar Equation Problems	353
	6.B.1 The Main Program	353
	6.B.2 The Derivative Function	357
	6.B.3 The MUSCL Function	358
	6.B.4 Initialization	360
6.C	R Code for Simulating 1D Euler Equations Problems	364
	6.C.1 The Main Routine	364
	6.C.2 Initialization	369
	6.C.3 The Derivative Function	370
	6.C.4 The MUSCL Function	373
	6.C.5 Postsimulation Calculations	374
References		377

7 Meshless Methods 380
7.1	Introduction	380
7.2	Radial Basis Functions (RBF)	381
	7.2.1 Positive Definite RBFs	382
	7.2.2 RBF with Compact Support (CSRBF)	382
7.3	Interpolation	384
	7.3.1 Interpolation Example: 1D	385

	7.3.2	Interpolation Example: 2D	387
	7.3.3	Larger Interpolation Example: 2D	389
	7.3.4	Interpolation Example: 3D	393
	7.3.5	Interpolation with Polynomial Precision	397
7.4	Differentiation	398
	7.4.1	Derivative Example: 1D	399
7.5	Local RBFs	401
	7.5.1	Allocating Stencil Nodes	403
	7.5.2	Choosing the Right Shape Parameter Value	404
7.6	Application to Partial Differential Equations	406
	7.6.1	Explicit Euler Integration	406
	7.6.2	Weighted Average Integration	407
	7.6.3	Method of Lines	408
	7.6.4	With Nonlinear Terms	408
	7.6.5	Initial Conditions (ICs) and Boundary Conditions (BCs)	409
	7.6.6	Stability Considerations	410
	7.6.7	Time-Dependent PDEs	410
	7.6.8	Time-Independent PDEs	434
7.A	Franke's Function	452
7.B	Halton Sequence	452
7.C	RBF Definitions	454
	References	455

8 Conservation Laws 457

8.1	Introduction	457
8.2	Korteweg–de Vries (KdV) Equation	459
	8.2.1 The *First* Conservation Law, u	459
	8.2.2 The *Second* Conservation Law, u^2	459
	8.2.3 The *Third* Conservation Law, $u^3 + \frac{1}{2}u_x^2$	460
	8.2.4 Another Conservation Law	460
	8.2.5 An *Infinity* of Conservation Laws	461
	8.2.6 KdV Equation: 2D	463
	8.2.7 KdV Equation with Variable Coefficients (vcKdV)	464
8.3	Conservation Laws for Other Evolutionary Equations	466
	8.3.1 Nonlinear Schrödinger Equation	466
	8.3.2 Boussinesq Equation	468
8.A	Symbolic Algebra Computer Source Code	468
	References	469

9 Case Study: Analysis of Golf Ball Flight 470

9.1	Introduction	470
9.2	Drag Force	472
9.3	Magnus Force	476
9.4	Gravitational Force	479
9.5	Golf Ball Construction	480
9.6	Ambient Conditions	480

	9.7	The Shot	483
	9.7.1	Golf Ball Compression	483
	9.7.2	Spin	484
	9.7.3	Launch Angle	484
	9.7.4	Bounce and Roll	485
	9.7.5	Shot Statistics	486
	9.8	Completing the Mathematical Description	487
	9.8.1	The Effect of Wind	488
	9.9	Computer Simulation	489
	9.9.1	Driver Shots	490
	9.9.2	Wood Shots	491
	9.9.3	Iron Shots	491
	9.9.4	Effect of Wind	492
	9.9.5	Effect of Differing Ambient Conditions	493
	9.9.6	Effect of Push/Pull and Inclined Golf Ball Spin Axis	495
	9.9.7	Drag/Lift Carry Test	497
	9.9.8	Drag Effect at Ground Level	497
	9.10	Computer Code	499
	9.10.1	Main Program	499
	9.10.2	Derivative Function	503
	9.10.3	Initial Conditions	505
	References		506

10 Case Study: Taylor–Sedov Blast Wave 508

10.1	Brief Background to the Problem	508
10.2	System Analysis	508
10.3	Some Useful Gas Law Relations	512
10.4	Shock Wave Conditions	514
10.5	Energy	515
10.6	Photographic Evidence	516
10.7	Trinity Site Conditions	518
10.8	Numerical Solution	519
10.9	Integration of PDEs	529
10.A	Appendix: Similarity Analysis	530
10.B	Appendix: Analytical Solution	531
	10.B.1 Closed-Form Solution	533
	10.B.2 Additional Complexity	537
	10.B.3 The Los Alamos Primer	537
References		537

11 Case Study: The Carbon Cycle 539

11.1	Introduction	539
11.2	The Model	539
	11.2.1 Atmosphere	542
	11.2.2 Oceans	543
	11.2.3 Air–Ocean Exchange	544
	11.2.4 Carbonate Chemistry	546

		11.2.5 Acidity of Surface Seawater	552
		11.2.6 Ocean Circulation	553
		11.2.7 Emission Profiles	554
		11.2.8 Earth's Radiant Energy Balance	557
		11.2.9 How the Atmosphere is Affected by Radiation	562
	11.3	Simulation Results	571
		11.3.1 Carbon Buildup in the Atmosphere	571
		11.3.2 Carbon Buildup in Surface Seawater and Accompanying Acidification	572
		11.3.3 Surface Temperature Changes	575
	11.A	Appendices	576
		11.A.1 Model Differential Equations	576
		11.A.2 Correlations for Chemical Equilibrium and Dissociation Constants	576
		11.A.3 Revelle and Uptake Factors	577
		11.A.4 Residence Time	579
		11.A.5 Mass Action	580
		11.A.6 The Electromagnetic Spectrum	580
	References		581

Appendix: A Mathematical Aide-Mémoire   585
Index   607

Color plates follow page 284

Preface

The language of science and engineering is largely mathematical, which, increasingly, requires solving problems that are described by ordinary differential equations (ODEs) and partial differential equations (PDEs). The primary focus of this book is numerical solutions to initial value problems (IVPs) and boundary value problems (BVPs) described by ODEs and PDEs. The solutions are implemented in computer code using the open source R language system.

The intended readership is senior undergraduates and postgraduate students in the subject areas of science, technology, engineering, and mathematics (STEM). The contents should also appeal to engineers and scientists in industry who need practical solutions to real-world problems. The emphasis is on understanding the basic principles of the methods discussed and how they can be implemented in computer code.

The aim of this book is to provide a set of software tools that implement numerical methods that can be applied to a broad spectrum of differential equation problems. Each chapter includes a set of references that provide additional information and insight into the methods and procedures employed. All chapters are more or less complete in themselves, except for a few references to other chapters. Thus each chapter can be studied independently.

It is assumed that the reader has a basic understanding of the R language, although the computer code is annotated to a level that should make understanding clear. Additional discussion is included in the text for more advanced language constructs. Some basic examples of the use of computer algebra systems are also included that make use of the R interface packages `Ryacas` and `rSymPy`.

R is a free, high-level software programming language and software environment that has traditionally been used for statistical computing and graphics. It has been widely used for many years by statisticians for data analysis being particularly effective in handling large data sets. However, recently, packages have been added to R to solve a wider range of numerical problems. In particular, the addition of package `deSolve` [Soe-10] opened up the language for solving differential equations by adding industrial strength integrators. The package `deSolve` is used extensively in the R examples provided in this book. For readers wishing to learn about R, *The Art of R Programming* [Mat-11] is a good introductory text, and *The R Book* [Cra-11] is a comprehensive description of the

R language. Both are highly recommended. The R language can be downloaded from http://cran.r-project.org/. To supplement the R installation, a free interactive development environment (IDE), *RStudio*, can be downloaded from https://www.rstudio.com/. RStudio provides a very user-friendly customizable user interface that greatly enhances programming productivity and graphical display of simulation results. For users familiar with Matlab, a comprehensive cross-reference between Matlab and R commands has been compiled by Dr. David Hiebeler of the University of Maine and can be downloaded from http://www.math.umaine.edu/~hiebeler.

We emphasize two areas of style usage that have been adopted in this book. The first is that the symbol used for *assignment* is <- (also known as the gets symbol), which is the R purist's form. However, it is acknowledged that = is used extensively in the R literature and that choosing between <- and = could be considered simply a matter of semantics, as there are only a few special situations in which = does not work properly. The second area of style usage relates to *global variables*, that is, those variables created outside of a function but that can be used inside a function. Global variables created using the *superassignment* symbol <<- can also be changed within a lower environment function using <<-. This means that, in general, they are allocated higher in the *environment hierarchy* than where they are used. Now, like all modern programming languages, variables in R are passed to functions by reference, but global variables are allowed. In this book, we tend to make great use of global variables, as such use leads to cleaner and less cluttered code. It also avoids having to process lists returned from functions where, in the rare situation, globals are actually changed within a function rather than just used. However, we do not use global variables wherever possible and pass values through function arguments where it is appropriate or where it makes the code clearer. For applications described in this book, the scope for global variables is generally between adjacent levels in the environment hierarchy and therefore is not readily prone to programming conflicts. A good discussion relating to the use of global variables can be found in [Mat-11, chap. 7].

OVERVIEW OF CONTENTS

Chapter 1 introduces the subject of ordinary differential equations, and basic integration methods are covered, including Euler, Runge–Kutta, variable step, extrapolation, BDFs, NDFs, and Adams. The Newton and Levenberg–Marquardt convergence methods are introduced in connection with implicit integration. A discussion relating to truncation error and verification of integration order is presented with a number of examples that use the concepts discussed. The chapter concludes with a brief discussion on stiffness.

Chapter 2 explores some of the theory and practical applications of ODE integrator stability analysis. The discussion includes topics on global order of accuracy, A-stable and zero-stable definitions, and the Dahlquist barrier theorems. Using the Dahlquist test problem, we investigate the stability of various integration methods, including Runge–Kutta, variable step, BDFs, NDFs, and Adams. For each method, the stability regions are plotted in the complex plane to illustrate the associated stability margins.

Chapter 3 discusses Cauchy problems in the context of initial value and/or boundary value problems. The basic concepts relating to PDEs and their classification are discussed, along with initial conditions (ICs) and the various types of boundary conditions (BCs). General ideas relating to discretization methods, mesh grid, stencils, upwinding, and the Courant–Friedrichs–Lewy number are introduced. This leads into a discussion on the method of lines (MOL), one of the major numerical schemes for solving PDEs. A good selection of 1D and 2D example MOL problems are solved. These examples, in both Cartesian and polar coordinates, have been chosen to bring out the main ideas and to show the variety of problems that can be solved using this method. Then follows a brief discussion of the following fully discrete methods: FTBS, implicit FTBS, FTCS, implicit BTCS, leapfrog, Beam–Warming, Lax–Friedrichs, and Lax–Wendroff. Solutions are compared for the different methods applied to sample problems. The ideas behind the finite volume method are presented in preparation for their use in high-resolution schemes, which are discussed subsequently in Chapter 6.

Chapter 4 discusses the concept of a system being *well-posed* and ways of using this idea to investigate system stability. The matrix stability method is outlined, and detailed stability calculations are included for semi-discrete PDE schemes. A number of examples are included that analyze PDEs, and results are presented in graphical form. The von Neumann stability method is discussed with the fundamental ideas developed using Fourier transforms. Various semi-discrete and fully discrete PDE systems are analyzed, with results presented graphically.

Chapter 5 introduces the idea of the *dispersion relation* and its relationship to wave number and wave frequency. The accuracy of numerical schemes is discussed in relation to numerical amplification and exact amplification factors. Also discussed is how these factors can be used to provide an indication of phase (dispersion) and amplitude (dissipation) errors. The chapter concludes with a discussion on phase and group velocities.

Chapter 6 introduces the idea of *high-resolution* schemes for the solution of PDEs. The Riemann problem is discussed, along with Godunov's method of providing an approximate solution. A discussion follows on the principle of total variation diminishing (TVD) and how Godunov's order barrier theorem places constraints on monotonic PDE solution methods. An introduction is then provided to two major methods that are employed widely to solve these types of problems. The first is the flux limiter method, whereby fifteen different flux limiter functions are presented. The second is the weighted essentially nonoscilliatory (WENO) method, whereby the associated weights and smoothness indicator calculations are presented. These methods are discussed within a finite volume and MOL framework. They are based on the monotone upstream-centered schemes for conservation laws (MUSCL) method, with the particular implementation being the Kurganov and Tadmor central scheme. A variety of 1D and 2D problems are solved using these methods, and the results are presented graphically. These include, advection, Burgers, Buckley–Everett, Euler equations, Sod's shock tube, Taylor–Sedov detonation, Woodward–Colella interacting blast wave, and frontogenesis: all computationally demanding PDE evolution problems.

Chapter 7 introduces the concept of *meshless methods* using radial basis functions (RBFs). These methods represent important tools for the numerical analyst and are becoming very popular for solving otherwise difficult problems. One of the main advantages of meshless methods is that they can be used on irregular grids and therefore are applicable to problem geometries of any shape. The ideas are presented from an introductory-basics level, with examples showing how the method is used to interpolate scattered data and also for solving partial differential equations. For a number of examples, the results are compared to analytical solutions to demonstrate the accuracy of the results obtained. The Halton sequence, which produces pseudo-random data, is introduced to demonstrate how the method readily handles irregular grids and/or scattered data. A number of globally and compactly supported RBFs are defined and their use illustrated with examples in 1D, 2D, and 3D. This chapter also includes discussion on the use of local RBFs, which allow the method to be used on very large problems. Local RBFs result in the system matrices becoming sparse, which facilitates the application of sparse matrix methods, which are provided by the R package `Matrix`.

Chapter 8 introduces the concept of *conservation laws* in the context of evolutionary PDEs. Under the assumption of certain decay conditions, it is shown how conserved quantities can be identified for particular PDEs. It is then shown that conserved quantities can be used to calculate associated constants of motion or invariants. These constants of motion are very useful in numerical analysis as they can be used to provide an indication of calculation accuracy. The ideas are discussed mainly in terms of the 1D Korteweg-de Vries (KdV) equation, where a number of the commonly known conservation laws applicable to this equation are derived. It is then shown using the Miura and Gardener transformations that, actually, the KdV equation possesses an infinity of conservation laws. The discussion is then extended to the 2D KdV equation and then to the KdV equation with variable coefficients (vcKdV). This chapter also includes conservation law discussions and example calculations in relation to the nonlinear Schrödinger (NLS) equation and the Boussinesq equation.

Chapter 9 is a case study into the analysis of a golf ball in flight. It has been the subject of many theoretical investigations, with some of the earliest being published in the late 1890s. This case study provides an in-depth study of this subject, with computer simulation results presented and compared to published performance data. The various forces acting on a golf ball in flight are discussed, namely, drag, Magnus, and gravitational. The differential equations that describe the golf ball flight are then derived. The effects of compression, spin, ambient conditions, wind, launch angle, bounce, and roll are all taken into account in the simulation calculations. The latest coefficients from the literature for drag and lift are used. The results are presented graphically and compared with measured statistics for club head speed, carry, and trajectory height for various classes of player, using different golf clubs. The effect of push, pull, fade, draw, slice, and hook shots are investigated, and the results are discussed. The magnitude of wind shear at ground level is calculated, and its effect on a static golf ball is investigated.

Chapter 10 is a case study into the problem of an intense explosion. In 1945, Sir Geoffrey Ingram Taylor was asked by the British *Military Application of Uranium Detonation* (MAUD) Committee to deduce information regarding the power of the first atomic explosion at the *Trinity site* in the New Mexico desert. He was able to estimate, using only public domain photographs of the blast, that the yield of the bomb was equivalent to between 16.8 and 23.7 kilotons of TNT. This case study discusses the subsequent seminal papers that Ingram published and traces the perceptive calculations that he made. A systems analysis starts with a form of the Euler equations, from which, using similarity analysis, certain important relationships are deduced. With the aid of a set of high-speed photographs of the detonation, and assuming spherical symmetry, the underlying characteristics of the blast are gradually revealed by a sequence of thermodynamic calculations. The thermodynamic gas laws required to unravel this puzzle are discussed, along with shock wave analysis using the Hugonoit–Rankine relations. Kinetic energy and heat energy integrals are evaluated, leading to a full description of the blast. This includes blast wave speed, pressure, and temperature over time. Closed-form analytical solutions to the problem, subsequently published by Sedov in 1959, are presented that provide a useful check on the accuracy of Ingram's numerical calculations. A similarity analysis is included, covering spherical, cylindrical, and planar blast situations.

Chapter 11 is a case study into the *global carbon cycle* and how increased concentrations in atmospheric carbon dioxide have implications for climate change. A simplified model is presented that considers four fossil-fuel emission scenarios based on the work of Caldeira and Wickett and how the atmosphere and oceans respond. Air–ocean interaction is modeled using the wind-driven gas–sea exchange relationship due to Wanninkhof, with the subsequent dispersion of gaseous CO_2 into $CO_2(aq)$, HCO_3^-, CO_3^{2-}, and H^+ in accordance with carbonate chemistry equilibria. It is shown how an increasing concentration of positive hydrogen ions causes seawater acidity to rise, that is, a fall in pH. Seawater buffering calculations are also introduced to demonstrate how the ocean's ability to absorb atmospheric CO_2 is diminished as seawater concentration of dissolved inorganic carbon increases. A discussion of solar and terrestrial radiation modeling is also included, along with calculations that show how increasing CO_2 concentrations in the atmosphere can lead to increases in the Earth's surface temperature, that is, to the so-called greenhouse effect.

All chapters include worked examples, many of which generate animations, along with annotated computer code, which is available for download. As an additional resource for some chapters, computer code is provided with the downloads for symbolic computer algebra analysis using Maple and Maxima/wxMaxima systems. Maxima is a free open-source program available for different operating systems that can be downloaded from http://maxima.sourceforge.net/. A GUI-based version, wxMaxima, can be downloaded from http://andrejv.github.io/wxmaxima/.

To maximize the benefit of studying numerical computing, the learning process should not be regarded as a passive pursuit. Rather, it should be regarded as a *hands-on* experimental activity. Because many problems are nonlinear, analytical or well-tried and tested solutions may not exist. In these situations it may be necessary to try various different options to find the most appropriate solution technique, with the best parameter values being deduced by trial and error. This book includes many and varied solutions to a wide range of problems, and it is hoped that the reader will find some that can be applied directly, or adapted, to the particular problem of interest.

To conclude, the major focus of this book is the numerical solution of initial value problems (IVPs) and boundary value problems (BVPs) described by ODEs and PDEs. The general approach and content complement the author's earlier books ([Gri-11] and [Sch-09]) and also the excellent book *Solving Differential Equations in R* [Soe-12].

ACKNOWLEDGMENTS

I would like to thank Professor William (Bill) E. Schiesser of Lehigh University for collaboration in the area of numerical solutions to differential equations over many years and for suggesting that I look into using the R language for solving ODEs and PDEs; Professor Alan M. Nathan of the University of Illinois for useful discussions on the flight of projectiles and for clarifying some concepts; and Professor Duncan Murdoch of the

University of Western Ontario for advice on using the R package `rgl` and for responding positively to requests for changes to the 3D graphics function `persp3d()`.

Graham W. Griffiths
Nayland, Suffolk, United Kingdom
June 2015
graham@griffiths1.com
www.pdecomp.net

REFERENCES

[Cra-11] Crawley, M. J. (2011), *The R Book*, John Wiley.

[Gri-11] Griffiths, G. W. and W. E. Schiesser (2011), *Traveling Wave Solutions of Partial Differential Equations: Numerical and Analytical Methods with Matlab and Maple*, Academic Press.

[Mat-11] Matloff, N. (2011), *The Art of R Programming*, No Starch Press.

[Sch-09] Schiesser, W. E. and G. W. Griffiths (2009), *A Compendium of Partial Differential Equation Models: Method of Lines Analysis with Matlab*, Cambridge University Press.

[Soe-10] Soetaert, K., T. Petzoldt and R. W. Setzer (2010), Solving Differential Equations in R: Package **deSolve**, *Journal of Statistical Software* **33**-9, 1–25.

[Soe-12] Soetaert, K., J. Cash and F. Mazzia (2012), *Solving Differential Equations in R*, Springer-Verlag.

1

ODE Integration Methods

1.1 INTRODUCTION

Before we start to solve *differential equations* (DEs) by *numerical integration*, we will briefly review some background topics prior to discussing a selection of the available methods. We will do this in a nonrigorous way, as this aspect is covered in detail in any textbook dealing with the *calculus*; for example, see [Joh-08] for an excellent basic introduction to differentiation and integration and [Kre-11] for a more in-depth and broader coverage of the calculus and its application. The latter book is considered by many science, technology, engineering, and mathematics (STEM) students to be their math bible.

When we first start to study the calculus, we learn a little about the historical context in which it was conceived and the priority dispute between Newton[1] and Leibniz[2] over who invented it. Today, this dispute is considered moot, and historians credit both with independent discovery. We are then introduced to the concept of a *limit*, whereby we evaluate a *continuous* function, $f(x)$, over the interval $[a, b]$—written as $f(x) \in C[a, b]$—as x moves closer to a particular value, say, $x = x_0$. This is, of course, only necessary when $f(x_0)$ is indeterminate, as otherwise we could just plug $x = x_0$ into $f(x)$. Later in the course, we learn advanced methods for dealing with indeterminate situations by obtaining limits that require the use of *derivatives*.

The study of limits leads naturally to the idea of a *derivative*, being equal in one-dimensional space to the *slope* of a *tangent* to the curve of our function $f(x)$. We write the derivative of $f(x)$ with respect to x as $\frac{df(x)}{dx}$ or, alternatively, as $f'(x)$, where the leading d in the numerator and denominator is an *operator* representing an *infinitesimal* change in $f(x)$ and x, respectively. Usually, our first attempt at obtaining an approximation to the tangent is to consider a straight line constructed between two points: one located at the point on the curve where we wish to evaluate the derivative, x_0, and the other a short

---

[1] Sir Isaac Newton, English mathematician and scientist (1642–1727), also famous for discovering the inverse law of gravity, which he published in 1687 in his famous book *Philosophi Naturalis Principia Mathematica* (Mathematical Principles of Natural Philosophy), commonly known as the *Principia*.
[2] Gottfried Wilhelm von Leibniz, German mathematician and philosopher (1646–1716). He was also linguist and wrote extensively on a wide range of subjects, including philosophy, politics, law, ethics, theology, history, and philology.

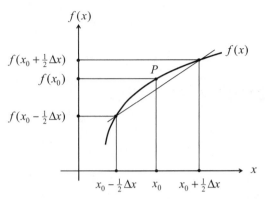

Figure 1.1. Diagram of a two-sided derivative construction. In the limit as $\Delta x \to 0$, the slope at point P will be equal to the derivative of $f(x)$ evaluated at $x = x_0$, that is, $\frac{df(x)}{dx}\big|_{x=x_0}$.

increment away. This is called a *single-sided* approximation, and the tangent slope, s, can be calculated in two different ways:

$$s_R = \frac{f(x_0 + \Delta x) - f(x_0)}{\Delta x}, \quad \text{right-sided}$$
$$s_L = \frac{f(x_0) - f(x_0 - \Delta x)}{\Delta x}, \quad \text{left-sided},$$
(1.1)

where the symbol Δx means a small increment in x.

A more accurate approximation is the *two-sided* approximation, where the points are located equidistantly either side of the point $f(x_0)$. The slope of this approximate tangent is given by

$$s = \frac{f(x_0 + \Delta x/2) - f(x_0 - \Delta x/2)}{\Delta x};$$
(1.2)

see Fig. (1.1).

Clearly the approximation of eqn. (1.2) will become increasingly more accurate as Δx becomes smaller and the points move closer to x_0. In the limit when $\Delta x = 0$, we obtain a value for the actual tangent slope at x_0 and, consequently, the derivative. This is why we were introduced to the concept of a limit prior to discussing derivatives. A two-sided derivative of $f(x)$ at $x = x_0$ can be defined as

$$\frac{df(x = x_0)}{dx} = \lim_{\Delta x \to 0} \frac{f(x_0 + \Delta x/2) - f(x_0 - \Delta x/2)}{\Delta x}, \quad f(x) \in C^1,$$
(1.3)

where the symbol C^1 means that this definition applies to a function $f(x)$ that is both continuous and has a continuous first derivative at $x = x_0$.

Figure 1.2 includes a plot showing how a two-sided derivative approximation converges rapidly as $\Delta x \to 0$. The function under consideration is $f(x) = x^{\frac{1}{3}}$, and the tangent slope is evaluated at $x = 0.5$.

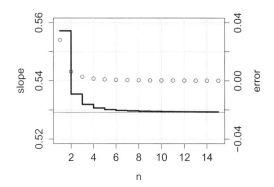

Figure 1.2. Plot of two-sided derivative approximate values of the function $f(x) = x^{\frac{1}{3}}$ evaluated as $x = 0.5$ (stepped) with corresponding errors (circles). Values for the denominator in eqn. (1.3) are $\Delta x = 0.5/i$, $i = 1, \ldots, 15$. The slope approximation converges rapidly toward the true value of 0.529 (thin horizontal line) as i increases.

The R code that produced Fig. (1.2) is shown in Listing 1.1.

```
# File: derivativeApprox.R
# Two-sided derivative approx. of f(x)=x^(1/3) evaluated at x=0.5
N <- 15; x <- seq(0,1,len=N); slope <- rep(0,N); err <- rep(0,N)
Nrange <- 1:N
x0 <- 0.5; x1 <- 0.25; x2 <- 0.75
Dx <- x0-x1 # equal spacing about x=0.5
s_anal <- 0.5291337 # derivative of f(x) at x=0.5
for(n in Nrange){
  dx <- Dx/n                  # Next spacing of x
  xa <- x0-dx; xb <- x0+dx    # x points for new slope
  fa <- xa^(1/3); fb <- xb^(1/3) # new function values
  slope[n] <- (fb-fa)/(xb-xa) # derivative approx.
}
err <- slope-s_anal           # error
txt <- sprintf("tangent slope=%5.3e at x0=%3.1f for dx=%5.3e\n",slope[N],x0,dx)
cat(txt) # print summary of final result
#
par(mar=c(4,5,2,5))
plot(Nrange,slope[Nrange],type="s",lwd=3, xlim=c(1,N),
     ylim=c(0.52,0.56), xlab="n",ylab="slope") # type="s" means step plot
grid(lty=1); abline(h=s_anal,col="red") # Analytical value
lines(Nrange,slope[Nrange],lwd=3,type="s")
#
par(new = T) # Plot error on second y-axis
plot(Nrange,err[Nrange], col = "blue", axes = F,
     xlab = NA, ylab = NA,ylim=c(-0.04,0.04))
axis(side = 4); mtext ( text ="error", side = 4, line = 3 )
```

Listing 1.1. File: derivativeApprox.R—Code to generate two-sided derivative approximations

Table 1.1. Basic derivative pairs

Function, $f(x)$	Derivative, $f'(x) = \dfrac{df(x)}{dx}$
x^{an}	nax^{an-1}
$\sin ax$	$a \cos ax$
$\tan ax$	$a \sec^2 ax$
$\exp ax$	$a \exp ax$
$\vdots$	$\vdots$
etc.	

Once we have grasped the concept of the derivative of a function, we then learn how to differentiate analytically a whole series of functions, including trigonometrical functions, quotients, powers, and so on. We also learn to differentiate a function $f(x)$ with respect to x multiple times, where the nth derivative is written as $\frac{d^n f(x)}{dx^n}$. From this point on, we are able to use a lookup table to obtain derivatives of standard functions, such as those listed in Table 1.1.

At this point, we are equipped to apply differentiation to solving real-world problems such as min/max of functions, obtaining velocity from distance traveled, and acceleration from velocity. We also learn more advanced methods of obtaining limits that require the use of derivatives, for example, *l'Hôpital's rule*.[3]

The next stage in learning the calculus is an introduction to the concept of *integration* of a function. We learn that *indefinite integration* is the opposite of differentiation (i.e., the *antiderivative*) and needs to include a constant of integration and that *definite integration* (also known as *quadrature*) is equal to the area under the curve of a function $f(x)$ from $x = x_a$ to $x = x_b$. We are shown how to obtain an approximation to the definite integral of our function $f(x)$ with respect to x, written as $\int_a^b f(x)\,dx$, by dividing the area into a set of rectangles and summing the areas of the individual rectangles. This is known as a *Riemann sum*.[4] A Riemann sum can be formed in three ways: left, center, and right (see Fig. 1.3).

The area of one rectangle at, say, $x = x_i$ is given by $A_i = (x_{i+1} - x_i) f(x_i)$. Therefore, using a left Riemann sum of n rectangles, the total area is given by $A = \sum_{i=0}^{n-1} A_i = \sum_{i=0}^{n-1} (x_{i+1} - x_i) f(x_i)$. Similar to the derivative discussed earlier, the approximation will become increasingly more accurate as n increases and x_i and x_{i+1} approach each other and the rectangle widths reduce. Using our knowledge of limits, we can now define a definite integral of $f(x)$ with respect to x from $x = x_a$ to $x = x_b$. This can be done using the left and right Riemann sums, as follows:

$$
\begin{aligned}
I_L &= \int_{x=x_a}^{x=x_b} f(x)\,dx = \lim_{n\to\infty} \sum_{i=0}^{n-1} (x_{i+1} - x_i) f(x_i), \quad x_a = x_1,\ x_b = x_n \\
I_R &= \int_{x=x_a}^{x=x_b} f(x)\,dx = \lim_{n\to\infty} \sum_{i=1}^{n} (x_{i+1} - x_i) f(x_i), \quad x_a = x_1,\ x_b = x_n,
\end{aligned}
\tag{1.4}
$$

---

[3] First published by French mathematician Guillaume François Antoine, Marquis de l'Hôpital (1661–1704). If $\lim_{x \to x_0} \frac{f(x)}{g(x)}$ is indeterminate, but $\lim_{x \to x_0} \frac{f'(x)}{g'(x)}$ exists, then the method states that the first limit also exists and is equal to the second limit.

[4] After German mathematician Georg Friedrich Bernhard Riemann (1826–1863).

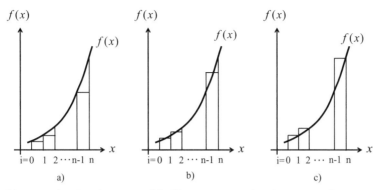

Figure 1.3. Diagram showing three possible Riemann sums for the area under the curve of function $f(x)$ between $x = x_0$ and $x = x_n$: (a) A *left* Riemann sum where the area $A = \sum_{i=0}^{n-1} f_i(x_{i+1} - x_i)$; (b) a *center* Riemann sum where the area $A = \sum_{i=0}^{n-1} (f_i + f_{i+1})(x_{i+1} - x_i)/2$; and (c) a *right* Riemann sum where the area $A = \sum_{i=1}^{n} f_i(x_{i+1} - x_i)$. Here we have used the symbol f_i to represents $f(x_i)$.

where I_L represents the *left* Riemann sum and I_R represents the *right* Riemann sum. Note that it is not a requirement that the x_i be spaced equally along the *x*-axis.

A more accurate approximation is obtained from the *center* Riemann sum, where the height of the rectangle at $f(x_i)$ is equal to the average height of the two values $f(x_i)$ and $f(x_{i+1})$. The integral approximation by this method is given by

$$I_C = \int_{x=x_a}^{x=x_b} f(x)\,dx = \lim_{n \to \infty} \sum_{i=0}^{n-1} \frac{1}{2}(x_{i+1} - x_i)(f(x_i) + f(x_{i+1})), \quad x_a = x_1, x_b = x_n \quad (1.5)$$

(see Fig. 1.1).

Figure 1.4 includes a plot of the center Riemann sum approximation to the definite integral of $f(x) = x^3$ evaluated between $x = 0$ and $x = 1$. The approximation converges toward the true integral value of 0.25 as the number of rectangles n increases. The center sum approximation requires 12 rectangles to reduce the error to below 1%, whereas left

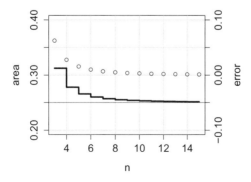

Figure 1.4. Plot of *center* Riemann sum approximate values of the integral of $f(x) = x^3$ evaluated between $x = 0$ and $x = 1$ (stepped) with corresponding errors (circles). The integral approximation converges to the true value of 0.25 (thin horizontal line) as n increases.

and right approximations would require more that 200 rectangles to achieve the same accuracy.

The R code that produced Fig. 1.4 is shown in Listing 1.2.

```
# File: RiemannSum.R
# Center Riemann sum of f(x)=x^3 evaluated from x=0 to x=1
N <- 15; Nrange <- seq(3,N,by=1)
area = rep (0,N)
I_anal <- 0.25           # Integration of f(x) between x=0 and x=1
for (n in Nrange ){      # calculate Riemann sum for 3 to N intervals
  x <- seq (0,1, len=n)  # Next sequence from x=0 to x=1
  dx <- x[2] -x [1]      # equal spacing
  f <- x^3               # evaluate f(x), x=0,...,1
  area[n] <- dx*(sum(f[2:(n-1)])+(f[1]+f[n])/2)
}
err <- area-I_anal       # error
txt <- sprintf("Center Riemann sum =%5.3e for n=%d rectangles \n", area[n],n)
cat(txt) # print summary of final result
# Plot results
par(mar=c(4,5,2,5))
plot ( Nrange , area [Nrange], type ="s", lwd =3, ylim =c (0.2 ,0.4),
       xlab ="n", ylab ="area") # type ="s" means step plot
grid ( lty =1); abline (h =0.25 , col =" red ") # Analytical value
#
par(new = T) # Plot error on second y=axis
plot(Nrange,err[Nrange], col = "blue", axes = F,
     xlab = NA, ylab = NA,ylim=c(-0.1,0.1))
axis(side = 4); mtext ( text ="error", side = 4, line = 3 )
```

Listing 1.2. File: `RiemannSum.R`—Code to generate Riemann sums

Once we have grasped the concept of integration, we learn how to integrate analytically a wide range of functions, including trigonometrical and logarithmic functions, quotients, and powers. As for derivatives, we are now able to use a lookup table to obtain integrals of standard functions, such as those listed in Table 1.2.

With our new knowledge of integration, we can now solve additional real-world problems to those we tackled using differentiation, such as distance traveled from velocity, volumes of revolution, areas between curves, and lengths of curves.

At this stage in our study, we have learned the basics of the calculus and how to use it to solve otherwise intractable problems. We are then equipped and ready to be exposed to the concept of *differential equations* (DEs), and this opens up a whole new field of study. Historically, differential equations arose from the analysis of physical system dynamics. Subsequently, the ideas have moved into other disciplines, such as finance and the social sciences.

Table 1.2. Basic integral pairs

Function, $f(x)$	Integral, $F(x) = \int f(x)\,dx$
x^{an}	$\frac{x^{an+1}}{an+1}$
$\sin ax$	$-\frac{1}{a}\cos ax$
$\tan ax$	$\frac{1}{a}\ln(|\sec ax|)$
$\exp ax$	$\frac{1}{a}\exp ax$
$\vdots$	$\vdots$
etc.	

Note: (a) An *indefinite integral* is equal to $F(x)+k$, where k represents a constant of integration. (b) A *definite integral* that is evaluated between the limits $x = x_a$ and $x = x_b$ is equal to $F(x_b) - F(x_a)$. This is known as the *fundamental theorem of calculus* and establishes a formal relationship between differentiation and integration.

A differential equation is an equation that can include constants, functions, and, of course, derivatives; hence use of the term *differential*. It is appropriate here to clarify the following terms that are used to classify differential equations:

- A DE must include one or more derivatives of any order, for example, first, second, or third and the order of a DE is determined by the highest-order derivative.
- A DE is called *linear* if all terms containing the dependent variable or its derivatives are of the first degree and do not contain higher powers or products.
- A DE is defined as *homogeneous* if all the terms contain the dependent variable and/or its derivative of any order. Thus a constant term destroys homogeneity and the DE becomes *nonhomogeneous*.

For example, the following represents a linear, nonhomogeneous, second-order differential equation:

$$d\frac{d^2 f(x)}{dx^2} + c\frac{df(x)}{dx} + b\,f(x) + a = 0, \tag{1.6}$$

where $a, b, c,$ and d are constants and $f(x)$ is an unknown function that we wish to determine, that is, the solution. However, for a sufficiently smooth function, there exists an infinite family of functions that will satisfy eqn. (1.6). Depending on the type and order of DE, to obtain a particular solution, $f(x)$, we also need to specify appropriate initial conditions (ICs) and/or boundary conditions (BCs). The number of these auxiliary conditions is equal to the *highest-order derivative with respect to the independent variable*. For example, a first-order initial value ODE with dependent variable $u(t)$ requires one IC at the initial value of t, that is, $u_0 = u(t_0)$. Similarly, the second-order ODE given in eqn. (1.6) requires one IC for the dependent variable at the initial value of x, that is, $f_0 = f(x_0)$, plus an initial value for the first derivative of the dependent variable, that is, $f'_0 = f'(x_0)$, and so on. These requirements are illustrated by example in subsequent sections and are also discussed in more detail for PDEs in Chapter 3.

Equation (1.6) is defined as an *ordinary differential equation* (ODE) as there is only one independent variable, x. If a DE includes more than one independent variable, say, x and t, then this DE is called a *partial differential equation* (PDE). For example, the following represents a second-order PDE, the so-called *diffusion equation*:

$$\frac{\partial u(x,t)}{\partial t} = D \frac{\partial^2 u(x,t)}{\partial x^2}, \qquad (1.7)$$

where D is the diffusion constant and the unknown function is represented by $u(x,t)$. Note that the derivatives are now called *partial derivatives* and the operator symbol has changed from d to ∂. A partial derivative is defined as

$$\frac{\partial u(x,t)}{\partial x} = \left. \frac{du(x,t)}{dx} \right|_t, \qquad (1.8)$$

which means that $\frac{\partial u(x,t)}{\partial x}$ is equal to $\frac{du(x,t)}{dx}$ with all independent variables, other than x, held constant—in this case, just t. Hence use of the term *partial*.

Partial derivatives still represent the slope of a tangent, but instead of the tangent being to a line, it is a tangent line on a surface. In this instance, the tangent line is aligned in the direction of the variable x. PDEs are discussed in more detail in Chapter 3.

The process of solving a differential equation involves integration, either directly or indirectly. Hence we refer to *integration of differential equations*. We usually start learning about differential equations by studying certain types of ODEs, some of which can be solved by substituting an *ansatz*[5] into the equation and solving for the unknowns. For example, letting $a = -1, b = c = 1, d = 0$, and $f(x) = y(x)$ in eqn. (1.6) yields the first-order linear nonhomogeneous ODE

$$\frac{dy}{dx} + y = 1, \quad y = y(x). \qquad (1.9)$$

In our DE course, we learned that for a linear nonhomogeneous ODE, the solution consists of two parts, a *homogeneous* solution, y_h, to the homogeneous part of the ODE and a *particular* solution, y_p, to the nonhomogeneous part, with the *total* solution being equal to $y = y_h + y_p$. First we substitute the ansatz $y = Ae^{mx}$ into the homogeneous part of eqn. (1.9) (by setting the nonhomogeneous right-hand side to zero), giving

$$Ame^{mx} + Ae^{mx} = 0. \qquad (1.10)$$

If we now divide through by Ae^{mx}, we obtain the *characteristic* equation

$$m + 1 = 0, \quad \therefore m = -1. \qquad (1.11)$$

Thus $y_h = Ae^{-x}$ is a *general* solution to the homogeneous part of eqn. (1.9), with A being an arbitrary constant.

---

[5] The term *ansatz* is used to describe an assumed form of solution, a trial solution, or an educated guess.

We now attempt to solve for y_p trying the ansatz $y_p = C \times 1$, a constant multiplied by the nonhomogeneous right-hand side of eqn. (1.9),[6] which yields

$$0 + C = 1. \tag{1.12}$$

Therefore $C = 1$ and $y = y_h + y_p = Ae^{-x} + 1$. However, if we know a value of y for a particular value of x, we can solve for A. This is known as the *particular* solution to eqn. (1.9) — because particular values are assigned to the constants A and m. Letting the known value be the initial condition $y = 0$ when $x = 0$ yields $A = -1$, and the overall particular solution satisfying the initial condition becomes

$$y = 1 - e^{-x}. \tag{1.13}$$

Of course, for this case, eqn. (1.9) is a *separable equation* and can be solved by a simple integration, that is,

$$\int \frac{1}{1-y} dy = \int dx, \tag{1.14}$$
$$\therefore -\ln(1-y) + k_1 = x + k_2, \quad \Rightarrow \quad y = 1 + Ae^{-x},$$

where $k = k_1 - k_2$ is the combined constant of integration and $A = -e^k$. Thus we have obtained the same general solution; and using the same initial condition, we obtain the same overall particular solution, $y = 1 - e^{-x}$.

Higher-order ODEs can, in general, be transformed into a set of coupled first-order ODEs. By "coupled," we mean that the dependent variables occur in more than one equation. Thus transformation often facilitates analysis and can simplify the process of obtaining a general or particular solution. Consider the following second-order ODE:

$$\frac{d^2 y}{dx^2} = a \frac{dy}{dx} + by + c, \quad y = y(x). \tag{1.15}$$

This equation can be transformed into two coupled first-order equations by letting $y_1 = y$ and $y_2 = \frac{dy_1}{dx}$, when we obtain, in matrix form,

$$\begin{bmatrix} \frac{dy_1}{dx} \\ \frac{dy_2}{dx} \end{bmatrix} = \begin{bmatrix} 0 & 1 \\ b & a \end{bmatrix} \begin{bmatrix} y_1 \\ y_2 \end{bmatrix} + \begin{bmatrix} 0 \\ c \end{bmatrix}. \tag{1.16}$$

In the same way, a third-order system can be transformed into a set of three coupled first-order equations, and so on for higher-order systems. This form of mathematical presentation is very useful for analytical analysis when eigenvalues can be obtained that provide insight into the stability and dynamic behavior of the system being investigated. This approach can also be applied to the solution of PDEs and is discussed in later chapters.

The analysis of *nonlinear* differential equations is beyond the scope of this brief introductory overview. A wide discussion on DEs in general may be found in [Zwi-98], and an

---

[6] If the homogeneous solution contains the nonhomogeneous part, then we try letting y_p equal the nonhomogeneous part multiplied by x^n, where n is the minimum integer that makes $y_h + y_p$ linearly independent. There are many tricks for solving nonhomogeneous ODEs, for example, see [Kre-11].

interesting book that discusses nonlinear DEs and provides an introduction to nonlinear dynamics is [Tel-06].

An amazing characteristic of DEs is that very similar equations crop up in many different disciplines, for example, rate of reaction and continuity in chemistry, celestial mechanics and diffusion in physics, natural modes of vibration and inertial systems in mechanical engineering, and resistance–capacitance–inductance circuits and electromagnetic waves in electrical engineering. Even in financial analysis, there is the *Black–Scholes* PDE model[7] for describing stochastic[8] processes involved in the financial instruments known as *derivative options*. These PDEs include convection- and diffusion-like terms. Consequently, from their wide application, it is clear that the solution of DEs is an immensely important area of applied mathematics.

For many problems, ODEs cannot be solved analytically by the preceding or similar methods. This can be due to the presence of nonlinear terms, parameters that vary over the solution domain, difficult boundary conditions, or just the sheer complexity of the problem. In these situations, we have to resort to *numerical methods*. The process of solving ODEs/PDEs numerically is called *numerical integration* and is performed by a *numerical integrator*, alternatively referred to as a *numerical solver*. The matrix form of eqn. (1.16) is particularly useful in numerical analysis as standard numerical solvers usually expect ODEs in this form. A numerical solution to a differential equation is one where the differentials are replaced by a set of algebraic equations and whose solution provides an approximation to the true solution. The job of the numerical analyst is to find an approximate solution that is sufficiently close to the true solution.

The idea of numerical integration is to use one or more past values to approximate the derivative of the dependent variable, and then to use this to project the solution forward, either in time or space. The new solution in turn becomes a past value that is used to project forward to the next solution value, and so on. In this step-by-step approach, a solution trajectory gradually evolves until the problem is solved to the satisfaction of the analyst. Of course, this is a rather simplistic overview, and many considerations need to be taken into account to obtain a good numerical solution. For example, in later sections, we discuss integration errors and methods for keeping the solution within desired accuracy bounds, and in the next chapter, we learn about methods to investigate solution stability.

Numerical integration methods is a vast subject that we can only discuss briefly here, and we shall only cover a small selection of the more common integrators to give some insight into the numerical solution of initial value problems. The integrator methods we consider are suitable for solving initial value problems described by ordinary differential equations (ODEs) of the form

$$\frac{dy}{dt} = f(y, t), \quad [y(t_0) = y_0], \quad y \in \mathbb{R}^m, t > 0. \tag{1.17}$$

---

[7] The so-called Black–Scholes model, shown here without discussion, was published by Fischer Black and Myron Scholes in their 1973 paper "The Pricing of Options and Corporate Liabilities," published in the *Journal of Political Economy*:

$$\frac{\partial V}{\partial t} + \frac{1}{2}\sigma^2 S^2 \frac{\partial^2 V}{\partial S^2} = rV - rS\frac{\partial V}{\partial S}.$$

[8] A stochastic process is a *random* process that is governed by its statistics. It can be considered to be the opposite of a *deterministic* process.

Although the integrators we discuss are straightforward to implement in a simplistic manner, many difficulties have to be overcome when programming a general purpose integrator. However, it is not our intention to develop specialist code here; rather, our intent is to give an explanation of the fundamentals to aid understanding of the methods. For those who wish to explore the subject further, to obtain a flavor of more advanced topics, a wide variety of integrator codes and formal test problems are readily available on the Internet. The *Test Set for Initial Value Problem Solvers* [Maz-08] and *GAMS*, the *Guide to Available Mathematical Software* [GAM-10], are both excellent free sources of such material.

1.2 EULER METHODS

1.2.1 Forward Euler

The *forward Euler* numerical integration method, also known as *explicit Euler*, is an explicit scheme based on a Taylor series expansion truncated after the second term. In general, the term *explicit* means that the next value of the dependent variable is calculated from one or more previously calculated values. However, for this particular method the next value at time t_n only requires knowledge of the previous value at t_{n-1}.

Expanding the Taylor series of $y(t+h)$ about $y(t)$ yields

$$y(t+h) = y(t) + hf(t, y) + \frac{h^2}{2!} f'(t, y) + \frac{h^3}{3!} f''(t, y) + \cdots \qquad (1.18)$$

If we now truncate this series after the second term, we obtain the following numerical approximation:

$$y(t_n + h) = y_n + hf(t_n, y_n) + \mathcal{O}(h^2), \qquad (1.19)$$

where h is the step size, and we write $t_n = t_0 + nh$ and y_n to represent the numerical approximation to $y(t = t_n)$, and so on. The term $\mathcal{O}(h^2)$ represents the *local truncation error*, that is, the difference between the numerical value and the exact or real value. The term h^2 indicates that the truncation error has *order* 2, and therefore the method is accurate to order 1. The reason for this is explained in Section 1.5.

We include forward Euler in our discussions because it is the simplest form of numerical integrator and it can be used to illustrate certain basic numerical integration concepts. In practice it is only used to obtain solutions in situations where small step sizes are acceptable (and do not lead to significant round-off errors) and stability issues do not arise. The stability aspects of this method are discussed in Chapter 2.

We can also arrive at an equivalent form of eqn. (1.19) by approximating the derivative by a finite difference, that is,

$$\frac{dy(t)}{dt} = f(t, y) \approx \frac{y(t+h) - y(t)}{h}, \qquad (1.20)$$

when, on rearranging and writing $y_n \approx y(t)$, and so on, we get

$$y(t+h) \approx y_{n+1} = y_n + hf(t_n, y_n). \qquad (1.21)$$

Equation (1.21) is equal to eqn. (1.19) less the truncation error term.

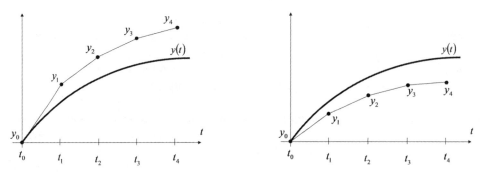

Figure 1.5. (left) Forward Euler integration example, with the error at t_n being given by err $= y_n - y(t_n)$. (right) Backward Euler example, with the error at t_n being given by err $= y_n - y(t_n)$.

An example of the forward Euler method in use is shown in Figure 1.5, where the current value at $t = t_n$ is extrapolated from point y_n to point y_{n+1} along a constant slope equal to the local derivative at y_n. This process then repeats.

1.2.2 Backward Euler

The *backward Euler* numerical integration method, also known as *implicit Euler*, is an *implicit* scheme similar to the forward Euler scheme, except that the derivative is evaluated at $t_{n+1} = t_n + h$ rather than at time t_n. Hence, the term *backward*. By *implicit*, here we mean that the independent variable t_{n+1} appears on both sides of the equation, which implies that the solution usually has to be solved *iteratively*. Thus, there is no explicit equation defining the required dependent variable $y(t_{n+1})$.

The numerical approximation therefore becomes

$$y(t_n + h) = y_n + hf(t_{n+1}, y_{n+1}) + \mathcal{O}(h^2), \tag{1.22}$$

where h is the step size.

Backward Euler is a simple form of numerical integrator but introduces additional complexity due to eqn. (1.22) being implicit and therefore requiring to be solved iteratively for most applications. It has the same limitations on accuracy as forward Euler but has a major advantage in that it is a very stable method. This aspect is covered in Chapter 2, where the stability characteristics of backward differentiation formula methods are discussed.

An example of the backward Euler method in use is shown in Figure 1.5, where the current value at $t = t_n$ is extrapolated from point y_n to point y_{n+1} along a constant slope equal to the local derivative at y_{n+1}. This process then repeats.

1.3 RUNGE–KUTTA METHODS

The Runge–Kutta family of integrators is used extensively to solve practical problems. They are called one-step methods because they only use current values of the dependent variables and do not depend upon past information. However, they do involve a number of stages in the calculation. Their advantage over simple one-step schemes, such as forward Euler is that they make a more accurate estimate of the derivative by using a *weighted combination of derivatives* calculated at different times.

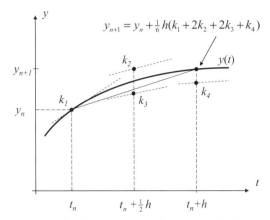

Figure 1.6. Example showing weighted average of k_i slopes used in Runge–Kutta order 4 scheme.

The general idea originated with Carl Runge [Run-95] and was extended by Karl Heun [Heu-00], who completed the analysis for methods up to third order. He also indicated the way forward to fourth order. However, it was Wilhelm Kutta [Kut-01], a German engineer, who finally gave a complete analysis of fourth-order methods. The order of a particular method is equal to the order of the local truncation error (see Section 1.5).

The now well-recognized general form of the *explicit* Runge–Kutta method is

$$y_{n+1} = y_n + h \sum_{i=1}^{s} b_i k_i, \qquad (1.23)$$

where h represents the time step and

$$\begin{aligned}
k_1 &= f(t_n, y_n), \\
k_2 &= f(t_n + hc_2, y_n + a_{21}k_1), \\
k_3 &= f(t_n + hc_3, y_n + a_{31}k_1 + a_{32}k_2), \\
&\vdots \\
k_s &= f(t_n + hc_s, y_n + a_{s1}k_1 + a_{s2}k_2 + \cdots + a_{ss-1}k_{s-1}).
\end{aligned} \qquad (1.24)$$

Refer to later discussion relating to coefficients a_{ij} and c_i for explicit methods.

Each of the k_i represents a derivative evaluated using a different t and y combination. The constants c_i represent weighting factors, so that the final derivative used in the step from y_n to y_{n+1} is a *weighted sum* of the separate individual derivative evaluations k_i. This process is shown in Figure 1.6 for a four-stage scheme.

Much work has been done since the pioneering days around the end of the nineteenth century, particularly in the derivation of coefficients, determination of necessary conditions for a s-stage method to achieve a particular order, and implicit versions.

The number of stages used in the method is specified by the constant s, and the constants a_{ij}, b_i, and c_i are given in the so-called *Butcher tableau*, as shown in Table 1.3.

For *explicit* methods we have the entries in the Butcher tableau $a_{ij} = 0$ for $j \geq i$ (matrix A is lower triangular); otherwise, the methods are *implicit*. We shall not consider implicit Runge–Kutta schemes. Specific entries in Table 1.3 are demonstrated subsequently, for example, Table 1.8 for the fourth-order method.

Table 1.3. The *Butcher tableau*

$$\begin{array}{c|cccc} c_1 & a_{11} & a_{12} & \cdots & a_{1s} \\ c_2 & a_{21} & a_{22} & \cdots & a_{2s} \\ \vdots & \vdots & \vdots & \ddots & \vdots \\ c_s & a_{s1} & a_{s2} & \cdots & a_{ss} \\ \hline & b_1 & b_2 & \cdots & b_s \end{array} = \begin{array}{c|c} c & A \\ \hline & b^T \end{array}$$

For consistency, the coefficients of Runge–Kutta methods are required to conform to at least the following conditions:

$$\sum_{j=1}^{i-1} a_{ij} = c_i, \ i = 2, \ldots, s$$
$$\sum_{j=1}^{s} b_j = 1. \tag{1.25}$$

In general it is not known how many stages are required to achieve a specified order for an explicit Runge–Kutta method. However, the result is known up to order 8 and is given in Table 1.4, which also shows for comparison the number of free parameters available in a s stage method [But-07].

In Table 1.4 the values for N^p represent the number of conditions that apply to achieve an order p method. As an example, the following eight conditions apply to an order 4, four-stage method:

$$\begin{aligned} b_1 + b_2 + b_3 + b_4 &= 1, \\ b_2 c_2 + b_3 c_3 + b_4 c_4 &= 1/2, \\ b_2 c_2^2 + b_3 c_3^2 + b_4 c_4^2 &= 1/3, \\ b_3 a_{32} c_2 + b_4 a_{42} c_2 + b_4 a_{43} c_3 &= 1/6, \\ b_2 c_2^3 + b_3 c_3^3 + b_4 c_4^3 &= 1/4, \\ b_3 c_3 a_{32} c_2 + b_4 c_4 a_{42} c_2 + b_4 c_4 a_{43} c_3 &= 1/8, \\ b_3 a_{32} c_2^2 + b_4 a_{42} c_2^2 + b_4 a_{43} c_3^2 &= 1/12, \\ b_4 a_{43} a_{32} c_2 &= 1/24. \end{aligned} \tag{1.26}$$

Table 1.4. Number of stages to achieve a specified order

Order, p	1	2	3	4	*	5	6	*	7	*	8
Order conditions, N^p	1	2	4	8	*	17	37	*	85	*	200
Stages, s	1	2	3	4	5	6	7	8	9	10	11
Free parameters, $M^s = s(s+1)/2$	1	3	6	10	15	21	28	36	45	55	66

* Values not valid for particular value of s.

Table 1.5. Order 1 tableau, forward Euler

0	
1	0
	1

These conditions arise directly from the fourth-order Runge–Kutta method and involve significant algebraic manipulation to arrive at this form. The additional conditions that apply to higher orders become extremely tedious to solve and will not be discussed here. For further details the reader is referred to the pioneering work of John Butcher [But-64, But-87]. Additional information on these aspects can also be obtained from the comprehensive paper by Enright et al. [Enr-94] and John Lambert's textbook [Lam-91].

1.3.1 RK Coefficients

The general scheme for the Runge–Kutta methods is given in eqns. (1.23) and (1.24), and the coefficients $a_{i,j}$ and b_j are given in the Butcher tableaux that follow for RK-1 to RK-4.

Runge–Kutta order 1

From Table 1.5, we see that the forward Euler scheme proceeds as follows:

$$k_1 = f(t_n, y_n), \tag{1.27}$$

with the next step given by

$$y_{n+1} = y_n + h k_1. \tag{1.28}$$

Runge–Kutta order 2

From Table 1.6 we see that for the mid-point rule, the scheme proceeds as follows:

$$\begin{aligned} k_1 &= f(t_n, y_n) \\ k_2 &= f\left(t_n + \tfrac{1}{2}h, y_n + \tfrac{1}{2}h k_1\right), \end{aligned} \tag{1.29}$$

with the next step given by

$$y_{n+1} = y_n + h k_2. \tag{1.30}$$

We also see that Heun's order 2 method proceeds as follows:

$$\begin{aligned} k_1 &= f(t_n, y_n) \\ k_2 &= f(t_n + h, y_n + h k_1), \end{aligned} \tag{1.31}$$

Table 1.6. Order 2 tableaux: (left) mid-point rule, (right) Heun's method

0	0			0	0	
$\tfrac{1}{2}$	$\tfrac{1}{2}$	0		1	1	0
	0	1			$\tfrac{1}{2}$	$\tfrac{1}{2}$

Table 1.7. Order 3 tableaux: (left) Kutta's scheme, (right) Heun's scheme

0				0			
$\frac{1}{2}$	$\frac{1}{2}$	0		$\frac{1}{3}$	$\frac{1}{3}$	0	
1	-1	2	0	$\frac{2}{3}$	0	$\frac{2}{3}$	0
	$\frac{1}{6}$	$\frac{2}{3}$	$\frac{1}{6}$		$\frac{1}{4}$	0	$\frac{3}{4}$

with the next step given by

$$y_{n+1} = y_n + \tfrac{1}{2}h\left(k_1 + k_2\right). \tag{1.32}$$

Runge–Kutta order 3

From Table 1.7 we see that Kutta's order 3 scheme proceeds as follows:

$$\begin{aligned} k_1 &= f\left(t_n, y_n\right), \\ k_2 &= f\left(t_n + \tfrac{1}{2}h, y_n + \tfrac{1}{2}hk_1\right), \\ k_3 &= f\left(t_n + h, y_n - hk_1 + 2hk_2\right), \end{aligned} \tag{1.33}$$

with the next step given by

$$y_{n+1} = y_n + \tfrac{1}{6}h\left(k_1 + 4k_2 + k_3\right). \tag{1.34}$$

We also see that Heun's order 3 method proceeds as follows:

$$\begin{aligned} k_1 &= f\left(t_n, y_n\right), \\ k_2 &= f\left(t_n + \tfrac{1}{3}h, y_n + \tfrac{1}{3}hk_1\right), \\ k_3 &= f\left(t_n + \tfrac{2}{3}h, y_n + \tfrac{2}{3}hk_2\right), \end{aligned} \tag{1.35}$$

with the next step given by

$$y_{n+1} = y_n + \tfrac{1}{4}h\left(k_1 + 3k_3\right). \tag{1.36}$$

Runge–Kutta order 4

From Table 1.8 we see that the order 4 scheme proceeds as follows:

$$\begin{aligned} k_1 &= f\left(t_n, y_n\right)r, \\ k_2 &= f\left(t_n + \tfrac{1}{2}h, y_n + \tfrac{1}{2}hk_1\right), \\ k_3 &= f\left(t_n + \tfrac{1}{2}h, y_n + \tfrac{1}{2}hk_2\right)r, \\ k_4 &= f\left(t_n + h, y_n + hk_3\right), \end{aligned} \tag{1.37}$$

with the next step given by

$$y_{n+1} = y_n + \tfrac{1}{6}h\left(k_1 + 2k_2 + 2k_3 + k_4\right). \tag{1.38}$$

The RK-4 method is one of the most popular general-purpose integrators as it is simple to implement and has good stability characteristics (see Chapter 2), and it generally gives good results for many problems. However, it is a fixed-step explicit integrator and therefore is not suitable for stiff problems. Also, the basic algorithm does not include error control, and therefore a suitable step size usually has to be determined by trial and error.

Table 1.8. The classic order 4 tableau

0	0			
$\frac{1}{2}$	$\frac{1}{2}$	0		
$\frac{1}{2}$	0	$\frac{1}{2}$	0	
1	0	0	1	0
	$\frac{1}{6}$	$\frac{1}{3}$	$\frac{1}{3}$	$\frac{1}{6}$

Consider the problem of eqn. (1.9), which we repeat for convenience:

$$\frac{dy}{dx} + y = 1 \quad y = y(x), \; y(0) = 0. \tag{1.39}$$

The results of solving this equation using the RK-4 method are shown in Fig. 1.7 for three different step sizes 0.1, 0.01, and 0.001. The steady state error is the result of the cumulative effect of the truncation errors that arise at each step. They reduce as the step size reduces, as expected (see discussion in Section 1.5). A variable step size algorithm with error control would improve matters as it would automatically reduce the step size for steep solution gradients and increase the step size as the gradient reduces. Thus, this approach gives the analyst more control over the quality of the simulation by being able to set available parameters, such as the maximum permitted error. Variable step size schemes also usually result in a more efficient overall computation owing to large

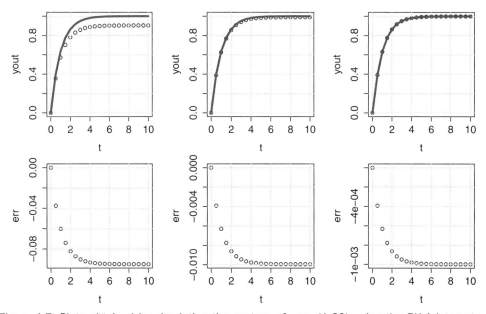

Figure 1.7. Plots obtained by simulating the system of eqn. (1.39) using the RK-4 integrator. The top row of plots show the analytical solution (solid line) with the simulation result at 21 equidistantly spaced times superimposed (circles). The bottom row shows the corresponding simulation errors. The results are shown left-to-right for step sizes $h = 0.1$, 0.01, and 0.001, respectively.

step sizes being possible as equilibrium conditions approach. This aspect is discussed in Section 1.3.2.

The R code that performs the simulation and produces the plots shown in Fig. 1.7 is included in Listing 1.3.

```
# File: main_rk4Test.R
cat("\014") # Clear console
rm(list = ls(all = TRUE)) # Delete workspace
#
require(compiler); enableJIT(3)
#
# 4th order Runge-Kutta, returns value of next step
rk4 <- function(U0,t,h,f){
  k1 <- f(t,     U0)
  k2 <- f(t+h/2, U0 + h/2*k1)
  k3 <- f(t+h/2, U0 + h/2*k2)
  k4 <- f(t+h,   U0 + h*k3)
  u  <- U0 + h/6*(k1+2*k2+2*k3+k4)
}
# derivative calculation - returns the derivative value at time=t
deriv_rk4Test <- function(t, u, parms) {
  ncall <<- ncall +1
  ut <- exp(-t)
  return(ut)
}
#
ptm0 <- proc.time()
print(Sys.time())
# Initial condition
y0 <- 0
#
####################################################
# Set simulation times
####################################################
nout=21          # No of output values for plotting
t0 <- 0          # start time
tf <- 10         # simulation end time
t <-rep(0,nout)  # vector of output times
####################################################
# ODE integration
####################################################
ncall <<- 0 # derivative calls (global)
h    <- c(0.1,0.01,0.001) # time step
nh   <- length(h)
#
par(mfcol=c(2,3),mar=(c(4,4,1,2)+0.1)) # define plot array
for(ih in 1:nh){ # set different step size parameters
  nsteps <- tf/(h[ih]*(nout-1))
```

```
tt        <- t0 # simulation start time
t[1]      <- t0 # first output time
yout      <- rep(0,nout); yout[1] <- y0
y1        <- y0 # starting point
for(iout in 2:nout){ # run simulation for ne step size
  # calculate next output
  # Take nsteps in time to next output
  if(iout <= nout){
    for(it in 1:nsteps){
      yn <- y1  # Advance y
      tt <- tt+h[ih] # Advance time
      y1 <- rk4(yn,tt,h[ih],deriv_rk4Test)
    }
  }
  t[iout] <- tt
  yout[iout] <- y1
  cat(sprintf("\n t = %7.4f, y = %7.4f",t[iout],yout[iout]))
  # Next output
}
ya <- 1-exp(-t)
err <- yout - ya
plot(t,yout,ylim=c(0,1));grid(lty=1)
lines(t,ya, col="red",lwd=3)
plot(t,err); grid(lty=1)
cat(sprintf("\nmax error = %e\n",max(abs(err))))
}
#######################################################
tFinish <- proc.time()-ptm0
cat(sprintf("\nCalculation time: %f\n",tFinish[3]))
```

Listing 1.3. File: `main_RK4Test.R`—Code to simulate system of eqn. (1.39) with different step sizes and to plot results

1.3.2 Variable Step Size Methods

By monitoring the error as integration proceeds, we are able to adjust the step size in order to keep the magnitude of the error within a desired bound. We will consider two methods that can be used effectively to monitor and thus control the error: *Richardson extrapolation* and *embedded solutions*.

1.3.2.1 Richardson Extrapolation

Richardson extrapolation, also known as the *deferred approach to the limit*, can be used on any numerical method of order p having a truncation error τ that takes the form

$$\begin{aligned}\tau &= c_{p+1}h^{p+1} + c_{p+2}h^{p+2} + c_{p+3}h^{p+3} + \cdots \\ &= c_{p+1}h^{p+1} + \mathcal{O}\left(h^{p+2}\right),\end{aligned} \quad (1.40)$$

where c_{p+1}, and so on, are constants.

This method is not very efficient and was used to control truncation error in integration schemes prior to the development of embedded solution methods. Basically, it involves comparing two values of the dependent variable computed from

1. a single-step solution based on a step size of $2h$
2. a solution consisting of two consecutive half-steps based on a step size of h

The difference between the two solutions can be used as an estimate of the local truncation error.

From Section 1.5 we see that in a Taylor series expansion for a single step of a p-order method the terms up to and including those of h^p are zero and the remaining terms are equal to the truncation error, τ, of the step. For the two-step solution we perform step 1 by $y(t_0) \to y(t_1)$ then step 2 by $y(t_1) \to y_2$, where $t_1 = t_0 + h$ and y_2 is our two-step integrated estimate of $y(t_0 + 2h)$. Therefore, assuming that the *localizing assumption* holds (see Section 1.5) and that there is no round-off error, we have

$$y(t_0 + 2h) - y_2 = \tau = \frac{h^{p+1}}{(p+1)!} y^{(p+1)} + \mathcal{O}\left(h^{p+2}\right)$$
$$= c_{p+1} h^{p+1} + \mathcal{O}\left(h^{p+2}\right), \tag{1.41}$$

where $y(t_0 + 2h)$ represents the real solution. Similarly, for the single-step solution $y(t_0) \to \hat{y}_2$, we have a truncation error given by

$$y(t_0 + 2h) - \hat{y}_2 = \hat{\tau} = \frac{(2h)^{p+1}}{(p+1)!} y^{(p+1)} + \mathcal{O}\left(h^{p+2}\right)$$
$$= 2^{p+1} c_{p+1} h^{p+1} + \mathcal{O}\left(h^{p+2}\right), \tag{1.42}$$

where $\hat{y}_2$ represents the single-step integrated estimate of the real solution at $t = t_0 + 2h$. On subtracting eqn. (1.41) from (1.42) we obtain an estimation of the local truncation error, ϵ, for the method, that is,

$$\epsilon = \frac{y_2 - \hat{y}_2}{2^{p+1} - 1} = c_{p+1} h^{p+1} + \mathcal{O}\left(h^{p+2}\right), \tag{1.43}$$

which means that the estimated local truncation error, ϵ, is proportional to h^{p+1} and is therefore of the same order of magnitude as defined for the p-order method. Thus, ϵ provides an estimate of the real local error, and by comparing it to the required tolerance, it can be used in a scheme to control error by appropriately indicating when the step size should be reduced or increased. First we determine the approximate rate of change of truncation error with step size by simply dividing the truncation error ϵ by the current time step h to get

$$r = \frac{\epsilon}{h}. \tag{1.44}$$

A new step size, h_{new}, is then calculated from

$$h_{new} = \eta \frac{\text{TOL}}{r}$$
$$\therefore h_{new} = \eta \frac{\text{TOL}}{\epsilon} h,$$

where η represents a *safety factor*, typically around 0.8 or 0.9, and TOL is the *tolerance or maximum allowable error*. The safety factor is included because the error estimate is not exact. If ϵ is smaller than TOL, the current step is accepted and the process continues with the new (larger) step size h_{new}. However, if ϵ is larger than TOL, the current step is rejected and the process is repeated with the new (smaller) step size h_{new} until $\epsilon <$ TOL.

However, although this method does work well in practice, it is rarely used in general-purpose integrators as it overestimates the error and is inefficient due to the number of stages that need to be evaluated for each step.

Richardson extrapolation can also be used to achieve a better approximation to a numerical value. For example, if we subtract eqn. (1.42) from eqn. (1.41) multiplied by 2^{p+1}, we obtain

$$\left(2^{p+1} - 1\right) y \left(t_0 + 2h\right) = 2^{p+1} y_2 - \hat{y}_2 + \mathcal{O}\left(h^{p+2}\right). \quad (1.45)$$

Therefore, on rearranging, we obtain y_2^*, an improved estimate for y_2,

$$y_2^* = \frac{2^{p+1} y_2 - \hat{y}_2}{2^{p+1} - 1} + \mathcal{O}\left(h^{p+2}\right), \quad (1.46)$$

which has order of accuracy $\mathcal{O}(h^{p+1})$, that is, 1 order higher than the order of accuracy $\mathcal{O}(h^p)$, of both y_2 and $\hat{y}_2$.

1.3.2.2 Embedded Solutions

Fehlberg [Feh-69a] reported a six-stage, fifth-order method that has a fourth-order method *embedded* within the scheme. The effect of this is that another combination of the six terms results in the fourth-order method. Thus, for six k_i evaluations, we obtain two values for the next step, that is, $\hat{y}_{n+1}$ for the fifth-order method and y_{n+1} for the fourth-order method (following Lambert [Lam-91], it is $\hat{y}_{n+1}$ that carries the solution of order $p+1$ and y_{n+1} the solution of order p). The importance of embedded solutions is that we have an efficient calculation for two values of the dependent variable that can be used to arrive at an estimate of the local error. Thus, for a p-order, s-stage scheme, the two *calculated* values for y_{n+1} will be equal to

$$y_{n+1} = y_n + h \sum_{i=1}^{s} b_i k_i + \mathcal{O}\left(h^p\right)$$

$$\hat{y}_{n+1} = y_n + h \sum_{i=1}^{s} \hat{b}_i k_i + \mathcal{O}\left(h^{p+1}\right), \quad (1.47)$$

and the error estimate at t_{n+1} is given by

$$\epsilon_{n+1} = \hat{y}_{n+1} - y_{n+1}$$
$$= h \sum_{i=1}^{s} \left(b_i - \hat{b}_i\right) k_i + \left(\mathcal{O}\left(h^{p+1}\right) - \mathcal{O}\left(h^p\right)\right) \quad (1.48)$$
$$\approx \mathcal{O}\left(h^p\right).$$

Note that the k_i are the same for the lower- and higher-order methods, that is,

$$k_i = f(t_n + hc_i, y_n + a_{i1}k_1 + a_{i2}k_2 + \cdots + a_{ii-1}k_{i-1}), \qquad (1.49)$$

which is the basis for an embedded method (i.e., the k_i do not have to be calculated for each order).

Recall that the order of a particular method is equal to 1 less than the order of the local truncation error.

If we now consider steps of different size, say, h_0 and h_1, having corresponding errors ϵ_0 and ϵ_1, they will have the similar relationships of the form $\epsilon_0 = ch_0^p$ and $\epsilon_1 = ch_1^p$, where c is unknown but considered approximately constant for step sizes h_0 and h_1 of similar magnitude. If we assume that $h > 0$ and neglect higher-order terms, then we can use the following approach to estimate an improved step size:

$$\frac{\epsilon_0}{\epsilon_{n1}} = \left(\frac{h_0}{h_1}\right)^p$$
$$\Downarrow \qquad\qquad (1.50)$$
$$h_0 = \eta h_1 \left|\frac{\epsilon_0}{\epsilon_1}\right|^{\frac{1}{p}},$$

where we have introduced η, a safety factor typically set between 0.8 and 0.9, because our error estimate is not exact. Therefore, if h_1 and ϵ_1 represent the current step size and current tolerance, and ϵ_0 represents a specified error tolerance, then h_0 is an estimate of the desired step size. Thus, by adjusting the step size according to whether the error is too high or too low, the integration process can proceed as quickly as the error will permit. The integration generally proceeds on the basis of the higher-order scheme y_{n+1}, which has usually been error tuned, rather than $\hat{y}_{n+1}$, although this does not apply in all cases (see Fehlberg method, discussed later).

R code for automatically adjusting the step size of a Runge–Kutta method is shown in Listing 1.4.

```
# File: RKvarStep.R
RKvarStep <- function(method,derivFun,y0,tspan,hmin,hmax,tol,maxEvals)
{
    #----------------------------------------------------------------
    # USAGE:
    #   c(nout,tout1,yout1) <- RKvarstep(method,deriv,y0,tspan,hmin,hmax,
    #                                    tol,maxEvals)
    #----------------------------------------------------------------
    # DESCRIPTION:
    #     Runge-Kutta embedded solution methods.
    #     Variable step size control to solve ODE of N dimensions:
    #         dx(t)/dt <- f(t,x(t))
    #     over the time interval t[t0:tf].
    #----------------------------------------------------------------
    # INPUTS:
    #     derivFun <- derivative function of the form:
    #                 function dy/dt <- f(t,y)
```

```
#       y0         <- N x 1 initial conditon vector, y <- y(t0)
#       tspan      <- M x 1 vector of times that solutions required.
#                     If just start and end times included, i.e. t0:tf
#                     then all solution points are included.
#       hmin       <- size of minimum step
#       hmax       <- size of maximum step
#       tol        <- upper bound on local truncation error
#       maxEvals   <- maximum number of scalar function evaluations
#-----------------------------------------------------------------
# OUTPUTS:
#       yout       <- M x N vector containing estimate of solution:
#                     y <- y(t), t <- tspan
#       tout       <- M x 1 vector of solution times
#-----------------------------------------------------------------
# FUNCTIONS CALLED:
#       RKdorpri54()
#       RKdorpri87()
#       RKfehlberg45()
#       RKcashKarp54()
#       RK23
#-----------------------------------------------------------------
# Initialize
sf      <- 0.9           # safety factor
M       <- length(tspan) # size of dependent variable vector
N       <- length(y0)
y1      <- y0            # initialise starting point
err0    <- tol           # initialise desired error
h       <- hmin          # initialise stepsize
stop    <- 0             # set stop flag to zero (GO)
#
if(method == 1){         # dorpri54
  integType <<- "RKdopri54"
  exponent <- 1/5
}else if(method == 2){   # dorpri87
  integType <<- "RKdopri87"
  exponent <- 1/8
}else if (method == 3){  # felberg45
  integType <<- "RKfehlberg45"
  exponent <- 1/5 # change to 1/4 for classic RKF45
}else if (method == 4){  # cashKarp
  integType <<- "RKcashKarp54"
  exponent <- 1/5
}else if (method == 5){  # rk23
  integType <<- "RK23"
  exponent <- 1/3
}
```

```
    y    <- y0
    err <- 0
    if (M==2){           # full output
      outFlg <- 1
      M1 <- maxEvals
    }else{               # specified output
      outFlg <- 0
      M1 <- M
    }
    tsi <- 0
    # Preallocation of arrays
    yout <- matrix(0,M1,N)
    tout <- rep(0,M1)
    # start main calculation
    for(i in 1:(M-1)){
      t  <- tspan[i]
      tf <- tspan[i+1]
      yout[i,] <- y
      tout[i]  <- t
      while(t < tf && stop == 0){
        err1   <- tol+1
        while(err1 > err0 && stop == 0){
          if (t+h > tf){
            h <- tf - t # Don't step past tf
          }
          integCommand <- sprintf("%s(y1,t,(t+h),derivFun)",integType)

          th <- t+ h
          # out <- eval(integType,c(y1,t,th,derivFun))
          out <- eval(parse(text=integCommand))
          #
          y[1:N]  <- out[1:N]
          err1    <- out[N+1]
          #
          if (ncall > maxEvals){
            stop <- 1
            cat(sprintf("\nINTEGRATION STOPPED AT t=%f, i=%d, err1=%g: Maximum
                        derivative evaluations exceeded!", t,i,err1))
            cat(sprintf("\n nevals <- %f, maxEvals <- %f \n",ncall, maxEvals));
          }
          err0 <- max(max(abs(y)),1)*tol
          # check if error acceptable - if so do update
          if (err1 <= err0){
            # set time step statistics
            # update time
            t <- t + h
```

```
      # update solution
      y1 <- y
      if (outFlg == 1){
        i<- i+1
        yout[i,] <- y
        tout[i]  <- t
      }
    }
  # desired step size
    hd <- sf*h*((abs(err0)/(abs(err1)+.Machine$double.eps))^exponent);
    if (hd < hmin){ # check limits
      h <- hmin
      stop <- 1
      cat(sprintf("\nINTEGRATION STOPPED AT t=%f, i=%d, err1=%g: Step size
              too small!",t,i,err1))
      cat(sprintf("\n  hd <- %g, hmin <- %g \n",hd,hmin))
    }else if (hd > hmax){
      h <- hmax
    }else {
      h <- hd
    }
  } # end while loop
  err <- err1
 } # end while loop
} # end for loop

if (outFlg == 0){
  # output only at specified times: N>2
  yout[M,] <- y
  tout[M] <- t
  nout    <- M
  tout1   <- tout[1:M]
  yout1   <- yout[1:M,]
}else{
  nout   <- i
  tout1 <- tout[1:nout]
  yout1 <- yout[1:nout,]
}
# Return solution
return(c(nout,tout1,yout1))
}
```

Listing 1.4. File: RKvarStep.R—Code for automatic step size control of Runge–Kutta method

Table 1.9. Modified Butcher tableau

c	A
	b^T
	$\hat{b}^T$
	ϵ^T

To present the coefficients for schemes that estimate local error, the Butcher tableau is modified to that shown in Table 1.9.

Runge–Kutta order 2/3

This method is used in the R ode23 integrator. The Butcher tableau is given in Table 1.10.

Where $e^T = \hat{b}^T - b^T$ and neglecting higher-order terms, the new *second*-order value for the dependent variable y and the *third*-order error estimate become

$$y_{n+1} = y_n + h \sum_{i=1}^{s} b_i k_i$$
$$\epsilon = h \sum_{i=1}^{s} \left(\hat{b}_i - b_i \right) k_i, \qquad (1.51)$$

and the k_i are given by eqn. (1.49).

R computer code for this method is similar to that given in Listing 1.5 for the RK–Felberg scheme and is included with the downloads for this book.

Runge–Kutta order 4/5 (Fehlberg) [Feh-69a])

The constants of the original 4/5 order (fourth-order with fifth-order error control) method reported by Fehlberg are given in Table 1.11.

Table 1.10. The Bogacki–Shampine order 2/3 tableau pair [Bog-89a]

0	0			
$\frac{1}{2}$	$\frac{1}{2}$	0		
$\frac{3}{4}$	0	$\frac{3}{4}$	0	
1	$\frac{2}{9}$	$\frac{1}{3}$	$\frac{4}{9}$	0
b^T	$\frac{2}{9}$	$\frac{1}{3}$	$\frac{4}{9}$	0
$\hat{b}^T$	$\frac{7}{24}$	$\frac{1}{4}$	$\frac{1}{3}$	$\frac{1}{8}$
e^T	$\frac{-5}{72}$	$\frac{1}{12}$	$\frac{1}{9}$	$\frac{-1}{8}$

Table 1.11. Order 4/5 Fehlberg tableau pair [Feh-69b]

0	0					
$\frac{1}{4}$	$\frac{1}{4}$	0				
$\frac{3}{8}$	$\frac{3}{32}$	$\frac{9}{32}$	0			
$\frac{12}{13}$	$\frac{1932}{2197}$	$\frac{-7200}{2197}$	$\frac{7296}{2197}$	0		
1	$\frac{439}{216}$	-8	$\frac{3680}{513}$	$\frac{-845}{4104}$	0	
$\frac{1}{2}$	$\frac{-8}{27}$	2	$\frac{-3544}{2565}$	$\frac{1859}{4104}$	$\frac{-11}{40}$	0
b^T	$\frac{25}{216}$	0	$\frac{1408}{2565}$	$\frac{2197}{4104}$	$\frac{-1}{5}$	0
$\hat{b}^T$	$\frac{16}{135}$	0	$\frac{6656}{12825}$	$\frac{28561}{56430}$	$\frac{-9}{50}$	$\frac{2}{55}$
e^T	$\frac{1}{360}$	0	$\frac{-128}{4275}$	$\frac{2197}{75240}$	$\frac{1}{50}$	$\frac{2}{55}$

where $e^T = \hat{b}^T - b^T$ and neglecting higher-order terms, the new *fourth*-order value for the dependent variable y and the *fifth*-order error estimate become

$$y_{n+1} = y_n + h\sum_{i=1}^{s} b_i k_i$$

$$\epsilon = h\sum_{i=1}^{s} \left(\hat{b}_i - b_i\right) k_i, \quad (1.52)$$

and the k_i are given by eqn. (1.49).

This scheme is one of the most popular methods and can also be run as a 5/4 order (fifth-order with fourth-order error control) method where the new fifth-order value for the dependent variable y becomes

$$\hat{y}_{n+1} = y_n + h\sum_{i=1}^{s} \hat{b}_i k_i. \quad (1.53)$$

The R code for the RK–Felhberg 4/5 scheme is given in Listing 1.5. We note the following points about this code.

Use of matrix operator: The R notation K%*%b_hat means multiply the matrix variable K by the vector variable b_hat.

Change to (5/4) scheme: It is a simple matter to convert to the fifth-*order* RK–Felhberg (5/4) scheme from the *classic* (4/5) scheme specified in Listing 1.5 by replacing the lines

```
yf <- y0 + h*K%*% b
y_tmp2 <- y0 + h*K%*%b_hat
```

with

```
yf      <- y0 + h*K%*%b_hat
y_tmp2  <- y0 + h*K%*%b
```

```
# RKfehlberg45.R
RKfehlberg45 <- function(y0,t0,tf,derivFun)
{
  # ----------------------------------------------------------------
  # USAGE:
  #   c(yf,err) <- RKfehlberg45 (y0,t0,tf,deriv)
  # ----------------------------------------------------------------
  # DESCRIPTION:
  #     Fifth order, 6 stage Runge-Kutta Fehlberg method with
  #     embedded fourth oder method.
  #     Ref: Fehlberg E., (1969). "Low-order classical Runge-Kutta
  #         formulas with step size control and their application to heat
  #         transfer problems". NASA Technical Report 315, (extract
  #         published in Computing, 6, 1970, 61-71).
  #     Performs a single step solution of ODE of N dimensions:
  #         dx(t)/dt <- f(t,x(t))
  #     from t<-t0 to t<-tf.
  # ----------------------------------------------------------------
  # INPUTS:
  #     y0       <- Mx1 initial conditon vector, y <- y(t0)
  #     t0       <- initial time
  #     tf       <- final time
  #     derivFun <- derivativef function of the form:
  #                 function dy/dt <- f(t,y)
  # ----------------------------------------------------------------
  # OUTPUTS:
  #     y        <- Nx1 vector containing estimate of solution:
  #                 y <- y(tf)
  #     err      <- max local truncation error from t<-t0 to t<-tf.
  # ----------------------------------------------------------------
  # FUNCTIONS CALLED:
  #     none
  # ----------------------------------------------------------------
  # PARAMETERS:
  #         Felberg 4/5 order RK constants
  m   <- 6
  a   <- c( 0, 0,     0, 0, 0, 0,
            1/4, 0,   0, 0, 0, 0,
            3/32, 9/32, 0, 0, 0, 0,
```

```
             1932/2197, -7200/2197, 7296/2197,    0,   0,   0,
             439/216,   -8,         3680/513,  -845/4104, 0, 0,
             -8/27,      2,        -3544/2565, 1859/4104, -11/40, 0)
  A     <- matrix(a,nrow=m,ncol=m,byrow=TRUE)
  b     <- c(25/216,  0, 1408/2565, 2197/4104,  -1/5, 0)
  b_hat <- c(16/135,  0, 6656/12825, 28561/56430, -9/50, 2/55)
  c     <- c(0, 1/4, 3/8, 12/13, 1, 1/2)
  # ---------------------------------------------------------------
  h     <- tf - t0
  M     <- length (y0)
  K     <- matrix(0,nrow=M,ncol=m)
  y_tmp <- rep(0,M)
  # Evaluate intermediate derivatives
  K[,1] <- derivFun(t0,y0)
  #
  y_tmp <- y0+ h*A[2,1]*K[,1]
  K[,2] <- derivFun(t0+h*c[2],y_tmp)
  #
  y_tmp <- y0+ h*(A[3,1]*K[,1]+A[3,2]*K[,2]);
  K[,3] <- derivFun(t0+h*c[3],y_tmp)
  #
  y_tmp <- y0+ h*(A[4,1]*K[,1]+A[4,2]*K[,2]+A[4,3]*K[,3]);
  K[,4] <- derivFun(t0+h*c[4],y_tmp)
  #
  y_tmp <- y0+ h*(A[5,1]*K[,1]+A[5,2]*K[,2]+A[5,3]*K[,3] +
                  A[5,4]*K[,4])
  K[,5] <- derivFun(t0+h*c[5],y_tmp)
  #
  y_tmp <- y0+ h*(A[6,1]*K[,1]+A[6,2]*K[,2]+A[6,3]*K[,3] +
                  A[6,4]*K[,4]+A[6,5]*K[,5])
  K[,6] <- derivFun(t0+h*c[6],y_tmp)
  # Compute yf
  yf <- y0 + h*K%*%b
  # Calculate intermediate variable y_tmp2
  y_tmp2 <- y0 + h*K%*%b_hat
  # Calculate step error O(h^5)
  err <- max(abs(yf-y_tmp2))
  return(c(yf,err))
}
```

Listing 1.5. File: RKfehlberg45.R—Code for the Fehlberg (4/5) Runge–Kutta method. Note the listing is for a stand-alone function. It is also included, along with other embedded RK integrators, in the RK library file: RKembedded_LIB.R.

Runge–Kutta order 5/4 (Dormand–Prince) [Dor-80]

The Dormand–Prince method is a seven-stage scheme but effectively uses only six function evaluations. This is because it has a numerical form whereby the the last row of A is the same as the vector b^T and therefore can be used in calculating the next time step. In other words, we have the equation for k_7

$$k_7 = f\left(t = t_n + h, y = y_n + h \sum_{j=1}^{6} a_{7j} k_j^{n+1}\right), \quad (1.54)$$

which uses an identical value for y to that used to evaluate the equation for y_{n+1}:

$$y_{n+1} = y_0 + f\left(t = t_n + h, y = y_n + h \sum_{j=1}^{6} b_j k_j^n\right). \quad (1.55)$$

The Dormand–Prince method is currently the method used in the R integrator ode45, and its coefficients are shown in Table 1.12.

Table 1.12. Order 5/4 Dormand–Prince tableau pair [Lam-91]

0	0						
$\frac{1}{5}$	$\frac{1}{5}$	0					
$\frac{3}{10}$	$\frac{3}{40}$	$\frac{9}{40}$	0				
$\frac{4}{5}$	$\frac{44}{45}$	$\frac{-56}{15}$	$\frac{32}{9}$	0			
$\frac{8}{9}$	$\frac{19372}{6561}$	$\frac{-25360}{2187}$	$\frac{64448}{6561}$	$\frac{-212}{729}$	0		
1	$\frac{9017}{3168}$	$\frac{-355}{33}$	$\frac{46732}{5247}$	$\frac{49}{176}$	$\frac{-5103}{18656}$	0	
1	$\frac{35}{384}$	0	$\frac{500}{1113}$	$\frac{125}{192}$	$\frac{-2187}{6784}$	$\frac{11}{84}$	0
b^T	$\frac{35}{384}$	0	$\frac{500}{1113}$	$\frac{125}{192}$	$\frac{-2187}{6784}$	$\frac{11}{84}$	0
$\hat{b}^T$	$\frac{5179}{57600}$	0	$\frac{7571}{16695}$	$\frac{393}{640}$	$\frac{-92097}{339200}$	$\frac{187}{2100}$	$\frac{1}{40}$
e^T	$\frac{71}{57600}$	0	$\frac{-71}{16695}$	$\frac{71}{1920}$	$\frac{-17253}{339200}$	$\frac{22}{525}$	$\frac{-1}{40}$

where $e^T = \hat{b}^T - b^T$ and neglecting higher-order terms, the new *fifth*-order value for the dependent variable y and the *fifth*-order error estimate become

$$\hat{y}_{n+1} = y_n + h \sum_{i=1}^{s} \hat{b}_i k_i$$

$$\epsilon = h \sum_{i=1}^{s} \left(\hat{b}_i - b_i\right) k_i, \quad (1.56)$$

where the k_i are given by eqn. (1.49).

Shortly after Dormand and Prince published their 5/4 order method, they also published a successful 13-stage eighth-order Runge–Kutta method with an embedded seventh-order method [Pri-81].

R computer code for the 5/4 method is similar to that given in Listing 1.5 for the RK–Felberg scheme. However, it is not presented here to save space but is included with the downloads. In addition, R code for the Dormand–Prince 8/7 method is also included.

Runge–Kutta order 5/4 (Cash–Karp) [Cas-90]
The Cash–Karp method is a fifth-order, six-stage Runge–Kutta scheme with embedded fourth-order method. Its coefficients are shown in Table 1.13.

Table 1.13. Order 5/4 Cash-Karp 5/4 tableau pair [Cas-90]

0	0					
$\frac{1}{5}$	$\frac{1}{5}$	0				
$\frac{3}{10}$	$\frac{3}{40}$	$\frac{9}{40}$	0			
$\frac{3}{5}$	$\frac{3}{10}$	$\frac{-9}{10}$	$\frac{6}{5}$	0		
1	$\frac{-11}{54}$	$\frac{5}{2}$	$\frac{-70}{27}$	$\frac{35}{27}$	0	
$\frac{7}{8}$	$\frac{1631}{55296}$	$\frac{175}{512}$	$\frac{575}{13824}$	$\frac{44275}{110592}$	$\frac{253}{4096}$	0
b^T	$\frac{2825}{27648}$	0	$\frac{18575}{48384}$	$\frac{13525}{55296}$	$\frac{277}{14336}$	$\frac{1}{4}$
$\hat{b}^T$	$\frac{37}{378}$	0	$\frac{250}{621}$	$\frac{125}{594}$	0	$\frac{512}{1771}$
e^T	$\frac{-277}{64512}$	0	$\frac{6925}{370944}$	$\frac{-6925}{202752}$	$\frac{-277}{14336}$	$\frac{277}{7084}$

where $e^T = \hat{b}^T - b^T$ and neglecting higher-order terms, the new *fifth*-order value for the dependent variable y and the *fourth*-order error estimate become, respectively,

$$\hat{y}_{n+1} = y_n + h \sum_{i=1}^{s} \hat{b}_i k_i$$
$$\epsilon = h \sum_{i=1}^{s} \left(\hat{b}_i - b_i\right) k_i, \qquad (1.57)$$

and the k_i are given by eqn. (1.49).

R computer code for this method is similar to that given in Listing 1.5 for the RK–Felberg scheme and is included with the downloads.

Variable step size test example
To illustrate the performance of the variable step code shown in Listing 1.4, we choose a reasonably complex function to integrate in order to test the ability of the variable step error control scheme to keep the error within the specified tolerance. It consists of the sum of two bell curves displaced from one another and is defined by

$$\frac{dy(t)}{dt} = f(t), \ y(t=0) = 0, \ t \geq 0$$
$$f(t) = 0.5e^{-(t-\mu_1)^2} + e^{-(t-\mu_2)^2}, \qquad (1.58)$$

with parameter values $\mu_1 = 7$ and $\mu_2 = 4$. The analytical solution is given by

$$y(t) = 0.4431134627 \operatorname{erf}(t - 7.0) + 0.8862269255 \operatorname{erf}(t - 4.0) + K, \qquad (1.59)$$

where the constant of integration $K = 1.329340375$.

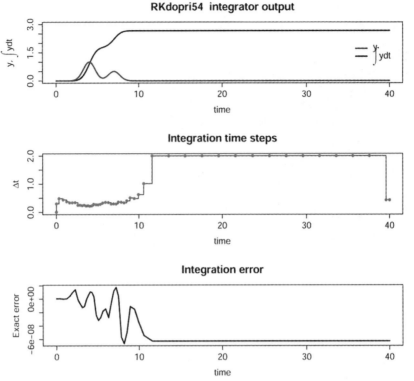

Figure 1.8. Results of the Runge–Kutta Dormand–Prince 5/4 variable step size integrator used to solve $dy(t)/dt = 0.5e^{-(t-\mu_1)^2} + e^{-(t-\mu_2)^2}$, $y(0) = 0$, $\mu_1 = 7$, and $\mu_2 = 4$. The integration solution times are indicated in the middle plot by small dots. As the solution becomes smooth, the integration time steps are increased automatically, while maintaining the error below the specified tolerance.

Note that the preceding integration is nontrivial and that attempts to obtain the analytical solution using the *computer algebra* packages Ryacas and rSymPy in R both failed. However, the commercial package *Maple* does find the correct solution, and the code is included with the downloads.

We select the Runge–Kutta Dormand–Prince 5/4 integration method and set the error tolerance to 10^{-7}. The result of the variable step integration is shown in Fig. 1.8 where it is apparent that the step size has been adjusted throughout the integration period. Smaller steps were taken where the derivative changes rapidly, then increasingly larger steps were taken as the derivative slope reduces. For this problem the integrator achieved a maximum error of 6.6262×10^{-8}, well within the required tolerance. However, for other nonlinear problems, this level of accuracy cannot be guaranteed, and some trial-and-error adjustment of the tolerance value may be necessary. The solution required 364 calls to the derivative function.

The variable step size algorithm can also be set to output results at specified intervals (see Fig. 1.9).

The R computer code for the main routine is shown in Listing 1.6. We note the following points about this program.

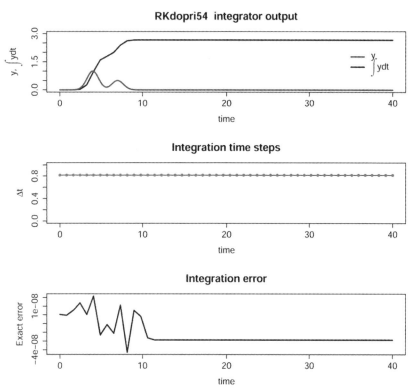

Figure 1.9. Results of the Runge–Kutta Dormand–Prince 5/4 variable step size integrator used to solve $dy(t)/dt = 0.5e^{-(t-\mu_1)^2} + e^{-(t-\mu_2)^2}$, $y(0) = 0$, $\mu_1 = 7$, and $\mu_2 = 4$. The *fixed* integration solution times are indicated in the middle plot by small dots. The solution outputs are provided at times specified by the variable tspan <- seq(t0,tf,length=nout) even though the actual integration steps change automatically within the fixed intervals to maintain the error below the specified tolerance. Note that the solution curve is not as smooth as in Fig. 1.8, due to the plot using fewer solution points.

Other integrator schemes: The test can be run with schemes other than the Dorman–Prince 5/4 method by changing the method from 1 to a value between 2 and 5 in the following code:

```
method <- 1  # 1 = dorpri54
             # 2 = dorpri87
             # 3 = felberg45
             # 4 = cashKarp54
             # 5 = RK23
```

Functions for all the preceding embedded RK methods are included in the embedded RK library contained in file RKembedded_LIB.R.

Solution output: The solution output times are set by the vector tspan. If the length of tspan is equal to 2, then all solutions are output. If tspan has length greater than 2, then solutions are only output at the times specified by tspan. Figure 1.8 shows the situation where they were set to be provided after each integration step by tspan <- c(t0,tf). In contrast, for Fig. 1.9, they were set to be provided at set intervals by tspan <- seq(t0,tf,length=nout).

Results for all schemes: The integrator performances for each method with the output set to occur after each integrator step are

```
dorpri54:    max err: 6.6262e-08, steps:  448
dorpri87:    max err: 1.8982e-09, steps:  650
felberg45:   max err: 6.4642e-07, steps:  465
cashKarp54:  max err: 1.2463e-07, steps:  438
RK23 :       max err: 1.1390e-05, steps: 1728
```

For the Dormand–Prince order 8/7 scheme, see downloads.

Plot scale: To ensure a scale that would clearly illustrate the evolution of time step changes in the middle plot of Fig. 1.8, the maximum time step value was set to hmax <- 2. This could actually be set to a much larger value without compromising accuracy.

```
# File: main_RKBellTest.R
rm(list = ls(all = TRUE)) # Delete workspace
library("rgl")
library("pracma") # for erf()
#
# Access routines
source("../integSRC/RKvarStep.R")
source("../integSRC/RKembedded_Lib.R") # load library of embedded RK methods
source("bellDeriv.R")
#
ptm = proc.time()
# Set parameters
ncall <<- 0 # global
nt      <- 50
t0      <- 0
tf      <- 40
y0      <- c(0.0) # c(0.0,0.0)
N       <- length(y0)
# Set integration method
method <- 1 # 1 <- dorpri54
            # 2 <- dorpri87
            # 3 <- felberg45
            # 4 <- cashKarp45
            # 5 <- RK23
            # 6 <- SHK_est
# Set solution times
```

```
# Set of outputs at fixed intervals
# tspan <- seq(t0,tf,length=nt)
#
# Outputs determined by integrator
tspan <- c(t0,tf)
# Set integration parameters
hmin<-1.0e-14; hmax <- 2
# hmin<- .Machine$double.eps*2; hmax <- 100
tol <-1.0e-07; mode <- 1; maxEvals<-5*10^3
hstart <- hmin
# calculate integration solution
out <- RKvarStep(method, bellDeriv, y0, tspan, hmin, hmax, tol, maxEvals);
#
nout <- out[1]
tout <- out[2:(nout+1)]
int_y <- out[nout+2:2:(nout+1)]
#
# extract step size information
#
stepInfo <- rep(0,nout)
for (i in 1:(nout-1)){
  stepInfo[i] <- tout[i+1]-tout[i]
}
stepInfo[nout] <- stepInfo[nout-1]

# Analytical solution of integral of y
int_ya <- 0.4431134627*erf(tout-mu1) + 0.8862269255*erf(tout-mu2) + 1.329340375;
# Error
err <- int_ya-int_y
# Print results
cat(sprintf("maximum integration error .....: %5.4 e\n",max (abs (err ))))
cat(sprintf("Number of derivative calls ....: %d\n",ncall));
#
ya <- bellDeriv(tout,1) #rep(0,m));
#Plot 2D images
op <- par(mfrow = c(3, 1))
par(mar = c(5,7,4,2))
plot(tout, int_y, type="l", lwd=2, lty=1, col = "black",
    xlim=c(0,tf),ylim=c(0,3),
    xlab="time", ylab=expression(paste(y,", ",integral(y*dt))),
    main=paste(integType," integrator output"))
lines(tout, ya, lwd=2, lty=1, col = "blue")
legend("right",lwd=c(2,2),lty=c(1,1),bty="n",
      legend=c(expression(paste(y)),
              expression(paste(integral(y*dt), "        "))),
      col=c("blue","black"))
```

```
#
plot(tout,stepInfo, type="s", lwd=1, lty=1, col = "black",
    xlab="time", ylab=expression(Delta*t));
points(tout,stepInfo, pch=16, col = "red",
    xlab="time", ylab=expression(Delta*t))
title("Integration time steps")
#
plot(tout,err, type="l", lwd=2, lty=1, col = "black",
    xlab="time", ylab="Exact error");
title("Integration error")
#
```

Listing 1.6. File: `main_RKBellTest.R`—Main routine for the Runge–Kutta Dormand–Prince 5/4 variable step size integrator test example

1.3.3 SHK: Sommeijer, Van Der Houwen, and Kok Method

The *Sommeijer, van der Houwen, and Kok* (SHK) method [Som-94] is a Runge–Kutta-like scheme. It is a recursive method that is particularly easy to program and is suitable for problems where the *derivative is not explicitly dependent upon time*, for example, $y'(t) = t\exp(mt)$. The order s of the method can be changed by simply varying the number of iterations, and it has minimal memory storage requirements. For the ODE problem of eqn. (1.17) the SHK method provides a numerical solution of the form

$$
\begin{aligned}
k_1 &= \frac{h}{s} f(y_n), \\
k_2 &= \frac{h}{s-1} f(y_n + k_1), \\
k_3 &= \frac{h}{s-2} f(y_n + k_2), \\
&\vdots \\
y_{n+1} &= y_n + \frac{h}{1} f(y_n + k_{s-1}).
\end{aligned}
\tag{1.60}
$$

Pseudo code that evaluates y_{n+1} for the SHK method is as follows:

⋮

```
k = 0
s = order
for (j=s to 1, step=-1)
    k = h*f(y(n) + k)/j
end
y(n+1) = y(n) + k
```

⋮

From an examination of eqn. (1.60) we see that the equation form is similar to Runge–Kutta methods and that it generates a s-order Taylor series about y_n. In fact, this is confirmed when we investigate the stability of the SHK method, as we see that it has exactly the same characteristics as the equivalent order RK method. This aspect is discussed further in Chapter 2.

R computer code for the SHK method is not presented here to save space but is included with the downloads.

1.4 LINEAR MULTISTEP METHODS (LMMS)

1.4.1 General

Linear multistep (or k-step) methods (LMMs) are generally applied in situations where stability rather than accuracy is the main requirement. They are used to solve initial value problems of the form of eqn. (1.17).

LMMs generally have very good stability characteristics and are usually written in the following *standard* mathematical form:

$$\sum_{j=0}^{k} \alpha_j y_{n+1-j} = \Delta t \sum_{j=0}^{k} \beta_j f\left(t_{n+1-j}, y_{n+1-j}\right) + \tau_n, \qquad (1.61)$$

where k represents the number of past steps used in a particular scheme, n is the solution sequence number, and α_j and β_j are constants defined for a particular scheme. The variable τ_n represents the local truncation error at $t = n\Delta t$ (see Section 1.5.1).

Equation (1.61) is presented in its most general form where the number of past values of y on the left, k, is equal to the number of derivatives, f, on the right. This need not be the case, and popular versions often only use one derivative value, for example, eqn. (1.63). The method requires that $\alpha_0 \neq 0$. When $\beta_0 \neq 0$, the method is implicit; otherwise, it is explicit. Equation (1.61) can be represented in alternative equivalent forms, some of which are discussed later.

With constant step size, order, and α_j and β_j coefficients, the stability and consistency properties of LMMs can be determined from the following first and second characteristic polynomials:

$$\begin{aligned}\rho(z) &= \alpha_0 z^k + \alpha_1 z^{k-1} + \alpha_2 z^{k-2} + \cdots + \alpha_k, \quad z \in \mathbb{C}; \\ \sigma(z) &= \beta_0 z^k + \beta_1 z^{k-1} + \beta_2 z^{k-2} + \cdots + \beta_k.\end{aligned} \qquad (1.62)$$

Stability aspects will be discussed in Chapter 2.

We will be mainly concerned with only a subset of LMMs, namely, *backward differentiation formula* (BDF) and *numerical differentiation* (NDF) methods. Adams–Bashforth and Adams–Moulton will also be briefly discussed. BDFs and NDFs include one derivative term, at $t = (n+1)\Delta t$, and are therefore implicit methods. Adams–Bashforth is an explicit method as it only uses one past value of the dependent variable and one or more past derivative terms. Adams–Moulton also uses one past value of the dependent variable and one or more derivative terms. However, it uses one derivative term at $t = (n+1)\Delta t$ and is therefore an implicit method.

Table 1.14. BDF coefficients for the *standard form* of eqn. (1.63)

k	1	2	3	4	5	6
$\widetilde{\alpha}_0$	1	1	1	1	1	1
$\widetilde{\alpha}_1$	-1	$-4/3$	$-18/11$	$-48/25$	$-300/137$	$-360/147$
$\widetilde{\alpha}_2$		$1/3$	$9/11$	$36/25$	$300/137$	$450/147$
$\widetilde{\alpha}_3$			$-2/11$	$-16/25$	$-200/137$	$-400/147$
$\widetilde{\alpha}_4$				$3/25$	$75/137$	$225/147$
$\widetilde{\alpha}_5$					$-12/137$	$-72/147$
$\widetilde{\alpha}_6$						$-10/147$
$\widetilde{\beta}_0$	1	$2/3$	$6/11$	$12/25$	$60/137$	$60/147$
p	1	2	3	4	5	6
$\widetilde{C}_{p+1}$	$-1/2$	$-2/9$	$-3/22$	$-12/125$	$-10/137$	$-20/343$

1.4.2 Backward Differentiation Formulas (BDFs)

In the literature, BDFs (also known as Gear's method [Gea-71]) are described in two equivalent mathematical forms. We shall refer to the following mathematical representation as the *standard form* for a BDF of order $p = k$:

$$\sum_{i=0}^{k} \widetilde{\alpha}_i y_{n+1-i} = \Delta t \widetilde{\beta}_0 f(t_{n+1}, y_{n+1}), \qquad (1.63)$$

where $\sum_{i=0}^{k} \widetilde{\alpha}_i = 0$ and we have dropped the truncation error term. We note that the results of eqn. (1.63) are not altered if all the coefficients are divided by any positive number. Thus, we are therefore able choose an arbitrary value for any one coefficient. If we choose $\widetilde{\alpha}_0 = 1$, that is divide through by $\widetilde{\alpha}_0$, we get the standard form, and the $\widetilde{\alpha}_i$ and $\widetilde{\beta}_0$ are constants have values as defined in Table 1.14. This table also includes the error constant for each scheme, $C_{p+1} = \frac{-\widetilde{\beta}_0}{k+1}$, which is equal to the constant of the leading term of the associated truncation error, τ_k.

Alternatively, if we divide the standard form eqn. (1.63) by $\widetilde{\beta}_0$, we obtain the following equivalent mathematical form:

$$\sum_{j=0}^{k} \alpha_j y_{n+1-j} = \beta_0 \Delta t f(t_{n+1}, y_{n+1}), \qquad (1.64)$$

where $\sum_{i=0}^{k} \alpha_i = 0$ and the α_j and β_0 coefficients are given in Table 1.15. We shall refer to this mathematical representation as the *normalized form*, which corresponds to the LMM equation (1.61). Note that the error constants $\widetilde{C}_{p+1}$ have also been divided by β_0, yielding $C_{p+1} = \frac{-1}{k+1}$.

The normalized form is a convenient alternative with which to work when developing a Newton convergence routine for a BDF integrator.

Using the *normalized* BDF coefficients from Table 1.15, we obtain the following expanded BDF forms.

BDF-1

The *BDF-1* method is the same as the *backward Euler* method. For the 1D ODE problem of eqn. (1.17) and taking constants from Table 1.14, we see from eqn. (1.63) that we obtain

Table 1.15. BDF coefficients for the *normalized form* of eqn. (1.64)

k	1	2	3	4	5	6
α_0	1	3/2	11/6	25/12	137/60	147/60
α_1	−1	−2	−3	−4	−5	−6
α_2		1/2	3/2	3	5	15/2
α_3			−1/3	−4/3	−10/3	−20/3
α_4				1/4	5/4	15/4
α_5					−1/5	−6/5
α_6						1/6
β_0	1	1	1	1	1	1
p	1	2	3	4	5	6
C_{p+1}	−1/2	−1/3	−1/4	−1/5	−1/6	−1/7

the following implicit first-order BDF equation:

$$\Delta t f(t_{n+1}, y_{n+1}) = y_{n+1} - y_n. \tag{1.65}$$

BDF-2

$$\Delta t f(t_{n+1}, y_{n+1}) = \frac{3}{2} y_{n+1} - 2y_n + \frac{1}{2} y_{n-1}. \tag{1.66}$$

BDF-3

$$\Delta t f(t_{n+1}, y_{n+1}) = \frac{11}{6} y_{n+1} - 3y_n + \frac{3}{2} y_{n-1} - \frac{1}{3} y_{n-2}. \tag{1.67}$$

BDF-4

$$\Delta t f(t_{n+1}, y_{n+1}) = \frac{25}{12} y_{n+1} - 4y_n + 3y_{n-1} - \frac{4}{3} y_{n-2} + \frac{1}{4} y_{n-3}. \tag{1.68}$$

BDF-5

$$\Delta t f(t_{n+1}, y_{n+1}) = \frac{137}{60} y_{n+1} - 5y_n + 5y_{n-1} - \frac{10}{3} y_{n-2} + \frac{5}{4} y_{n-3} - \frac{1}{5} y_{n-4}. \tag{1.69}$$

In each of the preceding cases it is necessary to iterate until y_{n+1} converges to within the required tolerance. Thus, for BDF-k, the sequence becomes

$$y_{n+1}^s = \Delta t \widetilde{\beta}_0 f\left(t_{n+1}, y_{n+1}^{s-1}\right) - \sum_{i=1}^{k} \widetilde{\alpha}_i y_{n+1-i}^{s-1},$$
$$\text{error} = y_{n+1}^s - y_{n+1}^{s-1}, \tag{1.70}$$

where $s = 1, 2, \ldots,$ is the iteration sequence number and convergence is achieved when

$$|\text{error}| < TOL. \tag{1.71}$$

The next solution estimate, y_{n+1}^{s+1}, could be obtained by use of the *direct substitution* method, that is, setting $y_{n+1}^{s-1} = y_{n+1}^s$ in eqn. (1.70) and repeating the calculation. However, this would usually be a slow process, and convergence may not be achieved within a reasonable number of iterations, if at all. Therefore, a more sophisticated method that provides accelerated convergence is usually adopted (see Section 1.4.4).

For the higher-order methods, past values have to be obtained to start the solution sequence. This can be achieved by using explicit integration methods or by using lower-order BDF methods. Example code for BDF-4 is included in Listing 1.8 and illustrates a process whereby the first two steps are taken with BDF-1, the third with BDF-2, the fourth with BDF-3, and the fifth and later steps with BDF-4. Note that the BDF-n method recommended iteration starting value calculation for each new time step (after the first) uses $n+1$ past values, even though the the iterative solution process only requires n past values (see Section 1.4.4.3).

1.4.2.1 Derivation of BDF Coefficients: Lagrange Polynomial Method

To derive numerical values for the coefficients $\sigma_p = \alpha_p/\beta_0$, $(p = 0, \ldots, M)$, we start by finding a polynomial of degree $\leq M$ that interpolates y at the data points corresponding to $t_0, t_1, \cdots t_M$. A suitable polynomial is the Lagrange interpolation polynomial

$$P_M(t) = \sum_{j=0}^{M} y_{n+1-j} p_j(t). \tag{1.72}$$

This consists of a weighted sum of subpolynomials $p(t)$, defined as

$$p_j(t) = \prod_{\substack{k=0,\\k\neq j}}^{M} \left(\frac{t-t_k}{t_j-t_k}\right), \tag{1.73}$$

A Lagrange interpolation function is provided in file LagrangeInterp.R as part of the downloads.

On differentiating the polynomial of eqn. (1.72) we obtain

$$\frac{dP_M(t)}{dt} = \sum_{k=0}^{M} y_{n+1-j} \frac{dp_j(t)}{dt}, \tag{1.74}$$

where the differentiated subpolynomials become

$$\frac{dp_j(t)}{dt} = \sum_{\substack{m=0\\m\neq j}}^{M} \left\{ \frac{\prod_{\substack{k=0\\k\neq j,m}}^{M}(t-t_k)}{\prod_{\substack{k=0\\k\neq j}}^{M}(t_j-t_k)} \right\}. \tag{1.75}$$

However, we are interested in the situation where the independent variable, t, is equally spaced by an amount $\Delta t = t_n - t_{n-1}$, when the Lagrange subpolynomials become

$$\frac{dp_j(t)}{dt} = \sum_{\substack{m=0 \\ m\neq j}}^{M} \left\{ \frac{\prod_{\substack{k=0 \\ k\neq j,m}}^{M} (M-k)\Delta t}{\prod_{\substack{k=0 \\ k\neq j}}^{M} (j-k)\Delta t} \right\} = \frac{1}{\Delta t}\sum_{\substack{m=0 \\ m\neq j}}^{M} \left\{ \frac{\prod_{\substack{k=0 \\ k\neq j,m}}^{M} (M-k)}{\prod_{\substack{k=0 \\ k\neq j}}^{M} (j-k)} \right\}. \tag{1.76}$$

But, from the initial value problem statement of eqn. (1.17) and eqn. (1.74), we get

$$P'_M(t) \approx y'_{n+1} = f(t_{n+1}, y_{n+1}), \tag{1.77}$$

from which it follows that

$$\sum_{j=0}^{M} y_{n+1-j} \frac{1}{\Delta t} \sum_{\substack{m=0 \\ m\neq j}}^{M} \left\{ \frac{\prod_{\substack{k=0 \\ k\neq j,m}}^{M} (M-k)}{\prod_{\substack{k=0 \\ k\neq j}}^{M} (j-k)} \right\} = f(t_{n+1}, y_{n+1}). \tag{1.78}$$

If we multiply eqn. (1.78) by Δt, we get

$$\sum_{j=0}^{M} \alpha_{M-j} y_{n+1-j} = \Delta t f(t_{n+1}, y_{n+1}), \tag{1.79}$$

which is equivalent to eqn. (1.64), where

$$\alpha_{M-j} = \sum_{\substack{m=0 \\ m\neq j}}^{M} \left\{ \frac{\prod_{\substack{k=0 \\ k\neq j,m}}^{M} (M-k)}{\prod_{\substack{k=0 \\ k\neq j}}^{M} (j-k)} \right\}. \tag{1.80}$$

We have now arrived at a point where we are able to calculate values for all coefficients. Equation (1.80) defines the coefficients for the BDF of eqn. (1.64) in normal form. Although this equation does look somewhat formidable, it is straightforward to evaluate the product and sum terms to obtain the coefficients σ_j. For example, if we let $M = 4$, then we get

$$\alpha_0 = \frac{6+8+12+24}{4 \times 3 \times 2 \times 1} = \frac{25}{12},$$

$$\alpha_1 = \frac{0+0+0+24}{3 \times 2 \times 1 \times -1} = -4,$$

$$\alpha_2 = \frac{0+0+0+12}{2 \times 1 \times -1 \times -2} = 3,$$

$$\alpha_3 = \frac{0+0+0+6}{1 \times -1 \times -2 \times -3} = -\frac{4}{3},$$

$$\alpha_4 = \frac{0+0+0+6}{-1 \times -2 \times -3 \times -4} = \frac{1}{4},$$

with $\beta_0 = 1$. Coefficients for BDF schemes up to order 6 in the form of eqn. (1.64) can be evaluated using R (see Listing 1.7). The results have been calculated for both and are listed in Tables 1.14 and 1.15.

```
# File: BDFcoeffs.R
# Calcuation of BDF coefficients
cat("\014") # Clear console
rm(list = ls(all = TRUE)) # Delete workspace
library("MASS")
M <- 4; # BDF Order
rho <- rep(0,M+1)
alpha_S <- rep(0,M+1)

for(j in 0:M){ # j <- alpha coefficient number
  p <- j+1;
  rho[p] <- 0;
  for(m in 0:M){
    num <- 1;den <- 1;
    for(k in 0:M){
      if(k != j && k != m){
        num <- num * (M-k);
      }
      if(k != j){
        den <- den * (j-k);
      }
    }
    if(m != j){
      rho[p] <- rho[p]+num/den;
    }
    #simplify(rho(j+1));
  }
}
```

```
for(i in 0:M){
  p <- i+1;
  alpha_S[p] <- rho[M+1-i]/rho[M+1];
}
cat(sprintf("Calcuation of BDF coefficients, Order=%d\n\n",M))
print("alpha coefficients - standard:"); print(fractions(alpha_S,5))
beta_S <- 1/rho[M+1]
print("Beta coefficient - standard:"); print(fractions(beta_S,5))
print("Error constant - standard,C(p+1)")
print(-fractions(beta_S/(M+1)))
#
print("alpha coefficients - normalized!");
alpha_N <- alpha_S * rho[M+1]
beta_N  <- 1
#
print(fractions(alpha_N,5))
print("Beta coefficient - normalized:"); print(fractions(beta_N,5))
print("Error constant - normalised,C(p+1)")
print(-fractions(beta_N/(M+1)))
```

Listing 1.7. File: BDFcoeffs.R—Code for calculating BDF coefficients

1.4.2.2 BDF Coefficients from Backward Differences

The BDF coefficients can also be obtained from *backward differences* by recasting eqn. (1.64) into the following form:

$$\sum_{m=1}^{k} \frac{1}{m} \nabla^m y_{n+1} = \Delta t f(t_{n+1}, y_{n+1}), \tag{1.81}$$

where we have set $\beta_0 = 1$ and

$$\begin{aligned}
\nabla^1 y_n &= y_n - y_{n-1}, \\
\nabla^2 y_n &= \nabla (y_n - y_{n-1}) = (y_n - y_{n-1}) - (y_{n-1} - y_{n-2}) = (y_n - 2y_{n-1} + y_{n-2}), \\
\nabla^3 y_n &= \nabla^2 (y_n - y_{n-1}) = \nabla (y_n - 2y_{n-1} + y_{n-2}) = (y_n - 3y_{n-1} + 3y_{n-2} - y_{n-3}), \\
&\vdots
\end{aligned} \tag{1.82}$$

etc.

For example, setting $k = 3$ in eqn. (1.81), we get

$$\begin{aligned}
&\tfrac{1}{1}(y_{n+1} - y_n) + \tfrac{1}{2}(y_{n+1} - 2y_n + y_{n-1}) \\
&\quad + \tfrac{1}{3}(y_{n+1} - 3y_n + 3y_{n-1} - y_{n-2}) = \Delta t f(t_{n+1}, y_{n+1}) \\
&\therefore \tfrac{11}{6} y_{n+1} - 3y_n + \tfrac{3}{2} y_{n-1} - \tfrac{1}{3} y_{n-2} = \Delta t f(t_{n+1}, y_{n+1}).
\end{aligned} \tag{1.83}$$

Thus, we have BDF coefficients that correspond to the values given in Table 1.15. By a similar approach, we can obtain BDF coefficients for any order.

BDF methods can also be used in a variable step scheme. However, it is more complex than for single-step methods because a new past history sequence has to be computed. In this situation, to change to a new step size h_{new}, previous solution values computed at a spacing of h_{old} need to be interpolated to obtain values at the new spacing. This variable step method is implemented in Listing 1.8, where the new values are obtained using *Lagrange* interpolation.

1.4.3 Numerical Differentiation Formulas (NDFs)

The NDF method is another LMM and is derived from the BDF method having one extra term. NDFs have smaller truncation errors than BDFs, leading to increased accuracy as fewer integration steps are required overall. However, higher-order schemes can be less stable than the corresponding order BDF. So the choice of which method to use becomes problem dependent.

Following generally the approach of Shampine and Reichelt [Sha-97], NDFs of order $p = k$ can be written as

$$\sum_{m=1}^{k} \frac{1}{m} \nabla^m y_{n+1} - \Delta t f(t_{n+1}, y_{n+1}) - \kappa \gamma_k \left(y_{n+1} - y_{n+1}^{(0)} \right) = 0, \qquad (1.84)$$

where we have dropped the truncation error term (see Section 1.5.1). ∇^m represents the mth backward-difference operator, and κ is a scalar parameter the values of which were selected to balance efficiency, in terms of step size, and stability angle. The parameter γ_k is defined as

$$\gamma_k = \sum_{j=1}^{k} \frac{1}{j}. \qquad (1.85)$$

It should be noted that the term $\sum_{m=1}^{k} \frac{1}{m} \nabla^m y_{n+1}$ in eqn. (1.84) is equivalent to the term $\sum_{j=0}^{k} \alpha_j y^{n+1-j}$ in eqn. (1.64) for normalized coefficients. Thus, if either $\kappa = 0$ or $\gamma = 0$, the NDFs reduce to BDFs.

Because NDFs, like BDFs, are implicit integrators, they have to be solved *iteratively*, usually using a Newton or similar method. For both methods the *recommended initial starting point* [Sha-97] for the iteration is

$$y_{y+1}^{(0)} = \sum_{j=0}^{k} \nabla^j y_n. \qquad (1.86)$$

This represents an *extrapolation* from k *past values* and provides an estimate from which to start the iteration process. Thus, the starting point is derived from one additional past value than that used in the iteration calculation. See further discussion in Section 1.4.4.3.

The bracketed last term in eqn. (1.84) is equal to

$$y_{n+1} - y_{n+1}^{(0)} = \nabla^{k+1} y_{n+1}. \qquad (1.87)$$

Table 1.16. The Klopfenstein–Shampine NDF coefficients with efficiency and A(α)-stability comparisons relative to BDFs [Sha-97]

order k	NDF coeff's κ_k	NDF coeff's γ_k	step ratio percent	stability angle BDF	stability angle NDF	percent change
1	-0.1850	1	26%	90^0	90^0	0%
2	$-1/9$	$\frac{3}{2}$	26%	90^0	90^0	0%
3	-0.0823	$\frac{11}{6}$	26%	86^0	80^0	-7%
4	-0.0415	$\frac{25}{12}$	12%	73^0	66^0	-10%
5	0	$\frac{137}{60}$	0%	51^0	51^0	0%

For $k = 2$, and substituting values for $y_{n+1}^{(0)}$ and $\nabla^{k+1}y_{n+1}$, we see the equality is confirmed, that is,

$$y_{n+1} - (3y_n - 3y_{n-1} + y_{n-2}) = (y_{n+1} - 3y_n + 3y_{n-1} - y_{n-2}). \quad (1.88)$$

Using the *normalized* BDF coefficients from Table 1.15 and values for κ and γ from Table 1.16, we obtain the following expanded NDF forms:

NDF-1

$$\Delta t f(t_{n+1}, y_{n+1}) = y_{n+1} - y_n \\ - \kappa_1 \gamma_1 (y_{n+1} - 2y_n + y_{n-1}). \quad (1.89)$$

NDF-2

$$\Delta t f(t_{n+1}, y_{n+1}) = \tfrac{3}{2} y_{n+1} - 2y_n + \tfrac{1}{2} y_{n-1} \\ + \kappa_2 \gamma_2 (y_{n+1} - 3y_n + 3y_{n-1} - y_{n-2}). \quad (1.90)$$

NDF-3

$$\Delta t f(t_{n+1}, y_{n+1}) = \frac{11}{6} y_{n+1} - 3y_n + \tfrac{3}{2} y_{n-1} - \tfrac{1}{3} y_{n-2} \\ + \kappa_3 \gamma_3 (y_{n+1} - 4y_n + 6y_{n-1} - 4y_{n-2} + y_{n-3}). \quad (1.91)$$

NDF-4

$$\Delta t f(t_{n+1}, y_{n+1}) = \frac{25}{12} y_{n+1} - 4y_n + 3y_{n-1} - \tfrac{4}{3} y_{n-2} + \tfrac{1}{4} y_{n-3} \\ + \kappa_4 \gamma_4 (y_{n+1} - 5y_n + 10y_{n-1} - 10y_{n-2} + 5y_{n-3} - y_{n-4}). \quad (1.92)$$

NDF-5

$$\Delta t f(t_{n+1}, y_{n+1}) = \frac{137}{60} y_{n+1} - 5y_n + 5y_{n-1} - \tfrac{10}{3} y_{n-2} + \tfrac{5}{4} y_{n-3} - \tfrac{1}{5} y_{n-4} \\ + \kappa_5 \gamma_5 (y_{n+1} - 6y_n + 15y_{n-1} - 20y_{n-2} + 15y_{n-3} - 6y_{n-4} + y_{n-5}). \quad (1.93)$$

Note that the details for NDF-5 are included only for completeness because, with $\kappa_5 = 0$ from Table 1.16, it is clear that NDF-5 is equivalent to BDF-5.

1.4.4 Convergence

BDFs and NDFs are k-step implicit schemes that we require to solve, and we have many methods at our disposal. We will look at two, namely, the *Newton* method and the *Levenberg–Marquardt* method. We will only illustrate solutions for the NDF method, as solutions for the BDF method follow by a straightforward simplification.

1.4.4.1 Newton's Method

We seek to solve eqn. (1.84) by iteration, so first we reformulate the equation to

$$g = \sum_{m=1}^{k} \frac{1}{m} \nabla^m y_{n+1} - \Delta t f(t_{n+1}, y_{n+1}) - \kappa \gamma_k \nabla^{k+1} y_{n+1} = 0. \tag{1.94}$$

We note that the leading terms of $\sum_{m=1}^{k} \frac{1}{m} \nabla^m y_{n+1}$ and $\nabla^{k+1} y_{n+1}$ are always y_{n+1}; therefore, on differentiation with respect to y_{n+1}, we obtain the Jacobian of g:

$$J = \frac{dg}{dy_{n+1}} = -\left[-(\alpha_0 - \kappa \gamma_k) + \Delta t \frac{df(t_{n+1}, y_{n+1})}{dy_{n+1}} \right], \tag{1.95}$$

where df/dy_{n+1} is the Jacobian of the derivative function f. We can now employ the standard Newton iteration scheme to solve eqn. (1.94) for y_{n+1}, that is,

$$\begin{aligned}\Delta y^s &= -J^{(s)-1} g^s \\ \therefore y^{s+1} &= y^s + \Delta y^s,\end{aligned} \tag{1.96}$$

where we have dropped the subscript $n+1$ and s represents the iteration number.

Equations (1.96) are recursive, with J^s and g^s being updated at each iteration until convergence is achieved in accordance with some criteria.

The solution for the BDFs follow by simply setting $\kappa \gamma_k = 0$ in eqns. (1.94) and (1.95).

1.4.4.2 Levenberg–Marquardt Method

The *Levenberg–Marquardt* method [Lev44, Mar-63] is a combination of *Newton* and *steepest descent* methods, that is,

$$\left(J^T J + \mu I \right) \Delta y = -J^T g, \tag{1.97}$$

where μ is a *weighting coefficient* or blending factor that determines the contribution of Newton and steepest descent to each solution step. For $\mu = 0$, eqn. (1.97) reduces to Newton's method, and for large μ, it reduces to the steepest descent method. Thus, it can be thought of as an interpolation between the two methods.

We proceed [Ans-97] in a similar way to the Newton method, by first obtaining the Jacobian J of eqn. (1.95). The L-M iterative scheme to solve eqn. (1.84) for y_{n+1} then becomes

$$\Delta y^s = -\left(J^{(s)T}J^s + \mu I\right)^{-1} J^{(s)T} g^s$$
$$\therefore y^{s+1} = y^s + \Delta y^s, \quad (1.98)$$

where again eqns. (1.98) are recursive, with J^s and g^s being updated at each iteration until convergence is achieved in accordance with some criteria.

The optimum value for μ is problem dependent and may need to change during the iteration procedure. A value of $\mu = 0.5$ seems to work well on relatively smooth problems. However, there are schemes aimed at solving computationally difficult problems that adjust μ according as to whether $|\Delta y^s|$ is increasing or decreasing. We will not discuss this aspect here.

The L-M method has advantages over straightforward Newton as it can find solutions in situations where J^s becomes singular.

As for the Newton method, the solutions for BDFs follow by simply setting $\kappa \gamma_k = 0$ in eqns. (1.94) and (1.95).

The R code for the variable-step BDF-4/NDF-4 methods is quite long, so only a selection of code is included in Listing 1.8, to give a flavor of the coding approach. A full listing that includes many comments is provided with the downloads.

```
# File:bdfNDFvs_4.R
bdfNDFvs_4 <- function(deriv,jac,tspan,y0,maxIter=20,maxEvals=5e5,
                TOL=1e-7,mode=0)
{
  # -----------------------------------------------------------------
  # USAGE:
  #  out <- ndfvs_4(deriv,jac,tspan,y0,maxiter,maxEvals,TOL, mode)
  # -----------------------------------------------------------------
  # DESCRIPTION:
  #     Variable step, general purpose BDF/NDF integrator
  #
.
.    <<<< code removed to save space >>>>
.
  stop <- 0
  for(iout in 1:Nt-1){
    if(stop == 1000){
      break
    }
.
.    <<<< code removed to save space >>>>
.
      while(stop == 0){
#        Converence loop
```

```
       .
       .    <<<< code removed to save space >>>>
       .
           }else if(totSteps4>4){
#            Used for fifth and later solution points
             if(contFlag4 == 0){
#              Starting point for NDF-4 iteration - 5 previous values
               ystart    <- 5*y0-10*y_old4[1,]+10*y_old4[2,]-5*y_old4[3,]+y_old4[4,]
               y         <- ystart # predicted value of y
               contFlag4 <- 1
             }
#            Jacobian matrix - update
             J <- jac(t,y)

#            g and f functions
             f <- deriv(t,y)
             g <- (h*f-(a40*y+a41*y0+a42*y_old4[1,]+a43*y_old4[2,]+a44*y_old4[3,]))
             if(mode == 0){ # BDF
                J <- h*J-a40*diag(neqn)
             }else{         # NDF
                nn <- K4*gamma4*(y-(5*y0-10*y_old4[1,]+10*y_old4[2,]-5*y_old4[3,]
                                    +y_old4[4,]))
                g  <- g+nn
                J  <- h*J-(a40-K4*gamma4)*diag(neqn)
             }
           }
           if(LM == 1){
#            Levenberg-Marquardt convergence Method
             JT <- (J)
             gJ <- JT%*%(g)
             G  <- JT*J +mu*diag(neqn)
#            Gaussian elimination
             dy <- -solve(G,gJ) # -(G\gJ)
           }else{
#            Newton convergence -Gaussian elimination
             dy <- -solve(J,(g)) # -(J\g)
           }
#          Update solution
           y <- y+dy
           stop <- 1
#          Check if the corrections are within the tolerance TOL
           err1 <- norm(dy/(y+.Machine$double.eps),"i") # Infinity norm
           if(err1>TOL){
#            Convergence not achieved; continue calculation
             niter <- niter+1
             stop  <- 0
```

```
        }
.
.   <<<< code removed to save space >>>>
.
#   Continue integration step
    } # end while loop - Integration step completed
#   Update past vales vector
    y_old4[5,] <- y_old4[4,]; y_old4[4,] <- y_old4[3,]
    y_old4[3,] <- y_old4[2,]; y_old4[2,] <- y_old4[1,]
    y_old4[1,] <- y0
    y0 <- y
    t <- t+h
    h_last[2] <- h_last[1]
    h_last[1] <- h
    if(t+h > tf){
        h <- tf - t
    }

#   Update history vectors
    y_hist <- rbind(t(y), y_hist)
    t_hist <- rbind(t, t_hist)
    if(outFlg == 0){
    # Check size of history vector and reduce if too large
      #[mh,nh] <- size(y_hist);
      mh <- nrow(y_hist); nh <- ncol(y_hist)
      if(mh>75){
#       delete rows above 25 - to reduce memory usage
        # y_hist[,26:length(y_hist)] <- []
        # t_hist[ 26:length(y_hist)] <- []
        y_hist[-c(26:length(y_hist)),]
        t_hist[-c(26:length(y_hist)),]
      }
    }

.   <<<< code removed to save space >>>>
.
    }
  }   # End while loop
#   fprintf('\n t <- #g, error <- #g, h <- #g, ncall <-
#d',t,err1,h_last(2), ncall);
    m <- nrow(y_hist); n <- ncol(y_hist)
    if(outFlg == 0){
      # Output only at specified times: Nt>2
      yout[iout+1,1:neqn] <- y_hist[1,1:neqn]
      tout[iout+1] <- t
    }
```

```
    } # End of for loop
    if(outFlg == 1){
        #[m,n] <- size(y_hist)
        m <- nrow(y_hist); n <- ncol(y_hist)
        yout <- matrix(y_hist[m:1,],nrow=m,ncol=n,byrow=FALSE)
        tout <- t_hist[m:1]
    }
    return(list(nout=length(tout),tout=tout,yout=yout))
}
```

Listing 1.8. File: bdfNDFvs_4.R—Part code of function bdfNDFvs_4() that implements the BDF-4 and NDF-4 methods. It calls functions deriv() and jac() passed as arguments from the appropriate main routine, and also the global function LagrangeInterp().

1.4.4.3 Iteration Starting Values

Because BDFs and NDFs are implicit integrators, they have to be *solved iteratively*, usually using a Newton-type method. As mentioned in Section 1.4.2, the higher-order methods require a set of past values to start the overall solution sequence. This can be achieved by using either explicit integration methods or lower-order BDF methods (as illustrated in Listing 1.8). However, a starting point also has to be found to initiate the iteration process at the start of each new time step. For both BDF and NDF methods the *recommended* nth *iteration starting point* is [Sha-97]

$$y_{n+1}^{(0)} = \sum_{j=0}^{k} \nabla^j y_n, \qquad (1.99)$$

where $\nabla^j y_n$ represents a *j*th-order backward-difference operator acting on y_n. Equation (1.99) is therefore an *extrapolation* from y_n to y_{n+1} using *k past values* and provides an estimate from which to start the iteration process. Thus, for each new time step (after the first), the iteration starting point is derived from one additional past value over and above that used in the subsequent iteration calculations.

Noting that $\nabla^0 y_n = y_n$, we obtain the following starting values:

$k = 1$

$$\begin{aligned} y_{n+1}^{(0)} &= y_n + (y_n - y_{n-1}) \\ &= 2y_n - y_{n-1}. \end{aligned} \qquad (1.100)$$

$k = 2$

$$\begin{aligned} y_{n+1}^{(0)} &= y_n + (y_n - y_{n-1}) + (y_n - 2y_{n-1} + y_{n-2}) \\ &= 3y_n - 3y_{n-1} + y_{n-2}. \end{aligned} \qquad (1.101)$$

$k = 3$

$$\begin{aligned} y_{n+1}^{(0)} &= y_n + (y_n - y_{n-1}) + (y_n - 2y_{n-1} + y_{n-2}) + (y_n - 3y_{n-1} + 3y_{n-2} - y_{n-3}) \\ &= 4y_n - 6y_{n-1} + 4y_{n-2} - y_{n-3}. \end{aligned} \qquad (1.102)$$

$k = 4$

$$y_{n+1}^{(0)} = y_n + (y_n - y_{n-1}) + (y_n - 2y_{n-1} + y_{n-2}) + (y_n - 3y_{n-1} + 3y_{n-2} - y_{n-3})$$
$$+ (y_n - 4y_{n-1} + 6y_{n-2} - 4y_{n-3} + y_{n-4}) \quad (1.103)$$
$$= 5y_n - 10y_{n-1} + 10y_{n-2} - 5y_{n-3} + y_{n-4}.$$

$k = 5$

$$y_{n+1}^{(0)} = y_n + (y_n - y_{n-1}) + (y_n - 2y_{n-1} + y_{n-2}) + (y_n - 3y_{n-1} + 3y_{n-2} - y_{n-3})$$
$$+ (y_n - 4y_{n-1} + 6y_{n-2} - 4y_{n-3} + y_{n-4})$$
$$+ (y_n - 5y_{n-1} + 10y_{n-2} - 10y_{n-3} + 5y_{n-4} - y_{n-5}) \quad (1.104)$$
$$= 6y_n - 15y_{n-1} + 20y_{n-2} - 15y_{n-3} + 6y_{n-4} - y_{n-5}.$$

Flame ball: Variable-step test example 1. To illustrate the performance of the BDF/NDF variable-step code shown in Listing 1.8, we choose to study the behavior of a simple combustion process that exhibits strong stiffness characteristics. The normalized system can be described by the ordinary differential equation [Rei-80] [Oma-91, 69–71]

$$\frac{dy(t)}{dt} = f(t), \quad y(t=0) = \epsilon, \quad 0 \le t \le 2/\epsilon$$
$$f(t) = y(t)^2 - y(t)^3, \quad (1.105)$$

where y represents the radius of a flame ball that grows rapidly until it reaches a critical size. It then remains at this size due to the amount of oxygen being consumed by combustion being just balanced by the amount available through the surface. Initially, in the interval $t \in [0, 1/\epsilon]$, the solution is positive, and hence the Jacobian, J, is greater than zero. This means that the process is unstable, although only just. However, in the interval $t \in [1/\epsilon, 2/\epsilon]$, the Jacobian is negative, and hence the process is stable. Nevertheless, the problem is relatively easy to solve numerically for $\epsilon \approx 0.01$. However, it becomes increasingly more difficult for smaller values of ϵ and, for $\epsilon \le 1.e\text{-}4$, it poses a real test for a numerical integrator. Further discussion of this aspect is given by Shampine [Sha-07] and Moler [Mol-03].

Before proceeding with the numerical integration, we note that the problem described by eqn. (1.105) is separable and, using a *partial fraction* expansion, can be integrated to give the following implicit solution in y:

$$\ln\left(\frac{y}{y-1}\right) - \frac{1}{y} = t + K, \quad (1.106)$$

where $K = \ln(\frac{\epsilon}{\epsilon - 1}) - \frac{1}{\epsilon}$ represents the constant of integration and $\epsilon = y$ is evaluated at $t = 0$. But this is not in a very useful form for our purposes owing to the singularity at $y = 1$ and the negative logarithm argument in K for small ϵ. However, eqn. (1.105) has the following explicit solution for y as a function of t:

$$y(t) = \left[W\left(a\,e^{a-t}\right) + 1\right]^{-1}, \quad a = 1/\epsilon - 1, \tag{1.107}$$

where $W()$ represents the Lambert function, which is a solution to the transcendental equation

$$W(z)\exp(W(z)) = z. \tag{1.108}$$

Some interesting historical background to the Lambert function $W()$ is given by Moler [Mol-03].

Note that eqn. (1.105) is another example of a nontrivial integration problem that can be readily solved numerically. Now, while the *computer algebra* packages rYacas and rSymPy in R can derive eqn. (1.106), both fail to find the solution given in eqn. (1.107). However, the commercial package Maple does find the correct solution, and the code is included with the downloads. Listings 1.9 and 1.10 include code for deriving eqn. (1.106) from eqn. (1.105) using the R packages rYacas and rSymPy, respectively. Refer to Appendices 1.A and 1.B for some information relating to these packages and their installation.

```
# File: intRyacas.R
cat("\014") # Clear console
rm(list = ls(all = TRUE)) # Delete workspace
#
library(Ryacas)
t <- Sym("t"); y <- Sym("y") # Specify symbolic variables
epsilon <- Sym("epsilon")
# Note: yacas fails to integrate: y^2-y^3 directly.
# So need to form partial fraction and integrate separately!
func1 <- function(y) {
  return(1/y - 1/(y-1))
}
func2 <- function(y) {
  return(1/y^2)
}
# Perform integration
F1 <- yacas(Integrate(func1(y), y)+Integrate(func2(y), y))
print(PrettyForm(F1))
# Evaluate K at t=0, when y=epsilon
K <- gsub("y", "epsilon", F1)
print(PrettyForm(K))
#
eqn <- sprintf("\n%s = t + %s\n",F1,K)
cat(eqn)
```

Listing 1.9. File: intRyacas.R—Code to derive the solution to eqn. (1.105), i.e., eqn. (1.106), using the Ryacas package. Note that it is necessary to provide Ryacas with the *partial fraction* form of the integrand and then to integrate terms separately.

```
# File: rSymPyInt.R
cat("\014") # Clear console
rm(list = ls(all = TRUE)) # Delete workspace
#
library(rSymPy)
# create a SymPy variable called y
sympy("var('y')"); sympy("var('K')")
sympy("var('t')"); sympy("var('epsilon')")
# Code to solve dydt=y^2-y^3, y(t=0)=epsilon
# by variable separation method
# Create a function f(y)
sympy("f=1/(y**2-y**3)")
# Now integrate the function
F <- sympy("F=integrate(f, y)")
cat(sprintf("F=%s\n",F))
#
# Now integrate the dt
G <- sympy("t=integrate(1, t)+K")
cat(sprintf("G=%s\n",G))
#
K <- sympy("K=limit(F, y, epsilon)")
cat(sprintf("K=%s\n",K))
# Final solution
t <- sympy("t=F-K")
cat(sprintf("t=%s\n",t))
#
y <-2; epsilon <- 1.01
TT <- eval(parse(text=t))
cat(sprintf("TT=%f\n",TT))
```

Listing 1.10. File: rSymPyInt.R—Code to derive the solution of eqn. (1.106) to eqn. (1.105) using the rSymPy package

The plots shown in Fig. 1.10 were generated by the R code in Listing 1.11 and illustrate analytical solutions for $\epsilon = 0.0001$ and $\epsilon = 0.01$. An additional plot for $\epsilon = 0.0001$ is also included, with an expanded time axis centered around $t = 10,000$.

Interestingly, it can be shown that the two curves are exactly similar and are related by the simple transformation

$$y_2(t; \epsilon_2) = y_1(t - \tau; \epsilon_1), \quad \tau = \ln \frac{(1 - \epsilon_2)\epsilon_1}{(1 - \epsilon_1)\epsilon_2} + \frac{\epsilon_1 - \epsilon_2}{\epsilon_1 \epsilon_2}. \quad (1.109)$$

Thus, for the preceding epsilon values, we find from eqn. (1.109) that $\tau = 9904.615121$.

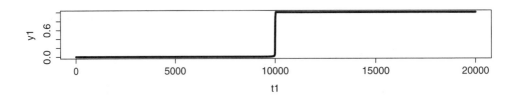
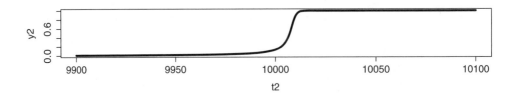
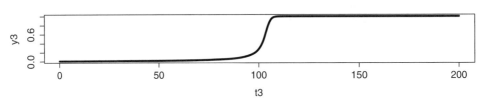

Figure 1.10. Plots of the analytical solution to eqn. (1.105) given by eqn. (1.107) for (top) $\epsilon = 0.0001$, (center) $\epsilon = 0.0001$ but with expanded time axis centered around $t = 10,000$, and (bottom) $\epsilon = 0.01$.

```
# File: flameBallAnalPlots.R
cat("\014") # Clear console
rm(list = ls(all = TRUE)) # Delete workspace
#
require(compiler)
enableJIT(3)
#
library(gsl) # Package containing Lambert function
#
N <- 1000 # Plot points
epsilon <- 0.0001
a    <- 1/epsilon-1
t1 <- seq(0,2/epsilon,len=N)
argLam <- pmin(a*exp(a-t1),.Machine$double.xmax/2) # Catch infinities
y1 <- (lambert_W0(argLam) + 1)^(-1) # Solution for epsilo=0.0001
#
epsilon <- 0.0001
b <- 1/epsilon-1
t2 <- seq(9900,10100,len=N)
y2 <- (lambert_W0(b*exp(b-t2)) + 1)^(-1) # y1 expanded around t=10,000
```

```
# #
epsilon <- 0.01
a <- 1/epsilon-1
t3 <- seq(0,2/epsilon,len=N)
y3 <- (lambert_W0(a*exp(a-t3)) + 1)^(-1) # Solution for epsilon=0.01
# Plot analytical solutions to the flame ball problem
par(mfrow=c(3,1))
plot(t1,y1,type="l",lwd=3)
plot(t2,y2,type="l",lwd=3)
plot(t3,y3,type="l",lwd=3)
```

Listing 1.11. File: flameBallAnalPlots.R—Code to plot the analytical solutions of eqn. (1.105) given by eqn. (1.107) with different values for ϵ

We now use the BDF/NDF code of Listing 1.8 to obtain a numerical solution to the ODE of eqn. (1.105) with $\epsilon = 0.0001$. The results are shown in Figs. 1.11 and 1.12.

The best (smallest error) results were found for NDF-4, which was also able to find a solution using fewer time steps. In Figure 1.11, it is seen that the time steps increased steadily after $t = 10,000$ until at the end of the simulation steps of 500 were being taken. The final time step was less than 500 to avoid overshooting the final time of $t = 20,000$. The numerical solution had a maximum error of 8.70672e-05 when compared to the analytical solution. This was achieved with 6,071 integration steps, which required 11,132 derivative evaluations. It is interesting to note that 90% of the time steps were needed between $t = 0$ and $t = 10,020$, with only 10% needed from $t = 10,020$ to $t = 20,000$. This is due to the stiff nature of the problem prior to the transition at around $t = 10,000$.

In Figure 1.12 the time scale has been expanded, from which we observe that the solution transition has the same shape as the transitions shown in Figure 1.10.

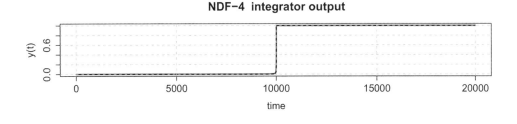

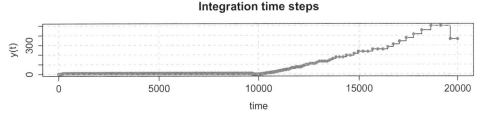

Figure 1.11. Plot of the numerical solution to eqn. (1.105) for $\epsilon_2 = 0.0001$. The upper plot shows the sharp transition of the solution at $t = 10000$. In the lower plot, the integration solution times are indicated by small dots, which show clearly how the time steps are being controlled by the integrator, with the maximum step reaching a value of 500.

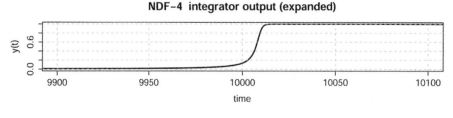
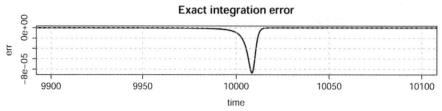

Figure 1.12. Plot of the numerical solution to eqn. (1.105) for $\epsilon = 0.0001$, but with expanded time axis. The lower plot shows the exact error, being the differerence between the analytical and numerical solutions. The error increases around the solution transition, peaking at |err|=8.70672e-05.

The R code for the *main* program that calls bdfNDFvs_4() is included in the downloads, along with other supporting routines.

Nonlinear pendulum: Variable-step test example 2. For a second example we illustrate the performance of the BDF/NDF integrator of Listing 1.8 when used to solve a problem with two state variables. We investigate the behavior of the nonlinear pendulum system detailed in Fig. 1.13. This system can be modeled by the ordinary differential equation (1.110) [deP-06], which becomes progressively more difficult to solve as the pendulum approaches equilibrium. Thus, we have chosen this particular ODE to integrate to test the ability of the variable-step error control to obtain a solution while keeping the convergence error within a specified tolerance.

$$\ddot{\phi} + \frac{\zeta}{I}\dot{\phi} + \frac{kd^2}{2I}\phi + \frac{\mu \operatorname{sgn}(\dot{\phi})}{I} + \frac{mgD\sin(\phi)}{2I}$$
$$= \frac{kd}{2I}\left(\sqrt{a^2 + b^2 - 2ab\cos(\omega t + \theta)} - (a-b)\right). \tag{1.110}$$

Single and double dots over a variable indicate differentiation with respect to time once and twice, respectively. Parameter values used in the the model, additional to those listed in Fig. 1.13, are the lumped mass $m = 1.47 \times 10^{-2}$ kg, gravitational acceleration $g = 9.81$ m/s^2, spring stiffness $k = 2.47$ N/m, inertia of the disc-lumped mass system $I = 1.738 \times 10^{-4}$ kg m^2, damping factor $\zeta = 2.368 \times 10^{-5}$ kg m^2/s, and coefficient of friction $\mu = 1.272 \times 10^{-4}$ Nm. The value for motor phase parameter θ is set for each individual test.

In their original analysis, de Paula et al. approximated the term sgn($\dot{\phi}$) in eqn. (1.110) by $\left(\frac{2}{\pi}\right)\arctan(q\dot{\phi})$, where q was set to a large number; they chose $q = 10^6$ [deP-06].

First we transform the second-order ODE of eqn. (1.110) into two first-order ODEs by letting $y_1 = \phi$ and $y_2 = \dot{\phi}$, then, on using the preceding approximation for sgn($\dot{\phi}$),

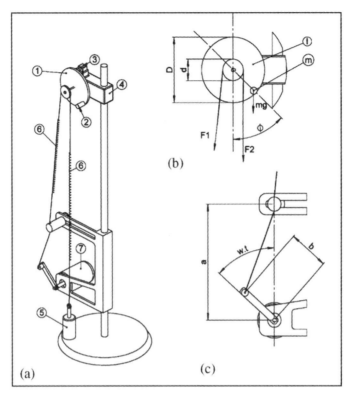

Figure 1.13. The pendulum apparatus of the BDF/NDF example, with dimensions $a = 1.6 \times 10^{-1}$ m, $b = 6.0 \times 10^{-2}$ m, $d = 4.8 \times 10^{-2}$ m, and $D = 9.5 \times 10^{-2}$ m. From [deP-06].

we obtain

$$\dot{y}_1 = y_2$$

$$\dot{y}_2 = -\frac{\zeta}{I} y_2 - \frac{kd^2}{2I} y_1 - \frac{2}{\pi} \frac{\mu}{I} \arctan(q\, y_2) - \frac{mgD \sin(y_1)}{2I} \quad (1.111)$$

$$+ \frac{kd}{2I} \left(\sqrt{a^2 + b^2 - 2ab \cos(\omega t + \theta)} - (a-b) \right).$$

We now consider four different *free response* ($\omega = 0$) situations where the corresponding initial conditions for state variables and motor phase parameter ($y_1, y_2 : \theta$) are ($2\pi, 0 : \pi$), ($5, \pi : \pi$), ($2\pi, -3\pi : \pi/2$), and ($0, 0 : 0$), respectively. The resulting transient plots are shown in Fig. 1.14 and the corresponding phase plots are shown in Fig. 1.15.

The *equilibrium points* associated with eqn. (1.111) are found by setting the time derivatives to zero, that is, $\dot{y}_1 = 0$ and $\dot{y}_2 = 0$, and solving for y_1 and y_2. Because $\dot{y}_1 = 0$, from eqn. (1.111) we must have $y_2 = 0$. Thus, we only need to find the equilibrium points for y_2. Using the earlier parameter values and $y_2 = 0$ yields

$$\dot{y}_2 = -34.10817031 - 16.37192175 y_1 - 39.41215478 \sin(y_1)$$
$$+ 341.0817031 \sqrt{0.02920 - 0.01920 \cos \theta}. \quad (1.112)$$

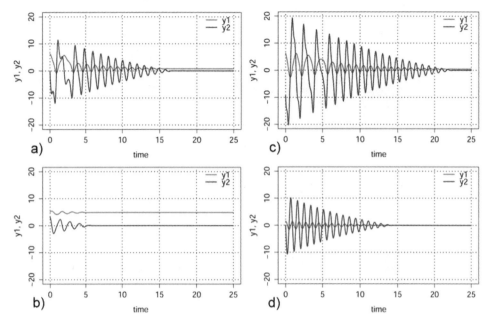

Figure 1.14. Transient plots of the BDF/NDF nonlinear pendulum example. The $(y_1, y_2, : \theta)$ initial conditions for each plot are (a) $(2\pi, 0 : \pi)$; (b) $(5, \pi : \pi)$; (c) $(2\pi, -3\pi : \pi/2)$, and (d) $(0, 0 : 0)$. (See color plate 1.14)

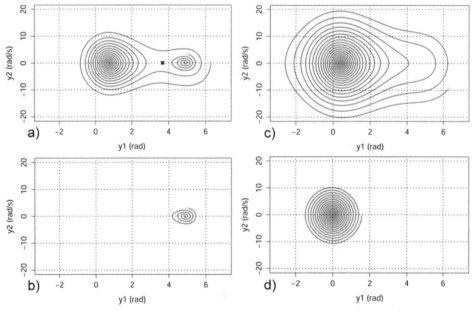

Figure 1.15. Phase plots of the BDF/NDF nonlinear pendulum example. The $(y_1, y_2, : \theta)$ initial conditions for each plot are (a) $(2\pi, 0 : \pi)$; (b) $(5, \pi : \pi)$; (c) $(2\pi, -3\pi : \pi/2)$, and (d) $(0, 0 : 0)$. The trajectory of plot (b) has been superimposed on plot (a) to emphasize how the choice of initial condition affects the ultimate equilibrium point. The small circular dots in each plot represent the initial conditions for y_1 and y_2, and the square in plot (a) represents the unstable equilibrium point at $(y_1 = 3.630396053, y_2 = 0)$. (See color plate 1.15)

We now set $\dot{y}_2 = 0$ and, for $\theta = \pi$, this yields three real solutions, two stable points [$(y_1 = 0.789988, y_2 = 0.0), (y_1 = 4.875392, y_2 = 0.0)$] and one unstable point ($y_1 = 3.630396, y_2 = 0$). Similarly, for $\theta = \pi/2$, this yields one real solution ($y_1 = 0.443559$, $y_2 = 0$). Finally, for $\theta = 0$, we obtain one solution ($y_1 = 0, y_2 = 0$). The plots in Figs. 1.14 and 1.15 show clearly the different *basins of attraction* and that each simulation terminates at the appropriate equilibrium point. The R code that solves for the equilibrium points of y_2 in eqn. (1.111) is given in Listing 1.12.

The paper by de Paula et al. extends the preceding analysis to cases where the pendulum response is forced, that is, $|\omega| > 0$. This leads to complex chaotic trajectories, and the interested reader can try setting ω to values between 3 and 7 [radians/s] and observe the various phase-plane plots generated for different initial conditions.

The R code for the *main* program that calls bdfNDFvs_4() is included with the downloads, along with other supporting routines. Options are included in the main program code to select the different initial conditions described here. Also, the main program is initially set to simulate the pendulum system with a *free response* by setting variable FR=0. However, three *non-free-response* options are provided in the main program that are selected by setting variable FR equal to 1, 2, or 3.

```
# File: pendulumEquilPoints.R
cat("\014") # Clear console
rm(list = ls(all = TRUE)) # Delete workspace
#
library(rootSolve)
equilibPts <- function(y1){
dy2dt <- -34.10817031-16.37192175*y1-39.41215478*sin(y1)+
         341.0817031*sqrt(0.29200e-1-0.01920*cos(theta))
}
# solve for pendulum equilibrium points
theta <- pi;   y11 <- uniroot.all(equilibPts, c(0,6))
print(y11)
theta <- pi/2; y12 <- uniroot.all(equilibPts, c(0,6))
theta <- 0;    y13 <- uniroot.all(equilibPts, c(0,6))
# Plots
y1a <- seq(0,6,len=100);
theta <- pi
plot(y1a,equilibPts(y1a),type="l",lwd=3,lty=1,ylim=c(-100,50),
     xlab="y1",ylab="dy2/dt")
grid(lty=1)
theta <- pi/2
lines(y1a,equilibPts(y1a),lwd=3,lty=2)
theta <- 0
lines(y1a,equilibPts(y1a),lwd=3,lty=3)
points(c(y11,y12,y13),c(0,0,0,0,0), pch=19, col="red")
```

Listing 1.12. File: pendulumEquilPoints.R—Code to calculate the equilibrium points for $\dot{y}_2$ in the nonlinear pendulum eqn. (1.111).

1.4.5 Adams Methods

The Adams methods are old methods that are still very popular. They form another subset of the LMMs. The *Adams–Bashforth* method is explicit, and the *Adams–Moulton* method is implicit.

Each can be used independently to solve initial value problems, and because they only require one past value, they can also be used together to form an effective *variable step, predictor-corrector* method. The predictor would be based on a k-step Adams–Bashforth method and provide a first estimate for y_{n+1}. The corrector would be based on a $(k-1)$-step Adams–Moulton method, which would use the y_{n+1} estimate and calculate a more accurate corrected value. More details on this aspect can be found in [Bud-93, Lam-91].

1.4.5.1 Adams–Bashforth Method

The k-step Adams–Bashforth method has order $p = k$ (AB-k) and is defined by the following equation, where symbols have the usual meaning:

$$y_{n+1} - y_n = h \sum_{j=1}^{k} \beta_j f_{n+1-j}. \tag{1.113}$$

It is a k-step method because it uses k past values to calculate the next step. Coefficients for the k-step Adams–Bashforth method can be derived in a similar manner to the derivation of BDF coefficients, and values for a constant step size method are given in Table 1.17.

We therefore see, for example, that the three-step, third-order Adams–Bashforth method AB-3 is given by

$$y_{n+1} = y_n + h\left(\frac{23}{12}f_n - \frac{16}{12}f_{n-1} + \frac{5}{12}f_{n-2}\right) + \frac{3}{8}h^4 f^{(iv)} + \mathcal{O}\left(h^5\right),$$

where the error terms have been included. Note that for $k > 1$, the Adams–Moulton error terms, C_{p+1}, are significantly smaller than the corresponding Adams–Bashforth error terms.

Table 1.17. Adams–Bashforth coefficients for eqn. (1.113)

k	1	2	3	4	5	6
β_1	1	$\frac{3}{2}$	$\frac{23}{12}$	$\frac{55}{24}$	$\frac{1901}{720}$	$\frac{4277}{1440}$
β_2		$-\frac{1}{2}$	$-\frac{16}{12}$	$-\frac{59}{24}$	$-\frac{2774}{720}$	$-\frac{7923}{1440}$
β_3			$\frac{5}{12}$	$\frac{37}{24}$	$\frac{2616}{720}$	$\frac{9982}{1440}$
β_4				$-\frac{9}{24}$	$-\frac{1274}{720}$	$-\frac{7298}{1440}$
β_5					$\frac{251}{720}$	$\frac{2877}{1440}$
β_6						$-\frac{475}{1440}$
p	1	2	3	4	5	6
C_{p+1}	$\frac{1}{2}$	$\frac{5}{12}$	$\frac{3}{8}$	$\frac{251}{720}$	$\frac{95}{288}$	$\frac{19087}{60480}$

Table 1.18. Adams–Moulton coefficients for eqn. (1.114)

k	1	2	3	4	5	6
β_0	1	$\frac{1}{2}$	$\frac{5}{12}$	$\frac{9}{24}$	$\frac{251}{720}$	$\frac{475}{1440}$
β_2		$\frac{1}{2}$	$\frac{8}{12}$	$\frac{19}{24}$	$\frac{646}{720}$	$\frac{1427}{1440}$
β_3			$-\frac{1}{12}$	$-\frac{5}{24}$	$-\frac{246}{720}$	$-\frac{798}{1440}$
β_4				$\frac{1}{24}$	$\frac{106}{720}$	$\frac{482}{1440}$
β_4					$-\frac{19}{720}$	$-\frac{173}{1440}$
β_5						$\frac{27}{1440}$
p	1	2	3	4	5	6
C_{p+1}	$-\frac{1}{2}$	$-\frac{1}{12}$	$-\frac{1}{24}$	$-\frac{19}{720}$	$-\frac{3}{160}$	$\frac{-863}{60480}$

1.4.5.2 Adams–Moulton Method

The $(k-1)$-step Adams–Moulton method has order $p = k$ (AM-k) and is defined by the following equation, where symbols have the usual meaning:

$$y_{n+1} - y_n = h \sum_{j=0}^{k-1} \beta_j f_{n+1-j}. \quad (1.114)$$

It is a $(k-1)$-step method because it uses $(k-1)$ past values to calculate the next step. Coefficients for the $(k-1)$-step Adams–Moulton method can be derived in a similar manner to the derivation of BDF coefficients, and values for a constant step size method are given in Table 1.18. Also, the implicit solution can be found using a similar iterative approach to that described for the BDF/NDF methods.

We therefore see, for example, that the two-step, third-order Adams–Moulton method AM-3 is given by

$$y_{n+1} = y_n + h\left(\frac{5}{12}f_{n+1} + \frac{8}{12}f_n - \frac{1}{12}f_{n-2}\right) - \frac{1}{24}h^4 f^{(iv)} + \mathcal{O}\left(h^5\right), \quad (1.115)$$

where the error terms have been included.

1.5 TRUNCATION ERROR AND ORDER OF INTEGRATION

Definition 1.5.1 (*Local truncation error* (LTE)). The *local truncation error of the method*, τ, is defined as being equal to the difference obtained when the true value is subtracted from the integrated value taken over one step. It is assumed that the integration step has been derived using *past exact values*. This is known as the *localizing assumption*. If the Taylor series expansion of the real value is subtracted from the Taylor series expansion of the integration method, the truncation error will be equal to the sum of those terms remaining (see eqn. (1.116)). A numerical method is said to be *consistent* if $\tau \to 0$ as the step size $h \to 0$.

Definition 1.5.2 (*Order of truncation error*). The *order of the truncation error* is defined as being equal to the lowest power of time step h, contained in the Taylor series terms of the truncation error. In the case of eqn. (1.116) it will be $p + 1$. Thus, if the truncation error is of order h^{p+1}, then this means that all error terms up to and including h^p are zero. Thus, the integration method is defined as having order p, that is, it is accurate to the pth term in a Taylor series expansion. The *"big oh"* of h, written as $\mathcal{O}(h)$, is used to signify the *rate* or *order of convergence* at which the error approaches zero as $h \to 0$. Thus, a function of $\mathcal{O}(h^{p+1})$ converges faster than a function of $\mathcal{O}(h^p)$.

Definition 1.5.3 (*Integration order of accuracy*). The *order of accuracy of an integration method* is defined as being equal to the order, less 1, of the first of the Taylor series terms contained in the truncation error. In the example of eqn. (1.116) the order will be p, and we write the order of accuracy as $\mathcal{O}(h^p)$. Care should be taken with operations on $\mathcal{O}(h^p)$ as it can represent different functions, for example,

- the result of adding of two functions that are $\mathcal{O}(h^p)$ and $\mathcal{O}(h^q)$, $p < q$ is $\mathcal{O}(h^p)$
- the result of multiplying two functions that are $\mathcal{O}(h^p)$ and $\mathcal{O}(h^q)$ is $\mathcal{O}(h^{p+q})$
- the result of multiplying a function of $\mathcal{O}(h^p)$ by a constant is still $\mathcal{O}(h^p)$
- the result of subtracting a function of $\mathcal{O}(h^p)$ from another of $\mathcal{O}(h^p)$ need not be equal to zero but is usually still equal to $\mathcal{O}(h^p)$; the exception is when we know *specifically* what each $\mathcal{O}(h^p)$ represents

Consider a numerical integration scheme that integrates an ordinary differential equation from t to $t + h$, starting from the exact value $y(t)$. The error contained in the integrated value $\hat{y}(t + h)$ can be determined by subtracting the Taylor expansion around the true value $y(t)$, from the Taylor series expansion of the numerical method result around $\hat{y}(t)$. Thus, we will have a situation where some of the lower-order expansion terms will be eliminated, and we will be left with a reduced series, that is,

$$\tau = \hat{y}(t+h) - y(t+h) = C_{p+1} h^{p+1} \frac{d^{p+1} y}{dt^{p+1}} + C_{p+2} h^{p+2} \frac{d^{p+2} y}{dt^{p+2}} + \cdots + \text{etc.} \quad (1.116)$$

Thus, the order of truncation error is $p + 1$, and the order of integration is p.
Additional discussion related specifically to LMMs is included in Section 1.5.1.

1.5.1 LMM Truncation Error

The local truncation error for an LMM is obtained by inserting the analytical solution into the particular scheme, that is,

$$\tau_k\left(t^{n+1}\right) = \sum_{j=0}^{k} \alpha_j y\left(t^{n+1-j}\right) - \Delta t \sum_{j=0}^{k} \beta_j f\left(t^{n+1-j}, y^{n+1-j}\right). \quad (1.117)$$

An estimate of the local truncation error is obtained using a Taylor expansion about $y = y(t^{n+1})$, from which we obtain

$$\tau_k\left(t^{n+1}\right) = \sum_{j=0}^{k} \alpha_j \left\{ y + (-j\Delta t)y' + \frac{1}{2!}(-j\Delta t)^2 y'' + \cdots + \frac{1}{(p+1)!}(-j\Delta t)^{p+1} y^{[p+1]} \right\}$$

$$- \Delta t \sum_{j=0}^{k} \beta_j \left\{ y' + (-j\Delta t)y'' + \cdots + \frac{1}{p!}(-j\Delta t)^p y^{[p+1]} \right\} + \mathcal{O}\left(\Delta t^{p+2}\right)$$

$$\Downarrow$$

$$= C_0 y + C_1 \Delta t y' + C_2 \Delta t^2 y'' + \cdots + C_{p+1} \Delta t^{p+1} y^{[p+1]} + \mathcal{O}\left(\Delta t^{p+2}\right). \quad (1.118)$$

The coefficients $C_0, \ldots, C_q$ in the preceding expression are given by [Gea-71, p118]

$$C_0 = \sum_{j=0}^{k} \alpha_j$$

$$\vdots \qquad\qquad (1.119)$$

$$C_q = \sum_{j=0}^{k} \left(\frac{-j^q}{q!} \alpha_j + \frac{-j^{q-1} \beta_j}{(q-1)!} \right), \quad q > 0.$$

1.5.1.1 BDF Error

We will illustrate the preceding approach by analyzing the BDF k-step scheme when eqn. (1.117) becomes

$$\tau_k\left(t^{n+1}\right) = \sum_{j=0}^{k} \alpha_j y^{n+1-j} - \beta_0 \Delta t f\left(t^{n+1}, y^{n+1}\right). \quad (1.120)$$

We now expand the RHS as a Taylor series to give

$$\tau_k\left(t^{n+1}\right) = \sum_{j=0}^{k} \alpha_j \left\{ y + (-j\Delta t)y' + \frac{1}{2!}(-j\Delta t)^2 y'' + \cdots + \frac{1}{(p+1)!}(-j\Delta t)^{p+1} y^{[p+1]} \right\}$$

$$- \Delta t \beta_0 \left\{ y' + (-\Delta t)y'' + \cdots + \frac{1}{p!}(-\Delta t)^p y^{[p+1]} \right\} + \mathcal{O}\left(\Delta t^{p+2}\right). \quad (1.121)$$

On grouping like terms we obtain

$$\tau_k\left(t^{n+1}\right) = y \sum_{j=0}^{k} \alpha_j + \Delta t y' \left(-\beta_0 + \sum_{j=0}^{k} -j\alpha_j \right)$$

$$+ \Delta t^2 y'' \left(\beta_0 + \sum_{j=0}^{k} \frac{1}{2!}(-1)^2 \alpha_j \right) + \cdots + \mathcal{O}\left(\Delta t^{p+2}\right) \qquad (1.122)$$

$$\Downarrow$$

$$= C_0 y + C_1 \Delta t y' + C_2 \Delta t^2 y'' + \cdots + C_{p+1} \Delta t^{p+1} y^{[p+1]} + \mathcal{O}\left(\Delta t^{p+2}\right),$$

where the coefficients $C_0, \ldots, C_q$ are equal to the following:

$$C_0 = \sum_{j=0}^{k} \alpha_j$$

$$C_1 = \sum_{j=0}^{k} -j\alpha_j$$

$$C_2 = \sum_{j=0}^{k} \frac{-j^2}{2!}\alpha_j \qquad (1.123)$$

$$\vdots$$

$$C_q = \sum_{j=0}^{k} \frac{-j^q}{q!}\alpha_j.$$

Note that the β_0 terms have disappeared. Other less compact forms for the coefficients C_q do include β_0. For example, the following is an equivalent form:

$$C_q = \sum_{j=0}^{k} \frac{(1-j)^q}{q!}\alpha_j - \frac{1}{(q-1)!}\beta_0. \qquad (1.124)$$

For consistency, in accordance with Definition 1.5.1, we must have $C_0 = C_1 = C_2 = \cdots = C_p = 0$. We see from eqn. (1.62) that $C_0 = \rho(1) = 0$, thus confirming the requirement for C_0. The requirement that $C_1, \ldots, C_p$ are all equal to zero can be confirmed by substituting the coefficients of either Table 1.14 or Table 1.15 into eqn. (1.123).

Thus, the local truncation error becomes

$$\begin{aligned} \tau_k &= C_{p+1}\Delta t^{p+1} y^{[p+1]} + \mathcal{O}\left(\Delta t^{p+2}\right) \\ &\approx C_{p+1}\Delta t^{p+1} \nabla^{p+1} y_{n+1}; \end{aligned} \qquad (1.125)$$

see Section 1.4.2.2.

The *error constant* for the BDF method of order p is found to be

$$\begin{aligned} \widetilde{C}_{p+1} &= -\frac{\beta_0}{k+1}, \quad \text{standard form} \\ C_{p+1} &= -\frac{1}{k+1}, \quad \text{normalised form}; \end{aligned} \qquad (1.126)$$

the appropriate values are given in Tables 1.14 and 1.15, respectively. Because of possible differences in error constant calculations, care has to be exercised when comparing the accuracy of different methods.

Finally, eqn. (1.125) can be used to estimate integration error in a variable-step BDF method.

1.5.1.2 NDF Error

We can extend the BDF local truncation error analysis to NDFs by including the appropriate extra term from eqn. (1.84) to eqn. (1.120), that is,

$$\begin{aligned}\tau_k\left(t^{n+1}\right) &= \sum_{m=1}^{k} \frac{1}{m}\nabla^m y_{n+1} - \kappa\gamma_k\left(y_{n+1} - y_{n+1}^{(0)}\right) - \beta_0 \Delta t f\left(t^{n+1}, y^{n+1}\right) \\ &= \sum_{j=0}^{k} \alpha_j y^{n+1-j} - \kappa\gamma_k\left(y_{n+1} - \sum_{j=0}^{k}\nabla^j y_n\right) - \beta_0 \Delta t f\left(t^{n+1}, y^{n+1}\right)\end{aligned} \quad (1.127)$$

On expanding the RHS as a Taylor series about $y = y\left(t^{n+1}\right)$ and grouping terms we get the same form of local truncation error as for the BDFs, that is,

$$\tau_k\left(t^{n+1}\right) = C_0 y + C_1 \Delta t y' + C_2 \Delta t^2 y'' + \cdots + C_{p+1}\Delta t^{p+1} y^{[p+1]} + \mathcal{O}\left(\Delta t^{p+2}\right), \quad (1.128)$$

where again, for consistency, in accordance with Definition 1.5.1 we must have: $C_0 = C_1 = C_2 = \cdots = C_p = 0$. Thus, the local truncation error becomes

$$\begin{aligned}\tau_k &= C_{p+1}\Delta t^{p+1} y^{[p+1]} + \mathcal{O}\left(\Delta t^{p+2}\right) \\ &\approx C_{p+1}\Delta t^{p+1} \nabla^{k+1} y_{n+1}.\end{aligned} \quad (1.129)$$

The *error constant* for the NDF method of order p is found to be

$$\begin{aligned}\tilde{C}_{p+1} &= -\beta_0 \left(\kappa\gamma_k + \frac{1}{k+1}\right), \quad \text{standard form} \\ C_{p+1} &= -\left(\kappa\gamma_k + \frac{1}{k+1}\right), \quad \text{normalised form.}\end{aligned} \quad (1.130)$$

Thus, from eqn. (1.130), it is clear that for correctly chosen *negative* values of $\kappa\gamma_k$, the absolute value of the local truncation error will be made smaller. However, a trade-off has to be made between a smaller truncation error and reduced stability. The constants in Table 1.16 were chosen to achieve good improvements in efficiency for a given error tolerance, while not reducing the $A(\alpha)$ stability angle excessively. For comparison purposes, C_{p+1} values for BDFs and NDFs are given in Table 1.19.

Again, we emphasize that because of possible differences in error constant calculations, care has to be exercised when comparing the accuracy of different methods.

Finally, as for the BDF method, eqn. (1.125) can be used to estimate integration error in a variable-step NDF method, but using C_{p+1} as defined by eqn. (1.130).

Table 1.19. BDF and NDF *normalized form* error constants

order, p	1	2	3	4	5
C_{p+1}, BDF	$-\frac{1}{2}$	$-\frac{1}{3}$	$-\frac{1}{4}$	$-\frac{1}{5}$	$-\frac{1}{6}$
C_{p+1}, NDF	$-\frac{63}{200}$	$-\frac{1}{6}$	$-\frac{5947}{60000}$	$-\frac{109}{960}$	$-\frac{1}{6}$

1.5.2 Verification of Integration Order

From eqn. (1.116) we see that for a p-order integration scheme, neglecting $p+2$ and higher terms in h, the truncation error is proportional to the step size raised to the power $p + 1$, that is,

$$\tau \propto h^{p+1}$$
$$\therefore \tau = K\, h^{p+1}, \tag{1.131}$$

where K is considered a constant for a particular method applied to a particular problem. Now, if we take one step of step size h_1 and r steps of size $h_2 = h_1/r$, we have the following relationships:

$$\tau_1 = Kh^{p+1}$$
$$\tau_2 = rK\left(\frac{h}{r}\right)^{p+1}. \tag{1.132}$$

Therefore, on dividing τ_1 by τ_2, rearranging, and taking logarithms, we obtain

$$p = \frac{\log \tau_2 - \log \tau_1}{\log r}. \tag{1.133}$$

We can use estimates for τ_1 and τ_2 using methods discussed in Section 1.3.2.2.

A practical scheme for verifying the integration order of a particular method involves obtaining solutions of a problem for various values of k and creating a log-log plot of the *maximum error* results. The slope of the graph will indicate the order p of the method. Figure 1.16 includes an example of such a plot created in R for the Runge–Kutta Dorman–Prince 5/4 method, based on the coefficients detailed in Table 1.12. It shows *estimated* and *analytical* error slopes calculated to be -5.307 and -5.385, respectively. The values used for the estimated error were obtained by the *embedded solutions method* outlined in Section 1.3.2.2. The integrator orders are equal to the absolute value of the respective slopes, that is, 5.307 and 5.385, and the analytical errors were obtained by comparing the integrated solution with the analytical solution.

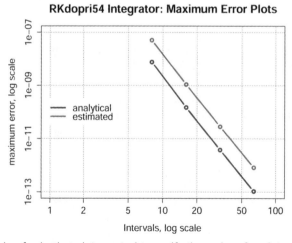

Figure 1.16. Example of a log-log plot created to verify the order of an integration method. The *estimated* (upper) and *analytical* (lower) error slopes are calculated to be -5.307 and -5.385, respectively. The slopes are evaluated from $(\log \text{error}_4 - \log \text{error}_1)/(\log \text{interval}_4 - \log \text{interval}_1)$.

The R code for generating Figure 1.16 is shown in Listing 1.13, where we integrate the test function $dx/dt = (A - x)$ that has an analytical solution $x = A[1 - \exp(-\lambda t)]$. The code can test the various RK integrators contained in RKembedded_Lib.R by setting the value of integType to between 1 and 7.

```
# File: RKerrorPlots.R
# Purpose: To generate error plots for RK integrators
rm(list = ls(all = TRUE)) # Delete workspace
library("rgl")
source("../integSRC/RKembedded_Lib.R") # load library of embedded RK methods
#
ptm = proc.time()
#
tend <- 1; # End time
# Test integration intervals
integInterval <- c(8, 16, 32, 64)
#
M          <- 4     # Number of interval sets to be calculated
pltFlg     <- 0     # 0<- plots off, 1<- plots on.
# Integrator RK types
integs <- c( "RK23","RKnystrom3","RKfehlberg45","RKcashKarp54",
             "RKdopri54","RKdopri87","SHK_est")
integType <- 5 # Set integrator type for test
# Problem parameters
lambda <- 1.0; A <- 1.0
#
# Derivative function
f <- function(t,x) lambda*(A - x) # dxdt
# Preallocate arrays
maxErr       <- rep(0,M)
maxEstErr    <- rep(0,M)
est_err_plot <- rep(0,M)
#
cat(sprintf("\nIntegrator Test: %s\n", integs[integType]))
for(idt in 1:M){
  N<-integInterval[idt];
  time<-seq(0,tend,length.out=N); # Gives N values but N-1 steps
  dt<-tend/(N-1);         # Important that this divisor is N-1
  x_plot   <- matrix(0,nrow=M, ncol=N)
  err_plot <- matrix(0,nrow=M, ncol=N)
  #   Initial condition
  x <- 0;
  #   Analytical solution
  x_anal <- A;
  #   Store base values
  x_tmp <- x; x_plot[,1] <-x;
  # Integrator command
```

```
    integCommand <- sprintf("%s(x,t,tf,f)",integs[integType])
    #  (N-1) steps in t
    for (j in 2:N){
      #     Runge Kutta methods
      t <- time[j-1]; tf <- time[j]
      # out <- RKdopri54(x,t,tf,f)
      out <- eval(parse(text=integCommand))
      #       Update dependant variable and estimated error
      x   <- out[1]; est_err <- out[2]
      #       Analytical solution and arays for plotting
      x_anal          <- A*(1.0-exp(-lambda*time[j]))
      x_plot[idt,j]   <- x
      err_plot[idt,j] <- x_plot[idt,j]-x_anal
      est_err_plot[j] <- est_err
    }
    #   Maximum errors
    maxErr[idt] <- max(abs(err_plot[idt,]));
    cat(sprintf("\n N <- %5d dt <- %10.3e  x <- %10.3e \n", N, dt, x))
    cat(sprintf("\n Maximum error: <- %0.3e\n", maxErr[idt]))
    #
    maxEstErr[idt] <- max(abs(est_err_plot))
    cat(sprintf("Estimated error: x <- %10.3e\n", maxEstErr[idt]))
    #
    if (pltFlg == 1){
      par(mfrow = c(1, 2));par(oma = c(0, 0, 3, 0))
      plot(time, x_plot[idt,],ylab="x"); grid(col="black")
      plot(time, err_plot[idt,],ylab="error"); grid(col="black")
      tStr <- sprintf("x and error plots for %d integration intervals",N)
      mtext(tStr,outer=TRUE)
    }
}# Next integration interval
# Calculate Integrator Order
intOrder <- abs((log10(maxErr[M])-log10(maxErr[1]))/
          (log10(integInterval[M])-log10(integInterval[1])))
cat(sprintf('\n Integrator Order (analytical) <- %6.3f',intOrder))
intEstOrder <- abs((log10(maxEstErr[M])-log10(maxEstErr[1]))/
             (log10(integInterval[M])-log10(integInterval[1])))
cat(sprintf("\n Integrator Order (estimated) <- %6.3f\n",intEstOrder))
# Plot errors;
par(mfrow = c(1, 1)); par(oma = c(0, 0, 0, 0))
titleTxt = sprintf("%s Integrator: Maximum Error Plots",integs[integType])
plot(integInterval, maxErr, type="b",lty=1,log="xy",col="blue",
     lwd=3,xlim=c(1e0,1e2), ylim=c(10^(-17),10^(-3)),
     xlab="Intervals, log scale",ylab="maximum error, log scale",
     main=titleTxt)
lines(integInterval, maxEstErr, type="b",lty=1,lwd=3,col="red")
```

```
abline(h=c(1e-17,1e-14,1e-11,1e-8,1e-5),v=c(1,2,5,10,20,50,100),
       col = "black", lty=3)
legend("left", lwd=c(3,3),bg="white",box.col="white",
       legend=c("analytical","estimated"),
       inset=0.01,col=c("blue","red"),bty="0")
```

Listing 1.13. File: RKerrorPlots.R—Code to generate the error plots of Fig. 1.16. The integrator function, RKdopri54(), was selected by setting variable integType = 5.

It is a simple matter to change the integrator specified in Listing 1.13 to a different type. For example, adding the line source("RKfehlberg45.R") and changing the statement integType <- "RKdopri54" to integType <- "RKfehlberg45" will result in the code generating error plots for the Rung–Kutta Felberg integrator of order 4/5. Similarly, other integrator methods with embedded solutions can also be specified (see Section 1.3.2.2). This capability is facilitated in R by the eval() function. The required command is first assembled programmatically as "text," which is achieved by the statement integCommand <- sprintf("%s(x,t,tf,f)",integType). The command is then executed by the statement out <- eval(parse(text=integCommand)).

1.6 STIFFNESS

The phenomenon of *stiffness* is related to the range of dynamic responses of different elements within a system described by differential equations. Stiffness is apparent in a system having *eigenvalues that differ greatly*, generally with ratios of orders of magnitude. This is known as the *stiffness ratio*. Thus, a stiff problem exhibits widely varying time scales; that is, some components of the solution move much more rapidly than others. Consequently, the integrator step size used to solve a stiff problem is usually *dictated by stability requirements* rather than by *accuracy requirements*. Also, in general, *explicit methods* cannot solve a stiff problem or, if they can, only with time steps that are impracticably small. Therefore, *implicit methods* are generally used to solve stiff problems.

The term *stiff differential equation(s)* is often used when describing a dynamic system exhibiting stiff characteristics.

For further discussion relating to stiffness, the reader is referred to [Gea-71, Mol-03, Sha-07].

1.7 HOW TO CHOOSE A NUMERICAL INTEGRATOR

There is no general *best* ODE integrator, and each particular problem must be assessed to determine the most appropriate solution method, often by trial and error. Here we consider a few different situations and discuss which ODE integrator may be suitable. This is not meant to cover all potential application scenarios but rather to provide some illustrative examples.

- *Embedded systems.* Where the code forms part of a much larger system that is often autonomous, and the size of computer code is important (not to be confused with the embedded Runge–Kutta systems of Section 1.3.2.2). However, computer code size is now becoming much less of a problem for this type of system as cost of memory has fallen drastically. Nevertheless, for some applications, keeping the code size as small as possible is a requirement. For these situations a fixed step method, such as one of the Runge–Kutta methods, is appropriate, depending on stability and accuracy considerations. There may have to be a trade-off between complexity of the method and acceptable step size, that is, between memory size and computational efficiency.

- *Real-time systems.* These are systems that control equipment to maintain a particular variable within a tight range or where a user interfaces with fast-moving graphics, for example, aircraft display. In these situations, computations often have to be completed in a predictable time that must be repeatable. Thus, some form of fixed step size method is usually employed, such as one of the explicit Adams or Runge–Kutta integrators. However, if accuracy is a particular concern, then a variable step size integrator may have to be employed. In this situation, extensive testing must be carried out prior to deployment to ensure that accuracy requirements will be met while maintaining real time.
- *Noncritical systems.* These are systems such as the "physics models" in computer games where it is only necessary to create an illusion of *perceived reality*. This type of application can usually achieve a suitable solution using a low-order explicit Runge–Kutta integrator. Possibly even the explicit Euler method may be deemed acceptable.
- *Stiff systems.* These are systems which have eigenvalues that differ by orders of magnitude and usually require some form of implicit integrator. For mildly stiff problems an Adams–Moulton or one of the BDF/NDF methods with a fixed step size may provide a suitable solution. As the system becomes more stiff, one of these methods together with a variable step size algorithm may be necessary to obtain an acceptable solution in a reasonable time. When the problem becomes very stiff, it may be necessary to use one of the specialist integrators, such as *lsodes* provided in the deSolve package.
- *Nonlinear systems.* For mild to strongly nonlinear systems a variable step size Runge–Kutta method is usually the preferred solution method. However, for highly nonlinear applications, again it may be necessary to use one of the specialist integrators, such as *lsodes* provided in the deSolve package.
- *Specialist systems.* Differential equations associated with certain applications sometimes require special integrators to obtain results that meet specific requirements. For example, PDEs with solutions that exhibit steep gradient phenomena are often solved using *high-resolution* methods, such as those discussed in Chapter 6. These seek to obtain *total variation deminishing* (TVD) solutions, for which a TDV Runge–Kutta method is often employed (again, see Chapter 6). Other applications, such as solutions to the *N-body problem* of celestial mechanics, often employ integrators specifically designed for the purpose; see [Fox-84] for some examples.

1.A INSTALLATION OF THE R PACKAGE Ryacas

The R code in Listing 1.9 provides a simple example of using the symbolic computer algebra system *yacas* using the R interface package Ryacas [Goe-14], which allows one to use yacas entirely from within R. It takes an R expression, an R one-line function, or a yacas string and returns an R expression or a variety of other formats. It can be used for symbolic mathematics, exact arithmetic, ASCII pretty printing, and R to TeX conversions. Although yacas does not have the full repertoire of a sophisticated commercial package like Maple, it does have a wide range of useful mathematical features. Also, it should be mentioned that there are syntax differences between various commands in Ryacas and R, for example, many Ryacas commands are capitalized, for example, sin(x) in R becomes Sin(x) in Ryacas.

However, before using Ryacas, the yacas program has to be installed, and this can be accomplished by running the following code [Goe-14, p9] from within R:

```
yacasInstall(url = "http://ryacas.googlecode.com/files/
                    yacas-1.0.63.zip",
overwrite = FALSE)
yacasFile(filename = c("yacas.exe", "scripts.dat", "R.ys"),
slash = c("\\", "/"))
```

The yacas package can also be used as a separate program, and full details are available at http://yacas.sourceforge.net/homepage.html

1.B INSTALLATION OF THE R PACKAGE rSymPy

The R package rSymPy is an R interface package giving R users access to the *SymPy* computer algebra system (CAS). It is written in python and runs on Jython (java-hosted python) from within R. It is available on CRAN, the Comprehensive R Archive Network.

The R code in Listing 1.10 provides a simple example of using the symbolic computer algebra package SymPy using the R interface package rSympy [Gro-14], which allows one to use SymPy entirely from within R. It takes an R expression in the form of a string and passes it to SymPy, and the result is returned to R as a string. The result can then be used in R calculations using the eval(parse(text=s)) command, where s represents a string returned from SymPy. The state of the SymPy session is maintained between calls.

To run rSymPy, it is essential that Java be installed on your machine and that it is the same version as used for R. To discover the version of Java on your machine, type the following line at the system command prompt (not in R):

```
java -version
```

You don't need python, Jython, or SymPy as they installed along with rSymPy. Installation is accomplished by typing the following line into R:

```
install.packages("rSymPy")
```

This installs rSymPy, which includes both SymPy and Jython, and also installs the rJava package if not already installed.

Again, SymPy does not have the full repertoire of a sophisticated commercial package like Maple but, it does includes features ranging from basic symbolic arithmetic to calculus, algebra, discrete mathematics, and quantum physics. It is also capable of formatting the result of the computations as LaTeX code. SymPy is free software and is licensed under a BSD license. The SymPy package can also be used as a separate program, and full details are available at http://docs.sympy.org/latest/index.html.

REFERENCES

[Ans-97] Anselmo, K. J. and W. E. Schiesser (1997), Some Experiences with a Differential Levenberg Marquardt Method, SIAM annual meeting, Stanford University, July 14–18.

[Bog-89a] Bogacki, P. and L. F. Shampine (1989), A 3(2) Pair of Runge-Kutta Formulas, *Applied Mathematics Letters* **2**, 1–9.

[Bog-89b] Bogacki, P. and L. F. Shampine (1989), An Efficient Runge-Kutta (4, 5) Pair, Report 89-20, Mathematics Department, SMU, Dallas, TX.

[Bud-93] Burden, R. L. and J. D. Faires (1993), *Numerical Analysis*, 5th ed., PWS.

[But-64] Butcher, J. (1964), On Runge-Kutta Processes of High Order, *Journal of the Australian Mathematical Society*, **4**, 179–193.

[But-87] Butcher, J. (1987), *The Numerical Analysis of Ordinary Differential Equations*, John Wiley.

[But-07] Butcher, J. (2007), Runge-Kutta Methods, *Scholarpedia*, http://www.scholarpedia.com.

[Cas-90] Cash, J. R. and A. H. Karp (1990), A Variable Order Runge-Kutta Method for Initial Value Problems with Rapidly Varying Right-Hand Sides, *ACM Transactions on Mathematical Software* **16**, 201–222.

[deP-06] de Paula, A. S., M. A. Savi, and F. H. I. Pereira-Pinto (2006), Chaos and Transient Chaos in an Experimental Nonlinear Pendulum, *Journal of Sound and Vibration* **294**-3, 585–595.

[Dor-80] Dormand, J. R. and P. J. Prince (1980), A Family of Embedded Runge-Kutta Formulae, *Journal of Computational and Applied Mathematics* **6**, 19–26.

[Enr-94] Enright, W. H., D. J. Higham, B. Owren, and P. W. Sharp (1994), A Survey of the Explicit RungeKutta Method, technical report 291/94, Department of Computer Science, University of Toronto (revised April 1995).

[Feh-69a] Fehlberg E. (1969), Klassiche Runge-Kutta Formeln fünfter and siebenter Ordnung mit Schrittweiten-Kontrolle, *Computing (Arch. Elektron. Rechnen)* **4**, 93–106.

[Feh-69b] Fehlberg E. (1969), Low-Order Classical Runge-Kutta Formulas with Step Size Control and Their Application to Heat Transfer Problems, NASA Technical Report 315 (extract published in *Computing* **6**, 1970, 61–71).

[Feh-70] Fehlberg E. (1970), Klassiche Runge-Kutta Formeln vierter und niedrigerer Ordnung mit Schrittweiten-Kontrolle und ihre Anwendung auf Wärmeleitungs probleme, *Computing* **6**, 61–71.

[Fox-84] Fox, K. (1984), Numerical Integration of the Equations of Motion of Celestial Mechanics, *Celestial Mechanics* **33**, 127–142.

[GAM-10] Gear, G. W. (1971), *GAMS, the Guide to Available Mathematical Software*, Mathematical and Computational Sciences Division, Information Technology Laboratory, National Institute of Standards and Technology. Available online at http://gams.nist.gov/.

[Gea-71] Gear, G. W. (1971), *Numerical Initial Value Problems in Ordinary Differential Equations*, Prentice Hall.

[Gea-90] Gear, C. W. and R. D. Skeel (1990), The Development of ODE Methods, in *A History of Scientific Computing*, Stephen G. Nash, editor, ACM Press, pp. 88–105.

[Goe-14] Goedman, G., and G. Grothendieck (2014), R Interface to the Yacas Computer Algebra System Ryacas. Available online at http://cran.r-project.org/web/packages/Ryacas/Ryacas.pdf.

[Gro-14] Grothendieck, G. (2014), R Interface to the Sympy Computer Algebra System rSymPy. Available online at http://cran.r-project.org/web/packages/rSymPy/rSymPy.pdf.

[Heu-00] Heun, K. (1900), Neue Methode zur approximativen Integration der Differentialgleichungen einer un-abhängigen Vëranderlichen, *Zeitschrift für Mathematik und Physik* **45**, 23–38.

[Joh-08] Johnson, D. (2008), *Teach Yourself VISUALLY Calculus*, John Wiley.

[Kre-11] Kreyszig, E. (2011), *Advanced Engineering Mathematics*, 10th ed., John Wiley.

[Kut-01] Kutta, M. W. (1901), *Beiträge zur näherungsweisen Integration totaler Differentialgleichungen*, PhD thesis, University of Munich, Germany.

[Lam-91] Lambert, J. D. (1991), *Numerical Methods for Ordinary Differential Systems*, John Wiley.

[Lev44] Levenberg, K. (1944), A Method for the Solution of Certain Non-linear Problems in Least Squares, *Quarterly Journal of Applied Mathematics*, **2**, 164–168.

[Mar-63] Marquardt, D. W. (1963), An Algorithm for Least-Squares Estimation of Nonlinear Parameters, *SIAM Journal Applied Mathematics*, **11**, 431–441.

[Maz-08] Mazzia, F. and C. Magherini (2008), Test Set for Initial Value Problem Solvers, release 2.4, Department of Mathematics, University of Bari and INdAM, Research Unit of Bari. Available online at http://www.dm.uniba.it/~testset.

[Mol-03] Moler, C. (2003), Stiff Differential Equations, *Matlab News & Notes*, May, Cleve's Corner. Available online at http://www.mathworks.com/company/newsletters/news_notes/clevescorner/may03_cleve.html.

[Oma-91] O'Malley, R. E., Jr. (1991), Singular Perturbation Methods for Ordinary Differential Equations. *Applied Mathematical Sciences* **89**, 69–71.

[Pri-81] Prince, P. J. and J. R. Dorman (1981), High Order Embedded Runge-Kutta Formulae. *Journal of Computational and Applied Mathematics* **7**, 67–75.

[Rei-80] Reiss, E. L. (1980), A New Asymptotic Method for Jump Phenomena, *SIAM Journal of Applied Mathematics* **39**, 440–455.

[Run-95] Runge, C. D. T. (1895), Uber die numerische Auflösung von Differentialgleichungen, *Mathematics Annals* **46**, 167–78.

[Sha-97] Shampine, L. F. and M. W. Reichelt (1997), The Matlab ODE Suite, *SIAM Journal of Scientific Computing* **18**, 1–22.

[Sha-07] Shampine, L. F. and S. Thompson (2007), Stiff Systems, *Scholarpedia*, **2**–3, 2855. Available online at http://www.scholarpedia.org/article/Stiff_systems.

[Som-94] Sommeijer, B. P., P. J. van der Houwen, and J. Kok (1994), Time Integration of Three-Dimensional Numerical Transport Models, *Applied Numerical Mathematics*, **16**, 201–225.

[Tel-06] Tél, T. and M. Gruiz (2006), *Chaotic Dynamics*, Cambridge University Press.

[Zwi-98] Zwillinger, D. (1998), *Handbook of Differential Equations*, Academic Press.

2

Stability Analysis of ODE Integrators

2.1 GENERAL

The integrator methods we will consider are those suitable for the numerical integration of ordinary differential equations (ODEs) of the form

$$\frac{dy(t)}{dt} = f(y,t), \quad [y(t_0) = y_0], \quad y \in \mathbb{R}^m, t > 0. \tag{2.1}$$

These methods have been described in Chapter 1.

First we include some definitions of terms that will be used in the analysis that follows [Gea-71, Lam-91, Ric-71].

Definition 2.1.1 (*Global order of accuracy*). A numerical integration method has global order of accuracy of k if

$$\max_{n \in (0, t/h)} |y^n - y(nh)| \leq \mathcal{O}(h^k), \quad \forall n > 0 \text{ as } h \to 0,$$

where h represents the time step size and n the step number. For any $f(y,t)$ that has k continuous derivatives with respect to y and t, methods with a higher k will tend to converge more rapidly than those with lower k.

Definition 2.1.2 (*A-stable*). A numerical integration method is *A*-stable if its *region of absolute stability*, S, contains the entire left half of the complex plane, that is,

$$S = \{\lambda_i h \in \mathbb{C} \ : \ \text{Re}\{\lambda_i h\} < 0, \ i = 1 \cdots m\},$$

where λ_i are the eigenvalues of $f(y,t)$. See Fig. 2.1.

Definition 2.1.3 ($A(\alpha)$ *stable*). A numerical integration method is $A(\alpha)$-stable if a subregion, S_α, of its *region of absolute stability*, S, lies in the left half of the complex plane entirely within a wedge, making an angle α with the real axis, that is,

$$S_\alpha = \{\lambda_i h \in \mathbb{C} \ : \ -\alpha < \pi - \arg\{\lambda_i h\} < \alpha, \ i = 1 \cdots m\},$$

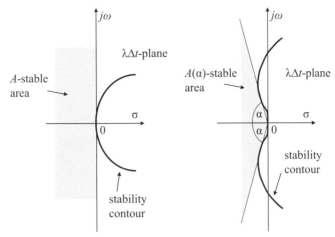

Figure 2.1. Illustration of (left) A-stability and (right) A(α)-stability.

where λ_i are the eigenvalues of $f(y, t)$. See Fig. 2.1. Consequently, an $A(\alpha)$-*stable method* has a weaker form of stability than an *A-stable* method.

Definition 2.1.4 (*Zero-stable*). A numerical integration method is *zero-stable* if it is well-posed. See also definition in Section 2.5.1.

2.1.1 Dahlquist Barrier Theorems

The first Dahlquist barrier theorem [Dah-56]:

- *No zero-stable linear k-step method can have order exceeding $k + 1$ when k is odd and $k + 2$ when k is even.*

The second Dahlquist barrier theorem [Dah-63]:

- *An explicit linear multistep method cannot be A-stable.*
- *The order of an A-stable linear multistep method cannot exceed 2.*
- *The second-order A-stable linear multistep method with smallest error constant is the* ***trapezoidal rule***.

For more details, refer also to Lambert [Lam-91].

2.2 DAHLQUIST TEST PROBLEM

Dahlquist [Dah-75] proposed the following ODE test problem to investigate the stability of numerical integrator schemes:

$$\frac{dy(t)}{dt} = \lambda y(t), \quad y(0) = y_0, \lambda \subset \mathbb{C}, \text{Re}(\lambda) \leq 0, t > 0, \quad (2.2)$$

where λ is the system eigenvalue and t usually represents time. We will use this equation subsequently to test the Runge–Kutta and backward differentiation formula schemes.

Equation (2.2) has the following analytical solution:

$$y(t) = y_0 e^{\lambda t}, \quad t \geq 0. \tag{2.3}$$

Application of a numerical integration scheme to eqn. (2.2) gives the approximation

$$y(nh) = G_f(\lambda h) y_n, \tag{2.4}$$

where again h represents the time step size and n the step number. G_f is called the *stability function* or *gain factor* of the method, and examples are given in the stability analysis of various integration schemes that follow.

2.3 EULER METHODS

2.3.1 Forward Euler

This method is the same as the first-order Runge–Kutta method (RK-1) and has the same stability characteristics (see Section 2.4.1).

2.3.2 Backward Euler

This method is the same as the first-order backward differentiation formula (BDF-1) and has the same stability characteristics (see Section 2.5.2.1).

2.4 RUNGE–KUTTA METHODS

Runge–Kutta methods have been described in Chapter 1, from which it is seen that they represent a family of one-step numerical integration schemes that require s evaluations of the derivative for each step. They are self-starting, and the order of accuracy depends on the particular scheme. Unless stated otherwise, from now on, we use h to represent time step size and n to represent step number.

Definition 2.4.1 The *region of absolute stability* of a numerical integration method, for the Dahlquist test problem $\dot{y} = \lambda y$, is the region, S, in the complex plane where

$$S = \left\{ \lambda h \in \mathbb{C} \; : \; \left\| y^n \right\| \leq \left\| y^{n-1} \right\|, \quad \forall n > 0 \right\}.$$

This definition allows the stability of Runge–Kutta methods to be analyzed in a straightforward manner.

2.4.1 RK-1: First-Order Runge–Kutta

This method is also known as the *forward Euler* method. For the 1D ODE problem of eqn. (2.1), the RK-1 method provides a numerical solution of the form

$$y^{n+1} = y^n + k_1, \tag{2.5}$$

where

$$k_1 = h f(y^n, t^n). \tag{2.6}$$

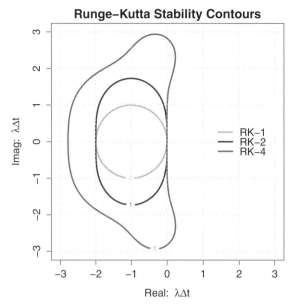

Figure 2.2. Runge–Kutta stability contours for RK-1 (inner) to RK-4 (outer) generated by the R code of Listing 2.1. Each scheme is stable for the region internal to its stability contour.

We now investigate the stability of this method using the *Dahlquist test problem*, eqn. (2.2). The RK-1 method results in the following recursive equation:

$$k_1 = h\lambda y$$
$$y^{n+1} = y^n \left[1 + h\lambda\right]. \tag{2.7}$$

For a stable system we must have the condition where the gain factor, G_f, is consistent with

$$|G_f| = \left|\frac{y^{n+1}}{y^n}\right| = |1 + z| \leq 1, \tag{2.8}$$

where $z = h\lambda$.

We can now determine the stability region for the RK-1 method by a *mapping* or *transformation* in the complex frequency domain. This entails mapping $z = h\lambda$ in the complex plane into its image in the corresponding $|G_f|$ plane. The *region of stability* for this method is $S = \{z \in \mathbb{C} : |G_f| \leq 1\}$. Establishing the boundary ∂S of the stability region is easily accomplished in R using the *contour function* to plot the contour for $|G_f| = 1$ (see Listing 2.1). The result is the Runge–Kutta stability plots for RK-1, RK-2, and RK-4 of Fig. 2.2, where the stable region for a particular method is represented by the interior of the corresponding contour. From this it is seen that the RK-1 method has a limited stability region and, consequently, can only be used for very stable systems.

```
# File: RKstabilityPlots.R
# Purpose: To plot stability contours for Runge-Kutta method
rm(list = ls(all = TRUE)) # Delete workspace
#
ptm = proc.time() # start process timer
```

```
#Specify x range and number of points
x0 <- -4.5; x1 <- 4.5
Nx <- 200 #301
# Specify y range and number of points
y0 <- -4; y1 <- 4
Ny <- 200 #301
# define complex domain
x <- seq(x0,x1,length=Nx)
y <- seq(y0,y1,length=Ny)
# Construct mesh
X <- outer(x,rep(1,Ny))
Y <- outer(rep(1,Nx),y)
# Calculate z
z <- X + 1i*Y
# Calculate gain factors
# Forward Euler (Runge-Kutta 1)
G1 <- 1 + z;
# Runge-Kutta 2
G2 <- 1 + z + 0.5*z^2
# Runge-Kutta 4
G4 <- 1 + z + 0.5*z^2 + (1/6)*z^3 + (1/24)*z^4
# Runge-Kutta RKF45 and RKCK45
G5 <- 1+z+(1/2)*z^2+(1/6)*z^3+(1/24)*z^4+(1/120)*z^5+(1/2080)*z^6
G6 <- 1+z+(1/2)*z^2+(1/6)*z^3+(1/24)*z^4+(10517/1228800)*z^5
      +(1771/1638400)*z^6
  # Calculate magnitude of G
G_mag1 <- abs(G1); G_mag2 <- abs(G2); G_mag4 <- abs(G4)
G_mag5 <- abs(G5); G_mag6 <- abs(G6)
# Plot contours of G_mag for RK-1 to RK-4
par(mar = c(5,5,2,5))
#plot.new()
contour(x,y,G_mag1,levels=1,col="green",lwd=3,asp=1,
        xlim=c(-3,3),ylim=c(-3,3),
        xlab=expression(paste("Real: ",lambda,Delta,"t")),
        ylab=expression(paste("Imag: ",lambda,Delta,"t")))
contour(x,y,G_mag2,levels=1,col="blue" ,lwd=3,add=TRUE)
contour(x,y,G_mag4,levels=1,col="red"  ,lwd=3,add=TRUE)
grid(col="lightgray",lwd=2)
title("Runge-Kutta Stability Contours");

legend("right",legend=c("RK-1","RK-2","RK-4"), inset=0.1,
      lwd=c(3,3,3),lty=c(1,1,1),bty="n", col=c("green","blue","red"))
#
#  Plot contours of G_mag for RK4 and RKF54
contour(x,y,G_mag4,levels=1,col="magenta",lwd=3,asp=1,
        xlim=c(-4.5,3.5),ylim=c(-4,4),
```

```
        xlab=expression(paste("Real: ",lambda,Delta,"t")),
        ylab=expression(paste("Imag: ",lambda,Delta,"t")))
contour(x,y,G_mag5,levels=1,col="black" ,lwd=3,add=TRUE);
grid(col="lightgray",lwd=2)
title("Comparison of RK4 and RKF54 Stability Contours");
legend("right",legend=c("RK-4","RKF54"), inset=0.05,
       lwd=c(3,3),lty=c(1,1),bty="n", col=c("magenta","black"))
# Plot contours of G_mag for RKF54 and RKCK45
contour(x,y,G_mag5,levels=1,col="red",lwd=3,asp=1,
        xlim=c(-4.5,3.5),ylim=c(-4,4),
        xlab=expression(paste("Real: ",lambda,Delta,"t")),
        ylab=expression(paste("Imag: ",lambda,Delta,"t")))
contour(x,y,G_mag6,levels=1,col="blue" ,lwd=3,add=TRUE);
grid(col="lightgray",lwd=2)
title("Comparison of RKF54 and RKCK45 Stability Contours")
legend("right",legend=c("RKF54","RKCK45"), inset=0.05,
       lwd=c(3,3),lty=c(1,1),bty="n", col=c("red","blue"))
cat(sprintf("time taken by process = %5.3f s\n",(proc.time()-ptm)[3]))
```

Listing 2.1. File: RKstabilityPlots.R—Code to plot Runge–Kutta stability contours

2.4.2 RK-2: Second-Order Runge–Kutta

This method is also known as *modified Euler* or *Heun's* method. For the 1D ODE problem of eqn. (2.1), the RK-2 method provides a numerical solution of the form

$$y^{n+1} = y^n + k_2, \qquad (2.9)$$

where

$$\begin{aligned} k_1 &= hf(y^n, t^n) \\ k_2 &= hf\left(y^n + \tfrac{1}{2}k_1,\ t^n + \tfrac{1}{2}h\right). \end{aligned} \qquad (2.10)$$

We can now investigate the stability of this numerical method as follows using the Dahlquist test problem of eqn. (2.2), when the RK-2 method results in the following recursive equation:

$$\begin{aligned} k_1 &= h\lambda y^n, \\ k_2 &= h\lambda\left(y + \tfrac{1}{2}h\lambda y^n\right), \\ y^{n+1} &= y^n \left[1 + h\lambda + \tfrac{1}{2}(h\lambda)^2\right]. \end{aligned} \qquad (2.11)$$

Again we define $z = h\lambda$. and for a stable system we must have the condition where the gain factor, G_f, is consistent with

$$|G_f| = \left|\frac{y^{n+1}}{y^n}\right| = \left|1 + z + \tfrac{1}{2}(z)^2\right| \le 1. \qquad (2.12)$$

We can now determine the stability region for the RK-2 method by plotting gain factor against $z = h\lambda$ in the complex frequency domain. The *region of stability* for this method is $S = \{z \in \mathbb{C} \ : \ |G_f| \le 1\}$.

The stability region is plotted in Fig. 2.2, where the stable region is interior to the contour. From this it is seen that the RK-2 method has a good region of stability and, due to its accuracy, being second order, and its simplicity, it finds application in many situations.

2.4.3 RK-4: Fourth-Order Runge–Kutta

For the 1D ODE problem of eqn. (2.1) the RK-4 method provides a numerical solution of the form

$$y^{n+1} = y^n + \tfrac{1}{6}\left(k_1 + 2k_2 + 2k_3 + k_4\right), \quad (2.13)$$

where

$$\begin{aligned}
k_1 &= hf\left(y^n, t^n\right), \\
k_2 &= hf\left(y^n + \tfrac{1}{2}k_1,\ t^n + \tfrac{1}{2}h\right), \\
k_3 &= hf\left(y^n + \tfrac{1}{2}k_2,\ t^n + \tfrac{1}{2}h\right), \\
k_4 &= hf\left(y^n + k_3,\ t^n + h\right).
\end{aligned} \quad (2.14)$$

We now investigate the stability of this numerical method as follows using the Dahlquist test problem of eqn. (2.2), when the RK-4 method results in the following recursive equation:

$$\begin{aligned}
k_1 &= h\lambda y, \\
k_2 &= h\lambda\left(y^n + \tfrac{1}{2}h\lambda y\right), \\
k_3 &= h\lambda\left(y^n + \tfrac{1}{2}h\lambda\left(y^n + \tfrac{1}{2}h\lambda y^n\right)\right), \\
k_4 &= h\lambda\left(y^n + h\lambda\left(y^n + \tfrac{1}{2}h\lambda\left(y^n + \tfrac{1}{2}h\lambda y\right)\right)\right), \\
y^{n+1} &= y^n\left[1 + \tfrac{1}{6}\left[6h\lambda + 3\left(h\lambda\right)^2 + \left(h\lambda\right)^3 + \tfrac{1}{4}\left(h\lambda\right)^4\right]\right].
\end{aligned} \quad (2.15)$$

For a stable system we must have the condition where the gain factor, G_f, is consistent with

$$|G_f| = \left|\frac{y^{n+1}}{y^n}\right| = \left|1 + z + \frac{1}{2!}(z)^2 + \frac{1}{3!}(z)^3 + \frac{1}{4!}(z)^4\right| \leq 1. \quad (2.16)$$

We can now determine the stability region for the RK-4 method by plotting a contour in the complex plane of gain factor for $|G_f| = 1$. This will define the *region of stability* for this method, that is, $S = \{z \in \mathbb{C} : |G_f| \leq 1\}$. The stability region is plotted in Fig. 2.2, where the stable region is interior to the contour. From this it is seen that the RK-4 method has a large region of stability and includes a significant part of the imaginary axis. Because of this, its high accuracy, and being fourth order, it is very widely used in industry for many and varied applications.

Note that the stability of Runge–Kutta schemes defined by a *Butcher tableau* (see Chapter 1) can be determined from [But-07]

$$R(z) = 1 + zb^T\left(I - zA\right)^{-1} - e, \quad (2.17)$$

where $e = [1, 1, \cdots, 1]^T$ is a vector of appropriate size and the stability region, S, in the complex plane is defined as

$$S = \{z \in \mathbb{C} : |R(z)| \leq 1\}. \quad (2.18)$$

Equation (2.17) generates the same stability terms, for example, eqn. (2.16) (see Listing 2.2 for the R code that solves eqn. (2.17) for various RK methods and also plots the corresponding contours).

```r
# File: butcherTableauGF.R
cat("\014") # Clear console
rm(list = ls(all = TRUE)) # Delete workspace
#
require(compiler);enableJIT(3)
library(pracma)
# Purpose: To generate Gain Factor from a Butcher tableau.
butcherGf <- function(A,b,z,N){
  n <-length(b)
  J <- diag(x = 1, nrow=n, ncol=n)
  R <- array(0,c(N,N))
  for(i in 1:N){
    for(j in 1:N){
      e <- rep(1,len=n)
      M <-inv(J-z[i,j]*A)
      Me <- M%*%e
      R[i,j] <- (1+z[i,j]*t(b)%*%Me)
    }
  }
  return(R)
}
# Compute contours
N <- 100 # number of points
# Specify x, y ranges and
x0 <- -5; x1 <- 3
y0 <- -4; y1 <- 4
# Construct mesh
x <- seq(x0,x1,len=N)
y <- seq(y0,y1,len=N)
X <- matrix(rep(x,each=N),nrow=N)
Y <- matrix(rep(y,each=N),nrow=N)
# Calculate z
z <- t(X) + 1i*Y
# Example: 4th order Runge-Kutta tableau
A1<-matrix(c(0,   0,   0, 0,
             1/2, 0,   0, 0,
             0,   1/2, 0, 0,
             0,   0,   1, 0),nrow=4,ncol=4,byrow=TRUE)
b14<-array(c(1/6, 1/3, 1/3, 1/6),c(4,1)) # column vector
# Example: 4-5th Fehlberg Runge-Kutta tableau
A2<-matrix(c(0,          0,          0,          0,    0, 0,
             1/4,        0,          0,          0,    0, 0,
             3/32,       9/32,       0,          0,    0, 0,
             1932/2197, -7200/2197,  7296/2197,  0,    0, 0,
```

```
                        439/216,         -8,      3680/513, -845/4104,     0,  0,
                        -8/27,            2,    -3544/2565, 1859/4104, -11/40, 0),
                     nrow=6,ncol=6,byrow=TRUE)
b24 <- array(c(25/216, 0, 1408/2565, 2197/4104, -1/5, 0 ),c(6,1)) # 4th
b25 <- array(c(16/135, 0, 6656/12825, 28561/56430, -9/50, 2/55),c(6,1)) # 5th
# Example: 4-5th Cash-Karp Runge-Kutta tableau
A3<-matrix(c(0,            0,       0,         0,       0, 0,
             1/5,          0,       0,         0,       0, 0,
             3/40,         9/40,    0,         0,       0, 0,
             3/10,        -9/10,    6/5,       0,       0, 0,
            -11/54,        5/2,    -70/27,    35/27,    0, 0,
             1631/55296,  175/512, 575/13824, 44275/110592, 253/4096, 0),
             nrow=6,ncol=6,byrow=TRUE)
b34 <- array(c(37/378,   0, 250/621,    125/594,    0,    512/1771),c(6,1)) # 4th
b35 <- array(c(2825/27648, 0, 18575/48384, 13525/55296, 277/14336, 1/4),
             c(6,1)) # 5th
#
case <- 1
#
if(case == 1){
  Gf <- butcherGf(A1,b14,z,N)
  titleStr = "RK 4th Order Stability Contour "
}else if(case == 2){
  Gf <- butcherGf(A2,b24,z,N)
  titleStr = "RKF 4th Order Stability Contour "
}else if(case == 3){
  Gf <- butcherGf(A2,b25,z,N)
  titleStr = "RKF 5th Order Stability Contour "
}else if(case == 4){
  Gf <- butcherGf(A3,b34,z,N)
  titleStr = "RKCK 4th Order Stability Contour "
}else if(case == 5){
  Gf <- butcherGf(A3,b35,z,N)
  titleStr = "RKCK 5th Order Stability Contour "
}
Gf_mag <- abs(Gf)
# Plot contours of G_mag
par (mar = c(5 ,5 ,2 ,5)) # Set plot margins
contour (x,y,Gf_mag,levels=1,col=" blue ",lwd=2, asp =1,
        xlab = expression ( paste (" Real : ", lambda ,Delta ,"t")),
        ylab = expression ( paste (" Imag : ", lambda ,Delta ,"t")))
grid (col =" lightgray ", lwd =2)
title (titleStr)
```

Listing 2.2. File: butcherTableauGF.R—Code to generate the gain factor, G_f from a Butcher tableau

2.4.4 RKF-54: Fehlberg Runge–Kutta

The RK-Fehlberg integrator [Feh-69] is traditionally presented as a fourth-order scheme (RKF-45) that includes the capability of providing a fifth-order error estimate that can be used for automatic error control. However, it can also be used as a fifth-order scheme (RKF-54) that uses the fourth-order result for error estimation purposes. We will analyze the stability of this scheme, which provides a numerical solution of the form

$$y^{n+1} = y^n + (33440k_1 + 146432k_3 + 142805k_4 - 50787k_5 + 10260k_6)/282150 \quad (2.19)$$

$$y^{n+1}_{est} = y^n + \frac{25}{216}k_1 + \frac{1408}{2565}k_3 + \frac{2197}{4104}k_4 - \frac{1}{5}k_5, \quad (2.20)$$

where

$$\begin{aligned}
k_1 &= hf(y^n, t^n), \\
k_2 &= hf\left(y^n + \tfrac{1}{4}k_1,\ t^n + \tfrac{1}{4}h\right), \\
k_3 &= hf\left(y^n + (3k_1 + 9k_2)/32,\ t^n + \tfrac{3}{8}h\right), \\
k_4 &= hf\left(y^n + (1932k_1 - 7200k_2 + 7296k_3)/2197,\ t^n + \tfrac{12}{13}h\right), \\
k_5 &= hf\left(y^n + (8341k_1 - 32832k_2 + 29440k_3 - 845k_4)/4104,\ t^n + h\right), \\
k_6 &= hf\left(y^n + (-6080k_1 + 41040k_2 - 28352k_3 + 9295k_4 - 5643k_5)/20520,\ t^n + \tfrac{1}{2}h\right).
\end{aligned} \quad (2.21)$$

Error control can be achieved by selecting a step size from

$$h_0 = \eta h_1 \left|\frac{\epsilon_0}{\epsilon_1}\right|^{\frac{1}{p}}, \quad (2.22)$$

where symbols and the algorithm are detailed in Chapter 1.

However, to investigate the stability of this numerical method, we only need to investigate the properties of eqn. (2.19). Again we use the Dahlquist test problem of eqn. (2.2), in the same way as for the RK-4 scheme. Therefore, from eqns. (2.19) and (2.21), we obtain the following recursive equation:

$$\begin{aligned}
k_1 &= z, \\
k_2 &= z(1 + k_1/4), \\
k_3 &= z(1 + (k_1 + 3k_2)3/32), \\
k_4 &= z(1 + (1932k_1 - 7200k_2 + 7296k_3)/2197), \\
k_5 &= z(1 + (8341k_1 - 32832k_2 + 29440k_3 - 845k_4)/4104), \\
k_6 &= z(1 + (-6080k_1 + 41040k_2 - 28352k_3 + 9295k_4 - 5643k_5)/20520), \\
y^{n+1} &= y^n \left[1 + (33440k_1 + 146432k_3 + 142805k_4 - 50787k_5 + 10260k_6)/282150\right],
\end{aligned} \quad (2.23)$$

where $z = h\lambda$.

Now, for a stable system, we must satisfy the condition whereby the magnitude of the gain factor is less than unity. Thus, on expanding and simplifying eqn. (2.23) the stability criterion becomes

$$|G_f| = \left|\frac{y^{n+1}}{y^n}\right| = \left|1 + z + \frac{1}{2}z^2 + \frac{1}{6}z^3 + \frac{1}{24}z^4 + \frac{1}{120}z^5 + \frac{1}{2080}z^6\right| \leq 1. \quad (2.24)$$

The R code that derives eqn. (2.24) using the package Ryacas is included in Listing 2.3.

```
# RK54_gainFactor_Ryacas.R
cat("\014") # Clear console
rm(list = ls(all = TRUE)) # Delete workspace
library(Ryacas) # R interface library to Yacas symbolic algebra system
# Calculation of gain factor for RKF54 (Felberg)
z   <- Sym("z")
k_1 <- z
k_2 <- z*(1 +       k_1/4)
k_3 <- z*(1 + (    k_1 +     3*k_2)*3/32)
k_4 <- z*(1 + ( 1932*k_1 - 7200*k_2 + 7296*k_3)/2197 )
k_5 <- z*(1 + ( 8341*k_1 - 32832*k_2 + 29440*k_3 - 845*k_4 )/4104)
k_6 <- z*(1 + (-6080*k_1 + 41040*k_2 - 28352*k_3 + 9295*k_4 - 5643*k_5)/20520)
Gf  <- Simplify((33440*k_1+146432*k_3+142805*k_4-50787*k_5+10260*k_6)/282150)
print(Gf)
# Test
z <- -1
zz <- eval(as.expression.Sym(Gf)) # Check if |Gf2(z)| <= 1
quit
```

Listing 2.3. File: RK54_gainFactor_Ryacas.R—Code to calculate the gain factor, G_f, of eqn. (2.24) for RKF-45

> The R code in Listing 2.3 provides a simple example of using the symbolic computer algebra system *yacas* via the interface the package Ryacas [Goe-14], which allows one to use yacas entirely from within R. It takes a R expression, an R one-line function, or a yacas string and returns an R expression or a variety of other formats. It can be used for various symbolic mathematical applications, and although yacas does not have the full repertoire of a sophisticated commercial package like *Maple*, it is a very effective system. A similar example using the R interface library rSymPy to the symbolic algebra system *SymPy* is included with the downloads (see file RK54_gainFactor_rSymPy.R).
>
> Refer also to appendices in Chapter 1 for additional information on installing Ryacas and rSymPy.

We can now determine the stability region for the RKF-54 method by plotting the gain factor G_f against $z = h\lambda$ in the complex frequency domain. The region of stability for this method is $S = \{\lambda h \in \mathbb{C} : |G_f| \leq 1\}$. The stability region associated the RKF-54 scheme is shown in Fig. 2.3 (plotted using the R code in Listing 2.1), where the stable region is interior to the contour. From this it is seen that the RKF-54 method has a larger region of stability than the RK-4 scheme. It also includes part of the right-half plane, which means it can handle some partially unstable problems. Because of its high accuracy, being

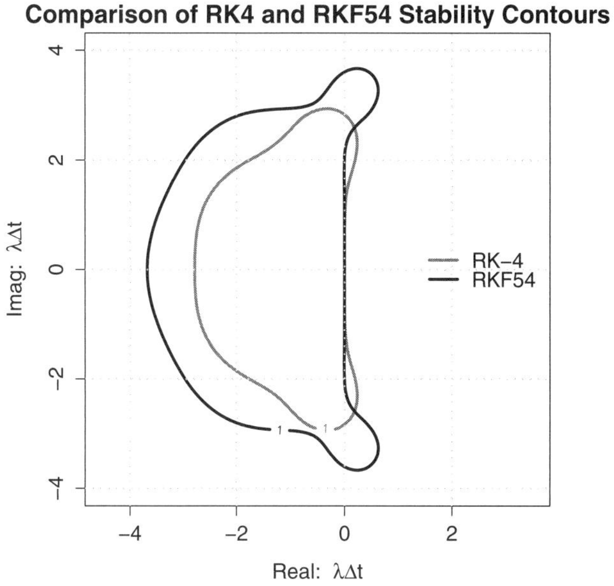

Figure 2.3. Comparison of RK-4 (inner) with RKF-54 (outer) stability regions generated by the R code of Listing 2.1.

fifth order, and its error estimating capability, it is often used for applications requiring automatic step size control. It is a simple matter to convert RKF-45 to RKF-54, as shown in Chapter 1. However, as a word of caution, we mention that it is not universally agreed that the RKF-54 scheme provides a superior solution to the RKF-45 scheme.

2.4.5 SHK: Sommeijer, van der Houwen, and Kok

The *Sommeijer, van der Houwen, and Kok* (SHK) method [Som-94] has a particularly easy recursive algorithm to program for problems where the derivative is not dependent on time. The order, s, can be changed by simply varying the number of iterations, and it has minimal memory storage requirements. For the 1D ODE problem of eqn. (2.1), the SHK method provides a numerical solution of the form

$$y^{n+1} = y^n + y_s^*, \qquad (2.25)$$

where the psudo code for evaluating y_s^* is

$$\begin{aligned} y_1^* &= y^n + \frac{h}{s} f(y^n) \\ y_2^* &= y^n + \frac{h}{s-1} f(y_1^*) \\ &\vdots \\ y_s^* &= y^n + \frac{h}{1} f(y_{s-1}). \end{aligned} \qquad (2.26)$$

We now investigate the stability of this numerical method using the Dahlquist test-problem of eqn. (2.2), when the *SHK* method with $s = 5$ results in the following recursive equation:

$$\begin{aligned}
y_1^* &= y^n + \frac{h}{5}\lambda y^n, \\
y_2^* &= y^n + \frac{h}{4}\lambda\left(y^n + \frac{h}{5}\lambda y^n\right), \\
y_3^* &= y^n + \frac{h\lambda}{3}\left(y^n + \frac{h}{4}\lambda\left(y^n + \frac{h}{5}\lambda y^n\right)\right), \\
y_4^* &= y^n + \frac{h\lambda}{2}\left(y^n + \frac{h\lambda}{3}\left(y^n + \frac{h}{4}\lambda\left(y^n + \frac{h}{5}\lambda y^n\right)\right)\right), \\
y_5^* &= y^n + \frac{h\lambda}{1}\left(y^n + \frac{h\lambda}{2}\left(y^n + \frac{h\lambda}{3}\left(y^n + \frac{h}{4}\lambda\left(y^n + \frac{h}{5}\lambda y^n\right)\right)\right)\right), \\
y^{n+1} &= y^n\left[1 + h\lambda + \frac{1}{2!}(h\lambda)^2 + \frac{1}{3!}(h\lambda)^3 + \frac{1}{4!}(h\lambda)^4 + \frac{1}{5!}(h\lambda)^5\right].
\end{aligned} \quad (2.27)$$

For a stable system we must have the condition where the gain factor

$$|G_f| = \left|\frac{y^{n+1}}{y^n}\right| = \left|1 + h\lambda + \frac{1}{2!}(h\lambda)^2 + \frac{1}{3!}(h\lambda)^3 + \frac{1}{4!}(h\lambda)^4 + \frac{1}{5!}(h\lambda)^5\right| \leq 1. \quad (2.28)$$

It is clear that each iteration adds yet another term to eqn. (2.28) and that if we had set $s = 4$, we would have exactly the same gain factor, eqn. (2.16), as for RK-4 and, hence, the same stability region.

Note that by inspection of the gain factors of the preceding examples, we can see that, up to order 4, the Runge–Kutta family of integrators construct Taylor series approximations of the integrand truncated to the term corresponding to the order of the particular scheme. The SHK scheme does the same, except that additionally, it constructs Taylor series approximations of the integrand for higher orders. However, as mentioned in Chapter 1, it is important to bear in mind that unlike the Runge–Kutta family, the SHK method is *not* suitable for integrating ODEs explicitly dependent on time.

We can now determine the stability region for the SHK-5 method by plotting gain factor against $z = h\lambda$ in the complex frequency domain. The region of stability for this method is $S = \{\lambda h \in \mathbb{C} : |G_f| \leq 1\}$. Plotting the stability region is easily accomplished in R using the contour function (see Listing 2.4). The resulting contour is the same as that shown for RK-54 in Fig. 2.3, where the stable region is interior to the contour. From this it is seen that the SHK-5 method has a large region of stability, which includes a significant part of the imaginary axis. It also includes part of the right-half plane, which means it can handle some partially unstable problems.

```
# File: SHKstabilityPlot5thOrd.R
# Purpose: To plot stability contour for 5th order SHK method
rm(list = ls(all = TRUE)) # Delete workspace
#
ptm = proc.time() # start process timer
#Specify x range and number of points
x0 <- -4; x1 <- 4
Nx <- 301
```

```
# Specify y range and number of points
y0 <- -4; y1 <- 4
Ny <- 301
# define complex domain
x <- seq(x0,x1,length=Nx)
y <- seq(y0,y1,length=Ny)
# Construct mesh
X <- outer(x,rep(1,Ny))
Y <- outer(rep(1,Nx),y)
# Calculate z
z <- X + 1i*Y
# Calculate gain factor
# 5th Order SHK
G5 <- 1 + z + 0.5*z^2 + (1/6)*z^3 + (1/24)*z^4 + (1/120)*z^5
# Calculate magnitude of G
G_mag5 <- abs(G5)
# Plot contours of G_mag
par(mar = c(5,5,2,5))
plot.new()
contour(x,y,G_mag5,levels=1,col="blue",lwd=2,asp=1,
        xlab=expression(paste("Real: ",lambda,Delta,"t")),
        ylab=expression(paste("Imag: ",lambda,Delta,"t")))
grid(col="lightgray",lwd=2)
title("SHK 5th Order Stability Contour")
#
cat(sprintf("time taken by process = %5.3f s\n",(proc.time()-ptm)[3]))
```

Listing 2.4. File: `SHKstabilityPlot5thOrd.R`—Code to generate the fifth-order SHK stability contour. The scheme is stable for the region internal to the contour

2.5 LINEAR MULTISTEP METHODS (LMMs)

2.5.1 General

General linear multistep (or k-step) methods (LMMs) have the following *standard* mathematical form:

$$\sum_{j=0}^{k} \alpha_j y^{n+1-j} = h \sum_{j=0}^{k} \beta_j f(t^{n+1-j}, y^{n+1-j}), \qquad (2.29)$$

where k represents the number of steps in a particular scheme and n the solution sequence number, and where α_j and β_j are constants defined for a particular scheme. The method requires that $\alpha_0 \neq 0$. When $\beta_0 \neq 0$, the method is implicit; otherwise, it is explicit. Equation (2.29) can be represented in alternative equivalent forms, some of which are discussed in the following.

With constant step size, order, and α_j and β_j coefficients, the stability and consistency properties of LMMs can be determined from the following *first* and *second characteristic polynomials* formed from the coefficients of eqn. (2.29), which are defined as

$$\rho(z) = \sum_{j=0}^{k} \alpha_j z^{k-j} = \alpha_0 z^k + \alpha_1 z^{k-1} + \alpha_2 z^{k-2} + \cdots + \alpha_k, \quad z \in \mathbb{C}$$
$$\sigma(z) = \sum_{j=0}^{k} \beta_j z^{k-j} = \beta_0 z^k + \beta_1 z^{k-1} + \beta_2 z^{k-2} + \cdots + \beta_k.$$
(2.30)

We now include some definitions that will assist the analysis that follows [Gea-71, Lam-91, Ric-71].

Definition 2.5.1 (*Consistency*). In the limit when $h \to 0$, a k-step LMM must be a consistent discretization of the ordinary differential equation, that is, the truncation error $\tau \to 0$ as the step size $h \to 0$ (see Chapter 1). This requirement is met iff

- $\sum_{j=0}^{k} \alpha_j = 0$. It is clear that this is a necessary requirement for a steady state equilibrium to exist. This is because, under steady state conditions, $y(t)$ is constant, that is, $y^{n+1} = y^n = \cdots =$ and so on, and $f(y(t)) = 0$; hence $\sum_{j=0}^{k} \alpha_j y^{n+j-1} = \sum_{j=0}^{k} \alpha_j = 0$.
- $\rho(1) = 0$ and $\rho'(1) = \sigma(1)$.

Additionally, a k-step method is consistent of order p iff

$$\rho(e^h) - h\sigma(e^h) = \mathcal{O}(h^{p+1}).$$

This implies that the *maximum absolute value of the truncation error* is less than or equal to a constant multiplied by the time step raised to the power p, that is,

$$\max_{t \in nh} \|\tau_p(t)\| \leq Ch^p.$$

Definition 2.5.2 (*Zero-stability*). In the limit when $h \to 0$, a LMM must not have solutions that can grow unbounded as $n \to \infty$. This requires that $\rho(z)$ satisfy the *root condition*, that is,

- all roots of $\rho(z) = 0$ have $|z| \leq 1$ and those roots where $|z| = 1$ are simple and not repeated

Definition 2.5.3 *The region of absolute stability* of a k-step method is defined by the contour Γ in the complex plane, given that $[\rho(z) - h\lambda\sigma(z)]$ satisfies the *root condition*. Thus, the *boundary inclusion locus*, Γ, is defined by

$$\Gamma = \left\{ h\lambda \in \mathbb{C} : h\lambda = \frac{\rho(z)}{\sigma(z)}, \quad z = e^{i\theta}, \theta \in [-\pi, +\pi] \right\}.$$

Definition 2.5.4 (*Convergence*). A LMM is convergent iff it is *consistent* and *stable* (*Dahlquist Equivalence Theorem*).

Definition 2.5.5 The *truncation error* of a LMM (see Chapter 1) is given by

$$\tau_k(t) = \sum_{j=0}^{k} \alpha_j \hat{y}(t+[k-j]h) - h\sum_{j=0}^{k} \beta_j f(\hat{y}(t+[k-j]h)),$$

where $\hat{y}(t)$ is the *exact solution* of the initial value problem. We can therefore represent the solution as

$$\hat{y}(t) = y^{n+1} + \tau_k(t) = y^{n+1} + C_{p+1}h^{p+1}\hat{y}^{[p+1]} + \mathcal{O}(h^{p+2}),$$

where C_{p+1} represents the *error constant* for the method of order p and $\hat{y}^{[p+1]}$ represents the $p+1$ derivative of $\hat{y}$ (see Chapter 1).

We will be concerned with only a subset of LMMs, namely, backward differentiation formula (BDF) and numerical differentiation formula (NDF) methods. BDFs and NDFs include one derivative term, at $t = (n+1)h$, and are therefore implicit methods.

2.5.2 Backward Differentiation Formulas (BDFs)

We will consider BDFs, which, in the literature, are defined in two equivalent mathematical forms. We shall refer to the following mathematical representation as the *standard form*:

$$\sum_{i=0}^{k} \widetilde{\alpha}_i y^{n+1-i} = h\widetilde{\beta}_0 f(t^{n+1}, y^{n+1}), \tag{2.31}$$

where $\sum_{i=0}^{k} \widetilde{\alpha}_i = 0$. We note that the results of eqn. (2.31) are not altered if all the coefficients are divided by any positive number. Thus, we are able choose an arbitrary value for any one coefficient. If we choose $\widetilde{\alpha}_0 = 1$, that is, divide through by $\widetilde{\alpha}_0$, we get the standard form, and the $\widetilde{\alpha}_i$ and $\widetilde{\beta}_0$ are constants that have values as defined in Table 2.1. This table also includes the error constant for each scheme, $C_{p+1} = \frac{-\widetilde{\beta}_0}{k+1}$, which is equal to the constant of the leading term of the associated truncation error, τ_k.

Table 2.1. BDF coefficients for the *standard form* of eqn. (2.31).

k	1	2	3	4	5	6
$\widetilde{\alpha}_0$	1	1	1	1	1	1
$\widetilde{\alpha}_1$	−1	−4/3	−18/11	−48/25	−300/137	−360/147
$\widetilde{\alpha}_2$		1/3	9/11	36/25	300/137	450/147
$\widetilde{\alpha}_3$			−2/11	−16/25	−200/137	−400/147
$\widetilde{\alpha}_4$				3/25	75/137	225/147
$\widetilde{\alpha}_5$					−12/137	−72/147
$\widetilde{\alpha}_6$						−10/147
$\widetilde{\beta}_0$	1	2/3	6/11	12/25	60/137	60/147
p	1	2	3	4	5	6
$\widetilde{C}_{p+1}$	−1/2	−2/9	−3/22	−12/125	−10/137	−20/343

Table 2.2. BDF coefficients for the *normalized form* of eqn. (2.32).

k	1	2	3	4	5	6
α_0	1	3/2	11/6	25/12	137/60	147/60
α_1	−1	−2	−3	−4	−5	−6
α_2		1/2	3/2	3	5	15/2
α_3			−1/3	−4/3	−10/3	−20/3
α_4				1/4	5/4	15/4
α_5					−1/5	−6/5
α_6						1/6
β_0	1	1	1	1	1	1
p	1	2	3	4	5	6
C_{p+1}	−1/2	−1/3	−1/4	−1/5	−1/6	−1/7

Alternatively, if we divide the standard form eqn. (2.31) by $\widetilde{\beta}_0$, we obtain the following equivalent mathematical form:

$$\sum_{j=0}^{k} \alpha_j y^{n+1-j} = \beta_0 h f\left(t^{n+1}, y^{n+1}\right), \quad (2.32)$$

where $\sum_{i=0}^{k} \alpha_i = 0$ and the α_j and β_0 coefficients are given in Table 2.2. We shall refer to this mathematical representation as the *normalized form*, which corresponds to the LMM eqn. (2.29). Note that the error constants $\widetilde{C}_{p+1}$ have also been divided by β_0, yielding $C_{p+1} = \frac{-1}{k+1}$.

Note that the normalized form is a convenient form with which to work when developing a Newton convergence routine for a BDF integrator.

2.5.2.1 BDF-1

The BDF-1 method is the same as the backward Euler method. For the 1D ODE problem of eqn. (2.1), and using the constants from Table (2.1), eqn. (2.31) yields the following recursive first-order BDF equation:

$$y^{n+1} = y^n + h f\left(y^{n+1}, t^n\right). \quad (2.33)$$

We can now investigate the stability of this numerical method as follows using the Dahlquist test-problem of eqn. (2.2), when the BDF-1 method results in the following recursive equation:

$$y^{n+1} = y^n \frac{1}{1 - h\lambda}. \quad (2.34)$$

For a stable system we must have the condition where the gain factor

$$\left|G_f\right| = \left|\frac{y^{n+1}}{y^n}\right| = \left|\frac{1}{1 - h\lambda}\right| \leq 1. \quad (2.35)$$

We could now determine the stability region for the BDF-1 method by plotting $\left|G_f\right| = 1$ in the complex frequency domain, as we did previously for the Runge–Kutta family of integrators, that is, for $S = \{\lambda h \in \mathbb{C} : |G_f| \leq 1\}$. However, for the BDF family of integrators, we adopt a different method of stability analysis, which is outlined in Section 2.5.2.2 for BDF-4 and which illustrates the general approach. Plotting the stability region is easily accomplished using the R plot function (see BDF stability Fig. 2.4, where the stable region is exterior to the BDF-1 contour). From this it is clear that the BDF-1 method is *A-stable*, that is, it is stable over the whole left half of the λh-plane. Moreover, it also includes a substantial part of the right-half plane. Consequently, it is very robust and can be used for applications that include the investigation of partially unstable systems.

2.5.2.2 BDF-4

Taking normalized constants from Table 2.2, we see from eqn. (2.31) that we obtain the following recursive fourth-order BDF equation:

$$y^{n+1} + \frac{1}{12}\left[-48y^n + 36y^{n-1} - 16y^{n-2} + 3y^{n-3}\right] = hf\left(y^{n+1}\right). \tag{2.36}$$

First we will investigate the stability of this method in a simple way as follows, using the Dahlquist test-problem of eqn. (2.2). The idea is to obtain λh as a function of powers of $e^{-\lambda h}$ and then plot this function in the complex plane for $|e^{-\lambda h}| = 1$. This is accomplished by setting $e^{-\lambda h} = e^{i\theta}$ and letting θ take values $0 \to 2\pi$. The result is a contour in the complex plane that defines the region of stability, $S = \{\lambda h \in \mathbb{C} : |G_f| \leq 1\}$.

From eqn. (2.3), the analytical solution of eqn. (2.2), we obtain

$$\begin{aligned}
y^{n+1} &= y_0 e^{(\lambda[n+1]h)} \\
&\Downarrow \\
y^n &= y_0 e^{(\lambda n h)} &&= y^{n+1} e^{(-\lambda h)} \\
y^{n-1} &= y_0 e^{(\lambda[n-1]h)} &&= y^{n+1} e^{(-2\lambda h)} \\
y^{n-2} &= y_0 e^{(\lambda[n-2]h)} &&= y^{n+1} e^{(-3\lambda h)} \\
y^{n-3} &= y_0 e^{(\lambda[n-3]h)} &&= y^{n+1} e^{(-4\lambda h)} \\
\frac{dy^{n+1}}{dt} &= \lambda e^{(\lambda[n+1]h)} &&= \lambda y^{n+1}.
\end{aligned} \tag{2.37}$$

Now, on letting $z = e^{(\lambda h)}$, we obtain

$$\begin{aligned}
y^n &= y^{n+1} z^{-1}, \\
y^{n-1} &= y^{n+1} z^{-2}, \\
y^{n-2} &= y^{n+1} z^{-3}, \\
y^{n-3} &= y^{n+1} z^{-4}.
\end{aligned} \tag{2.38}$$

Thus, eqn. (2.36) becomes

$$h\lambda y^{n+1} = \frac{1}{12} y^{n+1}\left(25 - 48z^{-1} + 36z^{-2} - 16z^{-3} + 3z^{-4}\right)$$

$$\therefore h\lambda = \frac{1}{12}\left(25 - 48z^{-1} + 36z^{-2} - 16z^{-3} + 3z^{-4}\right). \tag{2.39}$$

We can now perform a mapping of eqn. (2.39) in the complex frequency domain from the z-plane to the λh-plane. This is accomplished by setting $z = \exp(i\theta)$ and letting θ vary from 0 to 2π. The result is that as z moves around the unit circle in the z-plane, the stable interior of the unit circle will map to the stable exterior of the resulting contour in the λh-plane. This can be easily plotted in R (see Listing 2.6).

The resulting BDF-4 contour is shown in Fig. 2.4, where the region of absolute stability (exterior to the contour) has been plotted in accordance with Definition 2.5.3. It can be verified that the interior of the unit circle maps to the exterior of the contour by plotting a point inside the unit circle and observing the corresponding point in the λh-plane. Other BDF schemes are handled in a similar way, and the resulting contours for BDF-1, BDF-2, and BDF-3 are also included in Fig. 2.4.

Let us now revert to the preceding definitions so that we can treat the problem more rigorously. For BDF-4, eqns. (2.30) become

$$\begin{aligned} \rho(z) &= \tfrac{25}{12}z^4 - \tfrac{48}{12}z^3 + \tfrac{36}{12}z^2 - \tfrac{16}{12}z^1 + \tfrac{3}{12} \\ \sigma(z) &= z^4. \end{aligned} \tag{2.40}$$

We see from the coefficients of Table 2.1 that

$$\sum_{j=0}^{k} \widetilde{\alpha}_j = 0 \tag{2.41}$$

(equivalent to $\sum_{j=0}^{k} \alpha_j = 0$) and from eqn. (2.40) that

$$\rho(1) = 0 \quad \text{and} \quad \rho'(1) = \sigma(1) = \tfrac{12}{25}. \tag{2.42}$$

We could have alternatively used coefficients from Table 2.2. Thus, eqns. 2.41 and (2.42) establish that the method is *consistent* according to Definition 2.5.1. We can now investigate the *order* of the method by substituting $z = e^h$ into eqns. (2.39) to obtain

$$\begin{aligned} \rho\left(e^h\right) &= \tfrac{25}{12}e^{4h} - \tfrac{48}{12}e^{3h} + \tfrac{36}{12}e^{2h} - \tfrac{16}{12}e^h + \tfrac{3}{12} \\ h\sigma\left(e^h\right) &= he^{4h}. \end{aligned} \tag{2.43}$$

Note that this step is effectively the same process as taking Fourier transforms in the von Neumann analysis used to investigate the stability of numerical solutions to partial differential equations (see Chapter 4).

For ρ and σ we perform a Taylor series expansion on each of the e^{nh} terms to order 5:

$$\begin{aligned} e^h &= 1 + h + \tfrac{h^2}{2!} + \tfrac{h^3}{3!} + \tfrac{h^4}{4!} + \tfrac{h^5}{5!}, \\ e^{2h} &= 1 + 2h + \tfrac{4h^2}{2!} + \tfrac{8h^3}{3!} + \tfrac{16h^4}{4!} + \tfrac{32h^5}{5!}, \\ e^{3h} &= 1 + 3h + \tfrac{9h^2}{2!} + \tfrac{27h^3}{3!} + \tfrac{81h^4}{4!} + \tfrac{243h^5}{5!}, \\ e^{4h} &= 1 + 4h + \tfrac{16h^2}{2!} + \tfrac{64h^3}{3!} + \tfrac{256h^4}{4!} + \tfrac{1024h^5}{5!}. \end{aligned} \tag{2.44}$$

Substituting the expansions into eqns. (2.43), we find

$$\rho\left(e^h\right) - h\sigma\left(e^h\right) = -\tfrac{12}{25}h^5 - \tfrac{512}{25}h^6 + \cdots \tag{2.45}$$

Thus, terms involving h^4 and lower have all canceled. Hence, from Definition 2.5.1 we see that the lowest power of h is 5, which implies $p = 5$. Consequently, we have shown that the BDF-4 method has a *consistency* of order $p - 1 = 4$, as expected. The first term on the RHS of eqn. (2.45), that is, $-\frac{12}{125}h^5$, is equal to $-\frac{1}{5}\widetilde{\beta}_0 h^5 = \widetilde{C}_{p+1} h^5$, where $\widetilde{C}_{p+1}$ is the error constant for the standard form of BDF-4 (see Table 2.1). The terms involving h^p and higher constitute the truncation error τ_k (see Definition 2.5.5). For more details regarding the error constant, see Chapter 1.

We can now investigate the method's stability. Using the R package `rootSolve` (see Listing 2.5) we find that the roots of $\rho(z) = 0$ of eqn. (2.40) are

$$0.3815$$
$$0.2693 + 0.4920i$$
$$0.2693 - 0.4920i$$
$$1.0000$$

```
# File: BDF4_roots.R
cat("\014") # Clear console
rm(list = ls(all = TRUE)) # Delete workspace
# BDF-4 roots
library(rootSolve)
# Find complex roots - eqn. (39)
# Must supply coefficients in increasing order
zz <- polyroot(c(+3/12,-16/12,+36/12,-48/12,25/12))
print(zz)
plot(Re(zz), Im(zz), pch=16, cex=1.25,xlim=c(0,1),ylim=c(-0.6,0.6))
grid(lty=1)
```

Listing 2.5. File: BDF4_roots.R—Code to compute complex roots of eqn. (2.40) for BDF-4 and plot results

Thus, all the roots conform to $|z| \leq 1$ and those where $|z| = 1$ are simple and not repeated. Therefore, the *root condition* is satisfied and, according to Definition 2.5.2, the method is zero-stable; thus, the solution will not grow unbounded. Furthermore, because we have shown that the method is both consistent and stable, this means the method is also *convergent*! Note that BDFs of order 7 and higher do not satisfy the root condition.

Finally, we calculate the region of absolute stability. We do this by plotting the contour Γ in the complex plane, as defined in Definition 2.5.3, that is,

$$h\lambda = \frac{\rho(z)}{\sigma(z)} = \frac{\frac{25}{12}z^4 - \frac{48}{12}z^3 + \frac{36}{12}z^2 - \frac{16}{12}z + \frac{3}{12}}{z^4}, \quad z = e^{i\theta}, \quad \theta = [0, 2\pi]. \quad (2.46)$$

We see that eqn. (2.46) is equal to eqn. (2.39), which we arrived at heuristically by employing a Taylor series expansion. Hence, we get the same contour when plotted (see Fig. 2.4). To determine whether the stable region is interior or exterior to the contour Γ, plot θ equal to one of the roots of eqn. (2.39). If this point is exterior to Γ, then the region of absolute stability is external to Γ. Similarly, if the point is interior to Γ, then the stable region will be interior to Γ. In the case of BDFs, the region of absolute stability is external to Γ.

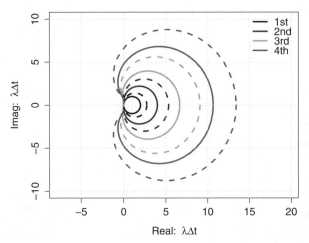

Figure 2.4. Stability contours Γ generated by the R code of Listing 2.6 for BDFs (solid) and NDFs (dotted). Stable regions are external to the contours, which are arranged from order 1 (inner) to order 4 (outer).

We have demonstrated that BDF-4 is both *consistent* and *stable*; therefore, in accordance with Definition 2.5.4, we can finally say the system is also *convergent*. R code for plotting the BDF stability contours is shown in Listing 2.6.

Control engineers will have recognized that we could have used Laplace transforms to transform eqn. (2.36) directly to

$$s\Delta t = \tfrac{1}{12}\left(25 - 48e^{-s\Delta t} + 36e^{-2s\Delta t} - 16e^{-3s\Delta t} + 3e^{-4s\Delta t}\right), \quad (2.47)$$

where $e^{-s\Delta t}$ represents in the complex frequency domain a time delay of Δt in the time domain (where we have used Δt rather than h). We could then obtain the same plot as earlier in the $s\Delta t$-plane by letting $s\Delta t$ on the RHS vary from $-\pi i$ to $+\pi i$. Practically, this has the effect of mapping the left-half plane in the complex frequency domain to the exterior of the contour in the $s\Delta t$-plane. However, theoretically, s should vary along a contour that encompasses the whole left-half plane. Fortunately, it is only necessary to vary s from $-\pi i$ to $+\pi i$ as the effect of sampling creates side-bands making the s-domain periodic along the imaginary axis, that is, $s = \sigma + i\omega = \sigma + i\left[\omega \pm 2n\pi\right], [n = 0, 1, \ldots, \infty]$.

We can also use z-transforms to further transform eqn. (2.47). By letting $z = e^{s\Delta t}$, we map from the s-plane to the z-plane and obtain

$$s\Delta t = \tfrac{1}{12}\left(25 - 48z^{-1} + 36z^{-2} - 16z^{-3} + 3z^{-4}\right), \quad (2.48)$$

which is equivalent to eqn. (2.39). Again, setting $z = e^{i\theta}$ and letting θ vary from $-\pi$ to $+\pi$ causes z to move round the unit circle in the z-plane when the stable interior of the unit circle will map to the stable exterior of the resulting contour in the $\lambda\Delta t$-plane. This is readily accomplished using R.

Table 2.3. The NDF coefficients with step size improvements and A(α) stability comparisons relative to BDFs [Sha-97]

Order k	NDF coeffs. κ_k	NDF coeffs. γ_k	Step size improvement (percent)	A(α) stability angle BDF	A(α) stability angle NDF	Percentage change
1	$-\frac{37}{200}$	1	26%	90°	90°	0%
2	$-\frac{1}{9}$	$\frac{3}{2}$	26%	90°	90°	0%
3	$-\frac{823}{10000}$	$\frac{11}{6}$	26%	86°	80°	−7%
4	$-\frac{83}{2000}$	$\frac{25}{12}$	12%	73°	66°	−10%
5	-0	$\frac{137}{60}$	0%	51°	51°	0%

2.5.3 Numerical Differentiation Formulas (NDFs)

The NDF method described in Chapter 3 can be written as

$$\sum_{m=1}^{k} \frac{1}{m} \nabla^m y_{n+1} - hf(t_{n+1}, y_{n+1}) - \kappa \gamma_k \left(y_{n+1} - y_{n+1}^{(0)} \right) = 0, \quad (2.49)$$

where ∇^m represents the mth backward difference operator, κ is a scalar parameter, and

$$\gamma_k = \sum_{j=1}^{k} \frac{1}{j}. \quad (2.50)$$

We can now investigate the stability of NDFs using the same approach that we used to investigate BDFs. Using the *normalized* BDF coefficients from Table 2.2 and values for κ and γ from Table 2.3, we will illustrate the method by application to NDF-4, which can be described by

$$hf(t_{n+1}, y_{n+1}) = \frac{25}{12} y_{n+1} - 4y_n + 3y_{n-1} - \frac{4}{3} y_{n-2} + \frac{1}{4} y_{n-3}$$
$$- \kappa_4 \gamma_4 \left(y_{n+1} - 5y_n + 10y_{n-1} - 10y_{n-2} + 5y_{n-3} - y_{n-4} \right). \quad (2.51)$$

Applying the Dahlquist test problem to eqn. (2.51), we arrive at the following result:

$$h\lambda = \frac{1}{12} \left(25 - 48z^{-1} + 36z^{-2} - 16z^{-3} + 3z^{-4} \right)$$
$$- \kappa_4 \gamma_4 (1 - 5z^{-1} + 10z^{-2} - 10z^{-3} + 5z^{-4} - z^{-5}). \quad (2.52)$$

Equation (2.52) is similar to eqn. (2.39), which we obtained for BDF-4. As for BDF-4, we can now perform a mapping in the complex frequency domain of eqn. (2.52) from the z-plane to the λh-plane by setting $z = \exp(i\theta)$ and letting θ vary from $-\pi$ to $+\pi$. The resulting contour is shown in Fig. 2.4, where the region of absolute stability (exterior to the contour Γ) has been plotted in accordance with Definition 2.5.3. Other NDF schemes are handled in the same way, and the resulting contours for NDF-1, NDF-2, and NDF-3 are also included in the figure, along with corresponding BDF contours for comparison. The R code for generating Fig. 2.4 and the stability angles listed in Table 2.3 is included in Listing 2.6.

```r
# File: BDF_NDFstabilityPlots.R
# Purpose: To plot stability contours for BDF and NDF methods
rm(list = ls(all = TRUE)) # Delete workspace
#
ptm = proc.time() # start process timer
#
theta <- seq(-pi, pi, length=100)
z <- exp(1i*theta)
# BDF 1st order
mu1 <-(1 - z^-1)
# NDF 1st order
K1<- -0.1850; gamma1<-1;
nu1 <- mu1 - K1*gamma1*(1 - 2*z^-1 + z^-2)
# BDF 2nd order
mu2 <-(3 - 4*z^-1 + z^-2)/2
# NDF 2nd order
K2<--1/9; gamma2<-3/2
nu2 <- mu2 - K2*gamma2*(1 - 3*z^-1 + 3*z^-2 - z^-3)
# BDF 3rd order
mu3 <-(11 - 18*z^-1 + 9*z^-2 - 2*z^-3)/6
# NDF 3rd order
K3<- -0.0823; gamma3<-11/6;
nu3 <- mu3 - K3*gamma3*(1 - 4*z^-1 + 6*z^-2 - 4*z^-3 + 1*z^-4)
# BDF 4th order
mu4 <-(25 - 48*z^-1 + 36*z^-2 - 16*z^-3 + 3*z^-4)/12
# NDF 4th order
K4<--0.0415; gamma4<-25/12;
nu4 <- mu4 - K4*gamma4*(1-5*z^-1 + 10*z^-2 - 10*z^-3 + 5*z^-4 - z^-5)
par(mar = c(5,5,2,5))
plot.new()
plot(mu1,col="black",type="l",lwd=3,xlim=c(-4,16),ylim=c(-10,10),asp=1,
     xlab=expression(paste("Real: ",lambda,Delta,"t")),
     ylab=expression(paste("Imag: ",lambda,Delta,"t")))
lines(mu2,col="blue"  ,lty=1,lwd=3)
lines(mu3,col="green",lty=1,lwd=3)
lines(mu4,col="red"  ,lty=1,lwd=3)
lines(nu1,col="black",lty=2,lwd=3)
lines(nu2,col="blue"  ,lty=2,lwd=3)
lines(nu3,col="green",lty=2,lwd=3)
lines(nu4,col="red"  ,lty=2,lwd=3)
#
grid(col="lightgray",lty=1,lwd=2)
title("Stability Contours for BDFs (solid) and NDFs (dashed)");
legend("topright",legend=c("1st ","2nd   ","3rd   ","4th   "),bg="white",
       lwd=c(3,3,3),lty=c(1,1,1),bty="n", col=c("black", "blue", "green","red"))
# Calculate alpha stability angles
```

```
# Note: alpha <= 90 deg.
# ============================
alpha_BDF1 <- min(90,180*(1-max(Arg(mu1))/pi))
alpha_NDF1 <- min(90,180*(1-max(Arg(nu1))/pi))
alpha_BDF2 <- min(90,180*(1-max(Arg(mu2))/pi))
alpha_NDF2 <- min(90,180*(1-max(Arg(nu2))/pi))
alpha_BDF3 <- min(90,180*(1-max(Arg(mu3))/pi))
alpha_NDF3 <- min(90,180*(1-max(Arg(nu3))/pi))
alpha_BDF4 <- min(90,180*(1-max(Arg(mu4))/pi))
alpha_NDF4 <- min(90,180*(1-max(Arg(nu4))/pi))
cat(sprintf('\n\nAlpha Stability Angles - degrees\n'))
cat(sprintf('-------------------------------\n'))
cat(sprintf('\nOrder:  1     2     3     4\n'))
cat(sprintf('===============================\n'))
cat(sprintf('BDF =%5.2f =%5.2f =%5.2f =%5.2f\n',
    alpha_BDF1, alpha_BDF2, alpha_BDF3, alpha_BDF4))
cat(sprintf('NDF =%5.2f =%5.2f =%5.2f =%5.2f\n',
    alpha_NDF1, alpha_NDF2, alpha_NDF3, alpha_NDF4))
cat(sprintf('-------------------------------\n\n'))
#
cat(sprintf("time taken by process = %5.3f s\n",(proc.time()-ptm)[3]))
```

Listing 2.6. File: BDF_NDFstabilityPlots.R—Code to plot BDF and NDF stability contours

The angle α, corresponding $A(\alpha)$-stability (see Table 2.3), is obtained from the argument of $h\lambda\left(e^{j\theta}\right), \theta \in (0, 2\pi)$, that is,

$$\alpha = \min\left[\frac{\pi}{2}, \pi - \sup_{\theta \in (0,2\pi)} \left(\arg\left\{h\lambda\left(e^{j\theta}\right)\right\}\right)\right]. \quad (2.53)$$

The leading term of the truncation error for a k-step NDF (normalized coefficients) is given by

$$-\left(\kappa\gamma + \frac{1}{k+1}\right)h^{k+1}y^{[k+1]} \approx -\left(\kappa\gamma + \frac{1}{k+1}\right)h^{k+1}\nabla^{k+1}y_{n+1}, \quad (2.54)$$

where $y^{[k+1]}$ represents the $(k+1)$ derivative of y.

Thus, from eqn. (2.54), it is clear that for correctly chosen *negative* values of $\kappa_k\gamma_k$, the truncation error will be made smaller. However, a trade-off has to be made between a smaller truncation error and reduced stability. The constants in Table 2.3 were chosen to *achieve good improvements in efficiency, while not reducing the stability angle excessively*.

2.5.4 Adams Methods

The stability of the Adams methods described in Chapter 3 can also be analyzed using the same methods as employed to analyze the BDF methods.

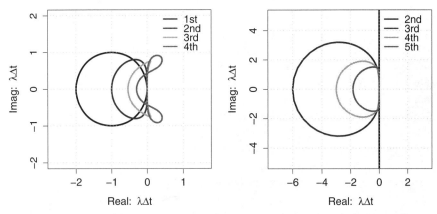

Figure 2.5. (left) Adams–Bashforth stability contours. (right) Adams–Moulton stability contours. Contours are arranged from low order (outer) to high order (inner). The stable region for AM-2 is the whole left-half plane; otherwise, the stable regions for both AB and AM are internal to the contours.

The *Adams–Bashforth k*-step method has order k and is defined as

$$y_{n+1} - y_n = h \sum_{j=1}^{k} \beta_j f_{n+1-j}. \quad (2.55)$$

The *Adams–Moulton* $(k-1)$-step method has order k and is defined as

$$y_{n+1} - y_n = h \sum_{j=0}^{k-1} \beta_j f_{n+1-j}. \quad (2.56)$$

Symbols in eqns. (2.55) and (2.56) have the usual meaning.

Applying the Dahlquist test-problem, eqn. (2.2), to eqn. (2.55), we arrive at the following result for the *third-order, three-step Adams–Bashforth* method:

$$h\lambda = \frac{(z-1)}{\left(\frac{23}{12} - \frac{16}{12}z^{-1} + \frac{5}{12}z^{-2}\right)} = \frac{\rho(z)}{\sigma(z)}. \quad (2.57)$$

We could also have arrived at eqn. (2.57) by use of eqn. (2.30).

We now perform a mapping in the complex frequency domain of eqn. (2.57) from the z-plane to the λh-plane by setting $z = \exp(i\theta)$ and letting θ vary from $-\pi$ to $+\pi$. The resulting contour for AB-3 is shown in Fig. 2.5, where the region of absolute stability (interior to the contour Γ) has been plotted in accordance with Definition 2.5.3. Other Adams–Bashforth schemes are handled in the same way, and the resulting contours for AB-1, AB-2, and AB-4 are also included in the figure.

In a similar manner, we apply the Dahlquist test-problem, eqn. (2.2), to eqn. (2.56), when we arrive at the following result for the *third-order, two-step Adams–Moulton* method:

$$h\lambda = \frac{(z-1)}{\left(\frac{5}{12}z + \frac{8}{12} - \frac{1}{12}z^{-1}\right)} = \frac{\rho(z)}{\sigma(z)}. \quad (2.58)$$

Again we perform a mapping in the complex frequency domain of eqn. (2.58) from the z-plane to the λh-plane by setting $z = \exp(i\theta)$ and letting θ vary from $-\pi$ to $+\pi$. The resulting contour Γ is shown in Fig. 2.5, where the region of absolute stability (LHS of λh-plane for $k = 2$ and interior to the contour for $k \geq 3$) has been plotted in accordance with Definition 2.5.3. Other Adams–Moulton schemes are handled in the same way, and the resulting contours for AM-2, AM-3, AM-4, and AM-5 are also included in Fig. 2.5.

The R code for generating Fig. 2.5 is included in Listing 2.7.

```
# File: AB_AMstabilityPlots.R
# Purpose: To plot stability contours for Adams methods
rm(list = ls(all = TRUE)) # Delete workspace
#
ptm = proc.time() # start process timer
#
theta <- seq(-pi, pi, length=100)
z <- exp(1i*theta)
# AB-1, 1st order
beta1<-1;
ab_mu1 <-(z - 1)/beta1;
# AM-1, 1st order
beta0<-1/2; beta1<-1/2;
am_mu1 <-(z-1)/(beta0*z+beta1);

# AB-2, 2nd order
beta1<-3/2; beta2<--1/2;
ab_mu2 <-(z-1)/(beta1 + beta2*z^(-1));
# # AM-2 2nd order
beta0<-5/12; beta1<-8/12; beta2<- -1/12;
am_mu2 <-(z-1)/(beta0*z + beta1 + beta2*z^(-1));
#
# AB-3 3rd order
beta1<-23/12; beta2<- -16/12; beta3<-5/12;
ab_mu3 <-(z-1)/(beta1 + beta2*z^-1 + beta3*z^(-2));
# AM-3 3rd order
beta0<-9/24; beta1<-19/24; beta2<- -5/24; beta3<-1/24;
am_mu3 <-(z-1)/(beta0*z + beta1 + beta2*z^-1 + beta3*z^(-2));
#
# AB-4 4th order
beta1<-55/24; beta2<- -59/24; beta3<-37/24; beta4<--9/24;
ab_mu4 <-(z-1)/(beta1 + beta2*z^-1 + beta3*z^-2 + beta4*z^(-3));
# AM-4 4th order
beta0<-251/720; beta1<-646/720; beta2<- -246/720; beta3<-106/720; beta4<- -19/
     720
am_mu4 <-(z-1)/(beta0*z + beta1 + beta2*z^-1 + beta3*z^(-2)+ beta4*z^(-3));
#
# Plot Adams-Bashforth stability contours
```

```
par(mar = c(5,5,2,5))
plot.new()
plot(ab_mu1,col="black",type="l",lwd=2,xlim=c(-2.5,0.5),ylim=c(-1.5,1.5),asp=1,
     xlab=expression(paste("Real: ",lambda,Delta,"t")),
     ylab=expression(paste("Imag: ",lambda,Delta,"t")))
lines(ab_mu2,col="blue" ,lty=1,lwd=2)
lines(ab_mu3,col="green",lty=1,lwd=2)
lines(ab_mu4,col="red"  ,lty=1,lwd=2)
#
grid(col="lightgray",lwd=2)
title("Stability Contours for Adams-Basford schemes");
legend("topright",legend=c("1st ","2nd   ","3rd    ","4th   "),
       lwd=c(2,2,2),lty=c(1,1,1),bty="n", col=c("black", "blue", "green","red"))
#
# Plot Adams-Moulton stability contours
plot(am_mu1,col="black",type="l",lwd=2,xlim=c(-7,3),ylim=c(-5,5),asp=1,
     xlab=expression(paste("Real: ",lambda,Delta,"t")),
     ylab=expression(paste("Imag: ",lambda,Delta,"t")))
lines(am_mu2,col="blue" ,lty=1,lwd=2)
lines(am_mu3,col="green",lty=1,lwd=2)
lines(am_mu4,col="red"  ,lty=1,lwd=2)
#
grid(col="lightgray",lwd=2)
title("Stability Contours for Adams-Moulton schemes");
legend("topright",legend=c("1st ","2nd   ","3rd    ","4th   "),
       lwd=c(2,2,2),lty=c(1,1,1),bty="n", col=c("black", "blue", "green","red"))
#
cat(sprintf("time taken by process = %5.3f s\n",(proc.time()-ptm)[3]))
```

Listing 2.7. File: AB_AMstabilityPlots.R—Code to plot Adams–Bashforth and Adams–Moulton stability contours

We will now check the convergence properties of AB-3 and AM-3.

From eqn. (2.30), we see that we have the following *first* and *second characteristic polynomials* for the third-order Adams–Bashforth method:

$$\rho(z) = z - 1$$
$$\sigma(z) = \frac{23}{12} - \frac{16}{12}z^{-1} + \frac{5}{12}z^{-2}. \tag{2.59}$$

From eqn. (2.59) we see that

$$\rho(1) = 0 \text{ and } \rho'(1) = \sigma(1) = 1. \tag{2.60}$$

This establishes that AB-3 is *consistent* in accordance with Definition 2.5.1. Furthermore, we see that the *first characteristic polynomial*, $\rho(1) = 0$, has a simple root at $z = 1$. Thus, all the roots conform to $|z| \leq 1$, and those where $|z| = 1$ are simple and not repeated. Therefore, the *root condition is satisfied*, and according to Definition 2.5.2, the method is *zero-stable* and the solution will not grow unbounded. Furthermore, because we have shown that the method is both *consistent* and *stable*, it means that this method is also *convergent*!

Similarly, we see that we have the following first and second characteristic polynomials for the third-order Adams–Moulton method:

$$\rho(z) = z - 1$$
$$\sigma(z) = \frac{5}{12}z + \frac{8}{12} - \frac{1}{12}z^{-1}. \qquad (2.61)$$

From equation (2.61) we see that

$$\rho(1) = 0 \quad \text{and} \quad \rho'(1) = \sigma(1) = 1. \qquad (2.62)$$

This establishes that AM-3 is also *consistent* in accordance with Definition 2.5.1. We also see that again the first characteristic polynomial, $\rho(1) = 0$, has a simple root at $z = 1$. Thus, all the roots conform to $|z| \leq 1$, and those where $|z| = 1$ are simple and not repeated. Therefore, the *root condition is satisfied*, and according to Definition 2.5.2, the method is *zero-stable* and the solution will not grow unbounded. Again, because we have shown that this method is both *consistent* and *stable*, it means that this method is also *convergent*!

REFERENCES

[But-07] Butcher, J. (2007), Runge-Kutta Methods, *Scholarpedia* **2–9**, 3147. Available online at http://www.scholarpedia.org/article/Runge-Kutta_methods.

[Dah-56] Dahlquist, G. (1956), Convergence and Stability in the Numerical Integration of Ordinary Differential Equations, *Mathematica Scandinavica* **4**, 33–53.

[Dah-63] Dahlquist, G. (1963), A Special Stability Problem for Linear Multistep Methods, *BIT Numerical Mathematics* **3**, 27–43.

[Dah-75] Dahlquist, G. (1975), Error Analysis for a Class of Methods for Stiff Nonlinear Initial Value Problems, *Procs. Numerical Analysis Conference*, Dundee, Lecture Notes in Math. 506, G.A. Watson (ed.), Springer Verlag, 1976, pp. 60–74.

[Feh-69] Fehlberg, E. (1969), Low-Order Classical Runge-Kutta Formulas with Stepsize Control and Their Application to Some Heat Transfer Problems, NASA Technical Report, TR R-315.

[Gea-71] Gear, C. W. (1971), Numerical Initial Value Problems in Ordinary Differential Equations, Prentice Hall.

[Goe-14] Goedman, G., and G. Grothendieck (2014), R Interface to the yacas Computer Algebra System Ryacas. Available online at http://cran.r-project.org/web/packages/Ryacas/Ryacas.pdf.

[Lam-91] Lambert, J. D. (1991), *Numerical Methods for Ordinary Differential Systems: The Initial Value Problem*, John Wiley.

[Ric-71] Richtmyer, R. D. and K. W. Morton (1967), *Difference Methods for Initial-Value Problems*, 2nd ed., John Wiley.

[Sha-94] Shampine, L. F. (1994), *Numerical Solution of Ordinary Differential Equations*, Chapman Hall.

[Sha-97] Shampine, L. F. and M. W. Reichelt (1997), The Matlab ODE Suite. *SIAM Journal on Scientific Computing* **1–18, 1**–22.

[Sch-91] Schiesser, W. E. (1991), *The Numerical Method of Lines: Integration of Partial Differential Equations*, Academic Press.

[Som-94] Sommeijer, B. P., P. J. van der Houwen and J. Kok (1994), Time Integration of Three-Dimensional Numerical Transport Models, *Applied Numerical Mathematics* **16**, 201–225.

[Wes-01] Wesseling, P. (2001), *Principles of Computational Fluid Dynamics, Springer*.

3
Numerical Solution of PDEs

3.1 SOME PDE BASICS

Our physical world is most generally described in scientific and engineering terms with respect to three-dimensional space and time, which is commonly referred to as *space-time*. An *evolution* equation is a *partial differential equation* (PDE) that describes mathematically the evolution over space and time of a physical system or process. PDEs are therefore among the most widely used forms of mathematics in many disciplines. As a consequence, methods for the solution of PDEs, such as the method of lines (MOL) [Sch-91], are of broad interest in science and engineering. Our main focus will be on MOL.

As a basic illustrative example of a PDE, we consider

$$\frac{\partial u}{\partial t} = D \frac{\partial^2 u}{\partial x^2}, \qquad (3.1)$$

where

- u is a dependent variable (dependent on x and t)
- t is an independent variable representing time
- x is an independent variable representing one dimension of three-dimensional space
- D real positive constant, explained later.

Note that eqn. (3.1) has two independent variables, x and t, which is the reason it is classified as a PDE (any differential equation with *more than one independent variable is a PDE*). A differential equation with only one independent variable is generally termed an *ordinary differential equation* (ODE); we will consider ODEs later as part of the MOL.

Equation (3.1) is termed the *diffusion equation* or *heat equation*. When applied to heat transfer, it is *Fourier's second law*; the dependent variable u is temperature and D is the *thermal diffusivity*. When eqn. (3.1) is applied to mass diffusion, it is *Fick's second law*; u is mass concentration and D is the *coefficient of diffusion* or the *diffusivity*.

The symbol $\frac{\partial u}{\partial t}$ is a partial derivative of u with respect to t (x is held constant when taking the derivative, which is why *partial* is used to describe this derivative). Equation (3.1) is *first order in t* because the highest order partial derivative in t is first order; it is *second order in x* because the highest order derivative in x is second order. Equation (3.1) is *linear* or *first degree* because all of the terms are to the first power (note that *order* and *degree* can be easily confused).

3.2 INITIAL AND BOUNDARY CONDITIONS

Before we can consider obtaining a solution to eqn. (3.1), we must specify some auxiliary conditions to complete the statement of the PDE problem. The number of required auxiliary conditions is determined by the *highest order derivative in each independent variable*. Because eqn. (3.1) is first order in t and second order in x, it requires one auxiliary condition in t and two auxiliary conditions in x. To have a complete *well-posed* problem, some additional conditions may have to be included, for example, that specify valid ranges for coefficients [Cra-47]. However, this is a more advanced topic and will not be developed further here.

The variable t is termed an *initial value variable* and therefore requires *one initial condition (IC)*. It is an initial value variable since it starts at an initial value, t_0, and moves forward over a *finite interval* $t_0 \leq t \leq t_f$ or a *semi-infinite interval* $t_0 \leq t \leq \infty$ without any additional conditions being imposed. Typically in a PDE application, the initial value variable is time, as in the case of eqn. (3.1).

The variable x is termed a *boundary value variable* and therefore requires *two boundary conditions (BCs)*. It is a boundary value variable since it varies over a *finite interval* $x_0 \leq x \leq x_f$, a *semi-infinite interval* $x_0 \leq x \leq \infty$ or a *fully infinite interval* $-\infty \leq x \leq \infty$, and at *two different values of x*, conditions are imposed on u in eqn. (3.1). Typically, the two values of x correspond to boundaries of a physical system, hence the name *boundary conditions*.

As examples of auxiliary conditions for eqn. (3.1),

- an IC could be
$$u(x, t = 0) = u_0(x), \tag{3.2}$$
where u_0 is a given function of x
- two BCs could be
$$u(x = a, t) = u_a(t), \quad x \in \partial\Omega \tag{3.3}$$
$$\frac{\partial u(x = x_b, t)}{\partial n} = f_b(u_b, t), \quad x \in \partial\Omega, \tag{3.4}$$

where u_a is a given *constant boundary* value of u at $x = a$ (possibly time varying). The symbol Ω represents the problem *spatial domain* and $\partial\Omega$ the *boundary* of Ω; n represents the outward normal to $\partial\Omega$ and f_b is a given *first derivative boundary* at $x = b$ (possibly time varying). Note that $\frac{\partial u}{\partial n}$ can also be written as $\frac{\partial \Omega}{\partial n}$ or $\partial\Omega_n$. This is illustrated in Fig. 3.1 for different domain types.

We now define the three types of boundary conditions:

1. If the dependent variable is specified as eqn. (3.3), that is, a constant, it is termed a *BC of the first type* or a *Dirichlet*[1] *BC*.

---

[1] The term was adopted in recognition of the work of *Johann Peter Gustav Lejeune Dirichlet* (1805–1859) [Che-05] on formalizing the *Theory of Fourier Series*. Dirichlet was a German mathematician and professor at the universities of Breslau, Berlin and Göttingen.

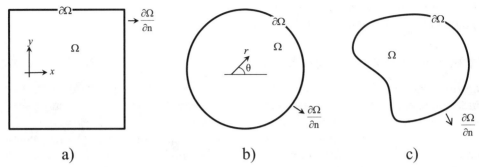

Figure 3.1. Examples of different 2D boundaries for (a) rectangular domain; (b) circular domain; and (c) irregular domain.

2. If the derivative of the dependent variable is specified, as in eqn. (3.4), it is termed a *BC of the second type* or a *Neumann*[2] *BC* (note that it does not have to be time varying).
3. If both the dependent variable and its derivative appear in the BC, it is termed a *BC of the third type* or a *Robin*[3] *BC*.

So in summary we have

Table 3.1. Summary of variables that have to be defined on the boundary for different types of BC

1st type BC Dirichlet	2nd type BC Neumann	3rd type BC Robin
u	$\dfrac{\partial u}{\partial n}$	$a_1 u + a_2 \dfrac{\partial u}{\partial n}$

Problems specified with Dirichlet BCs are known as *Dirichlet problems*; problems specified with Neumann BCs are known as *Neumann problems*; and problems specified with both Dirichlet and Neumann BCs are known as *mixed problems*.

For some problems the solution can be extended to an infinite domain by use of *periodic boundary conditions*. For example, for eqn. (3.1) if the function $u(x, t)$ satisfies the initial and boundary conditions

$$\begin{aligned}
u(x, 0) &= u_0(x), & x &\in (a, b), & &\text{IC,} \\
u(a, t) &= u(b, t), & t &> 0, & &\text{Dirichlet periodic BC,} \\
\frac{\partial u(a, t)}{\partial x} &= \frac{\partial u(b, t)}{\partial x}, & t &> 0, & &\text{Neumann periodic BC,}
\end{aligned} \quad (3.5)$$

---

[2] The term was adopted in recognition of the work of *Carl Gottfried Neumann* (1832–1925) on the *Dirichlet problem* [Che-05]. Neumann was a German mathematician and professor at the universities of Halle, Basel, Tübingen, and Leipzig. Carl Neumann was the son of German mineralogist, physicist, and mathematician *Franz Ernst Neumann* (1798–1895), who was professor of mineralogy and physics at Königsberg University.
[3] The term was apparently named after (*Victor*) *Gustave Robin* (1855–1897), who lectured in mathematical physics at the Sorbonne in Paris and worked in the area of thermodynamics [Gus-98a, Gus-98b, Gus-98c].

then spatially we can extend u periodically to $x \in (-\infty, \infty)$. The extended conditions satisfy the differential equation for all $x \in (-\infty, \infty)$; thus, because the function and derivative values match at the boundaries, it follows that the second derivative values will also match. This approach will be discussed further in examples that are presented subsequently.

A fourth type of boundary condition is the *absorbed* BC that is sometimes required for infinite domain problems, such as occur in electromagnetism. We shall not discuss this aspect further.

Under certain circumstances, boundary conditions may also be required at certain physical interfaces within the simulation region.

Discontinuities at the boundaries, produced, for example, by differences in initial and boundary conditions at the boundaries, can cause computational difficulties, particularly for hyperbolic problems.

To ensure that BCs are accounted for in the solution of a PDE, BCs have to be incorporated into the discretization process. This aspect is discussed in Section 3.8.2.

3.3 TYPES OF PDE SOLUTIONS

Equations (3.1), (3.2), (3.3), and (3.4) constitute a complete PDE problem, and we can now consider what we mean by a solution to this problem. Briefly, the solution of a PDE problem is a *function that defines the dependent variable as a function of the independent variables*, in this case $u(x, t)$. In other words, we seek a function that when substituted in the PDE and all of its auxiliary conditions satisfies simultaneously all of these equations.

The solution can be of two types:

1. If the solution is an actual mathematical function, it is termed an *analytical solution*. While analytical solutions are the *gold standard* for PDE solutions in the sense that they are exact, they are also generally difficult to derive mathematically for all but the simplest PDE problems (in much the same way that solutions to nonlinear algebraic equations generally cannot be derived mathematically, except for certain classes of nonlinear equations).
2. If the solution is in numerical form, for example, $u(x, t)$ tabulated numerically as a function of x and t, it is termed a *numerical solution*. Ideally, the numerical solution is simply a numerical evaluation of the analytical solution. But since an analytical solution is generally unavailable for realistic PDE problems in science and engineering, the *numerical solution is an approximation to the analytical solution*, and our expectation is that it represents the analytical solution with good accuracy. However, numerical solutions can be computed with modern-day computers for very complex problems, and they will generally have good accuracy (even though this cannot be established directly by comparison with the analytical solution since the latter is usually unknown). The focus of numerical solutions in this book is the MOL, with the emphasis on the *calculation of accurate numerical solutions*.

3.4 PDE SUBSCRIPT NOTATION

Before we go on to the general classes of PDEs that the MOL can handle, we briefly discuss an alternative notation for PDEs. Instead of writing the partial derivatives as in eqn. (3.1), we also adopt a *subscript notation* that is easier to state and bears a closer resemblance to the associated computer coding. For example, we can write eqn. (3.1) as

$$u_t = D u_{xx}, \qquad (3.6)$$

where, for example, u_t is subscript notation for $\frac{\partial u}{\partial t}$. In other words, a partial derivative is represented as the dependent variable with a subscript that defines the independent variable. For a derivative that is of order n, the independent variable is repeated n times, for example, for eqn. (3.1), u_{xx} represents $\frac{\partial^2 u}{\partial x^2}$. We will use both notations in this book, choosing the most appropriate for the discussion in hand.

3.5 A GENERAL PDE SYSTEM

Using the subscript notation, we can now consider some general PDEs. For example, a general PDE first order in t can be considered

$$\overline{u}_t = \overline{f}(\overline{x}, t, \overline{u}, \overline{u}_{\overline{x}}, \overline{u}_{\overline{xx}}, \cdots), \qquad (3.7)$$

where an overbar (overline) denotes a vector. For example, $\overline{u}$ denotes a vector of n dependent variables

$$\overline{u} = (u_1, u_2, \ldots, u_n)^T, \qquad (3.8)$$

that is, a system of n simultaneous PDEs. Similarly, $\overline{f}$ denotes a vector of n derivative functions

$$\overline{f} = (f_1, f_2, \ldots, f_n)^T, \qquad (3.9)$$

where T denotes the *transpose* operator, that is, the row vector is transposed to a column vector. Note also that $\overline{x}$ is a vector of spatial coordinates, so that, for example, in *Cartesian coordinates*, $\overline{x} = (x, y, z)^T$, while in *cylindrical coordinates*, $\overline{x} = (r, \theta, z)^T$. Thus, eqn. (3.7) can represent PDEs in one, two, and three spatial dimensions.

Because eqn. (3.7) is first order in t, it requires one initial condition

$$\overline{u}(\overline{x}, t = 0) = \overline{u}_0(\overline{x}, \overline{u}, \overline{u}_{\overline{x}}, \overline{u}_{\overline{xx}}, \cdots), \qquad (3.10)$$

where $\overline{u}_0$ is a vector of n initial condition functions

$$\overline{u}_0 = (u_{10}, u_{20}, \ldots, u_{n0})^T. \qquad (3.11)$$

The derivative vector $\overline{f}$ of eqn. (3.7) includes functions of various spatial derivatives, that is, $(\overline{u}, \overline{u}_{\overline{x}}, \overline{u}_{\overline{xx}}, \cdots)$, and therefore we cannot state a priori the required number of BCs. For example, if the highest-order derivative in $\overline{x}$ in all of the derivative functions is second order, then we require $2 \times n$ BCs. We state the general BC requirement of eqn. (3.7) as

$$\overline{f}_b(\overline{x}_b, \overline{u}, \overline{u}_{\overline{x}}, \overline{u}_{\overline{xx}}, \cdots, t) = 0, \qquad (3.12)$$

where the subscript b denotes *boundary*. The vector of boundary condition functions, $\overline{f}_b$, has a length (number of functions) determined by the highest-order derivative in $\overline{x}$ in each PDE in eqn. (3.7), as discussed previously.

3.6 CLASSIFICATION OF PDES

Equations (3.7), (3.10), and (3.12) constitute a general PDE system to which the MOL can be applied. Before proceeding to the details of how this might be done, we need to discuss the three basic forms of the PDEs as classified geometrically. First we will provide a descriptive form of these functions for the specification of the three geometric classes and then follow up with some background details of their geometric origins.

If the derivative functions in eqn. (3.7) contain *only first-order derivatives in* $\bar{x}$, the PDEs are classified as *first-order hyperbolic*. As an example, the equation

$$u_t + v u_x = 0 \tag{3.13}$$

is generally called the *linear advection equation*; in physical applications, v is a linear or flow velocity. Although eqn. (3.13) is possibly the simplest PDE, this simplicity is deceptive in the sense that it is can be very difficult to integrate numerically because it *propagates discontinuities*, which is a distinctive feature of first-order hyperbolic PDEs.

Equation (3.13) is termed a *conservation law* because it expresses conservation of *mass, energy, or momentum* under the conditions for which it is derived, that is, the *assumptions on which the equation is based*. Conservation laws are a bedrock of PDE mathematical models in science and engineering, and an extensive literature pertaining to their solution, both analytical and numerical, has evolved over many years (see Chapter 8).

An example of a *first-order hyperbolic system* (using the notation $u_1 \Rightarrow u, u_2 \Rightarrow v$) is

$$\begin{aligned} u_t &= v_x \\ v_t &= u_x. \end{aligned} \tag{3.14}$$

Equations (3.14) constitute a system of *two linear, constant coefficient, first-order hyperbolic PDEs*.

Differentiation and algebraic substitution can occasionally be used to eliminate some dependent variables in systems of PDEs. For example, if the first of eqns. (3.14) is differentiated with respect to t and the second is differentiated with respect to x, we obtain

$$\begin{aligned} u_{tt} &= v_{xt} \\ v_{tx} &= u_{xx}. \end{aligned} \tag{3.15}$$

We can then eliminate the mixed partial derivative between these two equations (assuming u and v are well behaved functions and that v_{xt} in the first equation equals v_{tx} in the second equation) to obtain

$$u_{tt} = u_{xx}. \tag{3.16}$$

Equation (3.16) is the *second-order hyperbolic wave equation*.

If the derivative functions in eqn. (3.7) contain *only second-order derivatives in* $\bar{x}$, the PDEs are classified as *parabolic*. Equation (3.1) is an example of a parabolic PDE.

Finally, if a PDE *contains no derivatives in t* (e.g., the LHS of eqn. (3.7) is zero), it is classified as *elliptic*. As an example,

$$u_{xx} + u_{yy} = 0 \tag{3.17}$$

is *Laplace's equation*, where x and y are independent spatial variables in Cartesian coordinates. Note that with no derivatives in t, elliptic PDEs require no ICs, that is, they are entirely boundary value PDEs.

PDEs with *mixed geometric characteristics* are possible and, in fact, are quite common in applications. For example, the PDE

$$u_t = -u_x + u_{xx} \tag{3.18}$$

is *hyperbolic-parabolic*. Because it frequently models convection (hyperbolic) through the term u_x and diffusion (parabolic) through the term u_{xx}, it is generally termed a *convection-diffusion equation*. If, additionally, it includes a function of the dependent variable u such as

$$u_t = -u_x + u_{xx} + f(u), \tag{3.19}$$

then it might be termed a *convection-diffusion-reaction equation* because $f(u)$ typically models the rate of a chemical reaction. If the function consists only of the independent variables, that is,

$$u_t = -u_x + u_{xx} + g(x, t), \tag{3.20}$$

the equation could be labeled an *inhomogeneous PDE*.

This discussion clearly indicates that PDE problems come in an infinite variety that depend on linearity, types of coefficients (constant, variable), coordinate system, geometric classification (hyperbolic, elliptic, parabolic), number of dependent variables (number of simultaneous PDEs), number of independent variables (number of dimensions), types of BCs, smoothness of the IC, and so on, so it might seem impossible to formulate numerical procedures with any generality that can address a relatively broad spectrum of PDEs. However, the MOL does actually provides a framework with considerable generality, although the success in applying it to a new PDE problem depends to some extent on the experience and inventiveness of the analyst. Thus, it should be appreciated that the MOL is not a single, straightforward, clearly defined approach to PDE problems but, rather, is a general concept (or philosophy) that requires specification of the details for each new PDE problem.

In subsequent sections we will illustrate the formulation of a MOL numerical algorithm with the caveat that this will not be a general discussion of MOL as it can be applied to any conceivable PDE problem.

Finally, we explain the preceding geometric nomenclature in relation to the most general second-order PDE with constant coefficients, that is,

$$a\frac{\partial^2 u}{\partial x^2} + b\frac{\partial^2 u}{\partial xy} + c\frac{\partial^2 u}{\partial y^2} + d\frac{\partial u}{\partial x} + e\frac{\partial u}{\partial y} + = 0. \tag{3.21}$$

Such equations are classified analogously with conic sections:

$$b^2 - 4ac \begin{cases} < 0 \Rightarrow \text{elliptic equation,} \\ = 0 \Rightarrow \text{parabolic equation,} \\ > 0 \Rightarrow \text{hyperbolic equation.} \end{cases} \tag{3.22}$$

The behavor of these equations and their method of solution can differ significantly. Examples include the following:

- **Elliptic**

$$\nabla^2 \phi = 4\pi G \rho; \quad \text{Poisson (gravitational potential)} \tag{3.23}$$

- **Parabolic**

$$\frac{\partial T}{\partial t} = \kappa \nabla^2 T; \quad \text{Diffusion (heat conduction)} \tag{3.24}$$

- **Hyperbolic**

$$\frac{\partial^2 p}{\partial t^2} + v^2 \nabla^2 p = 0; \quad \text{Wave (sound propagation)} \tag{3.25}$$

3.7 DISCRETIZATION

Discretization is a method, or process, whereby we construct an approximation to a continuous variable by using a set of *discrete* values together with an appropriate algorithm that provides additional information about the system being studied. For example, the *continuous* variable $u(x)$ may be approximated by a series of discrete values $u(x_1)$, $u(x_2), \ldots, u(x_i), \ldots, u(x_n)$ and a *difference* algorithm represented by the equation

$$u(x = x_j + a) \simeq u(x_j) + a \frac{u(x_{j+1}) - u(x_j)}{x_{j+1} - x_j}, \quad |a| < (x_{j+1} - x_j), \tag{3.26}$$

where $x_{j+1} > x_j$, etc. Equation (3.26) performs a linear interpolation of the discrete values of u to provide intermediate values. Now the fractional part of the second term on the right-hand side of eqn. (3.26) actually represents a first-order discrete/numerical approximation to the first derivative of $u(x)$ at $x = x_j$. So we could rewrite eqn. (3.26) as

$$u(x = x_j + a) \simeq u(x_j) + a \frac{du(x_j)}{dx}. \tag{3.27}$$

Higher-order approximations can be used to provide more accurate estimates of the numerical derivative, and a similar approach can be used to approximate higher derivatives. This aspect is discussed subsequently.

3.7.1 General Finite Difference Terminology

We will first define the following terminology that will allow difference operators to be used to derive FD methods [Hir-88].

Displacement operator E:

$$E u_i = u_{i+1}. \tag{3.28}$$

Forward difference operator δ^+:

$$\delta^+ u_i = u_{i+1} - u_i. \tag{3.29}$$

Backward difference operator δ^-:

$$\delta^- u_i = u_i - u_{i-1}. \tag{3.30}$$

Central difference operator $\bar{\delta}$:

$$\bar{\delta}u_i = \frac{1}{2}(u_{i+1} - u_{i-1}). \tag{3.31}$$

Differential operator D:

$$Du = u_x \equiv \frac{\partial u}{\partial x}. \tag{3.32}$$

By using these definitions, we can derive the following additional operators:

$$\begin{aligned}\delta^+ &= E - I \\ \delta^- &= I - E^{-1},\end{aligned} \tag{3.33}$$

where we have used the inverse displacement operator

$$E^{-1}u_i = u_{i-1}. \tag{3.34}$$

This leads to the following relationships:

$$\delta^- = E\delta^+ \tag{3.35}$$

and

$$\delta^+\delta^- = \delta^-\delta^+ = \delta^+ - \delta^- = \delta^2. \tag{3.36}$$

In general, n can be taken as positive or negative repeated operations, thus

$$E^n u_i = u_{i+n}, \tag{3.37}$$

$$\delta^{+2} = \delta^+\delta^+ = E^2 - 2E + 1, \tag{3.38}$$

$$\delta^{+3} = (E-1)^3 = E^3 - 3E^2 + 3E - 1. \tag{3.39}$$

Now consider the Taylor expansion

$$\begin{aligned}u(x + \Delta x) &= u(x) + \Delta x u_x(x) + \frac{\Delta x^2}{2!}u_{xx}(x) + \frac{\Delta x^3}{3!}u_{xxx}(x) + \cdots \\ &\Downarrow \\ Eu(x) &= \left(1 + \Delta xD + \frac{(\Delta xD)^2}{2!} + \frac{(\Delta xD)^3}{3!}u_{xxx}(x) + \cdots\right)u(x).\end{aligned} \tag{3.40}$$

This leads to the formal relationships

$$\begin{aligned}Eu(x) &= e^{\Delta xD}u(x), \\ &\Downarrow \\ E &= e^{\Delta xD}, \\ &\Downarrow \\ \Delta xD &= \ln E = \ln(1 + \delta^+).\end{aligned} \tag{3.41}$$

But, we have the series expansion $\ln(1+x) = x - \frac{x^2}{2} + \frac{x^3}{3} - \frac{x^4}{4} + \cdots$, which leads to

$$\Delta xD = \delta^+ - \frac{\delta^{+2}}{2} + \frac{\delta^{+3}}{3} - \frac{\delta^{+4}}{4} + \cdots \tag{3.42}$$

By truncating this last relationship, first derivative FD approximations can be obtained manually. For example, employing the earlier relationships, we obtain the *first derivative,*

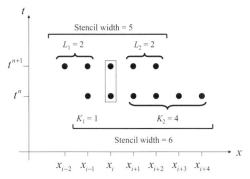

Figure 3.2. A typical stencil diagram for a 1D spatial grid [Lan-98].

backward difference approximation

$$\Delta x D = \ln E = -\ln(1 - \delta^-)$$
$$= \delta^- + \frac{\delta^{-2}}{2} + \frac{\delta^{-3}}{3} + \frac{\delta^{-4}}{4} + \cdots, \quad (3.43)$$

from which, to second-order accuracy, we obtain

$$(u_x)_i = D u_i = \frac{3u_i - 4u_{i-1} + u_{i-2}}{2\Delta x} + \frac{\Delta x^2}{3} u_{xxx}, \quad (3.44)$$

where the term $\frac{\Delta x^2}{3} u_{xxx}$ represents the leading term of the FD approximation *truncation error*. For additional FD approximation derivations using this approach, refer to [Hir-88, Chapter 4].

A more systematic approach is discussed in Section 3.8.2, where an R program is provided that automates the process of generating discretization schemes for approximating derivatives.

3.7.2 The Mesh

The term *mesh* or *grid* usually refers to a 2D or 3D computational arrangement representing points or cells in the spatial domain for which solutions will calculated. The mesh can be defined on any coordinate system, for example, *Cartesian* or *spherical/polar* coordinates. An example of a 2D mesh is shown in Fig. 3.4. For 1D applications the term is also used for fully discrete systems where time is represented on one axis and space on the other, as shown in Fig. 3.2.

- *Structured grids* are regular computational meshes used to discretize 2D/3D domains that can be referenced using two or three indices. This requires that the cells comprising the grid consist of nonoverlapping *quadrilaterals/cubes*—the cubes need not have equal sides.
- *Unstructured grids* are computational meshes used to discretize a 2D or 3D domain. They consist of a set of elements that are usually referenced using only one index. Elements are generally *triangles/tetrahedra* or *quadrilaterals/cubes*—the cubes need not have equal sides. They are used for problems that have irregular physical or computational boundaries.

- *Adaptive grids* are controlled by an algorithm that minimizes the computational effort by ensuring that a fine mesh is only applied to parts of the domain where needed, with a courser mesh applied to the remainder, for example, where steep gradients occur. The algorithm ensures that the fine mesh follows steep gradient phenomena or is introduced as and when steep gradient phenomena arise.

3.7.3 Nonuniform Grid Spacing

For some problems with a significant computational load, it can be advantageous to include a finer grid in areas of the solution domain where the rate of change of a dependent variable is known to be particularly high. This is known as the *nonuniform grid approach*. Although this approach can reduce the total calculations involved in a simulation, it comes at the price of complexity. The discretization has to be customized to the particular problem, and in the case of moving fronts, the discretization also has to change as the simulation proceeds. This latter approach is the *adaptive grid* approach mentioned previously.

3.7.4 The Courant–Friedrichs–Lewy Number

The *Courant–Friedrichs–Lewy number*, Co, also known as *CFL number* or *Courant number*, is a dimensionless quantity that is defined as the ratio of two velocities, that is,

$$\text{Co} = \frac{v \Delta t}{\Delta x} = \frac{v}{\Delta x / \Delta t} = \frac{\text{maximum local velocity}}{\text{speed of computational grid}}. \qquad (3.45)$$

As an example, v could be the local velocity of a system describe by a hyperbolic PDE such as eqn. (3.25), Δx the corresponding local computational grid spacing, and Δt the time step to be taken by a numerical integrator.

The *Courant number* can form part of a necessary condition to ensure stability, but it may not be a sufficient condition, particularly for nonlinear problems. Its application is discussed briefly later in this chapter, but more general examples of its use are provided in Chapter 4).

3.7.5 The Stencil

The *stencil*, also referred to as a *computational molecule*, is a convenient way of visualizing the numerical discretization of a mathematical problem. It describes the geometric layout of a set of grid points employed by a numerical approximation algorithm.

In an implicit scheme, the unknown solution at time level $n + 1$ typically depends upon itself and known solutions at earlier times. Thus, we have a situation such that

$$u_i^{n+1} = u\left(u_{i-K_1}^n, \ldots, +u_{i+K_2}^n; u_{i-L_1}^{n+1} + \ldots, u_{i+L_2}^{n+1}\right), \qquad (3.46)$$

where $(K_1, K_2, L_1, L_2) \in \mathbb{I} \geq 0$. This situation is illustrated in Fig. 3.2 for a typical stencil for a 1D spatial grid where $L_1 = 2, L_2 = 2, K_1 = 1$, and $K_2 = 4$.

As an example, suppose that an unknown implicit algorithm is of the following form:

$$u_i^{n+1} = u_i^n - \frac{\lambda}{2}\left(f(u_{i+1}^{n+1}) - f(u_{i-1}^{n+1})\right) + \frac{\lambda}{2}\left(f(u_{i+2}^n) - 4f(u_{i+1}^n) + 3f(u_i^n)\right). \qquad (3.47)$$

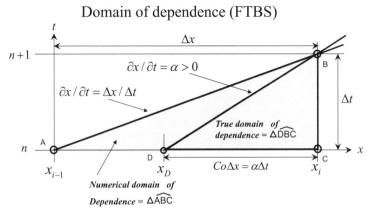

Figure 3.3. Domains of dependence for the *explicit FTBS* scheme.

Then we would have $L_1 = 0$, $L_2 = 2$, $K_1 = 1$, and $K_2 = 1$. For more detailed discussion on the subject of stencils, refer to [Lan-98].

The stencil concept is also useful for visualizing the *domain of dependence* of a particular discretization scheme. For an accurate stable scheme it is a requirement that the *numerical domain of dependence* should include the *real domain of dependence*. By this we mean that for a given spatial grid size Δx and time step Δt, the *characteristic speed c* should not exceed $\frac{\Delta x}{\Delta t}$. Thus, for a given c and Δx, the time step is constrained to $\Delta t < \frac{\Delta x}{c}$. Moreover, if we need to reduce Δx due to accuracy considerations, we will also need to make a corresponding reduction in Δt to ensure that the resulting solution is stable.

Consider now the explicit FTBS scheme of eqn. (3.116) applied to eqn. (3.115), which has characteristic speed α. In accordance with this, we see that the dimensionless group $\text{Co} = \frac{\alpha \Delta t}{\Delta x}$, and for stability we require that $|\text{Co}| < 1$. This is the *Courant–Friedricks–Lewy* (CFL) condition discussed previously, and the concept is illustrated in Fig. 3.3.

Further discussion on the subject of domain of dependence can be found in [Lev-02].

3.7.6 Upwinding

An *upwind* scheme is one where the discretization is biased in the direction of *propagation of the flow of material or information*, that is, in the direction of the *sign of the characteristic speed*. This means that if more points in the scheme are included on the upwind side than the downwind side, the scheme is considered to be an upwind scheme. This is in contrast to a centered scheme where there are an equal number of points on each of the upwind and downwind sides.

Upwinding is particularly relevant to the numerical solution of hyperbolic PDEs. For example, in eqn. (3.115), for $\alpha > 0$, where wave propagation would be left to right, a first-order upwind approximation to the spatial derivative at grid point i for a 1D system would be

$$\frac{\partial u_i}{\partial x} = \frac{1}{\Delta x}(u_i - u_{i-1}), \tag{3.48}$$

and a second-order upwind approximation example is

$$\frac{\partial u_i}{\partial x} = \frac{1}{2\Delta x}\left(3u_i - 4u_{i-1} + u_{i-2}\right). \tag{3.49}$$

However, for $\alpha < 0$, where wave propagation would be right to left, a first-order upwind approximation to the 1D spatial derivative would be

$$\frac{\partial u_i}{\partial x} = \frac{1}{\Delta x}\left(u_i - u_{i+1}\right), \tag{3.50}$$

and a second-order upwind approximation example is

$$\frac{\partial u_i}{\partial x} = \frac{1}{2\Delta x}\left(3u_i - 4u_{i+1} + u_{i+2}\right). \tag{3.51}$$

Thus, for positive α, the *upwind side* is to the *left* of grid point *i* and the *downwind side* is to the *right*. Similarly, for negative α, the *upwind side* is to the *right* of grid point *i* and the *downwind side* is to the *left*. The idea is readily extended to higher-order approximations.

Upwind schemes are generally more diffusive than centered schemes, hence they tend to be more stable.

3.8 METHOD OF LINES (MOL)

3.8.1 Introduction

The *method of lines* (MOL) is a semi-discrete method that approximates a system of partial differential equations (PDEs) by a system of ordinary differential equations (ODEs). This is achieved by discretizing all but one of the independent variables, usually time. The result is a set of ODEs that can then be integrated using a standard ODE solver. The advantage of this approach is that the ODE system is generally a simpler problem to solve than the original PDE system; additionally, all the power and ease of use of standard ODE integrators can be brought to bear on the problem. These include convenience of readily available well-proven computer code and library of explicit and implicit methods, along with automatic step-size control to maintain accuracy and stability.

To illustrate how the MOL works, consider the example of the *hyperbolic-parabolic* PDE given by eqn. (3.18), where the spatial derivatives have been discretized by a suitable FD scheme, including BCs. To complete the solution, the temporal derivative $\frac{du}{dt}$ has to be integrated; that is, for each element u_i in $\underline{u}$, we must solve

$$u_i = \int_{t=t_1}^{t=t_2}\left(\frac{du_i}{dt}\right)dt. \tag{3.52}$$

The time integration can be performed using any of the integrator methods described in Chapter 1 or by use of a numerical integrator from a standard library such as provided in R.

This method is simple to implement, and the discussion that follows, along with subsequent 1D and 2D examples, should make the solution process clear.

A full discussion of the MOL can be found in [Sch-91], [Ham-07], and [Sch-09].

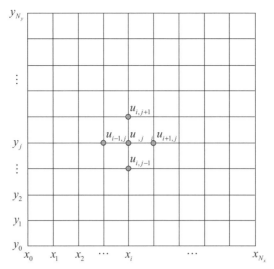

Figure 3.4. Typical 2D finite difference mesh with equally spaced grid points. The group of dots could represent a computational stencil (see Section 3.7.5).

3.8.2 Finite Difference Matrices

Finite difference methods work with *mesh point values* of the solution, that is,

$$u_{ij} \equiv u(x_i, y_j); \quad \begin{aligned} x_i &= i\Delta x, & y_j &= j\Delta y \\ i &= 0, \cdots, N_x, & j &= 0, \cdots, N_y. \end{aligned} \tag{3.53}$$

A typical 2D finite difference mesh is shown in Fig. 3.4.

Our discussion will first relate to the derivation of *finite difference* (FD) schemes and then to the application of these FD approximations to solve PDEs using the MOL.

The idea is to approximate a derivative by a numerical approximation equal to a function where the independent variables consist of a set of solution points adjacent to the point where the derivative is to be evaluated. For example, consider a 1D problem where the spatial variable y is discretized as shown earlier such that we now define $y_i = f(x_i)$. Then a m-point numerical FD approximation to the first derivative generally takes the form

$$\begin{aligned} \frac{\partial y_i}{\partial x} = \frac{\partial f(x_i)}{\partial x} &= \sum_{j=0}^{m-1} a_j y_{i+j-r} + \tau_i \\ &\simeq \sum_{j=0}^{m-1} a_j y_{i+j-r}, \quad 1 < m, \ 0 \le r < m, \end{aligned} \tag{3.54}$$

where r is the stencil offset from point i. For example, for a central scheme with odd m, we have $r = m/2 + 1/2$, and for even m, we have $r = m/2$.

The symbol τ_i represents a *truncation error* term that is dependent upon grid-point spacing, Δx, and which is discussed later. This method effectively represents a FD polynomial approximation to the spatial derivative, the order of which is given by $p = m - n$. There is another important interpretation of order, namely, the highest-order polynomial for which the FD approximation is exact.

To use a m-point FD approximation, we first need to identify the corresponding polynomial coefficient values, and we can evaluate these coefficients by use of a suitable approximating function. Methods based on the *Lagrange polynomial, Taylor series*, or *Newton divided differences* can all be used to equal effect. For convenience, we will assume that values of y are available at m equally spaced points and that the process can be considered smooth over the range of interest. The ideas are general and can be extended to different-order approximations and to higher derivatives. Nonequally spaced points can be treated in a similar manner but will not be discussed here. Methods for approximating nonsmooth processes, that is, those with *sharp gradients* or *shocks*, are discussed in Chapter 6.

There are a number of different methods for determining FD coefficients, but a procedure that works well, based on Lagrange polynomials, is as follows:

1. Obtain a Lagrange polynomial approximation $P(x)$ to the function $y = f(x)$, based on m equally spaced data points, that is, $x_0, x_1, x_2, \ldots, x_{m-1}$.
2. Differentiated $P(x)$ N times.
3. Define $x_1 = x_0 + \Delta x, x_2 = x_0 + 2\Delta x$, and so on
4. By setting x, in turn, equal to x_1, x_2, and so on, we obtain m sets of coefficients each valid for an approximation valid at a particular data point. The coefficients can be represented by a $m \times m$ matrix where the elements in row i constitute the coefficients associated with the FD approximation to the derivative at point i.

The associated *truncation error* terms are evaluated by use of Taylor series expansions and substituting them into the FD approximations.

The procedure is as follows:

1. Obtain a Taylor series expansion S_i for each u_i.
2. Form a weighted sum $\sum_{j=0}^{m-1} a_j y_{i+j-r}$ for each row of the FD coefficient matrix (see second of eqns. (3.54)).
3. In each row, various leading terms will cancel, leaving the actual derivative plus the truncation error terms.

The easiest and most straightforward way of implementing the preceding procedures is to employ a computer algebra program. Code written for the symbolic programs Maple and Maxima (open source) that includes both procedures and examples is included with the downloads. Output generated by these programs for a four-point, second-order FD approximation to a second-order derivative will be of the form

$$\frac{\partial^2 y(x=x_1)}{\partial x^2} = \frac{2y_1 - 5y_2 + 4y_3 - y_4}{\Delta x^2} + \tau_1,$$

$$\frac{\partial^2 y(x=x_2)}{\partial x^2} = \frac{y_1 - 2y_2 + y_3}{\Delta x^2} + \tau_2,$$

$$\frac{\partial^2 y(x=x_3)}{\partial x^2} = \frac{y_2 - 2y_3 + y_4}{\Delta x^2} + \tau_3,$$

$$\frac{\partial^2 y(x=x_4)}{\partial x^2} = \frac{-y_1 + 4y_2 - 5y_3 + y_4}{\Delta x^2} + \tau_4,$$

(3.55)

where the ith truncation errors τ_i is

$$\tau_i = C_i \frac{\partial^4 f(x = x_i)}{\partial x^4} \Delta x^2 + \mathcal{O}(\Delta x^3). \tag{3.56}$$

Thus the truncation is of order Δx^2, and for this particular example, $C_1 = -11/12$, $C_2 = +1/12$, $C_3 = +1/12$, and $C_4 = -11/12$. In general, for an m-point, nth derivative, the truncation error will be of order Δx^{m-n}, and the ith truncation error will be

$$\tau_i = C_i \frac{\partial^m f(x = x_i)}{\partial x^m} \Delta x^{m-n} + \mathcal{O}(\Delta x^{m-n+1}), \tag{3.57}$$

where C_i is a constant associated with the leading term of the ith truncation error.

Equation (3.55) can be written succinctly as

$$\frac{\partial^2 \underline{y}}{\partial x^2} = \frac{1}{\Delta x^2} \begin{bmatrix} 2 & -5 & 4 & -1 \\ 1 & -2 & 1 & 0 \\ 0 & 1 & -2 & 1 \\ -1 & 4 & -5 & 2 \end{bmatrix} \underline{y} + \underline{\tau}. \tag{3.58}$$

However, for practical situations where we have *many interior points*, the associated coefficients are defined by row 2 or 3, that is, the interior points are defined by row 2 or 3 shifted such that coefficient -2 for, say, $u_{xx}(x = x_k)$, is centered on the interior point being calculated, that is, k. Thus, for a grid with many interior points, the four-point stencil for a first-order approximation to a second derivative can be described in matrix form as

$$\frac{\partial^2 \underline{y}}{\partial x^2} = \frac{1}{\Delta x^2} \begin{bmatrix} 2 & -5 & 4 & -1 & & & & \\ 1 & -2 & 1 & & & & & \\ & 1 & -2 & 1 & & & & \\ & & \ddots & \ddots & \ddots & & & \\ & & & 1 & -2 & 1 & & \\ & & & & 1 & -2 & 1 & \\ & & & & -1 & 4 & -5 & 2 \end{bmatrix} \underline{y} + \underline{\tau}. \tag{3.59}$$

In a similar way, coefficients for FD approximations can be calculated for different derivatives based on differing numbers of grid points. For example, applying the same process, we obtain the following four-point stencil for a second-order approximation to a first derivative:

$$\frac{\partial \underline{y}}{\partial x} = \frac{1}{6\Delta x} \begin{bmatrix} -11 & 18 & -9 & 2 & & & & \\ -2 & -3 & 6 & -1 & & & & \\ & -2 & -3 & 6 & -1 & & & \\ & & \ddots & \ddots & \ddots & & & \\ & & & -2 & -3 & 6 & -1 & \\ & & & & 1 & -6 & 3 & 2 \\ & & & & -2 & 9 & -18 & 11 \end{bmatrix} \underline{y} + \underline{\tau}. \tag{3.60}$$

Note that when the FD stencil for a particular row is centered on the interior point being calculated, the approximation is a *central difference*; when it is shifted to the left, it is a *forward difference* approximation, and when it is shifted to the right, it is a *backward difference* approximation. In general, the magnitude of a central difference truncation error is less than the corresponding forward and backward difference truncation errors.

The *banded* structure of the derivative matrices means that *sparse matrix* methods can be used to speed up calculations for large problems, but we do not pursue this subject further here. However, this aspect is discussed in Chapter 7, where functions from the R package Matrix are employed to solve a sparse matrix PDE problem.

Refer to Appendix 3.A at the end of this chapter for additional FD matrix examples where truncation errors are also given.

We mention without detailed discussion that an alternative way of generating derivative matrix coefficients is by expanding $f(x) = \exp(i\omega x)$ locally around $h = 0$ and applying a *Padé* (rational function) approximation. Details of the method can be found in [For-98]. A computer program that implements this method in R using the interface package Ryacas to the symbolic algebra system Yakas is shown in Listing 3.1. Refer to Chapter 1 for more details on Ryacas and to R help for Ryacas for specific Yacas calls, as they do not use standard R syntax. This code, which can run slowly for stencils with more than eight points, also computes the order of approximation and a list of the truncation errors for each stencil location. Sample output showing derivative matrix, order, and truncation errors from this listing for a six-point second derivative matrix is given. The derivative matrix coefficients, divided by Δx^2, are used in function diffMatrix2nd6pt() contained in file deriv2nd6pt.R (see Appendix 3.B at the end of this chapter).

```
[1] "Starting Yacas!"

Coefficients for 6 point, 2nd derivative matrix

       [,1]   [,2]   [,3]  [,4]  [,5]   [,6]
[1,]  15/4  -77/6  107/6  -13  61/12  -5/6
[2,]   5/6   -5/4   -1/3  7/6  -1/2   1/12
[3,] -1/12    4/3   -5/2  4/3  -1/12    0
[4,]     0  -1/12    4/3 -5/2   4/3  -1/12
[5,]  1/12   -1/2    7/6 -1/3  -5/4   5/6
[6,]  -5/6  61/12    -13 107/6 -77/6  15/4

Numerical derivative scheme has accuracy of order: 4

truncation error at location 1: -137/180 * D(6)f(1)dx^4
truncation error at location 2:   13/180 * D(6)f(2)dx^4
truncation error at location 3:   -1/90  * D(6)f(3)dx^4
truncation error at location 4:   -1/90  * D(6)f(4)dx^4
truncation error at location 5:   13/180 * D(6)f(5)dx^4
truncation error at location 6: -137/180 * D(6)f(6)dx^4
```

```
# File: FDsTaylorExpansion.R
cat("\014") # Clear console
rm(list = ls(all = TRUE)) # Delete workspace
# Code to generate Finite difference coefficients, order and truncation errors
# Ref: Fornberg, B. (1998). Calculation of weights in finite difference
#       formulas.
#       SIAM Rev., 40(3), 685-691.
require(compiler)
enableJIT(3)
library(Ryacas)
library(MASS) # For fractions() function
m   <- 2   # Order of derivative matrix
N   <- 6   # number of stencil points
n   <- N-1 # order at which Taylor series to be truncated
var <- 'x' # variable name for Taylor function
if((N-m) < 1){
  print("ERROR: Number of points must be at least one\n");
  print("       greater than order of derivative !\n");
  print("       i.e. N >= m+1");
}else{
  res <- array(0,c((N),(N)))
  for(s in 0:n){ # iterate through each grid-point location
    f      <- sprintf('x^%s*Ln(x)^%s',s,m)
    call   <- sprintf('texp := Coef(Taylor( %s, 1., %d) %s,0 .. %d)', var,n,f,n)
    result <- yacas(call)
    res[(s+1),] <- c(unlist(eval(result[[1]])))
  }
  pref <- c("1st","2nd","3rd","4th","5th","6th","7th","8th","9th","10th","11th")
  cat(sprintf("\nCoefficients for %d point, %s derivative matrix\n\n",
      N,pref[m]))
  print(fractions(res))
  cat(sprintf("\nNumerical derivative scheme has accuracy of order: %d\n\n",
      N-m));
  #
  # Calculate truncation errors
  res2 <- array(0,c((N),(N+1)))
  for(s2 in 0:n){ # iterate through each grid-point location
    f2      <- sprintf('x^%s*Ln(x)^%s',s2,m)
    call2   <- sprintf('texp := Coef(Taylor( %s, 1., %d) %s,0 .. %d)', var,n+1,
      f2,n+1)
    result2 <- yacas(call2)
    res2[(s2+1),] <- c(unlist(eval(result2[[1]])))
    cat(sprintf("truncation error at location %d: %14s * D(%d)f(%d)dx^%d\n",
          (s2+1),fractions(-res2[(s2+1),(N+1)]),N,(s2+1),(N-m)))
  }
}
```

Listing 3.1. File: FDsTaylorExpansion.R—Code to derive FD coefficients using the symbolic algebra package Ryacas. This code also computes the order of approximation and a list of the truncation errors for each stencil location

Example
Consider the 1D PDE problem described by eqn. (3.18), that is,

$$u_t = -u_x + u_{xx}. \qquad (3.61)$$

This equation is transformed into the following ODE problem when we discretize the spatial derivatives using N grid points:

$$\frac{d\underline{u}}{dt} = -\frac{1}{2\Delta x} \begin{bmatrix} -3 & 4 & -1 & & & & \\ -1 & 0 & 1 & & & & \\ & -1 & 0 & 1 & & & \\ & & \ddots & \ddots & \ddots & & \\ & & & -1 & 0 & 1 & \\ & & & & -1 & 0 & 1 \\ & & & & 1 & -4 & 3 \end{bmatrix} \underline{u} + \frac{1}{\Delta x^2} \begin{bmatrix} 2 & -5 & 4 & -1 & & & \\ 1 & -2 & 1 & & & & \\ & 1 & -2 & 1 & & & \\ & & \ddots & \ddots & \ddots & & \\ & & & 1 & -2 & 1 & \\ & & & & 1 & -2 & 1 \\ & & & -1 & 4 & -5 & 2 \end{bmatrix} \underline{u} + \underline{\tau}^C, \qquad (3.62)$$

where we have used the symbol $\underline{u} = [u_1, u_2, \cdots, u_N]$ to represent a discrete approximation to the continuous variable u corresponding to points on the spatial domain $x = [x_1, x_2, \ldots, x_N]$.

Note that, for consistency, we have chosen to use second-order numerical approximations for both the first and second spatial derivatives. The effect of this means that the rate at which the combined truncation error $\underline{\tau}^C$ converges to zero is second order, that is, $\underline{\tau}^C \propto \mathcal{O}(\Delta x^2)$ as $\Delta x \to 0$. Refer also to the discussion on truncation error in Chapter 1.

The spatial discretization process is not yet complete, as we still have to account for boundary conditions. We will demonstrate this by assigning a *Neumann* BC to the left boundary ($x_1 = 0$) and a *Dirichlet* BC to the right boundary ($x_N = N\Delta x$).

The Neumann BC
Because our problem is subject to a *Neumann* BC, as described in Section 3.2, we need to ensure solutions are *consistent* with a BC that is given in the form of a first derivative [Sch-91]. This requires that the second derivative at the boundary be consistent with the first derivative at the same boundary. We will demonstrate this by example, where we use the following four-point approximations for the first derivative and second derivatives, which are derived from the appropriate coefficient matrices:

$$\frac{du_1}{dx} = \frac{-11u_1 + 18u_2 - 9u_3 + 2u_4}{6\Delta x} + \tau_1, \qquad (3.63)$$

$$\frac{d^2u_1}{dx^2} = \frac{2u_1 - 5u_2 + 4u_3 - u_4}{\Delta x^2} + \tau_2. \qquad (3.64)$$

Note that the first derivative and second derivative approximations are third order and second order, respectively, that is, $\tau_1 \propto \mathcal{O}(\Delta x^3)$ and $\tau_2 \propto \mathcal{O}(\Delta x^2)$ (see Appendix 3.A) at the end of this chapter. The reason for the order choices is explained subsequently.

We achieve the required consistency by solving the first equation for one point, say, u_i, and substituting the result back into the second equation. We then have a second derivative that is consistent with the first derivative. We can select i to be any point, $1 < i \leq 4$. Choosing $i = 4$, we proceed as follows.

Solving for u_4 gives

$$u_4 = \frac{11u_1 - 18u_2 + 9u_3}{2} + 3\Delta x \frac{du_1}{dx} + 3\tau_1 \Delta x. \tag{3.65}$$

Substituting eqn. (3.65) back into eqn. (3.64) yields

$$\frac{d^2 u_1}{dx^2} = \frac{2u_1 - 5u_2 + 4u_3 - \left(\dfrac{11u_1 - 18u_2 + 9u_3}{2} + 3\Delta x \dfrac{du_1}{dx} + 3\tau_1 \Delta x\right)}{\Delta x^2} + \tau_2 \tag{3.66}$$

$$= \frac{1}{2}\frac{-7u_1 + 8u_2 - u_3}{\Delta x^2} - \frac{3}{\Delta x}\frac{du_1}{dx} + \tau^C,$$

which we use to calculate a numerical approximation to the second derivative of u_1. We can now see the reason for selecting a third-order equation for the first derivative, namely, that the associated truncation error is divided by Δx; hence, reducing the order of approximation by 1 and making the overall approximation $\tau^C = \left(\frac{3}{\Delta x}\tau_1 + \tau_2\right) \propto \Delta x^2$, that is, second order.

Also, because $\frac{du_1}{dt}$ is specified, it does not need to be calculated numerically. However, for convenience, in practice it is usually calculated and then replaced by the appropriate Neumann BC value.

Thus, neglecting the truncation error, the numerical approximation to the first derivative becomes

$$\frac{du}{dx} = \frac{-1}{2\Delta x}\begin{bmatrix} 0 & 0 & 0 & & & & & \\ -1 & 0 & 1 & & & & & \\ & -1 & 0 & 1 & & & & \\ & & \ddots & \ddots & \ddots & & & \\ & & & & 1 & 0 & 1 & \\ & & & & & -1 & 0 & 1 \\ & & & & & 1 & -4 & 3 \end{bmatrix} \underline{u} - \begin{bmatrix} \frac{du_1}{dx} \\ 0 \\ 0 \\ \vdots \\ 0 \\ 0 \\ 0 \end{bmatrix}. \tag{3.67}$$

The Dirichlet BC
Because our problem is also subject to a *Dirichlet* BC, as described in Section 3.2, we therefore need to ensure solutions are *consistent* with a BC that is given in the form of an explicit value for u at the boundary, say, u_b. We do this by simply setting $u_N = u_b$, which also implies that $\frac{du_N}{dt} = 0$.

We are now in a position to specify the complete problem as

$$\frac{d\underline{u}}{dt} = \frac{-1}{2\Delta x}\begin{bmatrix} 0 & 0 & 0 & & & & & \\ -1 & 0 & 1 & & & & & \\ & -1 & 0 & 1 & & & & \\ & & \ddots & \ddots & \ddots & & & \\ & & & 1 & 0 & 1 & & \\ & & & & -1 & 0 & 1 & \\ & & & & & 1 & -4 & 3 \end{bmatrix}\underline{u} - \begin{bmatrix} \frac{du_1}{dx} \\ 0 \\ 0 \\ \vdots \\ 0 \\ 0 \\ 0 \end{bmatrix} \quad (3.68)$$

$$+ \frac{1}{\Delta x^2}\begin{bmatrix} \frac{7}{2} & \frac{8}{2} & -\frac{1}{2} & 0 & & & & \\ 1 & -2 & 1 & & & & & \\ & 1 & -2 & 1 & & & & \\ & & \ddots & \ddots & \ddots & & & \\ & & & 1 & -2 & 1 & & \\ & & & & 1 & -2 & 1 & \\ & & & & -1 & 4 & -5 & 2 \end{bmatrix}\underline{u} + \begin{bmatrix} -\left(\frac{3}{\Delta x}\right)\frac{du_1}{dx} \\ 0 \\ 0 \\ \vdots \\ 0 \\ 0 \\ 0 \end{bmatrix},$$

where $\frac{du_1}{dx}$ and u_N are specified inputs and the truncation error, τ^C, has been omitted. Thus, with the Dirichlet BC at the right-hand boundary, we have $u_N = u_b \Rightarrow \frac{du_N}{dt} = 0$, which effectively reduces the ODE vector by 1 as the time derivative at grid point N does not have to be calculated. Note that u_N can be time varying, in which case u_N is updated at each time step.

Other boundary condition combinations follow in a similar way, and this aspect will be illustrated by worked examples given in the following sections.

To implement the preceding approach in reusable computer code requires that the derivative matrices are generated from library functions. Because the matrices are constant, they are calculated once for the particular number of grid points and BCs. They can then be used repeatedly for calculating the spatial derivatives as the solution evolves over time. For this problem, the procedure for calculating the time derivative at each time step is as follows:

1. Replace the current value of u_N with the Dirichlet BC value u_b.
2. Calculate $\frac{du}{dx}$ based upon the first derivative matrix.
3. Replace $\frac{du_1}{dx}$ with the appropriate Neumann BC value.
4. Calculate $\frac{d^2u}{dx^2}$ based upon the second derivative matrix, modified for Neumann BC at $i = 1$.
5. Subtract three times the Neumann BC value from the second derivative of the first grid point, that is, $\frac{d^2u_1}{dx^2} = \frac{d^2u_1}{dx^2} - 3(\frac{du_1}{dx})$.
6. Assemble the time derivative $\frac{du}{dt} = -\frac{du}{dx} + \frac{d^2u}{dx^2}$.

These derivatives can be calculated using functions from the *derivative matrix library*, which is discussed in Appendix 3.B at the end of this chapter. We will now illustrate the

MOL by finding solutions to a number of 1D and 2D problems using functions from the derivative matrix library.

3.8.3 MOL 1D: Cartesian Coordinates

The application of MOL to 1D problems is generally straightforward, and we choose PDE examples that have quite different forms of solution. The ideas are expanded in the next section when a range of 2D examples is presented.

3.8.3.1 Korteweg–de Vries (KDV) Equation

The KDV equation in its canonical form is given by

$$\frac{\partial u}{\partial t} + 6u\frac{\partial u}{\partial x} + \frac{\partial^3 u}{\partial x^3} = 0, \quad u = u(x,t), \ t \geq t_0, \ x \in \mathbb{R}. \tag{3.69}$$

The equally valid form $\frac{\partial u}{\partial t} - 6u\frac{\partial u}{\partial x} + \frac{\partial^3 u}{\partial x^3} = 0$ is obtained by a simple change of variable, $u = -u$.

The KdV equation is very important in the modeling of shallow water waves and has historical significance. It is also applicable to many other scientific and mathematical areas [Dra-92, Dau-06] and represents an example of a system where *nonlinear* effects, due to the term $6u\frac{\partial u}{\partial x}$, and *dispersive* effects, due to the term $\frac{\partial^3 u}{\partial x^3}$, cancel each other out and enable *soliton* waves to propagate without dissipation. Solitons are solitary waves having the additional property that they can interact with other solitons such that they emerge following a collision without changing shape. However, collisions generally result in a small change in phase. They occur in the solutions of a large number of weakly nonlinear dispersive PDEs that describe physical processes and are observed to occur naturally in the corresponding real systems. Some example solitons include *hump* (similar to a Gaussian pulse), *kink* (an s-shaped curve), *breather* (an oscillating pulse), *compacton* (free from exponential tails (compact support)), and *peakon* (has a finite gradient discontinuity at its crest). The velocity of a soliton is related to its height and width. Taller, narrower solitons generally travel faster than shorter, wider solitons.

The occurrence of a solitary wave was first reported by John Scott-Russell, who observed such a phenomenon on the Edinburgh–Glasgow canal. He followed it on horseback for a distance of some two miles, noting that it progressed at a speed of approximately nine miles per hour. He then performed a series of experiments and published his empirical findings at a meeting of the British Association for the Advancement of Science [Sco-44]. At first the scientific establishment was skeptical about Scott-Russell's findings as the phenomenon could not be predicted by the then current theory of water waves. However, in 1895, Kortweg and de Vries published their analysis [Kor-95], including the KdV equation, showing that indeed a solitary wave was theoretically possible, in accordance with Scott-Russell's account. Previously, Boussinesq [Bou-72] and Lord Rayleigh (William Strutt) [Ray-76] had published earlier analyses that showed similar results but did not publish the actual KdV equation and thus were not recognized as having priority; so it was after Kortweg and de Vries that the KdV equation was named.

Few new findings were published during the next 60-plus years until Zabusky and Kruskal reported results of some numerical analysis they had carried out. Previously, Fermi, Pasta, and Ulam (FPU) published a seminal report [Fer-55] on a numerical model of a discrete nonlinear mass-spring system. Following on from the work of FPU, Zabusky and Kruskal formulated a continuum model of the nonlinear mass-spring system and considered the initial value problem of a variant of the KdV equation with periodic BCs. They observed the formation of solitons from a *cosine wave* initial condition and reported on their ability to interact with other solitons and emerge from collisions with their shape in tact, but with a slight phase change. In their paper detailing their findings [Zab-65], they also coined the term *soliton*. Subsequently, Lax gave a rigorous proof of these findings [Lax-68]. The publication of the Zabusky and Kruskal paper heralded in a new era in the analysis of nonlinear PDEs when many more evolutionary equations were found to exhibit soliton solutions. The field is still very active today, with many papers published each year.

In their investigation of the KdV equation, Zabusky and Kruskal [Zab-65] used the second-order fully explicit leapfrog finite difference scheme. However, here we use the semi-discrete MOL with sixth-order FD spatial discretization and the lsodes integrator from the R package deSolve.

The problem we solve numerically is the three-soliton solution to KdV eqn. (3.69) with zero *Dirichlet* boundary conditions. We use the following analytical solution $u_a(x, t_0)$ at $t_0 = -8$ as the initial condition, which is based on wave number values $k_1 = 1/2, k_2 = 3/4$, and $k_3 = 1$:

$$u_a = -[(225\tanh((1/2)t - (1/2)x)^2 + 1260\tanh(4t - x)\tanh(t/2 - x/2)$$
$$+ 1764\tanh(4t - x)^2 - 3564)\coth(27t/16 - 3x/4)^2$$
$$+ (-1440\tanh(4t - x)\tanh(t/2 - x/2)^2 - 4032\tanh(t/2 - x/2)\tanh(4t - x)^2$$
$$+ 4032\tanh(t/2 - x/2) + 1440\tanh(4t - x))\coth(27t/16 - 3x/4) \qquad (3.70)$$
$$+ (2304\tanh(4t - x)^2 - 1029)\tanh(t/2 - x/2)^2$$
$$- 1260\tanh(4t - x)\tanh(t/2 - x/2) + 300\tanh(4t - x)^2]$$
$$/ \left(8\left(-18\coth(27t/16 - 3x/4) + 7\tanh(t/2 - x/2) + 10\tanh(4t - x)\right)^2\right).$$

The analytical solution of eqn. (3.71) was derived using the *Bäcklund transformation* method [Gri-12].

Note, that for this problem, the soliton wave speeds are given by $c_i = 4k_i^2$ with the solitons moving from left to right along the x axis. Therefore, we have $c_1 = 1, c_2 = 2\frac{1}{4}, c_3 = 16$, where c_1 is associated with the short, slowest soliton and c_3 with the tall, fastest soliton. From Figs. 3.5 and 3.6, we observe the phenomena discussed earlier, that is, that the solitons emerge following a collision with their shape unchanged but with small changes in phase. Classic FD spatial discretizations are used with a seven-point, sixth-order first derivative and nine-point, sixth-order third derivative. It was necessary to use sixth-order discretizations to obtain the desired accuracy. By choosing discretizations of the same order the truncation errors will converge uniformly with cell spacing (as discussed in Section 3.8.2). This choice together with the lsodes integrator appear to give very good results. We can infer this qualitatively because we do not observe *trailing instabilities*, termed *radiation*, which would otherwise be introduced by very small inaccuracies or noise in the solution. For this particular problem, we can also check the accuracy against the theoretical solution. However, if the analytical solution is not known, there is another

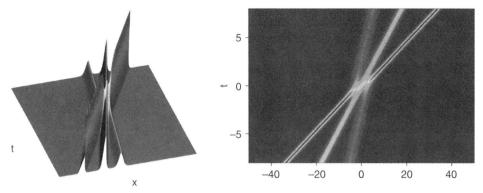

Figure 3.5. (left) Surface plot of a three-soliton solution to the KdV equation generated using the persp3d() function. (right) 2D space-time plot of the three-soliton collision generated using the image() function. The simulation employed 531 spatial grid points (horizontal axis) and 251 temporal grid points (vertical axis). The plots both show clearly how the taller faster solitons overtake the shorter slower solitons. The soliton phase changes are clearly visible (see main text). (See color plate 3.5)

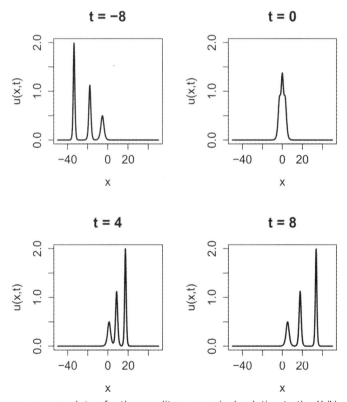

Figure 3.6. Time sequence plots of a three-soliton numerical solution to the KdV equation based on 531 spatial grid points. (top left) Initial condition from which the taller faster solititons move right to left and proceed to overtake the shorter slower solitons. (top right) All three solitons coinside at $x = 0$. (bottom left) Solitons having just emerged from the collision and progressively moving to the right. (bottom right) Soliton positions reversed and well clear of the collision.

Table 3.2. Summary of KdV simulation invariant results I_1, I_2, I_3 based on 531 spatial grid points and 251 temporal grid points. Also included in the right-hand column is the maximum absolute deviation of u from the analytical value u_a over all spatial grid points and over all time steps

Invariant	Analytical	Numerical		
		Max.	Min.	Max. err.
I_1	9.0000	9.0000	9.0000	0.0022
I_2	4.1250	4.1250	4.1250	0.0033
I_3	16.2379	16.2380	16.2373	0.0404

very good check using the fact that the KdV equation is associated with an infinite number of *conserved quantities* or *invariants* (see Chapter 8). Of these, three that have particular physical significance are as follows:

$$\text{Conservation of Mass:} \quad I_1 = \int_{-\infty}^{\infty} u(x,t)\, dx \quad (= 9.0000),$$

$$\text{Conservation of Momentum:} \quad I_2 = \int_{-\infty}^{\infty} \tfrac{1}{2} u(x,t)^2\, dx \quad (= 4.1250), \quad (3.71)$$

$$\text{Conservation of Energy:} \quad I_3 = \int_{-\infty}^{\infty} [2u(x,t)^3 - u(x,t)^2]\, dx \quad (= 16.2375).$$

Evaluating these invariants at each time step allows the analyst to judge how well the solution is evolving. By any standard, the solution obtained for this difficult problem by our numerical scheme should be considered a success as the invariants remain constant at their correct value throughout the simulation. The invariant data along with maximum errors is given in Table 3.2.

For a very good account of the historical background to the KdV equation the reader is referred to [De-11]; for an overview of the FPU work, see [Por-09]; and for further discussion relating to the KdV equation and solitons, and traveling waves in general, readers are referred to [Abl-11], [Deb-05], [Dra-92], [Joh-97], [Gri-11], and [Zab-10].

The R code for the main program that solves this problem is shown in Listing 3.2. This program calls functions for tasks that include initialization, derivative calculations, plotting results, and calculating invariants. These functions are contained in files KdV_anal.R, KdV_deriv.R, KdV_postSimCalcs.R, and KdV_invariants.R, respectively, which are all included with the downloads. The R code associated with KdV_deriv.R, and KdV_invariants.R is included in Listings 3.3 and 3.4.

```
# File: KdV_main.R
# ----------------------------------------------------------------
# 1D Korteweg-deVries (KdV) equation
#   ut + 6*u*ux+uxxx = 0
# Initial condition: u(x,t=0) = ua(x,t=0), where ua(x,t) is
# the analytical solution given by function KdV_anal()
# BCs: Dirichlet, u(t,xl)=ua(t,xl) and u(t,xu)=ua(t,xu)
# ----------------------------------------------------------------
```

```r
  rm(list = ls(all = TRUE)) # Delete workspace
  cat("\014")               # Clear console
#
#####################################################
# Required R library packages
#####################################################
  require(compiler)
  enableJIT(3)
  library("pracma")
  library("deSolve")
  library("rgl")
#####################################################
# Load sources for functions called
#####################################################
  source("deriv1st7pt.R")
  source("deriv3rd9pt.R")
  source("KdV_deriv.R")
  source("KdV_anal.R")
#
  ptm0 <- proc.time()
  print(Sys.time())
# Define spatial domain
  Nx <- 531
  xl <- -50; xu <- 50
# Grid
  dx <- (xu-xl)/(Nx-1)
  x  <- seq(from=xl,to=xu,by=dx)
#
#####################################################
# Define differential matrices
#####################################################
  diffMatX1st <- diffMatrix1st7pt(xl,xu,Nx)
  diffMatX3rd <- diffMatrix3rd9pt(xl,xu,Nx)
#####################################################
# ODE integration
#####################################################
# Set simulation times
#   t0   <- -20; tf <- 20
  t0   <- -8; tf <- 8
  t    <- c(t0, 0, 10, tf)
  nout <- 251
  t <- seq(t0,tf,length.out=nout)
# Initial condition
  uini <- KdV_anal(x,t0)
#
  ncall <<- 0
```

```
  out <- ode(method="lsodes",y=uini, times=t, func=KdV_deriv,
             rtol=1e-8,atol=1e-8, parms=NULL,
             maxsteps=50000, maxord=5, sparsetype = "sparseint")
####################################################
# Define array and plot data
####################################################
  U <- matrix(data=out[(nout+1):length(out)],ncol=Nx, nrow=nout)
  cat(sprintf("Derivative calls: %d\n",ncall))
#
  source("KdV_postSimCalcs.R") # Plot results
  source("KdV_invariants.R") # Calculate the invariants
#
  t1 <- proc.time()-ptm0
  cat(sprintf("Calculation time: %f\n",t1[3]))
```

Listing 3.2. File: KdV_main.R—Code for the *main program* that simulates KdV eqn. (3.69). This program calls functions for tasks that include initialization, derivative calculations, plotting results, and calculating invariants from files KdV_anal.R, KdV_deriv.R, KdV_postSimCalcs.R, and KdV_invariants.R, respectively

```
# File: KdV_deriv
KdV_deriv = function (t,u,parms) {
  u1 <- as.vector(u) # for martix multiplication
#   # Dirichlet BCs
  u1[1] <- 0;   u1[(Nx-3):Nx] <- 0 # BCs
  # Calculate 1D spatial 1st and 3rd derivatives
  ux   <- diffMatX1st %*% u1
  uxxx <- diffMatX3rd %*% u1
  # Calculate time derivative
  ut <- -uxxx-6*u*ux
  ut[1] <- 0; ut[Nx] <- 0 # BCs
  # Update ncall
  ncall <<- ncall+1
#
  return(list(c(ut)))
}
```

Listing 3.3. File: KdV_deriv.R—Code for the *derivative function* KdV_deriv(), called by the KdV main program of Listing 3.2

```
# File: KdV_invariants.R
# Invariants - Integrate by Simpson's rule
I1  <- rep(0,nout); I2  <- rep(0,nout); I3  <- rep(0,nout)
I1a <- rep(0,nout); I2a <- rep(0,nout); I3a <- rep(0,nout)
KdV_Err1 <-rep(0,nout);KdV_Err2 <-rep(0,nout);KdV_Err3 <-rep(0,nout)
```

```
for(i in 1:nout){
  # Invariant 1 - 3 soliton value: 9.0000
  u1     <- U[i,]
  I1[i]  <- dx*(1/3*(u1[1] + u1[Nx]) + 4/3*sum(u1[seq(2,(Nx-1),2)])
            + 2/3*sum(u1[seq(3,(Nx-2),2)]))
  u1a    <- KdV_anal(x,t[i])
  I1a[i] <- dx*(1/3*(u1a[1] + u1a[Nx]) + 4/3*sum(u1a[seq(2,(Nx-1),2)]) +
              2/3*sum(u1a[seq(3,(Nx-2),2)])) # analytical value
  # Invariant 2 - 3 soliton value: 4.1250
  u2     <- u1^2 / 2
  I2[i]  <- dx*(1/3*(u2[1] + u2[Nx]) + 4/3*sum(u2[seq(2,(Nx-1),2)])
            + 2/3*sum(u2[seq(3,(Nx-2),2)]))
  u2a    <- u1a^2 / 2
  I2a[i] <- dx*(1/3*(u2a[1] + u2a[Nx]) + 4/3*sum(u2a[seq(2,(Nx-1),2)])
            + 2/3*sum(u2a[seq(3,(Nx-2),2)])) # analytical value
  # Invariant 3 - 3 soliton value: 16.2379
  ux     <- diffMatX1st %*% U[i,]
  u3     <- 2*u1^3 - ux^2
  I3[i]  <- dx*(1/3*(u3[1] + u3[Nx]) + 4/3*sum(u3[seq(2,(Nx-1),2)])
            + 2/3*sum(u3[seq(3,(Nx-2),2)]))
  uxa    <- diffMatX1st %*% u1a
  u3a    <- 2*u1a^3 - uxa^2
  I3a[i] <- dx*(1/3*(u3a[1] + u3a[Nx]) + 4/3*sum(u3a[seq(2,(Nx-1),2)])
            + 2/3*sum(u3a[seq(3,(Nx-2),2)])) # analytical value
  KdV_Err1[i] <- max(abs(u1-u1a))
  KdV_Err2[i] <- max(abs(u2-u2a))
  KdV_Err3[i] <- max(abs(u3-u3a))
}
```

Listing 3.4. File: `KdV_invariants.R`—Code that calculates the *invariants* for KdV eqn. (3.69) called by the KdV main program of Listing 3.2

3.8.3.2 Corneal curvature[4]

Qualitative and quantitative models can provide a major advantage in the diagnosis and prediction of the outcome of therapies for different diseases and related health issues. In the case of vision, ophthalmologists and optometrists generally consider the cornea as one of the most important components of the eye that is crucial for normal vision [Plo-14]. For diagnostic purposes, a detailed knowledge of corneal topography and curvature is essential. To understand these features, a sufficient level of detail is achieved only by accurate measurements and correct mathematical models.

For corneal topography modeling, models based on various conical sections such as ellipsoids and paraboloids are well established and confirmed as accurate. This approach goes back to Helmholtz, who in 1867 proposed the ellipse as being suitable for a description of a corneal profile [Hel-24] . Models and associated numerical simulations of the

---

[4] This example is based on the paper "ODE/PDE Analysis of Corneal Curvature" by Łukasz Płociniczak et al. [Plo-14].

biomechanical shell that constitutes the cornea are now widely studied. These structural models are invaluable for understanding the fine details of many corneal features such as their age-dependent evolution and apex displacement caused by intra-ocular pressure. The greater complexity of biomechanical models provides enhanced insight of the model accuracy and physical interpretations.

Here we consider the numerical analysis of an intermediate (in accuracy and complexity) model between the conical sectional and structural mechanical approaches. The starting point for the analysis of corneal curvature is the static force balance

$$-T\nabla \cdot \left(\frac{\nabla h}{\left(\sqrt{1+|\nabla h|^2}\right)^3}\right) + kh = \frac{P}{\sqrt{1+|\nabla h|^2}}, \; h = h(x,y), \; (x,y) \in \mathbb{R}^2, \quad (3.72)$$

where h represents corneal surface height [m], (x,y) distance from axisymmetric center [m], T tension [N], P intraocular pressure [N/m], and k elastic constant [N/m^2].

We will consider a one-dimensional problem equation and, for computational convenience, we convert eqn. (3.72) to the nondimensional form

$$\frac{d^2h/dx^2}{(1+|dh/dx|^2)^{3/2}} - ah + \frac{b}{(1+|dh/dx|^2)^{1/2}} = 0, \; h = h(x), \; x \in \mathbb{R}, \quad (3.73)$$

where $a = R^2 k/T, b = RP/T$, and h, x have been rescaled by an R-typical corneal dimension such as its radius. From observation of the physical structure of the eye we can deduce that eqn. (3.73) is subject to a Neumann BC at $x = 0$ and a Dirichlet BC at $x = 1$, that is,

$$\frac{dh(x=0)}{dx} = 0, \; h(x=1) = 0. \quad (3.74)$$

The usual approach to solving the nonlinear time-independent PDEs is to form a suitable *functional*[5] and treat it as a *minimization* problem. Alternatively, a solution can be obtained using the method of *false transients*. This method treats the problem as being *dynamic* by replacing the right-hand zero in eqn. (3.73) with a *false time derivative* and integrating using the method of lines until a steady-state is reached. This transforms the problem from an ODE to a parabolic evolutionary PDE. Although, in general, the method is not the most efficient at finding a solution, it is a useful tool for the analyst and is in quite common usage. However, for this particular application, it offers a very straightforward approach and is more simple to apply than other methods. This is due, primarily, to the nonlinear denominators. As we are seeking only the steady state solution or equilibrium conditions, the dynamic response is not important providing that the method converges. For diffusion-type problems, this is generally not a problem, but this method can encounter difficulties for highly nonlinear problems, particularly those having multiple solutions. Some interesting complex examples of the false transients method applied to an aircraft tail plane and a flapping wing are given by Ly and Norrison [Ly-08].

---

[5] A *functional*, or *objective function*, takes a vector of input variables and returns a single scalar quantity representing the goodness or otherwise of a solution. It is often used in optimization algorithms where the objective is to minimize the functional value.

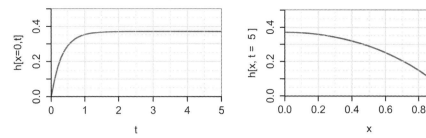

Figure 3.7. (left) Corneal height at $x = 0$ as the false transient method evolves eqn. (3.73) to steady state. (right) Corneal height profile when eqn. (3.73) has reached steady state and the complete solution is symmetrical about the origin, $x = 0$.

Thus, adopting the idea of false transients, we can define the problem mathematically on the 1D spatial domain $\Omega = [0, 1]$ with zero initial conditions and the boundary conditions of eqn. (3.74) imposed as follows:

$$\frac{\partial h}{\partial t} = \frac{\partial^2 h/\partial x^2}{(1+|\partial h/\partial x|^2)^{3/2}} - ah + \frac{b}{(1+|\partial h/\partial x|^2)^{1/2}}, \quad h = h(x,t), \ t \geq t_0, \ x \in \mathbb{R}, \tag{3.75}$$
$$h(x,0) = 0, \ \partial h(t,0)/\partial x = 0, \ h(t,1) = 0.$$

The simulation was run to $t = 5$ and carried out over 21 spatial grid points using a classical FD discretization. The first derivatives were approximated by a five-point scheme and the second derivative by a six-point scheme, both of which are fourth-order accurate. The results are shown in Fig. 3.7, with the height at the center being $h = 0.37$. Increasing the number of grid points to 201 and running for longer pseudo time gave the same results to six significant figures and had little effect on the computation time. This example clearly shows the utility of the false transient method for solving time-independent PDEs.

A full 2D numerical analysis of eqn. (3.72) using radial basis functions (RBFs), without any assumptions being made regarding radial symmetry, is given in [Gri-16a] and [Gri-16b].

The R code for solving this problem is shown in Listings 3.5 and 3.6.

```
# File: cornea_1D_main.R
# -------------------------------------------------
# 1D cornea curvature problem
#
#####################################################
# Delete workspace prior to new calculations
#####################################################
rm(list = ls(all = TRUE))
cat("\014") # Clear console
#
#####################################################
# Required R library packages
#####################################################
```

```
require(compiler)
enableJIT(3)
library("deSolve")
library("rgl")
#
#####################################################
# Load sources for functions called
#####################################################
source("deriv1st5pt.R")
source("deriv2nd6pt.R")
source("cornea_1D_deriv.R")
#
#####################################################
# Start timer
#####################################################
ptm = proc.time()
print(Sys.time())
#
#####################################################
# Declare array sizes
#####################################################
Nx   <- 21
#
#####################################################
# PDE parameters
#####################################################
# Note: xu rescaled by R, the cornea radius of
#       curvature, R - typically equal to 7mm
xl <- 0; xu <- 1; dx <- (xu-xl)/(Nx-1)
a  <- 1; b <- 1
#
#####################################################
# Define grid in x
#####################################################
x <- seq(from=xl,to=xu,by=dx)
#
#####################################################
# Define differentiation matrices
#####################################################
diffMat1X <- diffMatrix1st5pt(xl,xu,Nx)
diffMat2X <- diffMatrix2nd6pt(xl,xu,Nx,bc=c(1,0))
#####################################################
# Set parameters
#####################################################
tf <- 5; nout <- 51
t <- seq(from=0,to=tf,length.out=nout)
```

```
#
#####################################################
# Preallocate ICs for u1 and u2
#####################################################
uini <- rep(0,Nx)
#####################################################
# Initialze other variables
#####################################################
ncall <- 0
#
#####################################################
# Set ODE integrator and perform integration
#####################################################
out<-lsodes(y=c(uini), times=t, func=cornea_1D_deriv,
            sparsetype = "sparseint", ynames = FALSE,
            rtol=1e-12,atol=1e-12,maxord=5,parms=NULL)
#
#####################################################
# Create plot array
#####################################################
# Define plot data
u <- array(data=out[(nout+1):length(out)],c(nout,Nx))
cat(sprintf("u(max) = %f\n",max(u[nout,])))
#
source("cornea_1D_postSimCalcs.R")
#####################################################
# Print elapsed time
#####################################################
cat(sprintf("calculation time = %f\n",proc.time()-ptm)[3])
#
#####################################################
# End of program
#####################################################
```

Listing 3.5. File: cornea_1D_main.R—Code for the *main program* that simulates the cornea eqn. 3.75)

```
# File: cornea_1D_deriv.R
cornea_1D_deriv = function(t, u, parms) {
  #
  #####################################################
  # Set Dirichlet BC value
  #####################################################
  u[Nx]   <- 0
  #####################################################
```

```
# Calculate spatial derivatives
######################################################
#
ux    <- diffMat1X %*% u
uxx   <- diffMat2X %*% u
######################################################
# Set Neuman BC value
######################################################
ux[1] <- 0 # Neumann BC
######################################################
# Calculate time derivative
######################################################
#
den1 <- sqrt(1+ux^2)
den2 <- den1^3
#
ut <- uxx/den2 - a*u + b/den1
######################################################
# update number of derivative calls
######################################################
ncall <<- ncall+1
#
######################################################
# Return solution
return(list(c(ut)))
######################################################
}
```

Listing 3.6. File: `cornea_1D_deriv.R`—Code for the *derivative function* `cornea_1D_deriv()`, called from the cornea simulation main program of Listing 3.5

3.8.3.3 Convection–Diffusion–Reaction Equation

In this example we solve a problem similar to that represented by eqn. (3.18), but expanded to include a reaction term. Consider the following 1D PDE describing a *convective–diffusion–reaction* system:

$$\frac{\partial T}{\partial t} = v\frac{\partial T}{\partial x} + D\frac{\partial^2 T}{\partial x^2} + r, \quad T = T(x,t), \ t \geq t_0, \ x \in \mathbb{R}, \ r = r(T(x,t)), \quad (3.76)$$

where v represents bulk velocity of the medium, D a diffusion coefficient, and r a reaction term.

For this problem the reaction r is defined as

$$r(T) = \begin{cases} 0, & T < T_s, \ T_s = 0.25 \\ k(T_0 - T), & T \geq T_s, \ T_0 = 1, \end{cases} \quad (3.77)$$

where T_0 represents a reference temperature and T_s the *reaction strike temperature*, that is, the temperature at which the reaction starts. The parameter values used are $D = 0.2$,

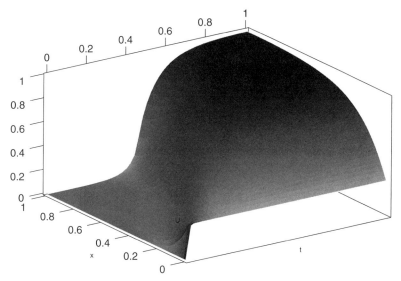

Figure 3.8. 3D surface plot of solution to the reactor problem of eqn. (3.76). The surface plot was generated using the `persp3d()` function.

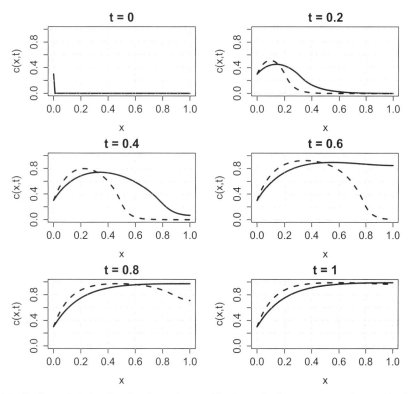

Figure 3.9. Series of plots showing how the profile through the reactor evolves with time. The calculations were based on 101 spatial grid points. Solid lines represent the solution with $D = 0.2$ and dashed lines with $D = 0.05$. The effect of reduced diffusion is clearly shown, with the profile taking significantly longer to reach steady state.

$k = 8$, and $v = 0.5$. The initial condition is $T(x, 0) = 0$ with Dirichlet boundary condition $T(0, t) = 0.3$ and Neumann boundary condition $T_x(1, t) = 0$. The simulation was carried out over the 1D spatial domain $x \in (0, 1)$ with 101 spatial grid points. The results of a classical FD discretization are shown in Figs. 3.8 and 3.9. The first derivative was approximated by a three-point scheme and the second derivative by a four-point scheme, which are both second-order accurate.

The R code for solving this problem is shown in Listings 3.7–3.9.

```
# File: convDiffReact_main.R
# -------------------------------------------------------------------------
# 1D Convection-diffusion-reaction example:
#     ut = D*uxx -v*ux + f(u), f(u) = k*(us-c)
# Initial condition: u(x,y,t=0) = 0
# BCs: u(x=0,t) = 0, Dirichlet
#     ux(x=1,t) = 0, Neumann
# -------------------------------------------------------------------------
####################################################
# Delete workspace prior to new calculations
####################################################
rm(list = ls(all = TRUE))
cat("\014") # Clear console
#
####################################################
# Required R library packages
####################################################
require(compiler)
enableJIT(3)
library("deSolve")
library("rgl")
library("shape")
#
####################################################
# Load sources for functions called
####################################################
source("deriv1st3pt.R")
source("deriv2nd4pt.R")
source("convDiffReact_deriv.R")
#
####################################################
# Start timer
####################################################
ptm0 = proc.time()
print(Sys.time())
#
####################################################
# Declare array sizes
```

```
#####################################################
Nx=101
#
#####################################################
# PDE parameters
#####################################################
xl <- 0; xu <- 1;
#
#####################################################
# Define grid in x, y
#####################################################
dx <- (xu-xl)/(Nx-1)
x  <- seq(from=xl,to=xu,by=dx)
#####################################################
# Define differential matrices
#####################################################
diffMat1X <- diffMatrix1st3pt(xl,xu,Nx)
diffMat2X <- diffMatrix2nd4pt(xl,xu,Nx,bc=c(0,1))
#####################################################
# Set reactor parameters
#####################################################
# u <- Concentration
v   <- 0.5  # Bulk velocity of medium
D   <- 0.2  # Diffusion coefficient
k   <- 8    # Reaction rate constant
ui  <- 0.25 # Concentration when reaction initiated
us  <- 1    # Saturation concentration
u1  <- 0.3  # Dirichlet BC at inlet
uxN <- 0    # Neumann BC at outlet
#
#####################################################
# Set time step parameters
tf <- 1; nout <- 51
t  <- seq(from=0,to=tf,length.out=nout)
#
#####################################################
# Set Initial Conditions
#####################################################
uini <- rep(0,Nx)
uini[1] <- u1 # feed concentration
#
#####################################################
# Initialze other variables
#####################################################
#
ncall <<- 0
```

```
#
####################################################
# ODE integration
####################################################
#
out<-lsodes(y=c(uini),times=t,func=convDiffReact_deriv,
            sparsetype ="sparseint", ynames = FALSE,
            rtol=1e-6,atol=1e-6,maxord=5,parms=NULL)
#
####################################################
# Create plot matrices
####################################################
#
U <- matrix(data=out[(nout+1):length(out)],ncol=Nx, nrow=nout)
#
t2 <- proc.time()-ptm0
cat(sprintf("Total time: %f\n",t2[3]))
#
source("convDiffReact_postSim.R")
####################################################
# End of program
####################################################
```

Listing 3.7. File: `convDiffReact_main.R`—Code for the *main program* that simulates the convection–diffusion–reaction system of eqn. (3.76)

```
# File: convDiffReact_deriv.R
convDiffReact_deriv <- function(t, u, parms) {
  # Function to calculate the time derivative of u
  ####################################################
  # Set Dirichlet BC for u[x=0]
  u[1] <- u1
  ####################################################
  # Calculate 2D spatial 2nd derivatives
  ####################################################
  ux    <- diffMat1X %*% u
  ux[Nx] <- uxN
  uxx   <- diffMat2X %*% u
  # Set Neumann BC for ux[x=1]
  uxx   <- diffBC2nd4pt(xy=1,xl,xu,uxx,bcL=0,bcU=uxN)
  #
  ####################################################
  # Calculate time derivative
  ####################################################
  #
```

```
  f    <- rep(0,Nx)
  rx   <- which(u >= ui) # Reaction initiates at u>=ui
  #
  f[rx] <- k*(us-u[rx]) # Reaction
  ut   <- (-v*ux + D*uxx + f)
  ut[1] <- 0;
  #
  # update number of derivative calls
  ncall <<- ncall+1
  #
  return(list(ut))
}
```

Listing 3.8. File: convDiffReact_deriv.R—Code for the *derivative function* convDiff-React_deriv(), called by the main program of Listing 3.7

```
####################################################
# Set animation flag
####################################################
#
animate <- FALSE # TRUE/FALSE
#
if (animate == TRUE){
  N_by=1
}else{
  N_by=10
}
cat(sprintf("\n Animation Set To: %s, N_by=%d \n",
            animate,N_by))
####################################################
#Define color scheme
####################################################
#
jet.colors = colorRampPalette(c(
  "#00007F", "blue", "#007FFF",
  "cyan", "#7FFF7F", "yellow",
  "#FF7F00", "red", "#7F0000"))

# Set palette
pal=jet.colors(100)
#pal=matlab.palette(100)
#
# Plot surface
open3d()
```

```r
bg3d("white")
# Set colour indices of each point for pers3d()
col.ind <- cut(U,100)#
persp3d(t,x,U, aspect=c(1, 1, 0.5), color = pal[col.ind],
        box=FALSE, smooth=FALSE)
UM <- matrix(data=c( 0.8178354, -0.5754114, 0.006800523, 0,
                     0.1779113,  0.2640698, 0.947952569, 0,
                    -0.5472587, -0.7740598, 0.318337947, 0,
                     0.0000000,  0.0000000, 0.000000000, 1),
            byrow=TRUE,nrow=4,ncol=4)
rgl.viewpoint(fov=0) # set before userMatrix
par3d(userMatrix=UM) # define 3d view
par3d(windowRect=c(20,100,820,900), zoom=1)
#
# Print to png file
# rgl.snapshot("convecDiffReacEqn_3D.png", fmt="png")
#
######################################################
#Plot 2D image
######################################################
# U2 <- U
# save(U2, file = "UD05.RData") # Save data
load("UD05.RData") # U2 Data from run with D=0.05
par(mfrow = c(3, 2))
#par(mar=c(2, 4, 4, 2) + 0.1, mgp = c(5, 1, 0))
par(mar=c(4, 4, 2, 2) + 0.1)
#par(oma = c(0,0,2,0))
for (i in seq(1,nout,by=10)){
  plot(x,U[i,], type="l", ylim=c(0,1), xlim=c(0,1),
       xlab="x", ylab="c(x,t)",lwd=2,
       main=paste("t =",t[i]))
  grid()
  lines(x,U2[i,],lty=2,lwd=2)
}
# mtext(side=3, outer=TRUE, cex=1.25, line=-1,
#       "Concentration profile")
par(mfrow = c(1, 1))
#animate=TRUE
if (animate == TRUE){
  png(file="example%02d.png", width=600, height=600)
  for (i in 1:nout){
    #plot.new() # Inserts a blank between plots
    persp(U[,,i], theta = -30, phi = 30,
          #xlim = range(x), ylim = range(y),
          #zlim = c(-0.3,1), ylab = "\n\n\nX",
          xlab = "\n\n\nY", zlab = "\n\n\nU",
```

```
        col = drapecol(U[,,i]), border=NA,
        main=paste("t =",t[i]), ticktype="detailed")
  }
  dev.off()
  # Create animated GIF file.
  # NOTE: Can also run the command, shown below in
  #       quotes, from command line prompt
  shell("convert -delay 20 example*.png AdvectionEx_1.gif",
        intern=TRUE)
  # Remove individual files
  file.remove(grep("example",list.files(),value=TRUE))
}
```

Listing 3.9. File: convDiffReact_postSim.R—code for the *postsimulation calculations* run after the main program of Listing 3.7

3.8.4 MOL 2D: Cartesian Coordinates

The application of MOL to 2D problems in Cartesian coordinates is a simple extension to the method used to solve 1D problems. The complexity of the problem increases due to influence of derivatives in both the *x* and *y* directions, and it is computationally more demanding as the total calculations increase by at least to the square of the number of grid points. The examples that follow demonstrate the utility and the ease of use of MOL to solve nontrivial problems.

3.8.4.1 Diffusion Equation

Diffusion[6] is a process where a physical quantity *diffuses* through the medium in which it is present. It is a naturally occurring process that takes the following general mathematical form of a parabolic PDE for an *isotropic* medium:

$$\frac{\partial u}{\partial t} = D \nabla u, \quad u = u(x,t),\ t \geq t_0,\ x \in \mathbb{R}^n. \tag{3.78}$$

As mentioned previously, the diffusion process occurs in a variety of different forms, for example, for heat transfer, it defines *Fourier's second law*, where *u* represents *temperature* and *D* *thermal diffusivity*; for mass transfer, it defines *Fick's second law*, where *u* represents *mass concentration* and *D* the *coefficient of diffusivity* or *mass diffusivity*.

In this example we consider the following 2D thermal problem with Dirichlet boundary conditions on the rectangular domain $\Omega = [0,1]^2$:

$$\rho C_p \frac{\partial T}{\partial t} = \left(k_x \frac{\partial^2 T}{\partial x^2} + k_y \frac{\partial y^2 T}{\partial^2}\right), \quad T = T(x,y,t),\ t \geq t_0,\ (x,y) \in \mathbb{R},$$

$$T(x,y,0) = \begin{cases} 100, & T(x,y) \in f(x,y), \\ 0, & \text{otherwise}, \end{cases} \tag{3.79}$$

$$\partial \Omega(x,y=0,$$

---
[6] From the Latin word *diffundere*, which means "to spread out."

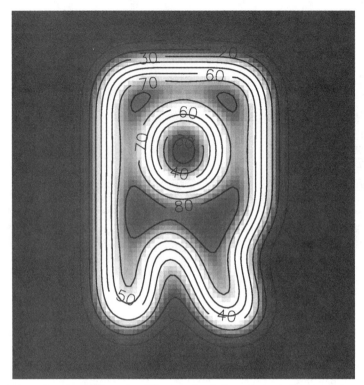

Figure 3.10. Image plot at $t = 90$ s of a diffusion simulation of steel plate based on a 100×100 grid. The letter "R" was initially at $100°C$ and the surrounding metal at $0°C$. The plot was generated using the image() function and contours were added using the contours() function. (See color plate 3.10)

where T represents the domain temperature. The medium consists of a thin sheet of steel measuring 1 m square with perfect insulation on both sides. It is initially at a temperature of $0°C$, except for the central area in the shape of the letter R, that is, $f(x, y)$, which is raised to a temperature of $100°C$. We assume that there are no temperature variations vertically through the sheet and that no convective or radiative heat transfer takes place. Also, we assume that the steel forms an isotropic medium with thermal conductivity $k_x = k_y = k = 43$W/m-C, density $\rho = 7800$kg/m^3, and specific heat $C_p = 0.473$J/kg-C. Thus, the thermal diffusivity constant becomes $D = k/\rho C_p = 1.66^{-5}$m^2/s. The Dirichlet BCs are defined in eqns. (3.79).

The calculations are performed on a 100×100 grid with the initial condition being loaded from an ASCII file at the start of the simulation. The spatial derivatives were approximated by six-point, fourth-order, second derivative matrices. A surface plot of the state at $t = 90°$ s is shown in Fig. 3.10, and an array of surface plots taken at various times is shown in Fig 3.11. Over a period of 500 s, the solution shows that the surface temperature falls to a maximum of $61°C$ on the R-shaped area from the initial uniform temperature of $100°C$.

The R code for the *main* program and *derivative* function for solving this problem is shown in Listings 3.10 and 3.11, respectively. Note that for this problem we need to set

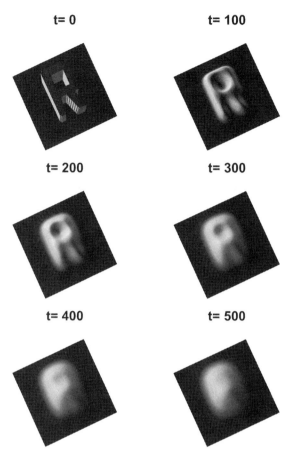

Figure 3.11. Surface plots showing how heat diffuses over time from the initially hot letter "R" at 100°C to the surrounding medium initially at 0°C. The surface plots were generated using the persp() function.

the variable case=0 in the main program. The results are plotted by additional R code, which can also be made to produce an animation by by setting the variable animation = TRUE. This additional code is in file diffusion2D_postSimCalcs.R, which is called from the main program (see example in Listing 3.15). The animation is created by assembling snapshots (frames) during the simulation. This process is performed using the program *ImageMagick*, which is a software suite to create, edit, compose, or convert images. It can read and write images in more than 200 formats and is freely available for download from http://www.imagemagick.org/script/index.php. Once installed, the R code calls ImageMagick to create the animation. All the associated R files are provided with the downloads.

```
# File: diffusion2D_main.R
# ----------------------------------------------------------------------------
# 2D Diffusion/Laplace/Poisson examples
```

```
#       ut = D*(uxx + uyy + f) (= 0)
# ----------------------------------------------------------------------
require(compiler)
enableJIT(3)
#
cat("\014") # Clear console
rm(list = ls(all = TRUE)) # Delete workspace
#
######################################################
# Required R library packages
######################################################
library("pracma")
library("deSolve")
library("rgl")
library("shape")
#
######################################################
# Load sources for functions called
######################################################
#source("deriv2nd4pt.R")
source("deriv2nd6pt.R")
# source("deriv2nd8pt.R")
# source("deriv2nd10pt.R")
source("diffusion2D_deriv.R")
#
ptm0 <- proc.time()
print(Sys.time())
# Declare array sizes
# Note for sizes must be the same as letter R data, see below!
Nx <- 100
Ny <- 100
#
# PDE parameters
xl <- 0; xu <- 1 # 1, pi, 1
yl <- 0; yu <- 1 # 1, pi, 1
#
# Spatial grid in x, y
x <- seq(from=xl,to=xu,by=(xu-xl)/(Nx-1))
y <- seq(from=yl,to=yu,by=(yu-yl)/(Ny-1))
#
######################################################
# Set initial condition uini
######################################################
uini <- array(0,c(Nx,Ny))
# Populate uini with initial condition
case <- 0
```

```
if(case == 0){
  simTitle <- "diffusion2D_letterR"
  letterR <- read.csv("letterR5.csv",
              header = FALSE, sep = ",") # 100x100
  # letterR <- read.csv("letterR5_2.csv",
  #             header = FALSE, sep = ",") # 200x200
  NlR <- dim(letterR)
  if(Nx != NlR[1] && Ny != NlR[2]){
    stop("Nx and Ny must be equal to size of letterR!")
  }
  # IC - rotate starting point
  uini <- as.matrix(t(apply(letterR,2,rev)))*100
  # Thermal diffusivity for steel
  D <- 1.166e-5
  tf <- 500
  BCx <- c(0,0); BCy <- c(0,0) # Dirichlet

}else if(case == 1){
  simTitle <- "Laplace2D"
  D <- 1
  tf <- 5
  BCx <- c(0,1); BCy <- c(0,0) # Dirichlet
}else if(case == 2){ # Iserles, p162
  simTitle <- "Poisson2D"
  g <-function(x,y){-(x^2+y^2)}
  f <- outer(x,y,g)
  D <- 1
  tf <- 8
  BCx <- c(0,1); BCy <- c(0,0) # Dirichlet/Neumann
}else if(case == 3){ #
  simTitle <- "Laplace2D"
  D <- 1
  tf <- 2
  BCx <- c(0,0); BCy <- c(0,0) # Dirichlet/Neumann
}
cat(sprintf("\n%s\n",simTitle))
#
#####################################################
# Define differential matrices
#####################################################
diffMat2X <- diffMatrix2nd6pt(xl,xu,Nx,bc=BCx)
diffMat2Y <- diffMatrix2nd6pt(yl,yu,Ny,bc=BCy)
#####################################################
# Set simulation times
#####################################################
nout=51
```

```
t=seq(from=0,to=tf,len=nout)
####################################################
# Define Dirichlet/Neumann BCs
####################################################
if(case == 0){            # Dirichlet
  BCyxl <- rep(0,Ny); BCyxu <- rep(0,Ny)
  BCxyl <- rep(0,Nx); BCxyu <- rep(0,Nx)
}else if(case == 1){      # Dirichlet
  BCyxl <- rep(0,Ny); BCyxl[1:(Ny/2)] <- 1
  BCyxu <- rep(0,Ny); BCyxu[(Ny/2):Ny] <- 1
  BCxyl <- rep(0,Nx); BCxyl[(Nx/2):Nx] <- 1
  BCxyu <- rep(0,Nx); BCxyu[1:(Nx/2)] <- 1
}else if(case == 2){ # Dirichlet/Neumann
  BCyxl <- sin(pi*y); BCux_xu <- pi*exp(pi)*sin(pi*y)+(y)^2
  BCxyl <- rep(0,Nx); BCxyu <- exp(pi*x)*sin(pi)+(1/2)*x^2
}else if(case == 3){ # Dirichlet
  BCyxl <- cos(x[1]*pi)*sinh(y*pi); BCyxu <- cos(x[Nx]*pi)*sinh(y*pi)
  BCxyl <- cos(x*pi)*sinh(y[1]*pi); BCxyu <- cos(x*pi)*sinh(y[Ny]*pi)
}
####################################################
# Apply Dirichlet BCs to IC                #
####################################################
if(case == 0 || case == 1 || case == 3){
  uini[1,] <- BCyxl; uini[Nx,] <- BCyxu
  uini[,1] <- BCxyl; uini[,Ny] <- BCxyu
}else if(case == 2){
  uini[1,] <- BCyxl;
  uini[,1] <- BCxyl; uini[,Ny] <- BCxyu
}
####################################################
# ODE integration
####################################################
ncall<<-0
out<-lsodes(y=c(uini), times=t, func=diffusion2D_deriv,
        sparsetype = "sparseint", ynames = FALSE,
        rtol=1e-6,atol=1e-6,maxord=5,parms=NULL)
####################################################
source("diffusion2D_postSimCalcs.R")
#
t1 <- proc.time()-ptm0
cat(sprintf("Calculation time: %f\n",t1[3]))
```

Listing 3.10. File: diffusion2D_main.R—Code for the *main program* that simulates diffusion eqn. (3.79), Laplace eqn. (3.81), and Poisson eqn. (3.83). Note that case=0 for *letter R* diffusion example, case=1 for *electrostatic potential* Laplace example, and case=2 for *Poisson* example.

```r
# File: diffusion2D_deriv.R
diffusion2D_deriv <- function(t, u, parms) {
    #
  # Reshape vector u -> Nx by Ny matrix
  u1 <- matrix(data=u,nrow=Nx, ncol=Ny, byrow=FALSE,
               dimnames=NULL)
  # Apply Dirichlet BCs
  if(case == 0 || case == 1 || case == 3){
    u1[1,] <- BCyxl; u1[Nx,] <- BCyxu
    u1[,1] <- BCxyl; u1[,Ny] <- BCxyu
  }else if(case == 2){
    u1[1,] <- BCyxl;
    u1[,1] <- BCxyl; u1[,Ny] <- BCxyu
  }
  #
  ######################################################
  # Calculate 2D spatial 2nd derivatives
  ######################################################
  u1_xx <-   diffMat2X %*% u1
  u1_yy <- t(diffMat2Y %*% t(u1))
  if(case == 2){# Set Neumann BCs
    u1_xx <- diffBC2nd6pt(xy=0,xl,xu,u1_xx,0, BCux_xu) # dudx on y
  }
  # Calculate time derivative
  ut = D*(u1_xx+u1_yy)
  if(case==2){ut <- ut + D*f}
  
  # Set time derivatives to zero for Dirichlet BCs
  if(case == 0 || case == 1 || case == 3){
    ut[1,] <- 0; ut[Nx,] <- 0
    ut[,1] <- 0; ut[,Ny] <- 0
  }else if(case == 2){
    ut[1,] <- 0;
    ut[,1] <- 0; ut[,Ny] <- 0
  }
  #
  # update number of derivative calls
  ncall <<- ncall+1
  #
  # Return solution
  return(list(ut))
}
```

Listing 3.11. File: diffusion2D_deriv.R—Code for the *derivative function* diffusion2D_deriv(), which is called from the diffusion main program of Listing 3.10

3.8.4.2 Laplace Equation

The *Laplace* equation represents the simplest form of an elliptical PDE. It is given by

$$\nabla u = 0, \quad u = u(x), \quad x \in \mathbb{R}^n. \tag{3.80}$$

Solutions to the Laplace equation are called *harmonic functions*[7] and will be unique if certain boundary conditions are satisfied. They can have no local *maxima* or *minima*, and extreme values must occur at the boundaries. Solutions will have similar form to those of the diffusion equation, evaluated at equilibrium or steady-state conditions.

A solution to Laplace's equation is uniquely determined if all boundaries are specified as being Dirichlet—the *Dirichlet problem*. However, any two solutions to Laplace's equation with the same Neumann boundaries specified—the *Neumann problem*—can differ only by a constant. Solutions to Laplace's equation with mixed boundary conditions will be unique. Also, because the problem is linear, the *superposition principle* applies, and therefore new solutions can be found by the addition of existing known solutions.

Consider the problem of an electrostatic potential distribution $V(x, y)$ inside a 2D square region, as shown in Fig. 3.12. Now, electrostatic potentials in a source-free region satisfy the Laplace equation; thus, adopting the idea of false transients, we can define the problem mathematically on the rectangular domain $\Omega = [0, 1]^2$ with Dirichlet boundary conditions from Fig. 3.12 imposed on $\partial \Omega$, as follows:

$$\frac{\partial V}{\partial t} = \frac{\partial^2 V}{\partial x^2} + \frac{\partial^2 V}{\partial^2 y}, \quad V = V(x, y, t), \ t \geq t_0, \ (x, y) \in \mathbb{R},$$

$$\partial \Omega(x, y) = \begin{cases} 1 : \frac{1}{2} < x \leq 1, \ y = 0, \\ 1 : 0 \leq x \leq \frac{1}{2}, \ y = 1, \\ 1 : 0 \leq y \leq \frac{1}{2}, \ x = 0, \\ 1 : \frac{1}{2} < y \leq 1, \ x = 1, \\ 0 : \text{otherwise.} \end{cases} \tag{3.81}$$

In this example we use the method of *false transients* discussed in Section 3.8.3.2. This method treats the problem as being *dynamic* by replacing the right-hand zero in eqn. (3.80) with a *false time derivative* and integrating using the MOL until a steady-state is reached. This transforms the problem from an elliptical PDE to a parabolic PDE. The problem was solved with initial condition $V(x, y, 0) = 0$ and a good solution was readily found using a four-point FD approximation to the second derivative on a 40 × 40 spatial grid and the standard ODE integrator ode45. The result is shown in Fig. 3.12, with electrostatic potential contours superimposed. A 3D surface plot is shown in Fig. 3.13.

The R code for the *main* program and *derivative* function that solve this problem is shown in Listings 3.10 and 3.11, respectively. Note that for this problem we need to set the variable case=1 in the main program. See Section 3.8.4.1 for details relating to the plotting of results.

---

[7] A harmonic function is twice continuously differentiable; i.e., for eqn. (3.80), we have within the solution domain $u \in C^2$.

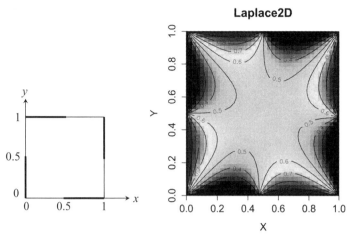

Figure 3.12. Laplace equation problem of eqn. (3.81). (left) 2D square domain where the potential of the highlighted boundary segments is 1 V and the nonhighlighted segments are at zero potential. (right) Solution showing electrostatic potential contours within the problem domain on a 40 × 40 grid. The plot was generated using the `image()` function and contours were added using the `contours()` function.

3.8.4.3 Poisson Equation

The *Poisson* equation is an elliptical PDE given by

$$0 = \nabla u + f, \quad u = u(x), \ f = f(x), \ x \in \mathbb{R}^n, \tag{3.82}$$

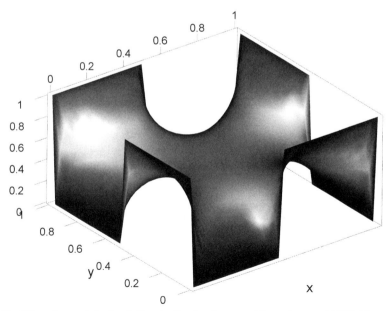

Figure 3.13. 3D surface plot of solution to the Laplace problem of eqn. (3.81). The segmented potentials result in a *castellation* effect, which is clearly visible. The surface plot was generated using the `persp3d()` function. (See color plate 3.13)

Table 3.3. Poisson equation solution errors for second- and fourth-order spatial discretization schemes with the domain divided into varying numbers of grid points

Number of grid points	Maximum absolute error	
	2nd-order approx.	4th-order approx.
10 × 10	0.16641	0.004232
20 × 20	0.04416	0.000216
30 × 30	0.02002	3.144e-05
40 × 40	0.01137	7.795e-06
50 × 50	0.00732	2.619e-06
80 × 80	0.00289	2.587e-07

and its solution is similar in form to that of the diffusion equation with a heat source, evaluated at equilibrium or steady state conditions. If the source function is zero, the problem reduces to the *Laplace* equation, $\nabla u = 0$.

We now consider the 2D problem of eqn. (3.83), adapted from [Ise-09, p162-3], that has a known analytical solution, u_a. It is defined on the rectangular domain $\Omega(x, y) = [0, 1]^2$ with three Dirichlet and one Neumann boundary conditions imposed on $\partial\Omega$, as appropriate, from u_a. We employ the false transient solution method by including an additional time derivative so that the problem takes the form of heat diffusion with source. Also, as we have transformed the problem to be time varying, the analytical solution now applies at $t \to \infty$:

$$\frac{\partial u}{\partial t} = \left(\frac{\partial^2 u}{\partial x^2} + \frac{\partial y^2 u}{\partial^2}\right) + f, \quad u = u(x, y, t), \; t \geq t_0, \; (x, y) \in \mathbb{R}, \tag{3.83}$$

$$u(x, y, 0) = 0, \; f(x, y) = -(x^2 + y^2), \; u_a(x, y) = \exp(\pi x)\sin(\pi y) + \tfrac{1}{2}(xy)^2.$$

We impose the following Dirichlet boundary conditions:

$$\partial\Omega(x, 0) = 0, \quad \partial\Omega(x, 1) = \exp(\pi x)\sin(\pi) + \tfrac{1}{2}y^2, \quad \partial\Omega(0, y) = \sin(\pi y), \tag{3.84}$$

along with the Neumann boundary condition

$$\partial\Omega_n(1, y) = \pi \exp(\pi)\sin(\pi y) + y^2. \tag{3.85}$$

Again, we find that the false transient method readily finds a good solution (see Fig. 3.14). The method employed a four-point FD approximation to the second derivative along with the standard ODE integrator lsodes with relative and absolute tolerances set to 1e-6. Using a 40 × 40 spatial grid, the resulting solution is very close to the analytical solution, with a maximum absolute error of 0.0114 ($\simeq 0.05\%$ of max u). Table 3.3 details how errors vary according to the discretization scheme and the number of grid points employed. These results are shown on a log-log plot in Fig. 3.15, where the *maximum absolute errors* follow closely the expected theoretical values, according to either Δx^2 or Δx^4, as discussed in Section 3.8.2.

We can determine the order of approximation achieved from the following equation using 10 and 80 grid point calculation results:

$$\text{order} = \frac{\log_{10} \tau_{10} - \log_{10} \tau_{80}}{\log_{10} N_{80} - \log_{10} N_{10}}, \tag{3.86}$$

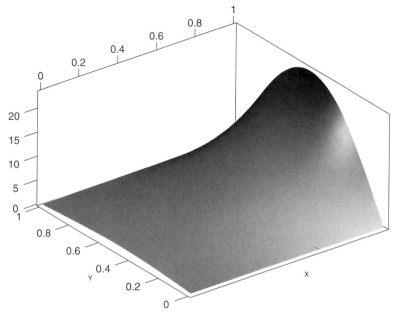

Figure 3.14. Surface plot of simulation on a 40 × 40 grid of the 2D Poisson eqn. (3.83). The surface plot was generated using the persp3d() function.

where τ_N represents the maximum absolute error using N grid points. From eqn. (3.86) we find that the four-point discretization scheme delivers an actual order of 1.95 and the six-point discretization an actual order of 4.67. It should be noted that the denominator of eqn. (3.86) is also equal to $(\log_{10} \Delta_{10} - \log_{10} \Delta_{80}) = (\log_{10} L/N_{10} - \log_{10} L/N_{80})$, where $L = 1$ represents domain lengths of x and y. Thus, we obtain the same order estimate from the calculation using number of grid points or grid spacing values. These results are very close to theoretical values of 2 and 4. However, if we continue to increase the

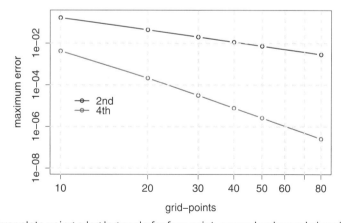

Figure 3.15. Error plots using a log-log scale for four-point, second-order and six-point, fourth-order spatial discretization schemes. The upper line shows second-order scheme error falling according to $\Delta x^{1.95}$. The lower line shows the error falling according to $\Delta^{4.67}$, where $\Delta = \Delta x = \Delta y$. These results are very close to theoretical value of Δ^2 and Δ^4.

number of grid points beyond 80, rounding errors will eventually start to come into play, and accuracy will begin to decline.

The R code for the *main* program and *derivative* function that solve this problem is shown in Listings 3.10 and 3.11, respectively. Note that for this problem we need to set the variable case=2 in the main program. See Section 3.8.4.1 for details relating to the plotting of results. The code to calculate the error plot of Fig. 3.15 is shown in Listing 3.12.

```
# File: PoissonErrPlots.R
# case=2 analytical solution and error calculation
ua <- array(0,c(Nx,Ny))
for(i in 1:Nx){
  for(j in 1:Ny){
    ua[i,j] <- exp(pi*x[i])*sin(pi*y[j])+(1/2)*(x[i]*y[j])^2
  }
}
err <- U[nout,,]-ua
maxErrLoc <- which.max(abs(err))
err_max <- err[maxErrLoc]
cat(sprintf("\nMaximum error = %e\n",err_max))
# Error for various grid-sizes
err_grid <- array(0,c(6,3))
# Number of grid points
err_grid[1:6,1] <- c(10,20,30,40,50,80)
# Errors for 2nd order discretization - calculated previously
err_grid[1:6,2] <- c(0.16641,0.04416,0.020018,0.011372,0.00732,0.002885)
# Errors for 4th order discretization - calculated previously
err_grid[1:6,3] <- c(0.0042432,0.0002159,3.1435e-5,7.7951e-6,2.6185e-6,
                   2.58704e-7)
plot(x=err_grid[,1],y=err_grid[,2],type="b",log="xy",
     ylim=c(0.1e-7,0.2),lwd=2,
     xlab="grid-points", ylab="maximum error",col="blue")
lines(x=err_grid[,1],y=err_grid[,3],type="b",lwd=2,col="red")
grid(lwd=2,lty=2)
legend(11,5e-5, legend=c("2nd","4th"), pch = 1,
       lwd=2,col=c("blue","red"),bg="white",box.col="white")
# Calculate order achieved
order1 <- (log10(err_grid[1,2])-log10(err_grid[6,2]))/
          (log10(err_grid[6,1])-log10(err_grid[1,1]))
order2 <- (log10(err_grid[1,3])-log10(err_grid[6,3]))/
  (log10(err_grid[6,1])-log10(err_grid[1,1]))
cat(sprintf("\n4-point, 2nd order scheme achieves order = %3.1f\n",order1))
cat(sprintf("\n6-point, 2nd order scheme achieves order = %3.1f\n",order2))
```

Listing 3.12. File: PoissonErrPlots.R—Code for calculating the *analytical* solution of eqn. (3.83) and the error plots of Fig. (3.15) for case=2 *Poisson* example

Figure 3.16. Surface plot of simulation on a 101 × 101 grid of the 2D advection of a rotating solid after one revolution, at $t = 3.1416$. The surface plot was generated using the `persp3d()` function.

3.8.4.4 Rotation of a Solid Body

We consider a standard 2D fluid mechanics problem where incompressible *rotational* flow is defined by the stream function

$$\psi(x, y) = x^2 + y^2, \tag{3.87}$$

with velocity field

$$\alpha_x(x, y) = 2y, \quad \alpha_y(x, y) = -2x, \tag{3.88}$$

that satisfies the continuity requirements

$$\frac{\partial \alpha_x}{\partial x} + \frac{\partial \alpha_y}{\partial y} = 0. \tag{3.89}$$

This is an advection problem that we define on the rectangular domain $\Omega = [-1, 1]^2$ and corresponding boundary $\partial \Omega$, as follows:

$$\frac{\partial u}{\partial t} + \alpha_x \frac{\partial u}{\partial x} + \alpha_y \frac{\partial u}{\partial y} = 0, \quad u = u(x, y, t), \ t \geq t_0, \ x \in \mathbb{R}, \ y \in \mathbb{R},$$
$$u(x, y, 0) = \exp\left[-\lambda \left((x_0 - x)^2 + (y_0 - y)^2\right)\right], \quad x_0 = 1/2, \ y_0 = 0, \tag{3.90}$$

where we impose Dirichlet boundary conditions $\partial \Omega = 0$. The initial condition $u(x, y, t = 0)$ is a Gaussian cone situated at the point (x_0, y_0).

From the velocity field it is clear that the velocity of any point rotating about the origin is equal to $v(r) = 2r = r \, d\theta/dt$. It therefore follows from simple integration that the time taken for one complete revolution is π. This means that all points on a radius vector will move in unison and take the same time to rotate through any angle θ, which corresponds to so-called *solid-body rotation*. It also follows from the definitions of α_x and α_y that the rotation will be clockwise, as shown in Fig. 3.17. Thus, we have a situation whereby a solid body rotates around the central point (0,0) without distortion in time $t = N\pi$ for any integer N, where N represents the number of complete revolutions made by the flow.

This is a good test for numerical solvers, and here we use the classical five-point approximation to the spatial first derivative. This fourth-order FD scheme along with the built-in ODE integrator ode45 handles this particular problem well with very little distortion (see Figs. 3.16, 3.17, and 3.18. Interestingly, the lsodes takes approximately 10 times longer to obtain the solution to within the same tolerances. The rotating solid completes

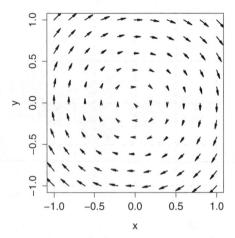

Figure 3.17. Velocity field plot generated using the `quiver()` function. The flow vectors at various points within the domain are represented by the length and direction of the corresponding arrows. To avoid an overcrowded situation, the plot was generated from a down-sampled set of data.

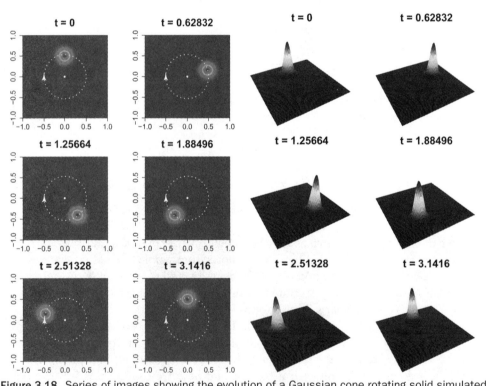

Figure 3.18. Series of images showing the evolution of a Gaussian cone rotating solid simulated on a 101 × 101 grid. The solid advects in a clockwise circular path with very little distortion. (left) Surface plots generated using the `persp()` function. (right) Plan view generated using the `image()` function.

one clockwise revolution, correctly returning to its original position at $t = 3.1416$. However, if the Gaussian cone is changed to a square solid, the situation changes drastically, with severe dissipation and instability becoming readily apparent after a very short time. We will see in Chapter 6 that this problem can be solved accurately using the WENO high-resolution scheme.

The R code for solving this problem is shown in Listings 3.13–3.15. An *animation* can be created by setting animation = TRUE in Listing 3.15—see discussion on imageMagick in Section 3.8.4.1.

The reader may like to experiment with different values for α_x and α_y to see the effects on rotation time and distortion of the Gaussian cone, if any. Also, the reader may like to see how the classic FD scheme handles a square solid by setting the initial condition parameter in the code appropriately (as mentioned earlier, not very well).

```
# File: advection2D_main.R
# -------------------------------------------------------------------
# 2D ADVECTION example: Rotating solid
#      ut     = alphaX*ux + alphaY*uy
# Initial condition: u(x,y,t=0) = 0
# BCs: u(x=0,y,t) = 0, u(x=1,y,t) = 0
#      ux(x,y=0,t) = 0, ux(x,y=1,t) = 0
# -------------------------------------------------------------------
require(compiler)
enableJIT(3)
##################################################
# Delete workspace prior to new calculations
##################################################
cat("\014") # Clear console
rm(list = ls(all = TRUE))
#
##################################################
# Required R library packages
##################################################
library("deSolve")
library("pracma")
library("rgl")
library("shape")
##################################################
# Load sources for functions called
##################################################
source("deriv1st5pt.R")
source("advection2D_deriv.R")
##################################################
# Start timer
##################################################
ptm0 = proc.time()
print(Sys.time())
```

```r
#
#####################################################
# Declare array sizes
#####################################################
Nx=100
Ny=100
#
#####################################################
# PDE parameters
#####################################################
xl <- -1; xu <- 1;
yl <- -1; yu <- 1;
#
#####################################################
# Define grid in x, y and velocity field
#####################################################
dx <- (xu-xl)/(Nx-1)
dy <- (yu-yl)/(Ny-1)
x  <- seq(from=xl,to=xu,by=dx)
y  <- seq(from=yl,to=yu,by=dy)
#
X <- outer(x,rep(1,Ny),"*")
Y <- outer(rep(1,Nx),y,"*")
#
alphaX <- +2*Y
alphaY <- -2*X
#####################################################
# Define differential matrices
#####################################################
diffMatX <- diffMatrix1st5pt(xl,xu,Nx)
diffMatY <- diffMatrix1st5pt(yl,yu,Ny)
#####################################################
# Set parameters
#####################################################
tf   <- 3.1416; nout <- 51
t    <- seq(from=0,to=tf,length.out=nout)
#
#####################################################
# Set Initial Conditions
#####################################################
case <- 0 # 0=Gaussian, 1=square
uini <- array(0,c(Nx,Ny))
lambda =50
offset_x <- 0.0
offset_y <- 0.5
for(i in 1:Nx){
```

```
  for(j in 1:Ny){
    # IC starting point
    if(case == 0){
      uini[i,j] =exp(-lambda*((offset_x-x[i])^2+(offset_y-y[j])^2))
    }else {
      if (x[i] > -0.25 && x[i] < 0.25 && y[j] > 0.1 && y[j]< 0.6){
        uini[i,j] = 1
      }
    }
  }
}
#
#####################################################
# Initialze other variables
#####################################################
#
ncall <<- 0
#
#####################################################
# ODE integration
#####################################################
# Note: lsodes takes approx. 10 times longer than ode45
# out<-lsodes(y=c(uini),times=t,func=advection2D_deriv,
#             sparsetype ="sparseint", ynames = FALSE,
#             rtol=1e-8,atol=1e-8,maxord=5,parms=NULL)
out<-ode(y=c(uini), times=t, func= advection2D_deriv,
         method = "ode45", ynames = FALSE, hini=0.001,
         rtol=1e-8,atol=1e-8,parms=NULL)
#####################################################
# Preallocate 2D plot array and plot results
#####################################################
U = array(0, c(nout, Nx, Ny))
#
source("advection2D_postSimCalcs.R")
#####################################################
# Print total elapsed time
#####################################################
t2 <- proc.time()-ptm0
cat(sprintf("Total time: %f\n",t2[3]))
#
#####################################################
# End of program
#####################################################
```

Listing 3.13. File: advection2D_main.R—Code for the *main program* that simulates the rotating solid-body example defined by eqn. (3.90)

```
# File: advection2D_deriv.R
advection2D_deriv <- function(t, u, parms) {
  ####################################################
  # Reshape vector u -> to a Nx by Ny matrix
  ####################################################
  u1 = array(u,c(Nx,Ny))
  # Set Dirichlet BCs for u[x=0,y] and u[x=1,y]
  #                       u[x,y=0] and u[y,x=1]
  u1[1,] <- 0; u1[Nx,] <- 0;
  u1[,1] <- 0; u1[,Ny] <- 0;
  #
  ####################################################
  # Calculate 2D spatial 1st derivatives
  ####################################################
  u1x <-   diffMatX %*% u1
  u1y <- t(diffMatY %*% t(u1))
  #
  ####################################################
  # Calculate time derivative
  ####################################################
  #
    ut <- -(alphaX*u1x + alphaY*u1y)
  #
  # update number of derivative calls
  ncall <- ncall+1
  #
  return(list(ut))
}
```

Listing 3.14. File: advection2D_deriv.R—Code for the *derivative* function that that is called by the rotating solid-body main program of Listing 3.13

```
# File: advection2D_postSimCalcs.R
####################################################
# Set animation flag
####################################################
animate <- FALSE # TRUE/FALSE
#
####################################################
# Create plot matrices
####################################################
if (animate == TRUE){
  N_by=1
```

```
}else{
  N_by=10
}
cat(sprintf("\n Animation Set To: %s, N_by=%d \n",
            animate,N_by))
for (i in seq(1,nout,by=N_by)){
  U[i,,]=matrix(data=out[i,2:(Nx*Ny+1)],ncol=Ny,
                nrow=Nx,byrow=FALSE,dimnames=NULL)
}
###################################################
#Define color scheme
###################################################
jet.colors = colorRampPalette(c(
  "#00007F", "blue", "#007FFF",
  "cyan", "#7FFF7F", "yellow",
  "#FF7F00", "red", "#7F0000"))
pal=jet.colors(100)
#
###################################################
#Plot 2D image
###################################################
par(mfrow = c(3, 2))
par(mar=c(2, 4, 4, 2) + 0.1, mgp = c(5, 1, 0))
par(oma = c(0,0,2,0))
for (i in seq(1,nout,by=10)){
  image(x,y,U[i,,], ylim=c(yl,yu), xlim=c(xl,xu),
        ylab="Y", xlab="X", col=pal, asp=1,
        main=paste("t =",t[i]),cex.main=1.5)
  if(case==0){rad <- 0.5}else{rad <- 0.35}
  plotcircle(r=rad, type="l", lwd=2,lty=3,arr.code=0,
             # from=0, to=2*pi,
             arrow = TRUE,arr.col="white",lcol="white")
  points(x=0,y=0,pch=20,col="white")
}
par(mfrow = c(1, 1))
###################################################
# Array of surface plots
###################################################
par(mfrow = c(3, 2))
par(mar=c(0, 0, 2, 0) + 0.1, mgp = c(3, 1, 0))
# par(oma = c(0,0,2,0))
for (i in seq(1,nout,by=10)){
  persp(x,y,U[i,,], theta = -30, phi = 30, expand=0.5,
        xlim = range(x), ylim = range(y), #zlim = c(-0.3,1),
        ylab = "x", xlab = "y",
        zlab = "U", col = drapecol(U[i,,]),border=NA,
```

```
            main=paste("t =",t[i]), #ticktype="detailed",
            box=FALSE, axes=FALSE,cex.main=1.5)
}
######################################################
# Plot velocity field
######################################################
par(mfrow = c(1, 1))
par(oma = c(0,0,2,0))
par(mar=c(4, 4, 0, 2) + 0.1)
#
plot(0,0,xlim=c(-1,1), ylim=c(-1,1),type="n",
     xlab="x", ylab="y",asp=1)
inds <- c(1,seq(10,100,10)) # down-sampled
quiver(X[inds,inds], Y[inds,inds], alphaX[inds,inds],
       alphaY[inds,inds],
       scale = 0.05, angle = 20, length = 0.1,lwd=2)
grid()
######################################################
# Plot 3D image
######################################################
open3d()
bg3d("white")
# Set colour indices of each point for pers3d()
col.ind <- cut(U[nout,,],100)#
persp3d(x,y,U[nout,,], aspect=c(1, 1, 0.5), color = pal[col.ind],
        ylab = "", xlab = "", zlab = "",
        box=FALSE,axes=FALSE,smooth=FALSE)
# Set display parameters
UM <- matrix(data=c( 0.7394126, -0.6731873, 0.009278561, 0,
                     0.1697968,  0.1998015, 0.965011239, 0,
                    -0.6514879, -0.7119666, 0.262041003, 0,
                     0.0000000,  0.0000000, 0.000000000, 1),
byrow=TRUE,nrow=4,ncol=4)
rgl.viewpoint(fov=0) # set before userMatrix
par3d(userMatrix=UM)
par3d(windowRect=c(20,100,820,900), zoom=1)
# rgl.snapshot("solidRotation3D.png", fmt="png") # print to file
######################################################
# Record animation if animate==TRUE
######################################################
#
if (animate == TRUE){
  png(file="example%03d.png", width=600, height=600)
  for (i in 1:nout){
    #plot.new() # Inserts a blank between plots
    persp(x,y,U[i,,], theta = -30, phi = 30, expand=0.5,
```

```
          xlim = range(x), ylim = range(y), zlim = c(-0.01,1.01),
          ylab = "\n\n\nX", xlab = "\n\n\nY", zlab = "\n\n\nU",
          col = drapecol(U[i,,]), border=NA,
          main=paste("t =",t[i]), ticktype="detailed")
  }
  dev.off()
  # Create animated GIF file.
  # NOTE: Can also run the command, shown below in
  #       quotes, from command line prompt
  shell("convert -delay 20 example*.png Advection2D_1.gif",
        intern=TRUE)
  # Remove individual files
  file.remove(grep("example",list.files(),value=TRUE))
}
```

Listing 3.15. File: advection2D_postSimCalcs.R—Code for the *postsimulation calculations* at the end of the rotating solid-body main program of Listing 3.13

3.8.4.5 Wave Equation

Linear and nonlinear waves occur in most physical systems, such as those that consist of liquid, acoustic, electromagnetic, or a solid medium. Further discussion and example applications on this subject can be found in [Gri-09].

The free surface shallow-water wave equations can be derived from the invisid Euler equations for incompressible flow. From *conservation of horizontal momentum* we obtain

$$\frac{\partial w}{\partial t} + w\frac{\partial w}{\partial x} + v\frac{\partial w}{\partial y} + g\frac{\partial \eta}{\partial x} = 0$$
$$\frac{\partial v}{\partial t} + w\frac{\partial v}{\partial x} + v\frac{\partial v}{\partial y} + g\frac{\partial \eta}{\partial y} = 0,$$
(3.91)

and from the *conservation of mass law* we obtain

$$\frac{\partial \eta}{\partial t} + \frac{\partial [(\eta+h)w]}{\partial x} + \frac{\partial [(\eta+h)v]}{\partial y} = 0,$$
(3.92)

where $\eta = \eta(x, y, t)$ represents variation of free water surface from equilibrium, $h = h(x, y)$ nominal water depth, $w = (w, x, y, t)$ fluid velocity in the x direction, $v = v(x, y, t)$ fluid velocity in the y direction, and g gravitational acceleration. Equations (3.91) and (3.92) are the so-called *shallow-water* equations, and a detailed derivation is given in most fluid dynamics textbooks, for example, [Joh-97].

We can linearize eqns. (3.91) and (3.92) by considering small disturbances about equilibrium conditions when

$$\eta = \eta_0 + \overline{\eta}, \quad w = w_0 + \overline{w}, \quad v = v_0 + \overline{v},$$
(3.93)

where subscripts 0 indicate values at equilibrium conditions and overbars indicate small perturbations. If we assume that $\eta_0 = 0$, $w_0 = 0$, and $v_0 = 0$, then, on substituting eqns. (3.93) into eqns. (3.91) and (3.92) and neglecting second-order effects, we obtain

$$\frac{\partial \eta}{\partial t} + \frac{\partial (wh)}{\partial x} + \frac{\partial (vh)}{\partial y} = 0,$$

$$\frac{\partial w}{\partial t} + g\frac{\partial \eta}{\partial x} = 0, \qquad (3.94)$$

$$\frac{\partial v}{\partial t} + g\frac{\partial \eta}{\partial y} = 0,$$

where the overbars have been dropped.

We can make a further simplification to obtain a single equation in η by cross-differentiation and eliminating w and v from the first of eqns. (3.94). This yields the classic 2D linear wave equation[8]

$$\frac{\partial^2 \eta}{\partial t^2} = c^2 \left(\frac{\partial^2 \eta}{\partial x^2} + \frac{\partial^2 \eta}{\partial y^2} \right), \qquad (3.95)$$

where $c = \sqrt{gh}$ represents the wave speed or celerity.[9] The advantage of this formulation is that the equation for η is *decoupled* from those for w and v and, therefore, can be solved independently.

Equation (3.95) will form the basis of the model for this example. The problem is defined on the rectangular domain $\Omega = [0, 1]^2$ with reflecting boundary conditions on $\partial \Omega$ as

$$\frac{\partial^2 u}{\partial t^2} = c^2 \left(\frac{\partial^2 u}{\partial x^2} + \frac{\partial^2 u}{\partial y^2} \right), \quad u = u(x, y, t), \ t \geq t_0, \ x \in \mathbb{R}, \ y \in \mathbb{R,}$$
$$u(x, y, 0) = \exp\left[-\lambda \left(x^2 + y^2\right)\right], \ \lambda = 50, \ c = 0.05, \qquad (3.96)$$

where u represents the height of shallow-water waves following the introduction of a disturbance — in this case, in the form of a Gaussian-shaped mass of water at location $(0.5, 0.5)$.

With regard to the reflective boundary conditions for eqn. (3.96), this is achieved by adding ghost cells at each boundary and including the following code statements: `u[,1]<-u[,2], u[,Ny+2]<-u[,Ny+1], u[1,]<-u[2,], u[Nx+2,]<-u[Nx+1,]`. Of course, the actual solution is only valid for `u[2:Nx+1:2:Ny+1]`, and it is these values that are plotted in Figs. 3.19 and 3.20.

Note that if we were modeling eqns. (3.91) and (3.92) instead of eqn. (3.96), then we would also need to ensure that fluid velocities reverse at each of the boundaries, that is, `w[,1<--w[,2], w[,Ny+2]<--w[,Ny+1]` and, similarly, `v[1,]<--v[2,], v[Nx+2,]<--v[Nx+1]`. This approach was used to model the 1D Woodward–Colella reflected boundary problem in Chapter 6.

To facilitate numerical integration of the temporal second derivative, $\frac{\partial^2 u}{\partial t^2}$, we adopt the usual approach and decompose eqn. (3.96) into an equivalent problem consisting of

---

[8] This result was published by Lagrange in 1871 [Lag-81].
[9] From the Latin *celeritās*, meaning "swiftness."

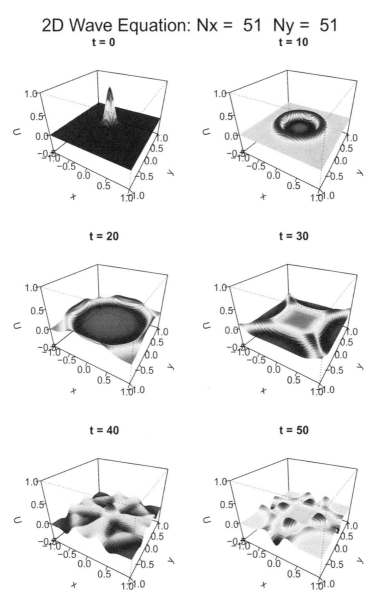

Figure 3.19. Series of images showing the evolution of waves in a square domain with reflective boundary conditions following a Gaussian-shaped disturbance. With the wave traveling at a velocity of $c = 0.05$, its peak reaches the boundary at $t = 20$, as expected. The surface plots were generated using the persp() function.

two first derivatives, that is,

$$\frac{\partial u_1}{\partial t} = u_2,$$
$$\frac{\partial u_2}{\partial t} = c^2 \left(\frac{\partial^2 u_1}{\partial x^2} + \frac{\partial^2 u_1}{\partial y^2} \right), \quad (3.97)$$

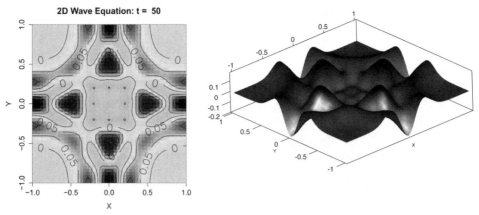

Figure 3.20. Solution to the wave equation problem of eqn. (3.96) at $t = 50$. The spatial domain is represented by a 51×51 grid. (left) Plot generated using the `image()` function with wave height contours added by the `contours()` function. (right) Surface plot of the same solution generated using the `persp3d()` function.

with the wave height solution being given by u_2. The simulation was performed on a 51×51 grid using a four-point FD approximation of the spatial domain, with integration performed by the `ode45` integrator from the `deSolve` package. The results are shown in Fig. 3.19, where a sequence of surface plots shows how the waves evolve following the introduction of a Gaussian disturbance. With the wave traveling at a velocity of $c = 0.05$, its main peak reaches the boundary at $t = 20$, as expected.

Figure 3.20 consists of a contour plot and 3D surface plot that show in more detail the final situation at $t = 50$.

The R code for solving this problem is shown in Listings 3.16–3.18. An *animation* can be created by setting `animation = TRUE` in Listing 3.18 (see discussion on imageMagick in Section 3.8.4.1).

The reader may like to experiment with different values for c to see the effects of faster or slower wave speeds. Also, the reader may like to see how the method handles differently shaped disturbances.

```
# File: wave2D_main.R
# --------------------------------------------------
# 2D Wave example: u_tt = C^2 * u_xx
# --------------------------------------------------
####################################################
# Delete workspace prior to new calculations
####################################################
cat("\014") # Clear console
rm(list = ls(all = TRUE))
####################################################
# Required R library packages
####################################################
require(compiler)
enableJIT(3)
library("deSolve")
```

```
library("rgl")
library("shape")
#################################################
# Load sources for functions called
#################################################
source("deriv2nd4pt.R")
source("deriv2nd6pt.R")
source("deriv2nd8pt.R")
source("deriv2nd10pt.R")
source("wave2D_deriv.R")
#################################################
# Start timer
#################################################
ptm0 <- proc.time()
print(Sys.time())
#################################################
# Declare array sizes
#################################################
Nx  <- 51;  Ny  <- 51
Nx1 <- Nx+2; Ny1 <- Ny+2 # Add ghost points
#################################################
# PDE parameters
#################################################
xl <- -1; xu <- 1; dx <- (xu-xl)/(Nx-1)
yl <- -1; yu <- 1; dy <- (yu-yl)/(Ny-1)
C=0.05; C2=C*C;
#################################################
# Define grid in x, y
#################################################
x=seq(from=xl,to=xu,length.out=Nx)
y=seq(from=yl,to=yu,length.out=Ny)
#################################################
# Define differential matrices
#################################################
# Allow for ghost points - used in derivative calc
diffMat2X <- diffMatrix2nd4pt(xl,(xu+2*dx),Nx1,c(0,0))
diffMat2Y <- diffMatrix2nd4pt(yl,(yu+2*dy),Ny1,c(0,0))
#################################################
# Set parameters
#################################################
tf <- 50; nout <- 51
t=seq(from=0,to=tf, length.out=nout)
#################################################
# Initial condition
#################################################
lambda =50
```

```
#fin  <- function(x,y){exp(-lambda*((0.5-x)^2+(0.5-y)^2))}
fin   <- function(x,y){exp(-lambda*((x)^2+(y)^2))}
u1ini <- outer(x,y,FUN=fin)
u2ini <- array(0,c(Nx,Ny))
uini  <- c(u1ini,u2ini)
####################################################
# ODE integration
####################################################
ncall <<- 0
# out<-lsodes(y=c(uini), times=t, func=wave2D_deriv,
#             sparsetype = "sparseint",
#             ynames = FALSE,
#             rtol=1e-12,atol=1e-12,maxord=5,parms=NULL)

out<-ode(method="ode45",y=c(uini), times=t, func=wave2D_deriv,
         ynames = FALSE,
         rtol=1e-12,atol=1e-12,parms=NULL)
####################################################
# Create plot matrices
####################################################
U = array(0, c(nout, Nx, Ny))
#
U[1:nout,,] <- out[1:nout,2:(Nx*Ny+1)]
####################################################
# Print elapsed time
####################################################
t1 <- proc.time()-ptm0
cat(sprintf("Calculation time: %f\n",t1[3]))
#
source("wave2D_postSimCalcs.R")
```

Listing 3.16. File: wave2D_main.R—Code for the *main* program that simulates the 2D wave example in Cartesian coordinates defined by eqn. (3.96)

```
# File: wave2D_deriv.R
wave2D_deriv <- function(t, u, parms) {
  ####################################################
  # Reshape vector u -> to two Nx by Ny matrices
  ####################################################
  u1 <- array(0,c(Nx1,Ny1))
  u2 <- array(0,c(Nx1,Ny1))
  u1[2:(Nx1-1),2:(Ny1-1)] <- u[1:(Nx*Ny)]
  u2[2:(Nx1-1),2:(Ny1-1)] <- u[(Nx*Ny+1):(2*Nx*Ny)]
  #
  ####################################################
  # Set Reflective BCs for:
  #     u[y,x=0], u[y,x=1], u[y=0,x] and u[y=1,x]
```

```
#####################################################
u1[1,]<-u1[2,]; u1[Nx1,]<-u1[(Nx1-1),]
u1[,1]<-u1[,2]; u1[,Ny1]<-u1[,(Ny1-1)]
u2[1,]<-u2[2,]; u2[Nx1,]<-u2[(Nx1-1),]
u2[,1]<-u2[,2]; u2[,Ny1]<-u2[,(Ny1-1)]
#####################################################
# Calculate 2D spatial 2nd derivatives
#####################################################
u1xx <-   diffMat2X %*% u1
u1yy <- t(diffMat2Y %*% t(u1))
#####################################################
# Calculate time derivative
#####################################################
u1t <- u2[2:(Nx1-1),2:(Ny1-1)]
u2t <- C2*(u1xx[2:(Nx1-1),2:(Ny1-1)] +
           u1yy[2:(Nx1-1),2:(Ny1-1)])
#####################################################
# update number of derivative calls
#####################################################
ncall <<- ncall+1
#####################################################
# Return solution
return(list(c(u1t,u2t)))
#####################################################
}
```

Listing 3.17. File: wave2D_deriv.R—Code for the *derivative* function that is called from the 2D wave main program of Listing 3.16

```
# File: wave2D_postSimCalcs.R
#####################################################
# Set animation flag
#####################################################
animate=FALSE
#####################################################
#Define color scheme
#####################################################
jet.colors = colorRampPalette(c("#00007F", "blue",
                                "#007FFF","cyan", "#7FFF7F",
                                "yellow", "#FF7F00",
                                "red", "#7F0000"))
pal=jet.colors(100)
#####################################################
#Plot 2D image
#####################################################
image(x,y,U[nout,,], ylim=c(yl,yu), xlim=c(xl,xu),
      ylab="Y", xlab="X", col=pal,
      main=paste("2D Wave Equation: t = ",tf))
```

```r
# Add some contours
contour(x,y,U[nout,,], labcex=1.5,add=TRUE)
#####################################################
# Plot 3D surface
#####################################################
open3d()
bg3d("white")
# Set colour indices of each point for pers3d()
col.ind <- cut(U[nout,,],100)#
persp3d(x,y,U[nout,,], aspect=c(1, 1, 0.25),
        color = pal[col.ind], smooth=TRUE,
        ylab="Y",xlab="X",zlab="",
        axes = TRUE, box = FALSE)
rgl.viewpoint(theta=-10,phi=-45,fov=0)
#
par3d(windowRect=c(20,100,820,900), zoom=1)
#
#####################################################
# Print to png file - uncomment if required
#####################################################
# rgl.snapshot("WaveEqn_3D.png", fmt="png")
#####################################################
# Array of surface plots
#####################################################
par(mfrow = c(3, 2))
par(mar=c(2, 4, 4, 2) + 0.1, mgp = c(5, 1, 0))
par(oma = c(0,0,2,0))
for (i in seq(1,nout,by=10)){
  persp(x,y,U[i,,], theta = 30, phi = 30, scale=FALSE,
        xlim = range(x),
        ylim = range(y),
        zlim = c(-0.5,1),#expand=0.5,
        ylab = "\n\n\ny", xlab = "\n\n\nx",
        zlab = "\n\n\nU",
        col = drapecol(U[i,,]),
        border=NA,
        main=paste("t =",t[i]), ticktype="detailed")
}
mtext(side=3, outer=TRUE, cex=1.25, line=-1,
      paste("2D Wave Equation: Nx = ",Nx," Ny = ", Ny))
#
par(mfrow = c(1, 1))
#
#####################################################
# Record animation if animate==TRUE
#####################################################
```

```
if (animate == TRUE){
  zRng <- range(U)
  png(file="example%02d.png", width=600, height=600)
  for (i in 1:nout){
    #plot.new() # Inserts a blank between plots
    persp(x,y,U[i,,], theta = 30, phi = 30, scale=FALSE,
          xlim = range(x), ylim = range(y),
          zlim = c(-0.5,1), #expand=0.5,
          ylab = "\n\n\nX",
          xlab = "\n\n\nY", zlab = "\n\n\nU",
          col = drapecol(U[i,,]),
          border=NA,
          main=paste("t =",t[i]), ticktype="detailed")
  }
  dev.off()
  # Create animated GIF file by call to imageMagick.
  # NOTE: Can also run the command, shown below in
  #       quotes, from command line prompt
  shell("convert -delay 20 example*.png wave2DEx_1.gif",
        intern=TRUE)
  # Remove individual files
  file.remove(grep("example",list.files(),value=TRUE))
}
```

Listing 3.18. File: wave2D_postSimCalcs.R—Code for the *postsimulation calculations* run at the end of the 2D wave main program of Listing 3.16

3.8.4.6 Frontogenesis

We now consider the dynamic *frontogenesis* problem. It is a standard test problem in the area of meteorology based upon analysis originally reported by Doswell [Dos-84] and requires the resolution of a complex weather front, for example, temperature, containing a vortex that evolves with time. For this idealized problem the wind field is taken to be a steady-state nondivergent vortex with a purely tangential wind field, v, which depends only on radius r from the origin. It is a smoothly varying vortex having continuous derivatives. The front slope is governed by the value of the front characterizing parameter δ, which for this problem is set to unity, resulting in a gentle transition from one region to the other. We explore a more difficult sharp front problem in Chapter 6, where the WENO method is used to resolve the vortex with $\delta = 10^{-6}$.

The nonlinear advection problem details are defined on the rectangular domain $\Omega = [-5, 5]^2$ and corresponding boundary $\partial\Omega$, as follows [Kal-11]:

$$\frac{\partial u}{\partial t} + \alpha_x \frac{\partial u}{\partial x} + \alpha_y \frac{\partial u}{\partial y} = 0, \quad u = u(x, y, t), \ t \geq t_0, \ x \in \mathbb{R}, \ y \in \mathbb{R},$$

$$u(x, y, 0) = \tanh\left(\frac{y}{\delta}\right),$$

$$\alpha_x = -y f(r), \quad \alpha_y = x f(r), \quad f(r) = \frac{1}{r}, \quad v(r) = \bar{v} \operatorname{sech}^2(r) \tanh(r),$$

$$r = \sqrt{x^2 + y^2}, \quad \bar{v} = 2.59807, \quad \delta = 1,$$

(3.98)

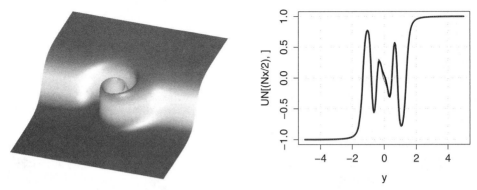

Figure 3.21. (left) Surface plot at $t = 4$ of solution to frontogenesis problem based on a 201×201 grid. The surface plot was generated using the persp3d() function. (right) Cross section through frontogenesis numerical solution (bold) with the almost indistinguishable analytical solution (thin) superimposed. This demonstrates the accuracy of the numerical solution.

where we impose Dirichlet boundary conditions $\partial\Omega(x, -5) = -1$, $\partial\Omega(x, 5) = 1$ and Neumann boundary conditions $\partial\Omega(-5, y) = 0$, $\partial\Omega(5, y) = 0$.

The analytical solution [Dav-85] is given by

$$u_a(x, y, t) = \tanh\left(\frac{y\cos(vt/r) - x\sin(vt/r)}{\delta}\right). \tag{3.99}$$

The vortex evolves from time $t = 0$ at $(x, y) = 0$ on a gentle front to form a complex rotational system until, at time $t = 4$, it straddles both sides of the front. This is a medium-difficulty problem for numerical solvers, and the classic nine-point FD scheme resolves the front reasonably well on a 201×201 grid, with the general shape exhibiting very little distortion. Figure 3.21 shows a surface plot together with the outline of a cross section through the numerical solution which compares very well to the analytical solution.

However, if δ is reduced to a very small value, the front becomes much sharper and the vortex arms will have narrow, flat tops rather than the rounded tops shown in Fig. 3.21. Under these conditions, classic FD schemes become unstable and cannot resolve the fine detail. Therefore, a different approach is needed (see Chapter 6), where the WENO method is applied to this problem. A sequence of plan views of the solutions is provided in Fig. 3.22, showing how the solution evolves over time.

The R graphic function persp3d() provides a particularly striking surface plot of the final state of the simulation. The reader may like to try different effects; for example, the lighting can be removed using the command clear3d(type="lights"). This leaves the scene in darkness, so additional lights need to be added with one or more commands light3d(theta=0, phi=0). The values of theta and phi can be set to any appropriate value, and more information can be found in the help facility for the package rgl. Up to eight light sources are supported.

The reader may also like to experiment with different values of $\bar{v}$ and different simulation run times to see how the vortex is affected or to plot the analytical solution of eqn. (3.99) with small values for δ when the arms become flat topped.

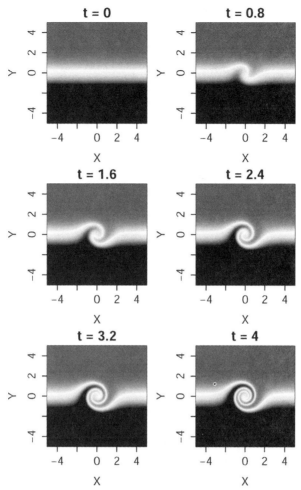

Figure 3.22. Array of 2D plots from simulation of frontogenesis at times from $t = 0$ to $t = 4$. The plots were generated using the persp() function.

The R code for solving this problem is shown in Listings 3.19–3.21.

```
# File: frontogenesis_main.R
# -----------------------------------------------------------------------------
# 2D frontogenesis example
# Ref: Kalise, Dante. "A Study of a WENO-TVD Finite Volume Scheme for the
#      Numerical Simulation of Atmospheric Advective and Convective Phenomena."
#      arXiv Preprint arXiv:1111.1712 (2011): 29. http://arxiv.org/abs/1111.1712.
# -----------------------------------------------------------------------------
cat("\014") # Clear console
rm(list = ls(all = TRUE)) # Delete workspace
#
require(compiler)
enableJIT(3)
```

```r
library("pracma")
library("deSolve")
library("rgl")
library("shape")

source("deriv1st9pt.R")
source("frontogenesis_deriv.R")
#
ptm <- proc.time()
print(Sys.time())
# Declare array sizes
Nx=201; Ny=201 # Equal grids
#
# PDE parameters
xl=-5; xu=5;
yl=-5; yu=5;
# Set times
tf=4; nout<-51
t=seq(from=0,to=tf,length.out=nout)
#
# Grid in x, y
x=seq(from=xl,to=xu,by=(xu-xl)/(Nx-1))
y=seq(from=yl,to=yu,by=(yu-yl)/(Ny-1))
#
####################################################
# Define differential matrices
####################################################
diffMatX <- diffMatrix1st9pt(xl,xu,Nx)
diffMatY <- diffMatrix1st9pt(yl,yu,Ny)
####################################################
# Define mesh
X <- outer(x,rep(1,Ny),"*")
Y <- outer(rep(1,Nx),y,"*")
#
r      <- sqrt(X^2+Y^2)
v0     <- 2.59807
delta  <- 1 #e-6
v      <- v0*sech(r)^2*tanh(r)
v_r    <- v/(1e-16+r)
alphaX <- v_r*Y
alphaY <- -v_r*X
Q      <- tanh((Y*cos(v_r*tf)-X*sin(v_r*tf))/delta) # Analytical soln
#
# Preallocate and define initial conditions
uini <- array(0,c(Nx,Ny))
uini = tanh(Y/delta)
```

```
#
# Initialze other variables
ncall=0
#
# ODE integration
out<-ode(method="ode45",y=c(uini), times=t, func=frontogenesis_deriv,
         ynames = FALSE,
         rtol=1e-12,atol=1e-12,parms=NULL)
# Create plot matrices#
U = array(0, c(nout, Nx, Ny))
#
for (i in seq(1,nout,by=10)){
  U[i,,]=matrix(data=out[i,2:(Nx*Ny+1)],ncol=Ny,
              nrow=Nx,byrow=FALSE,dimnames=NULL)
}
# Plot results
source("frontogenesis_postSimCalcs.R")
# Elapsed time
cat(sprintf("calculation time: %6.2f",proc.time()-ptm)[3])
```

Listing 3.19. File: frontogenesis_main.R—Code for the *main* program that simulates the frontogenesis example defined by eqn. (3.98)

```
# File: frontogenesis_deriv.R
frontogenesis_deriv = function(t, u, parms) {
  # Reshape vector u -> Nx by Ny matrix
  u <- matrix(data=u,nrow=Nx,ncol=Ny,byrow=FALSE,dimnames=NULL)
  #
  # Set Dirichlet BCs for u[y,x=0] and u[y,x=1]
  u[1,]=-1; u[Nx,]=1;
  ####################################################
  # Calculate 2D spatial 1st derivatives
  ####################################################
  ux <- (diffMatX %*% (u))
  uy <- t( diffMatY %*%t( u))
  # Set Neumann BCs for uy[y=0,x] and uy[y=1,x]
  ux[,1] = 0;
  uy[,Ny] = 0;
  ####################################################
  # Calculate time derivative
  ####################################################
  ut = -(alphaX*ux+alphaY*uy)
  ut[1,]=0; ut[Nx,]=0; # Dirichlet BCs
  ####################################################
  # update number of derivative calls
```

```
  ncall <<- ncall+1
  #
  # Return solution
  return(list(ut))
}
```

Listing 3.20. File: `frontogenesis_deriv.R`—Code for the *derivative function* that is called from the frontogenesis main program of Listing 3.19

```
# File: frontogenesis_postSimCalcs.R
# Plot results
UN<-U[nout,,]
#
plot(y,UN[(Nx/2),],type="l", lwd=3)
lines(y,Q[(Nx/2),],col="red")
grid()
#
#Define color scheme
jet.colors = colorRampPalette(c("#00007F", "blue", "#007FFF",
                                "cyan", "#7FFF7F", "yellow",
                                "#FF7F00", "red", "#7F0000"))
# Set palette
pal=jet.colors(100)
#Plot 2D image
op <- par(mfrow = c(3, 2))
par(mar=c(4,4,2,1))
for(i in seq(1,nout,by=10)){
  image(x,y,(U[i,,]), ylim=c(yl,yu), xlim=c(xl,xu), ylab="Y",
        xlab="X", col=pal, # "gray",
        asp=1, main=paste("t =",t[i]))
}
#
#####################################################
# Plot surface
#####################################################
open3d()
bg3d("white")
# Set colour indices of each point for pers3d()
col.ind <- cut(UN,100)#
persp3d(X,Y,UN, aspect=c(1, 1, 0.25), color = pal[col.ind],
        ylab = "", xlab = "", zlab = "",
        axes=FALSE,box=FALSE, smooth=TRUE)
UM <- matrix(data=c( 0.9051732,  0.4045764, -0.1303018, 0,
                    -0.1567654,  0.6027237,  0.7823988, 0,
                     0.3950762, -0.6877799,  0.6089932, 0,
                     0.0000000,  0.0000000,  0.0000000, 1),
```

```
          byrow=TRUE,nrow=4,ncol=4)
rgl.viewpoint(fov=0) # set before userMatrix
par3d(userMatrix=UM)
par3d(windowRect=c(20,100,820,900), zoom=1)
# Print to png file - uncomment if required
# rgl.snapshot("frontogensis3D.png", fmt="png") # print to file
#
```

Listing 3.21. File: `frontogenesis_postSimCalcs.R`—Code for the *postsimulation calculations*, run at the end of the frontogenesis main program of Listing 3.19

3.8.5 MOL 2D: Polar Coordinates

The application of MOL to 2D problems in *polar coordinates* is slightly more complex than problems specified in Cartesian coordinates. This is because inherent in many polar coordinate problems is the requirement to handle singularities at the origin. Also, graphical representation of the solution can be more difficult, as there tends to be less general-purpose graphing capability to handle polar coordinates provided in computer simulation programs.

Problems in polar coordinates are specified on a grid of the type shown in Fig. 3.23 and can be considered to be the same as *cylindrical coordinates*, except that there is zero variation in the z direction. The *del* and *Laplace* operators in *cylindrical* coordinates are shown operating on the scalar f:

$$\nabla f = \hat{\rho}\frac{\partial f}{\partial \rho} + \hat{\phi}\frac{\partial f}{\partial \phi} + \hat{z}\frac{\partial f}{\partial z} \quad \text{and} \quad \nabla^2 f = \frac{1}{\rho}\frac{\partial}{\partial \rho}\left(\rho\frac{\partial f}{\partial \rho}\right) + \frac{1}{\rho^2}\frac{\partial^2 f}{\partial \phi^2} + \frac{\partial^2 f}{\partial z^2} \quad (3.100)$$

$$f = f(\rho, \phi, z), \ \rho \in \mathbb{R}, \ \phi \in \mathbb{R}, \ z \in \mathbb{R}$$

respectively, where the ^ symbol implies a unit vector. Therefore, using r and θ in lieu of ρ and ϕ, and with zero variation in z, it follows that application of the del and Laplace operators to f in *polar* coordinates yields

$$\nabla f = \hat{r}\frac{\partial f}{\partial r} + \hat{\theta}\frac{\partial f}{\partial \theta} \quad \text{and} \quad \nabla^2 f = \frac{1}{r}\frac{\partial}{\partial r}\left(r\frac{\partial f}{\partial r}\right) + \frac{1}{r^2}\frac{\partial^2 f}{\partial \theta^2} \quad f = f(r, \theta), \ r \in \mathbb{R}, \ \theta \in \mathbb{R}. \quad (3.101)$$

Note that in eqns. (3.100) and (3.101), because f is a *scalar*, ∇f represents the *gradient* of f. However, if f were a *vector*, then we would form the *dot product* of ∇ and f, when $\nabla \cdot f$ would represent the *divergence* of f. The Laplace operator ∇^2 is also known as the *Laplacian*, which, when applied to f, is often represented by $\Delta f\ (=\nabla^2 f = \nabla \cdot \nabla f)$.

Now, for plotting purposes, we need to convert from polar to Cartesian coordinates, and this is achieved by the following transformation:

$$x = r\sin\theta \quad y = r\cos\theta. \quad (3.102)$$

Although this simple transformation works well, the result is a set of x and y vectors that are not monotonic. This can be a problem, as most surface-plotting routines require x and y to be vectors that are monotonically increasing or deceasing. The solution is to generate matrices X and Y instead of vectors x and y such that $u_{ij} = u(x = X_{ij}, y = Y_{ij})$ and use a surface-plotting routine that can handle data presented in this form—such as

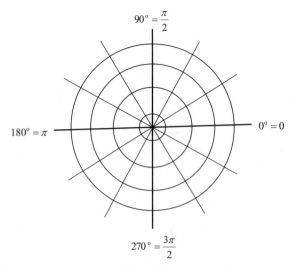

Figure 3.23. A simple representation of a finite difference grid in polar coordinates. The central point represents $r=0$, $\theta = 0, \cdots, 2\pi$, and the intersections of the circles and radial lines occur at points $r = i\Delta r$, $i = 2\cdots N_r$ and $\theta = j\Delta\theta$, $j = 1\cdots N_\theta$.

the R function `persp3d()`. The matrices are obtained from the the *outer* product (see the book's appendix) based on eqns. (3.102). Thus, we use the following data when plotting the solution:

$$X = r \otimes \sin\theta, \quad Y = r \otimes \cos\theta, \quad u = u(X, Y), \tag{3.103}$$

where now the coordinates are represented by the elements of the X and Y matrices. Or, equivalently, we could write

$$X_{ij} = r_i \sin\theta_j, \quad Y_{ij} = r_i \cos\theta_j, \quad u_{ij} = u(X_{ij}, Y_{ij}). \tag{3.104}$$

However, polar coordinates present an additional difficulty owing the Laplacian in eqns. (3.101) becoming singular at the origin. Nevertheless, the fact that in Cartesian coordinates the Laplacian is defined as

$$\nabla^2 f = \frac{\partial^2 f}{\partial x^2} + \frac{\partial^2 f}{\partial y^2}, \quad u = f(x, y), \; x \in \mathbb{R}, \; y \in \mathbb{R}, \tag{3.105}$$

which does not become singular at the origin, means that the singularity must be *removable*. We can therefore proceed by treating the origin as a separate case to the other points and approximate it in Cartesian coordinates. It should be noted that at this point, $f(r, \theta) = f(r = 0) = f_0$, that is, f does not depend upon θ and the grid surrounding the origin is made up of triangles instead of quadrilaterals. Adopting this approach and referring to Fig. 3.23, we observe that we can approximate $\nabla^2 f$ at the origin by a three-point, first-order finite difference equation of the form $\frac{\partial^2 f_i}{\partial r^2} \simeq (f_{i-1} - 2f_i + f_{i+1})/\Delta r^2$. Thus, for our problem, we could use

$$\nabla^2 f \simeq \frac{f(\Delta r, \pi) - 2f_0 + f(\Delta r, 0)}{\Delta r^2} + \frac{f(\Delta r, \pi/2) - 2f_0 + f(\Delta r, 3\pi/2)}{\Delta r^2}. \tag{3.106}$$

Note that the axis formed from radial vectors $f(r, \theta)$ and $f(r, 0)$ are orthogonal to the axis formed from $f(r, \pi/2)$ and $f(r, 3\pi/2)$, which is a necessary condition for this approach to work. Equally, we could approximate $\nabla^2 f$ at the origin by a finite difference equation based upon Cartesian coordinates formed from any other similar set of orthogonal radial vectors, that is, where again each coordinate would be formed from two opposing radial vectors. There is no preferred choice of coordinates, as they are all equally valid. Therefore we choose to take the average of all the possible pairs. Under these conditions, the second derivative approximation at the origin becomes

$$\nabla^2 f \simeq \frac{1}{\Delta r^2 (N_\theta/4)} \left(\sum_{j=1}^{N_\theta} f(\Delta r, j\Delta\theta) - f_0 \right). \tag{3.107}$$

Clearly, for the coordinates to be orthogonal, N_θ must be divisible by 4.

Higher-order approximations can be achieved in a similar way. For example, a five-point, third-order approximation of the form $\frac{\partial^2 f_i}{\partial r^2} \simeq \frac{1}{12}(-f_{i-2} + 16f_{i-1} - 30f_i + 16f_{i+1} - f_{i+2})/\Delta r^2$ would be

$$\nabla^2 f \simeq \frac{4}{\Delta r^2 N_\theta} \left(\frac{1}{12} \sum_{j=1}^{\frac{1}{2}N_\theta} -f(\Delta r, j\Delta\theta) + 16f(\Delta r, j\Delta\theta) - 30f_0 \right.$$
$$\left. + 16f(\Delta r, (j + \tfrac{1}{2}N_\theta)\Delta\theta) - f(\Delta r, (j + \tfrac{1}{2}N_\theta)\Delta\theta) \right) \tag{3.108}$$

$$\therefore \nabla^2 f \simeq \frac{4}{\Delta r^2 N_\theta} \left(\frac{1}{12} \sum_{j=1}^{N_\theta} -f(2\Delta r, i\Delta\theta) + 16f(\Delta r, j\Delta\theta) - 15f_0 \right).$$

Note that we underline the fact that this result, which is a Cartesian coordinate calculation, applies only at the origin. Therefore we have

$$\nabla^2 f(r=0, \theta) = \frac{\partial^2 f(x=0, y=0)}{\partial x^2} + \frac{\partial^2 f(x=0, y=0)}{\partial y^2}, \tag{3.109}$$

which is equal to eqn. (3.105).

A further complication with polar coordinates is that two *continuity conditions* need to be satisfied. The first continuity condition is due to the grid points at (r_1, θ_j) coalescing into a single point at the origin and is dealt with by the use of a Cartesian calculation for the derivative at $r = 0$, as discussed earlier. The second continuity condition is due to the grid points at (r_i, θ_1) and (r_i, θ_{N_θ}) being at the same locations within the (r, θ) domain. For consistency, this means that they must be equal. For a coherent finite difference scheme we need the approximations to straddle the $\theta \in (0, 2\pi)$ coordinates. We achieve this by extending the number of points along the θ coordinate by adding two ghost points: one ghost point before $\theta = 0$ and one after $\theta = 2\pi$. Thus, the number of grid points along the θ coordinate becomes $N_\theta + 2$, where θ_1 is the first ghost point and $\theta_{N_\theta+2}$ is the second ghost point. The ghost points are arranged so as to provide appropriate values for the finite difference equations that straddle these continuity points. This can be considered as being similar to the *periodic conditions* discussed briefly in Section 3.2. Thus, we need to make the following grid point assignments at each step prior to evaluating the Laplacian:

$$\begin{aligned}
f(r_i, \theta_{N_\theta+1}) &= f(r_i, \theta_2), & \text{continuity condition,} \\
f(r_i, \theta_{N_\theta+2}) &= f(r_i, \theta_3), & \text{ghost point,} \\
f(r_i, \theta_1) &= f(r_i, \theta_{N_\theta}). & \text{ghost point.}
\end{aligned} \tag{3.110}$$

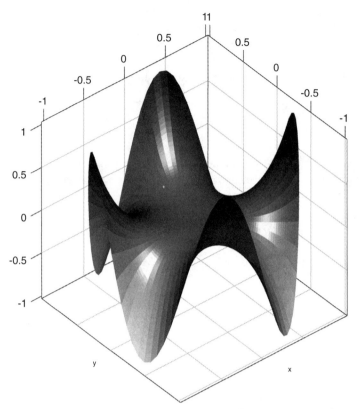

Figure 3.24. Surface plot of numerical solution to polar diffusion problem of eqns. (3.111) at $t = 1.5$. The surface plot was generated using the persp3d() function.

We emphasize that these assignments are made on the extended θ coordinate; and only following these assignments are the spatial derivatives calculated. However, the ghost points will then have served their purpose by allowing accurate numerical derivative estimates to be evaluated. Therefore, we revert back to the original number of θ grid points, N_θ, by ignoring the first and last grid points (the ghost points), when assembling ∇f and/or $\nabla^2 f$.

This method is illustrated in the following polar coordinate MOL examples.

3.8.5.1 Diffusion Equation

In this example we wish to find the steady state temperature distribution of the following 2D heat conduction problem, defined on the circular domain $\Omega = [0, 1] \times [0, 2\pi]$ and subject to Dirichlet boundary conditions

$$\frac{\partial T}{\partial t} = D \left[\frac{1}{r} \frac{\partial}{\partial r} \left(r \frac{\partial u}{\partial r} \right) + \frac{1}{r^2} \frac{\partial^2 u}{\partial \theta^2} \right], \quad T = T(r, \theta, t),\ t \geq t_0,\ (r, \theta) \in \mathbb{R}, \quad (3.111)$$

$$T(r, \theta, 0) = 0,\ \partial\Omega(r = 1, \theta) = \sin 4\theta,$$

where T represents the domain temperature. The medium consists of a thin sheet of heat-conducting material with perfect insulation on both sides. It is initially at a temperature of $T = 0°C$ when, at $t = 0$, the boundary temperature is raised to $T(r = 1, \theta) = \sin 4\theta °C$. We assume that there are no temperature variations vertically through the sheet and that no convective or radiative heat transfer takes place. Also, we assume that the material forms an isotropic medium having a thermal diffusivity constant equal to $D = 1 \text{ m}^2/\text{s}$. The Dirichlet BCs are defined in eqns. (3.111).

The standard FD method readily finds a good solution using a four-point FD approximation to the second spatial derivative, a three-point approximation to the first spatial derivative, and the lsodes ODE integrator. A surface plot of the numerical solution is shown in Fig. 3.24. Using 48 spatial grid points for the r coordinate and 96 for the θ cordinate, the method converges to a near-constant solution at $t = 1.5$. Doubling the run time to $t = 3$ makes no appreciable difference to the solution, with the maximum absolute difference being 0.0046, which is quite acceptable for most applications. We therefore conclude that we have found a satisfactory steady-state solution to the problem outlined in eqn. (3.111).

The R code for solving this problem is shown in Listings 3.22–3.24. Note that we need to set the variable case=0 in main program Listing 3.22.

```
# File: diffusionPolar2D_main.R
# -------------------------------------------------
# 2D Diffusion/Poisson polar co-ordinates examples:
#   u_t = D * (u_rr+u_r/r +u_pp/r^2) - f
# -------------------------------------------------
###################################################
# Delete workspace prior to new calculations
###################################################
cat("\014") # Clear console
rm(list = ls(all = TRUE))
#
###################################################
# Required R library packages
###################################################
require(compiler)
enableJIT(3)
library("deSolve")
library("rgl")
library("shape")
#
###################################################
# Load sources for functions called
###################################################
source("deriv1st5pt.R")
source("deriv2nd6pt.R")
source("diffusionPolar2D_deriv.R")
###################################################
# Start timer
###################################################
ptm = proc.time()
print(Sys.time())
```

```
#####################################################
# Declare array sizes
#####################################################
Nr   <- 48 # 24
Np   <- 96 # 32 #52 # Set to number divisable by 4
Np1  <- Np+2 # Add periodic ghost points at pl and pu
#####################################################
# PDE parameters
#####################################################
rl <- 0; ru <- 1.0; dr <- (ru-rl)/(Nr-1)
pl <- -pi; pu <- pi; dp <- (pu-pl)/(Np-1)
#####################################################
# Define grid in r, p
#####################################################
r <- seq(from=rl,to=ru,by=dr)
p <- seq(from=pl,to=pu,by=dp)
#####################################################
# Define differential matrices
#####################################################
diffMat2R <- diffMatrix2nd6pt(rl,ru,Nr,bc=c(0,0))
diffMat1R <- diffMatrix1st5pt(rl,ru,Nr)
diffMat2P <- diffMatrix2nd6pt(pl,pu,Np1,bc=c(0,0))
#####################################################
# Set parameters
#####################################################
tf <- 1.5; nout <- 51
t <- seq(from=0,to=tf,length.out=nout)
#
#####################################################
# Preallocate ICs for u1 and u2
#####################################################
u1ini <- matrix(data=0,nrow=Nr, ncol=Np, byrow=FALSE,
                dimnames=NULL)
#
#####################################################
# Set IC/BC conditions
#####################################################
#
case <- 1; cat(sprintf("\ncase = %d\n",case))
#
if (case == 0){     # For diffusion equation
  D <- 1
  u1ini[Nr,1:Np] <- 0
}else if (case == 1){ # For Poisson equation
  D <- 1
  f <- matrix(data=0,nrow=Nr, ncol=Np, byrow=FALSE)
```

```
#
  frho <- function(r,p){
    20-45*r-32*r*cos(p)+16*r*sin(p)+16*r^2-25*r^3+
    60*r^2*cos(p)-30*r^2*sin(p)}
  f <- outer(r,p,frho)
  fua <- function(r,p){
    r^2*(1-r)*((r*cos(p)-2)^2+(r*sin(p)+1)^2)}
  ua <- outer(r,p,fua)
  u1ini[Nr,] <- ua[Nr,]
}
uini <- c(u1ini)
#
#####################################################
# Set ODE integrator and perform integration
#####################################################
ncall <- 0
out<-lsodes(y=c(uini), times=t, func=diffusionPolar2D_deriv,
            sparsetype = "sparseint", ynames = FALSE,
            rtol=1e-6,atol=1e-6,maxord=5,parms=NULL)
#####################################################
# Create plot matrix
#####################################################
#
z0 <- array(0,dim=c(nout,Nr,Np))
#
z0[1:nout,,] <- out[1:nout,2:(Np*Nr+1)]
#
source("diffusionPolar2DpostSimCalcs.R")
#####################################################
# Print elapsed time
#####################################################
cat(sprintf("calculation time = %7.2f\n",(proc.time()-ptm)[3]))
#
#####################################################
# End of program
#####################################################
```

Listing 3.22. File: diffusionPolar2D_main.R—Code for the *main* program that simulates the diffusion system of eqn. (3.111) and the Poisson system of eqn. (3.112). Note: case=0 for *polar diffusion* example and case=1 for *polar Poisson* example.

```
# File: diffusionPolar2D_deriv.R
diffusionPolar2D_deriv = function(t, u, parms) {
  #
  #####################################################
```

```r
# Reshape vector u -> to two Nr by Np matrices
####################################################
u0 = matrix(data=u[1:(Nr*Np)],nrow=Nr, ncol=Np,
            byrow=FALSE, dimnames=NULL)
####################################################
# Set BC values
####################################################
if (case == 0){ # BC = Dirichlet
  u0[Nr,]  <- sin(4*p)
}else if (case == 1){ # BC = Dirichlet/Poisson Eqn
  u0[Nr,]  <- ua[Nr,]
}
####################################################
# Impose polar continuity condition
####################################################
u0[,Np] <- u0[,1] # Continuity
#
####################################################
# Extended domain matrix - with ghost points
####################################################
u1 <- matrix(data=0,nrow=Nr, ncol=Np1)
# Set values for actual domain
u1[,2:(Np1-1)] <-u0
####################################################
# Set ghost grid-point values
####################################################
# Set values for extended domain at p[1]
u1[,1] <- u1[,(Np1-2)]
# Set values for extended domain at p[Np]
u1[,Np1] <- u1[,3]
####################################################
# Calculate spatial derivatives
# and terms of the the Lapacian
####################################################
u1r   <- diffMat1R %*% u1
u1rr  <- diffMat2R %*% u1
u1pp  <- t(diffMat2P %*% t(u1 ))
u1pprr <- diffMat2R %*% u1pp
# Revert back to actual domain
u1rr0  <- u1rr[1:Nr,2:(Np1-1)]
u1r0   <- u1r[ 1:Nr,2:(Np1-1)]
u1pp0  <- u1pp[1:Nr,2:(Np1-1)]
# u1pprr0 <- u1pprr[1:Nr,2:(Np1-1)]
####################################################
# Preallocate arrays
####################################################
```

```r
  u1r_r   <- matrix(data=0,nrow=Nr,ncol=Np)
  u1pp_r2 <- matrix(data=0,nrow=Nr,ncol=Np)
  #
  # Terms of the Laplacian
  for (i in (2:Nr)){
    u1pp_r2[i,] <- u1pp0[i,]/(r[i]^2)
    u1r_r[i,]   <- u1r0[i,]/r[i]
  }
  # Regularize singularities by l'Hospital's rule
  # NOTE: Not needed as u1t[,1] calculated using
  # Cartesian coords
  #   u1pp_r2[1,] <- u1pprr0[1,]/2
  #   u1r_r[1,]   <- u1rr0[1,]
  # Impose continuity condition: u[,theta=pi]=u[,theta=0]
  u1pp_r2[,Np] <- u1pp_r2[,1]
  u1r_r[ ,Np]  <- u1r_r[,1]
  u1rr0[ ,Np]  <- u1rr0[,1]
  #
  ####################################################
  # Calculate time derivative
  ####################################################
  u1t <- D*( u1rr0 + u1r_r + u1pp_r2 )
  #
  u1_2 <- sum(u0[3,])
  u1_1 <- sum(u0[2,])
  u1_0 <- sum(u0[1,])
  u1t[1,] <- D*4*(u1_1-u1_0)/(Np*dr^2) # FD 3-pt scheme
  # u1t[1,] <- D*(4/12)*(-u1_2+16*u1_1-15*u1_0)/(Np*dr^2) # FD 5-pt scheme
  ####################################################
  # Subtract f to derivative term if Poisson eqn.
  ####################################################
  if (case == 1){
    u1t <- u1t - f
  }
  ####################################################
  # update number of derivative calls
  ####################################################
  ncall <<- ncall+1
  ####################################################
  # Return solution
  return(list(c(u1t)))
  ####################################################
}
```

Listing 3.23. File: diffusionPolar2D_deriv.R—Code for the *derivative* function that is called from the polar diffusion main program of Listing 3.22

```
# File: diffusionPolar2D_postSimCalcs.R
##################################################
# Define plot variables
z <- z0[nout,,]
zr <- c(range(z0[,1:Nr,1:Np]))
##################################################
# calculate error for case = 1
##################################################
if(case == 1){
  err <- z - ua #zAnal
  maxErrLoc <- which.max(abs(err))
  cat(sprintf("\nMaximum error = %e\n",err[maxErrLoc]))
}
##################################################
# Define polar plot variables
##################################################
fx <- function(r1,theta1){r1*sin(theta1)}
fy <- function(r1,theta1){r1*cos(theta1)}
x1 <- outer(r,p,fx)
y1 <- outer(r,p,fy)
##################################################
#Define color scheme
##################################################
jet.colors = colorRampPalette(c(
  "#00007F", "blue", "#007FFF",
  "cyan", "#7FFF7F", "yellow",
  "#FF7F00", "red", "#7F0000"))
# Set palette
pal <- jet.colors(100)
##################################################
# Plot surface
##################################################
open3d()
bg3d("white")
#
# Set colour indices of each point for pers3d()
col.ind <- cut(z,100)#
#objs <-
persp3d(x1,y1,z, aspect=c(1, 1, 1), color = pal[col.ind],
        xlab = "x", ylab = "y", zlab = "",
        zlim=zr, axes=TRUE,
        box=FALSE, smooth=TRUE)
um <- c( 0.7688994, -0.6393660, 0.00149996, 0,
         0.3885930,  0.4691809, 0.79300833, 0,
        -0.5077270, -0.6091616, 0.60920691, 0,
```

```
          0.0000000, 0.0000000, 0.00000000, 1)
UM <- matrix(data=um, byrow=TRUE,nrow=4,ncol=4)
rgl.viewpoint(fov=0) # set before userMatrix
par3d(userMatrix=UM)
par3d(windowRect=c(20,100,820,900), zoom=1)
grid3d(c("x+", "y+", "z"),col="gray",lty=2,lwd=1)
####################################################
# Print to png file - uncomment if required
# rgl.snapshot("diffusionEqn_polar_3D.png", fmt="png")
####################################################
```

Listing 3.24. File: diffusionPolar2D_postSimCalcs.R—Code for the *postsimulation calculations* run at the end of the polar diffusion main program of Listing 3.22

3.8.5.2 Poisson Equation

The *Poisson* equation is an elliptical PDE given by eqn. (3.82), and a Cartesian coordinates example was solved in Section 3.8.4.3.

We now consider the 2D problem of eqn. (3.112) that has a known analytical solution, u_a. It is defined on the circular domain $\Omega(r,\theta) = [0,1] \times [0,2\pi]$ with Dirichlet boundary conditions imposed on $\partial\Omega$ obtained from u_a. We employ the false transient solution method by including an additional time derivative so that the problem takes the form of heat diffusion with source. Also, as we have transformed the problem to be time varying, the analytical solution now applies at $t \to \infty$:

$$\frac{\partial^2 u}{\partial t^2} = c^2 \left[\frac{1}{r}\frac{\partial}{\partial r}\left(r\frac{\partial u}{\partial r}\right) + \frac{1}{r^2}\frac{\partial^2 u}{\partial \theta^2} \right] - f, \quad u = u(r,\theta,t), \ t \geq t_0, \ (r,\theta) \in \mathbb{R},$$

$$f(r,\theta) = 20 - 45r - 32r\cos\theta + 16r\sin\theta + 16r^2 - 25r^3 + 60r^2\cos\theta - 30r^2\sin\theta,$$

$$u(r,\theta,0) = 0, \ \partial\Omega(r=1,\theta) = 0, \ u_a(r,\theta) = r^2(1-r)\left[(r\cos\theta - 2)^2 + (r\sin\theta + 1)^2\right].$$

(3.112)

The problem is actually contrived, in as much as the analytical solution $u_a(x,t)$ was specified and the equation for $f(r,\theta)$ calculated by substituting into eqn. (3.82). The reason for this approach was to obtain a fairly complex problem to solve and also one which could be used to test the accuracy of the method. An additional bonus was that $u_a(r,\theta)$ and $f(r,\theta)$ both look interesting when plotted graphically.

The false transient method readily finds a good solution using a six-point FD approximation to the second derivatives, a five-point approximation to the first derivatives (including at the origin), and the lsodes ODE integrator. Surface plots of the solution and the function $f(r,\theta)$ are shown in Fig. 3.25. Using 48 spatial grid points for the r coordinate and 96 for the θ cordinate, the resulting solution is very close to the analytical solution with a maximum absolute error of 0.0034 ($\simeq$0.34% of max u). The errors converge according to the order of the discretization scheme and the number of grid points employed, similar to the Cartesian problem of Section 3.8.4.3. The interested reader may like to experiment with the discretization scheme and number of grid points to see the effect on the overall error.

The R code for solving this problem is shown in Listings 3.22–3.24. Note that for this problem we need to set the variable case=1 in the main program of Listing 3.22.

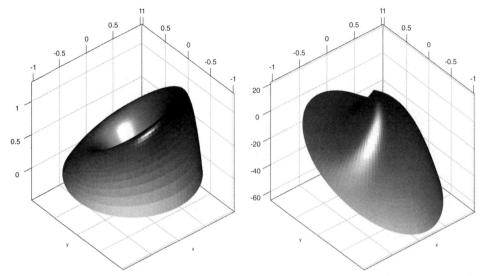

Figure 3.25. (left) Surface plot of the numerical solution $u(r, \theta, t)$ to the Poisson equation given in eqns. (3.112) at $t = 1.5$, based on 48 grid points for r and 96 grid points for θ. (right) Surface plot of the associated function $f(r, \theta)$. Both plots were generated using the persp3d() function.

3.8.5.3 Wave Equation

The classic 2D linear wave equation can also be simulated in polar coordinates, when eqn. (3.95) becomes

$$\frac{\partial^2 \eta}{\partial t^2} = c^2 \left[\frac{1}{r} \frac{\partial}{\partial r} \left(r \frac{\partial \eta}{\partial r} \right) + \frac{1}{r^2} \frac{\partial^2 \eta}{\partial \theta^2} \right], \tag{3.113}$$

where $c = \sqrt{gh}$ represents the wave speed or celerity, as for the Cartesian coordinates problem.

Equation (3.113) will form the basis of the model for this example. The problem is defined on the circular domain $\Omega = [0, 1] \times [0, 2\pi]$ with a reflecting boundary condition on $\partial \Omega$, as follows:

$$\frac{\partial^2 u}{\partial t^2} = c^2 \left[\frac{1}{r} \frac{\partial}{\partial r} \left(r \frac{\partial u}{\partial r} \right) + \frac{1}{r^2} \frac{\partial^2 u}{\partial \theta^2} \right], \quad u = u(r, \theta, t), \ t \geq t_0, \ r \in \mathbb{R}, \ \theta \in \mathbb{R},$$

$$u(r, \theta, 0) = \exp\left[-\lambda r^2 \left((\sin\theta - \delta x)^2 + (\cos\theta - \delta y)^2\right)\right], \tag{3.114}$$

$$\delta x = 0.0, \ \delta y = -0.5, \ c = 0.05,$$

where u represents the height of shallow-water waves subsequent to the introduction of a disturbance—in this case, in the form of a Gaussian-shaped mass of water at Cartesian coordinate location $x = 0.0, y = -0.5$. With regard to the reflective boundary condition for eqn. (3.114), this is achieved by adding ghost cells at the boundary $r = 1$.

The modeling approach used is as outlined previously with $\partial^2 u(r = 0, \theta) / \partial t^2$ calculated using a Cartesian coordinate approximation. The r coordinate is divided up into 48 grid points and the θ coordinate into 96 grid points. The extra grid points used for these calculations, compared to those of the Cartesian coordinate model, means that this simulation runs more slowly. Refer also to the discussion in Section 3.8.4.5 relating to reflective boundary conditions and the approach to calculating $\partial^2 u / \partial t^2$ from two first derivatives.

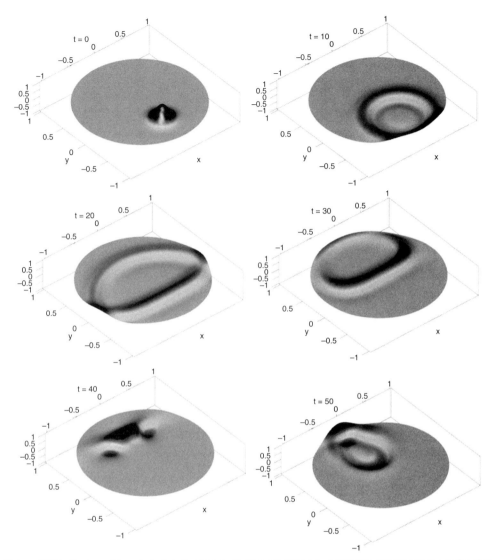

Figure 3.26. Series of images showing the evolution of waves in a circular domain with a reflective boundary condition at $r = 1$. Following a Gaussian shaped disturbance, the sequence proceeds right to left, top to bottom. With the wave traveling at a velocity of $c = 0.05$, its peak reaches the near boundary at $t = 10$ and the far boundary at $t = 30$, as expected. The surface plots were generated using the persp3d() function. (See color plate 3.26)

The results of the simulation are shown in Fig. 3.26, where a sequence of surface plots shows how the wave evolves following the introduction of a Gaussian-shaped disturbance. The simulation method generates a very good solution. With the wave traveling at a velocity of $c = 0.05$, its peak reaches the near boundary at $t = 10$ and the far boundary at $t = 30$, as expected.

The R code for solving this problem is shown in Listings 3.25–3.27. An *animation* can be created by setting animation = TRUE in Listing 3.27 (see discussion on imageMagick in Section 3.8.4.1).

```r
# File: wave2Dpolar_main.R
# -------------------------------------------------
# 2D Wave polar co-ordinates example:
#   u_tt = C^2 * (u_rr+u_r/r +u_pp/r^2)
# -------------------------------------------------
####################################################
# Delete workspace prior to new calculations
####################################################
cat("\014") # Clear console
rm(list = ls(all = TRUE))
####################################################
# Required R library packages
####################################################
require(compiler)
enableJIT(3)
library("deSolve")
library("rgl")
library("shape")
####################################################
# Load sources for functions called
####################################################
source("deriv1st5pt.R")
source("deriv2nd6pt.R")
source("wave2Dpolar_deriv.R")
source("p2c_2D.R")
####################################################
# Start timer
####################################################
ptm = proc.time()
print(Sys.time())
####################################################
# Declare array sizes
####################################################
Nr <- 48 # 24 # Must be divisible by 4
Np <- 96 # 48
#
Np1 <- Np+2 # Add periodic ghost points at pl and pu
Nr1 <- Nr+1 # Add reflective ghost point at r=ru
####################################################
# PDE parameters
####################################################
rl <- 0;   ru <- 1; dr <- (ru-rl)/(Nr-1)
pl <- -pi; pu <- pi; dp <- (pu-pl)/(Np-1)
C  <- 0.05; C2 <- C*C;
####################################################
# Define grid in x, y
####################################################
```

```
r <- seq(from=rl,to=ru,by=dr)
p <- seq(from=pl,to=pu,by=dp)
#####################################################
# Define differential matrices
#####################################################
diffMat2R <- diffMatrix2nd6pt(rl,(ru+dr),Nr1,bc=c(0,0))
diffMat1R <- diffMatrix1st5pt(rl,(ru+dr),Nr1)
diffMat2P <- diffMatrix2nd6pt(pl,(pu+2*dp),Np1,bc=c(0,0))
#####################################################
# Set parameters
#####################################################
tf <- 50; nout <- 51
t <- seq(from=0,to=tf,len=nout)
#####################################################
# Preallocate ICs for u1 and u2
#####################################################
u1ini <- matrix(data=0,nrow=Nr, ncol=Np, byrow=FALSE,
                dimnames=NULL)
u2ini <- u1ini
#####################################################
# Initial condition
#####################################################
lambda <- 50
offSet_x <- 0.0; offSet_y <- -0.5;
IC <- 1
for(i in 1:Nr){
  for(j in 1:Np){
    if(IC==1){
    # IC starting point
      u1ini[i,j] <- exp(-lambda*((r[i]*cos(p[j])-offSet_y)^2+
                      (r[i]*sin(p[j])-offSet_x)^2))
    }else if(IC == 2){

      u1ini[i,j] <- exp(-lambda*(((r[i])*cos(p[j])-offSet_y)^2 +
                      ((r[i])*sin(p[j])-offSet_x)^2))+
                    exp(-lambda*(((r[i])*cos(p[j])-3*offSet_y)^2+
                      ((r[i])*sin(p[j])-3*offSet_x)^2))
    }else{
      if(r[i]<0.1)
        u1ini[i,j] <- 1
    }
    # u1ini[i,j] <- i + 10*i
  }
}
uini <- c(u1ini,u2ini)
#####################################################
```

```
# ODE integration
######################################################
ncall <<- 0
out<-ode(y=c(uini), times=t, func=wave2Dpolar_deriv,
         method = "ode45", ynames = FALSE,
         rtol=1e-6,atol=1e-6,parms=NULL)
# Create plot matrix
z0 <- array(0,dim=c(nout,Nr,Np))
#
z0[1:nout,,] <- out[1:nout,2:(Np*Nr+1)]
#
# Plot results
source("wave2Dpolar_postSimCalcs.R")
######################################################
# Print elapsed time
######################################################
cat(sprintf("calculation time = %7.2f",(proc.time()-ptm)[3]))
#
######################################################
# End of program
######################################################
```

Listing 3.25. File: wave2Dpolar_main.R—Code for the *main* program that simulates the 2D polar wave eqn. (3.114)

```
# File: wave2Dpolar_deriv.R
wave2Dpolar_deriv = function(t, u, parms) {
  #
  ######################################################
  # Reshape vector u -> to two Nr by Np matrices
  ######################################################
  u10 = matrix(data=u[1:(Nr*Np)],nrow=Nr, ncol=Np,
               byrow=FALSE, dimnames=NULL)
  u20 = matrix(data=u[(Nr*Np+1):(2*Nr*Np)],nrow=Nr,
               ncol=Np, byrow=FALSE, dimnames=NULL)
  ######################################################
  # Set continuituy condition and reflective BCs
  ######################################################
  #
  u10[,Np] <- u10[,1] # Continuity condition
  u20[,Np] <- u20[,1] # Continuity condition
  #
  # Extended state matrix
  u1 <- matrix(data=0,nrow=Nr1, ncol=Np1)
  #
  # Assign values for actual domain
```

```
u1[1:Nr,2:(Np1-1)] <-u10
u1[Nr1,] <- u1[Nr,] # Reflective BCs
#
####################################################
# Set BCs (ghost grid-point values)
####################################################
# Set values for extended domain at p[1]
u1[,1] <- u1[,(Np1-2)]
# Set values for extended domain at p[Np]
u1[,Np1] <- u1[,3]
#
####################################################
# Calculate 2D spatial 2nd derivatives
# and terms of the the Lapacian
####################################################
#
u1r   <- diffMat1R %*% u1
u1rr  <- diffMat2R %*% u1
u1pp  <- t(diffMat2P %*% t(u1))
u1pprr <- diffMat2R %*% u1pp
####################################################
# Preallocate arrays
####################################################
#
u1r_r   <- matrix(data=0,nrow=Nr, ncol=Np)
u1pp_r2 <- matrix(data=0,nrow=Nr, ncol=Np)
#
# Revert back to actual domain
u1rr0  <- u1rr[1:Nr,2:(Np1-1)]
u1r0   <- u1r[ 1:Nr,2:(Np1-1)]
u1pp0  <- u1pp[1:Nr,2:(Np1-1)]
# u1pprr0 <- u1pprr[1:Nr,2:(Np1-1)]
#
# Terms of the Laplacian
for (i in (2:Nr)){
  u1pp_r2[i,1:(Np-1)] <- u1pp0[i,1:(Np-1)]/(r[i]^2)
  u1r_r[i,1:(Np-1)]   <- u1r0[i,1:(Np-1)]/r[i]
}
# By l'Hospital's rule (NOTE: Not needed as
# u1t[1,] calculated using Cartesian coords)
#   u1pp_r2[1,] <- u1pprr0[1,]/2
#   u1r_r[1,]   <- u1rr0[1,]
# Continuity condition
u1pp_r2[,Np] <- u1pp_r2[,1]
u1r_r[ ,Np] <- u1r_r[,1]
u1rr0[ ,Np] <- u1rr0[,1]
```

```
#
#####################################################
# Calculate time derivative
#####################################################
#
u1t <- u20
u2t <- C2*( u1rr0 + u1r_r + u1pp_r2 )
# Calculate ut at the origin
u1_2 <- sum(u10[3,])
u1_1 <- sum(u10[2,])
u1_0 <- sum(u10[1,])
# u2t[1,] <- C2*4*(u1_1-u1_0)/(Np*dr^2) # FD 3-pt scheme: Thomas, vol1, p198
u2t[1,] <- C2*(4/12)*(-u1_2+16*u1_1-15*u1_0)/(Np*dr^2) # FD 5-pt scheme
#####################################################
# update number of derivative calls
#####################################################
ncall <<- ncall+1
#
#####################################################
# Return solution
return(list(c(u1t,u2t)))
#####################################################
}
```

Listing 3.26. File: wave2Dpolar_deriv.R—Code for the *derivative* function that is called from the 2D polar wave main program Listing 3.25

```
# File: wave2Dpolar_postSimCalcs.R
#####################################################
# Set animation flag and plot results
#####################################################
animate <- FALSE # TRUE/FALSE
#####################################################
# Define polar plot variables
#####################################################
fx <- function(r1,theta1){r1*sin(theta1)}
fy <- function(r1,theta1){r1*cos(theta1)}
x1 <- outer(r,p,fx)
y1 <- outer(r,p,fy)
#####################################################
#Define color scheme
#####################################################
jet.colors = colorRampPalette(c("#00007F","blue","#007FFF",
      "cyan","#7FFF7F","yellow","#FF7F00","red","#7F0000"))
# Set palette
pal <- jet.colors(100)
```

```
#####################################################
# Plot surface
#####################################################
open3d()
bg3d("white")
#
z <- z0[nout,,]
# Set colour indices of each point for pers3d()
brks <- c(-0.7,seq(-0.3,0.3,len=97),0.7,1)
col.ind <- cut(z, breaks=brks,include.lowest = T)
persp3d(x1,y1,z, aspect=c(1, 1, 0.25), color = pal[col.ind],
        ylab = "y", xlab = "x", zlab = "",
        zlim=c(-1,1),box=FALSE, smooth=TRUE)
um <- c( 0.7688994, -0.6393660, 0.00149996, 0,
         0.3885930,  0.4691809, 0.79300833, 0,
        -0.5077270, -0.6091616, 0.60920691, 0,
         0.0000000,  0.0000000, 0.00000000, 1)
UM <- matrix(data=um, byrow=TRUE,nrow=4,ncol=4)
rgl.viewpoint(fov=0) # set before userMatrix
par3d(userMatrix=UM)
par3d(windowRect=c(20,100,820,900), zoom=1)
#####################################################
# Print to png file - uncomment if required
#####################################################
# rgl.snapshot("diffusionEqn_polar_2D.png", fmt="png")
#
if(animate == TRUE){
  for(i in seq(1,nout, by=1)){
    fileName <- sprintf("wave2dExample%03d.png",i)
    open3d(); bg3d("white")
    # New plot
    col.ind <- cut(z0[i,,],breaks=brks)#
    persp3d(x1,y1,z0[i,,], aspect=c(1, 1, 0.25),
            color = pal[col.ind],
            ylab = "y", xlab = "x", zlab = "",
            zlim=c(-1,1),
            box=FALSE, smooth=TRUE,
            main=paste("t =",t[i]), ticktype="detailed")
    rgl.viewpoint(fov=0) # set before userMatrix
    par3d(userMatrix=UM)
    par3d(windowRect=c(20,100,820,900), zoom=1)
    Sys.sleep(0.1)
    print(i)
    rgl.snapshot(fileName, fmt="png")
    rgl.close()
  }
```

```
    shell("convert -delay 20 wave2dExample*.png wave2dEx_polar_1.gif",
        intern=TRUE)
    # Remove individual files
    file.remove(grep("wave2dExample",list.files(),value=TRUE))
}
```

Listing 3.27. File: `wave2Dpolar_postSimCalcs.R`—Code for the *postsimulation calculations* run at the end of the 2D polar wave main program Listing 3.25

3.9 FULLY DISCRETE METHODS

3.9.1 Introduction

The main emphasis of this book is semi-discrete methods; thus, we will only discuss fully discrete methods briefly to mention a few of the standard schemes for solving hyperbolic PDEs similar to

$$\frac{\partial u(x,t)}{\partial t} + \alpha \frac{\partial u(x,t)}{\partial x} = 0; \quad t > 0, \quad u \in \mathbb{R}^N, \quad x \in \mathbb{R}^M, \tag{3.115}$$

where α represents *wave speed*.

Fully discrete methods discretize the problem in time and space where generally the time step is represented by a superscript n and, for a 2D system, the x, y spatial grid points by suffixes i, j, respectively. Thus, we have $t_n = n\Delta t, n = 0, 1, 2, \ldots, x_i = i\Delta x, i = 0, 1, 2\ldots, N_x$ and $y_j = j\Delta y, j = 0, 1, 2\ldots, N_y$, where N_x and N_y represent the number of grid points in the x and y coordinates, respectively (see Fig. 3.4).

3.9.2 Overview of Some Common Schemes

3.9.2.1 Forward in Time Backward in Space (FTBS)

The *FTBS* method is first-order accurate in both space and time for the hyperbolic system of eqn. (3.115). A 1D problem takes the following explicit form:

$$u_i^{n+1} = u_i^n - \frac{\alpha \Delta t}{\Delta x}\left(u_i^n - u_{i-1}^n\right), \quad \alpha > 0. \tag{3.116}$$

3.9.2.2 Forward in Time Backward in Space (FTBS), Implicit

The *FTBS* method is first-order accurate in time and in space for the hyperbolic system of eqn. (3.115). A 1D problem takes the following implicit form:

$$u_i^{n+1} = u_i^n - \frac{\alpha \Delta t}{\Delta x}\left(u_i^{n+1} - u_{i-1}^{n+1}\right), \quad \alpha > 0. \tag{3.117}$$

Rearranging, we get

$$u_i^{n+1} + \frac{\alpha \Delta t}{\Delta x}\left(u_i^{n+1} - u_{i-1}^{n+1}\right) = u_i^n, \tag{3.118}$$

which, on expansion and assuming periodic BCs, results in the following near bidiagonal problem:

$$\begin{bmatrix} 1+\text{Co} & & & & & & -\text{Co} \\ -\text{Co} & 1+\text{Co} & & & & & \\ & -\text{Co} & 1+\text{Co} & & & & \\ & & \ddots & & & & \\ & & & -\text{Co} & 1+\text{Co} & & \\ & & & & -\text{Co} & 1+\text{Co} & \\ & & & & & -\text{Co} & 1+\text{Co} \end{bmatrix} \begin{bmatrix} u_1^{n+1} \\ u_2^{n+1} \\ u_3^{n+1} \\ \vdots \\ u_{m-3}^{n+1} \\ u_{m-2}^{n+1} \\ u_{m-1}^{n+1} \end{bmatrix} = \begin{bmatrix} u_1^n \\ u_2^n \\ u_3^n \\ \vdots \\ u_m^{n1} \\ u_m^n \\ u_{m-1}^n \end{bmatrix}, \quad (3.119)$$

where $\text{Co} = \frac{\alpha \Delta t}{\Delta x}$ represents the Courant number and the problem can be solved using Gaussian elimination or some other linear equation solving method.

3.9.2.3 Forward in Time Central in Space (FTCS)

The *FTCS* method is first-order accurate in time and second-order accurate in space for the hyperbolic system of eqn. (3.115). A 1D problem takes the following explicit form:

$$u_i^{n+1} = u_i^n - \frac{1}{2}\frac{\alpha \Delta t}{\Delta x}\left(u_{i+1}^n - u_{i-1}^n\right), \quad \alpha > 0. \quad (3.120)$$

3.9.2.4 Backward in Time Central in Space (BTCS), Implicit

The *BTCS* method is first-order accurate in time and second-order accurate in space for the hyperbolic system of eqn. (3.115). A 1D problem takes the following implicit form:

$$u_i^{n+1} = u_i^n - \frac{1}{2}\frac{\alpha \Delta t}{\Delta x}\left(u_{i+1}^{n+1} - u_{i-1}^{n+1}\right), \quad \alpha > 0. \quad (3.121)$$

Rearranging, we get

$$u_i^{n+1} + \frac{1}{2}\frac{\alpha \Delta t}{\Delta x}\left(u_{i+1}^{n+1} - u_{i-1}^{n+1}\right) = u_i^n, \quad (3.122)$$

which, on expansion and assuming periodic BCs, results in the following near bidiagonal problem:

$$\begin{bmatrix} 1 & \text{Co}/2 & & & & & -\text{Co}/2 \\ -\text{Co}/2 & 1 & \text{Co}/2 & & & & \\ & -\text{Co}/2 & 1 & \text{Co}/2 & & & \\ & \ddots & \ddots & \ddots & & & \\ & & -\text{Co}/2 & 1 & \text{Co}/2 & & \\ & & & -\text{Co}/2 & 1 & \text{Co}/2 & \\ \text{Co}/2 & & & & -\text{Co}/2 & 1 \end{bmatrix} \begin{bmatrix} u_1^{n+1} \\ u_2^{n+1} \\ u_3^{n+1} \\ \vdots \\ u_{m-3}^{n+1} \\ u_{m-2}^{n+1} \\ u_{m-1}^{n+1} \end{bmatrix} = \begin{bmatrix} u_1^n \\ u_2^n \\ u_3^n \\ \vdots \\ u_m^n \\ u_m^n \\ u_{m-1}^n \end{bmatrix}, \quad (3.123)$$

where $\text{Co} = \frac{\alpha \Delta t}{\Delta x}$ represents the Courant number and the problem can be solved using Gaussian elimination or some other linear equation solving method.

3.9.2.5 Central in Time Central in Space (CTCS) or Leapfrog

The *CTCS* method is second-order accurate in both time and space for the hyperbolic system of eqn. (3.115). A 1D problem takes the following explicit form:

$$u_i^{n+1} = u_i^{n-1} - \frac{\alpha \Delta t}{\Delta x}\left(u_{i+1}^n - u_{i-1}^n\right). \qquad (3.124)$$

3.9.2.6 Beam-Warming (B-W)

The *beam-warming* method is second-order accurate in both space and time for the hyperbolic system of eqn. (3.115). A 1D problem takes the following explicit form:

$$u_i^{n+1} = u_i^n - \frac{1}{2}\frac{\alpha \Delta t}{\Delta x}\left(3u_i^n - 4u_{i-1}^n + u_{i-2}^n\right) + \frac{1}{2}\left(\frac{\alpha \Delta t}{\Delta x}\right)^2\left(u_i^n - 2u_{i-1}^n + u_{i-2}^n\right). \qquad (3.125)$$

3.9.2.7 Lax–Friedrichs (LxF)

The *Lax–Friedrich* method is second-order accurate in both space and time for the hyperbolic system of eqn. (3.115). A 1D problem takes the following explicit form:

$$u_i^{n+1} = \frac{1}{2}\left(u_{i+1}^n + u_{i-1}^n\right) - \frac{1}{2}\frac{\alpha \Delta t}{\Delta x}\left(u_{i+1}^n - u_{i-1}^n\right). \qquad (3.126)$$

3.9.2.8 Lax–Wendroff (LxW)

The *Lax–Wendroff* method, alternatively known as the *Cauchy–Kowalewski* method, is second-order accurate in both space and time for the hyperbolic system of eqn. (3.115). A 1D problem takes the following explicit form:

$$u_i^{n+1} = u_i^n - \frac{1}{2}\frac{\alpha \Delta t}{\Delta x}\left(u_{i+1}^n - u_{i-1}^n\right) + \frac{1}{2}\left(\frac{\alpha \Delta t}{\Delta x}\right)^2\left(u_{i+1}^n - 2u_i^n - u_{i-1}^n\right). \qquad (3.127)$$

For a 2D hyperbolic system the Lax–Wendruff scheme is significantly more complex than its 1D counterpart. The 2D scheme is given by [Lev-02]

$$\begin{aligned} u_{i,j}^{n+1} = u_{i,j}^n &- \frac{1}{2}\frac{\alpha_x \Delta t}{\Delta x}\left(u_{i+1,j}^n - u_{i-1,j}^n\right) - \frac{1}{2}\left(\frac{\alpha_y \Delta t}{\Delta y}\right)\left(u_{i,j+1}^n - u_{i,j-1}^n\right) \\ &+ \frac{1}{2}\frac{\alpha_x^2 \Delta t^2}{\Delta x^2}\left(u_{i+1,j}^n - 2u_{i,j}^n + u_{i-1,j}^n\right) + \frac{1}{2}\frac{\alpha_y^2 \Delta t^2}{\Delta y^2}\left(u_{i,j+1}^n - 2u_{i,j}^n + u_{i,j-1}^n\right) \\ &+ \frac{1}{8}\frac{\Delta t^2}{\Delta x \Delta y}(\alpha_x \alpha_y)\left[\left(u_{i+1,j+1}^n - u_{i-1,j+1}^n\right) - \left(u_{i+1,j-1}^n - u_{i-1,j-1}^n\right)\right], \end{aligned} \qquad (3.128)$$

where i and j represent spatial grid points in the x and y directions.

3.9.2.9 Crank–Nicholson

The *Crank–Nicholson* is an implicit method [Cra-47] that is historically famous in respect of the numerical solution of PDEs. This is because, when announced, it demonstrated some fundamentally new concepts in respect of computational stability. It is second-order accurate in both space and time and is often used to solve the following heat equation:

$$\frac{\partial u(x,t)}{\partial t} - \alpha \frac{\partial^2 u(x,t)}{\partial x^2} = 0, \quad u \in \mathbb{R}, \ x \in \mathbb{R}^m, \ m \in (1,2,3). \qquad (3.129)$$

The 1D heat problem takes the following implicit form:

$$u_i^{n+1} = u_i^n + \frac{1}{2}\frac{\alpha \Delta t}{(\Delta x)^2}\left[\left(u_{i+1}^{n+1} - 2u_i^{n+1} + u_{i-1}^{n+1}\right) + \left(u_{i+1}^n - 2u_i^n + u_{i-1}^n\right)\right]. \quad (3.130)$$

This equation (modified to take BCs into account) can be written more succinctly as

$$AT^{n+1} = BT^n, \quad (3.131)$$

where A and B are tridiagonal matrices with the three diagonals defined as

$$A_{i,i-1} = -\alpha, \quad A_{i,i} = 2\left(\frac{\Delta x^2}{\Delta t} + \alpha\right), \quad A_{i,i+1} = -\alpha, \quad (3.132)$$

$$B_{i,i-1} = \alpha, \quad B_{i,i} = 2\left(\frac{\Delta x^2}{\Delta t} - \alpha\right), \quad B_{i,i+1} = \alpha. \quad (3.133)$$

The solution to T is usually found by solving eqn. (3.131) by Gaussian elimination or some other similar linear equation solving method.

The 2D heat problem with $\Delta x = \Delta y$ takes the following implicit form:

$$\begin{aligned}u_{i,j}^{n+1} = u_{i,j}^n &+ \frac{1}{2}\frac{\alpha \Delta t}{(\Delta x)^2}\left[\left(u_{i+1,j}^{n+1} + u_{i-1,j}^{n+1} + u_{i,j+1}^{n+1} + u_{i,j-1}^{n+1} - 4u_{i,j}^{n+1}\right)\right.\\ &\left.+ \left(u_{i+1,j}^n + u_{i-1,j}^n + u_{i,j+1}^n + u_{i,j-1}^n - 4u_{i,j}^n\right)\right],\end{aligned} \quad (3.134)$$

where i and j represent spatial grid points in the x and y directions. It is computationally intensive, and other methods, such as *alternating-direction implicit* (ADI), are to be preferred. We will not discuss this method further here; readers are referred to [Tan-97] for further details.

3.9.3 Results from Simulating a Hyperbolic Equation

The various fully discrete methods described earlier have been used to simulate a 1D version of the hyperbolic eqn. (3.115), that is, the advection equation

$$\frac{\partial u(x,t)}{\partial t} + \alpha \frac{\partial u(x,t)}{\partial x} = 0; \quad t > 0, \quad u \in \mathbb{R}, \quad x \in \mathbb{R}. \quad (3.135)$$

For each method the system was subject to *square* and *cosine* pulse initial conditions. In addition, *periodic* boundary conditions were imposed so that the simulations could operate over long run-times. For stability considerations, we impose the Courant condition Co = $\left|\frac{\alpha \Delta t}{\Delta x}\right| < 1$.

Results from the simulations are shown in Figs. 3.27 and 3.28 with the exception of those from the explicit FTCS method, which has been omitted because it resulted in an unstable solution. From the results of these tests it is seen that none of the methods is ideal. They all suffer from either excessive dissipation or spurious oscillations. Increasing the number of grid points reduces dissipation but increases the frequency of spurious oscillations. However, even so, fully discrete methods are useful and do find application in solving some PDE problems. We will not pursue this subject further here; more discussion can be found in [Hir-88] and [Tan-97].

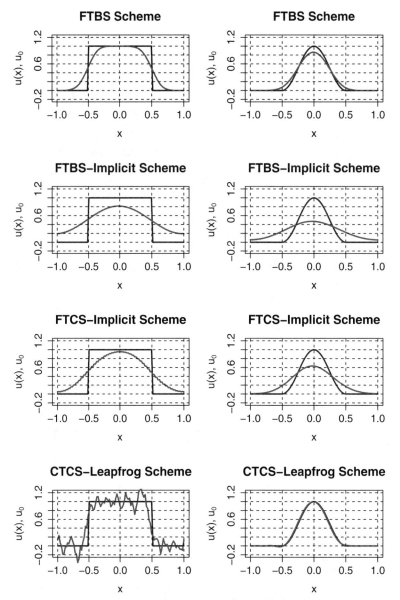

Figure 3.27. Step and cosine responses for fully discrete schemes applied to the advection equation $u_t + \alpha u_x = 0$, with $\alpha = 1$, Co = 0.8, simulation duration = 4 s, and periodic boundary conditions. The initial condition, u_0, is shown with the numerical solution, $u(x)$, superimposed. The simulation was performed over 100 grid points.

The R code for solving this problem is shown in Listings 3.28–3.30.

There are three parts to the code: the main program, the initialization functions, and the function that calculates the solutions. Following initialization, the main program uses two nested *for* loops to control the simulation. The outer loop steps through a list of the various methods to be, and the inner loop steps through a list of the the initial conditions to be used for each method. These lists can be easily modified to run

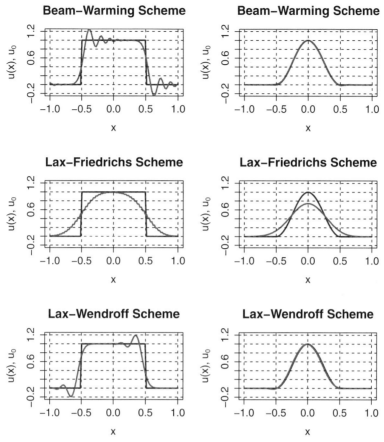

Figure 3.28. Step and cosine responses for fully discrete schemes applied to the advection equation $u_t + \alpha u_x = 0$, with $\alpha = 1$, Co=0.8, simulation duration = 4 s, and periodic boundary conditions. The initial condition, u_0, is shown with the numerical solution, $u(x)$, superimposed. The simulation was performed over 100 grid points.

different scenarios. The simulation calculation is performed by a call to the function fullyDiscrete_soln() following simulation initialization performed by a call to the function fullyDiscrete_init(). The detailed working of the program should be apparent from the comments included in the code.

```
# File: fullyDiscrete_main.R
# R Code used to simulate following Linear Hyperbolic PDE
#               Ut + alpha.Ux <- 0
# by fully discrete methods using periodic boundary conditions
rm(list = ls(all = TRUE)) # Delete workspace
library("rgl")
#library("gsl")
library("Matrix")
source("fullyDiscrete_init.R")
source("fullyDiscrete_soln.R")
```

```
# ADJUSTABLE PARAMETERS
# solType     - controlled in for loop
#  "0" = Forward in time backward in space (FTBS)
#  "1" = Forward in time backward in space (FTBS) - implicit
#  "2" = Backward in time central in space (FTCS) - implicit
#  "3" = Central in time central in space (CTCS) or Leapfrog
#  "4" = Forward in time central in space (FTCS) - UNSTABLE!
#  "5" = Beam-Warming
#  "6" = Lax-Friedrichs
#  "7" = Lax-Wendroff

# initWave - controlled in for loop
#  "0" = Step Pulse
#  "1" = Cosine Pulse
#  "2" = Triange Pulse
#  "3" = Forward Wedge Pulse
#  "4" = Backward Wedge Pulse
#  "5" = Gaussian Pulse
#  "6" = Ellipsoid Pulse
#  "7" = Composite Pulse Set
wavePeak <- 1.0   # Maximum height of wave
Xmax     <- 1.0   # End of spatial domain
Xmin     <- -Xmax # Start of spatial domain
M        <- 100   # Number of spatial grid points
alpha    <- 1.0   # Speed of wave in medium
                  # One loop every 2 [s]
x<-rep(0,M+1)
# INITIALISE VARS - CALCULATED
dx    <- (Xmax-Xmin)/(M-1)   # Segment length
x     <- seq(Xmin,Xmax,dx)   # Discretize spatial domain
Co    <- 0.8                 # Courant Number
dt    <- Co*dx/alpha         # time step
t0    <- 0.0                             # Set start time
tRun  <- 4.0                 # Set run duration
tEnd  <- t0 + tRun           # Set end time
figNum <- 0 # Initialize figure number

par(mfrow = c(4,2))
# Start simulation
for(iSim in c(0:3,5:7)){ # c(0:3,5:7)
  for(initWave in c(0,1)){ # Set the starting conditions
    u0 <- fullyDiscrete_init(x, initWave, Xmin, Xmax, wavePeak, M)
    solType <- iSim
    # cat(sprintfLAY SIMULATION DATA
    cat(sprintf(" \n"))
    cat(sprintf("=====================================================\n"))
```

```
cat(sprintf("            SIMULATION DATA              \n"))
if(solType == 0){
  strScheme <- "FTBS"
  cat(sprintf("  Forward in time forward in space - %s\n",strScheme))
}else if(solType == 1){
  strScheme <- "FTBS-Implicit"
  cat(sprintf("  Forward in time backward in space - %s\n",strScheme))
}else if(solType == 2){
  strScheme <- "FTCS-Implicit"
  cat(sprintf("  Forward in time central in space - %s\n",strScheme))
}else if(solType == 3){
  strScheme <- "CTCS-Leapfrog"
  cat(sprintf("  Central in time central in space - %s\n",strScheme))
}else if(solType == 4){
  strScheme <- "FTCS"
  cat(sprintf("  Forward in time central in space - %s\n",strScheme))
}else if(solType == 5){
  strScheme <- "Beam-Warming"
  cat(sprintf("   %s%s\n", strScheme,", Fully-Discrete method"))
}else if(solType == 6){
  strScheme <- "Lax-Friedrichs"
  cat(sprintf("   %s%s\n", strScheme,", Fully-Discrete method"))
}else if(solType == 7){
  strScheme <- "Lax-Wendroff"
  cat(sprintf("   %s%s\n", strScheme,", Fully-Discrete method"))
}
cat(sprintf("====================================================\n"))
cat(sprintf("Linear Hyperbolic Equation, ...: Ut + alpha.Ux = 0\n"))
cat(sprintf("Segment Length, ............, dx: %f %s\n",dx, " [m]"))
cat(sprintf("Number of grid points, ........, M: %d %s\n",M, " [-]"))
cat(sprintf("Total Length, .............., x: %f %s\n",(Xmax-Xmin), " [m]"))
cat(sprintf("Propogation Velocity, .., alpha: %f %s\n",alpha, " [m/s]"))
cat(sprintf("End Time, .............., tEnd: %f %s\n",tEnd, " [s]"))

if(initWave == 0){
  cat(sprintf("Initialised To ................: Step Pulse\n"))
}else if(initWave == 1){
  cat(sprintf("Initialised To ................: Cos Pulse\n"))
}else if(initWave == 2){
  cat(sprintf("Initialised To ................: Triangular Pulse\n"))
}else if(initWave == 3){
  cat(sprintf("Initialised To ................: Forward Wedge Pulse\n"))
}else if(initWave == 4){
  cat(sprintf("Initialised To ................: Backward Wedge Pulse\n"))
}else if(initWave == 5){
  cat(sprintf("Initialised To ................: Gaussian Pulse\n"))
```

```r
      }else if(initWave == 6){
        cat(sprintf("Initialised To ................: Ellipsoid Pulse\n"))
      }else if(initWave == 7){
        cat(sprintf("Initialised To ................: Composite Pulse set\n"))
      }
      # START SIMULATION
      U <- fullyDiscrete_soln(u0, t0, tEnd, dt, dx, alpha, M, solType)
#
      cat(sprintf("Time Step, ................, dt: %f\n",dt, " [s]"))
      cat(sprintf("Courant No, ...............,  Co: %f\n",Co, " [-]"))

      strT1 <- sprintf("%s Scheme", strScheme)
      plot(x,u0,col="blue",type="l",lwd=2,
           xlab="x",ylab=expression(paste("u(x), ",u[0])),
             ylim=c(-0.2,1.2),main=strT1)
      lines(x,U,col="red",lwd=2)
      grid(col="black",lty=2,lwd=1)

      status <- "OK"
      cat(sprintf("Simulation Completed, .........: %s\n", status))
      cat(sprintf("==================================================\n"))
#
  }
}
```

Listing 3.28. File: fullyDiscrete_main.R—Code for *main* program to solve PDE using *fully discrete* method

```r
# File: fullyDiscrete_init.R
# R code for initialising the spatial domain to simulate PDE problem's
fullyDiscrete_init <- function(xl, waveType, Xmin, Xmax, wavePeak, M)
{
  X <- Xmax-Xmin
  Z <- (Xmax + Xmin)/2
  if(waveType == 7){
    widthWave <- 0.1 # Fraction (0-1)
    x1 <- Z - widthWave*X/2
    x3 <- Z + widthWave*X/2
    x2 <- Z
    a <- 0.1
    delta <- 0.005
    beta <- 0.75*log(2)/(36*delta^2)
    triangeleOffset <- 0.1
    stepOffset      <- -0.3
    gaussOffset     <- -0.7
    ellipsoidOffset <- 0.5
```

```
  }else{
    widthWave <- 0.5 # Fraction (0-1)
    x1 <- Z - widthWave*X/2
    x3 <- Z + widthWave*X/2
    x2 <- Z
    a <- widthWave*X/2
    delta <- 0.005
    beta <- 0.025*log(2)/(36*delta^2)
    triangeleOffset <- 0
    stepOffset     <- 0
    gaussOffset    <- 0
    ellipsoidOffset <- 0
  }
  if(waveType == 0){  # Step Pulse
    ul<-stepPulse(xl, x1, x3, stepOffset, wavePeak, M)
  }else if(waveType == 1){   # Cosine Pulse
    ul<-cosPulse(xl, x1, x3, a, wavePeak, M)
  }else if(waveType == 2){   # Triangle Pulse
    ul<-triangle(xl, x1, x2, x3, triangeleOffset, wavePeak, M)
  }else if(waveType == 3){   # Forward Wedge
    ul<-forwardWedge(xl, x1, x3, wavePeak, M)
  }else if(waveType == 4){   # Backward Wedge
    ul<-backWedge(xl, x1, x3, wavePeak, M)
  }else if(waveType == 5){ # Gaussian Pulse
    ul<-gauss(xl, wavePeak, beta, gaussOffset, M)
  }else if(waveType == 6){ # Ellisoid Pulse
    ul<-ellipsoidPulse(xl, Xmin, Xmax, a, 1, x2, ellipsoidOffset)
  }else if(waveType == 7){ # Composite Pulse set
    ul1<-stepPulse(xl, x1, x3, stepOffset, wavePeak, M)
    ul2<-triangle(xl, x1, x2, x3, triangeleOffset, wavePeak, M)
    ul3<-ellipsoidPulse(xl, Xmin, Xmax, a, 1, ellipsoidOffset, 0)
    #    Add gauss pulse last!
    ul4<-gauss(xl, wavePeak, beta, gaussOffset, M)
    ul <- ul1+ul2+ul3+ul4
  }
return(ul)
}
stepPulse <- function(xl, x1, x3, offset, wavePeak, M)
{
  ul<-rep(0,M)
  for(J in 1:M){
    if(xl[J] > x1+offset && xl[J] < x3+offset){
      ul[J] <- wavePeak
    }else{
      ul[J] <- 0.0
    }
```

```r
  }
  return(ul)
}
gauss<- function(xl, wavePeak, beta, offset, M)
{
  ul<-rep(0,M)
  for(J in 1:M){
    ul[J] <- wavePeak*exp( -beta*(xl[J]-offset)^2)
  }
  return(ul)
}
ellipsoidPulse <- function(xl, Xmin, Xmax, a, b, x_off, y_off)
{
  t  <- seq(pi,0,-pi/1000) #pi,0,-pi/1000
  x1 <- x_off + a*cos(t)
  y1 <- y_off + b*sin(t)
  
  x2 <- seq(Xmin,x_off-a,length.out=101)
  y2 <- rep(0,100)
  
  x3 <- seq(x_off+a,Xmax,length.out=101)
  y3 <- rep(0,100)
  
  x4<-c(x2[1:100],x1,x3[2:101])
  y4<-c(y2,y1,y3)
  ul <- splinefun(x4,y4, method = "monoH.FC")(xl)
  return(ul)
}
backWedge <- function(xl, x1, x3, wavePeak, M)
{
  ul<-rep(0,M)
  for(J in 1:M){
    if(xl[J] < x1){
      ul[J] <- 0.0
    }else if(xl[J]>= x1 && xl[J] <= x3){
      ul[J] <- wavePeak*(xl[J]-x1)/(x3-x1)
    }else{
      ul[J] <- 0.0
    }
  }
  return(ul)
}
forwardWedge <- function(xl, x1, x3, wavePeak, M)
{
  ul<-rep(0,M)
  for(J in 1:M){
```

```r
    if(xl[J] < x1){
      ul[J] <- 0.0
    }else if(xl[J] >= x1 && xl[J] <= x3){
      ul[J] <- wavePeak*(x3-xl[J])/(x3-x1)
    }else{
      ul[J] <- 0.0
    }
  }
  return(ul)
}
triangle <- function(xl, x1, x2, x3, offset, wavePeak, M)
{
  ul<-rep(0,M)
  for(J in 1:M){
    if(xl[J] < x1+offset){
      ul[J] <- 0.0
    }else if(xl[J] >= x1+offset && xl[J] <= x2+offset){
      ul[J] <- wavePeak*(xl[J] - (x1+offset))/(x2-x1)
    }else if(xl[J] > x2+offset && xl[J] <= x3+offset){
      ul[J] <- wavePeak*(x3+offset - xl[J])/(x3-x2)
    }else{
      ul[J] <- 0.0
    }
  }
  return(ul)
}
cosPulse <- function(xl, x1, x3, widthWave, wavePeak, M)
{
  ul<-rep(0,M)
  for(J in 1:M){
    if(xl[J] > x1 && xl[J] < x3){
      ul[J] <- wavePeak*(1 - cos(pi*(xl[J] - x1)/(widthWave)))/2.0
    }else{
      ul[J] <- 0.0
    }
  }
  return(ul)
}
```

Listing 3.29. File: `fullyDiscrete_init.R`—Code functions for *initialization* of fully discrete PDE simulation

```r
# File: fullyDiscrete_soln.R
# R code that simulates the following PDE:
#             Ut + alpha.Ux <- 0, Linear Hyperbolic
# by various fully discrete methods
```

```r
fullyDiscrete_soln <- function(u0, t0, tEnd, dt, dx, alpha, M, solType)
{
  # START SIMULATION
  Co <- alpha*dt/dx # Courant Number. For stability must be < 1
  
  numSteps <- round(tEnd/dt) #+12
  Tend <- numSteps*dt
  t <- seq(t0,tEnd,numSteps)
  
  # Periodic boundary conditions
  # -------------------------------
  # Note: We use periodic BC's which means that grid points 1 and M
  #       are considered to be the same point. Grid point ip1 is
  #       equal to grid point i shifted left (i+1), and im1 is equal
  #       to grid point i shifted right (i-1). Similarly, ip2 and im2
  im1         <- rep(0,M); im2   <- rep(0,M)
  ip1         <- rep(0,M); ip2   <- rep(0,M)
  im1[2:M]    <- 1:(M-1); im1[1] <- (M-1)
  im2[3:M]    <- 1:(M-2); im2[1] <- (M-2); im2[2] <- (M-1)
  ip1[1:(M-1)] <- 2:M;    ip1[M] <- 2
  ip2[1:(M-2)] <- 3:M;    ip2[M] <- 3;    ip2[M-1] <- 2
  #
  # -------------------------------
  U  <- rep(0,M)
  U1 <- rep(0,M)
  U  <- u0
  U1 <- U
  if(solType == 1){ # FTBS
    V1 <- rep(1,M) + Co
    V2 <- rep(0,M)
    V3 <- rep(1,M)*(-Co)
    A <- bandSparse(M,M, k=list(0,1,-1), diagonals=list(V1,V2,V3),
                    symmetric = FALSE, giveCsparse = TRUE)
    A[1,M-1] <- -Co
    A[M,2]   <- Co
  }else if(solType == 2){ # FTCS
    V1 <- rep(1,M)
    V2 <- V1*Co/2
    V3 <- V1*(-Co/2)
    A <-bandSparse(M,M, k=list(0,1,-1), diagonals=list(V1,V2,V3),
                    symmetric = FALSE, giveCsparse = TRUE)
    A[1,M-1] <- -Co/2
    A[M,2]   <- Co/2
  }
  for(n in 1:numSteps){
    if(solType == 0){ # Forward in time backward in space (FTBS)
```

```
      #cat(sprintf("\n\n Co=%f \n",Co))
      #cat(print(U))
      U1 <- U - Co*(U-U[im1])
      #cat(sprintf("\n\n p2 \n"))
    }else if(solType == 1){ # Forwards in time backwards in space
                             (FTBS)-implicit
      U1 <- solve(A,U1)
      U1[M]<-U1[1]
    }else if(solType == 2){ # Backward in time central in space (BTCS)-implicit
      U1 <- solve(A,U1)
    }else if(solType == 3){ # Central in time central in space (CTCS)-Leapfrog
      if(n == 1){ # Use FTBS for first step
        Um1 <- U # save u_(n-1)
        U1  <- U - Co*(U-U[im1])
      }else{
        U1  <- Um1 - Co*(U[ip1]-U[im1])
        Um1 <- U # save u_(n-1)
      }
    }else if(solType == 4){ # Forward in time central in space (FTCS)
                            # UNSTABLE!
      U1 <- U - (1/2)*Co*(U[ip1]-U[im1])
    }else if(solType == 5){ # Beam-Warming
      U1 <- U - (1/2)*Co*(3*U-4*U[im1]+U[im2]) +
                             (1/2)*Co^2*(U-2*U[im1]+U[im2])
    }else if(solType == 6){ # Lax-Friedrichs
      U1<- (1/2)*(U[ip1]+U[im1]) - (1/2)*Co*(U[ip1]-U[im1])
    }else if(solType == 7){ # Lax-Wendroff
      U1<- U - (1/2)*Co*(U[ip1]-U[im1]) + (1/2)*Co^2 *
                                          ( U[ip1] - 2*U + U[im1])
    }
    U <- U1
  }
return(U1)
}
```

Listing 3.30. File: `fullyDiscrete_soln.R`—Code for a function that *computes* the fully discrete PDE solution

3.10 FINITE VOLUME METHOD[10]

3.10.1 General

The *finite volume method* is a method for representing and evaluating partial differential equations as algebraic equations. Similar to the finite difference method, values are

---

[10] This section is partly based on the author's contribution to Wikipedia [Wik-13].

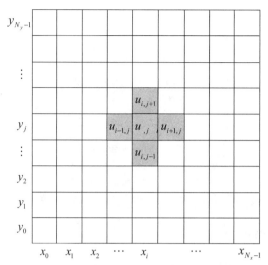

Figure 3.29. Typical 2D finite volume mesh with equally spaced cells. The group of shaded cells represents a computational stencil.

calculated at discrete places on a meshed geometry. *Finite volume* refers to the small volume surrounding each node point on a mesh. Although the method is referred to as a finite volume method, it actually works with *cell average values*, and the cells may represent the discretization of a 1D, 2D, or 3D domain. Thus, for the typical 2D finite volume mesh shown in Fig. 3.29, the i, jth cell average is given by

$$u_{ij} \equiv \frac{1}{\Delta x \Delta y} \int_{x_{i-\frac{1}{2}}}^{x_{i+\frac{1}{2}}} \int_{y_{j-\frac{1}{2}}}^{y_{j+\frac{1}{2}}} u(x, y)\, dxdy; \quad x_i = \left(i + \tfrac{1}{2}\right)\Delta x, \quad y_j = \left(j + \tfrac{1}{2}\right)\Delta y \qquad (3.136)$$
$$i = 0, \ldots, N_x - 1,\ j = 0, \ldots, N_y - 1.$$

In the finite volume method, volume integrals in a partial differential equation that contain a *divergence* term are converted to *surface integrals*, using the *divergence theorem*. These terms are then evaluated as fluxes at the surfaces of each finite volume. Because the flux entering a given volume is identical to that leaving the adjacent volume, these methods are *conservative*. Another advantage of the finite volume method is that it is easily formulated to allow for unstructured meshes. The method is used in many *computational fluid dynamics* packages.

3.10.2 Application to a 1D Conservative System

Consider a simple 1D conservative system defined by the following partial differential equation with appropriate boundary conditions:

$$\frac{\partial \rho}{\partial t} + \frac{\partial f}{\partial x} = 0, \quad t \geq 0. \qquad (3.137)$$

Here, $\rho = \rho(x, t)$ represents the *state* variable and $f = f(\rho(x, t))$ represents the *flux* or flow of ρ. Conventionally, positive f represents flow to the right, whereas negative f represents flow to the left. If we assume that eqn. (3.137) represents a flowing medium of constant area, we can subdivide the spatial domain, x, into *finite volumes* or *cells* with cell centers indexed as i. For a particular cell, i, we can define the *volume average* value of $\rho_i(t) = \rho(x,t)$ at time $t = t_1$ and $x \in [x_{i-\frac{1}{2}}, x_{i+\frac{1}{2}}]$ as

$$\overline{\rho}_i(t_1) = \frac{1}{x_{i+\frac{1}{2}} - x_{i-\frac{1}{2}}} \int_{x_{i-\frac{1}{2}}}^{x_{i+\frac{1}{2}}} \rho(x, t_1) \, dx, \tag{3.138}$$

and at time $t = t_2$ as

$$\overline{\rho}_i(t_2) = \frac{1}{x_{i+\frac{1}{2}} - x_{i-\frac{1}{2}}} \int_{x_{i-\frac{1}{2}}}^{x_{i+\frac{1}{2}}} \rho(x, t_2) \, dx, \tag{3.139}$$

where $x_{i-\frac{1}{2}}$ and $x_{i+\frac{1}{2}}$ represent locations of the upstream and downstream faces or edges, respectively, of the ith cell.

Integrating eqn. (3.137) in time, we have

$$\rho(x, t_2) = \rho(x, t_1) - \int_{t_1}^{t_2} f_x(\rho(x, t)) \, dt, \tag{3.140}$$

where $f_x = \frac{\partial f}{\partial x}$.

To obtain the volume average of $\rho(x, t)$ for the ith cell at time $t = t_2$, we integrate $\rho(x, t_2)$ over the cell volume, v_i, and divide the result by v_i, that is,

$$\overline{\rho}_i(t_2) = \frac{1}{v_i} \int_{v_i} \left(\rho(x, t_1) - \int_{t_1}^{t_2} f_x(\rho(x, t)) \, dt \right) dv. \tag{3.141}$$

We assume that f is well behaved and that we can reverse the order of integration for the second term. Also, recall that flow is normal to the unit area of the cell. Now, because in one dimension, $f_x := \nabla f$, we can apply the divergence theorem[11] and substitute for the volume integral of the divergence with the values of $f(x)$ at the cell edges $x_{i-\frac{1}{2}}$ and $x_{i+\frac{1}{2}}$ of the finite volume, as follows:

$$\overline{\rho}_i(t_2) = \frac{1}{\Delta x_i} \int_{x_{i-\frac{1}{2}}}^{x_{i+\frac{1}{2}}} \rho(x, t_1) \, dx - \frac{1}{\Delta x_i} \left(\int_{t_1}^{t_2} f_{i+\frac{1}{2}} \, dt - \int_{t_1}^{t_2} f_{i-\frac{1}{2}} \, dt \right),$$

$$\Downarrow \tag{3.142}$$

$$\overline{\rho}_i(t_2) = \overline{\rho}_i(t_1) - \frac{1}{\Delta x_i} \int_{t_1}^{t_2} \left(f_{i+\frac{1}{2}} - f_{i-\frac{1}{2}} \right) dt,$$

where the cell volume $\Delta x_i = (x_{i+\frac{1}{2}} - x_{i-\frac{1}{2}})$ and $f_{i\pm\frac{1}{2}} = f(\rho(x_{i\pm\frac{1}{2}}, t))$.

We can therefore derive a *semi-discrete* numerical scheme for the preceding problem with cell centers indexed as i and with cell edge fluxes indexed as $i \pm \frac{1}{2}$ by differentiating eqn. (3.142) with respect to time to obtain

$$\frac{d\overline{\rho}_i}{dt} + \frac{1}{\Delta x_i} \left[f_{i+\frac{1}{2}} - f_{i-\frac{1}{2}} \right] = 0. \tag{3.143}$$

It should be noted that eqn. (3.143) is *exact* and no approximations have been made during its derivation. For a practical scheme, the values for the edge fluxes, $f_{i\pm\frac{1}{2}}$, can be reconstructed by interpolation or extrapolation of the cell averages. This is discussed in Chapter 6.

---

[11] Recall that the divergence theorem is defined as $\iiint_v \nabla \cdot f \, dv = \oiint_S f \cdot dS$, where S represents the entire surface of the volume v.

3.10.3 Application to a General Conservation Law

We can also consider a general *conservation law* problem represented by the following PDE with appropriate boundary conditions:

$$\frac{\partial \mathbf{u}}{\partial t} + \nabla \cdot \mathbf{f}(\mathbf{u}) = \mathbf{0}. \tag{3.144}$$

Here, $\mathbf{u}$ represents a vector of states and $\mathbf{f}$ represents the corresponding *flux* vector. Again, we can subdivide the spatial domain into finite volumes or cells. For a particular cell, i, we take the volume integral over the total volume of the cell, v_i, which gives

$$\oiiint_{v_i} \left(\frac{\partial \mathbf{u}}{\partial t} + \nabla \cdot \mathbf{f}(\mathbf{u}) \right) dv = \mathbf{0}. \tag{3.145}$$

If we integrate the first term and divide by v_i to get the volume average, and then apply the divergence theorem to the second term, this yields

$$v_i \frac{d\bar{\mathbf{u}}_i}{dt} + \oiint_{S_i} \mathbf{f}(\mathbf{u}) \, dS = \mathbf{0}, \tag{3.146}$$

where S_i represents the total surface area of the cell. So, finally, we are able to present the general result equivalent to eqn. (3.143), that is,

$$\frac{d\bar{\mathbf{u}}_i}{dt} + \frac{1}{v_i} \oiint_{S_i} \mathbf{f}(\mathbf{u}) \, dS = \mathbf{0}. \tag{3.147}$$

Again, values for the edge fluxes can be reconstructed by interpolation or extrapolation of the cell averages. The actual numerical scheme will depend upon problem geometry and mesh construction. Various reconstruction methods, such as the *monotonic upwind scheme for conservation laws* (MUSCL) or *weighted essentially nonoscillatory* (WENO) scheme, are often used to achieve high-resolution solutions where shocks or discontinuities are present. These schemes are discussed in Chapter 6.

Finite volume schemes are inherently conservative as cell averages change through the edge fluxes. In other words, *one cell's loss is another cell's gain*!

3.11 INTERPRETATION OF RESULTS

It is important when performing a simulation that the analyst has confidence in the numerical results. This can be achieved by *verification* and *validation*.

3.11.1 Verification

Definition: The process of determining that a model implementation accurately represents the developer's conceptual description of the model and the solution of the model [AIA-98].

Verification has been aptly described as *"solving the equations right"* [Roa-98].

3.11.2 Validation

Definition: The process of determining the degree to which a model is an accurate representation of the real world from the perspective of the intended uses of the model [AIA-98].

Validation has been aptly described as *"solving the right equations"* [Roa-98].

3.11.3 Truncation Error

Truncation errors occur in general because we approximate a solution by a function with a finite number of terms. We discussed this in Section 3.8.2 for derivatives that are approximated by a truncated Taylor series.

3.A APPENDIX: DERIVATIVE MATRIX COEFFICIENTS

The derivative matrices given here for first, second, third, and fourth derivatives of various order have all been derived from the R code provided in Listing 3.1.

3.A.1 First Derivative Schemes

Three-point scheme, second-order accurate

$$\frac{\partial \mathbf{y}}{\partial x} = \frac{1}{2\Delta x} \begin{bmatrix} -3 & 4 & -1 \\ -1 & 0 & 1 \\ 1 & -4 & 3 \end{bmatrix} \times \begin{bmatrix} y_1 \\ y_2 \\ y_3 \end{bmatrix} + \begin{bmatrix} -\frac{\Delta x^2}{3} y^{(3)} + \mathcal{O}\left(\Delta x^3\right) \\ +\frac{\Delta x^2}{6} y^{(3)} + \mathcal{O}\left(\Delta x^3\right) \\ -\frac{\Delta x^2}{3} y^{(3)} + \mathcal{O}\left(\Delta x^3\right) \end{bmatrix} \quad (3.148)$$

Four-point scheme, third-order accurate

$$\frac{\partial \mathbf{y}}{\partial x} = \frac{1}{6\Delta x} \begin{bmatrix} -11 & 18 & -9 & 2 \\ -2 & -3 & 6 & -1 \\ 1 & -6 & 3 & 2 \\ -2 & 9 & -18 & 11 \end{bmatrix} \times \begin{bmatrix} y_1 \\ y_2 \\ y_3 \\ y_4 \end{bmatrix} + \begin{bmatrix} +\frac{\Delta x^3}{4} y^{(4)} + \mathcal{O}\left(\Delta x^4\right) \\ -\frac{\Delta x^3}{12} y^{(4)} + \mathcal{O}\left(\Delta x^4\right) \\ +\frac{\Delta x^3}{12} y^{(4)} + \mathcal{O}\left(\Delta x^4\right) \\ -\frac{\Delta x^3}{4} y^{(4)} + \mathcal{O}\left(\Delta x^4\right) \end{bmatrix} \quad (3.149)$$

Five-point scheme, fourth-order accurate

$$\frac{\partial \mathbf{y}}{\partial x} = \frac{1}{12\Delta x} \begin{bmatrix} -25 & 48 & -36 & 16 & -3 \\ -3 & -10 & 18 & -6 & 1 \\ 1 & -8 & 0 & 8 & -1 \\ -1 & 6 & -18 & 10 & 3 \\ 3 & -16 & 36 & -48 & 25 \end{bmatrix} \times \begin{bmatrix} y_1 \\ y_2 \\ y_3 \\ y_4 \\ y_5 \end{bmatrix} + \begin{bmatrix} -\frac{\Delta x^4}{5} y^{(5)} + \mathcal{O}\left(\Delta x^5\right) \\ +\frac{\Delta x^4}{20} y^{(5)} + \mathcal{O}\left(\Delta x^5\right) \\ -\frac{\Delta x^4}{30} y^{(5)} + \mathcal{O}\left(\Delta x^5\right) \\ +\frac{\Delta x^4}{20} y^{(5)} + \mathcal{O}\left(\Delta x^5\right) \\ -\frac{\Delta x^4}{5} y^{(5)} + \mathcal{O}\left(\Delta x^5\right) \end{bmatrix} \quad (3.150)$$

Seven-point scheme, sixth-order accurate

$$\frac{\partial \mathbf{y}}{\partial x} = \frac{1}{60\Delta x} \begin{bmatrix} -147 & 360 & -450 & 400 & -225 & 72 & -10 \\ -10 & -77 & 150 & -100 & 50 & -15 & 2 \\ 2 & -24 & -35 & 80 & -30 & 8 & -1 \\ -1 & 9 & -45 & 0 & 45 & -9 & 1 \\ 1 & -8 & 30 & -80 & 35 & 24 & -2 \\ -2 & 15 & -50 & 100 & -150 & 77 & 10 \\ 10 & -72 & 225 & -400 & -450 & -360 & 147 \end{bmatrix} \times \begin{bmatrix} y_1 \\ y_2 \\ y_3 \\ y_4 \\ y_5 \\ y_6 \\ y_7 \end{bmatrix}$$

$$+ \begin{bmatrix} -\frac{\Delta x^6}{7} y^{(7)} + \mathcal{O}(\Delta x^7) \\ +\frac{\Delta x^6}{42} y^{(7)} + \mathcal{O}(\Delta x^7) \\ -\frac{\Delta x^6}{105} y^{(7)} + \mathcal{O}(\Delta x^7) \\ +\frac{\Delta x^6}{140} y^{(7)} + \mathcal{O}(\Delta x^7) \\ -\frac{\Delta x^6}{105} y^{(7)} + \mathcal{O}(\Delta x^7) \\ +\frac{\Delta x^6}{42} y^{(7)} + \mathcal{O}(\Delta x^7) \\ -\frac{\Delta x^6}{7} y^{(7)} + \mathcal{O}(\Delta x^7) \end{bmatrix}, \quad (3.151)$$

Nine-point scheme, eighth-order accurate

$$\frac{\partial \mathbf{y}}{\partial x} = \frac{1}{840\Delta x} \begin{bmatrix} -2283 & 6720 & -11760 & 15680 & -14700 & 9408 & -3920 & 960 & -105 \\ -105 & -1338 & 2940 & -2940 & 2450 & -1470 & 588 & -140 & 15 \\ 15 & -240 & -798 & 1680 & -1050 & 560 & -210 & 48 & -5 \\ -5 & 60 & -420 & -378 & 1050 & -420 & 140 & -30 & 3 \\ 3 & -32 & 168 & -672 & 0 & 672 & -168 & 32 & -3 \\ -3 & 30 & -140 & 420 & -1050 & 378 & 420 & -60 & 5 \\ 5 & -48 & 210 & -560 & 1050 & -1680 & 798 & 240 & -15 \\ -15 & 140 & -588 & 1470 & -2450 & 2940 & -2940 & 1338 & 105 \\ 105 & -960 & 3920 & -9408 & 14700 & -15680 & 11760 & -6720 & 2283 \end{bmatrix} \times \begin{bmatrix} y_1 \\ y_2 \\ y_3 \\ y_4 \\ y_5 \\ y_6 \\ y_7 \\ y_8 \\ y_9 \end{bmatrix}$$

$$+ \begin{bmatrix} -\frac{\Delta x^8}{9} y^{(9)} + \mathcal{O}(\Delta x^9) \\ +\frac{\Delta x^8}{72} y^{(9)} + \mathcal{O}(\Delta x^9) \\ -\frac{\Delta x^8}{252} y^{(9)} + \mathcal{O}(\Delta x^9) \\ +\frac{\Delta x^8}{504} y^{(9)} + \mathcal{O}(\Delta x^9) \\ -\frac{\Delta x^8}{630} y^{(9)} + \mathcal{O}(\Delta x^9) \\ +\frac{\Delta x^8}{504} y^{(9)} + \mathcal{O}(\Delta x^9) \\ -\frac{\Delta x^8}{252} y^{(9)} + \mathcal{O}(\Delta x^9) \\ +\frac{\Delta x^8}{72} y^{(9)} + \mathcal{O}(\Delta x^9) \\ -\frac{\Delta x^8}{9} y^{(9)} + \mathcal{O}(\Delta x^9) \end{bmatrix},$$

$$(3.152)$$

3.A.2 Second Derivative Schemes

Four-point scheme, second-order accurate

$$\frac{\partial^2 \mathbf{y}}{dx^2} = \frac{1}{\Delta x^2} \begin{bmatrix} 2 & -5 & 4 & -1 \\ 1 & -2 & 1 & 0 \\ 0 & 1 & -2 & 1 \\ -1 & 4 & -5 & 2 \end{bmatrix} \times \begin{bmatrix} y_1 \\ y_2 \\ y_3 \\ y_4 \end{bmatrix} + \begin{bmatrix} -\frac{11}{12}\Delta x^2 y^{(4)} + \mathcal{O}(\Delta x^3) \\ +\frac{1}{12}\Delta x^2 y^{(4)} + \mathcal{O}(\Delta x^3) \\ +\frac{1}{12}\Delta x^2 y^{(4)} + \mathcal{O}(\Delta x^3) \\ -\frac{11}{12}\Delta x^2 y^{(4)} + \mathcal{O}(\Delta x^3) \end{bmatrix}. \quad (3.153)$$

Six-point scheme, fourth-order accurate

$$\frac{\partial^2 \mathbf{y}}{\partial x^2} = \frac{1}{12\Delta x^2} \begin{bmatrix} 45 & -154 & 214 & -156 & 61 & -10 \\ 10 & -15 & -4 & 14 & -6 & 1 \\ -1 & 16 & -30 & 16 & -1 & 0 \\ 0 & -1 & 16 & -30 & 16 & -1 \\ 1 & -6 & 14 & -4 & -15 & 10 \\ -10 & 61 & -156 & 214 & -154 & 45 \end{bmatrix} \times \begin{bmatrix} y_1 \\ y_2 \\ y_3 \\ y_4 \\ y_5 \\ y_6 \end{bmatrix} + \begin{bmatrix} -\frac{137}{180}\Delta x^4 y^{(6)} + \mathcal{O}(\Delta x^5) \\ +\frac{13}{180}\Delta x^4 y^{(6)} + \mathcal{O}(\Delta x^5) \\ -\frac{2}{180}\Delta x^4 y^{(6)} + \mathcal{O}(\Delta x^5) \\ -\frac{2}{180}\Delta x^4 y^{(6)} + \mathcal{O}(\Delta x^5) \\ +\frac{13}{180}\Delta x^4 y^{(6)} + \mathcal{O}(\Delta x^5) \\ -\frac{137}{180}\Delta x^4 y^{(6)} + \mathcal{O}(\Delta x^5) \end{bmatrix}$$

(3.154)

Eight-point scheme, sixth-order accurate

$$\frac{\partial^2 \mathbf{y}}{\partial x^2} = \frac{1}{180\Delta x^2} \begin{bmatrix} 938 & -4014 & 7911 & -9490 & 7380 & -3618 & 1019 & -126 \\ 126 & -70 & -486 & 855 & -670 & 324 & -90 & 11 \\ -11 & 214 & -378 & 130 & 85 & -54 & 16 & -2 \\ 2 & -27 & 270 & -490 & 270 & -27 & 2 & 0 \\ 0 & 2 & -27 & 270 & -490 & 270 & -27 & 2 \\ -2 & 16 & -54 & 85 & 130 & -378 & 214 & -11 \\ 11 & -90 & 324 & -670 & 855 & -486 & -70 & 126 \\ -126 & 1019 & -3618 & 7380 & -9490 & 7911 & -4014 & 938 \end{bmatrix} \times \begin{bmatrix} y_1 \\ y_2 \\ y_3 \\ y_4 \\ y_5 \\ y_6 \\ y_7 \\ y_8 \end{bmatrix}$$

$$+ \begin{bmatrix} -\frac{363}{560}\Delta x^6 y^{(8)} + \mathcal{O}(\Delta x^7) \\ +\frac{29}{560}\Delta x^6 y^{(8)} + \mathcal{O}(\Delta x^7) \\ -\frac{47}{5040}\Delta x^6 y^{(8)} + \mathcal{O}(\Delta x^7) \\ +\frac{1}{560}\Delta x^6 y^{(8)} + \mathcal{O}(\Delta x^7) \\ +\frac{1}{560}\Delta x^6 y^{(8)} + \mathcal{O}(\Delta x^7) \\ -\frac{47}{5040}\Delta x^6 y^{(8)} + \mathcal{O}(\Delta x^7) \\ +\frac{29}{560}\Delta x^6 y^{(8)} + \mathcal{O}(\Delta x^7) \\ -\frac{363}{560}\Delta x^6 y^{(8)} + \mathcal{O}(\Delta x^7) \end{bmatrix}.$$

(3.155)

Ten-point scheme, eighth-order accurate

$$\frac{\partial^2 \mathbf{y}}{\partial x^2} = \frac{1}{5040\Delta x^2} \begin{bmatrix} 32575 & -165924 & 422568 & -704368 & 818874 & -667800 & 375704 & -139248 & 30663 & -3044 \\ 3044 & 2135 & -28944 & 57288 & -65128 & 51786 & -28560 & 10424 & -2268 & 223 \\ -223 & 5274 & -7900 & -2184 & 10458 & -8932 & 4956 & -1800 & 389 & -38 \\ 38 & -603 & 6984 & -12460 & 5796 & 882 & -952 & 396 & -90 & 9 \\ -9 & 128 & -1008 & 8064 & -14350 & 8064 & -1008 & 128 & -9 & 0 \\ 0 & -9 & 128 & -1008 & 8064 & -14350 & 8064 & -1008 & 128 & -9 \\ 9 & -90 & 396 & -952 & 882 & 5796 & -12460 & 6984 & -603 & 38 \\ -38 & 389 & -1800 & 4956 & -8932 & 10458 & -2184 & -7900 & 5274 & -223 \\ 223 & -2268 & 10424 & -28560 & 51786 & -65128 & 57288 & -28944 & 2135 & 3044 \\ -3044 & 30663 & -139248 & 375704 & -667800 & 818874 & -704368 & 422568 & -165924 & 32575 \end{bmatrix} \times \begin{bmatrix} y_1 \\ y_2 \\ y_3 \\ y_4 \\ y_5 \\ y_6 \\ y_7 \\ y_8 \\ y_9 \\ y_{10} \end{bmatrix}$$

$$+ \begin{bmatrix} -\frac{7129}{12600}\Delta x^8 y^{(10)} + \mathcal{O}\left(\Delta x^9\right) \\ +\frac{481}{12600}\Delta x^8 y^{(10)} + \mathcal{O}\left(\Delta x^9\right) \\ -\frac{17}{2800}\Delta x^8 y^{(10)} + \mathcal{O}\left(\Delta x^9\right) \\ +\frac{37}{25200}\Delta x^8 y^{(10)} + \mathcal{O}\left(\Delta x^9\right) \\ -\frac{1}{3150}\Delta x^8 y^{(10)} + \mathcal{O}\left(\Delta x^9\right) \\ -\frac{1}{3150}\Delta x^8 y^{(10)} + \mathcal{O}\left(\Delta x^9\right) \\ +\frac{37}{25200}\Delta x^8 y^{(10)} + \mathcal{O}\left(\Delta x^9\right) \\ -\frac{17}{2800}\Delta x^8 y^{(10)} + \mathcal{O}\left(\Delta x^9\right) \\ +\frac{481}{12600}\Delta x^8 y^{(10)} + \mathcal{O}\left(\Delta x^9\right) \\ -\frac{7129}{12600}\Delta x^8 y^{(10)} + \mathcal{O}\left(\Delta x^9\right) \end{bmatrix}$$

(3.156)

3.A.3 Third Derivative Schemes

Seven-point scheme, fourth-order accurate

$$\frac{\partial^3 y}{dx^2} = \frac{1}{8\Delta x^3} \begin{bmatrix} -49 & 232 & -461 & 496 & -307 & 104 & -15 \\ -15 & 56 & -83 & 64 & -29 & 8 & -1 \\ -1 & -8 & 35 & -48 & 29 & -8 & 1 \\ 1 & -8 & 13 & 0 & -13 & 8 & -1 \\ -1 & 8 & -29 & 48 & -35 & 8 & 1 \\ 1 & -8 & 29 & -64 & 83 & -56 & 15 \\ 15 & -104 & 307 & -496 & 461 & -232 & 49 \end{bmatrix} \times \begin{bmatrix} y_1 \\ y_2 \\ y_3 \\ y_4 \\ y_5 \\ y_6 \\ y_7 \end{bmatrix}$$

$$+ \begin{bmatrix} -\frac{29}{15} \Delta x^4 y^{(7)} + \mathcal{O}\left(\Delta x^5\right) \\ -\frac{7}{120} \Delta x^4 y^{(7)} + \mathcal{O}\left(\Delta x^5\right) \\ +\frac{1}{15} \Delta x^4 y^{(7)} + \mathcal{O}\left(\Delta x^5\right) \\ -\frac{7}{120} \Delta x^4 y^{(7)} + \mathcal{O}\left(\Delta x^5\right) \\ +\frac{1}{15} \Delta x^4 y^{(7)} + \mathcal{O}\left(\Delta x^5\right) \\ -\frac{7}{120} \Delta x^4 y^{(7)} + \mathcal{O}\left(\Delta x^5\right) \\ -\frac{29}{15} \Delta x^4 y^{(7)} + \mathcal{O}\left(\Delta x^5\right) \end{bmatrix}.$$

(3.157)

Nine-point scheme, sixth-order accurate

$$\frac{\partial^3 y}{dx^3} = \frac{1}{240\Delta x^3} \begin{bmatrix} -2403 & 13960 & -36706 & 57384 & -58280 & 39128 & -16830 & 4216 & -469 \\ -469 & 1818 & -2924 & 2690 & -1710 & 814 & -268 & 54 & -5 \\ -5 & -424 & 1638 & -2504 & 2060 & -1080 & 394 & -88 & 9 \\ 9 & -86 & -100 & 882 & -1370 & 926 & -324 & 70 & -7 \\ -7 & 72 & -338 & 488 & 0 & -488 & 338 & -72 & 7 \\ 7 & -70 & 324 & -926 & 1370 & -882 & 100 & 86 & -9 \\ -9 & 88 & -394 & 1080 & -2060 & 2504 & -1638 & 424 & 5 \\ 5 & -54 & 268 & -814 & 1710 & -2690 & 2924 & -1818 & 469 \\ 490 & -4216 & 16830 & -39128 & 58280 & -57384 & 36706 & -13960 & 2403 \end{bmatrix} \times \begin{bmatrix} y_1 \\ y_2 \\ y_3 \\ y_4 \\ y_5 \\ y_6 \\ y_7 \\ y_8 \\ y_9 \end{bmatrix}$$

$$+ \begin{bmatrix} -\frac{29531}{15120} \Delta x^6 y^{(9)} + \mathcal{O}\left(\Delta x^7\right) \\ +\frac{1}{945} \Delta x^6 y^{(9)} + \mathcal{O}\left(\Delta x^7\right) \\ +\frac{331}{15120} \Delta x^6 y^{(9)} + \mathcal{O}\left(\Delta x^7\right) \\ -\frac{59}{3780} \Delta x^6 y^{(9)} + \mathcal{O}\left(\Delta x^7\right) \\ +\frac{41}{30240} \Delta x^6 y^{(9)} + \mathcal{O}\left(\Delta x^7\right) \\ -\frac{59}{3780} \Delta x^6 y^{(9)} + \mathcal{O}\left(\Delta x^7\right) \\ +\frac{331}{15120} \Delta x^6 y^{(9)} + \mathcal{O}\left(\Delta x^7\right) \\ +\frac{1}{945} \Delta x^6 y^{(9)} + \mathcal{O}\left(\Delta x^7\right) \\ -\frac{29531}{15120} \Delta x^6 y^{(9)} + \mathcal{O}\left(\Delta x^7\right) \end{bmatrix}.$$

(3.158)

3.A.4 Fourth Derivative Schemes

Ten-point scheme, sixth-order accurate

$$\frac{\partial^4 \mathbf{y}}{dx^4} = \frac{1}{240\Delta x^4} \begin{bmatrix} 4275 & -30668 & 99604 & -192624 & 244498 & -210920 & 123348 & -47024 & 10579 & -1068 \\ 1068 & -6405 & 17392 & -28556 & 31656 & -24638 & 13360 & -4812 & 1036 & -101 \\ 101 & 58 & -1860 & 5272 & -7346 & 6204 & -3428 & 1240 & -267 & 26 \\ -26 & 361 & -1112 & 1260 & -188 & -794 & 744 & -308 & 70 & -7 \\ 7 & -96 & 676 & -1952 & 2730 & -1952 & 676 & -96 & 7 & 0 \\ 0 & 7 & -96 & 676 & -1952 & 2730 & -1952 & 676 & -96 & 7 \\ -7 & 70 & -308 & 744 & -794 & -188 & 1260 & -1112 & 361 & -26 \\ 26 & -267 & 1240 & -3428 & 6204 & -7346 & 5272 & -1860 & 58 & 101 \\ -101 & 1036 & -4812 & 13360 & -24638 & 31656 & -28556 & 17392 & -6405 & 1068 \\ -1068 & 10579 & -47024 & 123348 & -210920 & 244498 & -192624 & 99604 & -30668 & 4275 \end{bmatrix} \times \begin{bmatrix} y_1 \\ y_2 \\ y_3 \\ y_4 \\ y_5 \\ y_6 \\ y_7 \\ y_8 \\ y_9 \\ y_{10} \end{bmatrix}$$

$$+ \begin{bmatrix} -\frac{4523}{945}\Delta x^6 y^{(10)} + \mathcal{O}\left(\Delta x^7\right) \\ -\frac{1271}{3780}\Delta x^6 y^{(10)} + \mathcal{O}\left(\Delta x^7\right) \\ +\frac{1279}{15120}\Delta x^6 y^{(10)} + \mathcal{O}\left(\Delta x^7\right) \\ -\frac{359}{15120}\Delta x^6 y^{(10)} + \mathcal{O}\left(\Delta x^7\right) \\ +\frac{41}{7560}\Delta x^6 y^{(10)} + \mathcal{O}\left(\Delta x^7\right) \\ +\frac{41}{7560}\Delta x^6 y^{(10)} + \mathcal{O}\left(\Delta x^7\right) \\ -\frac{359}{15120}\Delta x^6 y^{(10)} + \mathcal{O}\left(\Delta x^7\right) \\ +\frac{1279}{15120}\Delta x^6 y^{(10)} + \mathcal{O}\left(\Delta x^7\right) \\ -\frac{1271}{3780}\Delta x^6 y^{(10)} + \mathcal{O}\left(\Delta x^7\right) \\ -\frac{4523}{945}\Delta x^6 y^{(10)} + \mathcal{O}\left(\Delta x^7\right) \end{bmatrix} \qquad (3.159)$$

Table 3.4. List of spatial derivative matrix routines and associated file names

Function	Derivative	Order	Number of points	File Name
diffMatrix1st3pt()	1st	2nd	3	deriv1st3pt.R
diffMatrix1st5pt()	1st	4th	5	deriv1st5pt.R
diffMatrix1st7pt()	1st	6th	7	deriv1st7pt.R
diffMatrix1st9pt()	1st	8th	9	deriv1st9pt.R
diffMatrix2nd4pt()	2nd	2nd	4	deriv2nd4pt.R
diffMatrix2nd6pt()	2nd	4th	6	deriv2nd6pt.R
diffMatrix2nd8pt()	2nd	6th	8	deriv2nd8pt.R
diffMatrix2nd10pt()	2nd	8th	10	deriv2nd10pt.R
diffMatrix3rd5pt()	3rd	2nd	5	deriv3rd5pt.R
diffMatrix3rd9pt()	3rd	6th	9	deriv3rd9pt.R

3.B APPENDIX: DERIVATIVE MATRIX LIBRARY

The derivative matrix library is available with the downloads for this book. It includes the functions in Table 3.5 that calculate 1D or 2D derivative matrices.

In addition, the functions in Table 3.5 are provided to impose Neumann BCs at appropriate boundaries.

Table 3.5. List of functions provided to impose Neumann boundary conditions to numerical spatial derivatives and the associated file names

Function	Derivative	Order	Number of points	File name
diffBC2nd4pt()	2nd	2nd	4	deriv2nd4pt.R
diffBC2nd6pt()	2nd	4th	6	deriv2nd6pt.R
diffBC2nd8pt()	2nd	6th	8	deriv2nd8pt.R
diffBC2nd10pt()	2nd	8th	10	deriv2nd10pt.R

The preceding functions implement the general procedures/algorithms outlined in Section 3.8.2. An example showing how to use the these functions is given in Section 3.B.1.

R code for representative functions diffMatrix1st3pt(), diffMatrix2nd4pt() that create first and second derivative matrices are given in Listings 3.31 and 3.32, respectively. Function arguments are defined in the code along with other implementation details. The code for other functions follows a similar pattern but has been omitted to save space.

```
# File: deriv1st3pt.R
diffMatrix1st3pt<-function(zl,zu,N) {
  # Function uses a 3-point stencil to calculates
  # the 2nd order, 1D/2D, first derivative matrix
  #
  # Input Variables
  # zl = start of z spatial domain
  # zu = end of z spatial domain
  # N  = number of z spatial domain grid-points
  #
  # Grid spacing# Grid spacing
```

```r
    dz<-(zu-zl)/(N-1);
    #
    # Define dirivative matrix elements
    dz_11 <- c(-3, 4, -1)
    dz_MM <- c(-1, 0, 1)
    dz_N1 <- c( 1, -4, 3)
    #
    # Preallocate derivative matrix
    deriv2D <- matrix(data<-0,nrow<-N, ncol<-N)
    #
    mn <- dim(deriv2D)
    cat(sprintf("%d %d\n",mn[1],mn[2]))
    # Assign derivative matrix elements
    # Row 1
    deriv2D[1,1:3] <- dz_11
    # Rows 2:Nz-1
    for(i in 2:(N-1)){
        deriv2D[i,(i-1):(i+1)] <- dz_MM
    }
    # Row Nz
    deriv2D[N, (N-2):N] <- dz_N1
    #
    return(deriv2D/(2*dz))
}
```

Listing 3.31. File: `deriv1st3pt.R`—Code that generates a three-point, second-order first derivative matrix

```r
# File: deriv2nd4pt.R
diffMatrix2nd4pt<-function(zl,zu,N,bc=c(0,0)) {
    # Function uses a 4-point stencil to calculates
    # the 2nd order, 2D, 2nd derivative matrix
    #
    # Input Variables
    # zl = start of z spatial domain
    # zu = end of z spatial domain
    # N  = number of z spatial domain grid-points
    # bc = BC flag for z=zl; bc[1]=0/1, Dirichlet/Neumann
    # bc = BC flag for z=zu; bc[2]=0/1, Dirichlet/Neumann
    # Grid spacing
    dz<-(zu-zl)/(N-1)
    # Define derivative matrix elements
    if(bc[1]==0)
        dzz_11 = c( 2, -5, 4, -1)
    else{
        dzz_11 = c(-7, 8, -1, 0)/2 # used for Neumann BC
```

```r
    #Note: Additional term -3*dudz/dz to be added to
    #      dzz_11 by call to diffBC2nd4pt()
  }
  dzz_MM = c( 1, -2, 1)
  if(bc[2]==0)
    dzz_N1 = c(-1, 4, -5, 2)
  else{
    dzz_N1 = c(0, -1, +8, -7)/2 # used for Neumann BC
    #Note: Additional term +3*dudz/dz to be added to
    #      dzz_11 by call to diffBC2nd4pt()
  }
  # Preallocate derivative matrix
  deriv2D <- matrix(data<-0,nrow<-N, ncol<-N)
  #
  mn <- dim(deriv2D)
  cat(sprintf("%d %d\n",mn[1],mn[2]))
  # Assign derivative matrix elements
  # Row 1
  deriv2D[1,1:4] <- dzz_11
  # Rows 2:Nz-1
  for(i in 2:(N-1)){
    deriv2D[i,(i-1):(i+1)] <- dzz_MM
  }
  # Row Nz-1 and Nz
  deriv2D[N   , (N-3):N] <- dzz_N1
  #
  return(deriv2D/(dz^2))
}
diffBC2nd4pt<-function(xy,zl,zu,u_zz,bcL,bcU) {
  # Function apply Neumann BCs to the
  # 2D, 4pt, 2nd order, 2nd derivative matrix
  # This function applys "Neumann" BCs and is not called if all
  # BCs are of type "Dirichlet", i.e. BC arrays are both zero.
  # Also not called if the Neuman BCs are all zero.
  # Input Variables
  # xy   = Spatial derivative direction: 0=x, 1=y
  # zl   = start of spatial domain, z
  # zu   = end of spatial domain, z
  # u_zz = NxN derivative matrix
  # N    = number of spatial domain grid points
  # bcL  = BC array for z=zl; Neumann
  # bcR  = BC array for z=zu; Neumann
  #
  # Grid spacing
  m <- dim(u_zz)
  Nx <- m[1]; Ny <- m[2]
```

```
#    print(dim(u_zz))
#    print(length(bcL))
#    scan(quiet=TRUE)
  # Apply BCs
  if(xy ==0){ # Neumann BCs: dudx on y boundaries - at 1 and Ny
    dx<-(zu-zl)/(Nx-1)#
    u_zz[1,]  <- u_zz[1,]  - bcL*3/(dx)
    u_zz[Nx,] <- u_zz[Nx,] + bcU*3/(dx)
  }else{     # Neumann BCs: dudy on x boundaries - at 1 and Nx
    dy<-(zu-zl)/(Ny-1)#
    u_zz[,1]  <- u_zz[,1]  - bcL*3/(dy)
    u_zz[,Ny] <- u_zz[,Ny] + bcU*3/(dy)
  }
return(as.matrix(u_zz)) # Return updated derivative matrix
}
```

Listing 3.32. File: `deriv2nd4pt.R`—Code that generates a four-point, second-order second derivative matrix. This file also contains the function `diffBC2nd4pt` that can be used to impose Neumann boundary conditions on the derivative matrix.

3.B.1 Example

Here we provide examples of calculating 1D first and second derivatives. We will use the example 1D hyperbolic–parabolic PDE problem described by eqn. (3.61), that is,

$$\frac{\partial u}{\partial t} = -\frac{\partial u}{\partial x} + \frac{\partial^2 u}{\partial x^2}, \quad u = u(x,t), \ t \geq t_0, \ x \in (x_l, x_u), \tag{3.160}$$

which has a Neumann BC, $u_x = du_1/dx$, at $x = x_l$ and a Dirichlet BC, $u = u_b$, at $x = x_u$.

The discretization process transforms this equation into eqn. (3.68) from Section 3.8.2, which we repeat in the following for convenience:

$$\frac{d\underline{u}}{dt} = -\frac{1}{2\Delta x}\begin{bmatrix} 0 & 0 & 0 & & & & \\ -1 & 0 & 1 & & & & \\ & -1 & 0 & 1 & & & \\ & & \ddots & \ddots & \ddots & & \\ & & & 1 & 0 & 1 & \\ & & & & -1 & 0 & 1 \\ & & & & 1 & -4 & 3 \end{bmatrix}\underline{u} - \begin{bmatrix} \dfrac{du_1}{dx} \\ 0 \\ 0 \\ \vdots \\ 0 \\ 0 \\ 0 \end{bmatrix}$$

$$+\frac{1}{\Delta x^2}\begin{bmatrix} \dfrac{7}{2} & \dfrac{8}{2} & -\dfrac{1}{2} & 0 & & & \\ 1 & -2 & 1 & & & & \\ & 1 & -2 & 1 & & & \\ & & \ddots & \ddots & \ddots & & \\ & & & 1 & -2 & 1 & \\ & & & & 1 & -2 & 1 \\ & & & & -1 & 4 & -5 & 2 \end{bmatrix}\underline{u} + \begin{bmatrix} -\left(\dfrac{3}{\Delta x}\right)\dfrac{du_1}{dx} \\ 0 \\ 0 \\ \vdots \\ 0 \\ 0 \\ 0 \end{bmatrix}. \tag{3.161}$$

We proceed to code the problem of eqn. (3.161) as follows:

1. In the *main program*, load source files for second-order first and second derivative matrix functions,

   ```
   source("deriv1st3pt.R")
   source("deriv2nd4pt.R")
   ```

2. Still in the *main program*, create second-order first and second derivative matrices for u,

   ```
   diffMat1X <- diffMatrix1st3pt(xl,xu,N)
   diffMat2X <- diffMatrix2nd4pt(xl,xu,N,bc=c(1,0))
   ```

   where bc=c(1,0) defines (xl) as having a Neumann BC and (xu) as having a Dirichlet BC.
   Matrices diffMat1X and diffMat2X are global variables, and as such, they are created once but used many times in the derivative function.

3. In the *derivative function* (called by the *numerical integrator* at each time step), replace the current value of u_N with the Dirichlet BC value u_b, then calculate u_x and u_{xx}:

   ```
   u[N] <- ub
   ux <- diffMat1R %*% u
   uxx <- diffMat2R %*% u
   ```

   The terms %*% denote matrix/vector multiplication.

4. Still in the derivative function, replace $\frac{du}{dx}$ at $x = xl$ with the Neumann BC value u_{xb} and update u_{xx} with the same value:

   ```
   ux[1] <- uxb
   uxx <- diffBC2nd4pt(xy=0,xl,xu,uxx,uxb,0)
   ```

   The argument xy=0 denotes that the spatial derivative is in the *x* direction. For a 2D problem, xy=1 would denote that the spatial derivative is in the *y* direction. The last two arguments of the function diffBC2nd4pt() are used to update u_{xx} with Neumann BCs. The location of ubx denotes that u_{xx} is updated at $x = x_l$, as described earlier, and the location of 0 denotes no change to u_{xx} at $x = x_u$.

5. Finally, assemble the time derivative $\frac{du}{dt} = -\frac{du}{dx} + \frac{d^2u}{dx^2}$,

   ```
   ut <- -ux + uxx
   ```

See Section 3.8.3 for a number of 1D examples that illustrate this procedure.

The extension to 2D derivatives is straightforward. For example, if the preceding were a 2D problem, the procedure would generally be the same, except that derivatives in the *y* direction would need to be calculated; that is, we would change the code to

```
ux  <- diffMat1R %*% u
uy  <- t(diffMat1R %*% t(u))
uxx <- diffMat2R %*% u
uyy <- t(diffMat2R %*% t(u))
```

where u is now a 2 × 2 matrix and t denotes *matrix transpose*. A number of 2D examples are given in Sections 3.8.4 and 3.8.5.

REFERENCES

[Abl-11] Ablowitz, M. J. (2011), *Nonlinear Dispersive Waves: Asymptotic Analysis and Solitons*, Cambridge University Press.

[AIA-98] AIAA (1998), *Guide for the Verification and Validation of Computational Fluid Dynamics Simulations*, American Institute of Aeronautics and Astronautics.

[Bou-72] Boussinesq, J. (1872), Théorie des ondes et des remous qui se propagent le long d'un canal rectangulaire horizontal, en communiquant au liquide continu dans ce canal des vitesses sensiblement pareilles de la surface au fond, *Journal de Mathématiques Pures et Appliquées* **17**, 55–108.

[Che-05] Cheng, A. and D. T. Cheng (2005), Heritage and Early History of the Boundary Element Method, *Engineering Analysis with Boundary Elements* **29**, 268–302.

[Cra-47] Crank, J. and P. Nicholson (1947), A Practical Method for Evaluation of Solutions of Partial Differential Equations of the Heat Conduction Type, *Proceedings of the Cambridge Philosophical Society* **43**, 50–67.

[De-11] de Jager, E. M. (2011), On the Origin of the Korteweg-de Vries Equation, version 2, *arXiv e-print service*, available online at http://arxiv.org/pdf/math/0602661v2.pdf.

[Dau-06] Dauxois, T. and M. Peyrard (2006), *Physics of Solitons*, Cambridge University Press.

[Dav-85] Davies-Jones, R. (1985), Comments on "A kinematic analysis of frontogenesis associated with a nondivergent vortex," *Journal of the Atmospheric Sciences* **42**-19, 2073–2075. Available online at http://www.flame.org/~cdoswell/publications/Davies-Jones_85.pdf.

[Deb-05] Debnath, L. (2005), *Nonlinear Partial Differential Equations for Scientists and Engineers*, Birkhäuser.

[Dos-84] Doswell, C. A. I. (1984), A Kinematic Analysis of Frontogenesis Associated with a Nondiverent Vortex, *Journal of the Atmospheric Sciences* **41**-7, 1242–1248.

[Dra-92] Drazin, P. G. and R. S. Johnson (1992), *Solitons: An Introduction*, Cambridge University Press.

[Fer-55] Fermi, E., J. Pasta and S. Ulam (1955), Study of Nonlinear Problems I, Los Alamos Report LA 1940. Reproduced in *Nonlinear Wave Motion*, A. C. Newell, editor, American Mathematical Society, 1974.

[For-98] Fornberg, B. (1998), Calculation of Weights in Finite Difference Formulas, *SIAM Review* **40**-3, 685–691.

[Gri-12] Griffiths, G. W. (2012), *Bäcklund Transformation Method*, unpublished. Copy available from the author.

[Gri-16a] Griffiths, G. W., Ł. Płociniczak and W. E. Schiesser (2016a), Analysis of Cornea Curvature Using Radial Basis Functions—Part I: Methodology, *Journal of Computers in Biology and Medicine*, in press.

[Gri-16b] Griffiths, G. W., Ł. Płociniczak, and W. E. Schiesser (2016b), Analysis of Cornea Curvature Using Radial Basis Functions—Part II: Fitting to Data-Set, *Journal of Computers in Biology and Medicine*, in press.

[Gri-09] Griffiths, G. W. and W. E. Schiesser (2009), Linear and Nonlinear Waves, *Scholarpedia* **4**-7, 4308. Available online at http://www.scholarpedia.org/article/Linear_and_nonlinear_waves

[Gri-11] Griffiths, G. W. and W. E. Schiesser (2011), *Traveling Wave Solutions of Partial Differential Equations: Numerical and Analytical Methods with Matlab and Maple*, Academic Press.

[Gus-98a] Gustafson, K. (1998), Domain Decomposition, Operator Trigonometry, Robin Condition, *Contempory Mathematics* **218**, 432–437.

[Gus-98b] Gustafson, K. and T. Abe (1998), (Victor) Gustave Robin: 1855–1897, *The Mathematical Intelligencer* **20**, 47–53.

[Gus-98c] Gustafson, K. and T. Abe (1998), The Third Boundary Condition—Was It Robin's?, *The Mathematical Intelligencer* **20**, 63–71.

[Ham-07] Hamdi, S., W. E. Schiesser and G. W. Griffiths (2007), Method of Lines, *Scholarpedia* **2**-7, 2859. Available online at http://www.scholarpedia.org/article/Method_of_Lines.

[Hel-24] von Helmholtz, H. (1924), *Helmholtz's Treatise on Physiological Optics*, vol. 1, translated from the third German edition by J. P. C. Southall, Optical Society of America. The treatise was originally published in Leipzig in 1867.

[Hir-88] Hirsch, C. (1988), *Numerical Computation of Internal and External Flows*, vol. 1, John Wiley.

[Ise-09] Iserles, A. (2009), *A First Course in the Numerical Analysis of Differential Equations*, Cambridge University Press.

[Joh-97] Johnson, R. S. (1997), *A Modern Introduction to the Theory of Water Waves*, Cambridge University Press.

[Kal-11] Kalise, D. (2011), A Study of a WENO-TVD Finite Volume Scheme for the Numerical Simulation of Atmospheric, Advective and Convective Phenomena, *arXiv preprint arXiv:1111.1712, 29*. Available online at http://arxiv.org/abs/1111.1712.

[Kor-95] Korteweg, D. J. and G. de Vries (1895), On the Change of Form of Long Waves Advancing in a Rectangular Canal, and on a New Type of Long Stationary Waves, *Philosophical Magazine*, series 5 **39**, 422–443.

[Lag-81] Lagrange, J. L. (1781), Mémoire sur la Théorie du Mouvement des Fluides [Memoir on the Theory of Fluid Motion], *Nouveaux Mémoires de l'Acadaémie royale des Sciences et Belles-Lettres de Berlin, Reprinted in Oeuvres de Lagrange*, vol. IV, Gauthier-Villars, 1869.

[Lan-98] Laney, B. L. (1998), *Computational Gas Dynamics*, Cambridge University Press.

[Lax-68] Lax, P. (1968), Integrals of Nonlinear Equations of Evolution and Solitary Waves, *Communications on Pure Applied Mathematics*, **5**, 611–613.

[Lev-02] LeVeque, R. J. (2002), *Finite Volume Methods for Hyperbolic Problems*, Cambridge University Press.

[Lig-78] Lightfoot, Sir James (1978), *Waves in Fluids*, Cambridge University Press.

[Ly-08] Ly, E. and D. Norrison (2008), Application of Methods of False Transients to Generate Smooth Grids around a Body in Motion, *Anziam Journal*, **50**, C505–C515.

[Plo-14] Płociniczak, Ł., G. W. Griffiths and W. E. Schiesser (2014), ODE/PDE Analysis of Corneal Curvature, *Journal of Computers in Biology and Medicine*, **53**, 30–41.

[Por-09] Porter, M. A., N. J. Zabusky, B. Hu and D. K. Campbell (2009), Fermi, Pasta, Ulam and the Birth of Experimental Mathematics, *American Scientist* **97**-3, 214.

[Ray-76] Rayleigh, Lord (1876), On Waves, *Philosophical Magazine* **1** 257–279.

[Ric-67] Richtmyer, R. D. and K. W. Morton (1967), *Difference Methods for Initial Value Problems*, 2nd. ed. John Wiley.

[Roa-98] Roache, P. J. (1998), *Verification and Validation in Computational Science and Engineering*, Hermosa.

[Sch-91] Schiesser, W. E. (1991), *The Numerical Method of Lines: Integration of Partial Differential Equations*, Academic Press.

[Sch-09] Schiesser, W. E. and G. W. Griffiths (2009), *A Compendium of Partial Differential Equation Models: Method of Lines Analysis with Matlab*, Cambridge University Press.

[Sco-44] Scott-Russell, J. (1844), Report on Waves, in *14th Meeting of the British Association for the Advancement of Science*, John Murray, pp. 311–391.

[Shu-98] Shu, C.-W. (1998), Essentially Non-oscillatory and Weighted Essential Non-oscillatory Schemes for Hyperbolic Conservation Laws, in *Advanced Numerical Approximation of Nonlinear Hyperbolic Equations*, Lecture Notes in Mathematics, vol. 1697, B. Cockburn, C. Johnson, C.-W. Shu and E. Tadmor, editors, Springer, pp. 325–432.

[Tan-97] Tannehill, J. C., D. A. Anderson and R. H. Pletcher (1997), *Computational Fluid Mechanics and Heat Transfer*, Taylor and Francis.

[War-74] Warming, R. F. and B. J. Hyett (1974), The Modified Equation Approach to the Stability and Accuracy Analysis of Finite-Difference Methods, *Journal of Computational Physics* **14**, 159–179.

[Wes-01] Wesseling, P. (2001), *Principles of Computational Fluid Dynamics*, Springer.
[Wik-13] Wikipedia (2013), Finite Volume Method, available online at http://en.wikipedia.org/wiki/Finite-volume_method.
[Zab-65] Zabusky, N. J. and M. Kruskal (1965), Interaction of Solitons in a Collisionless Plasma and the Recurrence of Initial States, *Physical Review Letters* **15**-6, 240–243.
[Zab-10] Zabusky, N. J. and M. A. Porter (2010), Solitons, *Scholarpedia* **5**-8, 2068. Available online at http://www.scholarpedia.org/article/Solitons.

4

PDE Stability Analysis

4.1 INTRODUCTION

In this section we will discuss some ideas relating to the stability of systems described by partial differential equations. By stable solution, we generally mean a solution that is bounded, although we try to include a more precise definition.

First we will present an approach to assessing the intrinsic stability characteristics of basic PDEs, that is, before any numerical solution methods are applied. We will then deal separately with fully discrete systems and semi-discrete systems. Fully discrete systems are those where the PDE is discretized in both space and time, whereas semi-discrete systems are those where the PDE is discretized only in space to yield a system of ODEs. These ODEs can then be integrated by use of numerical integrators from standard libraries available for download over the Internet. We will only discuss stability of discretized systems in the context of solutions on regularly spaced grids.

Consider the following initial-boundary value problem consisting of a general, linear PDE of evolution in m space variables with *periodic* boundary conditions,

$$u_t(x,t) = \mathcal{P}\left(\frac{\partial}{\partial x}\right) u(x,t) + q(x,t), \quad x \in \mathbb{R}^m, \quad t > 0,$$
$$\text{ICs:} \quad u(0,t) = u_0(x), \quad q(0,t) = q_0(x) \quad (4.1)$$
$$\text{BCs:} \quad u(a,t) = u_a(x), \quad u(b,t) = u_b.$$

Here $\mathcal{P}(\frac{\partial}{\partial x})$ is a time-invariant, linear spatial differential operator of which the following are typical examples:

$$\mathcal{P}\left(\frac{\partial}{\partial x}\right) = \alpha_i \frac{\partial}{\partial x_i}; \quad \text{(advection)}$$

$$\mathcal{P}\left(\frac{\partial}{\partial x}\right) = \beta_{ii} \frac{\partial^2}{\partial x_i^2}; \quad \text{(diffusion)}$$

$$\mathcal{P}\left(\frac{\partial}{\partial x}\right) = \alpha_i \frac{\partial}{\partial x_i} + \alpha_j \frac{\partial}{\partial x_j} + \beta_{ii} \frac{\partial^2}{\partial x_i^2} + \beta_{ij} \frac{\partial^2}{\partial x_i \partial x_j}; \quad \text{(etc.)}$$

where α and β are constant coefficients.

We will consider the stability of semi-discrete (spatial discretization only) and fully discrete (both spatial and temporal discretization) initial value, homogeneous PDEs, that is, where $q = 0$. However, "space discretization cannot be said to be unstable by itself. It is only when it is coupled to a time integration that we can decide upon stability" [Hir-88, p. 415].

Once we have identified a potential discretization scheme, we need to investigate its stability characteristics to determine if it will provide a satisfactory solution when applied to our particular problem. A satisfactory solution will be one that converges to the correct solution. In this context, the *Lax Equivalence* theorem applies, which means that if the system is both *consistent* and *stable*, it will converge to the correct solution (refer to the book's appendix).

Recall from Chapter 1 that a *consistent scheme* is one whereby

$$\text{as } \Delta t \to 0, \quad \text{discrete solution} \to \text{analytical solution}$$

4.2 THE WELL-POSED PDE PROBLEM

Our discussion here is related primarily to solving partial differential equations numerically. Now, before embarking on the implementation of a numerical procedure, the analyst would usually like to have some idea as to the expected behavior of the system being modeled, ideally from an analytical solution. However, an analytical solution is not usually available; otherwise, we would not need a numerical solution. Nevertheless, we can usually carry out some basic analysis that may give insight into steady-state, long-term trend, bounds on key variables, reduced-order solutions for ideal or special conditions, and so on. One key characteristic that is useful to know is whether the fundamental system is *stable* or *well-posed*. This is particularly important because if our numerical solution produces seemingly unstable results, we need to know if this is fundamental to the problem or whether it has been introduced by the solution method we have selected to implement. For most situations involving simulation this is not a concern, as we would be dealing with a well-analyzed and documented system. But there are situations where real physical systems can be unstable, and we need to know these in advance. For a real system to become unstable, there usually needs to be some form of energy source, kinetic, potential, reaction, and so on, so this can provide a clue as to whether the system is likely to become unstable. If it is, then we may need to modify our computational approach so that we capture the essential behavior correctly—although a complete solution may not be possible. In a seminal paper [Tur-52], Alan Turing showed how chemical reaction–diffusion systems can become unstable under certain conditions and lead to the growth of stationary concentration patterns. He postulated that the emergence of such patterns may play a role in the biological creation of *form*. An example of this type of instability problem is given in the paper by Mosekilde and Jensen [Mos-95], where they discuss numerical simulation of the *Lengyel–Epstein* model.

Determining the stability of the basic system is not always easy or straightforward to achieve, but we do have some tools at our disposal to aid investigation. We will describe one of the few tools that can be used to investigate the intrinsic stability characteristics of partial differential equations, namely, the Fourier transform method.

We will only consider a continuous domain PDE system of equations (4.1) with $q = 0$. Because we are dealing with a linear system with constant coefficients and have specified periodic boundary conditions, we can apply the *Fourier transform* methods described in most advanced mathematical textbooks. Recall that the Fourier transform with respect to x is

$$\widehat{U}(\xi, t) = \mathcal{F}\{u(x, t)\}$$
$$= \int_{-\infty}^{+\infty} u(x, t) e^{-i\xi x} dx. \quad (4.2)$$

Some basic Fourier transform properties and pairs are given in Appendix 4.A.

We now differentiate with respect to t, and because the integral is with respect to x, the derivative can be taken under the integral. Thus, we obtain

$$\widehat{U}_t(\xi, t) = \int_{-\infty}^{+\infty} u_t(x, t) e^{-i\xi x} dx$$
$$= \int_{-\infty}^{+\infty} \mathcal{P}(u(x, t)) e^{-i\xi x} dx \quad (4.3)$$
$$= \widehat{\mathcal{P}}(i\xi) \widehat{U}(\xi, t),$$

where $\widehat{\mathcal{P}}(i\xi)$ represents the spatial differential operator in Fourier space and is known as the *symbol* of the differential operator $\mathcal{P}\left(\frac{\partial}{\partial x}\right)$. This equation represents an initial value ODE problem in time and, for each *wavenumber* $\xi = 2\pi/\lambda_L$, we have

$$\widehat{U}_t(\xi, t) = \widehat{\mathcal{P}}(i\xi) \widehat{U}(\xi, t), \quad t > 0, \quad \widehat{U}(\xi, 0) = \widehat{U}_0, \quad (4.4)$$

where λ_L= wavelength. Wavenumber is the spatial analog of the term freqency in the time-domain, and in our discussions it will be assumed to be a real quantity unless specified otherwise—for more details, see Section 4.4.1.1. Equation (4.4) represents an ODE with an initial condition $\widehat{U}_0$ equal to the Fourier transform of the original PDE initial condition, that is, $\widehat{U}_0 = \widehat{U}(\xi, 0) = \mathcal{F}\{u(x, 0) = f(x)\}$. It has a solution

$$\widehat{U}(\xi, t) = e^{\widehat{\mathcal{P}}(i\xi)t} \widehat{U}(\xi, 0), \quad t > 0, \quad (4.5)$$

from which it follows from the *Cauchy–Schwarz inequality* that

$$\left\| \widehat{U}(\xi, t) \right\| \leq \left\| e^{\widehat{\mathcal{P}}(i\xi)t} \right\| \left\| \widehat{U}(\xi, 0) \right\|, \quad (4.6)$$

where $\|\cdot\| = \|\cdot\|_2$ represents the *2-norm* (see the book's appendix).

Definition 4.2.1 Equation (4.4) is called *well-posed* [Kre-05, p-29] (and thus will not grow unbounded) if there are constants K and α, independent of ξ, such that

$$\left\| e^{\widehat{\mathcal{P}}(i\xi)t} \right\| \leq K e^{\alpha t}, \quad t > 0, \; \forall \xi \in \mathbb{R}^m. \quad (4.7)$$

By invoking *Parseval's theorem* of equality, $\left\| \widehat{U}(\xi, t) \right\|_2 = \left\| u(x, t) \right\|_2$, we are assured that we can investigate well-posedness in either the spatial domain or the complex frequency domain. Let $\lambda_i, i = 1, \ldots, m$ represent the eigenvalues of $\widehat{\mathcal{P}}(i\xi)$, then we can also define the *region of well-posedness* as being the set

$$D = \left\{ \lambda \in \mathbb{C}^m : \left\| \widehat{U}(\xi, t) \right\| \leq \left\| \widehat{U}(\xi, 0) \right\|, \forall \xi \in \mathbb{R}^m, t > 0 \right\}$$
$$= \left\{ \lambda \in \mathbb{C}^m : \left\| e^{\widehat{\mathcal{P}}(i\xi)t} \right\| \leq K e^{\alpha t}, \forall \xi \in \mathbb{R}^m, t > 0 \right\}. \quad (4.8)$$

For a 1D scalar PDE, the symbol $\widehat{\mathcal{P}}(i\xi)$ in eqn. (4.4) is equal to the system eigenvalue λ. Thus,

$$\widehat{U}(\xi,t) = e^{\lambda t}\widehat{U}(\xi,0). \tag{4.9}$$

Clearly, Definition 4.2.1 will hold (where $K=1$ and $\alpha = \lambda$), and this ODE will be well-posed for all $\xi \in \mathbb{R}$, providing Re$\{\lambda\} \leq 0$.

For a system of m PDEs, $\widehat{\mathcal{P}}(i\xi)$ represents a matrix and eqn. (4.4) represents a set of ODEs, again with an initial condition equal to the Fourier transform of the original PDE initial condition, that is, $\widehat{U}(\xi,0) = \mathcal{F}\{u(x,0) = f(x)\}$. It also has the solution

$$\widehat{U}(\xi,t) = e^{\widehat{\mathcal{P}}(i\xi)t}\widehat{U}(\xi,0), \quad t > 0. \tag{4.10}$$

However, in this case, $e^{\widehat{\mathcal{P}}(i\xi)t}$ is an $m \times m$ *matrix exponential* with the well-posedness of the system determined by the symbol $\widehat{\mathcal{P}}(i\xi)$. Now, where we have a single matrix $\widehat{\mathcal{P}}(i\xi) \in \mathbb{C}^{m \times m}$, it can be transformed to *Jordan canonical form* $J = R^{-1}\widehat{\mathcal{P}}(i\xi)R$ with eigenvalues $\lambda_i, i = (1,\ldots,m)$ [Fra-95], which yields

$$\begin{aligned}e^{\widehat{\mathcal{P}}(i\xi)t} &= e^{R^{-1}JRt} = e^{R^{-1}JtR} \\ &= R^{-1}e^{Jt}R \\ \therefore \left\|e^{\widehat{\mathcal{P}}(i\xi)t}\right\| &\leq \left\|R^{-1}\right\|\left\|e^{Jt}\right\|\|R\| = \left\|e^{Jt}\right\|\text{cond}(R),\end{aligned} \tag{4.11}$$

where we have again used the *Cauchy–Schwarz inequality* and cond($\cdot$) represents the matrix condition number (see the book's appendix).

It follows therefore that $\|e^{Jt}\|$ will stay bounded for all $t \geq 0$ iff the *spectral abscissa* $\phi(J) = \overset{\max}{i}\text{Re}\{\lambda_i\} < 0$ or, $\phi(J) = 0$ and the corresponding *Jordan block* J_r, of J, has dimension 1×1 [Kre-05, , pp. 44–45]. Note that if $\widehat{\mathcal{P}}(i\xi)$ has distinct eigenvectors $\mathbf{r}_i$ $(i=1,\ldots,m)$, then $R = \{\mathbf{r}_1 \cdots \mathbf{r}_m\}$ and diagonalization is guaranteed—true for hyperbolic systems.

Therefore, under these conditions, and noting that $\|e^{Jt}\| \leq e^{\phi(J)t}$, we can modify the requirement for well-posedness of eqn. (4.7) to

$$\begin{aligned}e^{\phi(J)t} &\leq Ke^{\alpha t}, \quad t > 0, \ \forall \xi \in \mathbb{R}^m \\ &\Downarrow \\ \phi(J) &\leq 0, \ \forall \xi \in \mathbb{R}^m, t > 0,\end{aligned} \tag{4.12}$$

and the modified region of stability then becomes

$$D = \{\lambda \in \mathbb{C}^m : \phi(J) \leq 0, \ \forall \xi \in \mathbb{R}^m, t > 0\}. \tag{4.13}$$

Thus, the problem of establishing well-posedness of a particular PDE is reduced to obtaining the Jordan cononical form J, of the symbol $\widehat{\mathcal{P}}(i\xi)$, and determining if the spectral abscissa is less than or equal to zero. For further details on the theory of well-posedness, refer to Kreiss and Lorenz [Kre-05, Chapter 2].

Example: Advection equation
We apply the preceding analysis to the scalar advection equation $u_t = -au_x$ with IC $u(x,0) = f(x)$ and BC periodic. The differential operator is given by

$$\mathcal{P}\left(\frac{\partial}{\partial x}\right)u(x,t) = -au_x, \tag{4.14}$$

and on taking Fourier transforms with respect to x (using the differentiation operator), we get

$$\mathcal{F}\{\mathcal{P}\left(\tfrac{\partial}{\partial x}\right) u(x,t)\} = \mathcal{F}\{-au_x\} \\ = -i\xi \widehat{U}(\xi,t). \tag{4.15}$$

Therefore, using the results from eqns. (4.4) and (4.5), we see that the advection equation in Fourier space becomes a homogeneous ODE, which has a standard solution

$$\widehat{U}_t(\xi,t) = -ia\xi \widehat{U}(\xi,t), \\ \Downarrow \\ \widehat{U}(\xi,t) = e^{-ia\xi t}\widehat{U}(\xi,0), \quad t > 0, \tag{4.16}$$

where ξ is a parameter and the initial condition is $\widehat{U}(\xi,0) = \mathcal{F}\{u(x,0) = f(x)\}$. Clearly, this ODE will be well-posed for all $\xi \in \mathbb{R}$, providing a is real, because, for example, Definition 4.2.1 is satisfied with $K = 1$, $\alpha \geq 1$ when we have

$$\left|e^{-ia\xi t}\right| \leq 1 \leq Ke^{\alpha t}. \tag{4.17}$$

We can also carry the analysis a little bit further by applying the inverse Fourier transform to eqn. (4.16) to obtain the time domain solution for the PDE, that is,

$$u(x,t) = \mathcal{F}^{-1}\{\widehat{U}(\xi,t)\} \\ = \tfrac{1}{2\pi}\int_{-\infty}^{+\infty}\widehat{U}(\xi,t)e^{i\xi x}d\xi \\ = \tfrac{1}{2\pi}\int_{-\infty}^{+\infty}\widehat{U}(\xi,0)e^{-ia\xi t}e^{i\xi x}d\xi \\ = \tfrac{1}{2\pi}\int_{-\infty}^{+\infty}\widehat{U}(\xi,0)e^{-i\xi(x-at)}d\xi \\ = f(x-at). \tag{4.18}$$

Thus, the initial condition propagates to the right with velocity a, which is the standard result.

Example: Heat equation
We apply the same analysis to the heat equation, $q_t = \mu q_{xx}$, with IC $q(x,0) = f(x)$ and BC periodic. The differential operator this time is given by

$$\mathcal{P}\left(\frac{\partial}{\partial x}\right) q(x,t) = \mu q_{xx}, \tag{4.19}$$

and on taking Fourier transforms with respect to x (using the differentiation operator), we get

$$\mathcal{F}\{\mathcal{P}\left(\frac{\partial}{\partial x}\right) q(x,t)\} = \mathcal{F}\{\mu q_{xx}\} \\ = -\mu\xi^2 \widehat{Q}(\xi,t). \tag{4.20}$$

Therefore, the heat equation in Fourier space becomes an ODE,

$$\widehat{Q}_t(\xi,t) = -\mu\xi^2 \widehat{Q}(\xi,t) \\ \Downarrow \\ \widehat{Q}(\xi,t) = e^{-\mu\xi^2 t}\widehat{Q}(\xi,0), \quad t > 0, \tag{4.21}$$

where again ξ is a parameter and the initial condition is $\widehat{Q}(\xi,0) = \mathcal{F}\{q(x,0) = f(x)\}$. Clearly, this ODE will be well posed for all $\xi \in \mathbb{R}$, providing μ is real and positive, because, for example, Definition 4.2.1 is satisfied with $K = 1$, $\alpha \geq 1$ when we have

$$\left|e^{-\mu\xi^2 t}\right| \leq 1 \leq Ke^{\alpha t}. \tag{4.22}$$

Again, Carrying the analysis a little bit further, we obtain the time domain solution for the PDE, that is,

$$\begin{aligned} q(x,t) &= \mathcal{F}^{-1}\left\{\widehat{Q}(\xi,t)\right\} \\ &= \frac{1}{2\pi}\int_{-\infty}^{+\infty}\widehat{Q}(\xi,t)e^{i\xi x}d\xi \\ &= \frac{1}{2\pi}\int_{-\infty}^{+\infty}\widehat{U}(\xi,0)e^{-\mu\xi^2 t}e^{i\xi x}d\xi. \end{aligned} \tag{4.23}$$

This is a difficult integral to evaluate, consisting of the product of an unknown initial condition function that could be simple or complex and an exponential with a squared exponent. But it is possible and is greatly helped if a symbolic mathematics program such as Maple can be brought to bear on the problem. Nevertheless, we can deduce from the solution to the ODE that the initial condition decays with a time constant $\tau = 1/(\mu\xi^2)$. This means that rate of decay is dependent upon wavenumber—faster for high wavenumbers, slower for low wavenumbers. Thus, rapid variations in the initial condition will attenuate more quickly than slow variations.

Example: Euler equations

We now apply the same analysis to the Euler equations given by $\frac{\partial u}{\partial t} + \frac{\partial F}{\partial x} = 0$, with IC $u(x,0) = f(x)$ and BC periodic. The main variables are defined as

$$\begin{aligned} u &= \begin{bmatrix} \rho \\ \rho u \\ E \end{bmatrix} = \begin{bmatrix} \rho \\ m \\ E \end{bmatrix}, \\ F &= \begin{bmatrix} \rho u \\ \rho u^2 + p \\ u(E+p) \end{bmatrix} = \begin{bmatrix} m \\ \frac{m^2}{\rho} \\ \frac{m}{\rho}(E+p) \end{bmatrix}, \\ E &= \rho\left(\tfrac{1}{2}u^2 + e\right), \end{aligned} \tag{4.24}$$

where

- e = specific internal energy [kJ/kg]
- ρ = fluid density [kg/m^2]
- u = fluid (advection) velocity [m/s]
- $m = \rho u$ = fluid momentum [kg s/m^2]
- p = pressure [Pa]

Now, because we are going to perform a Fourier analysis, which operates locally, we can linearize F to obtain

$$\frac{\partial u}{\partial t} = -A\frac{\partial u}{\partial x}, \tag{4.25}$$

where

$$A = \frac{\partial F}{\partial u} = \begin{bmatrix} 0 & 1 & 0 \\ \frac{\partial p}{\partial \rho} - \frac{m^2}{\rho^2} & 2\frac{m}{\rho} & 0 \\ \frac{m}{\rho}\frac{\partial p}{\partial \rho} - \frac{m}{\rho^2}(E+p)\frac{1}{\rho} & \frac{1}{\rho}(E+p) & \frac{m}{\rho} \end{bmatrix}. \qquad (4.26)$$

The differential operator for this system is given by

$$\mathcal{P}\left(\frac{\partial}{\partial x}\right) u(x,t) = -A u_x. \qquad (4.27)$$

On taking Fourier transforms with respect to x (using the differentiation operator), we get

$$\begin{aligned} \mathcal{F}\{\mathcal{P}\left(\tfrac{\partial}{\partial x}\right) u(x,t)\} &= \mathcal{F}\{-A u_x\} \\ &= -iA\xi \widehat{U}(\xi,t). \end{aligned} \qquad (4.28)$$

Therefore, the Euler equations in Fourier space become a set of ODEs,

$$\begin{aligned} \widehat{U}_t(\xi,t) &= -iA\xi \widehat{U}(\xi,t), \\ &\Downarrow \\ \widehat{U}(\xi,t) &= e^{-iA\xi t}\widehat{U}(\xi,0), \quad t > 0, \end{aligned} \qquad (4.29)$$

where again ξ is a parameter and the initial condition is equal to $\widehat{U}(\xi,0) = \mathcal{F}\{u(x,0) = f(x)\}$. We now have a 3×3 *matrix exponential* solution, and the stability of the system is determined by the eigenvalues of the symbol $\widehat{\mathcal{P}}(i\xi) = -iA\xi$. It can be shown that A has distinct eigenvectors and, consequently, has a full set of eigenvalues, that is,

$$\lambda = \begin{cases} u \\ u + \sqrt{\frac{\partial p}{\partial \rho}} \\ u - \sqrt{\frac{\partial p}{\partial \rho}} \end{cases}. \qquad (4.30)$$

A derivation of the eigenvalues for the Euler equations is given at the end of Chapter 6.

It follows, therefore, that the system will be well posed for all $\xi \in \mathbb{R}^3$ because both u, the advection velocity, and $\sqrt{\frac{\partial p}{\partial \rho}}$, the speed of sound in the medium, are real, and consequently the exponent of the exponential term in eqn. (4.29) will be purely imaginary. Thus Definition 4.2.1 will be satisfied; for example, with $K = 1$, $\alpha \geq 1$ we have

$$\left| e^{-i\left(u+\sqrt{\frac{\partial p}{\partial \rho}}\right)\xi t} \right| \leq Ke^{\alpha t}. \qquad (4.31)$$

4.3 MATRIX STABILITY METHOD

4.3.1 Semi-Discrete Systems

The first task is to discretize the spatial variable, and there are many ways in which this can be done. We will concentrate on some of the more popular schemes. Assuming a PDE defined by eqn. (4.1), once we have selected and applied the discretization scheme (taking into account the boundary conditions), we are left with a set of ODEs to solve by means of numerical integration, that is,

$$\frac{du}{dt} = Au + q, \quad u = [u_1, u_2, \ldots, u_m]^T, \quad (4.32)$$

where A is assumed to be a matrix with constant coefficients whose structure is specific to the discretization scheme employed, and q includes nonhomogeneous terms and boundary conditions. We begin our analysis by applying a perturbation to $u(t)$. On substituting $v(t) = u(t) + \varepsilon(t)$ into eqn. (4.32), where $v = [v_1, v_2, \ldots, v_m]^T$ and $\varepsilon = [\varepsilon_1, \varepsilon_2, \ldots, \varepsilon_m]^T$, $i = 1, \ldots, m$, we obtain

$$\frac{dv}{dt} = Av + q$$
$$\Downarrow$$
$$\frac{du + \varepsilon}{dt} = A(u + \varepsilon) \quad (4.33)$$
$$\Downarrow$$
$$\therefore \frac{d\varepsilon}{dt} = A\varepsilon.$$

The solution to this ODE is

$$\varepsilon(t) = \varepsilon(0)e^{At}. \quad (4.34)$$

This represents a *matrix exponential* solution. Now, for this equation to be stable, $\varepsilon(t)$ must not grow unbounded as $t \to \infty$. This requires that the eigenvalues $\lambda_i, i = 1, \ldots, m$ of A must all lie within the left half of the complex plane; that is, they must not have positive real parts. The *region of well-posedness* is therefore defined by the set

$$D = \{\lambda_i \in \mathbb{C} : \text{Re}\{\lambda_i\} \leq 0, i = (1, \ldots, m)\}. \quad (4.35)$$

A further requirement for *overall system stability* is that the eigenvalues must all be located within the integrator's stable region in the complex plane.

The matrix stability method identifies the eigenvalues of the discretized system, and if they are all located within the integrator stable region in the complex plane, the overall system will be *well-posed* or *stable*; otherwise, it is unstable. If we assume that the matrix A has full rank and, consequently, a full set of linearly independent *eigenvectors* $\mathbf{r}^i = [r_1, \ldots, r_m]^T, i = 1, \ldots, m$, it means that A can be *diagonalized* such that

$$\Lambda = RAR^{-1}, \quad (4.36)$$

where

$$\Lambda = \begin{bmatrix} \lambda_1 & & & \\ & \lambda_2 & & \\ & & \ddots & \\ & & & \lambda_m \end{bmatrix}, \quad (4.37)$$

$$R = \begin{bmatrix} \mathbf{r}^1 \cdots \mathbf{r}^m \end{bmatrix} \quad (4.38)$$

Thus, the solution reduces to $\varepsilon(t) = \varepsilon(0)R^{-1}e^{\Lambda t}R$. But the eigenvalues and eigenvectors still have to be found, and for a large problem size, this is not trivial. However, if we assume periodic BCs, then A becomes a *circulant* matrix where the row entries shift right at the next row down, that is,

$$A = \begin{bmatrix} a_1 & a_2 & a_3 & a_4 & \cdots & a_N \\ a_N & a_1 & a_2 & a_3 & \cdots & a_{N-1} \\ a_{N-1} & a_N & a_1 & a_2 & \cdots & a_{N-2} \\ \vdots & \vdots & \vdots & \vdots & \ddots & \vdots \\ a_2 & a_3 & a_4 & a_5 & \cdots & a_1 \end{bmatrix}. \quad (4.39)$$

Now, it transpires that circulant matrices have a very useful property whereby the eigenvalues are given by

$$\lambda_k = \sum_{j=1}^{N} a_j \exp(2\pi i(k-1)(j-1)/N), \quad i = \sqrt{-1}. \quad (4.40)$$

Thus, from a knowledge of the *first* row or column of A, we can immediately calculate Λ, which is a very much reduced problem

We now consider three examples.

4.3.1.1 Advection Equation with Upwind Spatial Discretization

For our first example, consider the advection equation

$$\frac{\partial u}{\partial t} = -a \frac{\partial u}{\partial x}, \quad u = u(x, t), \ t \geq t_0, \ x \in (x_l, x_u), \quad (4.41)$$

where $a > 0$ is the advection velocity. The spatial first derivative is discretized into N nodes by the first-order, two-point upwind scheme given in Chapter 3, but with Dirichlet boundary conditions imposed at the inlet (node 1) so that $u_1 = u_{in}$. The semi-discrete scheme therefore becomes

$$\frac{du_i}{dt} = -a \frac{(u_i - u_{i-1})}{\Delta x}, \quad i \in (2, \ldots, m), \ \Delta x = x_i - x_{i-1},$$
$$\frac{du_1}{dt} = 0, \quad (4.42)$$

which in matrix form is equivalent to eqn. (4.32) with $q = 0$, that is,

$$\frac{du}{dt} = Au,$$
$$\Downarrow$$
$$\frac{du}{dt} = \frac{-a}{\Delta x} \begin{bmatrix} 0 & 0 & 0 & & & & \\ -1 & 1 & 0 & & & & \\ & -1 & 1 & 0 & & & \\ & & \ddots & & & & \\ & & & & -1 & 1 & 0 \\ & & & & & -1 & 1 \end{bmatrix} \begin{bmatrix} u_{in} \\ u_2 \\ u_3 \\ \vdots \\ u_{m-1} \\ u_m \end{bmatrix}. \quad (4.43)$$

A simple check on the eigenvalues of the preceding sparse matrix using the R function eigen() reveals that they are all equal to $+1$, except for one equal to zero, which is associated with the Dirichlet BC (node 1). The system eigenvalues of A, therefore, become

$$\lambda_1 = 0, \quad \lambda_i = \frac{-a}{\Delta x}, \; i \in (2, \ldots, m-1). \tag{4.44}$$

Consequently, the system will be stable for $a > 0$ as no eigenvalues will have positive real parts, that is, $\text{Re}\{\lambda_i\} \leq 0$, and unstable for $a < 0$. All that remains is to select a suitable time integration scheme that has a stability region that includes $\lambda = \frac{-a}{\Delta x}$. R program `matStabAdvec_upwind.R`, which demonstrates this example, is included with the downloads.

For our second example we will demonstrate that it is simple to stabilize the system for $a < 0$ by applying a backward rather than forward space discretization. This time we apply a Dirichlet boundary condition at the opposite end of the spatial domain so that $u_m = u_{in}$ and node m becomes the inlet. The scheme is then changed to

$$\frac{du_i}{dt} = -a \frac{(u_i - u_{i+1})}{-\Delta x}, \; i \in (1, \ldots, m-1), \; -\Delta x = x_i - x_{i+1}$$
$$\frac{du_m}{dt} = 0, \tag{4.45}$$

which in matrix form becomes

$$\frac{du}{dt} = Au,$$
$$\Downarrow$$
$$\frac{du}{dt} = \frac{-a}{-\Delta x} \begin{bmatrix} 1 & -1 & & & & \\ 0 & 1 & -1 & & & \\ & 0 & 1 & -1 & & \\ & & & \ddots & & \\ & & & & 0 & 1 & -1 \\ & & & & & 0 & 0 \end{bmatrix} \begin{bmatrix} u_1 \\ u_2 \\ u_3 \\ \vdots \\ u_{m-1} \\ u_{in} \end{bmatrix}. \tag{4.46}$$

The eigenvalues of the sparse matrix are still all equal to $+1$, except for one equal to zero, which is associated with the Dirichlet BC (node m). However, as the grid spacing is now $-\Delta x$, the system eigenvalues of A becomes

$$\lambda_i = \frac{a}{\Delta x}, \; i \in (1, \ldots, m-1), \quad \lambda_m = 0. \tag{4.47}$$

Therefore, it follows that the system will be stable for $a < 0$, when $\text{Re}\{\lambda_i\} \leq 0$, and unstable for $a > 0$.

Note that in the preceding examples, if u_{in} is time varying, then it has to be updated at each time step.

4.3.1.2 Advection Equation with Central Spatial Discretization

For our third example, consider advection eqn. (4.41), again discretized into m nodes with periodic boundary conditions imposed. This time we use the following second-order, three-point central scheme given in Chapter 3, when the semi-discrete scheme becomes,

$$\frac{du_i}{dt} = -a\frac{(u_{i+1} - u_{i-1})}{2\Delta x}, \quad i \in (2, \ldots, m-1), \quad \Delta x = x_i - x_{i-1}$$
$$\frac{du_1}{dt} = -a\frac{(u_m - u_2)}{2\Delta x}, \tag{4.48}$$
$$\frac{du_m}{dt} = -a\frac{(u_{m-1} - u_1)}{2\Delta x},$$

which in matrix form we can write,

$$\frac{\partial u}{\partial x} = \frac{-a}{2\Delta x} \begin{bmatrix} 0 & -1 & 0 & 0 & \cdots & 1 \\ 1 & 0 & -1 & 0 & \cdots & 0 \\ 0 & 1 & 0 & -1 & \cdots & 0 \\ \vdots & \vdots & \vdots & \vdots & \ddots & \vdots \\ -1 & 0 & 0 & 0 & \cdots & 0 \end{bmatrix} \times \begin{bmatrix} u_1 \\ u_2 \\ u_3 \\ \vdots \\ u_m \end{bmatrix}, \tag{4.49}$$

where A is a *tridiagonal* matrix.

Now for a periodic tridiagonal matrix, the eigenvalues are known analytically [Lom-76] and are equal to

$$\lambda_j = b_2 + b_1 e^{-i2\pi j/m} + b_3 e^{i2\pi j/m}, \tag{4.50}$$

where we have used j as the index to avoid confusion with $i = \sqrt{-1}$. For the preceding matrix A, the tridiagonal bands are defined as $b_1 = 1, b_2 = 0, b_3 = -1$; therefore, using *Euler's formula*, we see that the system eigenvalues become

$$\lambda_j = \frac{-ia}{\Delta x} \sin(2\pi j/m), \quad j = 1, \ldots, m. \tag{4.51}$$

The system of eqn. (4.49) was analyzed using the matrix stability method and the results are shown in Fig. 4.1. A spatial domain of $x \in (-4, 4)$ was used, discretized into a grid of $m = 20$ nodes. A value for the *Courant–Fredrichs–Lewy* number, Co $= \Delta t |a|/\Delta x = 1$, was chosen so that $\Delta t = \Delta x/|a|$. The stability contour for the *forward Euler* integrator, defined on the *complex plane* as $\Gamma \in (-1 + \sin\theta, \cos\theta)$ (see Chapter 2), is overlaid on the plot, and for a stable system, all the scaled eigenvalues, $\lambda_j \Delta t$, must lie within this region. It is clear that this system is unstable, as all but one of the $\lambda_j \Delta t$ lie outside the integrator stability contour. The situation improves with Dirichlet BCs and/or reduced Δt, but some scaled eigenvalues always remain outside the integrator stability contour. Thus, we conclude that a three-point central scheme is not suitable for discretization of the spatial domain for purely advection problems. However, we will see in the next example that the addition of a diffusion term improves stability greatly.

The R code that generates the plots of Fig. 4.1 is contained in file matStabAdvec_central.R that is included with the downloads. The particular plot is selected by setting varable BC_flag=1 for *periodic* BCs or BC_flag=2 for *Dirichlet* BCs. See also discussion relating to the example in Section 4.3.1.3 and program listing for further details.

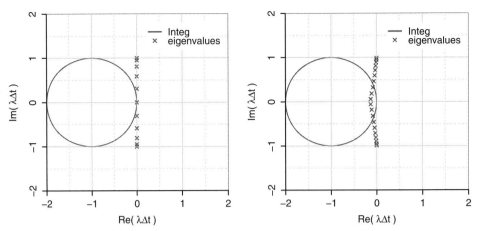

Figure 4.1. Complex domain showing *eigenvalues* for a three-point discretization of the first derivative of the *advection* eqn. (4.41). The stability contour for a *forward Euler* numerical integrator is overlaid. (left) With periodic boundary conditions. The system is unstable, as all but one of the scaled eigenvalues, $\lambda_k \Delta t$, lie outside the integrator stability region. (right) With a left-hand Dirichlet boundary condition, which moves the $\lambda_k \Delta t$ into the left-half plane (giving them negative real parts). Nevertheless, the system is unstable, as many $\lambda_k \Delta t$ still lie outside the integrator stability region.

4.3.1.3 Convection–Diffusion Equation

Consider the convection–diffusion equation

$$\frac{\partial u}{\partial t} = -v\frac{\partial u}{\partial x} + \mu\frac{\partial^2 u}{\partial x^2}, \quad u = u(x,t), \ t \geq t_0, \ x \in (x_l, x_u), \tag{4.52}$$

where $v > 0$ is the bulk medium velocity and $\mu > 0$ the diffusivity constant. The spatial domain is discretized into m nodes with the spatial first and second derivatives being approximated by three-point schemes.[1] With periodic boundary conditions imposed, the scheme becomes

$$\frac{\partial u}{\partial x} = \frac{-v}{2\Delta x} \begin{bmatrix} 0 & -1 & 0 & 0 & \cdots & 1 \\ 1 & 0 & -1 & 0 & \cdots & 0 \\ 0 & 1 & 0 & -1 & \cdots & 0 \\ \vdots & \vdots & \vdots & \vdots & \ddots & \vdots \\ -1 & 0 & 0 & 0 & \cdots & 0 \end{bmatrix} \times \begin{bmatrix} u_1 \\ u_2 \\ u_3 \\ \vdots \\ u_m \end{bmatrix}$$

$$+ \frac{\mu}{\Delta x^2} \begin{bmatrix} -2 & 1 & 0 & 0 & \cdots & 1 \\ 1 & -2 & 1 & 0 & \cdots & 0 \\ 0 & 1 & -2 & 1 & \cdots & 0 \\ \vdots & \vdots & \vdots & \vdots & \ddots & \vdots \\ 1 & 0 & 0 & 0 & \cdots & -2 \end{bmatrix} \times \begin{bmatrix} u_1 \\ u_2 \\ u_3 \\ \vdots \\ u_m \end{bmatrix} \tag{4.53}$$

---

[1] This is to simplify the discussion. For a practical situation, we would choose schemes of the same order for both spatial derivatives so that errors converge at the same rate with Δx, as described in Chapter 3.

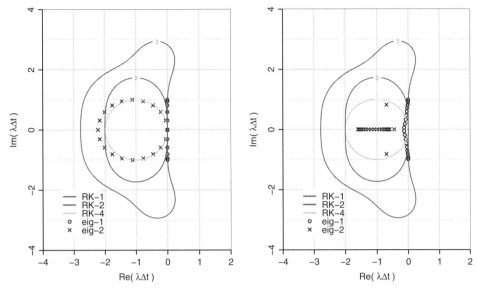

Figure 4.2. Complex domain showing stability plots for numerical integrators RK-1 (inner) to RK-4 (outer), with $\lambda_k \Delta t$ for the *convection–diffusion* eqn. (4.53) superimposed. Circles represent no diffusion, i.e., $\mu = 0$, and crosses diffusion with $\mu = 0.234$. (left) Periodic BCs. (right) Dirichlet BCs.

This reduces to

$$\frac{\partial u}{\partial t} = A \times \begin{bmatrix} u_1 \\ u_2 \\ u_3 \\ \vdots \\ u_m \end{bmatrix}, \qquad (4.54)$$

where A is a *tridiagonal* matrix with tridiagonal bands defined as

$$b_1 = -v/(2\Delta x) + \mu/(\Delta x^2), \quad b_2 = -2\mu/\Delta x^2, \quad b_3 = v/(2\Delta x) + \mu/(\Delta x^2). \quad (4.55)$$

We observe also that the matrix A is additionally a *circulant* matrix because rows 2 to m are cyclic permutations of the first row—the definition of a circulant matrix (see the appendix to the book). It follows, therefore, that the eigenvalues are known analytically from eqn. (4.40).

To calculate the eigenvalues, a 1D spatial domain of $x \in (-4, 4)$ is discretized into a grid of $m = 20$ nodes, and we choose the Péclet number[2] $Pe = (v\Delta x/\mu) = 1.8$, $Co = 1$, and $v = 1$. This gives a diffusivity of $\mu = 0.234$. The results of the three-point discretization schemes for the first and second spatial derivatives are shown in Fig. 4.2.

Consider the situation with $\mu = 0$, that is, no diffusion. From these figures we see that for *periodic* boundary conditions the system is *marginally stable* for integrator RK-4 and unstable for integrators RK-1 and RK-2. However, with a *Dirichlet* boundary imposed

---

[2] Equal to the ratio of advection to diffusion. Named after French physicist Jean Claude Eugène Péclet (1793–1857).

at the left boundary and *backward Euler* employed at the right boundary, the stability margin for RK-4 is improved due to the scaled eigenvalues, $\lambda_k \Delta t$, now having *negative real* parts. The improvement is not, however, sufficient to make RK-1 or RK-2 stable.

Now consider the situation with diffusion, where $\mu = 0.234$. From Fig. 4.2, we see that the system is stable using RK-4 for the periodic case, as all $\lambda_k \Delta t$ lie within the integrator stability region. But this case is unstable for integrators RK-1 and RK-2, as the $\lambda_k \Delta t$ lie outside their stability regions. However, from Fig. 4.2, again for $\mu = 0.234$, we see that the situation is improved by the imposition of a Dirichlet condition at the left boundary and use of backward Euler at the right boundary. In this case, all the $\lambda_k \Delta t$ lie within the RK-4, RK-2, and RK-1 integrator stability regions. Thus, under these conditions, the system would be stable for all the integrators under consideration.

The stability can be further improved for both the $\mu = 0$ and $\mu = 0.234$ cases by reducing the value for Co. Additionally, increasing the value of μ improves stability by increasing the eigenvalue negative real parts.

The R code for calculating the eigenvalues and plotting the results for this example is shown in Listings 4.1–4.3. The code is easily adapted to alternative discretization schemes and numerical integrators. The parameters Co and μ can be adjusted to experiment with system stability.

Listing 4.2 has been adapted from R code used in Chapter 2 for ODE stability calculations.

Listing 4.3 represents a short R program designed to be run after matStabConvDiff_main.R. It calculates eigenvalues for the *circulant* matrix A using eqn. (4.40) and obtains the same values as generated by the R function eigen().

```
# File: matStabConvDiff_main.R
# ----------------------------------------------------
# 1D convection-diffusion: u_t = v*u_x + D*u_xx
# Matrix method eigenvalue calculation
# ----------------------------------------------------
require(compiler)
enableJIT(3)
#
####################################################
# Delete workspace prior to new calculations
####################################################
cat("\014") # Clear console
rm(list = ls(all = TRUE))
#
ptm0 <- proc.time() # Start timer
print(Sys.time())
#
BC_flag <- 1 # 1 = periodic, 2= Dirichlet
calcTitle <- c("periodic", "Dirichlet")
cat(sprintf("\nStability plots for convection-diffusion eqn. with %s BCs!\n",
            calcTitle[BC_flag]))
# Declare array sizes
```

```
xl <- -4
xu <- 4
Nx <- 20 # number of points
x <- seq(xl,xu,len=Nx)

# Calculate node spacing
dx  <- x[2] - x[1]
dx2 <- dx*dx

# Set parameters
v  <- 1        # bulk velocity
Pe <- 1.8      # Peclet number
mu <- dx*v/Pe  # diffusivity

# Setup time step
CFL <- 1
dt <- CFL*dx/abs(v)

#
# Allocate matrices A and B
A <- array(0,c(Nx, Nx)) # Convection
B <- array(0,c(Nx, Nx)) # diffusion

# Construct A and B except for first and last row
for(i in 2:(Nx-1)){
        A[i,i-1] <-  v/(2*dx)
        A[i,i+1] <- -v/(2*dx)
        #
        B[i,i-1] =    mu/(dx2)
        B[i,i  ] = -2*mu/(dx2)
        B[i,i+1] =    mu/(dx2)
}

if(BC_flag == 1){ # Periodic bcs
  A[1 ,2  ] <-  -v/(2*dx)
  A[1 ,Nx ] <-   v/(2*dx)
  A[Nx,1  ] <-  -v/(2*dx)
  A[Nx,Nx-1] <- v/(2*dx)
  #
  B[1 ,1  ] <- -2*mu/(dx2)
  B[1 ,2  ] <-    mu/(dx2)
  B[1 ,Nx ] <-    mu/(dx2)
  #
  B[Nx,1  ] <-    mu/(dx2)
  B[Nx,Nx-1] <-   mu/(dx2)
  B[Nx,Nx ] <- -2*mu/(dx2)
```

```
}else{ # non-periodic bc's
  # Only the i+1 location need to be set in A.
  A[1,2] <-  -v/(2*dx)
  B[1,2] <- -2*mu/(2*dx2)
  B[1,3] <-    mu/(2*dx2)

  # Right hand boundary uses backwards difference
  A[Nx,Nx-1] <-   v/dx
  A[Nx,Nx  ] <-  -v/dx
  #
  B[Nx,Nx-2] <-    mu/dx2
  B[Nx,Nx-1] <- -2*mu/dx2
  B[Nx,Nx  ] <-    mu/dx2
}
AB <- A+B

# Calculate eigenvalues of A and A+B
lambdaA  <- eigen(A, only.values = TRUE)$values
lambdaAB <- eigen(AB,only.values = TRUE)$values

# Plot values for lambdaAB*dt
plot(Re(lambdaAB)*dt,Im(lambdaAB)*dt,
        type ="p",pch=4, xaxs="i",yaxs="i",
        xlim=c(-4,4),ylim=c(-4,4),col="black",lwd=2,
     xlab=expression(paste("Re( ",lambda*Delta*t," )")),
     ylab=expression(paste("Im( ",lambda*Delta*t," )")))
######
# Plot values for lambdaA*dt
theta <- seq(0,2*pi,len=101)
points(Re(lambdaA)*dt,Im(lambdaA)*dt,pch=1, col="black",lwd=2)
#
grid(lty=1,lwd=1,col="darkgray")
abline(h=c(-3,-1,1,3),lty=2,col="gray")
#
axis(side = 1, lwd = 2)
axis(side = 2, lwd = 2)
# Over plot integrator stability contours
source("RKcontours.R")
t1 <- proc.time()-ptm0
cat(sprintf("Calculation time: %f\n",t1[3]))
```

Listing 4.1. File: matStabConvDiff_main.R—Code for the main program to calculate the eigenvalues for the discretized convection–diffusion system of eqn. (4.52)

```r
# File RKcontours.R
# Construct mesh
Nx2 <- 100;Ny2 <- 100
x2  <- seq(xl,xu,len=Nx2)
y2  <- x2
X2<- matrix ( rep(x2, each=Nx2),nrow =Nx2);
Y2<- matrix ( rep(y2, each=Nx2),nrow =Nx2)
# Calculate z
z <- t(X2) + 1i*Y2 # Forward Euler (Runge - Kutta 1)
G1 <- 1 + z;
# Runge - Kutta 2
G2 <- 1 + z + 0.5* z ^2
# Runge - Kutta 4
G4 <- 1 + z + 0.5* z^2 + (1/6) *z^3 + (1/24) *z ^4;
# Runge - Kutta RKF45
#G5 <- 1+z +(1/2) *z ^2+(1/6) *z ^3+(1/24) *z ^4+(1/120) *z ^5+(1/2080) *z^6
# Calculate magnitude of G
G_mag1 <- abs(G1);
G_mag2 <- abs(G2);
G_mag4 <- abs(G4);
#G_mag5 <- abs(G5)
# Plot contours of G_mag for RK -1 to RK -4
contour (x2,y2,G_mag1 , levels =1, cex.lab=2,labels = "1", col =" green ",
         lwd =2, add= TRUE)
contour (x2,y2,G_mag2 , levels =1, cex.lab=2,labels = "2", col =" blue " ,
         lwd =2, add= TRUE )
contour (x2,y2,G_mag4 , levels =1, cex.lab=2,labels = "3", col =" red" ,
         lwd =2, add= TRUE )
```

Listing 4.2. File: RKcontours.R—Code for overlaying integrator stability contours on the eigenvalue plots generated by Listing 4.1. The code has been adapted from the ODE stability calculation R code used in Chapter 2 and is run at the end of the main program.

```r
# File: matConvDiffEigenCheck.R
# Eigenvalue calculation check for periodic BCs - to be run after
# the main program: matStabConvDiff_main.R
if(BC_flag != 1){
  cat(sprintf("\nERROR: Eigenvalue check only valid for periodic BCs!\n"))
}else{
  la <- rep(0,Nx); lab <- rep(0,Nx)
  #
  for(k in 1:Nx){
    la[k] <- 0; #lab[k] <- 0
    for(j in 1:Nx){
        la[k] <- la[k] + A[1,j] *exp((2*pi*1i*(j-1)/Nx)*(k-1))
```

```
      lab[k] <- lab[k] + AB[1,j]*exp((2*pi*1i*(j-1)/Nx)*(k-1))
    }
  }
  ea<-sort(Im(-la*dt)); eab<-sort(Im(-lab*dt))
  eA<-sort(Im(lambdaA*dt)); eAB<-sort(Im(lambdaAB*dt))
  #
  cat(sprintf("\nDifference between imaginary parts for lambda A\n"))
  cat(sprintf("calculated two different ways - see book text!\n\n"))
     print(eA-ea)
  #
  cat(sprintf("\nDifference between imaginary parts for lambda AB\n"))
  cat(sprintf("calculated two different ways - see book text!\n\n"))
  print(eAB-eab)
  # Uncomment to overplot points on exising plot
   points(Re(la)*dt,Im(la)*dt,pch=0,col="green")
   points(Re(lab)*dt,Im(lab)*dt,pch=0,col="green")
}
```

Listing 4.3. File: `matConvDiffEigenCheck.R`—Code to calculate the eigenvalues for a circulant matrix using eqn. (4.40) and to plot the results if required. The code is to be run for the *periodic* BCs case after the main program of file `matStabConvDiff_main.R` has completed.

For more details of the matrix stability method, readers are referred to Hirsh [Hir-88].

4.4 VON NEUMANN STABILITY METHOD

4.4.1 General

This method was developed by Von Neumann during World War II at Los Alamos and for some time was a closely guarded secret. It finally entered the public domain in 1947, when details were published by Crank and Nicholson [Cra-47]. The von Neumann method is generally applied to *linear* partial differential equations, discretized on a *uniform spatial grid*, $x \in (x_1, x_2, \ldots, x_j, \ldots, x_N)$, to determine the stability properties of a particular discretization scheme. The method also assumes periodic boundary conditions or the problem to be defined on an infinite grid, that is, an initial value problem with $N = \infty$.

The idea is to take the Fourier transform of the discretized partial differential equation, using the *shifting property* where spatial differentials occur, and then obtain an expression for the *gain factor*. The requirement for stability is that the modulus of the gain factor should be such that it does not result in an unbounded solution. Application of *Parseval's theorem* shows that if stability is established in the complex frequency, or Fourier domain, then this result applies equally to the spatial or grid domain. In other words, stability in the complex frequency domain implies stability in the spatial domain, and vice versa (see also Chapter 2).

4.4.1.1 Wavenumber

The term *wavenumber*, as mentioned earlier, essentially represents the spatial analog of the term freqency in the time domain. Other terms used in the literature include *circular*

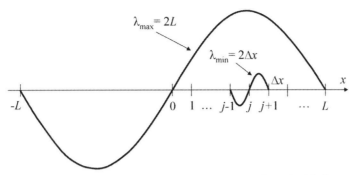

Figure 4.3. Relationship between grid spacing and wavelengths of resolvable frequencies [Hir-88, p. 285].

or *angular wavenumber* and *spatial frequency*. Also, different definitions of wavenumber are used in different branches of the physical sciences and different symbols are used; we shall define wavenumber as the real quantity

$$\xi = \frac{2\pi}{\lambda},$$

where λ represents wavelength.

We can see from Fig. 4.3 that grid spacing imposes a limit on the highest frequency that can be resolved by a discretization scheme. The wavelength of the highest frequency that can be resolved is $\lambda_{min} = 2\Delta x$, whereas the wavelength of the lowest frequency that can be resolved is $\lambda_{max} = 2L$, the length of the grid domain. Consequently, the limits of integration when performing spatial Fourier transforms only have to include up to the maximum wavenumber that can be resolved by the grid, that is, $\xi_{max} = \pm 2\pi/\lambda_{min}$. Outside of this range, the grid function is assumed to be periodic. Also, for finite spatial domains, there is a minimum wavenumber that can be resolved, that is, $\xi_{min} = \pm 2\pi/\lambda_{max}$.

Finally, when performing a Fourier transformation in the time domain, we obtain a *frequency spectrum*; similarly, when performing a Fourier transform in the spatial domain, we obtain a *wave number spectrum*.

4.4.2 Fully Discrete Systems

Consider the general, linear PDE of evolution in m space variables with periodic boundary conditions

$$u_t(x,t) = \mathcal{P}\left(\frac{\partial}{\partial x}\right) u(x,t), \quad x \in \mathbb{R}^m, \quad t > 0, \tag{4.56}$$

which is discretized in time and space as

$$u_j^{n+1} = P u_j^n, \tag{4.57}$$

where P is a space discretization operator. If we take the Fourier transform with respect to x, we get

$$\begin{aligned} \hat{U}_j^{n+1} &= \mathcal{F}\{Pu_j^n\} \\ &= \int_{-\infty}^{+\infty} e^{-i\xi x} P u_j^n \, dx \\ &= \hat{P}\left(e^{i\xi \Delta x}\right) \int_{-\infty}^{+\infty} e^{-i\xi x} u_j^n \, dx \\ \therefore \hat{U}_j^{n+1} &= \hat{P}\left(e^{i\xi \Delta x}\right) \hat{U}_j^n, \end{aligned} \quad (4.58)$$

where $\hat{P}(e^{i\xi \Delta x}) = G_f(\theta)$ $\{\theta = \xi \Delta x\}$ represents the result of the Fourier transform shift operator applied to P, and which is called the *gain factor* or *amplification ratio*. The gain factor also allows us to investigate additional properties of the discretization scheme, such as the dissipation error ε_D and the dispersion error ε_θ (see Chapter 5).

We now have a relationship between $\hat{U}_j^{n+1}$ and $\hat{U}_j^n$ that we can use to investigate the stability of the discretized system. From eqn. (4.58) it follows from the *Cauchy–Schwarz inequality* that

$$\left\| \hat{U}_j^{n+1} \right\| \leq \left\| G_f(\theta) \right\|^n \left\| \hat{U}_j^0 \right\|. \quad (4.59)$$

The requirement [Kre-05] for a well-behaved system to be stable and $\hat{U}_j^n$ not to grow unbounded is that there are real constants K and α independent of ξ and Δx such that

$$\left\| G_f(\theta) \right\| \leq K e^{\alpha \Delta t} = 1, \quad \forall |\theta| \leq \pi. \quad (4.60)$$

We can also define the *region of stability* as being the set

$$\begin{aligned} D &= \left\{ \lambda \in \mathbb{C}^m : \left\| \hat{U}_j^{n+1} \right\| \leq \left\| \hat{U}_j^n \right\|, \forall \theta \in \mathbb{R}^m, n > 0 \right\} \\ &= \left\{ \lambda \in \mathbb{C}^m : \left\| G_f(\theta) \right\| \leq K e^{\alpha \Delta t}, \forall \theta \in \mathbb{R}^m, n > 0 \right\}. \end{aligned} \quad (4.61)$$

Recall that the *spectral radius* $\rho(G_f(\theta)) \leq \|G_f(\theta)\|$, with equality being guaranteed for a *normal* matrix. Therefore, if $G_f(\theta)$ is a matrix, we can instead use $\rho(G_f(\theta))$, when the more restrictive requirement for stability becomes

$$\rho\left(G_f(\theta)\right) = \max_i |\lambda_i| \leq K e^{\alpha \Delta t} = 1, \quad \forall |\theta| \leq \pi \quad (4.62)$$

and the restricted *region of stability* is defined by the set

$$D = \left\{ \lambda \in \mathbb{C}^m : \rho\left(G_f(\theta)\right) \leq 1, \forall \theta \in \mathbb{R}^m, n > 0 \right\}. \quad (4.63)$$

By invoking *Parseval's theorem* of equality ($\|\hat{U}_j^n\|_2 = \|u_j^n\|_2$), we are assured that we can investigate stability in either the spatial (grid) domain or the complex frequency domain.

It should be noted that there are some problems where a component of the exact solution actually grows exponentially with increasing time, for instance, the heat conduction equation

$$\frac{\partial T}{\partial t} = \mu \frac{\partial^2 T}{\partial x^2} + q, \quad (4.64)$$

where the source term $q = kT$, $k > 0$. For these problems, the requirement for stability is

$$\left\| G_f(\theta) \right\| \leq K e^{\alpha \Delta t} = 1 + \mathcal{O}(\Delta t), \quad \forall |\theta| \leq \pi. \quad (4.65)$$

Nevertheless, for most practical problems, the condition of eq. (4.60) suffices and tends to err on the safe side, that is, it is more restrictive.

Although as an abstract concept, the preceding may seem rather complex, it turns out to be straightforward to implement for many practical schemes and can be applied in a mechanistic way. Recall that the Fourier transform of a function $f(x + \Delta x, t)$ is obtained by use of the shift operator to obtain $e^{i\xi \Delta x} \hat{F}(\xi, t)$, that is, simply the Fourier transform of $f(x, t) [= \hat{F}(\xi, t)]$ multiplied by $e^{i\xi \Delta x}$. Similarly, a shift by $p\Delta x$ equates to a multiplication by $e^{i\xi p \Delta x}$. Thus, we may summarize the method as follows:

1. Substitute $u_j^n = e^{i\xi j \Delta x} \hat{U}_j^n$ into the discretized PDE equation.
2. Cancel like terms to obtain a function in the form $\hat{U}_j^{n+1} = G_f(\theta) \hat{U}_j^n$, where $\theta = \xi \Delta x$.
3. Determine the limitations on Δx, Δt, and other parameters to ensure that $\|G_f(\theta)\|$ conforms with the stability requirement of eqn. (4.60).

For the more difficult discretization schemes, a suitable stability plot can easily be generated by use of a computer program. We illustrate a practical approach to the calculations in the following pages.

For more details on the von Neumann stability method, refer to Richtmyer and Morton [Ric-67] and Hirsch [Hir-88].

We will now consider the stability of various *fully discrete* numerical schemes applied to the one-dimensional scalar *hyperbolic* system

$$u_t + a u_x = 0, \tag{4.66}$$

with a constant. An expression for the gain factor will be calculated for each scheme (where we write $\text{Co} = \frac{a \Delta t}{\Delta x}$, the *Courant–Fredrichs–Lewy* number), and an associated computer-generated plot will be presented.

4.4.2.1 Forward in Time Backward in Space (FTBS)

After applying *FTBS* discretizations to the preceding hyperbolic system we get the following explicit scheme:

$$u_j^{n+1} = u_j^n - \frac{a \Delta t}{\Delta x} \left(u_j^n - u_{j-1}^n \right). \tag{4.67}$$

Step 1: Substituting $u_j^n = e^{i\xi j \Delta x} \hat{U}_j^n$ (equivalent to taking the Fourier transform with respect to x) yields

$$e^{i\xi j \Delta x} \hat{U}_j^{n+1} = e^{i\xi j \Delta x} \hat{U}_j^n - \text{Co} \left(e^{i\xi j \Delta x} \hat{U}_j^n - e^{i\xi (j-1) \Delta x} \hat{U}_j^n \right). \tag{4.68}$$

Step 2: Dividing both sides by $e^{i\xi j \Delta x}$, we obtain

$$\hat{U}_j^{n+1} = \left\{ 1 - \text{Co} \left(1 - e^{-i\xi \Delta x} \right) \right\} \hat{U}_j^n$$
$$\Downarrow$$
$$\hat{U}_j^{n+1} = G_f(\theta) \hat{U}_j^n, \tag{4.69}$$

where $\theta = \xi \Delta x$. Thus, we now have the following expression for *gain factor*:

$$G_f = \hat{P} \left(e^{i\xi \Delta x} \right) = 1 - \text{Co} \left(1 - e^{-i\theta} \right). \tag{4.70}$$

Step 3: Determine the limitations on $\theta = \xi \Delta x$. Now, using Euler's formula

$$e^{-i\theta} = \cos(\theta) - i\sin(\theta) \tag{4.71}$$

along with the half-angle trigonometric formulae, we find

$$G_f = 1 - 2\text{Co}\sin^2\left(\tfrac{\theta}{2}\right) - i\text{Co}\sin(\theta)$$

$$\therefore |G_f|^2 = 1 - 4\text{Co}\sin^2\left(\tfrac{\theta}{2}\right) + 4\text{Co}^2\sin^4\left(\tfrac{\theta}{2}\right) + 4\text{Co}^2\sin^2\left(\tfrac{\theta}{2}\right)\cos^2\left(\tfrac{\theta}{2}\right) \tag{4.72}$$

$$\Downarrow$$

$$|G_f| = \left|\sqrt{1 - 4\text{Co}(1-\text{Co})\sin^2\left(\tfrac{\theta}{2}\right)}\right|.$$

This system will be stable for all ξ providing $0 \leq \text{Co} \leq 1$ because, for example, eqn. (4.60) is satisfied with $K = 1$, $\alpha \geq 1$, when we have

$$\left|\sqrt{1 - 4\text{Co}(1-\text{Co})\sin^2\left(\tfrac{\theta}{2}\right)}\right| \leq 1 \leq Ke^{\alpha \Delta t}. \tag{4.73}$$

Thus, the stability requirement for the FTBS scheme is

$$0 \leq \text{Co} \leq 1. \tag{4.74}$$

This completes the fully discrete von Neumann stability analysis for the FTBS method.

Although the preceding is fairly straightforward to complete by hand, for more complex schemes, particularly involving multiple time steps, it can become extremely tedious and challenging. Consequently, a simple graphical method is to be preferred whereby the gain factor is plotted against Co for values of θ between 0 and 2π (see Listing 4.4 and the produced plots in Fig. 4.4). From the forward Euler figure it is seen that the graphical method confirms the analytical result.

```
# File: vonNeumannFullyDiscrete.R
# Von Neumann analysis of PDE fully-discrete Schemes
N <- 101
Co <- seq(-3, 3, len=N) # <- a*delta_t/delta_x
colNum <- c("red", "blue","green","mmagenta","black")
theta   <- pi/8
prtPlot <- 0 # 0=off, 1=on
cat(sprintf("print plot = %d\n", prtPlot))
# Plot contours of magmetude of Growth Factor - various schemes
# strT1 <- sprintf("PDE: ut + a.ux = 0 - ")
# strT2 <- sprintf("Gain plots for theta varying between 0 and 2*pi")
str1 <- rep(0,9)
#
for(testNo in 1:9){
#testNo <- 1
  for(m in 1:16){
    switch(testNo,
      "1"={# 1st Order - Forward Euler
        gf <- sqrt(1-4*Co*(1-Co)*sin(theta*m/2)^2)
```

```r
              str1[1] <- "Forward-Euler"},
       "2"={# 1st Order - Backwards Euler
              gf <- 1/(1 + Co*(1 - exp(-1i*theta*m)))
              str1[2] <- "Backwards-Euler"},
       "3"={# FTCS - 2nd Order in space
              gf <- 1 - Co*(exp(1i*theta*m) - exp(-1i*theta*m))/2
              str1[3] <- "FTCS" },
       "4"={# BTCS - 2nd Order in space
              # BTCS - Unconditionally stable - Leveque,1992 - pp99.
              gf <- 1 /(1 + Co*0.5*(exp(1i*theta*m) - exp(-1i*theta*m)))
              str1[4] <- "BTCS"},
       "5"={# Leapfrog - 2nd Order in space
              lambda1 <- -1i*Co*sin(theta*m)+
                   sqrt(as.complex(1-(Co*sin(theta*m))^2))
              lambda2 <- -1i*Co*sin(theta*m)-
                   sqrt(as.complex(1-(Co*sin(theta*m))^2))
              gf <- pmax(abs(lambda1), abs(lambda2))
              str1[5] <- "Leapfrog"},
       "6"={# Beam-Warming
              gf <- 1 - 0.5*Co*(3 - 4*exp(-1i*theta*m) + exp(-1i*2*theta*m) ) +
                   0.5*Co^2 *(1 - 2*exp(-1i*theta*m) + exp(-1i*2*theta*m) )
              str1[6] <- "Beam-Warming"},
       "7"={# Lax-Friedrichs
              gf <-  0.5*(exp(1i*theta*m) + exp(-1i*theta*m) ) -
                   0.5*Co*(exp(1i*theta*m) - exp(-1i*theta*m) )
              str1[7] <- "Lax-Friedrichs"},
       "8"={# Lax-Wendroff
              gf <- 1 - 1i*Co*sin(theta*m) - Co^2*(1-cos(theta*m))
              str1[8] <- "Lax-Wendroff"},
       "9"={gf <- 1 - Co*(3-4*exp(-1i*theta*m)+exp(-2*1i*theta*m))/2 # 2nd Order
              str1[9] <- "2nd Order"}, # Scheme: [3 -4 1]/2"
       {
         print('default')
       }
     )
# Calculate magnitude of gf - scheme unstable if gfmag > 1
     gfmag <- abs(gf)
     if(m==1){
       if(prtPlot==1){
       fileName <- sprintf("fullyDiscrete%s.png",str1[testNo])
       png(file=fileName,width = 719, height = 420, units = "pt",
           pointsize=24)}
       plot(Co, gfmag,type="l",xlim=c(-3,3),ylim=c(0,3),
            col="blue",lwd=2, xaxs="i",yaxs="i",xlab="Co",
            ylab=expression(paste("| ",G[f]," |")),
            main=str1[testNo])
```

```
        grid(lty=1,lwd=1,col="darkgray")
      }else{
        lines(Co, gfmag,col="blue",lwd=2)
      }
}
lines(Co,rep(1,N),col="black",lwd=6,lty=2)
axis(side = 1, lwd = 2)
axis(side = 2, lwd = 2)
if(prtPlot==1){dev.off()}
}
```

Listing 4.4. File: `vonNeumannFullyDiscrete.R`—Code to generate *fully discrete* gain factor plots shown in Fig. 4.4

The additional examples that follow take the same approach, and where appropriate, only the final results have been stated.

4.4.2.2 Forward in Time Backward in Space (FTBS), Implicit

We can also apply first-order *FTBS* discretizations to the hyperbolic system to get the following implicit scheme:

$$u_j^{n+1} = u_j^n - \frac{a\Delta t}{\Delta x}\left(u_j^{n+1} - u_{j-1}^{n+1}\right). \tag{4.75}$$

Applying the von Neumann procedure leads to the following stability requirement:

$$\left|G_f\right| = \left|\frac{1}{1+\text{Co}\left(1-e^{-i\theta}\right)}\right| \leq 1. \tag{4.76}$$

From Fig. 4.4 it is clear that the FTBS method is stable for Co ≤ -1 or Co ≥ 0.

4.4.2.3 Forward in Time Central in Space (FTCS)

After applying first-order central spatial and forward temporal discretizations to the hyperbolic system we get the following explicit scheme:

$$u_j^{n+1} = u_j^n - \frac{1}{2}\frac{a\Delta t}{\Delta x}\left(u_{j+1}^n - u_{j-1}^n\right). \tag{4.77}$$

Applying the von Neumann procedure leads to the following stability requirement:

$$\left|G_f\right| = \left|1 - \frac{\text{Co}}{2}\left(e^{i\theta} - e^{-i\theta}\right)\right| \leq 1. \tag{4.78}$$

From Fig. 4.4 it is clear that the FTCS method is unconditionally unstable.

4.4.2.4 Forward in Time Central in Space (FTCS), Implicit

After applying first-order central spatial and forward temporal discretizations to the above hyperbolic system we get the following implicit scheme:

$$u_j^{n+1} = u_j^n - \frac{1}{2}\frac{a\Delta t}{\Delta x}\left(u_{j+1}^{n+1} - u_{j-1}^{n+1}\right). \tag{4.79}$$

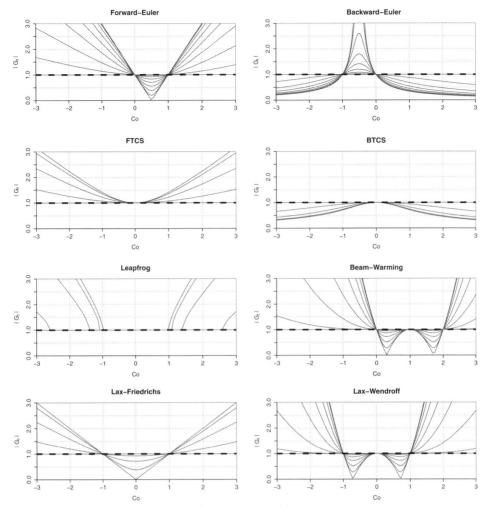

Figure 4.4. Gain diagrams for various discretization schemes. For each scheme the gain curves have been generated for a particular $\theta \in (0, \ldots, 2\pi)$. Schemes are stable for $|G_f| \leq 1$, i.e., below the dotted line.

Applying the von Neumann procedure leads to the following stability requirement:

$$|G_f| = \left| \frac{1}{1 + \frac{\text{Co}}{2}\left(e^{i\theta} - e^{-i\theta}\right)} \right| \leq 1. \tag{4.80}$$

From Fig. 4.4 it is clear that the implicit FTCS method is unconditionally stable.

4.4.2.5 Central in Time Central in Space (CTCS) or Leapfrog

After applying *CTCS* or *leapfrog* spatial and temporal discretizations to the hyperbolic system we get the following explicit scheme:

$$u_j^{n+1} = u_j^{n-1} - \frac{a\Delta t}{\Delta x}\left(u_{j+1}^n - u_{j-1}^n\right). \tag{4.81}$$

However, we cannot apply von Neumann stability analysis directly because of the u_j^{n-1} term. Therefore, we introduce additional variables $v_j^n = u_j^n$ and $w_j^n = u_j^{n-1}$ to get

$$\begin{bmatrix} v_j^{n+1} \\ w_j^{n+1} \end{bmatrix} = \begin{bmatrix} 0 & 1 \\ 1 & 0 \end{bmatrix} \begin{bmatrix} v_j^n \\ w_j^n \end{bmatrix} - \begin{bmatrix} \text{Co} & 0 \\ 0 & 0 \end{bmatrix} \begin{bmatrix} v_{j+1}^n \\ w_{j+1}^n \end{bmatrix} + \begin{bmatrix} \text{Co} & 0 \\ 0 & 0 \end{bmatrix} \begin{bmatrix} v_{j-1}^n \\ w_{j-1}^n \end{bmatrix}. \quad (4.82)$$

We are now able to apply the von Neumann procedure to get

$$\begin{aligned} \begin{bmatrix} \widehat{V}^{n+1} \\ \widehat{W}^{n+1} \end{bmatrix} &= \begin{bmatrix} 0 & 1 \\ 1 & 0 \end{bmatrix} \begin{bmatrix} \widehat{V}^n \\ \widehat{W}^n \end{bmatrix} - e^{i\theta} \begin{bmatrix} \text{Co} & 0 \\ 0 & 0 \end{bmatrix} \begin{bmatrix} \widehat{V}^n \\ \widehat{W}^n \end{bmatrix} + e^{-i\theta} \begin{bmatrix} \text{Co} & 0 \\ 0 & 0 \end{bmatrix} \begin{bmatrix} \widehat{V}^n \\ \widehat{W}^n \end{bmatrix} \\ &= \begin{bmatrix} \text{Co}\left(e^{-i\theta} - e^{i\theta}\right) & 1 \\ 1 & 0 \end{bmatrix} \begin{bmatrix} \widehat{V}^n \\ \widehat{W}^n \end{bmatrix} \\ &= \begin{bmatrix} -i2\text{Co}\sin(\theta) & 1 \\ 1 & 0 \end{bmatrix} \begin{bmatrix} \widehat{V}^n \\ \widehat{W}^n \end{bmatrix} \\ &= G_f \begin{bmatrix} \widehat{V}^n \\ \widehat{W}^n \end{bmatrix}. \end{aligned} \quad (4.83)$$

Because G_f is a matrix for this scheme, we need to obtain its *spectral radius*, that is, $\rho(G_f) = \max_k |\lambda_k| \le \|G_f\|$ (the equality only holds for *normal* G_f). The characteristic equation of G_f is

$$\lambda^2 + 2i\text{Co}\sin(\theta)\lambda - 1 = 0. \quad (4.84)$$

Thus, the eigenvalues of G_f are

$$\lambda = -i\text{Co}\sin(\theta) \pm \sqrt{1 - \text{Co}^2\sin^2(\theta)}. \quad (4.85)$$

Clearly, for $|\text{Co}| \le 1$, the square root term will be real resulting in $|\lambda_k| = \sqrt{\lambda_k \lambda_k^*} = 1$, $k \in (1, 2)$, where λ_k^* is the complex conjugate of λ_k. Hence, the system will be stable. However, for $|\text{Co}| > 1$, the square root term becomes purely imaginary for some values of θ when $\rho(G_f) = \max_{k=1,2} |\lambda_k| > 1$; hence, the system will be unstable. This situation is apparent from Fig. 4.4.

This example illustrates how to proceed when tackling discretization schemes that involve multiple time steps. A similar approach is also taken where the dependent variable is a vector quantity.

4.4.2.6 Beam-warming (B-W)
After applying B-W discretization to the hyperbolic system we get the following explicit scheme:

$$u_j^{n+1} = u_j^n - \frac{1}{2}\frac{a\Delta t}{\Delta x}\left(3u_j^n - 4u_{j-1}^n + u_{j-2}^n\right) + \frac{1}{2}\left(\frac{a\Delta t}{\Delta x}\right)^2 \left(u_j^n - 2u_{j-1}^n + u_{j-2}^n\right). \quad (4.86)$$

Applying the von Neumann procedure leads to the following stability requirement:

$$|G_f| = \left|1 - \frac{\text{Co}}{2}\left(3 - 4e^{-i\theta} + e^{-2i\theta}\right) + \frac{\text{Co}^2}{2}\left(1 - 2e^{-i\theta} + e^{-2i\theta}\right)\right| \leq 1. \quad (4.87)$$

From Fig. 4.4 it is clear that the B-W method is stable for $0 \leq \text{Co} \leq 2$.

4.4.2.7 Lax–Friedrichs (LxF)

After applying LxF discretization to the hyperbolic system we get the following explicit scheme:

$$u_j^{n+1} = \frac{1}{2}\left(u_{j-1}^n + u_{j+1}^n\right) - \frac{1}{2}\frac{a\Delta t}{\Delta x}\left(u_{j+1}^n - u_{j-1}^n\right). \quad (4.88)$$

Applying the von Neumann procedure leads to the following stability requirement:

$$\begin{aligned}|G_f| &= \left|\tfrac{1}{2}\left(e^{i\theta} + e^{-i\theta}\right) - \tfrac{\text{Co}}{2}\left(e^{i\theta} - e^{-i\theta}\right)\right| \leq 1 \\ &= \left|\cos(\theta) - i\text{Co}\sin(\theta)\right| \leq 1 \\ &= \sqrt{\cos^2(\theta) + \text{Co}^2\sin^2(\theta)} \leq 1.\end{aligned} \quad (4.89)$$

From Fig. 4.4 it is clear that the Lax–Friedrichs method is stable for $-1 \leq \text{Co} \leq 1$.

4.4.2.8 Lax–Wendroff (LxW)

After applying LxW discretization to the hyperbolic system we get the following explicit scheme:

$$u_j^{n+1} = u_j^n - \frac{1}{2}\frac{a\Delta t}{\Delta x}\left(u_{j+1}^n - u_{j-1}^n\right) + \frac{1}{2}\left(\frac{a\Delta t}{\Delta x}\right)^2\left(u_{j-1}^n - 2u_j^n + u_{j+1}^n\right). \quad (4.90)$$

Applying the von Neumann procedure leads to the following stability requirement:

$$\begin{aligned}|G_f| &= \left|1 - \tfrac{\text{Co}}{2}\left(e^{i\theta} - e^{-i\theta}\right) + \tfrac{\text{Co}^2}{2}\left(e^{i\theta} - 2 + e^{-i\theta}\right)\right| \leq 1 \\ &= \left|1 - i\text{Co}\sin(\theta) - \text{Co}^2(1 - \cos(\theta))\right| \leq 1.\end{aligned} \quad (4.91)$$

From Fig. 4.4 it is clear that the Lax–Wendroff method is stable for $-1 \leq \text{Co} \leq 1$.

4.4.2.9 Lax–Wendroff (LxW), 2D

We will now briefly consider the stability of the Lax–Wendroff numerical scheme applied to the 2D scalar *hyperbolic* system $u_t + a_x u_x + a_y u_y = 0$ with a_x and a_y constant.

For the preceding 2D hyperbolic system the Lax–Wendruff scheme is significantly more complex than its 1D counterpart. The 2D scheme is given by [LeV-02]

$$u_{j,k}^{n+1} = u_{j,k}^n - \frac{1}{2}\frac{a_x \Delta t}{\Delta x}\left(u_{j+1,k}^n - u_{j-1,k}^n\right) - \frac{1}{2}\left(\frac{a_y \Delta t}{\Delta y}\right)\left(u_{j,k+1}^n - u_{j,k-1}^n\right)$$

$$+ \frac{1}{2}\frac{a_x^2 \Delta t^2}{\Delta x^2}\left(u_{j+1,k}^n - 2u_{j,k}^n + u_{j-1,k}^n\right) + \frac{1}{2}\frac{a_y^2 \Delta t^2}{\Delta y^2}\left(u_{j,k+1}^n - 2u_{j,k}^n + u_{j,k-1}^n\right) \quad (4.92)$$

$$+ \frac{1}{8}\frac{\Delta t^2}{\Delta x \Delta y}(a_x a_y)\left[\left(u_{j+1,k+1}^n - u_{j-1,k+1}^n\right) - \left(u_{j+1,k-1}^n - u_{j-1,k-1}^n\right)\right].$$

To investigate stability for this case, we need to take the 2D Fourier transform, and this leads to the following *gain factor*:

$$G_f e^{i(j\theta + k\alpha)} = e^{i(j\theta + k\alpha)} - \frac{\mathrm{Co}_x}{2}\left(e^{i([j+1]\theta + k\alpha)} - e^{i([j-1]\theta + k\alpha)}\right)$$

$$- \frac{\mathrm{Co}_y}{2}\left(e^{i(j\theta + [k+1]\alpha)} - e^{i(j\theta + [k-1]\alpha)}\right)$$

$$+ \frac{\mathrm{Co}_x^2}{2}\left(e^{i([j+1]\theta + k\alpha)} - 2e^{i(j\theta + k\alpha)} + e^{i([j-1]\theta + k\alpha)}\right) \quad (4.93)$$

$$+ \frac{\mathrm{Co}_y^2}{2}\left(e^{i(j\theta + [k+1]\alpha)} - 2e^{i(j\theta + k\alpha)} + e^{i(j\theta + [k-1]\alpha)}\right)$$

$$+ \frac{\mathrm{Co}_x \mathrm{Co}_y}{8}\left[\left(e^{i([j+1]\theta + [k+1]\alpha)} - e^{i([j-1]\theta + [k+1]\alpha)}\right)\right.$$

$$\left. - \left(e^{i([j+1]\theta + [k-1]\alpha)} - e^{i([j-1]\theta + [k-1]\alpha)}\right)\right].$$

Converting to trigonometrical expressions, we get

$$G_f = 1 - \mathrm{Co}_x i \sin(\theta) - \mathrm{Co}_y i \sin(\alpha)$$

$$+ \frac{\mathrm{Co}_x^2}{1}(\cos(\theta) - 1) + \frac{\mathrm{Co}_y^2}{1}(\cos(\alpha) - 1) \quad (4.94)$$

$$+ \frac{\mathrm{Co}_x \mathrm{Co}_y}{4}[i \sin(\theta) \sin(\alpha)].$$

For a stable system the requirement is that $G_f \leq 1$, and this condition will be met if $\sqrt{\mathrm{Co}_x^2 + \mathrm{Co}_y^2} \leq \frac{1}{\sqrt{2}}$. For more details regarding the stability analysis of multidimension schemes, the reader is referred to Hirsch [Hir-88].

4.4.2.10 Crank–Nicholson

The *Crank–Nicholson* method when applied to the 1D heat equation $u_t = \mu u_{xx}$, with i.c. $u(x, 0) = f(x)$ and b.c. periodic, takes the following implicit form:

$$u_j^{n+1} = u_j^n + \frac{1}{2}\frac{\mu \Delta t}{(\Delta x)^2}\left[u_{j+1}^{n+1} + u_{j+1}^n - 2\left(u_j^{n+1} + u_j^n\right) + u_{j-1}^{n+1} + u_{j-1}^{n1}\right]. \quad (4.95)$$

Following the approach outlined in Section 4.4.2, we first substitute $u_j^n = e^{i\xi j\Delta x}\hat{U}_j^n$ into the discretized PDE eqn. (4.95) and rearrange to obtain the gain factor

$$G_f = \frac{\hat{U}^{n+1}}{\hat{U}^n} = \frac{1 - 2k\sin^2(\theta/2)}{1 + 2k\sin^2(\theta/2)}, \quad (4.96)$$

where $k = \frac{1}{2}\frac{\mu\Delta t}{(\Delta x)^2}$, $\theta = \xi j\Delta x$, and we have used the trignometrical half-angle formula.

Clearly, $|G_f| \leq 1$ for all values of θ providing that $\mu > 0$. Therefore, under these conditions, the Crank–Nicholson method is unconditionally stable, and the spatial grid spacing and time step may be chosen to achieve an acceptable truncation error.

4.4.3 Semi-Discrete Systems

The approach will be generally similar to that applied to the fully discrete schemes, except that semi-discrete schemes only apply discretization to the *spatial domain*, and this will change the analysis. However, we will arrive at similar requirements for stability, although, as mentioned earlier, "space discretization cannot be said to be unstable by itself. It is only when it is coupled to a time integration that we can decide upon stability" [Hir-88, p. 415]. Thus, we are essentially investigating the well-posedness of the scheme.

The idea is to take the Fourier transform of the semi-discretized partial differential equation, using the *shifting property* where spatial differentials occur. We then have a system of ordinary differential equations in the complex frequency domain whose stability properties will be determined by its eigenvalues. The requirement for stability is that the eigenvalues should not have positive real parts. Application of *Parseval's theorem* shows that if stability is established in the complex frequency, or Fourier domain, then this result applies equally to the spatial or grid domain. In other words, stability in the complex frequency domain implies stability in the time domain, and vice versa.

Consider the general, linear PDE of evolution in m space variables with periodic boundary conditions and suitable initial conditions

$$u_t(x,t) = \mathcal{P}\left(\frac{\partial}{\partial x}\right)u(x,t), \quad x \in \mathbb{R}^m, \quad t > 0, \quad (4.97)$$

semi-discretized as follows:

$$\frac{du_j(t)}{dt} = Pu_j(t), \quad u_j(0) = u(j\Delta x, 0), \quad (4.98)$$

where P is a space discretization operator. If we take the Fourier transform with respect to x, we get

$$\begin{aligned}
\frac{d\hat{U}_j(\xi,t)}{dt} &= \mathcal{F}\{Pu_j(t)\} \\
&= \int_{-\infty}^{+\infty} e^{-i\xi x}Pu_j(t)\,dx \\
&= \hat{P}\left(e^{i\xi\Delta x}\right)\int_{-\infty}^{+\infty} e^{-i\xi x}u_j(t)\,dx \\
\therefore \frac{d\hat{U}_j(\xi,t)}{dt} &= \hat{P}\left(e^{i\xi\Delta x}\right)\hat{U}_j,
\end{aligned} \quad (4.99)$$

where $\hat{P}\left(e^{i\xi\Delta x}\right) = G_f(\theta)\ \{\theta = \xi\Delta x\}$ represents the result of the Fourier transform shift operator applied to P, which is called the *gain factor* or *amplification ratio*. In practice, as for the fully discrete case, the limits of integration only have to include wavenumbers that can be resolved by the grid ($\pm\pi/\Delta x$). Outside of this range, the grid function is assumed to be periodic.

We now have an ODE relationship for $\hat{U}_j$ that we can use to investigate the stability of the discretized system. It is clear from linear system theory that the solution to eqn. (4.99) is

$$\hat{U}_j(\xi, t) = e^{G_f(\theta)t}\hat{U}_j(\xi, 0), \qquad (4.100)$$

where $\hat{U}_j(\xi, 0) = \mathcal{F}\{u_j(0)\}$. It therfore follows from the *Cauchy–Schwarz inequality* that

$$\left\|\hat{U}_j(\xi, t)\right\| \leq \left\|e^{G_f(\theta)t}\right\| \left\|\hat{U}_j(\xi, 0)\right\|. \qquad (4.101)$$

Note that eqn. (4.100) also implies that $\hat{U}_j(\xi, [n+1]\Delta t) = e^{G_f(\theta)\Delta t}\hat{U}_j(\xi, n\Delta t)$. Now, the requirement [Kre-05] for the system to be well behaved and stable, that is, for $\hat{U}_j$ not to grow unbounded, is that there are real constants K and α independent of ξ and Δx, such that

$$\left\|e^{G_f(\theta)}\right\| \leq Ke^\alpha = 1, \quad \forall |\theta| \leq \pi. \qquad (4.102)$$

We can also define the *region of well-posedness* as being the set

$$D = \left\{\hat{P}\left(e^{i\xi\Delta x}\right) \in \mathbb{C}^{m \times m} : \left\|\hat{U}_j(\xi, t)\right\| \leq \left\|\hat{U}_j(\xi, 0)\right\|, \forall \xi \in \mathbb{R}^m, t > 0\right\},$$

$$\equiv \left\{G_f(\theta) \in \mathbb{C}^{m \times m} : \left\|e^{G_f(\theta)}\right\| \leq 1, \forall \theta \in \mathbb{R}^m, t > 0\right\}.$$

For a 1D scalar PDE, $G_f(\theta)$ in eqn. (4.102) is equal to the system eigenvalue λ. Thus, eqn. (4.100) becomes $\hat{U}(\xi, t) = e^{\lambda t}\hat{U}(\xi, 0)$. Clearly, eqn. (4.102) will hold (where $K = 1$ and $\alpha = \lambda$) and the ODE of eqn. (4.99) be well-posed for all ξ, providing Re $\{\lambda\} \leq 0$.

Now, where we have a single matrix $G_f(\theta) \in \mathbb{C}^{m \times m}$, it can be transformed to *Jordan canonical form* $J = RG_f(\theta)R^{-1}$ with eigenvalues λ_i, $i = (1, \ldots, m)$ [Fra-95]. Thus, utilizing the matrix exponential characteristic $e^{At} = R^{-1}e^{\Lambda t}R$ (see the appendix to the book), we proceed as follows:

$$e^{G_f(\theta)t} = e^{R^{-1}JRt} = e^{R^{-1}JtR}$$
$$= R^{-1}e^{Jt}R \qquad (4.103)$$
$$\therefore \left\|e^{G_f(\theta)t}\right\| \leq \left\|R^{-1}\right\| \left\|e^{Jt}\right\| \|R\| = \left\|e^{Jt}\right\| \text{cond}(R).$$

It therefore follows that $\left\|e^{Jt}\right\|$ will stay bounded for all $t \geq 0$, providing that either of the following conditions exists for the *spectral abscissa*, $\phi(J) = \max_k \text{Re}\{\lambda_k\}$:

- $\phi(J) < 0$ or
- $\phi(J) = 0$ (when the corresponding *Jordan block* J_r, of J, has dimension 1×1 [Kre-05, p. 44–45]).

Note that if $\hat{\mathcal{P}}(i\xi)$ has distinct eigenvectors $\mathbf{r}_i$ $i = (1, \ldots, m)$, then $R = \{\mathbf{r}_1 \cdots \mathbf{r}_m\}$ and diagonalization is guaranteed—true for hyperbolic systems. Therefore, we have

$\left\| e^{G_f(\theta)t} \right\| \leq \left\| e^{Jt} \right\| \text{cond}(R) = e^{\phi(J)t} \text{cond}(R)$, and we can modify the requirement for the well-posedness of eqn. (4.102) to[3]

$$e^{\phi(J)t} \leq Ke^{\alpha t}, \quad t > 0, \; \forall \xi \in \mathbb{R}^m$$

$$\Downarrow \tag{4.104}$$

$$\phi(J) \leq 0, \; \forall \xi \in \mathbb{R}^m, t > 0,$$

and the modified region of stability then becomes

$$D = \left\{ \lambda \in \mathbb{C}^m \; : \; \phi(J) \leq 0, \; \forall \xi \in \mathbb{R}^m, t > 0 \right\}. \tag{4.105}$$

Thus, the problem of establishing well-posedness of a semi-discrete discretization scheme is reduced to obtaining the Jordan cononical form, J, of the symbol $\widehat{\mathcal{P}}(i\xi)$, and determining if the spectral abscissa is less than or equal to zero. For further details on the theory of well-posedness, refer to Kreiss and Lorenz [Kre-05, Chapter 2].

By invoking *Parseval's theorem* of equality, $\left\| \hat{U}_j(\xi, t) \right\|_2 = \left\| u_j(t) \right\|_2$, we are assured that we can investigate stability in either the spatial (grid) domain or the complex frequency domain.

We reiterate the point made for the fully discrete case that there are some problems where a component of the exact solution actually grows exponentially with increasing time, for instance, the heat conduction equation

$$\frac{\partial T}{\partial t} = \mu \frac{\partial^2 T}{\partial x^2} + q, \tag{4.106}$$

where the source term $q = kT$, $k > 0$. For these problems, the requirement for stability is

$$\left\| e^{G_f(\theta)} \right\| \leq Ke^{\alpha} = 1 + \mathcal{O}(\Delta t), \quad \forall |\theta| \leq \pi. \tag{4.107}$$

However, for most practical problems, this condition is rarely used as the condition of eqn. (4.102) or (4.104) suffices. For more details, refer to Richtmyer and Morton [Ric-67] and Hirsch [Hir-88].

We may summarize the method as follows:

1. Substitute $u_j(t) = e^{i\xi j \Delta x} \hat{U}_j(\xi, t)$ into the semi-discretized PDE equation.
2. Cancel like terms to obtain a function in the form $\frac{d\hat{U}_j(\xi,t)}{dt} = G_f(\theta) \hat{U}_j(\xi, t)$, where $G_f(\theta) = \hat{P}(e^{i\xi \Delta x})$ and $\theta = \xi \Delta x$. From linear theory we obtain the analytical solution $\hat{U}_j(\xi, t) = e^{G_f(\theta)t} \hat{U}_j(\xi, 0)$.
3. Determine the limitations on Δx and other parameters such that $\phi(J) = \max_i \text{Re}\{\lambda_i\} \leq 0$.

For the more difficult discretization schemes, a stability plot can easily be generated by use of a computer program. We illustrate a practical approach to the calculations in the following examples.

---

[3] Because the constant on the RHS of eqn. (4.104) is sufficiently large and arbitry, we can choose it to be equal to $K \times \text{cond}(R)$ and divide through by $\text{cond}(R)$ to obtain the desired result.

We will now consider the stability or well-posedness of *semi-discrete* numerical schemes applied to the scalar advection system $u_t + au_x = 0$ with a constant. Semi-discrete schemes only apply discretization to the *spatial domain*. An expression for the gain factor will be calculated for each scheme. The approach will be generally similar to that applied to the fully discrete schemes.

4.4.3.1 First-Order Upwind, 1D

For the advection system after spatial discretization we get

$$\frac{du_j(t)}{dt} = -\frac{a}{\Delta x}\left(u(t)_j - u(t)_{j-1}\right). \tag{4.108}$$

Step 1: Substituting $u_j = e^{i\xi j \Delta x}\hat{U}_j(\xi, t)$ (equivalent to taking the Fourier transform with respect to x) yields

$$e^{i\xi j \Delta x}\frac{d\hat{U}_j(\xi, t)}{dt} = -\frac{a}{\Delta x}\left(e^{i\xi j \Delta x}\hat{U}_j(\xi, t) - e^{i\xi(j-1)\Delta x}\hat{U}_j(\xi, t)\right). \tag{4.109}$$

Step 2: Dividing both sides by $e^{i\xi j \Delta x}$, we obtain

$$\begin{aligned}\frac{d\hat{U}_j(\xi, t)}{dt} &= -\frac{a}{\Delta x}\left(1 - e^{-i\xi \Delta x}\right)\hat{U}_j \\ &= G_f(\theta)\hat{U}_j \\ &\Downarrow \\ \hat{U}_j(\xi, t) &= e^{G_f(\zeta,\theta)t}\hat{U}_j(\xi, 0),\end{aligned} \tag{4.110}$$

where $\theta = \xi \Delta x$ and $\zeta = \frac{a}{\Delta x}$. Thus, from the arguments subsequent to eqn. (4.101), we now have the following expression that defines the *gain factor*:

$$e^{G_f(\zeta,\theta)} = e^{-\zeta\left(1-e^{-i\theta}\right)}. \tag{4.111}$$

Step 3: Determine the limitations on $\zeta = a/\Delta x$. For the system to be well-posed and stable we require either of the following equivalent conditions to be true:

$$\text{Re}\left\{G_f(\zeta,\theta)\right\} \leq 0 \quad \text{or} \quad \left|e^{G_f(\zeta,\theta)}\right| \leq 1.$$

We have used R to generate plots for both criteria (see Listing 4.5. We see from both plots of Fig. 4.5 that the requirement of eqn. (4.101) is satisfied for this scheme when we have

$$\zeta \geq 0 \quad \Rightarrow \quad a \geq 0. \tag{4.112}$$

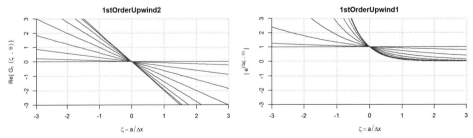

Figure 4.5. Gain factor plot for first-order semi-discrete scheme. The gain curves have been generated for a particular $\theta \in (0, \ldots, 2\pi)$. (left) Stable for $\text{Re}\{G_f\} \leq 0$. (right) Stable for $|e^{G_f}| \leq 1$.

Thus, the scheme is well-posed when applied to the 1D advection PDE. Or, from eqn. (4.102) and subsequent discussion, if the eigenvalues of the discretized system do not have positive real parts. This completes the semi-discrete von Neumann stability analysis for the first-order upwind scheme method.

```r
# Von Neumann Analysis of PDE Discretisation Schemes
N <- 101
zeta <- seq(-3, 3, len=N) # <- a*delta_t/delta_x

colNum <- c("red", "blue","green","mmagenta","black")
theta    <- pi/8
prtPlot <- 0 # 0=off, 1=on
cat(sprintf("print plot = %d\n", prtPlot))
# Plot contours of magmetude of Growth Factor-various schemes
strT1 <- sprintf("PDE: ut + a.ux = 0 - ")
strT2 <- sprintf("Gain plots for theta varying between 0 and 2*pi")
str1  <- rep(0,9)
#
for(testNo in 1:4){
#testNo <- 1
  for(m in 1:16){ # angle = m*zeta
    switch(testNo,
      "1"={# 1st order upwind
          gf <- -zeta*(1 - exp(-1i*theta*m))
          abs_eGf <- abs(exp(gf)) # Stable if |exp(Gf)| <= 1
          str1[1] <- "1stOrderUpwind1"
          yLabel <- expression(paste("| ",e^{(G[f]*list(zeta*phantom(0),
             ~theta))}," |"))},
      "2"={# 1st order upwind
          gf <- -zeta*(1 - exp(-1i*theta*m))
          abs_eGf <- Re(gf) # Stable if Re{ Gf } <= 0
          str1[2] <- "1stOrderUpwind2"
          yLabel <- expression(paste("Re{ ",G[f]*phantom(0)(list(zeta*phantom(0),
             ~theta))," }"))},
      "3"={# 2nd order upwind
          gf <- -zeta*(3 - 4*cos(-theta*m) + cos(-2*theta*m))/2 # 2nd Order
          abs_eGf <- abs(exp(gf)) # Stable if |exp(Gf)| <= 1
          str1[3] <- "2ndOrderUpwind1"
          yLabel <- expression(paste("| ",e^{(G[f]*list(zeta*phantom(0),
             ~theta))}," |")) },
      "4"={# 2nd Order in upwind
          gf <- -zeta*(3 - 4*cos(-theta*m) + cos(-2*theta*m))/2 # 2nd Order
          abs_eGf <- Re(gf) # Stable if Re{ Gf } <= 0
          str1[4] <- "2ndOrderUpwind2"
          yLabel <- expression(paste("Re{ ",G[f]*phantom(0)(list(zeta*phantom(0),
             ~theta))," }"))},
```

```
            {# Default
                print('ERROR: Wrong testNo!')
            }
        )
    # Calculate magnitude of gf - Note: scheme unstable if gfmag > 1
        if(m==1){
          if(prtPlot ==1){
          fileName <- sprintf("semiDiscrete%s.png",str1[testNo])
          png(file=fileName,width = 719, height = 420, units = "pt",
                pointsize=24)}
          par(mai=c(2,2,1,0.25))
          plot(zeta, abs_eGf,type="l",xlim=c(-3,3),ylim=c(-3,3),
                col="blue",lwd=2, xaxs="i",yaxs="i",
                xlab=expression(paste(zeta==a/Delta*x)),
                ylab=yLabel,
                main=str1[testNo])
          grid(lty=1,lwd=1,col="darkgray")
        }else{
          lines(zeta, abs_eGf,col="blue",lwd=2)
        }
    }
    #lines(zeta,rep(1,N),col="black",lwd=6,lty=2)
    axis(side = 1, lwd = 2)
    axis(side = 2, lwd = 2)
    if(prtPlot ==1){dev.off()}
    }
```

Listing 4.5. File: `vonNeumannSemiDiscrete.R`—Code to generate *semi-discrete* gain factor plots shown in Figs. 4.5 and 4.6

4.4.3.2 Second-Order Upwind, 1D

For the advection system we can apply a second-order upwind spatial discretization [Sch-91] to get

$$\frac{du_j(t)}{dt} = -\frac{a}{\Delta x}\left(3u(t)_j - 4u(t)_{j-1} + u(t)_{j-2}\right)/2. \tag{4.113}$$

Step 1: Substituting $u_j = e^{i\xi j\Delta x}\hat{U}_j(\xi,t)$ (equivalent to taking the Fourier transform with respect to x) yields

$$e^{i\xi j\Delta x}\frac{d\hat{U}_j(\xi,t)}{dt} = -\frac{a}{\Delta x}\left(3e^{i\xi j\Delta x}\hat{U}_j(\xi,t) - 4e^{i\xi(j-1)\Delta x}\hat{U}_j(\xi,t) + e^{i\xi(j-2)\Delta x}\hat{U}_j(\xi,t)\right)/2. \tag{4.114}$$

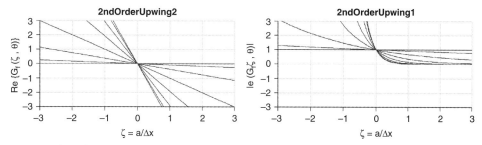

Figure 4.6. Gain factor plot for second-order, upwind, semi-discrete scheme. The gain curves have each been generated for a particular $\theta \in (0, \ldots, 2\pi)$; (left) stable for $\text{Re}\{G_f\} \leq 0$; (right) stable for $\left|e^{G_f}\right| \leq 1$.

Step 2: Dividing both sides by $e^{i\xi j \Delta x}$, we obtain

$$\frac{d\hat{U}_j(\xi, t)}{dt} = -\frac{a}{\Delta x}\left(3 - 4e^{-i\xi \Delta x} + e^{-i2\xi \Delta x}\right)\hat{U}_j/2 \qquad (4.115)$$
$$\Downarrow$$
$$\hat{U}_j(\xi, t) = e^{G_f(\zeta, \theta)t}\hat{U}_j(\xi, 0),$$

where $\theta = \xi \Delta x$ and $\zeta = \frac{a}{\Delta x}$. Thus, from the arguments subsequent to eqn. (4.101), we now have the following expression, which defines the *gain factor*:

$$e^{G_f(\zeta, \theta)} = e^{-\zeta\left(3 - 4e^{-i\theta} + e^{-i2\theta}\right)/2}. \qquad (4.116)$$

Step 3: Determine the limitations on $\zeta = a/\Delta x$. For the system to be well-posed and stable we again require either of the following equivalent conditions to be true:

$$\text{Re}\{G_f(\zeta, \theta)\} \leq 0 \quad \text{or} \quad \left|e^{G_f(\zeta, \theta)}\right| \leq 1. \qquad (4.117)$$

We have used R to generate plots for both criteria (see Listing 4.5). We see from both plot (a) and plot (b) of Fig. 4.6 that the requirement of eqn. (4.101) is satisfied for this scheme when we have

$$\zeta \geq 0 \quad \Rightarrow \quad a \geq 0. \qquad (4.118)$$

Thus, the scheme is also well-posed when applied to the 1D advection PDE, or, from eqn. (4.102) and subsequent discussion, if the eigenvalues of the discretized system do not have positive real parts. This completes the semi-discrete von Neumann stability analysis for the second-order upwind scheme method.

4.4.3.3 First-Order Upwind, 2D

After spatial discretization of a 2D instance of the advection system we obtain

$$\frac{du_{k,j}(t)}{dt} = -\frac{a}{\Delta x}\left(u(t)_{k,j} - u(t)_{k-1,j}\right) - \frac{a}{\Delta y}\left(u(t)_{k,j} - u(t)_{k,j-1}\right). \qquad (4.119)$$

We now proceed in the same way as for the 1D case, except that we are now dealing with the two-dimensional, $U(x, y, t)$. Thus, we have the following:

Step 1: Substituting $u_{i,j} = e^{i\xi k \Delta x} e^{i\xi j \Delta y} \hat{U}_{k,j}(\xi, \eta, t)$, which is equivalent to taking the Fourier transform with respect to x and y, yields

$$e^{i\xi k \Delta x} e^{i\xi j \Delta y} \frac{d\hat{U}_{k,j}(\xi,\eta,t)}{dt} = -\frac{ae^{i\xi j \Delta y}}{\Delta x}\left(e^{i\xi k \Delta x}\hat{U}_{k,j}(\xi,\eta,t) - e^{i\xi (k-1)\Delta x}\hat{U}_{k,j}(\xi,\eta,t)\right)$$
$$-\frac{ae^{i\xi k \Delta x}}{\Delta y}\left(e^{i\xi j \Delta y}\hat{U}_{k,j}(\xi,\eta,t) - e^{i\xi (j-1)\Delta y}\hat{U}_{k,j}(\xi,\eta,t)\right). \quad (4.120)$$

Step 2: Dividing both sides by $e^{i\xi k \Delta x} e^{i\xi j \Delta y}$, we obtain

$$\frac{d\hat{U}_{k,j}(\xi,\eta,t)}{dt} = -\left[\frac{a}{\Delta x}\left(1 - e^{-i\xi \Delta x}\right) + \frac{a}{\Delta y}\left(1 - e^{-i\eta \Delta y}\right)\right]\hat{U}_{k,j}$$
$$= G_f \hat{U}_{k,j}$$
$$\Downarrow \qquad (4.121)$$
$$\hat{U}_{k,j}(\xi,\eta,t) = e^{G_f t}\hat{U}_{k,j}(\xi,\eta,0)$$
$$\Downarrow$$
$$\hat{U}_{k,j}(\xi,\eta,(n+1)\Delta t) = e^{G_f \Delta t}\hat{U}_{k,j}(\xi,\eta,n\Delta t),$$

where $G_f = G_f(\theta, \phi, \zeta_{\Delta x}, \zeta_{\Delta y})$ and $\theta = \xi \Delta x$, $\phi = \eta \Delta y$, $\zeta_{\Delta x} = \frac{a}{\Delta x}$ and $\zeta_{\Delta y} = \frac{a}{\Delta y}$. Thus, from the arguments subsequent to eqn. (4.101), we now have the following expression, which defines the *gain factor*:

$$e^{G_f} = e^{-\left[\zeta_{\Delta x}\left(1-e^{-i\theta}\right) + \zeta_{\Delta y}\left(1-e^{-i\phi}\right)\right]}. \quad (4.122)$$

Step 3: Determine the limitations on $\zeta_{\Delta x}$ and $\zeta_{\Delta y}$. For the system to be well-posed and stable we require either of the following equivalent conditions to be true:

$$\text{Re}\left\{G_f\right\} \leq 0 \quad \text{or} \quad \left|e^{G_f}\right| \leq 1.$$

By inspection of eqn. (4.122) we see that the requirement of eqn. (4.101) is satisfied for this scheme when we have

$$\zeta_{\Delta x} \geq 0 \text{ and } \zeta_{\Delta y} \geq 0 \quad \Rightarrow \quad a \geq 0. \quad (4.123)$$

Thus, the scheme is well-posed when applied to the 2D advection PDE. Or, from eqn. (4.102) and subsequent discussion, the scheme is stable if the eigenvalues of the discretized system do not have positive real parts.

This completes the semi-discrete von Neumann stability analysis for the 2D first-order upwind scheme method.

4.5 UNSTRUCTURED GRIDS

Professor M. B. Giles has analyzed the stability of PDE systems on unstructured grids and offers the following comment:

> Fourier analysis is the standard method for analyzing the stability of discretisations of an initial value pde on a regular structured grid. For each point in the computational grid, a linear model problem is constructed on an infinite grid with uniform grid spacing and coefficients matching those of the chosen point. This model problem has Fourier eigenmodes whose stability is relatively easily analysed. If they are stable at all points in the grid, and the discretisation of the boundary conditions is also stable (which can be analysed using Godunov/Ryabenkii or GKS stability theory) then for most applications the overall discretisation is stable, in the sense of Lax). The Lax Equivalence theorem then applies, that if the

discretisation is consistent (for a dense subset of sufficiently smooth initial conditions) then the discrete solution will approach the analytic solution for all initial conditions as the grid spacing and time step are reduced to zero. However, engineering applications of CFD are increasingly using finite volume and finite element methods based on unstructured grids. For these, Fourier stability analysis is not applicable, and one must instead consider the full discrete matrix that arises from the combined spatial and temporal discretization of the pde and associated boundary conditions.

Readers interested in pursuing this aspect further are referred to the insightful paper by Giles [Gil-97].

4.A FOURIER TRANSFORMS

$$x(t) = \frac{1}{\sqrt{2\pi}} \int_{-\infty}^{\infty} X(\omega) e^{j\omega t} d\omega, \quad j = \sqrt{-1} \tag{4.124}$$

$$X(\omega) = \frac{1}{\sqrt{2\pi}} \int_{-\infty}^{\infty} x(t) e^{-j\omega t} dt \tag{4.125}$$

Parseval's theorem: general statement

$$\int_{-\infty}^{+\infty} |x(t)|^2 dt = \int_{-\infty}^{+\infty} |X(\omega)|^2 d\omega \tag{4.126}$$

Table 4.1. Fourier transform properties

Property	Function $f(t)$	Fourier transform $F(\omega)$
	$x(t)$	$X(\omega)$
	$y(t)$	$Y(\omega)$
Linearity	$ax(t) + by(t)$	$aX(\omega) + bY(\omega)$
Time shift	$x(t - t_0)$	$e^{-j\omega t_0} X(\omega)$
Frequency shifting	$e^{j\omega_0 t} x(t)$	$X((\omega - \omega_0))$
Conjugation	$x^*(t)$	$X^*(-\omega)$
Time reversal	$x(-t)$	$X(-\omega)$
Scaling	$x(at)$	$\frac{1}{|a|} X\left(\frac{\omega}{a}\right)$
Convolution in time domain	$x(t) * y(t)$	$X(\omega) Y(\omega)$
Multiplication in time domain	$x(t) y(t)$	$\frac{1}{2\pi} X(\omega) * Y(\omega)$
nth time derivative	$\frac{d^n}{dt^n} x(t)$	$(j\omega)^n X(\omega)$
Integration	$\int_{-\infty}^{t} x(t) dt$	$\frac{1}{j\omega} X(\omega) + \pi X(0) \delta(\omega)$
nth frequency derivative	$t^n x(t)$	$(j)^n \frac{d^n}{d\omega^n} X(\omega)$

Table 4.2. Fourier transform pairs

Function, $f(t)$	Fourier transform, $F(\omega)$
$\sum_{k=-\infty}^{+\infty} a_k e^{jk\omega_0 t}$	$2\pi \sum_{k=-\infty}^{+\infty} a_k \delta(\omega - k\omega_0)$
$e^{j\omega_0 t}$	$2\pi \delta(\omega - \omega_0)$
$\cos \omega_0 t$	$\pi[\delta(\omega - \omega_0) + \delta(\omega + \omega_0)]$
$\sin \omega_0 t$	$\dfrac{\pi}{j}[\delta(\omega - \omega_0) - \delta(\omega + \omega_0)]$
$x(t) = 1$	$2\pi \delta(\omega)$
Periodic square wave $x(t) = \begin{cases} 1, & |t| < T_1 \\ 0, & T_1 < |t| \leq \frac{T}{2} \end{cases}$ and $x(t+T) = x(t)$	$\sum_{k=-\infty}^{+\infty} \dfrac{2 \sin k\omega_0 T_1}{k} \delta(\omega - k\omega_0)$
$\sum_{n=-\infty}^{+\infty} \delta(t - nT)$	$\dfrac{2\pi}{T} \sum_{k=-\infty}^{+\infty} \delta\left(\omega - \dfrac{2\pi k}{T}\right)$
$x(t) \begin{cases} 1, & |t| < T_1 \\ 0, & |t| > T_1 \end{cases}$	$\dfrac{2 \sin \omega T_1}{\omega}$
$\dfrac{\sin Wt}{\pi t}$	$X(\omega) = \begin{cases} 1, & |\omega| < W \\ 0, & |\omega| > W \end{cases}$
$\delta(t)$	1
$u(t)$	$\dfrac{1}{j\omega} + \pi \delta(\omega)$
$\delta(t - t_0)$	$e^{-j\omega t_0}$
$e^{-at} u(t),\ a \in \mathbb{R} > 0$	$\dfrac{1}{a + j\omega}$
$t e^{-at} u(t),\ a \in \mathbb{R} > 0$	$\dfrac{1}{(a + j\omega)^2}$
$\dfrac{t^{n-1}}{(n-1)!} e^{-at} u(t),\ a \in \mathbb{R} > 0$	$\dfrac{1}{(a + j\omega)^n}$

REFERENCES

[And-95] Anderson, J. D. (1995), *Computational Fluid Dynamics: The Basics with Applications*, McGraw-Hill.

[Cra-47] Crank, J. and P. Nicholson (1947), A Practical Method for Evaluation of Solutions of Partial Differential Equations of the Heat Conduction Type, *Proceedings of the Cambridge Philosophical, Society* **43**, 50–67.

[Gil-97] Giles, M. B. (1997), On the Stability and Convergence of Discretisations of Initial Value PDE's, *IMA Journal of Numerical Analysis* **17**, 563–576.

[Fra-95] Fraleigh, J. B. and R. A. Beauregard (1995), *Linear Algebra*, 3rd ed., Addison-Wesley.
[Hir-88] Hirsch, C. (1988), *Numerical Computation of Internal and External Flows*, vol. 1, John Wiley.
[Koe-05] Koev, P. (2005), *Numerical Methods for Partial Differential Equations*, math 336 *lecture notes*, Massachusetts Institute of Technology, Spring.
[Kre-97] Kreiss, H. O. and S. H. Lui (1997), Nonlinear Stability of Time Dependent Differential Equations, *UCLA Computational and Applied Mathematics Report* 97–43.
[Kre-05] Kreiss, H. O. and J. Lorenz (2005), *Initial-Boundary Value Problems and the Navier-Stokes Equations*, SIAM.
[LeV-02] LeVeque, R. J. (2002), *Finite Volume Methods for Hyperbolic Problems*, Cambridge University Press.
[LeV-06] LeVeque, R. J. (2006), *AMath 585-6 Lecture Notes*, University of Washington.
[Lom-76] Lomax, H. (1976), Recent Progress in Numerical Techniques for Flow Simulation, *AIAA Journal* **14**, 512–518.
[Mos-95] Mosekilde, E. and O. Jensen (1995), Simulation of Turing Patterns in a Chemical Reaction-Diffusion System, *SAMS* **18–19**, 45–54.
[Ric-67] Richtmyer, R. D. and K. W. Morton (1967), *Difference Methods for Initial Value Problems*, 2nd ed., John Wiley.
[Sch-91] Schiesser, W. E. (1991), *The Numerical Method of Lines*, Academic Press.
[Tan-97] Tannehill, J. C., D. A. Anderson, and R. H. Pletcher (1997), *Computational Fluid Mechanics and Heat Transfer*, Taylor and Francis.
[Tur-52] Turing, A. (1952), The Chemical Basis of Morphogenesis, *Philosophical Transactions of the Royal Society* **237**, 37.

5

Dissipation and Dispersion

5.1 INTRODUCTION

The accuracy of a discretization scheme can be determined by comparing the *numeric amplification factor* $G_{numeric}$ with the *exact amplification factor* G_{exact} over one time step. Errors in magnitude are termed *dissipation*, and errors in phase are called *dispersion*. These terms are clarified subsequently. The term *amplification factor* is used to represent the change in magnitude of a solution over time. It can be calculated in either the time domain, by considering solution harmonics, or in the complex frequency domain, by taking Fourier transforms; we will work with solution harmonics. This analysis is intended to serve as an introduction to basic concepts and will therefore only consider systems described by linear PDEs with periodic boundary conditions and discretized over a uniform mesh.

For further reading, refer to [Hir-88, Chapter 8], [Lig-78, Chapter 3], [Tan-97, Chapter 4], and [Wes-01, Chapters 8 and 9].

5.2 DISPERSION RELATION

Physical waves that propagate in a particular medium will, in general, exhibit a specific group velocity as well as a specific phase velocity. This is because, within a particular medium, there is a fixed relationship between the wavenumber, ξ, and the frequency, ω, of waves. Thus, frequency and wavenumber are not independent quantities and are related by a functional relationship, known as the *dispersion relation*.[1] The term *frequency* is used here to describe the rate of oscillation of a wave in *time*, whereas the term *wave number* is used to describe the rate of oscillation in *space*.

We will demonstrate the process of obtaining the dispersion relation by example, using the advection equation

$$u_t + au_x = 0. \tag{5.1}$$

---

[1] The dispersion relation characterizes the mathematical relationship between temporal and spacial variations. For example, for a harmonic wave of the form $A \sin\left[\frac{2\pi}{\lambda}(x - at)\right]$ we have wavenumber $\xi = \frac{2\pi}{\lambda}$ and frequency $\omega = \frac{2\pi a}{\lambda}$, where λ represents wavelength, giving the dispersal relation $\omega(\xi) = a\xi$.

Generally, each wavenumber, ξ, corresponds to s frequencies, where s is the order of the PDE with respect to t. Now any linear PDE with constant coefficients admits a solution of the form

$$u(x,t) = u_0 e^{i(\xi x - \omega t)}. \tag{5.2}$$

Because we are considering linear systems, the principle of superposition applies, and eqn. (5.2) can be considered to be a frequency component or harmonic of the Fourier series representation of a specific solution to the wave equation. On inserting this solution into a PDE we obtain the so-called dispersion relation between ω and ξ, that is,

$$\omega = \omega(\xi), \tag{5.3}$$

and each PDE will have its own distinct form. For example, we obtain the specific dispersion relation for the advection equation by substituting eqn. (5.2) into eqn. (5.1) to obtain

$$-i\omega u_0 e^{i(\xi x - \omega t)} = -ia\xi u_0 e^{i(\xi x - \omega t)}$$
$$\Downarrow \tag{5.4}$$
$$\omega = a\xi.$$

This confirms that ω and ξ cannot be determined independently for the advection equation, and therefore eqn. (5.2) becomes

$$u(x,t) = u_0 e^{i\xi(x - at)}. \tag{5.5}$$

The physical meaning of eqn. (5.5) is that the initial value $u(x,0) = u_0 e^{i\xi x}$ is propagated from left to right, unchanged, at velocity a. Thus, there is no dissipation or attenuation and no dispersion.

Note that for a particular numerical scheme, if the imaginary part of $\omega(\xi)$ is equal to zero for all ξ, the scheme will be nondissipative. Similarly, if the imaginary part of $\omega(\xi)$ is greater than zero for all $\xi \neq 0$, the scheme will be dissipative.

A similar approach can be used to establish the dispersion relation for systems described by other forms of PDE.

5.3 AMPLIFICATION FACTOR

As mentioned earlier, the accuracy of a scheme can be determined by comparing the *numeric amplification factor* $G_{numeric}$ with the *exact amplification factor* G_{exact} over one time step. We can determine the "exact" amplification factor by considering the change that takes place in the "exact" solution during a single time step. For example, taking the advection of eqn. (5.1) and assuming a solution of the form of eqn. (5.2), we have

$$G_{exact} = \frac{u(x, t + \Delta t)}{u(x,t)} = \frac{u_0 e^{i\xi(x - a(t+\Delta t))}}{u_0 e^{i\xi(x-at)}} \tag{5.6}$$
$$\therefore G_{exact} = e^{-ia\xi \Delta t}.$$

We can also represent eqn. (5.6) in the form

$$G_{exact} = |G_{exact}| e^{i\Phi_{exact}}, \tag{5.7}$$

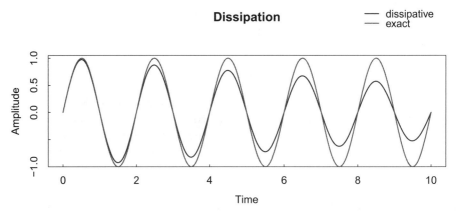

Figure 5.1. Illustration of pure dissipation effect on a single sinusoid where the system amplitude reduces over time but stays in phase.

where Φ represents the *phase*, given by

$$\Phi_{exact} = \angle G = \tan^{-1}\left(\frac{\text{Im}\{G_{exact}\}}{\text{Re}\{G_{exact}\}}\right). \tag{5.8}$$

Thus, using *Euler's identity*,[2]

$$|G_{exact}| = 1$$

and (5.9)

$$\Phi_{exact} = \tan^{-1}(\tan(-a\xi\Delta t)) = -a\xi\Delta t.$$

Values of $|\Phi| \to 0$ are associated with low frequencies and values of $|\Phi| \to \pi$ are associated with high frequencies.

Note that if $\text{Im}\{G\} = 0$, the scheme does not introduce a phase error and is purely dissipative. Conversely, if $\text{Re}\{G\} = 1$, the scheme does not introduce an amplitude error and is purely dispersive.

Now that we have an expression for G_{exact} in a useful form, we are in a position to calculate the relative numerical dissipation and dispersion effects due to the discretization process.

5.4 DISSIPATION

A situation where waves of different frequencies are damped by different amounts is called *dissipation* (see Fig. 5.1). Generally, this results in the higher-frequency components being damped more than lower-frequency components. The effect of dissipation, therefore, is that sharp gradients, discontinuities, or *shocks* in the solution tend to be smeared out, thus losing resolution (see Fig. 5.2). Fortunately, in recent years, various *high-resolution* schemes have been developed to obviate this effect to enable shocks to be captured with a high degree of accuracy, albeit at the expense of complexity. Examples of particularly effective schemes are based upon *flux/slope limiters* [Wes-01] and

---

[2] If $G = e^{i\theta} = \cos\theta + i\sin\theta$, then $|G| = \cos^2(\theta) + \sin^2(\theta) = 1$ and $\angle G = \dfrac{\sin\theta}{\cos\theta} = \tan\theta$.

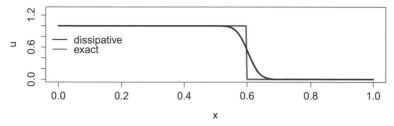

Figure 5.2. Effect of numerical dissipation on a step wave applied to pde $u_t + u_x = 0$.

WENO methods [Shu-98]. Dissipation can be introduced by numerical discretization of a partial differential equation that models a nondissipative process. Generally, dissipation improves stability and, in some numerical schemes, it is introduced deliberately by adding a diffusion term to aid stability of the resulting solution. Dissipation, whether real or numerically induced, tends to cause waves to lose energy.

The dissipation error resulting from discretization can be determined by comparing the magnitude of the numeric amplification factor $|G_{numeric}|$, with respect to the magnitude of the exact amplification factor $|G_{exact}|$, over one time step. The *relative numerical diffusion error* or *relative numerical dissipation error* compares real physical dissipation with the anomalous dissipation that results from numerical discretization. It is defined as

$$\varepsilon_D = \frac{|G_{numeric}|}{|G_{exact}|}, \qquad (5.10)$$

and the *total dissipation error* resulting from n steps will be

$$\varepsilon_{Dtotal} = (|G_{numeric}|^n - |G_{exact}|^n) u_0. \qquad (5.11)$$

If, for a given discretization scheme, $\varepsilon_D > 1$ for a given value of Co=$\frac{a\Delta t}{\Delta x}$ (the Courant–Friedrichs–Lewy number), this scheme will be unstable, and some type of modification will be necessary. This usually takes the form of reducing the time step.

As mentioned, if the imaginary part of $\omega(\xi)$ is zero for a particular discretization, then the scheme is nondissipative.

5.5 DISPERSION

A situation where waves of different frequencies move at different speeds without a change in amplitude is called *dispersion* (see Fig. 5.3). Alternatively, the Fourier components of a wave can be considered to disperse relative to each other. It therefore follows that a *dispersive wave*, if it is composed of different harmonics, is deformed while traveling. However, the energy contained within the wave is not lost and travels at the *group velocity* (discussed subsequently). Generally, this results in higher-frequency components traveling at slower speeds than the lower-frequency components. The effect of dispersion, therefore, is that often spurious oscillations or wiggles occur in solutions that include sharp gradients, discontinuities, or shocks. This usually manifested by high-frequency oscillations trailing the the particular effect (see Fig. 5.4).

The degree of dispersion can be determined by comparing the phase of the numeric amplification factor $|G_{numeric}|$ with the phase of the exact amplification factor $|G_{exact}|$ over

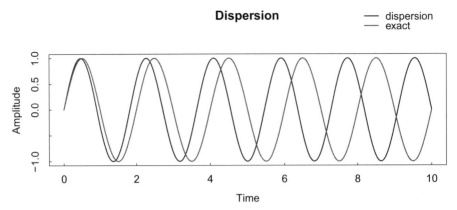

Figure 5.3. Illustration of pure dispersion effect on a single sinusoid where the system phase lags over time, but amplitude is unaffected.

one time step. Dispersion represents phase shift and results from the imaginary part of the amplification factor. The relative numerical dispersion error compares real physical dispersion with the anomalous dispersion that results from numerical discretization. It is defined as

$$\varepsilon_P = \frac{\Phi_{numeric}}{\Phi_{exact}}, \quad (5.12)$$

where $\Phi = \angle G = \tan^{-1}\left(\frac{\text{Im}\{G\}}{\text{Re}\{G\}}\right)$. The total phase error resulting from n steps will be

$$\varepsilon_{Ptotal} = n\left(\Phi_{numeric} - \Phi_{exact}\right). \quad (5.13)$$

If $\varepsilon_P > 1$, this is termed a *leading phase* error. This means that the Fourier component of the numerical solution has a wave speed *greater* than that of the exact solution. Similarly, if $\varepsilon_P < 1$, this is termed a *lagging phase* error. This means that the Fourier component of the numerical solution has a wave speed *less* than that of the exact solution.

Again, high-resolution schemes can all but eliminate this effect, but at the expense of complexity. Although many physical processes are modeled by PDEs that are nondispersive, when numerical discretization is applied to analyze them, the result is often to add some dispersion, even though this may be quite small.

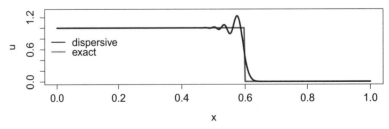

Figure 5.4. Effect of numerical dispersion on a step wave applied to pde $u_t + u_x = 0$.

5.6 DISSIPATION AND DISPERSION ERRORS

We now discuss the calculation of dissipative and dispersive errors, which we will illustrate by example. From Section (5.3) the gain for the exact advection equation,

$$u_t = -au_x, \qquad (5.14)$$

was calculated to be $e^{-ia\xi\Delta t}$ (see eqn. (5.6)). Also, from eqns. (5.9), the gain in magnitude and phase were found to be equal to

$$|G_{exact}| = 1$$
$$\Phi_{exact} = -a\xi\Delta t. \qquad (5.15)$$

Values of $|\Phi| \to 0$ are associated with low frequencies and values of $|\Phi| \to \pi$ are associated with high frequencies. As the real part of G_{exact} is equal to unity, the advection PDE is nondissipative.

These expressions for G_{exact} and Φ_{exact} will now be used to calculate the relative numerical dissipation and dispersion errors due to discretization.

5.6.1 The 1D Advection Equation, Semi-Discrete Upwind

We will now apply the semi-discrete upwind discretization scheme to the 1D scalar hyperbolic system of eqn. (5.14), which yields

$$\frac{du_j^n}{dt} = -\frac{a}{\Delta x}\left(u_j^n - u_{j-1}^n\right). \qquad (5.16)$$

From Chapter 4 we found that application of the von Neumann procedure leads to the following expression for the numerical amplification factor derived from the discretized spatial derivative:

$$\hat{U}_j(\xi, (n+1)\Delta t) = e^{G_f \Delta t}\hat{U}_j(\xi, n\Delta t)$$
$$= e^{-\frac{a\Delta t}{\Delta x}(1 - e^{-i\theta})}\hat{U}_j(\xi, n\Delta t), \qquad (5.17)$$

where $\hat{U}(\xi, n\Delta t)$ is the Fourier transform of u_j^n. This amplification factor is due solely to the discretized spatial derivative, as the time derivative is solved analytically in the Fourier domain. On converting to trigonometrical form we obtain,

$$\hat{U}_j(\xi, (n+1)\Delta t) = e^{-\frac{a\Delta t}{\Delta x}(1 - \cos\theta + i\sin\theta)}\hat{U}_j(\xi, n\Delta t). \qquad (5.18)$$

By application of *Parseval's theorem* (see the book's appendix), it can be shown that the amplification factor e^{G_f} applies equally to the time domain as well as to the Fourier domain.

We are seeking to compare the numerical amplification with the exact amplification over a single time step Δt, to ascertain the relative errors introduced by discretization.

First we need to identify the magnitude and phase components of the numerical amplification factor, which are

$$|G_{numeric}| = \left|e^{G_f \Delta t}\right|$$
$$= e^{\frac{-a\Delta t}{\Delta x}(1-\cos\theta)} \quad (5.19)$$

and

$$\Phi_{numeric} = \angle e^{G_f \Delta t}$$
$$= -\frac{a\Delta t}{\Delta x}\sin\theta. \quad (5.20)$$

Now, using this expression for the exact and numerical amplification factors, the relative dissipation error is determined as follows:

$$\varepsilon_D = \frac{|G_{numeric}|}{|G_{exact}|} = \frac{e^{-\frac{a\Delta t}{\Delta x}(1-\cos(\theta))}}{1} \quad (5.21)$$
$$= e^{-Co(1-\cos(\theta))},$$

and, similarly, the *relative dispersion* error is determined as

$$\varepsilon_P = \frac{\Phi_{numeric}}{\Phi_{exact}} = \frac{-\frac{a\Delta t}{\Delta x}\sin(\theta)}{-a\xi\Delta t} \quad (5.22)$$
$$= \frac{\sin(\theta)}{\theta},$$

where $\xi\Delta x = \theta$. As the real part of $G_{numeric}$ is not equal to unity, the scheme is dissipative; that is, it introduces dissipation.

Note that for the advection equation, the phase error, ε_P, will generally be independent of Co when standard type semi-discrete schemes, such as this scheme, are used.

Polar plots of ε_D and ε_P are shown in Fig. 5.5. To achieve this on a Cartesian grid, we use the x and y components of ε_D, which we calculate as follows:

$$\varepsilon_D(x) = cos(\theta) \times \varepsilon_D(\theta)$$
$$\varepsilon_D(y) = sin(\theta) \times \varepsilon_D(\theta), \quad (5.23)$$

and similarly for the Cartesian (x, y) coordinates of ε_P.

5.6.2 The 1D Advection Equation, Semi-Discrete Second-Order Upwind

Applying a second-order upwind discretization scheme to the spatial derivative of the 1D scalar advection eqn. (5.14) yields

$$\frac{du_j^n}{dt} = -\frac{a}{2\Delta x}\left(3u_j^n - 4u_{j-1}^n + u_{j-2}^n\right). \quad (5.24)$$

From Chapter 4 we found that application of the von Neumann procedure leads to the following expression for the numerical amplification factor derived from the discretized spatial derivative:

$$\hat{U}_j(\xi, (n+1)\Delta t) = e^{G_f \Delta t} \hat{U}_j(\xi, n\Delta t)$$
$$= e^{-\frac{a\Delta t}{\Delta x}(3 - 4e^{-i\theta} + e^{-i2\theta})} \hat{U}_j(\xi, n\Delta t), \quad (5.25)$$

where, again, $\hat{U}(\xi, n\Delta t)$ is the Fourier transform of u_j^n and, by application of Parseval's theorem, the amplification factor e^{G_f} also applies to the time domain. Again, this amplification factor is due solely to the discretized spatial derivative, as the time derivative is solved analytically in the Fourier domain.

We are seeking to compare the numerical amplification factor over a single time step Δt with the exact amplification factor; therefore, converting to trigonometrical functions and rearranging, we obtain

$$|G_{numeric}| = \left|e^{G_f \Delta t}\right|$$
$$= e^{\frac{-a\Delta t}{2\Delta x}(3 - 4\cos\theta + \cos 2\theta)} \quad (5.26)$$

and

$$\Phi_{numeric} = \angle e^{G_f \Delta t}$$
$$= -\frac{a\Delta t}{2\Delta x}(4\sin\theta - \sin 2\theta). \quad (5.27)$$

Now, using the previous expressions for the exact amplification factor, eqns. (5.15), we can calculate the relative dissipation error

$$\varepsilon_D = \frac{|G_{numeric}|}{|G_{exact}|}$$
$$= \frac{e^{\frac{-a\Delta t}{2\Delta x}(3 - 4\cos\theta + \cos 2\theta)}}{1} \quad (5.28)$$

and the relative dispersion error

$$\varepsilon_P = \frac{\Phi_{numeric}}{\Phi_{exact}}$$
$$= \frac{-\frac{a\Delta t}{2\Delta x}(4\sin(\theta) - \sin(2\theta))}{-a\xi \Delta t} \quad (5.29)$$
$$= \left(\frac{4\sin(\theta) - \sin(2\theta)}{2\theta}\right).$$

As the *real part* of $G_{numeric}$ is not equal to unity, the scheme is *dissipative*.

Note that for this scheme, ε_P is also independent of Co.

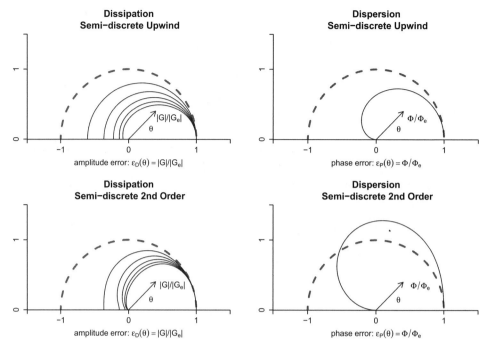

Figure 5.5. Relative dissipation and dispersion errors for advection PDE resulting from semi-discretization with values for Co=(0.25, 0.5, 0.75, 1.0, 1.25) corresponding to inner through outer curves. Dotted unit semicircles represent zero error solution. Each curve represents the locus of the radial vector as it moves through the angle $\theta = 0, \ldots, \pi$ radians. The plots have been generated from the R code in Listing 5.1 with testNo=9 for the *semi-discrete upwind* plots and testNo=10 for the *semi-discrete second-order* plots.

Polar plots of ε_D and ε_P are shown in Fig. 5.5, and the associated R code is given in Listing 5.1.

```
# File: dissipationDispersion.R
# Plots stability contours for PDE
#
cat("\014") # Clear console
rm(list = ls(all = TRUE)) # Delete workspace
library(shape)
# Plot relative dissipation and dispersion errors
# for range of Co numbers
#
# Specify y range and number of points
theta0 <- 0    # -pi;
theta1 <- pi
N      <- 101
# Linear vector of points
theta <- seq(theta0,theta1,len=N);
# Vector of Courant, Friedichs, Lewy values for plots
```

```r
Co <- c(0.25, 0.5, 0.75, 1.0, 1.25)
colNum <- c("red","blue","green","magenta","black")
M <- length(Co)
ev <- array(0,c(N,M))
prtPlot <- 0 # 0<-off, 1<-on
cat(sprintf("print plot = %d\n", prtPlot))
testNo <- 9
# Calculate eigenvalues for each Co
for(k in 1:M){
  switch(testNo,
  "1"={ # FTFS !!
     ev[,k] <- 1-Co[k]*(1 - exp(-1i*theta))
     str1 <- "FTFS"
     },
  "2"={ # BTFS
     ev[,k] <- 1 /(1 + Co[k]*(1 - exp(-1i*theta)))
     str1 <- "BTFS"
     },
  "3"={ # FTCS - 2nd Order in space
     ev[,k] <- 1 - Co[k]*(exp(i*theta) - exp(-1i*theta))/2
     str1 <- "FTCS"
     },
  "4"={ # BTCS - 2nd Order in space
       # BTCS - Unconditionally stable - Leveque,1992 - pp99.
     ev[,k] <- 1 /(1 + Co[k]*0.5*(exp(1i*theta) - exp(-1i*theta)))
     str1 <- "BTCS"
     },
  "5"={ # Leapfrog - 2nd Order in space
     lambda1 <- -i*Co[k]*sin(theta) + sqrt(1-(Co[k]*sin(theta))^2)
     lambda2 <- -i*Co[k]*sin(theta) - sqrt(1-(Co[k]*sin(theta))^2)
#          if (abs(lambda1) > abs(lambda2) )
         ev[,k] <- lambda1
#          else
#              ev[,k] <- lambda2;
#          end
     str1 <- "Leapfrog"
     },
  "6"={ # Beam-Warming
     ev[,k] <- 1 - 0.5*Co[k]*(3 - 4*exp(-1i*theta) + exp(-i*2*theta) )
              + 0.5*Co[k]^2 *(1 - 2*exp(-1i*theta) + exp(-i*2*theta) )
     str1 <- "Beam-Warming"
     },
  "7"={ # Lax Wendroff
     ev[,k] <- 1 - 1i*Co[k]*sin(theta) - Co[k]^2*(1-cos(theta))
     str1 <- "Lax-Wendroff"
     },
```

```r
    "8"={ # Lax-Friedrichs
       ev[,k] <- cos(theta) -1i*Co[k]*sin(theta)
       str1 <- "Lax-Friedrichs"
       },
    "9"={ # Semi-discrete Upwind
       ev[,k] <- exp(-Co[k]*(1-exp(-1i*theta)))
       str1 <- "Semi-discrete Upwind"
       },
    "10"={ # Semi-discrete 2nd Order
       ev[,k] <- exp( -Co[k]*(3 - 4*exp(-1i*theta) + exp(-1i*2*theta))/2 )
       str1 <- "Semi-discrete 2nd Order"
       },
       {
         break;
       }
         )
}
cat(sprintf("Test Number = %d\n", testNo))
# Plot contours of relative dissipation and dispersion
#      - various schemes
############################################################
par(mfrow=c(1,2),mar=c(7,4,4,2))
titleStr1 <- sprintf("Dissipation\n %s",str1)
for(k in 1:M){
  absEV <- abs(ev[,k]);
  angEV <- Arg(ev[,k]);
  angEV <- -angEV/(Co[k]*theta)
# #     tmp2 <- asin( Co[k].*sin(tmp2 ))/
# #           (-Co[k].*theta) # For leap-frog
# NOTE: Polar coordinates converted to Cartesian coordinates by
# multiplying by cos(theta) to get "x" and sin(theta) to get "y".
  if(k==1){
    plot( cos(theta)*absEV, sin(theta)*absEV, type="l", col="blue",
        xlim=c(-1.5,1.5), ylim=c(0,1.5),main=titleStr1,asp=1,ylab="",
        xaxs="i",yaxs="i",xaxt="n",yaxt="n",lwd=2,bty="n",
        xlab=expression(paste("amplitude error: ",
            epsilon[D](theta)=="|",G,"|/","|",G[e],"|")))
#   Following equiv. to polar(theta, absEV)
    plotcircle(mid=c(0,0), r=1,from=0,to=pi,lcol="red",lwd=5,lty=2)
    #lines(c(0,0.4), c(0,0.4),col="blue",lwd=2)
    arrows(0,0,0.4,0.4,length=0.15,col="blue",lwd=2)
    text(x=0.33,y=0.15,labels=expression(theta))
    text(x=0.60,y=0.30,labels=expression(paste("|",G,"|/","|",G[e],"|")))
    axis(side = 1, lwd = 2,at=c(-1.5,-1.0,0,1.0,1.5),labels=c(NA,-1.0,0,1.0,NA))
    axis(side = 2, lwd = 2,at=c(0,1.0,1.5),labels=c(0,1.0,NA))
  }else{
```

```
    lines( cos(theta)*absEV, sin(theta)*absEV, type="l", col="blue",lwd=2)
  }
# grid(lty=1,lwd=1,col="darkgray")
}
#
plot.new
titleStr2 <- sprintf("Dispersion\n %s",str1)
for(k in 1:M){
  absEV <- abs(ev[,k]);
  angEV <- Arg(ev[,k]);
  angEV <- -angEV/(Co[k]*theta)
# #     tmp2 <- asin( Co[k].*sin(tmp2 ))/
# #         (-Co[k].*theta) # For leap-frog
  if(k==1){
    plot( cos(theta)*angEV, sin(theta)*angEV, type="l", col="blue",
         xlim=c(-1.5,1.5), ylim=c(0,1.5),main=titleStr2,asp=1,ylab="",
         xaxs="i",yaxs="i",xaxt="n",yaxt="n",lwd=2,bty="n",
         xlab=expression(paste("phase error: ",
                               epsilon[P](theta)==Phi/Phi[e])))
#    polar( theta(1,], absEV(1,])
    plotcircle(mid=c(0,0),r=1,from=0,to=pi,lcol="red",lwd=5,lty=2)
    #lines(c(0,0.4), c(0,0.4),col="blue",lwd=2)
    arrows(0,0,0.4,0.4,length=0.15,col="blue",lwd=2)
    text(x=0.33,y=0.15,labels=expression(theta))
    text(x=0.60,y=0.30,labels=expression(paste(Phi/Phi[e])))
    axis(side = 1, lwd = 2,at=c(-1.5,-1.0,0,1.0,1.5),labels=c(NA,-1.0,0,1.0,NA))
    axis(side = 2, lwd = 2,at=c(0,1.0,1.5),labels=c(0,1.0,NA))
  }else{
    lines( cos(theta)*angEV, sin(theta)*angEV, type="l", col="blue",lwd=2)
  }
}
# grid(lty=1,lwd=1,col="darkgray")
```

Listing 5.1. File: dissipationDispersion.R—Code to generate the relative dissipation and dispersion error plots shown in Figure (5.5)

5.6.3 The 1D Advection Equation, Fully Discrete Upwind

Applying the upwind discretization scheme to both the temporal and spatial derivatives of the 1D scalar advection eqn. (5.14) yields

$$u_j^{n+1} = u_j^n - \frac{a\Delta t}{\Delta x}\left(u_j^n - u_{j-1}^n\right). \tag{5.30}$$

From Chapter 4 we found that application of the *von Neumann procedure* leads to the following expression for the numerical amplification factor:

$$G_f = 1 - 2\text{Co}\sin^2\left(\frac{\theta}{2}\right) - i\text{Co}\sin(\theta), \tag{5.31}$$

from which we obtain

$$|G_{numeric}| = |G_f| = \sqrt{1 - 4\text{Co}(1 - \text{Co})\sin^2\left(\frac{\theta}{2}\right)} \quad (5.32)$$

and

$$\Phi_{numeric} = \angle G_f = \arctan\left(\frac{-\text{Co}\sin(\theta)}{1 - \text{Co}(1 - \cos(\theta))}\right), \quad (5.33)$$

where $\text{Co} = \frac{a\Delta t}{\Delta x}$. Now, using the previous expressions for the exact amplification factor, eqns. (5.15), we can calculate the relative dissipation error as

$$\varepsilon_D = \frac{|G_{numeric}|}{|G_{exact}|} = \frac{\sqrt{1 - 4\text{Co}(1 - \text{Co})\sin^2\left(\frac{\theta}{2}\right)}}{1} \quad (5.34)$$

and the relative dispersion error as

$$\varepsilon_P = \frac{\Phi_{numeric}}{\Phi_{exact}} = \left(\frac{\arctan\left(\frac{-\text{Co}\sin(\theta)}{1 - \text{Co}(1 - \cos(\theta))}\right)}{-\text{Co}\,\theta}\right). \quad (5.35)$$

Because the *real part* of $G_{numeric}$ is not equal to unity, the scheme is *dissipative*.

Polar plots of ε_D and ε_P are shown in Fig. 5.6.

5.6.4 The 1D Advection Equation, Fully Discrete Lax–Friedrichs (LxF)

Applying the LxF discretization scheme to the 1D scalar advection eqn. (5.14) yields

$$u_j^{n+1} = \frac{1}{2}\left(u_{j-1}^n + u_{j+1}^n\right) - \frac{1}{2}\frac{a\Delta t}{\Delta x}\left(u_{j+1}^n - u_{j-1}^n\right). \quad (5.36)$$

From Chapter 4 we see that application of the von Neumann procedure leads to the expression for amplification factor

$$G_f = \cos(\theta) - i\,\text{Co}\sin(\theta), \quad (5.37)$$

from which we get

$$|G_{numeric}| = |G_f| = \sqrt{\cos^2(\theta) + \text{Co}^2\sin^2(\theta)} \quad (5.38)$$

and

$$\Phi_{numeric} = \angle G_f = \tan^{-1}(-\text{Co}\tan(\theta)), \quad (5.39)$$

where $\text{Co} = \frac{a\Delta t}{\Delta x}$.

Now, using the previous expressions for the exact amplification factor, eqns. (5.15), we can calculate the relative dissipation error,

$$\varepsilon_D = \frac{|G_{numeric}|}{|G_{exact}|} = \frac{\sqrt{\cos^2(\theta) - \text{Co}^2\sin^2(\theta)}}{1}, \quad (5.40)$$

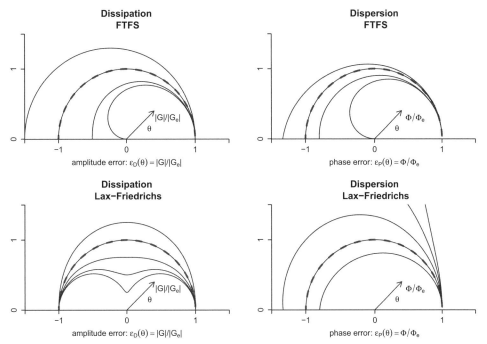

Figure 5.6. Relative dissipation and dispersion errors for advection PDE resulting from full-discretization with values for Co=(0.25, 0.5, 0.75, 1.0, 1.25) corresponding to inner through outer curves. Dotted unit semi-circles represent zero error solution. Each curve represents the locus of the radial vector as it moves through the angle $\theta = 0, \ldots, \pi$ radians. The plots have been generated from the R code in Listing 5.1 with testNo=1 for the FTFS plots and testNo=8 for the Lax–Friedrichs plots.

and the relative dispersion error,

$$\varepsilon_P = \frac{\Phi_{numeric}}{\Phi_{exact}} = \left(\frac{\tan^{-1}(-\text{Co}\tan(\theta))}{-\text{Co}\,\theta} \right). \quad (5.41)$$

As the *real part* of G_{exact} is not equal to unity, the scheme is *dissipative*.

Polar plots of ε_D and ε_P are shown in Fig. 5.6.

5.7 GROUP AND PHASE VELOCITIES

The term *group velocity* refers to the speed of a *wave packet* consisting of a low-frequency signal modulated (or multiplied) by a higher-frequency wave. The result is a low-frequency wave, consisting of a fundamental plus harmonics (a wave packet), that moves along a continuum with group velocity c_g. *Wave energy* and *information signals* propagate at this velocity, which is defined as being equal to the derivative of the real part of the frequency with respect to wave number, that is,

$$c_g = \frac{d\,\text{Re}\{\omega\}}{d\xi}. \quad (5.42)$$

If there are a number of spatial dimensions, then the group velocity is equal to the gradient of frequency with respect to the wavenumber vector, that is, $c_g = \nabla \text{Re}\{\omega(\xi)\}$.

The complementary term to group velocity is *phase velocity*, and this refers to the speed of propagation of an individual frequency component of the wave. It is defined as being equal to the ratio of frequency to wavenumber, that is,

$$c_p = \text{Re}\left\{\frac{\omega}{\xi}\right\}. \tag{5.43}$$

It can also be viewed as the speed at which a particular phase of a wave propagates, for example, the speed of propagation of a wave crest. In one wave period T, the crest advances one wavelength λ, therefore the phase velocity is also given by $c_p = \lambda/T$. We see that this second form is equal to eqn. (5.43) due to the following relationships: wavenumber $\xi = \frac{2\pi}{\lambda}$ and frequency $\omega = 2\pi f$, where $f = \frac{1}{T}$.

Note that for a nondispersive wave, $c_g = c_p$.

5.7.1 Exact Relationships for the Basic PDE

We will now consider the 1D scalar hyperbolic system

$$u_t = -au_x. \tag{5.44}$$

We found in Section 5.2 that by assuming a solution to the advection equation of the form $u(x, t) = u_0 e^{i(\xi x - \omega t)}$, the associated dispersion relation was

$$\omega = a\xi. \tag{5.45}$$

Thus, we are now able to use this result together with the definitions of eqns. (5.42) and (5.43) to determine the group and phase, that is,

$$\begin{aligned} c_g &= \frac{d\text{Re}\{\omega\}}{d\xi} = a \\ c_p &= \text{Re}\left\{\frac{\omega}{\xi}\right\} = a. \end{aligned} \tag{5.46}$$

We see that c_p and c_g are equal; this means that the advection wave is not subjected to either dissipation or dispersion, that is, it propagates a wave undistorted.

The group and phase velocities can be obtained for other PDEs in a similar way.

5.7.2 Semi-Discrete, First-Order Upwind Discretization

If the preceding advection eqn. (5.44) is discretized in space using a first-order upwind scheme, we obtain for the jth step

$$\frac{du_j}{dt} = -a\frac{(u_j - u_{j-1})}{\Delta x}. \tag{5.47}$$

Again, we analyze this scheme by letting $u(x_j, t) = u_0 e^{i(\xi x - \omega t)}$, which on substitution into eqn. (5.47) and writing x for x_j and $x - \Delta x$ for x_{j-1} we obtain the dispersion relation, as follows:

$$-i\omega u_0 e^{i(\xi x - \omega t)} = -a\left(1 - e^{-i\xi \Delta x}\right) u_0 e^{i(\xi x - \omega t)}/\Delta x,$$

$$\therefore \omega = -\frac{ia}{\Delta x}\left(1 - e^{-i\xi \Delta x}\right),$$

$$= -i\frac{a}{\Delta x}\left(1 - [\cos(\xi \Delta x) - i\sin(\xi \Delta x)]\right), \quad (5.48)$$

$$\therefore \omega = \frac{a}{\Delta x}\left(\sin(\xi \Delta x) - i[1 - \cos(\xi \Delta x)]\right).$$

From this result we are able to determine the group and phase velocities, that is,

$$c_g = \frac{d\mathrm{Re}\{\omega\}}{d\xi} = a\cos(\theta)$$

$$c_p = \mathrm{Re}\left\{\frac{\omega}{\xi}\right\} = a\frac{\sin(\theta)}{\theta} = a\,\mathrm{sinc}(\theta), \quad (5.49)$$

where $\theta = \xi \Delta x$.

Now, comparing with the exact PDE, we see that the group and phase velocities have both changed and that they are **not** equal. This means that the process of discretization has introduced some errors. These errors will manifest themselves as dissipation and dispersion effects ε_D and ε_P, respectively. But, for small θ, which corresponds to low wavenumbers, c_p and c_g will both be almost equal to the advection velocity a, that is, they will be approximately equal to those of the exact PDE. Therefore, we have found that, providing θ is kept small, semi-discrete forward upwind discretization of the advection equation will introduce very little dissipation and/or dispersion at low wavenumbers. However, as θ increases, the errors will increase and the solution will diverge from the exact PDE solution.

5.7.3 Semi-Discrete Leapfrog Discretization

Applying the leapfrog numerical discretization scheme to the spatial domain yields

$$\frac{du_j}{dt} = -a\frac{\left(u_{j+1}^n - u_{j-1}^n\right)}{2\Delta x}. \quad (5.50)$$

We again analyze this scheme by letting $u(x_j, t) = u_0 e^{i(\xi x - \omega t)}$, which, on substitution into eqn. (5.50) and writing $(x - \Delta x)$ for (x_{j-1}) and $(x + \Delta x)$ for (x_{j+1}), we obtain the dispersion relation as follows:

$$-i\omega u_0 e^{i(\xi x - \omega t)} = -a\left(\frac{e^{i(\xi \Delta x)} - e^{-i(\xi \Delta x)}}{2\Delta x}\right) u_0 e^{i(\xi x - \omega t)} \quad (5.51)$$

$$\therefore \omega = \frac{a}{\Delta x}\sin(\xi \Delta x). \quad (5.52)$$

Thus, the group and phase velocities are

$$c_g = \frac{d\operatorname{Re}\{\omega\}}{d\xi} = a\cos(\theta)$$

$$c_p = \operatorname{Re}\left\{\frac{\omega}{\xi}\right\} = a\frac{\sin(\theta)}{\theta} = a\operatorname{sinc}(\theta). \tag{5.53}$$

We see again that the phase velocity and group velocity have changed from those of the exact PDE but are actually the same as for the forward upwind discretization. However, the dispersion relation is different.

5.7.4 Fully Discrete Leapfrog Discretization

In this example we will apply the leapfrog numerical discretization scheme to both the spatial domain and the time domain of eqn. (5.44), which yields

$$\frac{\left(u_j^{n+1} - u_j^{n-1}\right)}{2\Delta t} = -a\frac{\left(u_{j+1}^n - u_{j-1}^n\right)}{2\Delta x}. \tag{5.54}$$

Again we let $u(x_j, t_n) = u_0 e^{i(\xi x - \omega t)}$, which on substitution into eqn. (5.54) and on writing $(t - \Delta t)$ for (t_{n-1}), $(t + \Delta x)$ for (t_{n+1}), $(x - \Delta x)$ for (x_{j-1}), and $(x + \Delta x)$ for (x_{j+1}), we obtain the dispersion relation, as follows:

$$\left(e^{-i\omega\Delta t} - e^{i\omega\Delta t}\right) u_0 e^{i(\xi x - \omega t)} = -\frac{a\Delta t}{\Delta x}\left(e^{i(\xi\Delta x)} - e^{-i(\xi\Delta x)}\right) u_0 e^{i(\xi x - \omega t)},$$

$$-\sin(\omega\Delta t) = -\operatorname{Co}\sin(\theta), \tag{5.55}$$

$$\therefore \omega = \frac{1}{\Delta t}\sin^{-1}(\operatorname{Co}\sin(\theta)).$$

Thus, as for previous examples, we are now able to determine the phase and group velocities, that is,

$$c_p = \operatorname{Re}\left\{\frac{\omega}{\xi}\right\} = \frac{a}{\operatorname{Co}\theta}\sin^{-1}(-\operatorname{Co}\sin(\theta))$$

$$c_g = \frac{d\operatorname{Re}\{\omega\}}{d\xi} = -a\frac{\cos(\theta)}{\cos(\omega\Delta t)}. \tag{5.56}$$

To obtain c_p, we have multiplied both sides of eqn. (5.55) by ω/ξ and rearranged, whereas, to obtain c_g, we have used implicit differentiation. We see that the phase velocity and group velocity have now changed again. However, from eqns. (5.55), we see that for small θ and small $\omega\Delta t$, which corresponds to low wavenumbers and low frequencies, the dispersion relation is $\omega \simeq a\xi$. Consequently, under these conditions, c_p and c_g will both be equal to a, that is, the same as for the exact and semi-discrete PDEs. This means that the fully discrete advection equation will also be neither dissipative nor dispersive at low wavenumbers and low frequencies. However, again, as θ and/or $\omega\Delta t$ increase, they will diverge from the exact PDE values.

Plots of the dispersion relations are included in Fig. 5.7 for the three preceding examples, where we have set the advection velocity to $a = 1$. The associated R code is shown in Listing 5.2.

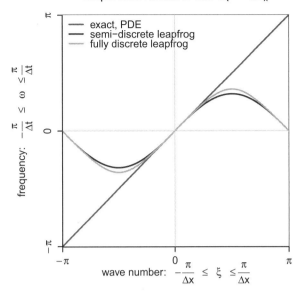

Figure 5.7. Dispersion relations for the advection equation for Co = 0.75; $a = 1$; $\Delta x = 1$. The plots show how the nondissipative, nondispersive characteristic of the *exact* PDE is modified by discretization.

```
# File: dispersionRelations.R
# Dispersion relation Analysis of PDE Discretisation Schemes
cat("\014") # Clear console
rm(list = ls(all = TRUE)) # Delete workspace
N<-301;
Co <- 0.75;
a<-1;dx<-1;
dt<-dx*Co/a;
L<-pi/dx;
M<-4; # Mode/harmonic number (M<-1 -> fundamental)
xi <- seq(-L,L,len=N);

w_basic <- a*xi; # Exact PDE

w_semi <- (a/dx)*sin(xi*dx); # semi-discrete 1st order upwind

w_full <- 1/(dt)*asin( Co*sin(xi*dx ) );
par(mar=c(7,6,4,2))
plot(xi,w_basic,type="l",col="red", asp=1,
    xaxs="i",yaxs="i",xaxt="n",yaxt="n",lwd=3,
    xlab=expression(paste("wave number: ",-frac(pi,Delta*x),
      phantom(0)<=phantom(0),xi,phantom(0)<=frac(pi,Delta*x))),
    ylab=expression(paste("frequency: ",-frac(pi,Delta*t),
      phantom(0)<=phantom(0),omega,phantom(0)<=frac(pi,Delta*t))))
```

```
lines(xi,w_semi,col="blue",lwd=3);
lines(xi,w_full,col="green",lwd=3);
legend("topleft",legend=c("exact, PDE", "semi-discrete leap-frog",
    "fully discrete leap-frog"),
      lwd=c(3,3,3), lty = c(1,1,1),
      col=c("red","blue","green"),bty="n")
axis(side=1,lwd=2,at=c(-pi,0,pi),labels=c(expression(-pi),0,expression(pi)))
axis(side=2,lwd=2,at=c(-pi,0,pi),labels=c(expression(-pi),0,expression(pi)))
# daspect([2 4 1]);
title(main=expression(paste("Dispersion relations for: ",u[t] == - au[x])))
abline(h=0, col="darkgray")
abline(v=0, col="darkgray")
```

Listing 5.2. File: dispersionRelations.R—Code to generate the dispersion relation plots shown in Fig. 5.7

5.8 MODIFIED PDES

We can also examine the accuracy of numerical schemes by trying to see what physical processes are represented and, neglecting boundary condition, would actually be solved by the discretization process. We can do this by expanding the individual terms of the scheme in a Taylor series about a particular grid point to try to ascertain what physical process is actually being modeled. Equations derived this way are called *modified equations* [War-74].

We will illustrate this approach by example.

Example: First-order upwind semi-discretization
Assuming a regular spaced grid, let us examine a first-order spatial derivative subject to first-order upwind discretization, that is,

$$\frac{\partial u_j}{\partial x} \cong \frac{u_j - u_{j-1}}{\Delta x}. \tag{5.57}$$

Expanding the term u_{j-1} as a Taylor series yields

$$u_{j-1} = u_j - \Delta x \frac{\partial u_j}{\partial x} + \frac{\Delta x^2}{2!} \frac{\partial^2 u_j}{\partial x^2} - \frac{\Delta x^3}{3!} \frac{\partial^3 u_j}{\partial x^3} + \cdots,$$

$$\therefore \frac{u_j - u_{j-1}}{\Delta x} = \frac{\partial u_j}{\partial x} + \frac{\Delta x}{2!} \frac{\partial^2 u_j}{\partial x^2} + O\left(\Delta x^2\right). \tag{5.58}$$

We can now investigate the equivalent advection equation as modified by the discretization process. Substituting eqn. (5.58) into the semi-discrete advection eqn. (5.57) gives the *modified equation*, that is,

$$\frac{\partial u}{\partial t} + a \frac{\partial u}{\partial x} = \frac{a \Delta x}{2} \frac{\partial^2 u}{\partial x^2}, \tag{5.59}$$

where we have truncated the Taylor series after the term containing the second derivative.

Thus, we see that the effect of first-order upwind discretization on the advection equation is to add a second-order spatial derivative, which is equivalent to adding an artificial diffusion or viscosity term. This has the undesirable effect of smoothing out any sharp gradients and the desirable effect of improving stability. Equation (5.59) demonstrates that this scheme discretizes the *original* advection equation to first-order accuracy in space.

Example: Central semi-discretization
Again assuming a regular spaced grid, let us examine a first-order spatial derivative subject to first-order central discretization, that is,

$$\frac{\partial u_j}{\partial x} \simeq \frac{u_{j+1} - u_{j-1}}{2\Delta x}. \tag{5.60}$$

Expanding the term u_{j-1} as a Taylor series at u_j yields

$$u_{j-1} = u_j - \Delta x \frac{\partial u_j}{\partial x} + \frac{\Delta x^2}{2!} \frac{\partial^2 u_j}{\partial x^2} - \frac{\Delta x^3}{3!} \frac{\partial^3 u_j}{\partial x^3} + \cdots,$$

$$u_{j+1} = u_j + \Delta x \frac{\partial u_j}{\partial x} + \frac{\Delta x^2}{2!} \frac{\partial^2 u_j}{\partial x^2} + \frac{\Delta x^3}{3!} \frac{\partial^3 u_j}{\partial x^3} + \cdots, \tag{5.61}$$

$$\therefore \frac{u_{j+} - u_{j-1}}{2\Delta x} = \frac{\partial u_j}{\partial x} + \frac{\Delta x^2}{3!} \frac{\partial^3 u_j}{\partial x^3} + \frac{\Delta x^4}{5!} \frac{\partial^5 u_j}{\partial x^5} + O\left(\Delta x^5\right).$$

We can now investigate the equivalent advection equation as modified by the discretization process. Substituting eqn. (5.61) into the semi-discrete advection equation gives the *modified equation*, that is,

$$\frac{\partial u}{\partial t} + a \frac{\partial u}{\partial x} = \frac{a \Delta x^2}{6} \frac{\partial^3 u}{\partial x^3}, \tag{5.62}$$

where we have truncated the Taylor series after the term containing the third derivative.

Equation (5.62) demonstrates that this scheme discretizes the *original* advection equation to second-order accuracy in space.

Note that the even terms of eqn. (5.61) have canceled. Thus, we see that the effect of second-order central discretization on the advection equation is to add a third-order spatial derivative, which is similar to adding an artificial surface tension term to fluid flow type problems. This has the undesirable effect of adding dispersion, that is, phase error, which can result in spurious oscillations in the solution.

Example: FTCS full discretization
Assuming a regularly spaced time interval, let us examine a first-order time derivative subject to first-order upwind discretization, that is,

$$\frac{\partial u}{\partial t} \cong \frac{u_j^{n+1} - u_j^n}{\Delta t}. \tag{5.63}$$

Expanding the term u_j^{n+1} as a Taylor series at u_j^n yields

$$u_j^{n+1} = u_j^n + \Delta t \frac{\partial u_j^n}{\partial t} + \frac{\Delta x^2}{2!} \frac{\partial^2 u_j^n}{\partial t^2} + \frac{\Delta x^3}{3!} \frac{\partial^3 u_j}{\partial x^3} + \cdots,$$

$$\therefore \frac{u_j^{n+1} - u_j^n}{\Delta t} = \frac{\partial u_j^n}{\partial t} + \frac{\Delta x}{2!} \frac{\partial^2 u_j^n}{\partial t^2} + O\left(\Delta t^2\right). \tag{5.64}$$

We can now examine the equivalent advection equation as modified by this discretization process. Substituting eqn. (5.64) into the fully discrete advection for the time derivative, along with eqn. (5.61) for the spatial derivative, gives the *modified equation*, that is,

$$\frac{\partial u}{\partial t} + a\frac{\partial u}{\partial x} = -\frac{\Delta x}{2!}\frac{\partial^2 u}{\partial t^2} + \frac{a\Delta x^2}{6}\frac{\partial^3 u}{\partial x^3}, \qquad (5.65)$$

where we have truncated the Taylor series for u_j^{n+1} after the term containing the second derivative. The Taylor series for the spatial derivative has been truncated after the term containing the third derivative, as for the previous example.

Equation (5.65) demonstrates that this scheme discretizes the *original* advection equation to first-order accuracy in time and second-order accuracy in space.

Thus, we see that the effect of first-order temporal upwind discretization on the advection equation is to add a second-order time derivative, that is, diffusion term. This is in addition to the third-order spatial derivative that introduces dispersion. The overall effect is not very good at all: as we have seen in Chapter 4, the scheme is *unconditionally unstable*!

We have demonstrated here a general approach to obtaining the modified PDE equations. Similar methods to these can be applied to more complex PDEs of higher order. However, the Taylor analysis, although straightforward, can become a lengthy, tedious process, particularly when mixed derivatives are involved. For more details, refer to [Tan-97].

REFERENCES

[Hir-88] Hirsch, C. (1988), *Numerical Computation of Internal and External Flows, Vol. 1*, John Wiley.
[Lev-02] LeVeque, R. J. (2002), *Finite Volume Methods for Hyperbolic Problems*, Cambridge University Press.
[Lig-78] Lightfoot, Sir James (1978), *Waves in Fluids*, Cambridge University Press.
[Ric-67] Richtmyer, R. D. and K. W Morton (1967), *Difference Methods for Initial Value Problems*, 2nd ed., John Wiley.
[Shu-98] Shu, C.-W. (1998), Essentially Non-oscillatory and Weighted Essential Non-oscillatory Schemes for Hyperbolic Conservation Laws, in *Advanced Numerical Approximation of Nonlinear Hyperbolic Equations*, Cockburn, B., C. Johnson, C.-W. Shu, and E. Tadmor, editors, Lecture Notes in Mathematics 1697, Springer, pp. 325–432.
[Tan-97] Tannehill, J. C., D. A. Anderson and R. H. Pletcher (1997), *Computational Fluid Mechanics and Heat Transfer*, Taylor and Francis.
[War-74] Warming, R. F. and B. J. Hyett (1974), The Modified Equation Approach to the Stability and Accuracy Analysis of Finite-Difference Methods, *J. Comput. Phys.* **14**, 159–179.
[Wes-01] Wesseling, P. (2001), *Principles of Computational Fluid Dynamics*, Springer.

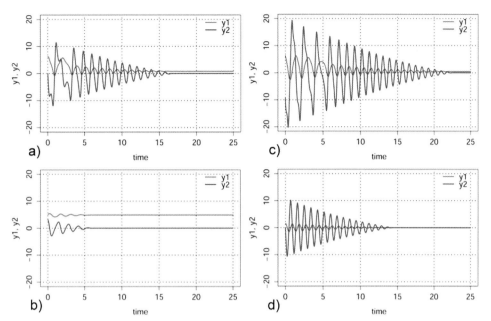

Plate 1.14. Transient plots of the BDF/NDF nonlinear pendulum example. The $(y_1, y_2, : \theta)$ initial conditions for each plot are (a) $(2\pi, 0 : \pi)$; (b) $(5, \pi : \pi)$; (c) $(2\pi, -3\pi : \pi/2)$, and (d) $(0, 0 : 0)$.

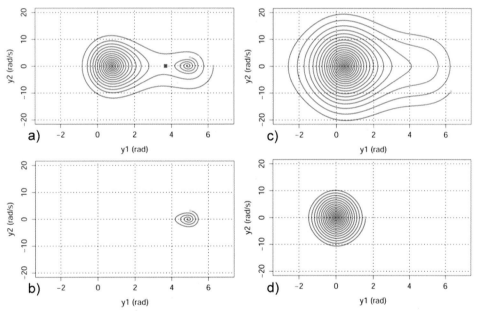

Plate 1.15. Phase plots of the BDF/NDF nonlinear pendulum example. The $(y_1, y_2, : \theta)$ initial conditions for each plot are (a) $(2\pi, 0 : \pi)$; (b) $(5, \pi : \pi)$; (c) $(2\pi, -3\pi : \pi/2)$, and (d) $(0, 0 : 0)$. The trajectory of plot (b) has been superimposed on plot (a) to emphasize how the choice of initial condition affects the ultimate equilibrium point. The small circular dots in each plot represent the initial conditions for y_1 and y_2, and the square in plot (a) represents the unstable equilibrium point at $(y_1 = 3.630396053, y_2 = 0)$.

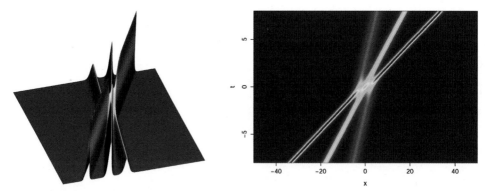

Plate 3.5. (left) Surface plot of a three-soliton solution to the KdV equation generated using the `persp3d()` function. (right) 2D space-time plot of the three-soliton collision generated using the `image()` function. The simulation employed 531 spatial grid points (horizontal axis) and 251 temporal grid points (vertical axis). The plots both show clearly how the taller faster solitons overtake the shorter slower solitons. The soliton phase changes are clearly visible (see main text).

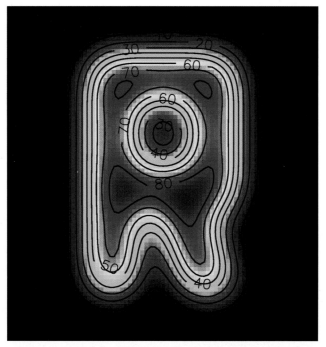

Plate 3.10. Image plot at $t = 90$ s of a diffusion simulation of steel plate based on a 100×100 grid. The letter "R" was initially at $100°$C and the surrounding metal at $0°$C. The plot was generated using the `image()` function and contours were added using the `contours()` function.

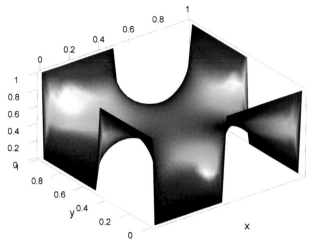

Plate 3.13. 3D surface plot of solution to the Laplace problem of eqn. (3.81). The segmented potentials result in a *castellation* effect, which is clearly visible. The surface plot was generated using the `persp3d()` function.

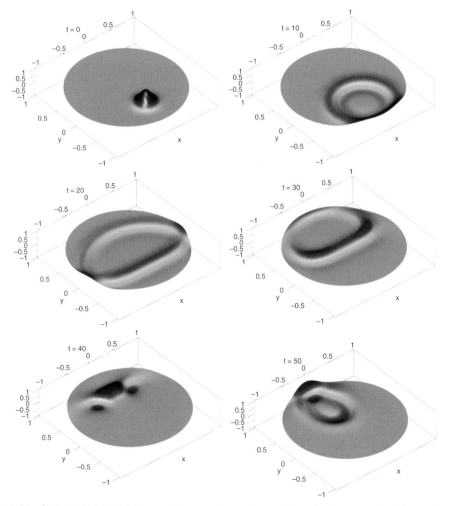

Plate 3.26. Series of images showing the evolution of waves in a circular domain with a reflective boundary condition at $r = 1$. Following a Gaussian shaped disturbance, the sequence proceeds right to left, top to bottom. With the wave traveling at a velocity of $c = 0.05$, its peak reaches the near boundary at $t = 10$ and the far boundary at $t = 30$, as expected. The surface plots were generated using the `persp3d()` function.

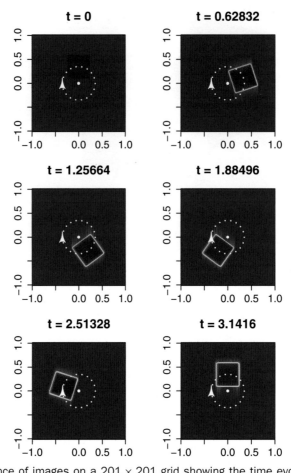

Plate 6.33. Sequence of images on a 201 × 201 grid showing the time evolution of a rotating-solid. The center of the solid follows the dotted circle of radius $r = 0.35$. Plots show clearly that after some initial dissipation the square then advects without noticeable distortion. Solution based upon the Kuganov and Tadmor central scheme with WENO5 reconstruction.

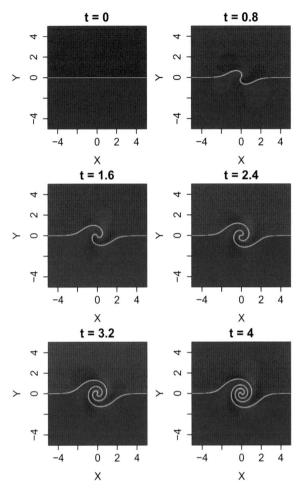

Plate 6.35. Array of 2D plots of high-resolution simulation of frontogenesis showing time evolution from $t = 0$ to $t = 4$. Solution based upon the Kuganov and Tadmor central scheme with WENO5 reconstruction.

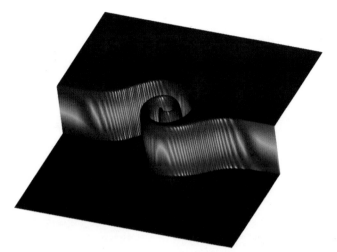

Plate 6.36. Surface plot of high-resolution simulation of the *frontogenesis* problem at $t = 4$.

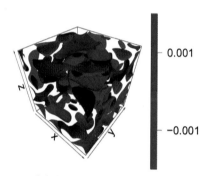

Plate 7.11. Iso-surface contours of the errors associated with the 3D interpolation problem.

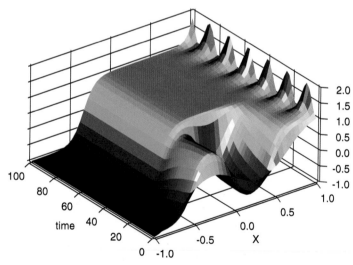

Plate 7.16. Solution $u(x, t)$ obtained from a meshless method simulation of the Allen–Cahn eqn. (7.84) from $t = 0$ to 100 on a grid of 41 nodes over the domain $x \in (-1, 1)$.

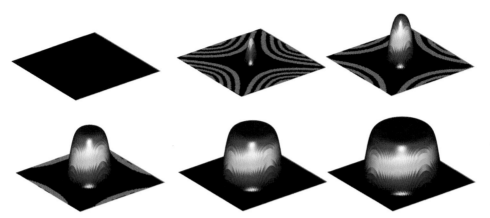

Plate 7.20. Sequence of *local* meshless method solutions $u(x, y, t)$ of the 2D Fisher–Kolmogorov equation (7.92). From left to right, top to bottom, results of simulation at times $t = 0, 0.0004, 0.0008, 0.0012, 0.0016$, and 0.002.

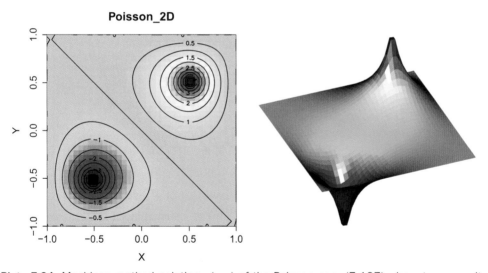

Plate 7.24. Meshless method solution $u(x, y)$ of the Poisson eqn. (7.107) where two opposite electrostatic charges are located within a square domain with Dirichlet boundary conditions. (left) 2D plot with contours. (right) Surface plot.

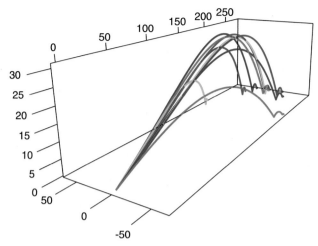

Plate 9.18. Perspective view showing the typical range of golf shots depicted in Fig. 9.17. Distances are shown in yards.

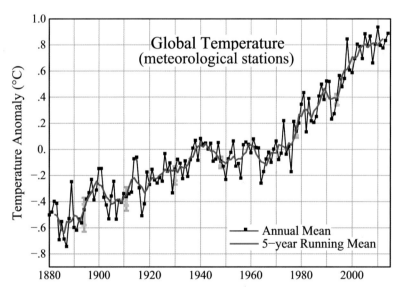

Plate 11.11. Global annual-mean surface air temperature change, with the base period 1951–1980, derived from the meteorological station network (this is an update of Plate 6(b) in [Han-01]). Uncertainty bars (95% confidence limits) are shown for both the annual and five-year means. They account only for incomplete spatial sampling of data. Source: NASA, Goddard Space Flight Center, available online at http://data.giss.nasa.gov/gistemp/graphs_v3/.

6

High-Resolution Schemes[1]

6.1 INTRODUCTION

High-resolution schemes are used in the numerical solution of partial differential equations (PDEs) where high accuracy is required and the solution exhibits *shocks, discontinuities*, or *steep gradients*. They generally have the following properties:

- Second- or higher-order spatial accuracy is obtained in smooth parts of the solution.
- Solutions are free from spurious oscillations or wiggles.
- High accuracy is obtained around shocks and discontinuities.
- The number of mesh points containing the wave is small compared with a first-order scheme with similar accuracy.

High-resolution schemes often use *flux/slope limiters* or *WENO methods* to modify the gradient around shocks or discontinuities so as to ensure a TVD solution. Particularly successful high-resolution schemes include the *MUSCL scheme* and the *Kurganov and Tadmore scheme*, which employ *state extrapolation* to achieve high accuracy. A typical approach to high resolution calculation is shown in Fig. 6.1.

We discuss various approaches to high-resolution PDE solutions in the following sections.

6.2 THE RIEMANN PROBLEM

The *Riemann problem* has uniform initial conditions on an infinite spatial domain, except for a single jump discontinuity [Lan-98]. For example, a 1D hyperbolic PDE, the Riemann problem could be defined as

$$\frac{\partial u}{\partial t} + \frac{\partial f(u)}{\partial x}, \quad t \geq t_0, \ x \in \mathbb{R}, \quad u(x, t_0) = \begin{cases} u_L, & x < x_0, \\ u_R, & x > x_0, \end{cases} \quad (6.1)$$

where u_L and u_R are constants. This idea is readily extended to vector quantities in multiple dimensions.

---

[1] This chapter draws upon some of the author's contributions to Wikipedia.

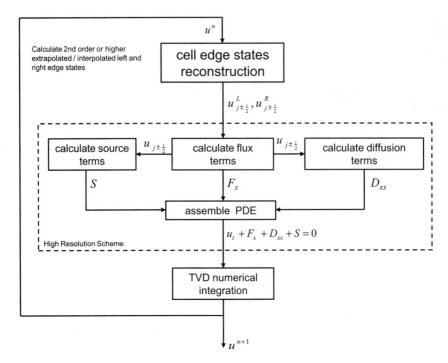

Figure 6.1. Typical high-resolution scheme based on cell edge reconstruction.

Godunov showed that it is not possible to construct a linear approximate solution to the Riemann problem having an order greater than 1 that is nonoscillatory (see Section 6.4.1). In subsequent sections we will discuss some nonlinear approximations to the Riemann problem that can provide nonoscillatory solutions.

6.3 TOTAL VARIATION DIMINISHING (TVD) METHODS

In systems described by *partial differential equations*, such as the 1D *hyperbolic advection* equation

$$\frac{\partial u}{\partial t} + a \frac{\partial u}{\partial x} = 0, \quad u = u(x, t), \ t \geq t_0, \ x \in \mathbb{R}, \tag{6.2}$$

the *total variation* (TV) is given by

$$TV = \int \left| \frac{\partial u}{\partial x} \right| dx, \tag{6.3}$$

and the total variation for the discrete case is

$$TV = \sum_j \left| u_{j+1} - u_j \right|. \tag{6.4}$$

A numerical method is said to be *total variation diminishing* (TVD) if

$$TV\left(u^{n+1}\right) \leq TV\left(u^n\right). \tag{6.5}$$

A system is said to be *monotonicity preserving* if the following properties are maintained as a function of t:

- No new local extrema can be created within the solution spatial domain.
- The value of a local minimum is nondecreasing, and the value of a local maximum is nonincreasing.

For *physically realizable* systems where there is *energy dissipation* of some kind, the total variation generally does not increase with time. Harten [Har-83] proved the following properties for a numerical scheme:

- A *monotone scheme* is TVD.
- A TVD scheme is *monotonicity preserving*.

Monotone schemes are attractive for solving engineering and scientific problems because they tend not to provide nonphysical solutions, although this is not guaranteed. The TVD condition should be thought of as a nonlinear stability condition.

Godunov's theorem proves that only first-order linear systems preserve monotonicity and are therefore TVD. Higher-order linear schemes, although more accurate for smooth solutions, are not TVD and tend to introduce spurious oscillations (wiggles) where discontinuities or shocks arise. To overcome these draw backs, various *high-resolution, nonlinear* techniques have been developed, often using *flux/slope limiters*.

6.3.1 TVD Numerical Integration

Some standard integration methods can degrade TVD schemes, and consequently a number of TVD integrators have been developed to avoid this problem. The schemes are designed to solve a system of ODEs in N dimensional space,

$$u_t = \mathcal{L}(u), \quad u = u(x,t), \ t \geq t_0, \ x \in \mathbb{R}^N, \tag{6.6}$$

with appropriate initial conditions. This system is assumed to result typically from a method of lines approximation to a hyperbolic conservation law of the form

$$u_t = -f(u)_x, \tag{6.7}$$

where the spatial derivative $f(u)_x$ is approximated by a discrete TVD scheme, denoted by $\mathcal{L}(u)$, and has the property of eqn. (6.5) subject to a *Courant–Friedrichs–Lewy* (CFL) condition restriction on the time step Δt such that

$$\Delta t \leq c \Delta t_1, \tag{6.8}$$

where c is termed the CFL coefficient for the high-order time discretization, CFL $= |a|\Delta t_1/\Delta x \leq 1$, and $|a|$ represents the maximum wave speed of the system.

A general Runge–Kutta method for integrating eqn. (6.7) is written in the form

$$u^{(i)} = \sum_{k=0}^{i-1} \left(\alpha_{i,k} u^{(k)} + \Delta t \beta_{i,k} \mathcal{L}(u^{(k)}) \right), \quad i = 1, \ldots, m,$$

$$u^0 = u^n, \quad u^{(m)} = u^{n+1}. \tag{6.9}$$

Thus, if all the coefficients are nonnegative ($\alpha_{i,k} \geq 0, \beta_{i,k} \geq 0$), then eqn. (6.9) is simply a convex combination of Euler forward operators, with Δt replaced by $\frac{\beta_{i,k}}{\alpha_{i,k}} \Delta t$. For consistency we must have $\sum_{k=0}^{i-1} \alpha_{i,k} = 1$. Therefore, it follows that for the scheme in eqn. (6.9) to be TVD, it is required that c in eqn. (6.8) must be equal to

$$c = \left\{ \min_{i,k} \frac{\alpha_{i,k}}{\beta_{i,k}} : \alpha_{i,k} \geq 0, \beta_{i,k} \geq 0 \right\}. \tag{6.10}$$

Second-third-, and fourth-order schemes have been analyzed in [Got-98], where the utility of such schemes are demonstrated. The second- and third-order schemes are simple to implement, as illustrated in what follows. However, the fourth-order scheme is more complex, requiring problem-specific residue minimization evaluations, and will not be discussed further here.

Second-order scheme

$$\begin{aligned} u^{(1)} &= u^n + \Delta t \mathcal{L}(u^n) \\ u^{n+1} &= \tfrac{1}{2} u^n + \tfrac{1}{2} u^{(1)} + \tfrac{1}{2} \Delta t \mathcal{L}(u^{(1)}). \end{aligned} \tag{6.11}$$

Third-order scheme

$$\begin{aligned} u^{(1)} &= u^n + \Delta t \mathcal{L}(u^n), \\ u^{(2)} &= \tfrac{3}{4} u^n + \tfrac{1}{4} u^{(1)} + \tfrac{1}{4} \Delta t \mathcal{L}(u^{(1)}), \\ u^{n+1} &= \tfrac{1}{3} u^n + \tfrac{2}{3} u^{(2)} + \tfrac{2}{3} \Delta t \mathcal{L}(u^{(2)}). \end{aligned} \tag{6.12}$$

The third-order scheme of eqns. (6.12) is a popular and practical method for use in high-resolution applications. R code implementions are included in Listings 6.12 and 6.16.

6.4 GODUNOV METHOD

Godunov's scheme is a *conservative numerical scheme*, suggested by S. K. Godunov for solving partial differential equations [God-54, God-59]. In this method, the conservative variables are considered as *piecewise constant* over the mesh cells at each time step, and the time evolution is determined by the exact solution of the *Riemann problem* (shock tube) at the intercell boundaries.

Following *Hirsch* [Hir-90], the scheme involves three distinct steps to obtain the solution at $t = (n+1)\Delta t$ from the known solution at $t = n\Delta t$, as follows:

Step 1: Define piecewise constant approximation of the solution at $t = (n+1)\Delta t$. Because the piecewise constant approximation is an average of the solution over the cell size Δt, the spatial error is of order Δx, and hence the resulting scheme will be first-order accurate in space. Note that this approximation corresponds to a *finite volume method* representation whereby the discrete values represent averages of the state variables over the cells. Exact relations for the averaged cell values can be obtained from the integral conservation laws (see Chapter 7).

Step 2: Obtain the solution for the local Riemann problem at the cell interfaces. This is the only physical step of the whole procedure. The discontinuities at the interfaces are resolved in a superposition of waves satisfying locally the conservation equations. The original Godunov method is based upon the exact solution of the Riemann problems. However, approximate solutions can be applied as an alternative.

Step 3: Average the state variables after a time interval Δt. The state variables obtained after step 2 are averaged over each cell defining a new piecewise constant approximation resulting from the wave propagation during the time interval Δt. To be consistent, the time interval Δt should be limited such that the waves emanating from an interface do not interact with waves created at the adjacent interfaces. Otherwise, the situation inside a cell would be influenced by interacting Riemann problems. This leads to the CFL condition $|a_{max}|\Delta t < \Delta x/2$, where $|a_{max}|$ is the maximum wave speed obtained from the cell eigenvalue(s) of the local *Jacobian* matrix.

The first and third steps are solely of a numerical nature and can be considered as a projection stage, independent of the second, physical step, the *evolution stage*. Therefore, they can be modified without influencing the physical input, for instance, by replacing the piecewise constant approximation by a piecewise linear variation inside each cell, leading to the definition of second-order space-accurate schemes, such as the *monotone upstream-centered schemes for conservation laws* (MUSCL) (see Section 6.6).

Mathematically, the Godunov scheme for the advection eqn. (6.2) can be described as a Riemann solver applied to a finite volume representation. Now the integral representation of this conservation law is

$$\int_{x-1/2}^{x+1/2} \int_{t^n}^{t^{n+1}} (\partial_t u + a \partial_x u) \, dx \, dt = 0. \tag{6.13}$$

Thus, using the exact relationships

$$u_i^n = \int_{x_{i-1/2}}^{x_{i+1/2}} u(x, t^n) \, dx \tag{6.14}$$

and

$$u_{i+1/2}^{n+1/2} = \frac{1}{\Delta t} \int_{t^{n-1/2}}^{t^{n+1/2}} u(x_{i+1/2}, t) \, dt, \tag{6.15}$$

the finite volume representation becomes

$$\frac{u_i^{n+1} - u_i^n}{\Delta t} + a \frac{u_{i+1/2}^{n+1/2} - u_{i-1/2}^{n+1/2}}{\Delta x}. \tag{6.16}$$

The values for u_i^n and $u_{i\pm 1/2}^{n\pm 1/2}$ can be obtained by solving the exact Riemann problem with piecewise initial data—the *first-order Godunov method*. This results in a first-order upwind difference scheme.

Instead of a piecewise constant approximation of the edge values, a piecewise linear approximation can be used and the Riemann problem solved—the *second-order*

Godunov method. However, this can introduce new nonreal extrema for solutions involving shocks or steep gradient phenomena and may necessitate extremely small grid spacing to achieve required accuracy. This can be computationally demanding. Extending the idea to systems of nonlinear conservation equations increases the complexity further and will not be pursued here. Alternatively, to avoid these difficulties, an approximate Riemann solution can be found using a high-resolution TVD method, such as the MUSCL scheme with appropriate edge reconstruction (see Section 6.6).

6.4.1 Godunov's Theorem

Sergei K. Godunov's work in the area of applied and numerical mathematics has had a major impact on science and engineering, particularly in the development of methodologies used in computational fluid dynamics (CFD) and other computational fields. One of his major contributions was to prove the theorem that bears his name [God-54, God-59]. *Godunov's theorem*, also known as *Godunov's order barrier theorem*, states that

> linear numerical schemes for solving *partial differential equations* (PDEs), having the property of not generating new extrema (*monotone scheme*), can be at most first-order accurate.

The theorem was originally proved by Godunov as a PhD student at Moscow State University and has been extremely important in the development of the theory of *high-resolution schemes* for the numerical solution of PDEs.

The Theorem

We generally follow Wesseling [Wes-01].

Aside

Assume a continuum problem described by a *partial differential equation* is to be computed using a numerical scheme based upon a uniform computational grid and a one-step, constant-step-size, M grid point integration algorithm, either implicit or explicit. Then if $x_j = j\Delta x$ and $t^n = n\Delta t$, such a scheme can be described by

$$\sum_m^M \beta_m \varphi_{j+m}^{n+1} = \sum_m^M \alpha_m \varphi_{j+m}^n. \tag{6.17}$$

It is assumed that β_m determines φ_j^{n+1} uniquely. Now, because the preceding equation represents a linear relationship between φ_j^n and φ_j^{n+1}, we can perform a linear transformation to obtain the following equivalent form:

$$\varphi_j^{n+1} = \sum_m^M \gamma_m \varphi_{j+m}^n. \tag{6.18}$$

Theorem 6.1 *(Monotonicity preserving).* The scheme of eqn. (6.18) is monotonicity preserving if and only if

$$\gamma_m \geq 0, \quad \forall m. \tag{6.19}$$

Proof. Godunov (1959). □

Case 1 (sufficient condition)

Assume that eqn. (6.19) applies and that φ_j^n is monotonically increasing with j.

Then, because $\varphi_j^n \leq \varphi_{j+1}^n \leq \cdots \leq \varphi_{j+m}^n$, it follows that $\varphi_j^{n+1} \leq \varphi_{j+1}^{n+1} \leq \cdots \leq \varphi_{j+m}^{n+1}$ because

$$\varphi_j^{n+1} - \varphi_{j-1}^{n+1} = \sum_{m}^{M} \gamma_m \left(\varphi_{j+m}^n - \varphi_{j+m-1}^n \right) \geq 0. \tag{6.20}$$

This means that monotonicity is preserved for this case.

Case 2 (necessary condition)

For the same monotonically increasing φ_j^n, assume that $\gamma_p < 0$ for some p, and choose

$$\varphi_i^n = 0, \quad i < k; \quad \varphi_i^n = 1, \quad i \geq k. \tag{6.21}$$

Then, from eqn. (6.18), we get

$$\varphi_j^{n+1} - \varphi_{j-1}^{n+1} = \sum_{m}^{M} \gamma_m \left(\varphi_{j+m}^n - \varphi_{j+m-1}^n \right) = \begin{cases} 0, & (j+m \neq k) \\ \gamma_m, & (j+m = k). \end{cases} \tag{6.22}$$

Now choose $j = k - p$, to give

$$\varphi_{k-p}^{n+1} - \varphi_{k-p-1}^{n+1} = \gamma_p \left(\varphi_k^n - \varphi_{k-1}^n \right) < 0, \tag{6.23}$$

which implies that φ_j^{n+1} is *not* increasing, and we have a contradiction. Thus, monotonicity is *not* preserved for $\gamma_p < 0$, which completes the proof.

Theorem 6.2 *(Godunov's Order Barrier Theorem)*. Linear one-step second-order accurate numerical schemes for the convection equation

$$\frac{\partial \varphi}{\partial t} + c \frac{\partial \varphi}{\partial x} = 0, \quad \varphi = \varphi(x, t), \ t > 0, \ x \in \mathbb{R} \tag{6.24}$$

cannot be monotonicity preserving unless

$$\sigma = |c| \frac{\Delta t}{\Delta x} \in \mathbb{N}, \tag{6.25}$$

where σ is the signed *Courant–Friedrichs–Lewy* (CFL) condition number.

Note that Godunov defines a second-order accurate numerical scheme as

that which will produce the exact solution if $\varphi(x, 0)$ is a polynomial of second degree.

Proof. Godunov (1959). □

Assume a numerical scheme of the form described by eqn. (6.18) and choose

$$\varphi(x, 0) = \left(\frac{x}{\Delta x} - \frac{1}{2} \right)^2 - \frac{1}{4}, \quad \varphi_j^0 = \left(j - \frac{1}{2} \right)^2 - \frac{1}{4}. \tag{6.26}$$

The exact solution is

$$\varphi(x, t) = \left(\frac{x - ct}{\Delta x} - \frac{1}{2} \right)^2 - \frac{1}{4}. \tag{6.27}$$

If we assume the scheme to be at least second-order accurate, it should produce the following solution exactly:

$$\varphi_j^1 = \left(j - \sigma - \frac{1}{2}\right)^2 - \frac{1}{4}, \quad \varphi_j^0 = \left(j - \frac{1}{2}\right)^2 - \frac{1}{4}. \tag{6.28}$$

Substituting eqn. (6.28) into eqn. (6.18) gives

$$\left(j - \sigma - \frac{1}{2}\right)^2 - \frac{1}{4} = \sum_{m}^{M} \gamma_m \left\{ \left(j + m - \frac{1}{2}\right)^2 - \frac{1}{4} \right\}. \tag{6.29}$$

Suppose that the scheme **IS** monotonicity preserving; then according to Theorem 6.1, $\gamma_m \geq 0$.

Now, it is clear from eqn. (6.29) that

$$\left(j - \sigma - \frac{1}{2}\right)^2 - \frac{1}{4} \geq 0, \quad \forall j. \tag{6.30}$$

Assume $\sigma > 0, \sigma \notin \mathbb{N}$, and choose j such that $j > \sigma > (j-1)$. This implies that $(j - \sigma) > 0$ and $(j - \sigma - 1) < 0$.

It therefore follows that

$$\left(j - \sigma - \frac{1}{2}\right)^2 - \frac{1}{4} = (j - \sigma)(j - \sigma - 1) < 0, \tag{6.31}$$

which contradicts eqn. (6.30) and completes the proof.

The exceptional situation whereby $\sigma = |c|\frac{\Delta t}{\Delta x} \in \mathbb{N}$ is only of theoretical interest, since this cannot be realized with variable coefficients. Also, integer CFL numbers greater than unity would not be feasible for practical problems.

6.5 FLUX LIMITER METHOD

Flux limiters are used in numerical schemes to solve problems in science and engineering, particularly fluid dynamics, that are described by *partial differential equations* (PDEs). Their main purpose is to avoid the spurious oscillations (wiggles) that would otherwise occur with high-order spatial discretization (as predicted by Godunov [God-54, God-59]) due to shocks, discontinuities, or steep gradients in the solution domain. They can be used directly on *finite difference schemes* for simple applications, such as the 1D advection PDE of eq. (6.2). However, for more complex systems involving conservation laws, *finite volume high-resolution schemes* are employed that are also designed to avoid entropy violations.

A popular high-resolution scheme is the *MUSCL scheme* (MUSCL stands for *monotone upstream-centered schemes for conservation laws*; see Section 6.6). The term was introduced in a seminal paper by Bram van Leer [vanL-79]. In this paper he constructed the first high-order, *total variation diminishing* (TVD) scheme where he obtained second-order spatial accuracy. The concept of TVD was introduced by Ami Harten [Har-83] and relates to a *monotone* numerical scheme. A system is said to be *monotonicity preserving* if the following properties are maintained over time:

- *No new local extrema are created* within the solution spatial domain,
- The value of a *local minimum is nondecreasing*, and the value of a *local maximum is nonincreasing*.

Use of flux limiters, together with an appropriate high-resolution scheme, makes the resulting PDE numerical solutions *total variation diminishing*.

Flux limiters are also referred to as *slope limiters* because both have the same mathematical form, and both have the effect of limiting the solution gradient near shocks or discontinuities. In general, the term *flux limiter* is used when the limiter acts on system *fluxes*, and *slope limiter* is used when the limiter acts on system *states*.

6.5.1 How Limiters Work

The main idea behind the construction of flux limiter schemes is to limit the spatial derivatives to realistic values—for scientific and engineering problems this usually means physically realizable values. They are used in high-resolution schemes for solving problems described by PDEs and only come into operation when sharp wave fronts are present. For smoothly changing waves, the flux limiters do not operate and the spatial derivatives can be represented by higher-order approximations without introducing nonreal oscillations.

Consider the 1D *semi-discrete scheme*

$$\frac{du_i}{dt} + \frac{1}{\Delta x_i}\left[F\left(u_{i+\frac{1}{2}}\right) - F\left(u_{i-\frac{1}{2}}\right)\right] = 0, \tag{6.32}$$

where, for a *finite volume scheme*, $F(u_{i+\frac{1}{2}})$ and $F(u_{i-\frac{1}{2}})$ represent edge fluxes for the ith cell. Similarly, for a *finite difference scheme*, they represent flux values on the grid at point $x = x_{i+\frac{1}{2}}$ and point $x = x_{i-\frac{1}{2}}$. If these fluxes can be represented by *low*, and *high*-resolution schemes, then a flux limiter can switch between these schemes depending upon the gradients close to the particular cell, as follows:

$$F\left(u_{i+\frac{1}{2}}\right) = f^{low}_{i+\frac{1}{2}} - \phi\left(r_i\right)\left(f^{low}_{i+\frac{1}{2}} - f^{high}_{i+\frac{1}{2}}\right)$$
$$F\left(u_{i-\frac{1}{2}}\right) = f^{low}_{i-\frac{1}{2}} - \phi\left(r_{i-1}\right)\left(f^{low}_{i-\frac{1}{2}} - f^{high}_{i-\frac{1}{2}}\right), \tag{6.33}$$

where f^{low} is the low-resolution flux, f^{high} is the high-resolution flux, $\phi\left(r\right)$ is the flux limiter function, and r represents the ratio of successive gradients on the solution mesh, that is,

$$r_i = \frac{u_i - u_{i-1}}{u_{i+1} - u_i}. \tag{6.34}$$

The limiter function is constrained to be greater than or equal to zero, $r \geq 0$. Therefore, when the limiter is equal to zero (sharp gradient, opposite slopes, or zero gradient), the flux is represented by a *low-resolution scheme*. Similarly, when the limiter is equal to 1 (smooth solution), it is represented by a *high-resolution scheme*. The various limiters listed as follows have differing switching characteristics and are selected to suit the particular problem and numerical solution scheme. No particular limiter has been found to work well for all problems, and a particular choice is usually made on a trial-and-error basis.

6.5.2 Limiter Functions

The following are common forms of flux/slope limiter function, $\phi(r)$:

HCUS* [Wat-95]

$$\phi_{hc}(r) = \frac{1.5(r+|r|)}{(r+2)}; \quad \lim_{r\to\infty} \phi_{hc}(r) = 3$$

HQUICK* [Wat-95],

$$\phi_{hq}(r) = \frac{2(r+|r|)}{(r+3)}; \quad \lim_{r\to\infty} \phi_{hq}(r) = 4$$

Koren [Kor-93]

$$\phi_{kn}(r) = \max[0, \min(2r, (1+2r)/3, 2)]; \quad \lim_{r\to\infty} \phi_{kn}(r) = 2$$

minmod[#] [Roe-86]

$$\phi_{mm}(r) = \max[0, \min(1, r)]; \quad \lim_{r\to\infty} \phi_{mm}(r) = 1$$

monotonized central (MC)[#] [vanL-77]

$$\phi_{mc}(r) = \max[0, \min(2, 2r, (1+r)/2)]; \quad \lim_{r\to\infty} \phi_{mc}(r) = 2$$

Osher [Cha-83]

$$\phi_{os}(r) = \max[0, \min(r, \beta)], \quad (1 \le \beta \le 2); \quad \lim_{r\to\infty} \phi_{os}(r) = \beta$$

ospre[#] [Wat-95]

$$\phi_{op}(r) = \frac{1.5(r^2+r)}{(r^2+r+1)}; \quad \lim_{r\to\infty} \phi_{op}(r) = 1.5$$

smart* [Gas-88]

$$\phi_{sm}(r) = \max[0, \min(2r, (0.25+0.75r), 4)]; \quad \lim_{r\to\infty} \phi_{sm}(r) = 4$$

superbee[#] [Roe-86]

$$\phi_{sb}(r) = \max[0, \min(2r, 1), \min(r, 2)]; \quad \lim_{r\to\infty} \phi_{sb}(r) = 2$$

Sweby[#] [Swe-84]

$$\phi_{sw}(r) = \max[0, \min(\beta r, 1), (r, \beta)], \quad \min(1 \le \beta \le 2); \quad \lim_{r\to\infty} \phi_{sw}(r) = \beta$$

UMIST [Lie-94]

$$\phi_{um}(r) = \max[0, \min(2r, (0.25+0.75r), (0.75+0.25r), 2)]; \quad \lim_{r\to\infty} \phi_{um}(r) = 2$$

van Albada 1[#] [vanA-82]

$$\phi_{va1}(r) = \frac{r^2+r}{r^2+1}; \quad \lim_{r\to\infty} \phi_{va1}(r) = 1$$

van Albada 2[*] Alternative form used on high spatial order schemes [Ker-03]

$$\phi_{va2}(r) = \frac{2r}{r^2+1}; \quad \lim_{r \to \infty} \phi_{va2}(r) = 0$$

van Leer 1[#] [vanL-74]

$$\phi_{vl1}(r) = \min(2r, \min((1+r)/2, 2)); \quad \lim_{r \to \infty} \phi_{vl1}(r) = 2$$

van Leer 2[#] [vanL-74]

$$\phi_{vl2}(r) = \frac{r + |r|}{1 + |r|}; \quad \lim_{r \to \infty} \phi_{vl2}(r) = 2$$

[*] Limiter is *not second-order TVD*!
[#] Limiter is *symmetric* and exhibits the following symmetry property:

$$\frac{\phi(r)}{r} = \phi\left(\frac{1}{r}\right)$$

This is a desirable property as it ensures that the limiting actions for forward and backward gradients operate in the same way.

Unless indicated to the contrary, the preceding limiter functions are second-order *total variation diminishing* (TVD). This means that they are designed such that they pass through a certain region of the solution, known as the TVD region, to guarantee stability of the scheme. Second-order TVD limiters satisfy at least the following criteria:

- $r \leq \phi(r) \leq 2r$, $(0 \leq r \leq 1)$
- $1 \leq \phi(r) \leq r$, $(1 \leq r \leq 2)$
- $1 \leq \phi(r) \leq 2$, $(r > 2)$
- $\phi(1) = 1$.

The admissible limiter region for second-order TVD schemes is shown in the *Sweby Diagram* (Fig. 6.2), and plots showing limiter functions overlaid onto the TVD region are shown in Fig. 6.3, where plots for the Osher and Sweby limiters have been generated using $\beta = 1.5$. The plots were generated by the R program in file `limiterPlots.R` that calls the function `limiter()` in Listing 6.1 and is included with the downloads.

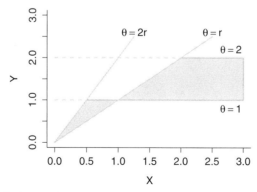

Figure 6.2. Admissible limiter region (shaded) for second-order TVD schemes [Swe-84].

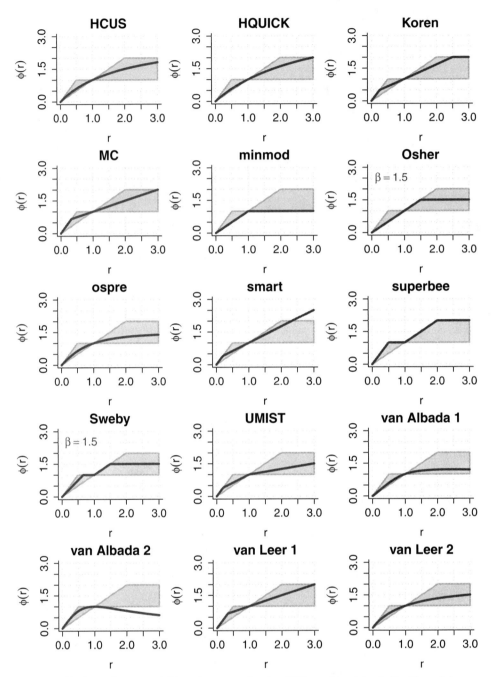

Figure 6.3. Limiter plots overlaid onto second-order TVD region (shaded). The plots were generated by the R program in file `limiterPlots.R` that calls the function `limiter()` in Listing 6.1.

```r
# Calculations for various Flux Limiters
limiter <- function(r, m, limiterType, Beta){
  # Flux Limiter Functions
  #
  m   <- length(r)
  psi <- rep(0,m);
  if(limiterType == 1){
    # None
  }else if(limiterType ==2){
    psi <- pmax( 0, 1.5*(r + abs(r))/(r + 2)) # HCUS
  }else if(limiterType ==3){
    psi <- pmax( pmax(0, 2*(r+abs(r)) / (r+3))) # HQUICK
  }else if(limiterType ==4){
    psi <- pmax(0, pmin(2*r, pmin(2*r/3 + 1/3, 2))) # Koren
  }else if(limiterType ==5){
    psi <- pmax(0, pmin(2*r,pmin((1+r)/2, 2))) # MC
  }else if(limiterType ==6){
    psi <- pmax(0, pmin( 1, r)) # minmod
  }else if(limiterType ==7){
    psi <- pmax(0, pmin(Beta, r) ) # Osher
  }else if(limiterType ==8){
    r2 <- r^2
    psi <- pmax(0, (3/2)*(r2+r)/(r2+r+1)) # ospry
  }else if(limiterType ==9){
    psi <- pmax( 0, pmin( pmin( 0.25+0.75*r, 4), 2*r)) # Smart
  }else if(limiterType ==10){
    psi <- pmax(0, pmin( 2*r, 1), pmin( r, 2)) # SuperBee
  }else if(limiterType ==11){
    psi <- pmax( pmax(0, pmin(Beta*r, 1)), pmin(r, Beta)) # Sweeby
  }else if(limiterType ==12){
    psi <- pmax(0, pmin( 2*r, pmin(0.25 + 0.75*r,
         pmin( 0.75 + 0.25*r, 2)))) # UMIST bounded QUICK
  }else if(limiterType ==13){
    r2 <- r^2
    psi <- pmax( pmax(0, (r2 + r)/(r2 + 1))) # Albada 1
  }else if(limiterType ==14){
    psi <- pmax( pmax(0, 2*r/(r^2 + 1))) # Albada 2 (for high spatial order schemes)
  }else if(limiterType ==15){
    psi <- pmax(0, pmin(2*r, pmin((1 + r)/2, 2))) # van Leer 1
  }else if(limiterType ==16){
    psi <- pmax(0, (r + abs(r))/(1 + abs(r)) )# van Leer 2
  }
  return(psi)
}
```

Listing 6.1. File: Limiter.R—Code for function `limiter()` that calculates various limiter responses

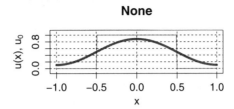

Figure 6.4. 1D advective equation $u_t + u_x = 0$, with step wave propagating to the right at $t = 10$. Finite volume method performed on a grid of 201 cells with periodic BCs and CLF=0.45. Shows the analytical solution with simulation result (superimposed) based upon a *first-order upwind* spatial discretization scheme. Time integration was performed using the *third-order TVD RK scheme*.

6.6 MONOTONE UPSTREAM-CENTERED SCHEMES FOR CONSERVATION LAWS (MUSCL)

The term MUSCL was introduced in a seminal paper by Bram van Leer (van Leer, 1979). In this paper he constructed the first *high-order, total variation diminishing* (TVD) scheme where he obtained second-order spatial accuracy. It is a finite volume method that provides high accuracy numerical solutions to partial differential equations, which can involve solutions that exhibit shocks, discontinuities, or large gradients.

The idea is to replace the piecewise constant approximation of Godunov's scheme by reconstructed states, derived from cell-averaged states obtained from the previous time step. For each cell, slope limited, reconstructed left and right states are obtained and used to calculate fluxes at the cell boundaries (edges). These fluxes can, in turn, be used as input to a Riemann solver, following which the solutions are averaged and used to advance the solution in time. Alternatively, the fluxes can be used in Riemann-free-solver schemes, such as the *Kurganov and Tadmor scheme* outlined in Section 6.6.2.

6.6.1 Linear Reconstruction

We will introduce the fundamentals of the MUSCL scheme by considering the following simple first-order, scalar, 1D system, which is assumed to have a wave propagating in the positive x direction:

$$u_t + F_x(u) = 0. \quad u = u(x, t), \ t \geq t_0, \ x \in \mathbb{R}, \tag{6.35}$$

where u represents a state variable and F represents a *flux* variable.

The basic scheme of Godunov uses piecewise constant approximations for each cell and results in a first-order upwind discretization of the preceding problem with cell centers indexed as i. A semi-discrete scheme can be defined as follows:

$$\frac{du_i}{dt} + \frac{1}{\Delta x_i}[F(u_{i+1}) - F(u_i)] = 0. \tag{6.36}$$

This basic scheme is not able to handle shocks or sharp discontinuities due to excessive dissipation, which causes the solution to become smeared. An example of this effect is shown in Fig. 6.4, which illustrates a 1D advective equation with a step wave propagating to the right. The simulation was carried out with a mesh of 201 cells and used a third-order TVD RK time integrator. The excessive dissipation introduced by this method is clearly evident.

To provide higher resolution of discontinuities, Godunov's scheme can be extended to use piecewise linear approximations of each cell, which results in a *central difference* scheme that is *second-order* accurate in space. The piecewise linear approximations are obtained from

$$u(x) = u_i + \frac{(x - x_i)}{(x_{i+1} - x_i)} (u_{i+1} - u_i), \quad x \in [x_i, x_{i+1}]. \tag{6.37}$$

The corresponding R code for this linear extrapolation is given in Listing 6.2.

```
# File: linearExtrapolation.R
linearExtrapolation <- function(u, limiterType, Beta){
# Slope-limited linear extrapolation
  N  <- length(u)
  rL <- rep(0,N)
  rR <- rep(0,N)

  BC <- 0
  if(BC == 0){
    # Shifted u arrays for Dirichlet BCs
    up1 <- c(u[2:N],u[N])
    up2 <- c(u[3:N],u[N],u[N])
    um1 <- c(u[1],u[1:(N-1)])
    um2 <- c(u[1],u[1],u[1:(N-2)])
  }else if(BC == 1){
    # Shifted u arrays for periodic BCs
    up1 <- c(u[2:N],u[1])
    up2 <- c(u[3:N],u[1],u[2])
    um1 <- c(u[N],u[1:(N-1)])
    um2 <- c(u[N-1],u[N],u[1:(N-2)])
  }
  # Check which denominators of rL > e-6
  rLGT0    <- which(abs(up1-u)>1e-6) # Avoids devide by zero
  rL[rLGT0] <- (u[rLGT0] - um1[rLGT0])/(up1[rLGT0]-u[rLGT0])
  # Check which denominators of rR > e-6
  rRGT0    <- which(abs(u-um1)>1e-6) # Avoids divide by zero
  rR[rRGT0] <- (up1[rRGT0] - u[rRGT0])/(u[rRGT0]-um1[rRGT0])

  uL   <- u + 0.5*Limiter(rL, N, limiterType, Beta)*(up1-u) # u(j+1/2)
  uRm1 <- u + 0.5*Limiter(rR, N, limiterType, Beta)*(um1-u) # u(j-1/2)
  uR   <- c(uRm1[2:N], uRm1[1])

  return(c(uL,uR))
}
```

Listing 6.2. File: `linearExtrapolation.R`—Code for the function `linearExtrapolation()` that performs slope-limited extrapolation

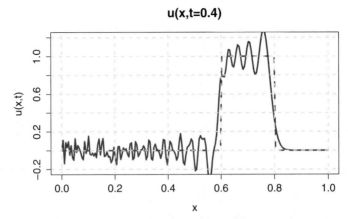

Figure 6.5. 1D advective equation $u_t + u_x = 0$, with step wave propagating to the right. Shows the analytical solution *dashed*, with simulation result superimposed *solid*, based upon a second-order, central difference spatial discretization scheme.

Thus, evaluating fluxes at the cell edges, we get the following semi-discrete scheme:

$$\frac{du_i}{dt} + \frac{1}{\Delta x_i}\left[F\left(u_{i+\frac{1}{2}}\right) - F\left(u_{i-\frac{1}{2}}\right)\right] = 0, \qquad (6.38)$$

where $u_{i+\frac{1}{2}}$ and $u_{i-\frac{1}{2}}$ are the piecewise approximate values of cell edge variables, that is,

$$u_{i+\frac{1}{2}} = 0.5\left(u_i + u_{i+1}\right) \qquad (6.39)$$

$$u_{i-\frac{1}{2}} = 0.5\left(u_{i-1} + u_i\right). \qquad (6.40)$$

Although the preceding second-order scheme provides greater accuracy for smooth solutions, it is not a *total variation diminishing* (TVD) scheme and introduces spurious oscillations into the solution where discontinuities or shocks are present. A fairly extreme example of this effect is shown in Fig. 6.5, which illustrates the 1D advective equation $u_t + u_x = 0$, with a unit step-pulse wave propagating to the right. This loss of accuracy is to be expected due to Godunov's theorem discussed in Section 6.4.1. This *finite difference* simulation was carried out on a mesh of 200 grid points and used the default numerical integrator ode() from the R package deSolve, with relative and absolute error tolerances both set to 10^{-8}. The spatial derivative was approximated using a three-point central scheme (see Chapter 3 for more discussion on spatial derivative approximations). Note that a biased upwind scheme would improve matters but would not totally eliminate the spurious oscillations. The R code for the main program that generated the plots in Fig. 6.5 is shown in Listing 6.3, and the associated derivative function is shown in Listing 6.4.

```
# File: advec2ndCent_main.R
# 1D advection equation
#    ut + a*ux = 0
#    Initial condition: u(x,t=0) = step pulse
#    Dirichlet BCs
#    Analytical solution: The BC value traveling left to right
#                         at velocity a
```

```r
# -----------------------------------------------------------------------------
# Delete workspace prior to new calculations
  cat("\014") # Clear console
  rm(list = ls(all = TRUE))
# Library of R ODE solvers
  library("deSolve")
#
# Include spatial differentiator
  source("advec2ndCent_deriv.R")
#
# PDE parameters
  a <- 1; xl=0; xu=1;
#
# Spatial grid in x
  n <- 200
  x <- seq(from=xl,to=xu,by=(xu-xl)/(n-1))
  # Grid spacing
  dx <- (xu-xl)/(n-1)
#
# Initial condition
# u(x,t=0) <- 0
  uini=rep(0,n)
  for(i in 1:n){
    if(x[i] >= 0.2 && x[i] <= 0.4){
      uini[i] <- 1
    }
  }
# Analytical solution for comparison purposes
  uend <- rep(0,n)
  for(i in 1:n){
    if(x[i] >= 0.6 && x[i] <= 0.8){
      uend[i] <- 1
    }
  }
#
  nout=101
# t interval
  times = seq(0,0.4, length.out=nout)
#
# ODE integration
  out <- ode(y = uini, times = times, func = advec2ndCent_deriv,
             rtol = 1e-8, atol = 1e-8,parms = NULL)

# u(x=0,t) not returned by ODE routine advec2ndCent(); so defined here
  out[,2] <- 0
#
```

```
# Plot ODE Analytical and numerical solutions
  plot(x=x,y=out[nout,2:(n+1)],type="l",xlab="x",ylab="u(x,t)",ylim=c(-0.2,1.2),
       col="blue", ,lwd=3, main="u(x,t=0.4)")
  lines(x=x,y=uend,type="l",lty=2,lwd=3,col="red")
  grid(lwd=2,lty=2)
```

Listing 6.3. File: advec2ndCent_main.R—Code for main program that that generates plots shown in Fig. 6.5

```
# File: advec2ndCent_deriv.R
advec2ndCent_deriv <- function(t,u,params){
  u[1] <- 0; u[n] <- 0 # Dirichlet BC
# Declare arrays
  ux <- rep(0,n)
# Second order central difference spatial derivative approximation
  ux[1] <- (-3*u[ 1]+4*u[ 2]-1*u[ 3])
  for(i in 2:(n-1)){
    ux[i] <- (-1*u[i-1] +0*u[i]+1*u[i+1])}
  ux[n] <- (1*u[n-2]-4*u[n-1]+3*u[n])
  ux <- ux/(2*dx)
# Time derivative
  ut <- -a*ux
#
# Dirichlet BC
  ut[1] <- 0; ut[n] <- 0
#
  list(c(ut))
}
```

Listing 6.4. File: advec2ndCent_deriv.R—Code for derivative function advec2ndCent() that calculates a second-order, three-point spatial derivative approximation. It is called from the main program of Listing 6.3

MUSCL-based numerical schemes extend the idea of using a linear piecewise approximation to each cell by using *slope limited* left and right *extrapolated states* (see Fig. 6.6). This results in the following high-resolution, TVD discretization scheme:

$$\frac{du_i}{dt} + \frac{1}{\Delta x_i}\left[F\left(u^*_{i+\frac{1}{2}}\right) - F\left(u^*_{i-\frac{1}{2}}\right)\right] = 0, \tag{6.41}$$

which, alternatively, can be written in the more succinct form

$$\frac{du_i}{dt} + \frac{1}{\Delta x_i}\left[F^*_{i+\frac{1}{2}} - F^*_{i-\frac{1}{2}}\right] = 0. \tag{6.42}$$

The numerical fluxes $F^*_{i\pm\frac{1}{2}}$ correspond to a nonlinear combination of first- and second-order approximations to the continuous flux function.

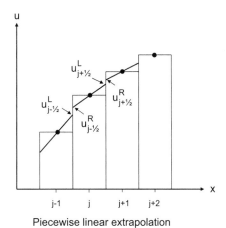

Piecewise linear extrapolation

Figure 6.6. An example of MUSCL type left- and right-state linear extrapolation.

The symbols $u^*_{i+\frac{1}{2}}$ and $u^*_{i-\frac{1}{2}}$ represent scheme-dependent functions (of the slope-limited extrapolated cell edge variables), that is,

$$u^*_{i+\frac{1}{2}} = u^*_{i+\frac{1}{2}}\left(u^L_{i+\frac{1}{2}}, u^R_{i+\frac{1}{2}}\right), u^*_{i-\frac{1}{2}} = u^*_{i-\frac{1}{2}}\left(u^L_{i-\frac{1}{2}}, u^R_{i-\frac{1}{2}}\right), \qquad (6.43)$$

and

$$\begin{aligned} u^L_{i+\frac{1}{2}} &= u_i + 0.5\phi\left(r_i\right)\left(u_{i+1} - u_i\right), \\ u^R_{i+\frac{1}{2}} &= u_{i+1} - 0.5\phi\left(r_{i+1}\right)\left(u_{i+2} - u_{i+1}\right), \end{aligned} \qquad (6.44)$$

$$\begin{aligned} u^L_{i-\frac{1}{2}} &= u_{i-1} + 0.5\phi\left(r_{i-1}\right)\left(u_i - u_{i-1}\right), \\ u^R_{i-\frac{1}{2}} &= u - 0.5\phi\left(r_i\right)\left(u_{i+1} - u_i\right), \end{aligned} \qquad (6.45)$$

$$r_i = \frac{u_i - u_{i-1}}{u_{i+1} - u_i}. \qquad (6.46)$$

The function $\phi(r_i)$ is a limiter function that limits the slope of the piecewise approximations to ensure the solution is TVD, thereby avoiding the spurious oscillations that would otherwise occur around discontinuities or shocks (see Section 6.5). The limiter is equal to zero when $r \leq 0$ and is equal to unity when $r = 1$. Thus, the accuracy of a TVD discretization degrades to first order at local extrema but tends to second order over smooth parts of the domain.

The algorithm is straightforward to implement. Once a suitable scheme for $F^*_{i+\frac{1}{2}}$ has been chosen, such as the *Kurganov and Tadmor scheme* (see later), the solution can proceed using standard integration techniques such as method of lines (MOL).

6.6.2 Kurganov and Tadmor Central Scheme

The Kurganov and Tadmor central scheme is a Riemann-solver-free, second-order, *high-resolution* scheme that uses MUSCL reconstruction. It is straightforward to implement and can be used on scalar and vector problems. The algorithm is based upon *central differences* with comparable performance to Riemann-type solvers when used to obtain solutions for PDEs describing systems that exhibit high-gradient phenomena. The forerunner to this scheme was the fully discrete NT scheme due to Nessyahu and Tadmor [Nes-90]. The Kurganov and Tadmore scheme can be implemented as a fully discrete or semi-discrete scheme. Here we consider the semi-discrete scheme.

The calculation proceeds as follows:

$$F^*_{i-\frac{1}{2}} = \frac{1}{2}\left\{\left[F(u^R_{i-\frac{1}{2}}) + F(u^L_{i-\frac{1}{2}})\right] - a_{i-\frac{1}{2}}\left[u^R_{i-\frac{1}{2}} - u^L_{i-\frac{1}{2}}\right]\right\}$$
$$F^*_{i+\frac{1}{2}} = \frac{1}{2}\left\{\left[F(u^R_{i+\frac{1}{2}}) + F(u^L_{i+\frac{1}{2}})\right] - a_{i+\frac{1}{2}}\left[u^R_{i+\frac{1}{2}} - u^L_{i+\frac{1}{2}}\right]\right\}, \quad (6.47)$$

where the *local propagation speed*, $a_{i\pm\frac{1}{2}}$, is the maximum absolute value of the eigenvalue of the Jacobian of $F(u(x,t))$ over cells i, $i \pm 1$, given by

$$a_{i\pm\frac{1}{2}}(t) = \max\left[\rho\left(\frac{\partial F^R(u_i(t))}{\partial u}\right), \rho\left(\frac{\partial F^L(u_{i+1}(t))}{\partial u}\right)\right], \quad (6.48)$$

and ρ represents the *spectral radius* of $\partial F(u(t))/\partial u$. Beyond these CFL-related speeds, no characteristic information is required [Kur-00a].

This flux calculation is sometimes referred to as *local Lax–Friedrichs* (LLF) flux or *Rusanov* flux [Lax-54] [Rus-61] [Tor-99] [Lev-02].

The scheme can readily include diffusion terms, if they are present. For example, if the preceding 1D scalar problem is extended to include a diffusion term, we get

$$u_t + F_x(u) = Q_{xx}(u, u_x), \quad (6.49)$$

for which Kurganov and Tadmor propose the following central difference approximation,

$$\frac{du_i}{dt} = -\frac{1}{\Delta x_i}\left[F^*_{i+\frac{1}{2}} - F^*_{i-\frac{1}{2}}\right] + \frac{1}{\Delta x_i}\left[P_{i+\frac{1}{2}} - P_{i-\frac{1}{2}}\right], \quad (6.50)$$

where

$$P_{i+\frac{1}{2}} = \frac{1}{2}\left[Q\left(u_i, \frac{u_{i+1} - u_i}{\Delta x_i}\right) + Q\left(u_{i+1}, \frac{u_{i+1} - u_i}{\Delta x_i}\right)\right] \quad (6.51)$$

$$P_{i-\frac{1}{2}} = \frac{1}{2}\left[Q\left(u_{i-1}, \frac{u_i - u_{i-1}}{\Delta x_{i-1}}\right) + Q\left(u_i, \frac{u_i - u_{i-1}}{\Delta x_{i-1}}\right)\right]. \quad (6.52)$$

Full details of the algorithm (*fully* and *semi-discrete* versions) and its derivation can be found in the original paper by Kurganov and Tadmor [Kur-00a], along with a number of 1D and 2D examples. Additional information is also available in an earlier related paper [Nes-90].

Note that although this scheme was originally presented by Kurganov and Tadmor as a second-order scheme based upon *linear extrapolation*, a later paper by Kurganov and Levy [Kur-00b] demonstrates that it can also form the basis of a third-order scheme, using *parabolic reconstruction*. Some 1D advective and Euler equation examples of using linear and parabolic reconstruction are provided in the following sections. It will also be shown subsequently that the Kurganov and Tadmor scheme can also be used with *WENO reconstruction* to good effect.

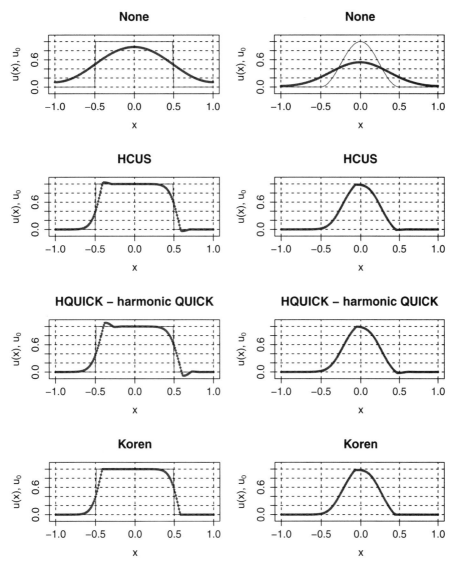

Figure 6.7. The 1D advective equation $u_t + u_x = 0$, with waves propagating to the right at $t = 10$: (left) step pulse and (right) cosine pulse. Finite volume method performed on a grid of 200 cells with periodic BCs and CLF=0.45. Shows the analytical solution along with simulation results superimposed, based upon the Kurganov and Tadmor central scheme with linear extrapolation reconstruction for different limiter cases. Time integration was performed using the third-order TVD RK scheme.

6.6.2.1 Scalar Problems
Example: Advection of step and cosine pulses
The effectiveness of using a high-resolution scheme is shown in Figs. 6.7–6.10), which illustrate solutions to the 1D advective equation $u_t + u_x = 0$ with step and cosine pulse initial conditions. The corresponding R code for the advection equation flux is given in Listing 6.5. The main program and subsidiary functions are discussed in Appendix 6.B at the end of this chapter.

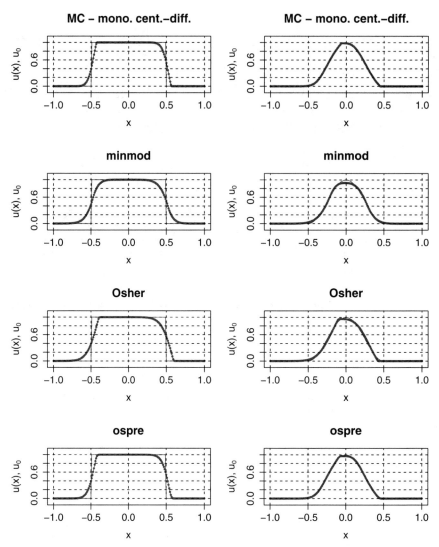

Figure 6.8. The 1D advective equation $u_t + u_x = 0$, with waves propagating to the right at $t = 10$: (left) step pulse and (right) cosine pulse. Finite volume method performed on a grid of 200 cells with periodic BCs and CLF=0.45. Shows the analytical solution (thin) along with simulation results (thick) based upon the Kurganov and Tadmor central scheme with linear extrapolation reconstruction and different limiter cases. Time integration was performed using the third-order TVD RK scheme.

```
# File: advection.R
advection <- function(u, alpha){
  f <- alpha*u
return(f)
}
```

Listing 6.5. File: advection.R—Code for function advection() that calculates the flux for the advection equation

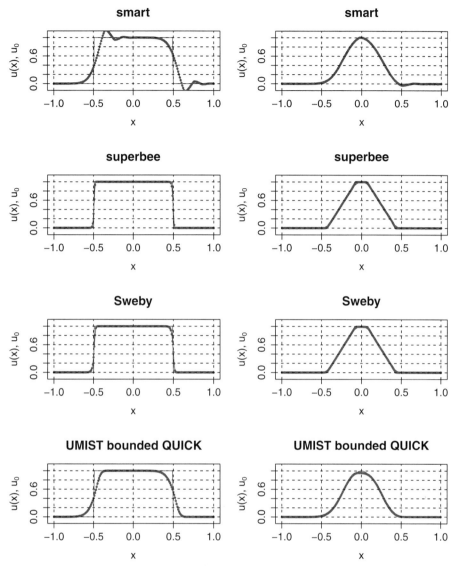

Figure 6.9. The 1D advective equation $u_t + u_x = 0$, with waves propagating to the right at $t = 10$; (left) step pulse and (right) cosine pulse. Finite volume method performed on a grid of 200 cells with periodic BCs and CLF=0.45. Shows the analytical solution with simulation results superimposed, based upon the Kurganov and Tadmor central scheme with linear extrapolation reconstruction for different limiter cases. Time integration was performed using the third-order TVD RK scheme.

However, although the results are a great improvement on standard MOL solutions, it will be seen that none of the limiters gives an ideal solution to the advection equation with step and cosine pulse waves propagating to the right. The superbee limiter gives very good results for the step pulse but causes unrealistic sharpening of the smooth cosine pulse. The Koren limiter gives good results for the cosine pulse and reasonable results for the step pulse, although the shape is asymmetrical. Some limiters result in excessive

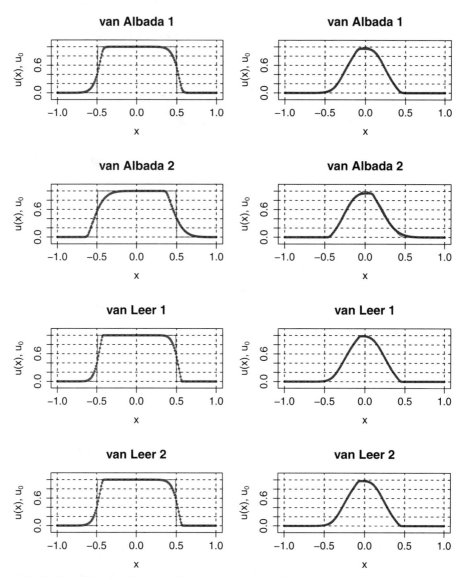

Figure 6.10. The 1D advective equation $u_t + u_x = 0$, with waves propagating to the right at $t = 10$: (left) step pulse and (right) cosine pulse. Finite volume method performed on a grid of 200 cells with periodic BCs and CLF=0.45. Shows the analytical solution with simulation results superimposed, based upon the Kurganov and Tadmor central scheme with linear extrapolation reconstruction for different limiter cases. Time integration was performed using the third-order TVD RK scheme.

dissipation, whereas others result in overshoots in the presence of sharp gradients. Thus, for each application, the analyst will need to select a limiter based upon trial and error.

The simulations were carried out on a mesh of 201 cells, using the Kurganov and Tadmor central scheme with a third-order TVDC RK time integration. These results contrast extremely well against the preceding first-order upwind and second-order central difference results. This scheme also provides good results when applied to sets of equations—see later results for this scheme applied to the Euler equations. However, care has to be taken in choosing an appropriate limiter because, for example, the superbee limiter can cause unrealistic sharpening for some smooth wave solutions.

Example: Burgers equation subject to cosine pulse
Burgers equation[2] is an important partial differential equation that occurs in various branches of fluid mechanics and applied mathematics. The general 1D form of Burgers equation (also known as the *viscous* Burgers equation) is

$$\frac{\partial u}{\partial t} + u\frac{\partial u}{\partial x} = v\frac{\partial^2 u}{\partial x^2}, \quad u = u(x,t),\ t \geq t_0,\ x \in \mathbb{R}, \quad (6.53)$$

where u represents velocity and v viscosity. Note that the problem (eqn. (6.53) plus IC) can be transformed using the *Cole–Hopf transformation* (see Chapter 9) to an equivalent *heat equation* problem, which, providing the transformation is well defined, can be solved analytically. Subsequently, inverting the heat equation solution leads to a solution to eqn. (6.53).

For the situation where $v = 0$, eqn. (6.53) reduces to the *inviscid* Burgers equation, which we represent in hyperbolic conservative form as

$$\frac{\partial u}{\partial t} + \frac{\partial(\frac{1}{2}u^2)}{\partial x} = 0. \quad (6.54)$$

It is this form of Burgers equation that we will consider. The R code is given in Listing 6.6.

```
# File: Burgers.R
Burgers <- function(u, alpha){
  f <- alpha*u*u/2
return(f)
}
```

Listing 6.6. File: Burgers.R—Code for function Burgers() that calculates the flux for the invisid Burgers equation

For many initial conditions, a shock condition arises in the form of a rarefaction wave, as the solution evolves. Therefore, the inviscid Burgers equation with appropriate IC is a good test for high-resolution schemes. Figure 6.11 shows the solution to eqn. (6.54) subject to a *cosine pulse* initial condition for $t \in (0, 0.8)$. The main program and subsidiary functions are discussed in Appendix 6.B at the end of this chapter.

It is seen that the solution is well resolved and that a shock gradually develops, culminating in a rarefaction wave that proceeds right to left.

Example: Buckley–Leverett equation subject to step pulse
The *Buckley–Leverett equation* describes immiscible displacement in a reservoir of porous medium containing two distinct phases, oil and water. The fluids are assumed to be incompressible and capillary pressure losses are assumed negligible. The 1D saturation

---
[2] Named after Dutch physicist Johannes Martinus Burgers (1895–1981).

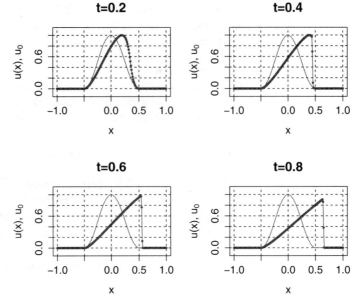

Figure 6.11. A solution to the 1D invisid Burgers equation shown at increasing times from $t = 0.2$ to $t = 0.8$, along with the cosine pulse initial condition. The finite volume method was employed on a grid of 200 cells with Dirichlet BCs and CLF=0.45. The simulation was based upon the Kurganov and Tadmor central scheme with linear extrapolation reconstruction and the Koren limiter. Time integration was performed using the third-order TVD RK scheme.

equation in conservative form is [Buc-42]

$$\frac{\partial u}{\partial t} = \frac{\partial f(u)}{\partial x}, \quad u = u(x,t), \ t \geq t_0, \ x \in \mathbb{R},$$

$$f(u) = \frac{u^2}{u^2 + c(1-u)^2}, \quad c > 0,$$

(6.55)

where $0 \leq u \leq 1$ represents water saturation and c the water over oil viscosity ratio. The saturation of a liquid is defined as the total volume of liquid in the system divided by the pore volume.

The corresponding R code is given in Listing 6.7.

```
# File: BuckeyLeverett.R
BuckeyLeverett <- function(u){
  c <- 0.25
  f <-u^2/(u^2+c*(1-u)^2)
return(f)
}
```

Listing 6.7. File: `BuckeyLeverett.R`—Code for the function `BuckeyLeverett()` that calculates the flux for the Buckey–Leverett equation

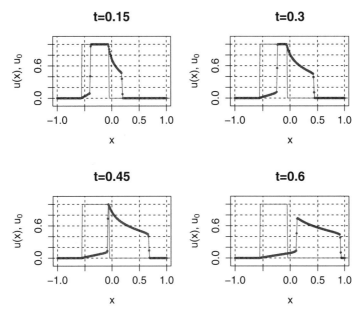

Figure 6.12. The 1D Buckley–Everett equation solution shown at increasing times from $t = 0.15$ to $t = 0.6$, overlaid with the step pulse initial condition. Finite volume method performed on a grid of 201 cells with Dirichlet BCs and CLF=0.45. The simulation was based upon the Kurganov and Tadmor central scheme with linear extrapolation reconstruction and the Koren limiter. Time integration was performed using the third-order TVD RK scheme.

Figure 6.12 shows the solution to eqn. (6.55) for the parameter $c = 0.25$ and $t \in (0, 0.6)$.

This problem has a nonconvex flux term (see Fig. 6.13), which puts an additional constraint on the solution Courant number. It requires that, at each time step,

$$\text{Co} = \max \left| \frac{\Delta t}{\Delta x} f'(u(x)) \right|, \qquad (6.56)$$

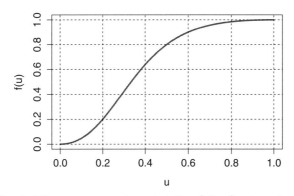

Figure 6.13. Plot showing the nonconvex characteristic of the flux term in the Buckley–Everett equation.

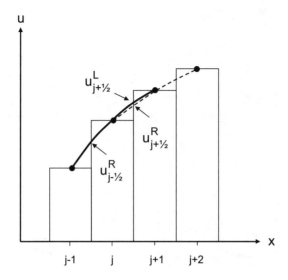

Piecewise parabolic reconstruction

Figure 6.14. An example of MUSCL-type state parabolic reconstruction.

where u is taken over the entire range of u. To avoid calculating Co at each time step, the simulation was run with a constant value Co = 0.45. The main program and subsidiary functions are discussed in Appendix 6.B at the end of this chapter.

For further discussion on the theory and practical issues relating to Buckley–Leverett equation solutions, refer to [Lev-02] and [Lan-98].

6.6.3 Piecewise Parabolic Reconstruction

It is possible to extend the idea of linear extrapolation to higher-order reconstruction, and an example is shown in Fig. 6.14. However, for this case, the left and right states are estimated by interpolation of a second-order, upwind-biased difference equation. The result is a parabolic reconstruction scheme that is third-order accurate in space.

We follow the approach of Kermani [Ker-03] and present a third-order upwind-biased scheme, where the symbols $u^*_{i+\frac{1}{2}}$ and $u^*_{i-\frac{1}{2}}$ again represent scheme-dependent functions (of the limited reconstructed cell edge variables). But for this case they are based upon parabolic-reconstructed states, that is,

$$u^*_{i+\frac{1}{2}} = f\left(u^L_{i+\frac{1}{2}}, u^R_{i+\frac{1}{2}}\right), \quad u^*_{i-\frac{1}{2}} = f\left(u^L_{i-\frac{1}{2}}, u^R_{i-\frac{1}{2}}\right), \tag{6.57}$$

and

$$u^L_{i+\frac{1}{2}} = u_i + \frac{\phi(r_i)}{4}\left[(1-\kappa)\delta u_{i-\frac{1}{2}} + (1+\kappa)\delta u_{i+\frac{1}{2}}\right], \tag{6.58}$$

$$u^R_{i+\frac{1}{2}} = u_{i+1} - \frac{\phi(r_{i+1})}{4}\left[(1-\kappa)\delta u_{i+\frac{3}{2}} + (1+\kappa)\delta u_{i+\frac{1}{2}}\right], \tag{6.59}$$

$$u^L_{i-\frac{1}{2}} = u_{i-1} + \frac{\phi(r_{i-1})}{4}\left[(1-\kappa)\delta u_{i-\frac{3}{2}} + (1+\kappa)\delta u_{i-\frac{1}{2}}\right], \tag{6.60}$$

$$u^R_{i-\frac{1}{2}} = u_i - \frac{\phi(r_i)}{4}\left[(1-\kappa)\delta u_{i+\frac{1}{2}} + (1+\kappa)\delta u_{i-\frac{1}{2}}\right], \tag{6.61}$$

where

$$\delta u_{i+\frac{1}{2}} = (u_{i+1} - u_i), \quad \delta u_{i-\frac{1}{2}} = (u_i - u_{i-1}), \tag{6.62}$$

$$\delta u_{i+\frac{3}{2}} = (u_{i+2} - u_{i+1}), \quad \delta u_{i-\frac{3}{2}} = (u_{i-1} - u_{i-2}), \tag{6.63}$$

$\kappa = 1/3$, and the limiter function $\phi(r)$ can be selected from any of those described in Section 6.5.2.

Parabolic reconstruction is straightforward to implement and can be used with the Kurganov and Tadmor scheme in lieu of the linear extrapolation. This has the effect of raising the spatial solution of the KT scheme to third-order. It performs well when solving the Euler equations (see Section 6.6.4). However, although this increase in spatial order has certain advantages over second-order schemes for problems exhibiting smooth solutions, for shocks, it is more dissipative.

Some 1D scalar example applications using parabolic extrapolation are discussed later.

The corresponding R code for parabolic reconstruction is given in Listing 6.8. The main program and subsidiary functions are discussed in Appendix 6.B at the end of this chapter.

```
# File: parabolicReconstruction.R
# Calculates slope-limited parabolic reconstructed derivative of f(U), wrt x
parabolicReconstruction <- function(u,limtype,Beta){
  scheme <- 1
  #
  N <- length(u)
  r <- rep(0,N)
  if(BC == 0){
    # Shifted u arrays for Dirichlet BCs
    up1 <- c(u[2:N],u[N])
    up2 <- c(u[3:N],u[N],u[N])
    um1 <- c(u[1],u[1:(N-1)])
  }else if(BC == 1){
    # Shifted u arrays for periodic BCs
    up1 <- c(u[2:N],u[1])
    up2 <- c(u[3:N],u[1],u[2])
    um1 <- c(u[N],u[1:(N-1)])
  }

  du    <- up1 - u
  du_p1 <- c(du[2:N],du[1])
  du_m1 <- c(du[N],du[1:(N-1)])
```

```
# Check which r > e-6
rGT0      <- which(abs(up1-u)>1e-6) # avoids divide by zero
r[rGT0] <- (u[rGT0] - um1[rGT0])/(up1[rGT0]-u[rGT0])
phi       <- Limiter(r, N, limiterType, Beta)
phi_p1 <- c(phi[2:M],phi[1])

# Parabolic - (Kermani, et al, 2003)
uL <- u   + 0.25*phi  *(0.666*du_m1 + 1.333*du)
uR <- up1 - 0.25*phi_p1*(0.666*du_p1 + 1.333*du)

return(c(uL,uR))
}
```

Listing 6.8. File: parabolicReconstruction.R—Code for function parabolicReconstruction() that calculates slope-limited parabolic reconstruction of flux

6.6.3.1 Scalar Problems

The main program and subsidiary functions for the following parabolic reconstruction examples are discussed in Appendix 6.B at the end of this chapter.

Example: Advection of step and cosine pulses

Compare the solution to the 1D advective problem ($u_t + u_x = 0$) shown in Fig. 6.9, obtained using linear extrapolation and superbee limiter, to the solution shown in Fig. 6.15 obtained using parabolic reconstruction and the alternative form of van Albada limiter:

$$\phi_{va}(r) = \frac{2r}{1+r^2}. \tag{6.64}$$

The step result is slightly more smeared but with no oscillations, and the cosine pulse result is comparable.

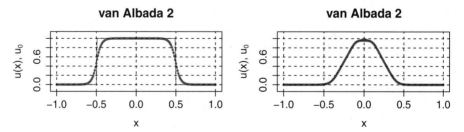

Figure 6.15. The 1D advective equation $u_t + u_x = 0$, with step and cosine waves propagating to the right at $t = 10$: (left) step pulse and (right) cosine pulse. Finite volume method performed on a grid of 201 cells with periodic BCs and CLF=0.45. Shows the analytical solution (thin) along with simulation results (thick) based upon the Kurganov and Tadmor central scheme with parabolic reconstruction and van Albada 2 limiter. Time integration was performed using the third-order TVD RK scheme.

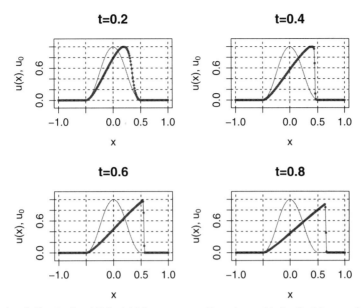

Figure 6.16. A solution to the 1D *invisid Burgers equation* shown (dotted) at increasing times from $t = 0.2$ to $t = 0.8$, along with the cosine pulse initial condition (solid). The finite volume method was employed on a grid of 200 cells with Dirichlet BCs and CLF=0.45. The simulation was based upon the Kurganov and Tadmor central scheme with parabolic reconstruction and the van Albada 2 limiter. Time integration was performed using the third-order TVD RK scheme.

The limiter of eqn. (6.64) was used to avoid the spurious oscillations experienced when using some other limiters; in particular, the superbee limiter gave very poor performance when used with parabolic extrapolation.

Example: Burgers equation subject to cosine pulse
Compare the solution to the 1D Burgers problem ($u_t + 2uu_x = 0$) shown in Fig. 6.11, obtained using linear extrapolation and superbee limiter, to the solution to the same problem shown in Fig. 6.16 obtained using parabolic reconstruction and the alternative form of van Albada limiter. We observe very little difference between the solutions obtained by the two methods for this particular problem.

Example: Buckley–Leverett equation subject to step pulse
The solution to the 1D Buckley–Leverett problem ($u_t + (\frac{u^2}{u^2+c(1-u)^2})_x = 0$) shown in Fig. 6.12, obtained using linear extrapolation and Koren limiter, is comparable to the solution to the same problem shown in Fig. 6.17 obtained using parabolic reconstruction and the Sweby limiter. So again, we observe that there is very little difference between the two methods for this particular problem.

6.6.4 Solutions to the Euler Equations

We consider 1D (planar) cases as discussed in Appendix 6.A at the end of this chapter where, in conservative vector form, the general Euler equations are

$$\frac{\partial U}{\partial t} + \frac{\partial f}{\partial x} = 0, \quad t \geq t_0, \ x \in \mathbb{R}, \tag{6.65}$$

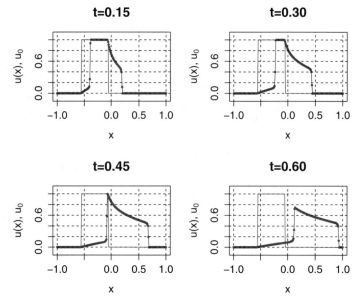

Figure 6.17. The 1D Buckley–Everett equation solution shown at increasing times from $t = 0.15$ to $t = 0.6$, along with the step pulse initial condition. Finite volume method performed on a grid of 201 cells with Dirichlet BCs and CLF=0.45. The simulation was based upon the Kurganov and Tadmor central scheme with parabolic reconstruction and the Sweby limiter. Time integration was performed using the third-order TVD RK scheme.

where

$$U = \begin{bmatrix} \rho \\ \rho u \\ E \end{bmatrix} \quad f = \begin{bmatrix} \rho u \\ p + \rho u^2 \\ u(E + p) \end{bmatrix}, \quad (\rho, u, p, E) \in \mathbb{R}, \tag{6.66}$$

and U is a vector of states and f is a vector of fluxes.

These equations represent conservation of *mass, momentum,* and *energy*. There are thus three equations and four unknowns: ρ (density), u (fluid velocity), p (pressure), and E (total energy). The *total energy*, E, and *entropy*, S, are given by

$$E = \rho e + \frac{1}{2}\rho u^2, \quad S = \log_{10}(p/\rho^\gamma), \tag{6.67}$$

respectively, where e represents specific internal energy.

To close the system, an *equation of state* is required. One that suits our purpose is

$$p = \rho(\gamma - 1)e \tag{6.68}$$

for a polytropic process, where γ is equal to the ratio of specific heats $[c_p/c_v]$ for the fluid.

We can now proceed, as shown in the simple 1D scalar example, by obtaining limited left and right extrapolated states for each state variable. Thus, for density, we obtain

$$\rho^*_{i+\frac{1}{2}} = \rho^*_{i+\frac{1}{2}}\left(\rho^L_{i+\frac{1}{2}}, \rho^R_{i+\frac{1}{2}}\right), \rho^*_{i-\frac{1}{2}} = \rho^*_{i-\frac{1}{2}}\left(\rho^L_{i-\frac{1}{2}}, \rho^R_{i-\frac{1}{2}}\right), \tag{6.69}$$

Table 6.1. Left and right region parameters for Sod's shock tube problem: density, pressure, and velocity

	Left	Right
$\rho_{L,R}$	1	0.125
$p_{L,R}$	1	0.1
$u_{L,R}$	0	0

where for linear extrapolation,

$$\rho^L_{i+\frac{1}{2}} = \rho_i + 0.5\phi(r_i)(\rho_{i+1} - \rho_i), \rho^R_{i+\frac{1}{2}} = \rho_{i+1} - 0.5\phi(r_{i+1})(\rho_{i+2} - \rho_{i+1}) \quad (6.70)$$

$$\rho^L_{i-\frac{1}{2}} = \rho_{i-1} + 0.5\phi(r_{i-1})(\rho_i - \rho_{i-1}), \rho^R_{i-\frac{1}{2}} = \rho_i - 0.5\phi(r_i)(\rho_{i+1} - \rho_i). \quad (6.71)$$

Similarly, we obtain limited left and right extrapolated states for momentum $m = \rho u$ and total energy E. The extrapolated states for velocity u are calculated from momentum, and for pressure p from the equation of state.

Having obtained the limited extrapolated states, we then proceed to construct the edge fluxes using these values. With the edge fluxes known, we can now construct the *semi-discrete scheme*, that is,

$$\frac{dU_i}{dt} = -\frac{1}{\Delta x_i}\left[F^*_{i+\frac{1}{2}} - F^*_{i-\frac{1}{2}}\right], \quad (6.72)$$

according to eqns. (6.47). The solution can now proceed by integration using standard numerical techniques.

The preceding illustrates the basic idea of the MUSCL scheme for systems involving vector quantities.

6.6.4.1 Sod's Shock Tube Problem

A *shock tube* consists of a tube with two regions separated by a diaphragm, with each region occupied by an ideal gas initially at rest. The left region is initialized to a higher pressure than the right region so that when the diaphragm is removed, the fluid on the left moves to the right at a supersonic velocity. As the system evolves, a contact discontinuity moves to the right, with a shock wave ahead of it, whereas a rarefaction wave moves to the left. The problem has an analytical solution, so it is ideal for testing the ability of fluid dynamic codes to resolve problems that exhibit rarefaction waves, contact discontinuities, and shocks and how well the *Rankine–Hugoniot conditions* are satisfied (see Chapter 10).

We consider the G. A. Sod's *shock tube problem* [Sod-78] with a tube of length=1 and divided into two equal regions with parameters as defined in Table 6.1.

Example: Linear extrapolation with ospre limiter
Figure 6.18 shows a second-order solution to Sod's problem using the preceding high-resolution Kurganov and Tadmor central scheme (KT) with linear extrapolation and the ospre limiter.

The closeness of the numerical solution to the analytical solution demonstrates clearly the effectiveness of the MUSCL approach to solving the Euler equations with good resolution. The simulation was carried out on a mesh of 200 cells using the KT algorithm

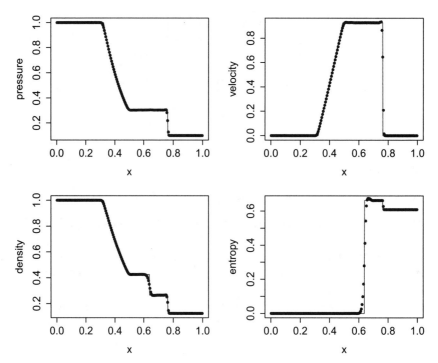

Figure 6.18. High-resolution simulation of Euler equations based on Sod's shock tube problem. The analytical solutions (solid) are shown along with simulated solutions (dotted) based upon the Kuganov and Tadmor central scheme with linear extrapolation reconstruction and ospre limiter. The time duration was 0.15 and $\lambda = \Delta t / \Delta x = 0.5$.

and ospre limiter. Time integration was performed by the third-order TVD Runge–Kutta integrator described in Section 6.3.1. Additional grid points would improve resolution of the solution.

The R code for the `main program` and supporting functions that simulate the Sod problem are given in Appendix 6.C at the end of this chapter, along with code that defines the `initial conditions`. Additionally, the R code that generated the plots in Fig. 6.18 is shown in Listing 6.20. This postsimulation code first calculates the exact solution based on the Rankine–Hugoniot conditions, then for comparison purposes, these data are plotted together with the results from the example simulation. This same code was also used to generate the Sod problem plots for the parabolic reconstruction example that follows and the WENO3 reconstruction example detailed in Section 6.7.12.1.

Example: Parabolic extrapolation with van Albada limiter

We now consider the previous Sod tube example with the same parameters, but this time with parabolic extrapolation and the van Albada 2 limiter.

Figure 6.19 shows the third-order solution again using the preceding high-resolution Kurganov and Tadmor central scheme.

The numerical solution is fairly close to the analytical solution, illustrating the effectiveness of the MUSCL approach to solving the Euler equations with good resolution. However, the solution is noticeably less accurate than that achieved by the linear extrapolation method. The simulation was carried out on a mesh of 200 cells using the KT

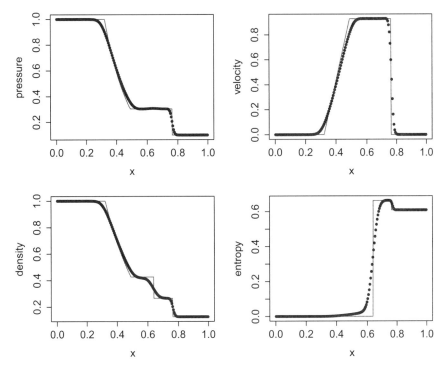

Figure 6.19. High-resolution simulation of Euler equations based on Sod's shock tube problem. Shows the analytical solutions (solid) along with simulated (third-order) solutions (dotted) based upon the Kuganov and Tadmor central scheme with parabolic reconstruction and van Albada 2 limiter.

algorithm with parabolic extrapolation. The *van Albada 2 limiter*,

$$\phi_{va}(r) = \frac{2r}{1+r^2}, \qquad (6.73)$$

was used as this gave the best accuracy, although other limiters did give similar results. Time integration was again performed by the third-order TVD RK integrator. Additional grid points improve resolution of the solution.

Various alternative high-resolution flux limiter schemes have been developed that solve the Euler equations with good accuracy, and the interested reader can find further information on these methods in the references detailed in Section 6.8.

6.6.4.2 Woodward–Colella Interacting Blast Wave Problem

We consider the 1D *Woodward–Colella interacting blast wave problem* with domain length $= 1$, which is divided into three regions with parameters as defined in Table 6.2.

The test problem was originally introduced by Woodward [Woo-82] to illustrate the strong relationship between the accuracy of the overall flow solution and the thinness of discontinuities on the computational grid. Woodward and Colella demonstrated that it involves multiple interactions of strong shocks and rarefactions with each other and with contact discontinuities [Woo-84]. The initial condition consists of three constant states of a polytropic gas, with $\gamma = 1.4$, which is at rest between reflecting walls. Two strong blast waves develop and collide, producing a new contact discontinuity. From each end, shocks

Table 6.2. Left center and right region parameters for the Woodward–Colella interacting blast wave problem: density, pressure, and velocity

	Left ($x \leq 0.1$)	Center ($0.1 < x < 0.9$)	Right ($x \geq 0.9$)
$\rho_{L,C,R}$	1	1	1
$p_{L,C,R}$	1000	0.01	100
$u_{L,C,R}$	0	0	0

move toward the center and rarefaction waves move outward and collide with the walls. The reflections produce regions of nearly constant pressure and density near the walls, which gradually decline over time.

Example: Linear extrapolation with Sweby limiter
Figure 6.20 shows the results of a simulation performed on a mesh of 500 cells, using the high-resolution Kurganov and Tadmor central scheme with linear reconstruction and

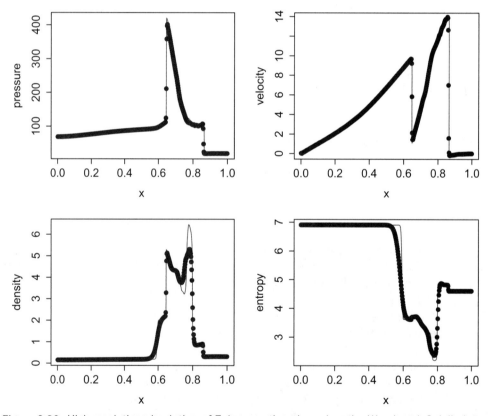

Figure 6.20. High-resolution simulation of Euler equations based on the Woodward–Colella interacting blast wave problem. The reference solutions (solid) are shown along with simulated solutions (dotted) based upon the Kuganov and Tadmor central scheme with linear extrapolation reconstruction and Sweby limiter. The time duration was $t = 0.0.38$ and $\lambda = \Delta t / \Delta x = 0.004$.

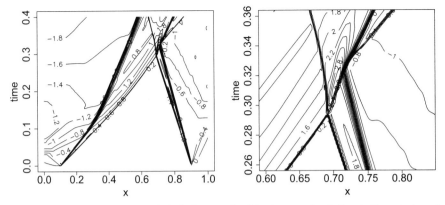

Figure 6.21. Plot of the Woodward–Colella interacting blast wave simulation in the *x-t* plane based on a mesh of 500 cells. (left) Log ρ plot with 20 contours and (right) close-up near collision.

Sweby limiter. Time integration was performed using the third-order TVD RK scheme and the time duration was 0.038 and $\lambda = \Delta t/\Delta x = 0.004$. A solution based on a very fine mesh of 4000 cells is also included as a reference. The closeness of the numerical solution to the reference solution provides a further demonstration of the effectiveness of the MUSCL approach to solving the Euler equations with good resolution.

Figure 6.21 shows a plot in the *x-t* plane of $\log \rho$ for the same simulation with 20 contours, where time has been subsampled to reduce the image processing time. A similar plot based on the reference solution would show the shocks being much sharper.

The R code for the `main program` and supporting functions that simulate the Woodward–Collela problem are given in Appendix 6.C at the end of this chapter, along with code that defines the `initial conditions`. Additionally, the R code that generated the plots in Figs. 6.20 and 6.21 is shown in Listing 6.21. This postsimulation code first loads the previously calculated 4000-point reference data solution, then for comparison purposes these data are plotted together with the results from the example simulation. This same code was also used to generate the Woodward–Colella plots for the parabolic reconstruction example that follows and the WENO3 reconstruction example detailed in Sections 6.7.12.2. Additional grid points would improve resolution of the simulation, as seen from the reference solution.

Example: Parabolic reconstruction with Albada 2 limiter

Figure 6.22 shows the results of a simulation performed on a mesh of 500 cells, using the high-resolution Kurganov and Tadmor central scheme with parabolic reconstruction and van Albada 2 limiter. Time integration was performed using the third-order TVD RK scheme and the time duration was 0.038 and $\lambda = \Delta t/\Delta x = 0.004$. A solution based on a very fine mesh of 4000 cells is also included as a reference. The solution exhibits generally the correct shape for pressure and velocity, but density and entropy solutions tend to be smeared out. We therefore conclude that parabolic reconstruction is not suitable for solving the difficult Woodward–Collela problem, even though additional grid points would improve the resolution. Therefore, for this problem, linear extrapolation is the preferred choice for flux limiter-based solution methods. However, we will see in Section 6.7.12.2 that the WENO method does provide a good alternative.

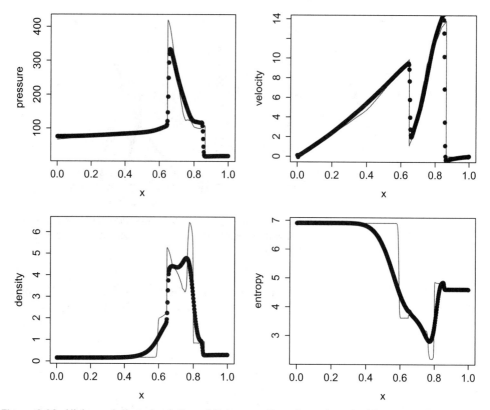

Figure 6.22. High-resolution simulation of Euler equations based on the Woodward–Colella interacting blast wave problem. The reference solutions (solid) are shown along with simulated solutions (dotted) based upon the Kuganov and Tadmor central scheme with parabolic reconstruction and van Albada 2 limiter. The time duration was $t = 0.0.38$ and $\lambda = \Delta t / \Delta x = 0.004$.

Figure 6.23 shows a plot in the x-t plane of $\log \rho$ for the same simulation with 20 contours, where time has been subsampled to reduce the image processing time. Again, we see the smearing effect present in Fig. 6.22.

6.6.4.3 Taylor–Sedov Blast Wave Problem

The *Taylor–Sedov* problem is a historically interesting problem as it was first investigated in detail following detonation of the first atomic bomb in 1945 at the Trinity site in the New Mexico desert. It involves a point-blast detonation which results in a very sharp shock wave that proceeds radially outward at great velocity from the point of detonation. This quickly moving shock wave poses a very stringent test for high-resolution numerical schemes of their ability to resolve the solution accurately. It also has the added advantage that the analytical solution [Sed-59] to the problem is known. The similarity relationship

$$R = S \left(\frac{E_0}{\rho_0} t^2 \right)^{\frac{1}{\nu+2}} \tag{6.74}$$

enables the shock wave to be located at time t. This information together with a set of appropriate shock conditions enables a full evolutionary solution to be obtained. In eqn. (6.74), R represents the radius of the shock wave at time t, E_0 the initial energy of the explosion, $\rho_0 = 1.225$ kg/m^3 and $p_0 = 10^5$ Pa the ambient density, and pressure prior to

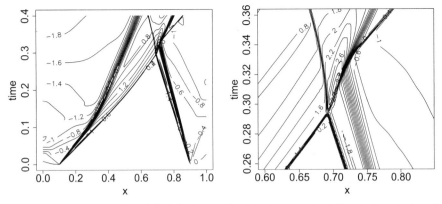

Figure 6.23. Plot of the Woodward–Colella interacting blast wave simulation in the x-t plane based on a mesh of 500 cells. (left) Log ρ plot with 20 contours and (right) close-up near collision.

the explosion, and $\nu = 1$ for planar geometry (2 cylindrical, 3 spherical) and $S = 0.9756$ (1.0040 cylindrical, 1.0328 spherical). Values for pressure, density, and velocity are then obtained from Table 10.3 Chapter 10 also provides a detailed analysis of the problem and some historical background.

We consider the 1D planar Taylor–Sedov blast wave problem with domain length = 200 and an initial detonation energy value of $E_0 = 2.5$ GJ located in the central two cells of the problem domain. This is a much less powerful explosion than the Trinity atomic bomb but, nevertheless, provides a good test for the numerical scheme.

> Ideally, we would use a *well-balanced* solution method, such that the flux and any source terms are balanced exactly at the steady state [Xin-13]. However, this is not necessary, as our simulation will only be performed for a very short period following detonation.

Example: Linear extrapolation with Koren limiter

Figure 6.24 shows the results of a simulation performed on a mesh of 400 cells, using the high-resolution Kurganov and Tadmor central scheme with linear extrapolation and Koren limiter. Time integration was performed using the third-order TVD RK scheme.

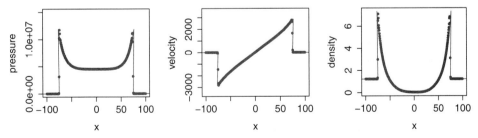

Figure 6.24. High-resolution simulation of Euler equations based on the Taylor–Sedov problem. The analytical solutions (solid) are shown along with simulated solutions (dotted) based upon the Kuganov and Tadmor central scheme with linear extrpolation reconstruction and Koren limiter. The time duration was 0.015 and $\lambda = \Delta t / \Delta x = 0.00001$.

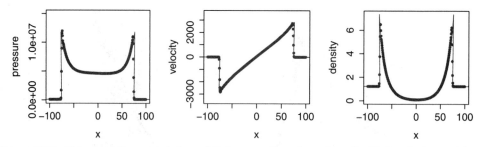

Figure 6.25. High-resolution simulation of Euler equations based on the Taylor–Sedov problem. The analytical solutions (solid) are shown along with simulated solutions (dotted) based upon the Kuganov and Tadmor central scheme with parabolic reconstruction and van Albada 2 limiter. The time duration was 0.015 and $\lambda = \Delta t / \Delta x = 0.00001$.

The time duration was 0.015 s and, for stability reasons, a value for $\lambda = \Delta t / \Delta x = 0.00001$ was needed which represents a Co $\simeq 0.12$. An analytical solution is also plotted as a reference.

The closeness of the numerical solution to the reference solution provides a further demonstration of the effectiveness of the MUSCL approach to solving the Euler equations with good resolution.

The R code for the main program and supporting functions that simulate the Taylor–Sedov problem are given in Appendix 6.C at the end of this chapter, along with code that defines the initial conditions. Additionally, the R code that generated the plots in Fig. 6.24 is shown in Listing 6.22. This postsimulation code first performs some prliminary calculations, followed by the loading of 39-point reference data that enable the exact solution to be calculated using the method outlined in Chapter 12. Then, for comparison purposes, these data are plotted together with the results from the example simulation. This same code was also used to generate the Taylor–Sedov plots for the parabolic reconstruction example that follows and the WENO5 reconstruction example detailed in Section 6.7.12.3. Additional grid points would improve resolution of the solution.

Example: Parabolic extrapolation with Albada 2 limiter
Figure 6.25 shows the results of a simulation performed on a mesh of 400 cells, using the high-resolution Kurganov and Tadmor central scheme with parabolic reconstruction and van Albada 2 limiter. Time integration was performed using the third-order TVD RK scheme. The time duration was 0.015 s and, for stability reasons, a value for $\lambda = \Delta t / \Delta x = 0.00001$ was needed, which represents a Co $\simeq 0.12$. An analytical solution is also plotted as a reference.

The closeness of the numerical solution to the reference solution provides a further demonstration of the effectiveness of the MUSCL approach to solving the Euler equations with good resolution. However, we observe that for this particular problem the solution based on linear extrapolation with Koren limiter provided slightly improved resolution over this solution based on parabolic extrapolation.

6.7 WEIGHTED ESSENTIALLY NONOSCILLATORY (WENO) METHOD

Weighted essentially nonoscillatory (WENO) schemes use a weighted set of polynomials to provide an approximation of a variable within a computational cell. First, we need to clarify what is meant by *nonoscillatory* in this context. The term would appear to imply that the solutions do not contain spurious oscillations or wiggles, but this is certainly *not* the case! WENO schemes are derived from the *essentially nonoscilliatory* (ENO) concept described in the seminal paper by Harten and Osher [Har-87], in which the nonoscillatory property is defined as

> [an] approximation, which is non-oscillatory in the sense that the number of extrema of the discrete solution is not increasing in time.

Thus, the term is almost synonymous with the TVD concept discussed in Section 6.3.

To demonstrate the method, we solve hyperbolic conservation laws by use of interpolation polynomials. To arrive at the required solution, it is necessary that these polynomials provide approximations of the spatial derivative at cell boundaries, or edges. The derivation of these polynomials and the associated weights are described in what follows.

6.7.1 Polynomial Reconstruction: Finite Volume Approach

Here we derive the interpolation polynomial from neighboring cell averages using the finite volume approach, as described in Chapter 3. Following [Shu-98], we consider the 1D WENO reconstruction method based on a mesh with interval $I_i \equiv [x_{i-\frac{1}{2}}, x_{i+\frac{1}{2}}]$, where $x_i \equiv \frac{1}{2}(x_{i-\frac{1}{2}} + x_{i+\frac{1}{2}})$ and $\Delta x_i \equiv x_{i+\frac{1}{2}} - x_{i-\frac{1}{2}}, i = 1, 2, \ldots, N$. We wish to determine the spatial derivative of a flux variable, $v(x)$, at grid point i.

Define the average or mean value of $v(x)$ over cell i as

$$\bar{v}_i = \frac{1}{\Delta x_i} \int_{x_{i-\frac{1}{2}}}^{x_{i+\frac{1}{2}}} v(x)\,dx, \quad i = 1, 2, \ldots, N. \tag{6.75}$$

Approximate the flux by the polynomial

$$p_i(x) = v(x) + \mathcal{O}(\Delta x^k), \quad x \in I_i, \ i = 1, 2, \ldots, N. \tag{6.76}$$

Then the right and left edge flux values are given by

$$v^+_{i+\frac{1}{2}} = p_i(x_{i+\frac{1}{2}}), \ v^-_{i-\frac{1}{2}} = p_i(x_{i-\frac{1}{2}}), \ i = 1, 2, \ldots, N, \tag{6.77}$$

and therefore

$$v^+_{i+\frac{1}{2}} = v(x_{i+\frac{1}{2}}) + \mathcal{O}(\Delta x^k), \ v^-_{i-\frac{1}{2}} = v(x_{i-\frac{1}{2}}) + \mathcal{O}(\Delta x^k). \tag{6.78}$$

The polynomial $p(x)$ could be replaced by other simple functions.

For a given cell I_i and a desired order of accuracy, k, choose a stencil $S(i)$, based on r cells to the left and s cells to the right and I_i itself, such that $r, s \geq 0$ and $r + s + 1 = k$. Thus,

$$S(i) = \{I_{i-r}, \ldots, I_{i+s}\}. \tag{6.79}$$

There will be a unique polynomial of degree at most $k - 1 = r + s$ denoted by $p(x)$ whose cell average in each of the in $S(i)$ agrees with $v(x)$, that is,

$$\frac{1}{\Delta x_i} \int_{x_{i-\frac{1}{2}}}^{x_{i+\frac{1}{2}}} p(\xi)\,d\xi = \bar{v}_i, \quad i = 1, 2, \ldots, N. \tag{6.80}$$

The polynomial $p(x)$ is the kth-order approximation we seek to find! We also need to obtain approximations to $v(x)$ at the cell boundaries, $v^+_{i+\frac{1}{2}}$ and $v^-_{i-\frac{1}{2}}$.

It follows that, because the mapping from given cell averages to $v^+_{i+\frac{1}{2}}$ and $v^-_{i-\frac{1}{2}}$ are linear, there must exist constants c_{rj} and $\tilde{c}_{rj}$ such that

$$v^+_{i+\frac{1}{2}} = \sum_{j=0}^{k-1} c_{rj} \bar{v}_{i-r+j}$$
$$v^-_{i-\frac{1}{2}} = \sum_{j=0}^{k-1} \tilde{c}_{rj} \bar{v}_{i-r+j}.$$
(6.81)

If we identify the left shift r not with the cell I_i, but with the point of reconstruction, $x_{i-\frac{1}{2}}$, that is, with the stencil $S(i)^3$, then we can drop the superscript $\pm$, as it is clear that

$$c_{rj} = \tilde{c}_{rj}.$$
(6.82)

Note that this does not mean that $v^+_{i+\frac{1}{2}} = v^-_{i-\frac{1}{2}}$:

$$\therefore v_{i+\frac{1}{2}} = v(x_{i+\frac{1}{2}}) + \mathcal{O}(\Delta x^k).$$
(6.83)

Now, consider the primative function

$$V(x) = \int_{-\infty}^{x} v(\xi)\,d\xi,$$
(6.84)

where the lower limit, $-\infty$, is unimportant and can be replaced by any fixed number. It follows that

$$V(x_{i+\frac{1}{2}}) = \sum_{j=-\infty}^{i} \int_{x_{j-\frac{1}{2}}}^{x_{j+\frac{1}{2}}} v(\xi)\,d\xi$$
$$= \sum_{j=-\infty}^{i} \bar{v}_j \Delta x_j.$$
(6.85)

Thus, from a knowledge of cell averages, $\bar{v}_i$, we are also able to obtain values for the primative function $V(x)$ at the cell boundaries exactly.

Let us denote the unique polynomial, of degree k at most, that interpolates $V(x_{i+\frac{1}{2}})$ at the $k+1$ points

$$x_{i-r-\frac{1}{2}}, \ldots, x_{-+s+\frac{1}{2}}$$
(6.86)

by $P(x)$ and denote its derivative by $p(x)$, that is,

$$p(x) = P'(x).$$
(6.87)

---

[3] This is a similar situation to the parabolic extrapolation shown in Fig. 6.14, where superscripts [+/−] correspond to superscripts [L/R], respectively.

Then we have

$$\frac{1}{\Delta x_i} \int_{x_{j-\frac{1}{2}}}^{x_{j+\frac{1}{2}}} p(\xi)\,d\xi = \frac{1}{\Delta x_i} \int_{x_{j-\frac{1}{2}}}^{x_{j+\frac{1}{2}}} P'(\xi)\,d\xi$$

$$= \frac{1}{\Delta x_i}\left(P(x_{j+\frac{1}{2}}) - P(x_{j-\frac{1}{2}})\right)$$

$$= \frac{1}{\Delta x_i}\left(V(x_{j+\frac{1}{2}}) - V(x_{j-\frac{1}{2}})\right) \quad (6.88)$$

$$= \frac{1}{\Delta x_i}\left(\int_{-\infty}^{x_{j+\frac{1}{2}}} v(\xi)\,d\xi - \int_{-\infty}^{x_{j-\frac{1}{2}}} v(\xi)\,d\xi\right)$$

$$= \frac{1}{\Delta x_i} \int_{x_{j-\frac{1}{2}}}^{x_{ij+\frac{1}{2}}} v(\xi)\,d\xi$$

$$= \bar{v}_j, \quad j = i-r, \dots, i+s,$$

where the final equality holds because $P(x)$ interpolates $V(x)$ at points $x_{j-\frac{1}{2}}$ and $x_{j+\frac{1}{2}}$ for $j = i-r, \dots, i+s$. This implies that $p(x)$ is the polynomial we seek. Also, this implies that within the cell, we have

$$P'(x) = V'(x) + \mathcal{O}\left(\Delta x^k\right), \quad x \in I_i. \quad (6.89)$$

6.7.2 Polynomial Coefficients

Now consider the Lagrange form of the interpolation polynomial, that is,

$$P(x) = \sum_{m=0}^{k} V\left(x_{i-r+m-\frac{1}{2}}\right) \prod_{\substack{\ell=0 \\ \ell \neq m}}^{k} \frac{x - x_{i-r+\ell-\frac{1}{2}}}{x_{i-r+m-\frac{1}{2}} - x_{i-r+\ell-\frac{1}{2}}}. \quad (6.90)$$

To ease the algebra, we subtract $V(x_{i-r-\frac{1}{2}})$ and use the fact that at $x = x_b$,

$$\sum_{m=0}^{k} \prod_{\substack{\ell=0 \\ \ell \neq m}}^{k} \frac{x - x_{i-r-\frac{1}{2}}}{x_{i-r+m}} = 1, \quad (6.91)$$

where b denotes boundary. Therefore we have

$$P(x) - V\left(x_{i-r-\frac{1}{2}}\right)$$

$$= \sum_{m=0}^{k} \left(V\left(x_{i-r+m-\frac{1}{2}}\right) - V\left(x_{i-r-\frac{1}{2}}\right)\right) \prod_{\substack{\ell=0 \\ \ell \neq m}}^{k} \frac{x - x_{i-r+\ell-\frac{1}{2}}}{x_{i-r+m-\frac{1}{2}} - x_{i-r+\ell-\frac{1}{2}}}. \quad (6.92)$$

After taking the derivative of both sides and noting that $V(x_{i-r+m-\frac{1}{2}}) - V(x_{i-r-\frac{1}{2}}) = \sum_{j=0}^{m-1} \bar{v}_{i-r+j} \Delta x_{i-r+j}$, we obtain

$$p(x) = \sum_{m=0}^{k} \sum_{j=0}^{m-1} \bar{v}_{i-r+j} \Delta x_{i-r+j} \left\{ \frac{\sum_{\ell=0, \ell \neq m}^{k} \prod_{q=0, q \neq m,\ell}^{k} \left(x - x_{i-r+q-\frac{1}{2}}\right)}{\prod_{\ell=0, \ell \neq m}^{k} \left(x_{i-r+m-\frac{1}{2}} - x_{i-r+\ell-\frac{1}{2}}\right)} \right\}. \quad (6.93)$$

We can now evaluate at $x = x_{j+\frac{1}{2}}$ to finally obtain

$$v_{j+\frac{1}{2}} = p(x_{j+\frac{1}{2}}) = \sum_{j=0}^{k-1} \left\{ \bar{v}_{i-r+j} \Delta x_{i-r+j} \sum_{m=j+1}^{k} \frac{\sum_{\ell=0, \ell \neq m}^{k} \prod_{q=0, q \neq m,\ell}^{k} \left(x - x_{i-r+q-\frac{1}{2}}\right)}{\prod_{\ell=0, \ell \neq m}^{k} \left(x_{i-r+m-\frac{1}{2}} - x_{i-r+\ell-\frac{1}{2}}\right)} \right\}. \quad (6.94)$$

Thus, we see that the constants c_{rj} in

$$v_{i+\frac{1}{2}} = \sum_{j=0}^{k-1} c_{rj} \bar{v}_{i-r+j} \quad (6.95)$$

are given by

$$c_{rj} = \Delta x_{i-r+j} \sum_{m=j+1}^{k} \frac{\sum_{\ell=0, \ell \neq m}^{k} \prod_{q=0, q \neq m,\ell}^{k} \left(x - x_{i-r+q-\frac{1}{2}}\right)}{\prod_{\ell=0, \ell \neq m}^{k} \left(x_{i-r+m-\frac{1}{2}} - x_{i-r+\ell-\frac{1}{2}}\right)}. \quad (6.96)$$

The constants c_{rj} obtained from eqn. (6.96) can be used in a *nonuniform grid* system and should be precomputed for each specific problem. However, for a *uniform grid*, $\Delta x_i = \Delta x$, and because the product terms in the numerator will have one less term than the denominator, the Δx terms will cancel. For this situation, eqn. (6.96) takes the much simplified form

$$c_{rj} = \sum_{m=j+1}^{k} \frac{\sum_{\ell=0, \ell \neq m}^{k} \prod_{q=0, q \neq m,\ell}^{k} (r - q + 1)}{\prod_{\ell=0, \ell \neq m}^{k} (m - \ell)} \quad (6.97)$$

and does not depend upon j or Δx. Consequently, these constants will apply generally.

A function WENO_CR_coeffs (k) that calculates c_{rj} according to eqn. (6.97) is given in Listing 6.9 and a corresponding test program is given in Listing 6.10.

```r
# File: WENO_crj_coeffs.R
WENO_crj_coeffs <- function(k){
  # Calculation is in accordance with:
  # Chi-Wang Shu(1997), "Essentailly Non-Oscillatory and Weighted Essentially
  # Non-Oscilliatory Schemes for Hyperbolic Conservation Laws, NASA/CR-97-206253,
  # Icase report 97-65. Also appears in in "Advanced Numerical Approximation
  # of Nonlinear Hyperbolic Equations", B. Cockburn, C. Johnson, C.-W. Shu
  # and E. Tadmor (Editor: A. Quarteroni), Lecture Notes in Mathematics,
  # volume 1697, Springer, 1998, pp.325-432.
  
  # NOTE:
  # 1) Program calculates a (k x k) matrix of Langrange polynomial coefficients
  #    as provided in:
  #    Balsara, Dinshaw S. and Chi-Wang Shu (1999), "Monotonicity Preserving
  #    Weighted Essentially Non-oscillatory Schemes with Increasingly High Order
  #    of Accuracy", JCP, vol 160, 405-452.
  # 2) The (k+1 x k) matix of coefficients given in Shu(1997) above can be
  #    generated by setting max interations of the outer loop to: M=k-1
  
  # k <- Order of interpolation
  # C <- Matrix of interpolation coefficients
  
  L <- k-1 # Inner Loop - max iterations
  M <- k-2 # Outer loop - max iterations
  # Set M <- k-1 to generate a k+1 x k matix of coefficients
  #       <- k-2 to generate a k x k matrix of coefficients
  # See Note 2) above!
  
  crj <- matrix(0,nrow=k,ncol=k)
  #print(crj)
  
  # for r=-1:k-1, # Gives k+1 rows and k columns
  for(r in -1:M){ # Outer Loop
    R <- r+2
    for(j in 0:(k-1)){ # Inner Loop
      S <- k-j # j+1
      crj[R,S] <- 0
      # Start outer summation
      for(m in (j+1):k){
        N <- 0;
        # Start inner summation
        for(p in 0:k){
          Z <- 0
```

```
          # Start of numerator product
          if(p != m){
            Z <- 1
            for(q in 0:k){
              if(q != m && q != p){
                Z <- Z*(r-q+1) #
              }
            }
          }
          # End of numerator product
          N <- N + Z
        }
        # End inner summation
        D <- 1
        # Start denominator product
        for(p in 0:k){
          if(p != m){
            D <- D*(m-p) #
          }
        }
        # End demominator product
        crj[R,S] <- crj[R,S] + N/D
        # End outer summation
      }
    }
  }
  # print(fractions(crj))
  return( crj )
}
```

Listing 6.9. File: WENO_crj_coeffs.R—Code for the function WENO_crj_coeffs() that calculates WENO c_{rj} coefficients in accordance with eqn. (6.97)

```
# File: WENO_coeffs_Test.R
library("MASS")
library("pracma")
source("WENO_crj_coeffs.R")
#
N   <- 3                 # Order of sub-polynomials
crj <-WENO_crj_coeffs(N) # A matrix, NxN
print(fractions(crj))    # Make coeffs rational
crj2 <- rot90(crj,2)     # B Matrix (mirror of A)
print(fractions(crj2))   # Make coeffs rational
```

Listing 6.10. File: WENO_coeffs_Test.R—Code of program to test function crj_coeffs() in Listing 6.9

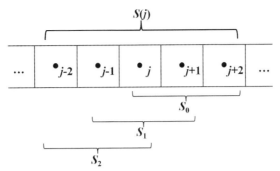

Figure 6.26. A fifth-order WENO scheme comprising the five-cell stencil S(j), constructed from overlapping third-order stencils S_0, S_1, and S_3, which are used to form a weighted sum. Note that coefficients c_{rj} described in Section 6.7.2 belong to stencil S_r.

6.7.3 Polynomial Reconstruction: Finite Difference Reconstruction

WENO reconstruction can also be performed based on an interpolation polynomial derived from point values. However, unlike the finite volume or cell average approach, the finite difference or point value approach is based on the essential requirement that the grid be uniform, that is, $\Delta x_i = \Delta x$. Otherwise, it can be shown that the spatial derivative approximation cannot be higher than second order.

This approach is not applied widely and will not be discussed further here. The interested reader is referred to [Shu-97, Section 2.1.2] for more details.

6.7.4 WENO Reconstruction

We consider a finite volume reconstruction scheme as illustrated by the fifth-order WENO stencil diagram shown in Fig. 6.26.

The idea of the WENO method is to construct a number of low-order overlapping polynomials that each apply over a substencil. Then, when formed into a weighted combination, they result in a high-order polynomial that approximates over the whole stencil, as illustrated in Fig. 6.26. For smooth solutions, the high-order polynomial provides an approximation $\hat{u}_{j+1/2}$, at the cell edge with truncation error of order equal to the higher-order polynomial degree. However, a key characteristic of the method is that when used with a corresponding approximation to $\hat{u}_{j-1/2}$, to calculate the spatial derivative

$$\frac{\partial u_j}{\partial x} = \frac{\hat{u}_{j+1/2} - \hat{u}_{j-1/2}}{\Delta x}, \qquad (6.98)$$

the accuracy is still equal to the degree of the higher-order polynomial, even though the calculation involves division by Δx. This is due to cancellation of low-order truncation terms.

For example, for a smooth solution, a five-cell stencil with three second-order sub-polynomials can be used to provide approximate values of u at the jth cell edges, that is, $\hat{u}_{j\pm 1/2}$. The approximations each have truncation error $C_3 \Delta x^3 + \mathcal{O}(\Delta x^4)$. We will see in Section 6.7.5 that when an appropriately weighted combination of the three sub-polynomials is constructed, this results in a fifth-degree polynomial and, for smooth solutions, the truncation error will be equal to $C_5 \Delta x^5 + \mathcal{O}(\Delta x^6)$. The resulting fifth-order polynomials are centered on the jth cell and can be used to provide approximate solution values $\hat{u}_{j\pm 1/2}$, at the edges of the jth cell. When these edge values are used to provide an approximation to $\partial u_j / \partial x$, the lowest-order truncation terms cancel, resulting in a truncation error of $\mathcal{O}\left(\Delta x^6\right)/\Delta x = \mathcal{O}\left(\Delta x^5\right)$, which is fifth-order accurate. However, for nonsmooth solutions, the WENO method employs smoothing indicators β, which are used to modify the weights d, so that when combined, the approximation eliminates or reduces the influence of low-order polynomials that straddle cells where the solution has a high spatial derivative. Thus, by this method, the WENO algorithm effectively elects when to use the optimal stencil and when to use a smaller stencil so as to avoid interpolation across steep gradients that can cause the solution to exhibit spurious oscillations. The resulting solution in the vicinity of shocks or steep gradients is therefore degraded and has a lower-order approximation than where the solution is smooth.

This process will be clarified further in subsequent sections.

6.7.5 Alternative Calculation for Substencil Coefficients

To provide further insight into the process of finding WENO coefficients, we will illustrate an alternative method for obtaining coefficients for three third-order polynomials, which are then used to form a weighted sum that yields a fifth-order WENO polynomial. The process provides a set of coefficients the same as calculated by eqn. (6.97). Here, we generally follow [Hen-05] but reconstruct *state* variables rather than *flux* variables. For problems with *diffusion* or *higher-order terms* and/or *source terms*, this approach enables a set of appropriate coherent approximations to the additional terms to be constructed directly from the reconstructed state variables.

The idea is to approximate the state variable $u(x)$ by a set of three second-degree polynomials, with as yet unknown coefficients, which are then used to find the mean value $\bar{u}$ for each cell. We do this by integrating each polynomial over cell length, Δx, and then dividing by Δx, as follows.

Let the state variable u be approximated by the second-degree polynomial $\hat{u}$, that is,

$$u(x) \simeq \hat{u}(x) = a_{k,0} + a_{k,1} x + a_{k,2} x^2. \tag{6.99}$$

Therefore, the mean state variable value for the cell centered at $x = X$ and width Δx is given by

$$\bar{u}(X) = \frac{1}{\Delta x} \int_{X-\Delta x/2}^{X+\Delta x/2} \hat{u}(x)\, dx$$

$$= a_{k,0} + a_{k,1} X + a_{k,2} \left(X^2 + \frac{\Delta x^2}{12} \right). \tag{6.100}$$

Evaluation at three consecutive node points is achieved by letting $X = X_j + (0 - k)\Delta x$, $X = X_j + (1 - k)\Delta x$, and $X = X_j + (2 - k)\Delta x$, where we let $X_j = 0$ to obtain

$$\bar{u}_{j+0-k} = a_{k,0} + a_{k,1}(0-k)\Delta x - a_{k,2}\left[(0-k)^2 + \tfrac{1}{12}\right]\Delta x^2,$$

$$\bar{u}_{j+1-k} = a_{k,0} + a_{k,1}(1-k)\Delta x - a_{k,2}\left[(1-k)^2 + \tfrac{1}{12}\right]\Delta x^2, \tag{6.101}$$

$$\bar{u}_{j+2-k} = a_{k,0} + a_{k,1}(2-k)\Delta x - a_{k,2}\left[(2-k)^2 + \tfrac{1}{12}\right]\Delta x^2.$$

We now have three equations and three unknown coefficients $a_{k,0}$, $a_{k,1}$, and $a_{k,2}$, which we can solve for in terms of $\bar{u}_{(\cdot)}$, as follows.

For $k = 0$ we obtain

$$a_{0,0} = \frac{23\bar{u}_j + 2\bar{u}_{j+1} - \bar{u}_{j+2}}{24},$$

$$a_{0,1} = \frac{1}{2}\frac{-3\bar{u}_j + 4\bar{u}_{j+1} - \bar{u}_{j+2}}{\Delta x}, \qquad (6.102)$$

$$a_{0,2} = \frac{1}{2}\frac{\bar{u}_j - 2\bar{u}_{j+1} + \bar{u}_{j+2}}{\Delta x^2};$$

for $k = 1$

$$a_{1,0} = \frac{-\bar{u}_{j-1} + 26\bar{u}_j - \bar{u}_{j+1}}{24},$$

$$a_{1,1} = \frac{1}{2}\frac{-\bar{u}_{j-1} + \bar{u}_{j+1}}{\Delta x}, \qquad (6.103)$$

$$a_{1,2} = \frac{1}{2}\frac{\bar{u}_{j-1} - 2\bar{u}_j + \bar{u}_{j+1}}{\Delta x^2};$$

and for $k = 2$

$$a_{2,0} = \frac{-\bar{u}_{j-2} + 2\bar{u}_{j-1} + 23\bar{u}_j}{24},$$

$$a_{2,1} = \frac{1}{2}\frac{\bar{u}_{j-2} - 4\bar{u}_{j-1} + 3\bar{u}_j}{\Delta x}, \qquad (6.104)$$

$$a_{2,2} = \frac{1}{2}\frac{\bar{u}_{j-2} - 2\bar{u}_{j-1} + \bar{u}_j}{\Delta x^2}.$$

Thus, on substituting these coefficients in turn into eqn. (6.99), we obtain approximations $\hat{u}_{(\cdot)}$ to the function $u_{(\cdot)}$ over each three-point stencil, that is,

$$u_j(x) \simeq \hat{u}_j(x) = \left(\frac{23\bar{u}_j + 2\bar{u}_{j+1} - \bar{u}_{j+2}}{24}\right) + \left(\frac{1}{2}\frac{-3\bar{u}_j + 4\bar{u}_{j+1} - \bar{u}_{j+2}}{\Delta x}\right)x$$

$$+ \left(\frac{1}{2}\frac{\bar{u}_j - 2\bar{u}_{j+1} + \bar{u}_{j+2}}{\Delta x^2}\right)x^2, \qquad (6.105)$$

$$u_{j-1}(x) \simeq \hat{u}_{j-1}(x) = \left(\frac{-\bar{u}_{j-1} + 26\bar{u}_j - \bar{u}_{j+1}}{24}\right) + \left(\frac{1}{2}\frac{-\bar{u}_{j-1} + \bar{u}_{j+1}}{\Delta x}\right)x$$

$$+ \left(\frac{1}{2}\frac{\bar{u}_{j-1} - 2\bar{u}_j + \bar{u}_j}{\Delta x^2}\right)x^2, \qquad (6.106)$$

$$u_{j-2}(x) \simeq \hat{u}_{j-2}(x) = \left(\frac{-\bar{u}_{j-2} + 2\bar{u}_{j-1} + 23\bar{u}_j}{24}\right) + \left(\frac{1}{2}\frac{\bar{u}_{j-2} + -4\bar{u}_{j-1} + 3\bar{u}_j}{\Delta x}\right)x$$

$$+ \left(\frac{1}{2}\frac{\bar{u}_{j-2} - 2\bar{u}_{j-1} + \bar{u}_j}{\Delta x^2}\right)x^2. \qquad (6.107)$$

Therefore, recalling that r represents the number of cells in the stencil $S(j)$ to the left of location I_j, the cell edge values for $\hat{u}^r_{j+\frac{1}{2}} = \hat{u}\left(x + \left(\frac{1}{2} - r\right)\Delta x\right)$ centered around $x_j = 0$ are

$$\hat{u}^0_{j+\frac{1}{2}} = \tfrac{1}{6}(2\bar{u}_j + 5\bar{u}_{j+1} - 1\bar{u}_{j+2}),$$

$$\hat{u}^1_{j+\frac{1}{2}} = \tfrac{1}{6}(-\bar{u}_{j-1} + 5\bar{u}_j + 2\bar{u}_{j+1}), \qquad (6.108)$$

$$\hat{u}^2_{j+\frac{1}{2}} = \tfrac{1}{6}(2\bar{u}_{j-2} - 7\bar{u}_{j-1} + 11\bar{u}_j).$$

The corresponding cell edge values for $\hat{u}^r_{j-\frac{1}{2}} = \hat{u}(x - \Delta x/2)$ are obtained by a simple decrement of the index j, giving

$$\hat{u}^0_{j-\frac{1}{2}} = \tfrac{1}{6}(2\bar{u}_{j-1} + 5\bar{u}_j - 1\bar{u}_{j+1}),$$

$$\hat{u}^1_{j-\frac{1}{2}} = \tfrac{1}{6}(-\bar{u}_{j-2} + 5\bar{u}_{j-1} + 2\bar{u}_j), \qquad (6.109)$$

$$\hat{u}^2_{j-\frac{1}{2}} = \tfrac{1}{6}(2\bar{u}_{j-3} - 7\bar{u}_{j-2} + 11\bar{u}_{j-1}).$$

It should be noted that these are *approximations to the actual state variable values*, u, NOT approximations to the mean state variable values, $\bar{u}$, even though they are calculated from cell mean state variable value approximations (refer to the discussion on the finite volume method in Chapter 3).

By appealing to a corresponding Taylor series expansion of u, we see that these three-point stencils each provide a *third-order approximation* to the state variable value at the right-hand edge of cell j, that is,

$$u_{j+\frac{1}{2}} = \hat{u}^0_{j+\frac{1}{2}} + \frac{1}{12}\frac{d^3u}{dx^3}\Delta x^3 + \mathcal{O}(\Delta x^4),$$

$$u_{j+\frac{1}{2}} = \hat{u}^1_{j+\frac{1}{2}} - \frac{1}{12}\frac{d^3u}{dx^3}\Delta x^3 + \mathcal{O}(\Delta x^4),$$

$$u_{j+\frac{1}{2}} = \hat{u}^2_{j+\frac{1}{2}} - \frac{1}{12}\frac{d^3u}{dx^3}\Delta x^3 + \mathcal{O}(\Delta x^4).$$

Also, we obtain a third-order approximation to the spatial derivative, even though we have to divide by Δx. This is because the lowest-order truncation error terms cancel, for example, for stencil 1,

$$\frac{du}{dx} = \frac{u_{j+\frac{1}{2}} - u_{j-\frac{1}{2}}}{\Delta x}$$

$$= \frac{\left(\hat{u}^1_{j+\frac{1}{2}} - \frac{1}{12}\frac{d^3u}{dx^3}\Delta x^3 + \mathcal{O}(\Delta x^4)\right) - \left(\hat{u}^1_{j-\frac{1}{2}} - \frac{1}{12}\frac{d^3u}{dx^3}\Delta x^3 + \mathcal{O}(\Delta x^4)\right)}{\Delta x} \qquad (6.110)$$

$$= \frac{\hat{u}^1_{j+\frac{1}{2}} - \hat{u}^1_{j-\frac{1}{2}}}{\Delta x} + \mathcal{O}(\Delta x^3).$$

Recall that $\mathcal{O}(\Delta x^3) - \mathcal{O}(\Delta x^3)$ is still equal to $\mathcal{O}(\Delta x^3)$ (see discussion on truncation errors in Chapter 3). Thus, for stencil 1, we have

$$\left.\frac{du}{dx}\right|_{x=x_j} \simeq \frac{1}{6}\frac{(\overline{u}_{j-2} - 6\overline{u}_{j-1} + 3\overline{u}_j + 2\overline{u}_{j+1})}{\Delta x}, \qquad (6.111)$$

which is third-order accurate.

The three stencils can be combined as a weighted sum to form a *centered five-point stencil*, that is,

$$\begin{aligned}\hat{u}_{j+\frac{1}{2}} &= d_0 \hat{u}^0_{j+\frac{1}{2}} + d_1 \hat{u}^1_{j+\frac{1}{2}} + d_2 \hat{u}^2_{j+\frac{1}{2}} \\ &= \frac{1}{60}\left(2\overline{u}_{j-2} - 13\overline{u}_{j-1} + 47\overline{u}_j + 27\overline{u}_{j+1} - 3\overline{u}_{j+2}\right),\end{aligned} \qquad (6.112)$$

where the so-called *ideal weights* are equal to

$$d_0 = \frac{1}{10}, \quad d_1 = \frac{6}{10}, \quad d_1 = \frac{3}{10}. \qquad (6.113)$$

Note that d_r is always positive and, to ensure consistency,

$$\sum_{r=0}^{k-1} d_r = 1. \qquad (6.114)$$

Equation (6.112) increases to *fifth order* the approximation of the state value at the right-hand edge of cell j, that is, $u_{j+1/2} = \hat{u}_{j+1/2} + \mathcal{O}(\Delta x^5)$. We obtain a similar approximation to the state value at the left-hand edge of cell j, that is, $u_{j-1/2} = \hat{u}_{j-1/2} + \mathcal{O}(\Delta x^5)$, by a simple decrement of j in eqn. (6.112) to give

$$\hat{u}_{j-\frac{1}{2}} = \frac{1}{60}\left(2\overline{u}_{j-3} - 13\overline{u}_{j-2} + 47\overline{u}_{j-1} + 27\overline{u}_j - 3\overline{u}_{j+1}\right). \qquad (6.115)$$

The weights are modified according to the smoothness of the solution (see Sections 6.7.6 and 6.7.7) and the result is the five-point WENO reconstruction scheme, commonly referred to as WENO5.

The analysis here has a one-point upwind bias in the optimum linear stencil, which makes it suitable for wind blowing from left to right. For wind blowing from right to left the procedure is modified symmetrically with respect to $x_{j+1/2}$.

Finally, it should be noted that for a uniform grid $\Delta x_j = \Delta x$, the expression for $\hat{u}_{j+\frac{1}{2}}$ does not depend upon j or Δx.

The preceding substencil coefficient calculations are quite tedious to perform by hand. So, as an extra resource that readers may find useful, we include with the downloads code written for symbolic algebra programs Maple and Maxima (open source) that derives substencil coefficients for the Nth-order stencil.

6.7.6 Weights

The weights d_r associated with sub-stencil r, discussed in Section 6.7.4 for $\hat{u}_{j+\frac{1}{2}}$, are modified according to the smoothness indicators β_r to arrive at the modified weights ω_r, as follows:

$$\alpha_r = \frac{d_r}{(\epsilon + \beta_r)^2} \tag{6.116}$$

$$\omega_r = \frac{\alpha_r}{\sum_{s=0}^{2} \alpha_s}. \tag{6.117}$$

For stability and consistency we require that

$$\omega_r \geq 0, \quad \sum_{r=0}^{k-1} \omega_r = 1. \tag{6.118}$$

A method for improving the weight calculations, particularly around critical points, is reported in [Hen-05] and forms the basis of the so-called WENO5M scheme.

6.7.7 Smoothness Indicators

The weights used in eqn. (6.112) are modified to reflect the smoothness or otherwise of the solution at cell j. This is achieved by applying smoothness indicators, defined as follows:

$$\beta_r = \sum_{\ell=1}^{k} \Delta x^{2\ell-1} \int_{x_{j-\frac{1}{2}}}^{x_{j+\frac{1}{2}}} \left(\frac{\partial^\ell \hat{u}_j^r(x)}{\partial x^\ell} \right)^2 dx. \tag{6.119}$$

The right-hand side of eqn. (6.119) is the sum of the L_2 norms for all the derivatives of the interpolation polynomial $\hat{u}_j^r(x)$ over the interval $(x_{j-\frac{1}{2}}, x_{j+\frac{1}{2}})$. The factor $\Delta x^{2\ell-1}$ is introduced to remove any Δx dependency in the derivatives, to preserve self-similarity when used for hyperbolic PDEs.

6.7.8 Calculation of Smoothness Indicator Coefficients

We continue the alternative method of Section 6.7.5 to obtain the smoothing indicator coefficients for the three-point stencils.

From eqn. (6.99) we have

$$\hat{u}(x) = a_{k,0} + a_{k,1}x + a_{k,2}x^2, \tag{6.120}$$

with coefficients defined by eqns. (6.102), (6.103), and (6.104). Differentiating $\ell = 1$ and $\ell = 2$ times, we obtain

$$\frac{d\hat{u}}{dx} = a_{k,1} + 2a_{k,2}x$$

$$\frac{d^2\hat{u}}{dx^2} = 2a_{k,2}. \tag{6.121}$$

Equation (6.119) therefore becomes

$$\beta_r = \int_{x_{j-\frac{1}{2}}}^{x_{j+\frac{1}{2}}} \left\{ \Delta x \left(a_{k,1} + 2a_{k,2}x \right)^2 + \Delta x^3 \left(2a_{k,2} \right)^2 \right\} dx$$

$$= a_{k,1}^2 \Delta x^2 + \frac{13}{3} a_{k,2}^2 \Delta x^4. \tag{6.122}$$

On substituting values for $a_{k,1}$ and $a_{k,2}$ from eqns. (6.102), (6.103), and (6.104) we obtain the smoothness indicators for edge value $\hat{u}_{j+\frac{1}{2}}$:

$$\beta_0 = \frac{13}{12}\left(\bar{u}_j - 2\bar{u}_{j+1} + \bar{u}_{j+2}\right)^2 + \frac{1}{4}\left(\bar{u}_j - 4\bar{u}_{j+1} + 3\bar{u}_{j+2}\right)^2.$$

$$\beta_1 = \frac{13}{12}\left(\bar{u}_{j-1} - 2\bar{u}_j + \bar{u}_{j+1}\right)^2 + \frac{1}{4}\left(\bar{u}_{j-1} - \bar{u}_{j+1}\right)^2. \quad (6.123)$$

$$\beta_2 = \frac{13}{12}\left(\bar{u}_{j-2} - 2\bar{u}_{j-1} + \bar{u}_j\right)^2 + \frac{1}{4}\left(3\bar{u}_{j-2} - 4\bar{u}_{j-1} + \bar{u}_j\right)^2.$$

These calculations for the smoothness coefficients are quite straightforward to perform by hand. Nevertheless, as an extra resource, we include with the downloads code written for symbolic algebra programs Maple and Maxima (open source) that derives eqns. (6.123).

6.7.9 Flux Splitting

For conservation laws, where waves may travel in positive and negative directions, some form of flux splitting may be necessary, such as the Lax–Friedrichs method employed in Section 6.6.2. A number of other flux-splitting schemes have been reported in the literature, for example, Steger–Warming flux splitting and Van Leer splitting, which we will not discuss here (for further discussion, see [Liu-03]).

The objective of flux splitting is to modify schemes designed for unidirectional wave propagation such that they are able to handle problems with solutions that exhibit waves traveling in different directions. For the finite volume WENO scheme implementation, we employ the Lax–Friedrichs flux-splitting method as described in Section 6.6.2 and recommended in [Shu-03]. Thus, we approximate the flux value at each edge of the jth cell by

$$\begin{aligned} F_{j+1/2} &= \tfrac{1}{2}\left[f^L_{j+1/2} + f^R_{j+1/2} - a_{j+1/2}\left(u^R_{j+1/2} - u^L_{j+1/2}\right)\right] \\ F_{j-1/2} &= \tfrac{1}{2}\left[f^L_{j-1/2} + f^R_{j-1/2} - a_{j-1/2}\left(u^R_{j-1/2} - u^L_{j-1/2}\right)\right], \end{aligned} \quad (6.124)$$

and the spatial derivative of flux for the jth cell is approximated by

$$\frac{\partial f(u_j)}{\partial t} = -\frac{1}{2}[F_{j+1/2} - F_{j-1/2}]. \quad (6.125)$$

This scheme is employed in Section 6.7.10.

6.7.10 Implementation of a WENO Finite Volume Scheme

We describe a procedure for implementing the WENO finite volume method for application to the 1D conservative problem

$$\frac{\partial u}{\partial t} + \frac{\partial f(u)}{\partial x} = 0, \quad u = u(x,t),\ x \in \mathbb{R}. \quad (6.126)$$

Recall that when we use the terms *left* and *right*, we refer the left side of the jth cell edge at $[j+1/2]$ and the right side of the jth cell edge at $[j-1/2]$. Thus, u^L_j is the reconstructed *left side* value of $\bar{u}$ at $x = (j + 1/2)\Delta x$ and u_j^R is the reconstructed right side of the jth cell edge at $x = (j - 1/2)\Delta x$ (see Fig. 6.14).

Step 1: Smoothing indicators

The smoothness indicators for $k = 3$ are

$$\beta_0 = \frac{13}{12}(u_i - 2u_{i+1} + u_{i+2})^2 + \frac{1}{4}(3u_i - 4u_{i+1} + u_{i+2})^2,$$

$$\beta_1 = \frac{13}{12}(u_{i-1} - 2u_i + u_{i+1})^2 + \frac{1}{4}(u_{i-1} + u_{i+1})^2, \quad (6.127)$$

$$\beta_2 = \frac{13}{12}(u_{i-2} - 2u_{i-1} + u_i)^2 + \frac{1}{4}(3u_{i-2} - 4u_{i-1} + u_i)^2.$$

These are the same as given in eqn. (6.123) but with the overbar dropped. Note that they apply to both *left* and *right* side reconstruction calculations.

Step 2: Stencil weights

The so-called *optimum weights* for $k = 3$ are

$$\begin{aligned} d_0^L &= 3/10, & d_1^L &= 3/5, & d_2^L &= 1/10, \; : \text{Left edges} \\ d_0^R &= 1/10, & d_1^R &= 3/5, & d_2^R &= 3/10, \; : \text{Right edges.} \end{aligned} \quad (6.128)$$

The modified *left* side weights for $k = 3$ are given by

$$\omega_0^L = \frac{\alpha_0^L}{\sum_{s=0}^{2}\alpha_s^L}, \quad \omega_1^L = \frac{\alpha_1^L}{\sum_{s=0}^{2}\alpha_s^L}, \quad \omega_2^L = \frac{\alpha_2^L}{\sum_{s=0}^{2}\alpha_s^L}, \quad (6.129)$$

and the modified *right* side weights by

$$\omega_0^R = \frac{\alpha_0^R}{\sum_{s=0}^{2}\alpha_s^L}, \quad \omega_1^R = \frac{\alpha_1^R}{\sum_{s=0}^{2}\alpha_s^L}, \quad \omega_2^R = \frac{\alpha_2^R}{\sum_{s=0}^{2}\alpha_s^L}, \quad (6.130)$$

with

$$\alpha_0^L = \frac{d_0^L}{(\epsilon + \beta_0)^2}, \quad \alpha_1^L = \frac{d_1^L}{(\epsilon + \beta_1)^2}, \quad \alpha_2^L = \frac{d_2^L}{(\epsilon + \beta_2)^2} \quad (6.131)$$

and

$$\alpha_0^R = \frac{d_0^R}{(\epsilon + \beta_0)^2}, \quad \alpha_1^R = \frac{d_1^R}{(\epsilon + \beta_1)^2}, \quad \alpha_2^R = \frac{d_2^R}{(\epsilon + \beta_2)^2}. \quad (6.132)$$

Step 3: Subpolynomials

The subpolynomial coefficients c_{rj} for $k = 3$ obtained from eqn. (6.97) are given in Table 6.3.

The subpolynomials for *left* side reconstructions are calculated from

$$\begin{aligned} p_0^L &= c_{0,0}u_j + c_{0,1}u_{j+1} + c_{0,2}u_{j+2}, \\ p_1^L &= c_{1,0}u_{j-1} + c_{1,1}u_j + c_{1,2}u_{j+1}, \\ p_2^L &= c_{2,0}u_{j-2} + c_{2,1}u_{j-1} + c_{2,2}u_j, \end{aligned} \quad (6.133)$$

Table 6.3. CRJ coefficients calculated from eqn. (6.97)

k	r	j = 0	j = 1	j = 3
	−1	11/6	−7/6	1/3
	0	1/3	5/6	−1/6
3	1	−1/6	5/6	1/3
	2	1/3	−7/6	11/6

Note: The 3 × 3 matrix formed from the three rows $r = [-1, 0, 1]$ form the mirror matrix of the matrix formed from rows $r = [0, 1, 2]$.

and the subpolynomials for *right* side reconstructions are calculated from

$$p_0^R = c_{-1,0} u_j + c_{-1,1} u_{j+1} + c_{-1,2} u_{j+2},$$
$$p_1^R = c_{00} u_{j-1} + c_{01} u_j + c_{0,2} u_{j+1}, \quad (6.134)$$
$$p_2^R = c_{10} u_{j-2} + c_{11} u_{j-1} + c_{1,2} u_j.$$

Step 4: Reconstructed edge values
The *left* side edge values $u_{j+1/2}^L$ are calculated from the weighted sum of the subpolynomials

$$u_{j+1/2}^L = \omega_0^L p_0^L + \omega_1^L p_1^L + \omega_2^L p_2^L, \quad (6.135)$$

and, similarly, the *right* side edge values $u_{j-1/2}^L$ are calculated from

$$u_{j-1/2}^R = \omega_0^R p_0^R + \omega_1^R p_1^R + \omega_2^R p_2^R. \quad (6.136)$$

We simply increment the j index of $u_{j-1/2}^R$ to obtain $u_{j+1/2}^R$ and now refer to $u_{j+1/2}^L$ and $u_{j+1/2}^R$ as being the values of u on the left and right sides of the cell edge at $x = (j + 1/2)\Delta x$. We now have values for $u_{j+1/2}^L$ and $u_{j+1/2}^R$ for all j, from which we calculate the corresponding fluxes $f_{j+1/2}^L = f(u_{j+1/2}^L)$ and $f_{j+1/2}^R = f(u_{j+1/2}^R)$.

Step 5: Flux splitting
We now apply the Lax–Friedrichs flux-splitting method described in Section 6.7.9 to obtain

$$F_{i+1/2} = \tfrac{1}{2}[f_{j+1/2}^L + f_{j+1/2}^R - a_{j+1/2}(u_{j+1/2}^R - u_{j+1/2}^L)]$$
$$F_{i-1/2} = \tfrac{1}{2}[f_{j-1/2}^L + f_{j-1/2}^R - a_{j-1/2}(u_{j-1/2}^R - u_{j-1/2}^L)], \quad (6.137)$$

where $a_{j\pm1/2} = \max_u |\partial f(u_{j\pm1/2})/\partial x|$ is a constant representing the maximum absolute value of *local wave speed* at the *cell edges* of the corresponding stencil. The flux $F_{i-1/2}$ does not have to be calculated separately, as it is obtained by simply decrementing the index of $F_{i+1/2}$. An estimate of the Jacobian of $f(u_{j+1/2})$ can be obtained from

$$\frac{\partial f(u_{j+1/2})}{\partial x} = \frac{f(u_{j+1}) - f(u_j)}{(u_{i+1} - u_i)}. \quad (6.138)$$

Finally, the WENO approximation to eqn. (6.126) becomes

$$\frac{\partial u}{\partial t} = -\frac{1}{\Delta x}(F_{j+1/2} - F_{j-1/2}), \qquad (6.139)$$

which can be integrated using any suitable numerical integrator, such as the third-order TVD Runge-Kutta scheme described in Section 6.3.1.

See [Luo-13] for a similar approach.

R computer code for a function that performs these calculations for a WENO5 reconstruction is given in Listing 6.11.

```
# File WENO5thOrdReconstruction.R
WENO5thOrdReconstruction <- function(u){ # Fifth-order WENO reconstruction
  epsilon <- 10^-10
  p <- 2 # default value <- 2 - smaller values give sharper result
  N <- length(u)
  uR <- rep(0,N) # Preallocate array
  # Preallocate arrays
  pL0 <- rep(0,N);pL1 <- rep(0,N);pL2 <- rep(0,N)
  pR0 <- rep(0,N);pR1 <- rep(0,N);pR2 <- rep(0,N)

  #BC <- 0
  if(BC == 0){
    # Shifted u arrays for Dirichlet BCs
    up1 <- c(u[2:N],u[N])
    up2 <- c(u[3:N],u[N],u[N])
    um1 <- c(u[1],u[1:(N-1)])
    um2 <- c(u[1],u[1],u[1:(N-2)])
  }else if(BC == 1){
    # Shifted u arrays for periodic BCs
    up1 <- c(u[2:N],u[1])
    up2 <- c(u[3:N],u[1],u[2])
    um1 <- c(u[N],u[1:(N-1)])
    um2 <- c(u[N-1],u[N],u[1:(N-2)])
  }
  # crj constants for (r-1)th degree Reconstruction Polynomials
  # Left edge reconstruction, rows crj0 ... crj2
  # Right edge reconstruction, rows crjm ... crj1
  cm <- c( 11, -7,  2 )/6
  c0 <- c(  2,  5, -1 )/6
  c1 <- c( -1,  5,  2 )/6
  c2 <- c(  2, -7, 11 )/6

  # Linear Weights for Smoothing Indicators
  dL0 <- 3/10; dL1 <- 3/5; dL2 <- 1/10 # Left edges
  dR0 <- 1/10; dR1 <- 3/5; dR2 <- 3/10 # Right edges
  # Smoothing Indicators Left and right edges
  Beta0 <- (13/12)*(u - 2*up1 + up2)^2 +
           (1/4)*(3*u - 4*up1 + up2)^2
```

```
Beta1 <- (13/12)*(um1 - 2*u + up1)^2 +
        (1/4)*( um1 -      up1)^2
Beta2 <- (13/12)*(um2 - 2*um1 + u )^2 +
        (1/4)*( um2 - 4*um1 + 3*u )^2

alphaL_0 <- dL0/(epsilon + Beta0)^p
alphaL_1 <- dL1/(epsilon + Beta1)^p
alphaL_2 <- dL2/(epsilon + Beta2)^p

alphaR_0 <- dR0/(epsilon + Beta0)^p
alphaR_1 <- dR1/(epsilon + Beta1)^p
alphaR_2 <- dR2/(epsilon + Beta2)^p

sum_alphaL <- alphaL_0 + alphaL_1 + alphaL_2
sum_alphaR <- alphaR_0 + alphaR_1 + alphaR_2

omegaL0 <- alphaL_0/sum_alphaL
omegaL1 <- alphaL_1/sum_alphaL
omegaL2 <- alphaL_2/sum_alphaL

omegaR0 <- alphaR_0/sum_alphaR
omegaR1 <- alphaR_1/sum_alphaR
omegaR2 <- alphaR_2/sum_alphaR

# Left edges
pL0 <- c0[1]*u   + c0[2]*up1 + c0[3]*up2
pL1 <- c1[1]*um1 + c1[2]*u   + c1[3]*up1
pL2 <- c2[1]*um2 + c2[2]*um1 + c2[3]*u
# Right edges
pR0 <- cm[1]*u   + cm[2]*up1 + cm[3]*up2
pR1 <- c0[1]*um1 + c0[2]*u   + c0[3]*up1
pR2 <- c1[1]*um2 + c1[2]*um1 + c1[3]*u
# Left edge values at [j+1/2]
uL  <- omegaL0*pL0 + omegaL1*pL1 + omegaL2*pL2
# Right edge values at [j-1/2]
uRm <- omegaR0*pR0 + omegaR1*pR1 + omegaR2*pR2
# Increment uRm to get Right edge values at [j+1/2]
uR[1:(N-1)] <- uRm[2:N]
uR[N]       <- uRm[1]
# Return Left and Right edge vectors at [j+1/2]
return(c(uL,uR))
}
```

Listing 6.11. File: WENO5thOrdReconstruction.R—Code for function WENO5thOrdReconstruction() that performs a fifth-order WENO reconstructuction

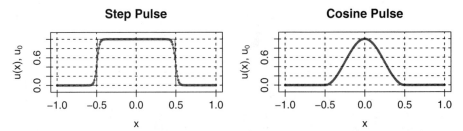

Figure 6.27. The 1D advective equation $u_t + u_x = 0$, with waves propagating to the right at $t = 10$: (left) step pulse and (right) cosine pulse. Finite volume method performed on a grid of 201 cells with periodic BCs and CLF=0.45. Shows the analytical solution (solid) along with simulation results (dotted) based upon the Kurganov and Tadmor central scheme with WENO5 reconstruction. Time integration was performed using the third-order TVD RK scheme.

Additional code for different order WENO reconstructions are provided with the downloads for this book but are omitted here due to space restrictions.

6.7.11 Scalar Problems

The main program and subsidiary functions for the following WENO examples are discussed in Appendix 6.B at the end of this chapter.

Example: Advection of step and cosine pulses
This test provides a comparison with the flux limiter scheme examples of Sections 6.6.2.1 and 6.6.3.1. The resolution of step pulse compares well with most of the flux limiter solutions, except those of the superbee and Sweby, which provide very sharp resolution. However, with respect to the cosine pulse, the WENO5 solution is superior in resolution and in terms of symmetry.

The computer code for this example is given in Listing 6.12, which should be run with variable strRecon set to 5 (WENO fifth order) and strInit set to "Step Pulse". Then the simulation should be run again with variable strInit set to "Cosine Pulse".

Example: Advection of a composite waveform
As a further, more demanding example, we use the following composite pulse wave function described by Balsara and Shu [Bal-00]. The problem includes a number of differently shaped pulses that provide a difficult advection test for numerical codes to solve.

The composite waveform includes the following pulse shapes, arranged from left to right: (1) a combination of Gaussians, (2) a square wave, (3) a sharply peaked triangle, and (4) a half-ellipse. These pulses are defined by eqns. (6.140) and (6.141):

$$\begin{aligned} u_0(x) &= \tfrac{1}{6}[G(x,\beta,z-\delta) + G(x,\beta,z+\delta) + G(x,\beta,z)], & -0.8 \leq x \leq -0.6, \\ &= 1, & -0.4 \leq x \leq -0.2, \\ &= 1 - |10(x-0.1)|, & 0.0 \leq x \leq 0.2, \\ &= \tfrac{1}{6}[F(x,\alpha,a-\delta) + F(x,\alpha,a+\delta) + F(x,\alpha,a)], & 0.4 \leq x \leq 0.6, \end{aligned} \quad (6.140)$$

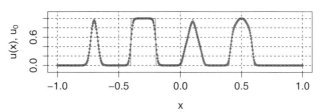

Figure 6.28. Solution to the 1D advective equation $u_t + u_x = 0$, with composite wave (see text) propagating to the right at $t = 10$. Finite volume method performed on a grid of 500 cells with periodic BCs and CLF=0.45. Shows the analytical solution (solid) along with simulation results (dotted) based upon the Kurganov and Tadmor central scheme with WENO5 reconstruction. Time integration was performed using the third-order TVD RK scheme with Co = 0.45.

where

$$G(x, \beta, z) = e^{-\beta(x-z)^2}$$
$$F(x, \alpha, z) = \sqrt{\max(1 - \alpha^2(x-a)^2)}. \quad (6.141)$$

The constants are equal to

$$a = 0.5, \quad z = -0.7, \quad \delta = 0.005, \quad \alpha = 10, \quad \beta = \frac{\log_{10} 2}{36\delta^2}. \quad (6.142)$$

This is a particularly difficult test problem because the composite wave consists of non-smooth functions and smooth functions that become discontinuous. The Gaussians differ from the triangle in that their profile has an inflection in the second derivative. A good numerical method should be able to advect information with a high degree of fidelity and be able to preserve the specific features of the analytical solution. It is seen from Fig. 6.28 that the WENO5 scheme provides good resolution of the advected composite pulse waveform.

The computer code for this example is given in Listing 6.15 which should be run with variable `strInit` set to "Composite Pulse Set".

As an exercise, the reader should run the flux limiter simulations of Sections 6.6.2.1 and 6.6.3.1 with the composite waveform as an IC and compare the result with Fig. 6.28. It will be found that the flux limiter schemes for this problem generally perform very poorly compared with the WENO5 method for the same number of grid points.

Overall, the WENO5 method is generally to be preferred over flux limiter methods, particularly for advection problems that exhibit a combination of smooth and sharply changing solutions.

6.7.12 Euler Equation Problems

We now introduce some example problems featuring the Euler equations described in Section 6.6.4 and Appendix 6.A at the end of this chapter.

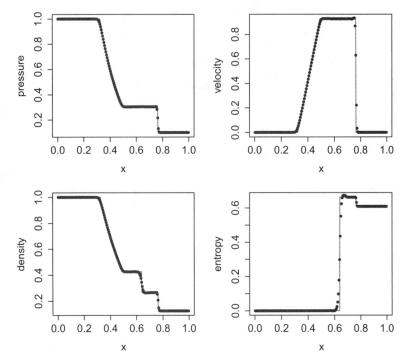

Figure 6.29. High-resolution simulation of Euler equations based on Sod's shock tube problem. The analytical solutions (solid) are shown along with simulated solutions (dotted) based upon the Kuganov and Tadmor central scheme with WENO3 reconstruction. Time integration was performed using the third-order TVD RK scheme and the time duration was 0.15 and $\lambda = \Delta t / \Delta x = 0.5$.

6.7.12.1 Sod's Shock Tube

Figure 6.29 shows a third-order solution to Sod's problem, detailed previously in Section 6.6.4.1 with parameters defined in Table 6.1, using the high-resolution Kurganov and Tadmor central scheme with WENO3 reconstruction.

The closeness of the numerical solution to the analytical solution demonstrates clearly the effectiveness of the MUSCL approach to solving the Euler equations with good resolution. The simulation was carried out on a mesh of 200 cells, and time integration was performed by a third-order TVD Runge–Kutta integrator. This solution resembles very closely the second- and third-order solutions obtained using linear and parabolic reconstructions (see Section 6.6.4.1.

The R code for the `main program` and supporting functions that simulate the Sod problem are given in Appendix 6.C at the end of this chapter, along with code that defines the `initial conditions`. Additionally, the R code that generated the plots in Fig. 6.29 is shown in Listing 6.20. This postsimulation code first calculates the exact solution based on the *Rankine-Hugoniot jump conditions*, then for comparison purposes, the data are plotted together with the results from the example simulation. This same code was also used to generate the Sod problem plots for the linear and parabolic reconstruction examples discussed in Section 6.6.4.1. Additional grid points improve the resolution of the solution.

Table 6.4. Left, center, and right region parameters for the Woodward–Colella interacting blast wave problem

	Left ($x \leq -0.1$)	Center ($0.1 < x < 0.9$)	Right ($x \geq 0.9$)
$\rho_{L,C,R}$	1	1	1
$p_{L,C,R}$	1000	0.01	100
$u_{L,C,R}$	0	0	0

6.7.12.2 Woodward–Colella Interacting Blast Wave

We consider the 1D Woodward–Colella interacting blast wave problem with domain length=1, which is divided into three regions with parameters as defined in Table 6.4.

The test problem details are described in Section 6.6.4.2.

Figure 6.30 plots the results of a simulation performed on a mesh of 500 cells, using the high-resolution Kurganov and Tadmor central scheme with WENO3 reconstruction. Time integration was performed using the third-order TVD RK scheme and the time duration was 0.038 and $\lambda = \Delta t / \Delta x = 0.004$. A solution based on a very fine mesh of 4000 cells is also included as a reference. The closeness of the numerical solution to the

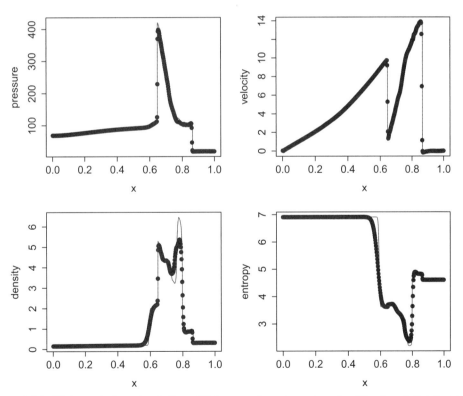

Figure 6.30. High resolution simulation of Euler equations based on the Woodward–Colella interacting blast wave problem. The reference solutions (solid) are shown along with simulated solutions (dotted) based upon the Kuganov and Tadmor central scheme with WENO3 reconstruction. The time duration was 0.0.38 and $\lambda = \Delta t / \Delta x = 0.004$.

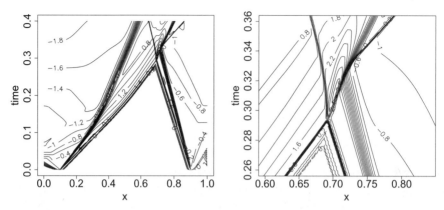

Figure 6.31. Plot of the Woodward–Colella interacting blast wave simulation in the x-t plane based on a mesh of 500 cells. (left) Log ρ plot with 20 contours and (right) close-up near collision.

reference solution provides a further demonstration of the effectiveness of the MUSCL approach to solving the Euler equations with good resolution. However, although for this particularly demanding test, the WENO3 solution is reasonable and captures the salient features of the reference solution, we observe from Fig. 6.20 that a solution based on linear extrapolation and a Sweby limiter provides better resolution.

Figure 6.31 shows a plot of the same simulation in the x-t plane with 20 contours, where time has been subsampled to reduce the computational effort. A similar plot based on the reference solution would show the shocks being much sharper.

The R code for the `main program` and supporting functions that simulate the Woodward–Collela problem are given in Appendix 6.C at the end of this chapter, along with code that defines the `initial conditions`. Additionally, the R code that generated the plots in Figs. 6.30 and 6.31 is shown in Listing 6.21. This postsimulation code first loads the previously calculated 4000-point reference data solution, then for comparison purposes these data are plotted together with the results from the example simulation. This same code was also used to generate the Woodward–Colella plots for the parabolic reconstruction example detailed in Section 6.6.4.2. Additional grid points improve the resolution of the solution, as seen from the reference solution.

6.7.12.3 Taylor–Sedov Blast Wave

The problem is described in Section 6.6.4.3.

We consider the 1D planar Taylor–Sedov blast wave problem with domain length=200 and an initial detonation energy value of $E_0 = 2.5$ GJ located in the central two cells of the problem domain. This is a much less powerful explosion than the Trinity atomic bomb but, nevertheless, provides a good test for the numerical scheme.

Figure 6.32 shows the results of a simulation performed on a mesh of 400 cells, using the above high resolution Kurganov and Tadmor central scheme with WENO5 reconstruction. Time integration was performed using the third-order TVD RK scheme. The time duration was 0.015s and, for stability reasons, a value for $\lambda = \Delta t/\Delta x = 0.00001$ was needed which represents a Courant number of Co $\simeq 0.12$. An analytical solution is also plotted as a reference. The closeness of the numerical solution to the reference solution provides a further demonstration of the effectiveness of the MUSCL approach to solving

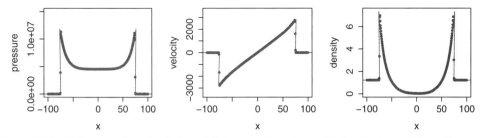

Figure 6.32. High-resolution simulation of Euler equations based Taylor–Sedov problem. The analytical solutions (solid) are shown along with simulated solutions(dotted) based upon the Kuganov and Tadmor central scheme with WENO5 reconstruction. The time duration was 0.015 and $\lambda = \Delta t/\Delta x = 0.00001$.

the Euler equations with good resolution. For this particular problem, the WENO5 solution is observed to give similar resolution to the solution based on linear extrapolation with Koren limiter (Fig. 6.24), but gives slightly improved resolution over the solution based on parabolic extrapolation with van Albada 2 limiter (Fig. 6.25).

The R code for the `main program` and supporting functions that simulate the Taylor–Sedov problem are given in Appendix 6.C at the end of this chapter, along with code that defines the `initial conditions`. Additionally, the R code that generated the plots in Fig. 6.24 is shown in Listing (6.22). This postsimulation code first performs some prliminary calculations, followed by the loading of 39-point reference data that enable the exact solution to be calculated using the method outlined in Chapter 12. Then, for comparison purposes, these data are plotted together with the results from the example simulation. This same code was also used to generate the Taylor–Sedov plots for the linear and parabolic reconstruction examples detailed in Section 6.6.4.3. Additional grid points improve the resolution of the solution.

6.7.13 2D Examples

In this section we illustrate how the WENO method can be applied to 2D problems. The examples have been selected because they pose serious difficulties for classic finite difference or finite volume schemes, but high-resolution solutions are obtained in a straightforward way using a 2D WENO scheme.

6.7.13.1 Rotation of a Solid Body

We now revisit the 2D fluid mechanics problem solved in Chapter 3, where we noted that the classic finite difference method was capable of resolving the solution well for a Gaussian cone, but failed when the rotating body was square shaped. We repeat the mathematical description for this advection problem for convenience.

Rotational flow is defined by the stream function

$$\psi(x, y) = x^2 + y^2, \tag{6.143}$$

with a resulting velocity field

$$\alpha_x(x, y) = 2y, \quad \alpha_y(x, y) = -2x \tag{6.144}$$

that satisfies the continuity requirements ($\partial \alpha_x/\partial x + \partial \alpha_y/\partial y = 0$).

The advection problem on the rectangular domain $\Omega = [-1, 1]^2$ and corresponding boundary $\partial \Omega$, is defined as follows:

$$\frac{\partial u}{\partial t} + \alpha_x \frac{\partial u}{\partial x} + \alpha_y \frac{\partial u}{\partial y} = 0, \quad u = u(x, y, t), \ t \geq t_0, \ x \in \mathbb{R}, \ y \in \mathbb{R},$$

$$u(x, y, 0) = \begin{cases} 1, & -0.25 \geq x \leq 0.25 \ \text{AND} \ 0.1 \geq y \leq 0.6 \\ 0, & \text{otherwise} \end{cases}, \quad (6.145)$$

$$\alpha_x = 2y, \quad \alpha_y = -2x,$$

where we impose Dirichlet boundary conditions $\partial \Omega = 0$.

We saw previously that the time taken for one complete revolution is $t = \pi$ and that all points on a radius vector will move in unison, taking the same time to rotate through any angle θ. A solid body therefore rotates around the central point (0,0) without distortion in time $t = N\pi$ for any integer N, where N represents the number of complete revolutions made by the flow.

This a difficult problem for numerical solvers. However, although there is some initial dissipation, the 2D WENO5 scheme handles the problem well with negligible dissipation subsequently (see Figs. 6.34 and 6.33). The rotating solid completes one clockwise revolution, correctly returning to its original position at $t = 3.1416$.

The reader may like to experiment with different values for α_x and α_y to see the effects on rotation time and distortion of the square, if any.

The R code for solving this problem is omitted to save space, as it follows the same general format as used for the rotating solid problem in Chapter 3, but with the WENO calculations described in Section 6.7.10. However, it is provided with the downloads.

6.7.13.2 Doswell Frontogenesis

We now consider a modified version of the dynamic frontogenesis problem discussed in Chapter 3 which increases the difficulty for numerical solvers. The problem is amended by decreasing the front characterizing parameter from $\delta = 1$ to $\delta = 10^{-6}$, which modifies the front from one that transitions smoothly to one that changes abruptly.

This advection problem is specified on the rectangular domain $\Omega = [-5, 5]^2$ and corresponding boundary $\partial \Omega$, which are repeated again for convenience:

$$\frac{\partial u}{\partial t} + \alpha_x \frac{\partial u}{\partial x} + \alpha_y \frac{\partial u}{\partial y} = 0, \quad u = u(x, y, t), \ t \geq t_0, \ x \in \mathbb{R}, \ y \in \mathbb{R},$$

$$u(x, y, 0) = \tanh\left(\frac{y}{\delta}\right), \quad (6.146)$$

$$\alpha_x = -y f(r), \quad \alpha_y = x f(r), \quad f(r) = \frac{1}{r}, \quad v(r) = \bar{v} \, \text{sech}^2(r) \tanh(r),$$

$$r = \sqrt{x^2 + y^2}, \quad \bar{v} = 2.59807, \quad \delta = 10^{-6},$$

where we impose Dirichlet boundary conditions $\partial \Omega(x, -1) = 0$, $\partial \Omega(x, 1) = 0$, and Neumann boundary conditions $\frac{\partial \Omega}{\partial n}(-1, y) = 0$, $\frac{\partial \Omega}{\partial n}(1, y) = 0$.

The analytical solution is given by

$$u_a(x, y, t) = \tanh\left(\frac{y \cos(vt/r) - x \sin(vt/r)}{\delta}\right). \quad (6.147)$$

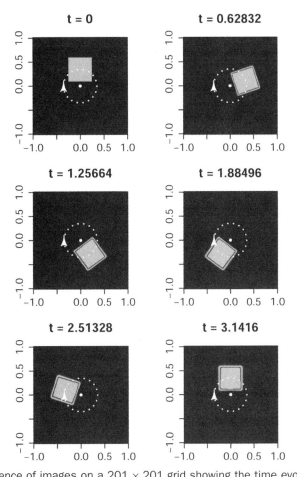

Figure 6.33. Sequence of images on a 201 × 201 grid showing the time evolution of a rotating-solid. The center of the solid follows the dotted circle of radius $r = 0.35$. Plots show clearly that after some initial dissipation the square then advects without noticeable distortion. Solution based upon the Kuganov and Tadmor central scheme with WENO5 reconstruction. (See color plate 6.33)

Figure 6.34. Surface plot of high-resolution simulation of 2D advection of the rotating-solid problem after one revolution, at $t = 3.1416$.

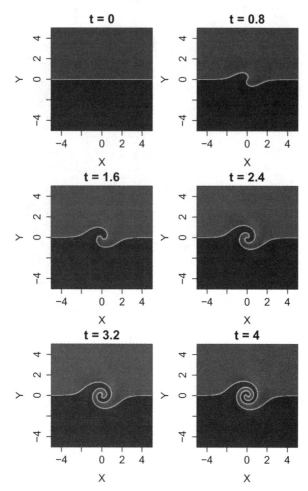

Figure 6.35. Array of 2D plots of high-resolution simulation of frontogenesis showing time evolution from $t = 0$ to $t = 4$. Solution based upon the Kuganov and Tadmor central scheme with WENO5 reconstruction. (See color plate 6.35)

This is another difficult problem for numerical solvers and, again, the 2D WENO5 scheme resolves the sharp front solution well, with the general shape exhibiting very little distortion (see Figs. 6.35 and 6.36). The vortex evolves from zero at $t = 0$ on a very sharp front to form a complex rotational system that straddles both sides of the front at $t = 4$. However, the vortex arms should have narrow flat tops rather than the rounded tops shown in Fig. 6.36. This is due to the small number of grid points spanning the vortex arms. It can be improved by including a finer overall grid or by refining the the grid around the votex area [Abo-08].

The reader may like to experiment with different values of $\bar{v}$ to see how the vortex is affected and also to try different simulation run times to see how the vortex evolves.

The R code for solving this problem is omitted to save space, as it follows the same general format as used for the frontogenesis problem in Chapter 3, but with the WENO calculations described in Section 6.7.10. However, it is provided with the downloads.

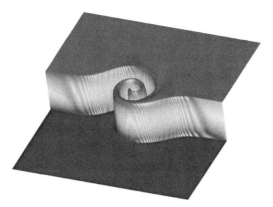

Figure 6.36. Surface plot of high-resolution simulation of the *frontogenesis* problem at $t = 4$. (See color plate 6.36)

6.8 FURTHER READING

For further discussion relating to the theory and application of high-resolution schemes in general, the reader is referred to [Dri-05], [Hir-90], [Lan-98], [Lev-02], [Tor-99], [Tan-97], and [Wes-01].

For further discussion relating to the theory and application of WENO schemes, the reader is referred to [Shu-98] [Bal-00], [Lat-05], and [Tit-03].

6.A EIGENVALUES OF EULER EQUATIONS

For the Euler equations in Cartesian coordinates without heat transfer and without body force we have [Lan-98, p. 14]

$$\frac{\partial U}{\partial t} + \frac{\partial f}{\partial x} = 0, \quad U = \begin{bmatrix} \rho \\ \rho u \\ \rho e_T \end{bmatrix}, f = \begin{bmatrix} \rho u \\ \rho u^2 + p \\ (\rho e_T + p) u \end{bmatrix}, u = u(x, t), t \geq t_0, x \in \mathbb{R}, \quad (6.148)$$

where

- U = vector of *conserved quantities*
- f = *flux* vector
- ρ = density, [kg/m^3]
- u = velocity, [m/s]
- p = pressure, [Pa]
- e_T = $(e + \frac{1}{2}u^2)$, total energy per unit mass, [J/kg]
- e = internal energy per unit mass, [J/kg]

Alternatively, the Euler equations for 1D, 2D, and 3D geometrically symmetric problems can be reduced to the more simple 1D case, as follows:

$$\frac{\partial U}{\partial t} + \frac{\partial f}{\partial r} = S, \quad S = -\begin{bmatrix} \frac{(\nu - 1)}{r} \rho u \\ \frac{(\nu - 1)}{r} \rho u^2 \\ \frac{(\nu - 1)}{r} (\rho e_T + p) u \end{bmatrix}, \quad (6.149)$$

where r represents the symmetry axis, $\nu = 1$ for the planar case, $\nu = 2$ for the cylindrical case, and $\nu = 3$ for the spherically symmetric case. Other variables are as previously defined. Equations (6.149) are presented in conservative form with source term S as this facilitates solution by the finite volume method.

We now make use of the thermodynamic equation of state for a polytropic gas:

$$p/\rho = (\gamma - 1)e = (\gamma - 1)\left(e_T - \tfrac{1}{2}u^2\right), \qquad (6.150)$$

where γ [-] represents the ratio of specific heats. Defining conserved variables $\rho u = m$ and $\rho e_T = E$, the flux vector becomes

$$f = \begin{bmatrix} m \\ \left(\tfrac{1}{2}(\gamma - 1)(-m^2 + 2\rho E) + m^2\right)/\rho \\ \tfrac{1}{2}m\left((1-\gamma)m^2 + 2\gamma\rho E\right)/\rho^2 \end{bmatrix}. \qquad (6.151)$$

Taking the Jacobian of f with respect to the conserved quantities ρ, m, and E yields the matrix $A = \partial f/\partial U$, that is,

$$A(U) = \begin{bmatrix} 0 & 1 & 0 \\ \tfrac{1}{2}\dfrac{m^2(\gamma-3)}{\rho^2} & -\dfrac{m}{\rho}(\gamma-3) & (\gamma-1) \\ \dfrac{m}{\rho^3}\left(m^2(\gamma-1) - \gamma\rho E\right) & \tfrac{1}{2}\dfrac{(-3m^2(\gamma-1) + 2\gamma\rho E)}{\rho^2} & \gamma\dfrac{m}{\rho} \end{bmatrix}. \qquad (6.152)$$

We now convert back to the primitive variables, ρ, u, and e_T to obtain

$$A(U) = \begin{bmatrix} 0 & 1 & 0 \\ \tfrac{1}{2}u^2(\gamma-3) & -u(\gamma-3) & (\gamma-1) \\ -\gamma e_T + (\gamma-1)u^3 & \gamma e_T - \tfrac{3}{2}u^2(\gamma-1) & \gamma u \end{bmatrix}, \qquad (6.153)$$

when we find the eigenvalues of the matrix A to be

$$\lambda_1 = u + 2\gamma(\gamma-1)(2e - u^2), \quad \lambda_2 = u, \quad \lambda_3 = u - 2\gamma(\gamma-1)(2e - u^2). \qquad (6.154)$$

These can be simplified to

$$\lambda_1 = u + c, \quad \lambda_2 = u, \quad \lambda_3 = u - c \qquad (6.155)$$

by using the following relationships for internal energy and speed of sound, respectively:

$$e = \dfrac{p}{(\gamma-1)\rho} + \tfrac{1}{2}u^2, \quad c = \sqrt{\gamma p/\rho}. \qquad (6.156)$$

These calculations can be quite tedious by hand, and it is easier to employ a computer algebra program. Code written for the symbolic programs Maple and Maxima (open source) that derives eqns. (6.153) and (6.155) are included with the downloads.

Using the relationships

$$\begin{aligned} h &= e + p/\rho \\ h_T &= e_T + p/\rho, \end{aligned} \qquad (6.157)$$

where

h = enthalpy per unit mass [J/kg]
h_T = total enthalpy per unit mass [J/kg]

we obtain from eqn. (6.150)

$$\gamma u e_T = u h_T - \frac{1}{2}(\gamma - 1)u^3. \tag{6.158}$$

Therefore, eqn. (6.153) can be written in the alternative form

$$A(U) = \begin{bmatrix} 0 & 1 & 0 \\ \frac{1}{2}u^2(\gamma - 3) & -u(\gamma - 3) & (\gamma - 1) \\ -uh_T + \frac{1}{2}(\gamma - 1)u^3 & h_T - (\gamma - 1)u^2 & \gamma u \end{bmatrix}. \tag{6.159}$$

6.B R CODE FOR SIMULATING 1D SCALAR EQUATION PROBLEMS

This section details the primary R routines that were used to in the preceding scalar examples. Other subsidiary functions have been detailed previously.

6.B.1 The Main Program

The main program is where the following simulation configuration options are set:

- initial condition type
- limiter type (for linear or parabolic reconstruction only)
- reconstruction type
- numerical integrator
- scalar PDE to be simulated
- boundary conditions, appropriate to PDE
- various parameters related to the initial condition
- plot requirements

The various options available are detailed in the code and should be made in accordance with the particular test simulation discussion.

The *Initial condition* type(s) and *Limiter* type(s) are set in the two `for` loops (around the middle of the code), to facilitate running multiple simulations consecutively that have different configurations. These results can then be used for comparison purposes.

The *Reconstruction* type is set by variable `reconType` with linear and parabolic being used with limiters and the other options being for the WENO methods.

The numerical *integrator* type is set by variable `integType` where the choice is between *TVD-RK* and *SHK*.

The simulation *run time* is controlled by the value of `tEnd` and should be selected so that for advection simulations waves propagate integer multiples of the spatial domain with periodic BCs. This will enable the analytical solution to be compared to the simulation results. However, for Burgers equation and Buckley–Leverett equation simulations the BCs should be set to Dirichlet and the value of tEnd should be selected so that waves do not propagate beyond the domain boundaries.

```r
# File: scalar1D_main.R
require(compiler)
enableJIT(3)
rm(list = ls(all = TRUE)) # Delete workspace
#library("rgl")
#library("Matrix")
library("pracma")
source("initialScalar.R")
source("linearExtrapolation.R")
source("parabolicReconstruction.R")
source("WENO3rdOrdReconstruction.R")
source("WENO3rdOrdReconstructionKL.R")
source("WENO5thOrdReconstruction.R")
source("Limiter.R")
source("advection.R")
source("Burgers.R")
source("BuckeyLeverett.R")
source("derivScalar.R")
source("MUSCL_KTscalar.R")
# R Code used to simulate conservation PDEs, i.e.
#               Ut + f(U)x <- 0
# with Dirichlet or periodic boundary conditions
#
ptm = proc.time()

strInit    <- c("Step Pulse","Cosine Pulse","Triange Pulse",
                "Forward Wedge Pulse","Backward Wedge Pulse",
                "stepUp","stepDown","Composite Pulse Set")
strLimiter <- c("None","HCUS",
                "HQUICK - harmonic QUICK","Koren",
                "MC - mono. cent.-diff.",
                "minmod","Osher","ospre","smart",
                "superbee","Sweby","UMIST bounded QUICK",
                "van Albada 1","van Albada 2",
                "van Leer 1","van Leer 2")
strRecon   <- c("Linear extrapolation","Parabolic",
                "WENO-3rd order","WENOKL-3rd order","WENO-5th order")
strInteg   <- c("TVD RK, 3rd Order","SHK")
strEqn     <- c("Advection","Burgers","Buckley-Leverett")
strBCs     <- c("Dirichlet","Periodic")

wavePeak  <- 1.0 # Maximum height of wave
Xmax      <- 1   # End of spatial domain
```

```
Xmin       <- -Xmax # Start of spatial domain
N          <- 200   # Number of spatial cells
M          <- N     # Number of Cell Interfaces (for periodic BCs)
alpha      <- +1    # Wave speed, +ve right to left, -ve left to right
                    # One loop every (Xmax-xmin)/alpha [s]
Beta       <- 1.5   # Limiter constant
reconType  <- 2     # Reconstruction method
solType       <- 1  # Solution method
integType  <- 1     # Integration method
eqnType    <- 3     # Problem equation
BC         <- 0     # BC Type: 0=Dirichlet (Burgers/B-L eqn.)
                    #          1=periodic (other)
# INITIALISE VARS - CALCULATED
dx    <- (Xmax-Xmin)/(N)              # Segment length
x     <- seq(Xmin,Xmax,length.out=N) #dx) # Discretize spatial domain
Co       <- 0.45                      # Courant Number
dt    <- Co*dx/abs(alpha) # time step
t0    <- 0.0                          # Set start time
tRun  <- 0.45          # Set run duration
tEnd  <- t0 + tRun     # Set end time
# ADJUSTABLE PARAMETERS - INPUTS

par(mfrow = c(1,1))
# Start simulation
for(iLim in c(4)){ # c(1:16)
  for(initWave in c(1)){
    u0 <- init(x, initWave, Xmin, Xmax, wavePeak, eqnType) # Set the starting
       conditions

    limiterType <- iLim
    # cat(sprintfLAY SIMULATION DATA
    cat(sprintf(" \n"))
    cat(sprintf("==================================================\n"))
    cat(sprintf("             SIMULATION DATA              \n"))
    cat(sprintf("==================================================\n"))
    cat(sprintf("Problem Equation, ............: %s\n",strEqn[eqnType]))
    if(solType == 1){
      cat(sprintf("Reconstruction, ..............: %s\n", strRecon[reconType]))
    }
    if(reconType < 3){
      cat(sprintf("Limiter,......................: %s\n", strLimiter[limiterType]))
    }
    cat(sprintf("Integrator type, .............: %s\n", strInteg[integType]))
    cat(sprintf("Segment Length, ..........., dx: %f %s\n",dx, " [m]"))
    cat(sprintf("Number of segments, ........, M: %d %s\n",M, " [-]"))
    cat(sprintf("Total Length, .............., x: %f %s\n",(Xmax-Xmin), " [m]"))
```

```r
cat(sprintf("Propogation Velocity, .., alpha: %f %s\n",alpha, " [m/s]"))
if(BC == 0){
  cat(sprintf("Boundary conditions, ..........: %s\n","Dirichlet"))
}else if(BC == 1){
  cat(sprintf("Boundary conditions, ..........: %s\n","Periodic"))
}
cat(sprintf("End Time, ..............., tEnd: %f %s\n",tEnd, " [s]"))

cat(sprintf("Initialised To ................: %s\n",strInit[initWave]))

# START SIMULATION
#######################################################
Co <- alpha*dt/dx  # Courant Number. For stability must be < 1

numSteps <- ceiling(tEnd/dt) #+12
t <- seq(t0,tEnd,length.out=numSteps)
Dt <- (tEnd-t0)/(numSteps-1)

U       <- matrix(0,nrow=numSteps,ncol=(M+1))
U[1,2:(M+1)] <- u0
U0      <- U[1,2:(M+1)]
U1      <- U0
shkOrd <- 4
for(it in 2:numSteps){
  t1 <- t[(it-1)]
  if(integType == 1){
    # 3rd Ord TVD RK Calcs
    # Ref: C-W Shu and O. Osher, Efficient implementation of
    # essentially non-oscilliatory shock capturing schemes,
    # J. Comp. Phys. 77, 1988, 439-471.
    f1 = derivScalar(U0, alpha, Beta, dx, limiterType, reconType, M, eqnType)
    U1 = U0 + f1 * Dt
    f2 = derivScalar(U1, alpha, Beta, dx, limiterType, reconType, M, eqnType)
    U2 = (3/4)*U0 + (1/4)*U1 + (1/4)*f2 * Dt
    f3 = derivScalar(U2, alpha, Beta, dx, limiterType, reconType, M, eqnType)
    U3 = (1/3)*U0 + (2/3)*U2 + (2/3)*f3 * Dt
    U[it,1]    <- t[it]
    U[it,2:(M+1)] <- U3
    U0         <- U3
  }else if(integType == 2){
    # End TVD RK Calcs
    for(s in seq(shkOrd,1,-1)){
      f <- derivScalar(U1, alpha, Beta, dx, limiterType, reconType, M, eqnType)
      U1 <- U0+ f*Dt/s
    }
    U[it,1]    <- t[it]
```

```
        U[it,2:(M+1)] <- U1
        U0  <- U1;
      }
    }
    cat(sprintf("Time Step, ..............., dt: %f\n",dt, " [s]"))
    cat(sprintf("Courant No, ..............., Co: %f\n",Co, " [-]"))

    if(reconType<3){
      strT1 <- sprintf("%s", strLimiter[limiterType])
    }else{
      strT1 <- sprintf("%s", strInit[initWave])
    }
    plot(x,u0,col="red",type="l",lwd=1,
        xlab="x",ylab=expression(paste("u(x), ",u[0])),
          ylim=c(-0.1,1.1),main=strT1)
    lines(x,U[numSteps,(1:N+1)],col="blue", lwd=1)
    points(x,U[numSteps,(1:N+1)],col="blue", pch=20,cex=0.75)
    grid(col="black",lty=2,lwd=1)
    # abline(h=(c(-1:6)/5),v=c(-1:8),lty=2,lwd=1)
    status <- "OK"
    cat(sprintf("Simulation Completed, .........: %s\n", status))
    cat(sprintf("==================================================\n"))
  }
}
#
# Elapsed time
cat(sprintf("calculation time = %5.2f\n",(proc.time()-ptm)[3]))
```

Listing 6.12. File: scalar1D_main.R—Code for the *scalar* examples main program. This program calls initialScalar(), where the simulation configuration is defined. It also calls various other functions, the files of which are all included with the downloads

6.B.2 The Derivative Function

The scalar1D_deriv() function calculates the spatial derivative $\partial u/\partial t = -\partial f/\partial x$ by calling the appropriate reconstruction method to obtain left and right edge values u_L and u_R. This is followed by a call to MUSCL_KTscalar() to obtain $\partial f/\partial x$ which is negated and returned as the time derivative to the numerical integrator.

```
# File: scalar1D_deriv.R
# File: scalar1D_deriv.R
scalar1D_deriv <- function(u0, alpha, Beta, dx, limiterType, reconType,
                    m, eqnType){

  a    <- rep(0,m)
  flux <- rep(0,m)
```

```
    dfdx <- rep(0,m)

    # Reconstructions for (j+1/2)
    if(reconType == 1){ # Left and right linear state extrapolation
      uLuR <- linearExtrapolation(u0,limiterType,Beta)
      uL   <- uLuR[1:m]
      uR   <- uLuR[(m+1):(2*m)]
    }else if(reconType == 2){ # Left and right state parabolic interpolation
      uLuR <- parabolicReconstruction(u0,limiterType,Beta)
      uL   <- uLuR[1:m]
      uR   <- uLuR[(m+1):(2*m)]
    }else if(reconType == 3){ # Left and right state 3rd Order WENO
      uLuR <- WENO3rdOrdReconstruction(u0)
      uL   <- uLuR[1:m]
      uR   <- uLuR[(m+1):(2*m)]
    }else if(reconType == 4){ # Left and right state 3rd Order WENOKL
      uLuR <- WENO3rdOrdReconstructionKL(u0)
      uL   <- uLuR[1:m]
      uR   <- uLuR[(m+1):(2*m)]
    }else if(reconType == 5){ # Left and right state 5th Order WENO
      uLuR <- WENO5thOrdReconstruction(u0)
      uL   <- uLuR[1:m]
      uR   <- uLuR[(m+1):(2*m)]
    }
    dfdx <- MUSCL_KTscalar(alpha, uL, uR, dx, m, eqnType)
    return( - dfdx )
}
```

Listing 6.13. File: scalar1D_deriv—Code for the derivative function scalar1D_deriv(). This function calls one of the reconstruction functions extrapolation(), parabolicReconstruction(), WENO3rdOrdReconstruction(), WENO3rdOrdReconstructionKL(), or WENO5thOrdReconstruction() and also MUSCL_KTscalar()

6.B.3 The MUSCL Function

The MUSCL_KTscalar() function uses the left and right edge vales, u_L and u_R, to calculate MUSCL fluxes based on the Kurganov and Tadmore algorithm. These are then used to calculate an approximation to the appropriate spatial derivative $\partial f/\partial x$ for the PDE being simulated. This spatial derivative is then passed back to the derivative function derivScalar() which calculates the time derivative for use by the numerical integrator.

```
# File: MUSCL_KTscalar.R
# Kurganov and Tadmore Semi-Discrete' method,
# for solving conservation PDE's
#      Ut + f(U)x <- 0
# This routine calculates the flux limited spatial derivative of f(U), wrt x
```

```
MUSCL_KTscalar <- function(alpha, uL, uR, dx, m, eqnType){

  a    <- rep(0,m)
  flux <- rep(0,m)
  dfdx <- rep(0,m)

  if(eqnType == 1){
    fL <- advection(uL, alpha)
    fR <- advection(uR, alpha)
  }else if(eqnType == 2){
    fL <- Burgers(uL, alpha)
    fR <- Burgers(uR, alpha)
  }else if(eqnType==3){
    fL <- BuckeyLeverett(uL)
    fR <- BuckeyLeverett(uR)
  }
  dU <- uL-uR # At edge (j+1/2)

  sml <- 0.01
  if(abs(uL[2]-uL[1]) > sml && abs(uR[1]-uR[m]) > sml){
    a[1] <- max(abs((fL[2]-fL[1])/(uL[2]-uL[1])),
                abs((fR[1]-fR[m])/(uR[1]-uR[m])))
  }else if(eqnType == 0){
    a[1] <- abs(alpha) # advection
  }else if(eqnType == 1){
    a[1] <- abs(uL[1]*alpha) # Burgers
  }else{
    a[1] <- abs(2*uL[1]^2/(2*uL[1]^2+(1-uL[1])^2)) # Buckley-Everett (c=2)
  }

  for(j in 2:(m-1)){
    if(abs(uL[(j+1)]-uL[j]) > sml && abs(uR[(j)]-uR[(j-1)]) > sml){
      a[j] <- pmax(abs((fL[(j+1)]-fL[j])/(uL[j+1]-uL[j])),
                   abs((fR[(j)]-fR[(j-1)])/(uR[j]-uR[(j-1)])))
    }else if(eqnType == 1){
      a[j] <- abs(alpha) # advection
    }else if(eqnType == 2){
      a[j] <- abs(uL[j]*alpha) # Burgers
    }else{
      a[j] <- abs(2*uL[j]^2/(2*uL[j]^2+(1-uL[j])^2)) # Buckley-Everett (c=2)
    }
  }
  if(abs(uL[1]-uL[m]) > sml && abs(uR[m]-uR[(m-1)]) > sml){
    a[m] <- pmax(abs((fL[1]-fL[m])/(uL[1]-uL[m])),
                 abs((fR[m]-fR[(m-1)])/(uR[m]-uR[(m-1)])))
```

```
        }else if(eqnType == 1){
          a[m] <- abs(alpha) # advection
        }else if(eqnType == 2){
          a[m] <- abs(uL[m]*alpha) # Burgers
        }else{
          a[m] <- abs(2*uL[m]^2/(2*uL[m]^2+(1-uL[m])^2)) # Buckley-Everett (c=2)
        }
        flux[1:m] <- ( (fL[1:m] + fR[1:m]) + a[1:m]*dU[1:m] )/2.0

        if(BC == 0){       # Dirichlet
          dfdx[1]   <- 0
        }else if(BC == 1){ # Periodic
        dfdx[1]   <- (flux[1] - flux[m])/dx
        }
        dfdx[2:m] <- (flux[2:m] - flux[1:(m-1)])/dx

        return(dfdx)
}
```

Listing 6.14. File: MUSCL_KTscalar.R—Code for function MUSCL_KTScalar(). This function calls advection(), Burgers(), or BuckleyLeverett()

6.B.4 Initialization

The initScalar() function is where the scalar simulations are initialized. It contains various subsidiary functions that set the state variable starting values over the entire spatial domain, according to the particular initial waveform as configured in the main program.

```
# File: scalar1D_init.R
# Initialise the spatial domain for simulating PDE problem's
scalar1D_init <- function(xl, waveType, Xmin, Xmax, wavePeak, eqnType)
{
  X <- Xmax-Xmin
  Z <- (Xmax + Xmin)/2
  W <- length(xl)
  #cat(sprintf("%d",W))

  if(waveType == 8){
    widthWave <- 0.1 # Fraction (0-1)
    x1 <- Z - widthWave*X/2
    x3 <- Z + widthWave*X/2
    x2 <- Z
    a <- 0.1
    triangeleOffset <- 0.1
    stepOffset      <- -0.3
```

```r
  }else{
    if(eqnType == 1){
      widthWave <- 0.5 # advection eqn. - Fraction (0-1)
    }else{
      widthWave <- 0.25 # Burgers/B-L eqn. - Fraction (0-1)
    }
    x1 <- Z - widthWave*X/2
    x3 <- Z + widthWave*X/2
    x2 <- Z
    triangeleOffset <- 0
    if(eqnType == 1){
      stepOffset     <- 0    # advection eqn.
    }else{
      stepOffset     <- -0.3 # Burgers/B-L eqn.
    }
    stepUpOffset   <- -0.5
    stepDownOffset <- -0.5
  }
  if(waveType == 1){  # Step Pulse
    ul<-stepPulse(xl, x1, x3, stepOffset, wavePeak, W)
  }else if(waveType == 2){   # Cosine Pulse
    ul<-cosPulse(xl, x1, x3, wavePeak, W)
  }else if(waveType == 3){   # Triangle Pulse
    ul<-triangle(xl, x1, x2, x3, triangeleOffset, wavePeak, W)
  }else if(waveType == 4){   # Forward Wedge
    ul<-forwardWedge(xl, x1, x3, wavePeak, W)
  }else if(waveType == 5){   # Backward Wedge
    ul<-backWedge(xl, x2, offset, wavePeak, W)
  }else if(waveType == 6){ # stepUp
    ul<-stepUp(xl, x2, stepUpOffset, wavePeak, W)
  }else if(waveType == 7){ # stepDown
    ul<-stepDown(xl, x2, stepDownOffset, wavePeak, W)
  }else if(waveType == 8){ # Composite Pulse set
    ul1<-stepPulse(xl, x1, x3, stepOffset, wavePeak, W)
    ul2<-triangle(xl, x1, x2, x3, triangeleOffset, wavePeak, W)
    ul3<-ellipsoidPulse(xl, wavePeak, W)
    ul4<-gauss(xl, wavePeak, W)
    ul <- ul1+ul2+ul3+ul4
  }
return(ul)
}
stepPulse <- function(xl, x1, x3, offset, wavePeak, M)
{
  ul<-rep(0,M)
  for(J in 1:M){
    if(xl[J] >= (x1+offset) && xl[J] <= (x3+offset)){
```

```
      ul[J] <- wavePeak
    }
  }
  return(ul)
}
stepDown <- function(xl, x2, offset, wavePeak, M)
{
  print(x2)
  ul<-rep(0,M)
  for(J in 1:M){
    if((xl[J]-offset) <= x2){
      ul[J] <- wavePeak
    }else{
      ul[J] <- 0
    }
  }
  return(ul)
}
stepUp <- function(xl, x2, offset,wavePeak, M)
{
  print(x2)
  ul<-rep(0,M)
  for(J in 1:M){
    if((xl[J]-offset) <= x2){
      ul[J] <- 0
    }else{
      ul[J] <- wavePeak
    }
  }
  return(ul)
}
gauss<- function(xl, wavePeak, M)
{
  delta <- 0.005
  z     <- -0.7
  beta  <- log(2)/(36*delta^2)
  ul<-rep(0,M)
  for(J in 1:M){
    if(xl[J] >= -0.8 && xl[J] <= -0.6){
      ul[J] <- (1/6)*wavePeak*(exp( -beta*(xl[J]-(z-delta))^2) +
                          exp( -beta*(xl[J]-z)^2) +
                          4*exp( -beta*(xl[J]-(z+delta))^2))
    }
  }
  return(ul)
}
```

```
ellipsoidPulse <- function(xl, wavePeak, M)
{
  a     <- 0.5
  delta <- 0.005
  alpha <- 10
  ul<-rep(0,M)
  for(J in 1:M){
    if(xl[J] >= 0.4 && xl[J] <= 0.6){
      ul[J] <- (1/6)*wavePeak*(sqrt(max(1 - alpha^2*(xl[J]-(a-delta))^2,0) ) +
                  sqrt(max(1 - alpha^2*(xl[J]-(a+delta))^2,0) ) +
                  4*sqrt(max(1 - alpha^2*(xl[J]-a)^2,0) ))
    }
  }
  return(ul)
}
backWedge <- function(xl, x1, x3, wavePeak, M)
{
  ul<-rep(0,M)
  for(J in 1:M){
    if(xl[J] < x1){
      ul[J] <- 0.0
    }else if(xl[J]>= x1 && xl[J] <= x3){
      ul[J] <- wavePeak*(xl[J]-x1)/(x3-x1)
    }else{
      ul[J] <- 0.0
    }
  }
  return(ul)
}
forwardWedge <- function(xl, x1, x3, wavePeak, M)
{
  ul<-rep(0,M)
  for(J in 1:M){
    if(xl[J] < x1){
      ul[J] <- 0.0
    }else if(xl[J] >= x1 && xl[J] <= x3){
      ul[J] <- wavePeak*(x3-xl[J])/(x3-x1)
    }else{
      ul[J] <- 0.0
    }
  }
  return(ul)
}
triangle <- function(xl, x1, x2, x3, offset, wavePeak, M)
{
  ul<-rep(0,M)
```

```
    for(J in 1:M){
      if(xl[J] >= (x1+offset) && xl[J] <= (x3+offset)){
        ul[J] <- wavePeak*(1 - abs((2/(x3-x1))*(xl[J]-offset)))
      }
    }
    return(ul)
}
cosPulse <- function(xl, x1, x3, wavePeak, M)
{
  ul<-rep(0,M)
  for(J in 1:M){
    if(xl[J] > x1 && xl[J] < x3){
      ul[J] <- wavePeak*(1 - cos(2*pi*(xl[J] - x1)/(x3-x1)))/2.0
    }else{
      ul[J] <- 0.0
    }
  }
  return(ul)
}
```

Listing 6.15. File: scalar1D_init.R—Code for function scalar1D_init() called by the main program of Listing 6.12. This function calls various subsidiary functions as determined by the main program configuration

6.C R CODE FOR SIMULATING 1D EULER EQUATIONS PROBLEMS

This appendix details the primary R routines that were used to generate solutions to the 1D examples discussed in Sections 6.6.4 and 6.7.12. Other subsidiary functions have been detailed previously.

6.C.1 The Main Routine

The main program is where the following simulation configuration options are set:

- initialization
- limiter type (for linear or parabolic reconstruction only)
- reconstruction type
- numerical integrator
- boundary conditions
- various parameters related to the initial condition
- plot requirements

The various options available are detailed in the code and should be made in accordance with the particular test simulation discussion.

The *initial conditions* and the *boundary conditions* are set automatically in the initialization code where the *test case* option is set.

The *Reconstruction type* is set by variable reconType with linear and parabolic being used with limiters and the other options being for the WENO methods.

The numerical integrator type is set by variable integType where the choice is between *TVD-RK* and *SHK*

```
# File: Euler_main.R
require(compiler)
enableJIT(3)
rm(list = ls(all = TRUE)) # Delete workspace
# MUSCL solution for for one-dimensional Euler equations

library("rgl")
source("Euler_deriv.R")
source("MUSCL_KTEuler.R")
source("linearExtrapolation.R")
source("parabolicReconstruction2.R")
source("WENO3rdOrdReconstructionKL.R")
source("WENO5thOrdReconstruction.R")
source("Limiter.R")
ptm <- proc.time()
# ....................Config Start..........................
#
reconType <- 1 # 1 <- Linear Extrapolation
               # 2 <- Parabolic Reconstruction
               # 3 <- 3rd Order WENO Reconstruction
               # 4 <- 5th Order WENO Reconstruction
limiterType <- 8 # " 1" <- "None"
             # " 2" <- "HCUS"
             # " 3" <- "HQUICK - harmonic QUICK"
             # " 4" <- "Koren"
             # " 5" <- "MC- monotonised central-difference"
             # " 6" <- "minmod"
             # " 7" <- "Osher"
             # " 8" <- "ospre"
             # " 9" <- "smart"
             # "10" <- "superBee"
             # "11" <- "Sweby"
             # "12" <- "UMIST bounded QUICK"
             # "13" <- "van Albada 1"
             # "14" <- "van Albada 2"
             # "15" <- "van Leer 1"
             # "16" <- "van Leer 2"
Beta <- 1.5   # Parameter for Osher and Sweeby Limiters
             # "1.0" <- non-oscillatory,
             # "2.0" <- least dissipative
             # "1.5" <- default
```

```
intMethod <- 1  # Parameter for Integration Method
            #     "0" <- SHK Runge Kutta,
            #     "1" <- 3rd Order TDV Runge Kutta - use for Sedov-Taylor

if( reconType > 1){
   limiterType <- 1
}

gamma <- 1.4      # Ratio of specific heats
N <- 500          # Number of grid cells
#
source("Euler_initial.R")
# ....................Config Ends...........................

dx <- Length/N            # Cell size
dt <- lambda*dx           # Time step
n  <- floor(tend/dt)      # Number of time-steps
#
#              Definition of cell numbering
#       x=0                                          x=L
# grid  |---o---|---o---|---o--- ... ---o-- ... --|---o---|
#  cells: 1       2       3              i             N

if(testcase == "Taylor-Sedov blast wave problem 1"){
  xcenter <- dx*(1:N) - dx/2 - Length/2 # Location of cell centers
  xcenterAbs <- abs(xcenter)
}else{
  xcenter <- dx*(1:N) - dx/2            # Location of cell centers
}

press0 <- rep(0,N)              # Preallocation of pressure,
rho0   <- rep(0,N); u0 <- rep(0,N)       # density, velocity,

if(testcase == "Woodward-Colella blast wave problem"){
  for(j in 1:length(xcenter)){         # Initial conditions
    if(xcenter[j] < 0.1*Length){
       press0[j] <- pleft; rho0[j] <- rholeft; u0[j] <- uleft
    }else if(xcenter[j] >= 0.1*Length && xcenter[j] <= 0.9*Length){
       press0[j] <- pmid; rho0[j] <- rhomid; u0[j] <- umid
    }else{
       press0[j] <- pright; rho0[j] <- rhoright; u0[j] <- uright
    }
    BC <- 1
  }
}else if(testcase == "Taylor-Sedov blast wave problem 1"){
  for(j in 1:length(xcenter)){         # Initial conditions
```

```
    Nj <- round(N/2)
    if(j < Nj){
        press0[j] <- pleft; rho0[j] <- rholeft; u0[j] <- uleft;
    }else if(j >= (Nj) && j <= (Nj+1)){
        press0[j] <- pmid
        rho0[j] <- rhomid; u0[j] <- umid
    }else{
        press0[j] <- pright; rho0[j] <- rhoright; u0[j] <- uright
    }
    BC <- 0
  }
}else{
  for(j in 1:length(xcenter)){         # Initial conditions
    if(xcenter[j] < 0.5*Length){
        press0[j] <- pleft; rho0[j] <- rholeft; u0[j] <- uleft
    }else{
        press0[j] <- pright; rho0[j] <- rhoright; u0[j] <- uright
    }
  }
  BC <- 0
}

# eT <- rho*(total energy/vol)
m0     <- rho0*u0                       # m <- Momentum
eT0    <- (0.5*m0*u0 + (1/(gamma-1))*press0)
eTleft <- eT0[1]; eTright <- eT0[j]

t <- 0
ncall <<- 1
U0 <- c(rho0,m0,eT0)
#U1 <- U0
U  <- matrix(0,nrow=n,ncol=(3*N))
shkOrd <- 4 # Set order of SHK integration
cat(sprintf("n=%d\n\n",n))
for(it in 1:n){
  t <- t + dt
  if(it%%10 == 0){
    cat(proc.time()-ptm)
    cat(sprintf(", t=%e\n",t))
  }

  # K and T scheme
  if(intMethod == 0){ # SHK integration
    for(s in seq(shkOrd,1,-1)){
      f <- derivEuler(U0, Beta, dx, limiterType, reconType, N)
      U1 <- U0+ f*dt/s
```

```
        }
    U[it,] <- U1
    U0    <- U1
  }else{
    # 3rd order TVD Runge-Kutta
    f1 <- derivEuler(U0, Beta, dx, limiterType, reconType, N)
    #print(f1)
    U1 <- U0 + f1 * dt
    #print(U1)
    f2 <- derivEuler(U1, Beta, dx, limiterType, reconType, N)
    U2 <- (3/4)*U0 + (1/4)*U1 + (1/4)*f2 * dt
    f3 <- derivEuler(U2, Beta, dx, limiterType, reconType, N)
    U3 <- (1/3)*U0 + (2/3)*U2 + (2/3)*f3 * dt
    U[it,] <- U3
    U0    <- U3
  }
}
rho     <- U[it,1:N]
m       <- U[it,(N+1):(2*N)]
u       <- m/rho
eT      <- U[it,(2*N+1):(3*N)]
press   <- (gamma-1)*pmax(1.e-6,(eT - 0.5*m*u));
vs      <- sqrt(gamma*press/rho) # speed of sound
mach    <- u/vs
entropy <- log(press/rho^gamma)
#
# Code for saving/loading reference solutions
# save(xcenter4k,press4k,u4k,rho4k,entropy4k,mach4k,eT4k,
#      file="WoodwardColella4k_x.data",ascii=TRUE)
# load("WoodwardColella_4k.data") # for comparison
# save(xcenter,press,u,rho,entropy,mach,eT,
#      file="SedovTaylor2_501.data",ascii=TRUE)
# load("SedovTaylor2_501.data") # for comparison

par(mfrow = c(3,2))
# Start plots
plot(xcenter,press,col="blue",type="l",pch=20,lwd=3, xlab="x",
     ylab="pressure")#,ylim=c(0,500))
plot(xcenter,u,col="blue",type="l",pch=20,lwd=3, xlab="x",
     ylab="velocity")
plot(xcenter,rho,col="blue",type="l",pch=20,lwd=3, xlab="x",
     ylab="density")#,ylim=c(0,7))
plot(xcenter,entropy,col="blue",type="l",pch=20,lwd=3, xlab="x",
     ylab="entropy")#,ylim=c(2,7))
plot(xcenter,mach,col="blue",type="l",pch=20,lwd=3, xlab="x",
     ylab="Mach number")#,ylim=c(0,2.5))
```

```r
plot(xcenter,eT,col="blue",type="l",pch=20,lwd=3, xlab="x",
    ylab="total energy")#,ylim=c(0,1200))
#
if(testcase == "Shocktube problem of G.A. Sod"){
  source("SodProbPostSimCalcs.R")
}else if(testcase == "Woodward-Colella blast wave problem"){
  source("Woodward-ColellaPostSimCalcs.R")
}else if(testcase == "Taylor-Sedov blast wave problem 1"){
  source("SedovTaylorPostSimCalcs.R")
}

# Elapsed time
totalTime<-proc.time()-ptm
cat(sprintf("\nTotal elapsed time: %6.2f sec.",totalTime[3]))
```

Listing 6.16. File: Euler_main.R—Code for the Euler *main* program. This program calls functions Euler_initial() and Euler_deriv(), which are detailed in Listings 6.17 and 6.18

6.C.2 Initialization

The in-line R code of Listing 6.17 is where the Euler equation simulations are initialized. The value of variable initEuler determines which test case will be used to initialize the state variable starting values over the entire spatial domain. Other variables are also initialized here for subsequent use in the *main* program of Listing 6.16.

```r
# File: Euler_initial.R
# Specification of Riemann problems for the Euler eqns.
# Set initEuler as necessary
initEuler <- 1
#
if(initEuler == 1){
  testcase <- "Shocktube problem of G.A. Sod"; #, JCP 27:1, 1978
  pleft <- 1.0; pright <- 0.1; rholeft <- 1.0; rhoright <- 0.125; Length <- 1;
  uleft <- 0; uright <- 0; tend <- 0.15; lambda <- 0.5; # lambda <- dt/dx
}else if (initEuler == 2){
  testcase <- "Woodward-Colella blast wave problem";
  # NOTE: Needs reflective boundary conditions!
  # There is no analytical solution for this problem and a hi-res solution
  # based on 4000 cells is generally taken to be a reference solution.
  pleft <- 1000.0; pmid <- 0.01; pright <- 100.0;
  rholeft <- 1.0; rhomid <- 1.0; rhoright <- 1.0; Length <- 1;
  uleft <- 0.0; umid <- 0.0; uright <- 0.0; tend <- 0.038; #tend <-0.038;
  lambda <- 0.004; # lambda <- dt/dx;
  # WENO 3rd order, TVD RK, N=500 gives good results
  # N=4000 gives very good results! (Long run!)
```

```
}else if (initEuler == 3){
  testcase <- "Sedov-Taylor blast wave problem 1"; # 2-cell pulse
  rholeft <- 1.225; rhomid <- 1.225; rhoright <- 1.225; Length <- 200;
  pleft <- 1e5; pmid <- 1e9; pright <- pleft; E0 <- 2*pmid*Length/N/(gamma-1)
  uleft <- 0.0; umid <- 0.0; uright <- 0.0; tend <- 0.015 #tend<-0.015, L<-200
  lambda <- 0.00001; # dt/dx; # UNITS: eT [J/kg], E0 [J/m^nu] see book text
  # WENO5 (175 sec) or linear extrap and Koren Limiter with TVD RK gives good
      results.
  # Parabolic reconstruction does not give acceptable results.
}
```

Listing 6.17. File: Euler_initial.R—Code for the initialization of program data and called by the *main* program detailed in Listing 6.16

6.C.3 The Derivative Function

The Euler_deriv() function of Listing 6.18 calculates the spatial derivatives of the conservative Euler equations $\partial u/\partial t = -\partial f/\partial x$, by calling the appropriate reconstruction method to obtain left and right edge values $U_1^L, U_1^R, U_2^L, U_2^R$ and U_3^L, U_3^R (see Section 6.6.4). This is followed by some housekeeping thermodynamic calculations and three calls to MUSCL_KTEuler() to obtain $\partial f/\partial x$, which is negated and returned as the time derivative to the numerical integrator.

Note that the Woodward–Colella problem requires *reflective* BCs, and this is implemented by first checking the testcase and then assigning the first and last grid numbers, as shown.

```
if(testcase == "Woodward-Colella blast wave problem"){
  # Reflective BC's
  rhostar[1] <- rhostar[2];rhostar[m] <- rhostar[(m-1)]
  mstar[1]   <- -mstar[2]; mstar[m]   <- -mstar[(m-1)]
  eTstar[1]  <- eTstar[2]; eTstar[m]  <- eTstar[(m-1)]
}
```

```
# File: Euler_deriv.R
# Euler derivatives with left and right reconsruction. Called by Euler_main()
Euler_deriv <- function(U0, Beta, dx, limiterType, reconType, m){
  ncall <<- ncall+1
  a    <- rep(0,m)
  flux <- rep(0,m)
  dfdx <- rep(0,m)
# Extract state variables
  rhostar  <- U0[1:m]
```

```
    mstar    <- U0[(m+1):(2*m)]
    eTstar   <- U0[(2*m+1):(3*m)]
  if(testcase == "Woodward-Colella blast wave problem"){
    # Reflective BC's
    rhostar[1] <- rhostar[2]; rhostar[m] <- rhostar[(m-1)]
    mstar[1]   <- -mstar[2];  mstar[m]   <- -mstar[(m-1)]
    eTstar[1]  <- eTstar[2];  eTstar[m]  <- eTstar[(m-1)]
  }else{
    # Dirichlet BC's
    rhostar[1] <- rholeft;       rhostar[m] <- rhoright
    mstar[1]   <- rholeft*uleft; mstar[m]   <- rhoright*uright
    eTstar[1]  <- eTleft;        eTstar[m]  <- eTright
  }
  if(reconType == 1){
    ULU1R <- linearExtrapolation(rhostar,limiterType, Beta) # Left and right
    U1L   <- ULU1R[1:m]                        # Linear extrapolated
    U1R   <- ULU1R[(m+1):(2*m)]                       # states
    #
    ULU2R <- linearExtrapolation(mstar,limiterType, Beta)
    U2L   <- ULU2R[1:m]
    U2R   <- ULU2R[(m+1):(2*m)]
    #
    ULU3R <- linearExtrapolation(eTstar,limiterType, Beta)
    U3L   <- ULU3R[1:m]
    U3R   <- ULU3R[(m+1):(2*m)]
    #
  }else if(reconType == 2){
    ULU1R <- parabolicReconstruction2(rhostar,limiterType, Beta) # Left and right
    U1L   <- ULU1R[1:m]                        # parabolic interpolated
    U1R   <- ULU1R[(m+1):(2*m)]                       # states
    #
    ULU2R <- parabolicReconstruction2(mstar,limiterType, Beta)
    U2L   <- ULU2R[1:m]
    U2R   <- ULU2R[(m+1):(2*m)]
    #
    ULU3R <- parabolicReconstruction2(eTstar,limiterType, Beta)
    U3L   <- ULU3R[1:m]
    U3R   <- ULU3R[(m+1):(2*m)]
    #
  }else if(reconType == 3){
    ULU1R <- WENO3rdOrdReconstructionKL(rhostar) # Left and right
    U1L   <- ULU1R[1:m]             # 3rd Order CWENO reconstructed
    U1R   <- ULU1R[(m+1):(2*m)]              # states
    #
    ULU2R <- WENO3rdOrdReconstructionKL(mstar)
    U2L   <- ULU2R[1:m]
```

```
  U2R    <- ULU2R[(m+1):(2*m)]
  #
  ULU3R <- WENO3rdOrdReconstructionKL(eTstar)
  U3L    <- ULU3R[1:m]
  U3R    <- ULU3R[(m+1):(2*m)]
  #
}else if(reconType == 4){
  ULU1R <- WENO5thOrdReconstruction(rhostar) # Left and right
  U1L    <- ULU1R[1:m]           # 5th Order WENO Reconstruction
  U1R    <- ULU1R[(m+1):(2*m)]              # states
  #
  ULU2R <- WENO5thOrdReconstruction(mstar)
  U2L    <- ULU2R[1:m]
  U2R    <- ULU2R[(m+1):(2*m)]
  #
  ULU3R <- WENO5thOrdReconstruction(eTstar)
  U3L    <- ULU3R[1:m]
  U3R    <- ULU3R[(m+1):(2*m)]
  #
}
  EL <- U3L/U1L # specific internal energy
  ER <- U3R/U1R
  uL <- U2L/U1L # velocity
  uR <- U2R/U1R
  tmp <- pmax(1.e-6,(EL - 0.5*uL^2))
  cL <- sqrt(gamma*(gamma-1)*tmp) # sonic speed
  cR <- sqrt(gamma*(gamma-1)*tmp)
  pL <- (gamma-1)*U1L*tmp # pressure
  pR <- (gamma-1)*U1R*tmp
  HL <- EL + pL/U1L # specific enthalpy
  HR <- ER + pR/U1R
# MUSCL fluxes
  dFlux1 <- MUSCL_KTEULER(cL, cR, uL, uR, U1L, U1R, U2L, U2R, U3L, U3R,
                          pL, pR, HL, HR, 1)
  dFlux2 <- MUSCL_KTEULER(cL, cR, uL, uR, U1L, U1R, U2L, U2R, U3L, U3R,
                          pL, pR, HL, HR, 2)
  dFlux3 <- MUSCL_KTEULER(cL, cR, uL, uR, U1L, U1R, U2L, U2R, U3L, U3R,
                          pL, pR, HL, HR, 3)
  dfdx <- c(dFlux1,dFlux2,dFlux3)/dx
  # print(dfdx)
  return( - dfdx )
}
```

Listing 6.18. File: Euler_deriv.R—Code for the derivative function Euler_deriv(). This function is called by the selected numerical integrator

6.C.4 The MUSCL Function

The MUSCL_KTEuler() function uses the left and right edge vales, $U1_L, U1_R, U2_L, U2_R$ and $U3_L, U3_R$, to calculate MUSCL fluxes based on the Kurganov and Tadmore algorithm. These are then used to calculate an approximation to the corresponding three spatial derivatives $\partial f_{1,2,3}/\partial x$ for the test case being simulated. These spatial derivatives are then passed back to the derivative function derivEuler(), which calculates the appropriate time derivatives for use by the numerical integrator.

```r
# File: MUSCL_KTEuler.R
# R Code for 'Kurganov and Tadmore Semi-Discrete' method,
# for solving the Euler PDE's
# This routine calculates the spatial derivative of f(U), wrt "x"
MUSCL_KTEULER <- function(cL, cR, uL, uR, U1L, U1R, U2L, U2R,
                U3L, U3R, pL, pR, HL, HR, fluxNum ){
  N  <- length(U1L)
  aL <- rep(0,N)
  aR <- rep(0,N)
  df <- rep(0,N)
  sourceTerm <- 0 # For sedov-Taylor blast problem
  nu         <- 0 # 0=planar, 1=cylindrical, 2=spherical
  if(fluxNum == 1){
      fL   <- U2L
      fR   <- U2R
      dU   <- U1R-U1L  # d(rho)
      dU[1] <- 0.0
      if(sourceTerm == 1){
        S <- (nu/xcenterAbs)*(fL+fR)/2 # nu=2 => 3-dimensional problem
      }
  }else if(fluxNum == 2){
      fL   <- U2L*uL + pL
      fR   <- U2R*uR + pR
      dU   <- U2R-U2L  # d(m)
      dU[1] <- 0.0
      if(sourceTerm == 1){
        S <- (nu/xcenterAbs)*(fL+fR-(pL+pR))/2
      }
  }else{
      fL   <- U2L*HL
      fR   <- U2R*HR
      dU   <- U3R-U3L  # d(rho*E)
      dU[1] <- 0.0
      if(sourceTerm == 1){
        S <- (nu/xcenterAbs)*(fL+fR)/2
      }
  }
```

```r
  # Calculate the characteristic speed
  r <- 3 # zones of influence upwind (increasing r
         # allows more cells to be taken into account)
  m <- r - 1
  for(j in r:(N-r)){
    for(i in -m:m){
      # aL[j] <- max(aL[j], cL[j] + U1L[j], cL[(j-i)] + U1L[(j-i)])
      aL[j] <- max(aL[j], cL[j+i])
    }
  }
  for(j in r:(N-r)){
    for(i in -m:m){
      # aR[j] <- max(aR[j], cR[j] + U1R[j], cR[(j-i)] + U1R[(j-i)])
      aR[j] <- max(aR[j], cR[j+i])
    }
  }
  a <- pmax(aL, aR)
  flux <- 0.5 * ( fL + fR ) - 0.5*a*dU
  
  df[2:(N-1)] <- flux[2:(N-1)] - flux[1:(N-2)]
  df[1] <- df[2]; df[N] <- df[(N-1)]
  
  if( sourceTerm == 1){
    df[2:(N-1)] <- df[2:(N-1)] + 0.5*(S[2:(N-1)]+S[1:(N-2)])*dx
  }
  return(df)
}
```

Listing 6.19. File: MUSCL_KTEuler.R—Code for the MUSCL function MUSCL_KTEuler(). This function implements the Kurganov and Tadmore semi-discrete scheme, and is called by derivEuler()

6.C.5 Postsimulation Calculations

The postsimulation calculations are performed for comparison purposes and allow reference solutions to be overlaid on plots of the simulation results. In the case of the *Sod* and *Taylor–Sedov* examples, the reference solutions are exact, whereas in the case of the *Woodward–Colella interacting blast wave* example, the reference solution is obtained by use of a very dense spatial grid.

6.C.5.1 Sod's Shock Tube

```r
# File: Sod_PostSimCalcs.R
# Calculate exact solution according to the Rankine-Hugoniot jump conditions
source("Sod_RiemannSoln.R")
par(mfrow = c(3,2))
# Start points
```

```
plot(xx,pexact,col="red",type="l",lwd=1, xlab="x",ylab="pressure")
points(xcenter,press,col="blue",pch=20,lwd=1, xlab="x",ylab="pressure")
plot(xx,uexact,col="red",type="l",lwd=1, xlab="x",ylab="velocity")
points(xcenter,u,col="blue",pch=20,lwd=1, xlab="x",ylab="velocity")
plot(xx,rhoexact,col="red",type="l",lwd=1, xlab="x",ylab="density")
points(xcenter,rho,col="blue",pch=20,lwd=1, xlab="x",ylab="density")
plot(xx,entroexact,col="red",type="l",lwd=1, xlab="x",ylab="entropy")
points(xcenter,entropy,col="blue",pch=20,lwd=1, xlab="x",ylab="entropy")
plot(xx,machexact,col="red",type="l",xlab="x",ylab="Mach number")
points(xcenter,mach,col="blue",pch=20,lwd=1, xlab="x",ylab="Mach number")
plot(xx,eTexact,col="red",type="l",lwd=1, xlab="x",ylab="total energy")
points(xcenter,eT,col="blue",pch=20,lwd=1, xlab="x",ylab="total energy")
```

Listing 6.20. File: Sod_PostSimCalcs.R—Code to be run at the end of the *main* program of Listing 6.16 for the *Sod shock tube* case. It generates the plots shown in Figs. 6.18, 6.19, and 6.29

6.C.5.2 Woodward–Colella

```
# File: Woodward-ColellaPostSimCalcs.R
# Woodward-Colella post simulation calcs
load("WoodwardColella4k_x.data") # 4000-point data for comparison
# Start plots
par(mfrow = c(3,2))
plot(xcenter4k,press4k,col="red",type="l",lwd=1, xlab="x",ylab="pressure")
points(xcenter,press,col="blue",pch=20,lwd=3, xlab="x",ylab="pressure")
plot(xcenter4k,u4k,col="red",type="l",lwd=1, xlab="x",ylab="velocity")
points(xcenter,u,col="blue",pch=20,lwd=3, xlab="x",ylab="velocity")
plot(xcenter4k,rho4k,col="red",type="l",lwd=1, xlab="x",ylab="density")
points(xcenter,rho,col="blue",pch=20,lwd=3, xlab="x",ylab="density")
plot(xcenter4k,entropy4k,col="red",type="l",lwd=1, xlab="x",ylab="entropy")
points(xcenter,entropy,col="blue",pch=20,lwd=3, xlab="x",ylab="entropy")
plot(xcenter4k,mach4k,col="red",type="l",lwd=1, xlab="x",ylab="Mach number")
points(xcenter,mach,col="blue",pch=20,lwd=3, xlab="x",ylab="Mach number")
plot(xcenter4k,eT4k,col="red",type="l",lwd=1, xlab="x",ylab="total energy")
points(xcenter,eT,col="blue",pch=20,lwd=3, xlab="x",ylab="total energy")
# x-t plot
xtPlt <- 1
if(xtPlt == 1){
  rho2   <- t(U[seq(1,n,by=10),1:N]) # sub-sample time
  rho2log <- log(rho2)
  t2 <-seq(0,0.4,length.out=ncol(rho2log))
  par(mfrow = c(1,2))
  contour(xcenter,t2,rho2log,nlevels=20,
          xlab="x",ylab="time")
```

```
contour(xcenter,t2,rho2log,nlevels=20,
        xlab="x",ylab="time",
        xlim=c(0.6,0.84), ylim=c(0.26,0.36))
}
```

Listing 6.21. File: `Woodward-ColellaPostSimCalcs.R`—Code to be run at the end of the *main* program of Listing 6.16 for the *Woodward-Colella interacting blast wave* case. It generates the plots shown in Figs. 6.20, 6.22, and 6.30

6.C.5.3 Taylor–Sedov

```
# File: SedovTaylorPostSimCalcs.R
# Taylor-Sedov post simulation calcs
# Ref: Sedov-59, p212-4 and p222. Note: gamma = 1.4
N2    <- ceiling(N/2)
rho1  <- rholeft # density at ambient conditions
S_pl  <- 0.9756 # = alpha_planar^(-1/nu+2)=1.077^(-1/3)
r2    <- (S_pl*E0/rho1)^(1/3)*tend^(2/3) # Position of shock wave, p213
rho2  <- rho1*(gamma+1)/(gamma-1)       # Density at r2,       p212
c2    <- (2/3)*r2/tend                   # Speed of sound at r2, p214
v2    <- (2/(gamma+1))*c2                # Shock velocity at r2, p212
p2    <- (2/(gamma+1))*rho1*c2^2         # Pressure at r2,      p212
# load nu=1 data adapted from Sedov-59, p222
load("SedovTaylorPlanar_39pts.data")
# Note: rxr2z is ratio r/r2 from Sedov table, p222
#       pxp2z is ratio p/p2 from Sedov table, p222
#       vxv2z is ratio v/v2 from Sedov table, p222
N <- length(rxr2z)
rxr2zr2 <- rxr2z*r2
dr <- rxr2zr2[1]-rxr2zr2[2]
# Prepare data adding an extra 2 points at each end to aid plotting
rx <- c(-Length,rxr2zr2[1]-dr,rxr2zr2,rxr2zr2[N]+dr,Length)
px <- c(pleft,pleft,pxp2z*p2,pright,pright)
vx <- c(uleft,uleft,vxv2z*v2,uright,uright)
rhox <- c(rholeft,rholeft,rhoxrho2z*rho2,rhoright,rhoright)
# Start plots
par(mfrow = c(1,3))
plot(rx,px,col="red",type="l",lwd=1,
     xlim=Length*c(-1,1)/2,ylim=c(0,1.5e7),
     xlab="x",ylab="pressure")
points(xcenter,press,col="blue",pch=20,lwd=1)
plot(rx,vx,col="red",type="l",lwd=1,
     xlim=Length*c(-1,1)/2,ylim=c(-3500,3500),
     xlab="x",ylab="velocity")
points(xcenter,u,col="blue",pch=20,lwd=1)
```

```
plot(rx,rhox,col="red",type="l",lwd=1,
    xlim=Length*c(-1,1)/2,
    xlab="x",ylab="density")
points(xcenter,rho,col="blue",pch=20,lwd=1)
```

Listing 6.22. File: `SedovTaylorPostSimCalcs.R`—Code to be run at the end of the *main* program of Listing 6.16 for the *Taylor–Sedov* case. It generates the plots shown in Figs. 6.24, 6.25, and 6.32

REFERENCES

[Abo-08] Aboiyar, T. (2008), *Non-Oscillatory Finite Volume Methods for Conservation Laws on Unstructured Grids*, D.Phil. thesis, University of Leicester.

[Bal-00] Balsara, D. and C. W. Shu (2000), Monotonicity Preserving Weighted Essentially Non-oscillatory Schemes with Increasingly High Order of Accuracy, *Journal of Computational Physics* **160**, 405–452.

[Buc-42] Buckley, S. E. and M. C. Leverett (1942), Mechanism of Fluid Displacement in Sands, *Transactions of the AIME* **146**, 187–196.

[Cha-83] Chatkravathy, S. R. and S. Osher (1983), High Resolution Applications of the Osher Upwind Scheme for the Euler Equations, *AIAA Paper 83-1943, Proc. AIAA 6th Comutational Fluid Dynamics Conference*, 363–373.

[Dav-85] Davies-Jones, R. (1985), Comments on "A kinematic analysis of frontogenesis associated with a nondivergent vortex," *Journal of Atmopsheric Sciences* **42**-19, 2073–2075. Retrieved from http://www.flame.org/~cdoswell/publications/Davies-Jones_85.pdf.

[Dos-84] Doswell, C. A. I. (1984), A Kinematic Analysis of Frontogenesis Associated with a Nondiverent Vortex, *Journal of Atmospheric Sciences* **41**-7, 1242–1248.

[Dri-05] Drikakis, D. and W. Rider (2005), *High-Resolution Methods for Incompressible and Low-Speed Flows*, Springer.

[Gas-88] Gaskell, P. H. and A. K. C. Lau (1988), Curvature-Compensated Convective Transport: SMART, a New Boundedness-Preserving Transport Algorithm, *International Journal for Numerical Methods in Fluids* **8**, 617.

[God-54] Godunov, S. K. (1954), *Different Methods for Shock Waves*, PhD dissertation, Moscow State University.

[God-59] Godunov, S. K. (1959), A Difference Scheme for Numerical Solution of Discontinuous Solution of Hydrodynamic Equations, *Math. Sbornik* **47**, 271–306, translated US Joint Publ. Res. Service, JPRS 7226, 1969.

[Got-98] Gottlieb, S. and C. W. Shu (1998), Total Variation Diminishing Runge-Kutta Schemes, *Mathematics of Computation* **67**-221, 73–86.

[Har-83] Harten, A. (1983), High Resolution Schemes for Hyperbolic Conservation Laws, *Journal of Computational Physics* **49**, 357–393.

[Har-87] Harten, A. and S. Osher (1987), Uniformly High-Order Accurate Nonoscillatory Schemes. I, *SIAM Journal on Numerical Analysis* **24**-2, 279–309.

[Hen-05] Henrick, A. K., T. D. Aslam and J. M. Powers (2005), Mapped Weighted Essentially Non-oscillatory Schemes: Achieving Optimal Order Near Critical Points, *Journal of Computational Physics* **207**, 542–567.

[Hir-90] Hirsch, C. (1990), *Numerical Computation of Internal and External Flows, Volume 2: Computational Methods for Inviscid and Viscous Flows*, John Wiley.

[Kal-11] Kalise, D. (2011), A Study of a WENO-TVD Finite Volume Scheme for the Numerical Simulation of Atmospheric Advective and Convective Phenomena, *arXiv preprint arXiv:1111.1712, 29*. Retrieved from http://arxiv.org/abs/1111.1712.

[Ker-03] Kermani, M. J., A. G. Gerber and J. M. Stockie (2003), Thermodynamically Based Moisture Prediction Using Roe's Scheme, *4th Conference of Iranian AeroSpace Society, Amir Kabir University of Technology, Tehran, Iran*, January 27–29.

[Kor-93] Koren, B. (1993), A Robust Upwind Discretization Method for Advection, Diffusion and Source Terms, in *Numerical Methods for Advection-Diffusion Problems*, C. B. Vreugdenhil and B. Koren editors, Vieweg, Braunschweig, 117.

[Kur-00a] Kurganov, A. and E. Tadmor (2000), New High-Resolution Central Schemes for Non-linear Conservation Laws and Convection-Diffusion Equations, *Journal of Computational Physics* **160**, 214–282. Available online at http://www.cscamm.umd.edu/centpack/publications/files/KT_semi-discrete.JCP00-centpack.pdf.

[Kur-00b] Kurganov, A. and D. Levy (2000), A Third-Order Semidiscrete Central Scheme for Conservation Laws and Convection-Diffusion Equations, *SIAM Journal of Scientific Computing* **22**, 1461–1488. Available online at http://www.cscamm.umd.edu/centpack/publications/files/Kur-Lev3rdsemidiscrete.SINUM00-centpack.pdf.

[Lan-98] Laney, C. B. (1998), *Computational Gas Dynamics*, Cambridge University Press.

[Lat-05] Latini, M. and O. Schilling (2005), *Weighted Essentially Non-Oscillatory Simulations and Modeling of Complex Hydrodynamic Flows. Part 1: Regular Shock Refraction*, University of California, Lawrence Livermore National Laboratory report under contract W-7405-Eng-48, February 3.

[Lax-54] Lax, P. D. (1954), Weak Solutions of Non-linear Hyperbolic Equations and Their Numerical Computation, *Communications on Pure and Applied Mathematics* **VII**, 159–193.

[Lev-02] Leveque, R. J. (2002), *Finite Volume Methods for Hyperbolic Problems*, Cambridge University Press.

[Lie-94] Lien, F. S. and M. A. Leschziner (1994), Upstream Monotonic Interpolation for Scalar Transport with Application to Complex Turbulent Flows, *International Journal in Numerical Methods in Fluids* **19**, 527.

[Liu-03] Liu, J.-G. and C. Wang (2003), Positivity Property of Second-Order Flux-Splitting Schemes for the Compressible Euler Equations, *Discrete and Continuous Dynamical Systems, Series B* **3**-2, 201–228.

[Luo-13] Luo, J. and K. Xu (2013), A high-Order Multidimensional Gas-Kinetic Scheme for Hydrodynamic Equations, *Science China—Technological Sciences* **56**-10, 2370–2384.

[Nes-90] Nessyahu, H. and E. Tadmor (1990), Non-oscillatory Central Differencing for Hyperbolic Conservation Laws, *Journal of Computational Physics* **87**, 408–463. Available online at http://www.cscamm.umd.edu/centpack/publications/files/NT2.JCP90-centpack.pdf.

[Roe-86] Roe, P. L. (1986), Characteristic-Based Schemes for the Euler Equations, *Annual Review of Fluid Mechanics* **18**, 337.

[Rus-61] Rusanov, V. V. (1961), Calculation of Intersection of Non-Steady Shock Waves with Obstacles, *Journal of Computational Mathematics and Physics USSR* **1**, 267–279.

[Sed-59] Sedov, L. I. (1959), *Similarity and Dimensional Methods in Mechanics*, Academic Press.

[Shu-97] Shu, C. W. (1997), *Essentailly Non-Oscillatory and Weighted Essentially Non-Oscillatory Schemes for Hyperbolic Conservation Laws*, NASA/CR-97-206253, ICASE report 97-65, Hampton, VA.

[Shu-98] Shu, C. W. (1998), Essentially Non-oscillatory and Weighted Essential Non-oscillatory Schemes for Hyperbolic Conservation Laws, in *Advanced Numerical Approximation of Nonlinear Hyperbolic Equations*, Lecture Notes in Mathematics, vol. 1697. B. Cockburn, C. Johnson, C.-W. Shu, and E. Tadmor, editors, Springer, pp. 325–432.

[Shu-03] Shu, C. W. (2003), High-Order Finite Difference and Finite Volume WENO Schemes and Discontinuous Galerkin Methods for CFD, *International Journal of Computational Fluid Dynamics* **17**-2, 107–118.

[Sod-78] Sod, G. A. (1978), A Numerical Study of a Converging Cylindrical Shock, *Journal of Fluid Mechanics* **83**, 785–794.

[Swe-84]   Sweby, P. K. (1984), High Resolution Schemes Using Flux-Limiters for Hyperbolic Conservation Laws, *SIAM Journal of Numerical Analysis* **21**, 995–1011.
[Tan-97]   Tannehill, J. C., D. A. Anderson and R. H. Pletcher (1997), *Computational Fluid Mechanics and Heat Transfer*, 2nd ed., Taylor and Francis.
[Tit-03]   Titarev, V. and E. F. Toro (2003), *On the Use of TVD Fluxes in ENO and WENO Schemes*, Technical Report UTM 635, Matematica, University of Trento, January.
[Tor-99]   Toro, E. F. (1999), *Riemann Solvers and Numerical Methods for Fluid Dynamics*, Springer.
[vanA-82] van Albada, G. D., B. Van Leer and W. W. Roberts (1982), A Comparative Study of Computational Methods in Cosmic Gas Dynamics, *Astronomy and Astrophysics* **108**, 76.
[vanL-74] van Leer, B. (1974), Towards the Ultimate Conservative Difference Scheme II. Monotonicity and Conservation Combined in a Second Order Scheme, *Journal of Computational Physics* **14**, 361–370.
[vanL-77] van Leer, B. (1977), Towards the Ultimate Conservative Difference Scheme III. Upstream-Centered Finite-Difference Schemes for Ideal Compressible Flow, *Journal of Computational Physics* **23**, 263–275.
[vanL-79] van Leer, B. (1979), Towards the Ultimate Conservative Difference Scheme V, *Journal of Computational Physics* **32**, 101.
[Wat-95]   Waterson, N. P. and H. Deconinck (1995), A Unified Approach to the Design and Application of Bounded Higher-Order Convection Schemes, *VKI Preprint* **21**.
[Wes-01]   Wesseling, P. (2001), *Principles of Computational Fluid Dynamics*, Springer.
[Woo-82]   Woodward, P. (1982), Trade-Offs in Designing Explicit Hydrodynamics Schemes for Vector Computers, in *Parallel Computation*, G. Rodrigue, editor, Academic Press, pp. 153–171.
[Woo-84]   Woodward, P. and P. Colella (1984), The Numerical Simulation of Two-dimensional Fluid Flow with Strong Shocks, *Journal of Computational Physics* **173**, 115–173.
[Xin-13]   Xing, Y. and C.-W. Shu (2013), High Order Well-Balanced WENO Scheme for the Gas Dynamics Equations under Gravitational Fields, *Journal of Scientific Computing* **54**-2/3, 645–662.
[Zho-95]   Zhou, G. (1995), *Numerical Simulations of Physical Discontinuities in Single and Multi-fluid Flows for Arbitrary Mach Numbers*, PhD thesis, Chalmers University of Technology.

7

Meshless Methods

7.1 INTRODUCTION

In this chapter we provide an introduction to *meshless methods*. It will, of necessity, be brief as this is a very wide subject that is still developing and that has many application areas. Here the main focus of our discussion will be *interpolation of scattered data* and the numerical solution of partial differential equations (PDEs) using *radial basis functions* (RBFs). A *basis* set is a set of elements that are used to describe a particular system of interest. The elements referred to here are a set of *linearly independent basis functions* (with *radial symmetry* or that are *radially invariant*) that apply to a *function space*. The field of radial basis functions (RBFs) was initiated by Hardy [Har-71, Har-90], who proposed the now widely used multiquadratic RBF for use in a multidimensional scattered interpolation method.

Every continuous function can generally be represented as a linear combination of *basis functions* (also called *blending functions*). In other words, if we wish to describe a function mathematically over a continuous domain, we can form a representation from a weighted sum of appropriate basis functions. For example, a representation of a periodic function $f(t)$ can be formed using a weighted sum of sines and cosines. This is called a *Fourier series*,[1] and the set of Fourier series functions has $\{\sin(n\omega t) \cup \cos(n\omega t) : n \in \mathbb{Z}\}$ as its basis, where ω is the *fundamental frequency* of the process represented by $f(t)$ and $n\omega$ the nth *harmonic*. This definition of the basis is applicable because each Fourier series can be formed from a linear combination of these basis functions.

For problems having complex solutions it may be advantageous to restrict the influence of distant RBFs. Consequently, the class of *compactly supported RBFs* has been developed whereby their value is truncated at a certain distance from the associated collocation point (see Section 7.2.2). As problems increase in size, the computation can be made more efficient by employing *local RBF* methods such that the solution at a particular collocation point, say, x_i, is approximated by a weighted set of local RBFs which are close neighbors, rather than by all the RBFs within the solution domain (see Section 7.5).

---

[1] The *Fourier series* was discovered by French mathematician and physicist Joseph Fourier (1768–1830) during his investigations into the flow of heat.

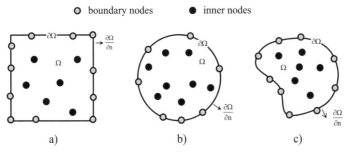

Figure 7.1. Examples of different 2D domains with *scattered data* for (a) rectangular domain; (b) circular domain; and (c) irregular domain.

In general the meshless method is very effective for appropriate problems and may be applied to a wide range of applications, including scattered data interpolation and the solution of partial differential equations. It does not restrict the domain shape, which means that the shape of the domain can be chosen to simplify the process of obtaining a solution for a particular application (see Fig. 7.1).

Therefore, because it does not rely on any domain-based constraints, the domain shape is not dictated by the the numerical solution method. Consequently, it does not suffer from coordinate distortions or singularities. In addition, it can handle large problems in multiple dimensions and can be efficiently implemented on a solution domain containing thousands of collocation points. These attributes are clearly illustrated in the paper by Fusilier and Wright [Fus-13].

Time-dependent PDEs that are purely *advective* can cause difficulty for RBF methods as their differentiation matrices may have eigenvalues with positive real parts, which will result in an unstable solution. However, for PDEs containing *diffusive* terms, stability can usually be assured by choosing a suitable integration algorithm along with an appropriate time step. Thus, RBF methods can be used effectively to solve *advection-diffusion* problems (see [Pla-06], [Sar-08], and [Sar-12] for further discussion).

7.2 RADIAL BASIS FUNCTIONS (RBF)

Radial basis functions are employed in methods that seek to approximate functions that are not normally known in advance. RBFs are usually finite linear combinations of translates of radially symmetric basis functions, say, $\varphi(\|\cdot\|)$, where $\|\cdot\|$ represents the *Euclidean norm* [Buh-03]. Radial symmetry means that the function only depends upon the Euclidean distance of the argument of φ from the origin, and any rotations do not affect the function value. The field of radial basis functions (RBFs) was instigated by Hardy (1971, 1990), who proposed the now widely used multiquadratic RBF for use in a multidimensional scattered interpolation method.

The simplest form for an RBF is $\varphi(r) = r$ where $r = \|x\|$ and $x \in \mathbb{R}^s, s \in (1,2,3)$. One of the most common in use is the *multiquadratic* RBF, $\varphi(r) = \sqrt{(\epsilon r)^2 + 1}$, where ϵ represents a *shape parameter* that is used to control the overall shape of the RBF, discussed subsequently. There are many RBFs in common use and examples are given in Appendix 7.C at the end of this chapter.

Linear combinations of RBFs are used to form *trial functions* that are not fixed in advance but, rather, are data dependent. These trial functions can be used to approximate scattered data in N-dimensional space. They are very efficient for high-dimensional reconstruction problems as they are multivariate but reduce to a scalar function of their argument (see Section 7.3).

Our discussions will be confined to reconstruction of *real* functions, but the methods are equally applicable to the reconstruction of *complex* functions.

7.2.1 Positive Definite RBFs

Meshless methods usually require the solution of set of linear equations of the form $Ac = f$, where the entries of A are of the form given in eqn. (7.5). Thus, it is necessary for matrix A to be *nonsingular*, that is, $\det(A) \neq 0$. We will also refer to the matrix A as a RBF. Although this is not strictly correct, it conforms with usual practice and, hopefully, the context will make clear to which we refer.

An important class of matrices are the so-called *positive definite* matrices, which have positive eigenvalues and are therefore nonsingular. It follows, therefore, that if matrix A above is positive definite, a solution will exist to eqn. (7.5). Readers are referred to [Gan-60, Gan-77, Gol-96] for general discussion on positive definite matrices and to [Fas-07] for discussion on positive definite matrices related to RBFs and their application.

Clearly, for meshless methods, it is advantageous to employ radial functions that give rise to positive definite matrices. The individual functions that make up RBFs are usually referred to as *positive definite radial functions*, and an example of such a function is the *multiquadratic* function $\varphi(r) = \sqrt{\epsilon^2 r^2 + 1}$. This and other positive definite RBFs are described in Appendix 7.C at the end of this chapter, along with some of their associated derivatives.

7.2.2 RBF with Compact Support (CSRBF)

We define the *support* of a function as the set of points where the function is not zero-valued, and define a function with *compact support* as one which has *bounded* support. Thus, a function such as $f(x) = \frac{x}{1+x^2}$ vanishes at infinity, but its support is not compact; rather, its support is *global*.

For a RBF, say, $\varphi(r)$, to have *compact support*, it must be greater than or equal to zero for all values of r, where r must be bounded. This requires that for a RBF, that would become negative for some value of r, to have compact support, it must be truncated at this value. Having compact support automatically ensures that an RBF is strictly positive definite and therefore the matrix A in eqn. (7.5) is invertible.

It can be shown that RBFs with compact support are not valid for all dimensions [Buh-03, Chapter 6], [Fas-07, Chapters 3 and 11]. Therefore, they are commonly represented as

$$\varphi_{s,k}(r), \tag{7.1}$$

where the symbol s represents the maximum applicable dimension and $2k$ the smoothness, C^{2k}. Thus, $\varphi_{3,2}$ means that (1) φ is strictly positive definite in $\mathbb{R}^3$ and therefore can be used on 1D, 2D, and 3D domains without loss of compact support and that (2) it is a

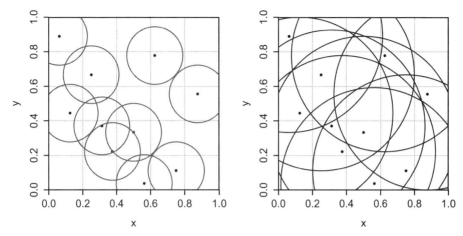

Figure 7.2. Examples of a domain with 10 Halton centers (left) Small overlapping circles which indicates that the CSRBF has small compact support. This situation results in a discontinuous reconstruction function that will not provide accurate interpolation (right) Large overlapping circles which will result in increased interpolation accuracy, albeit at the expense of a more computationally demanding dense reconstruction function. For practical situations, compact support must be chosen to provide a trade-off between accuracy and computational load.

continuous function with continuous derivatives up to and including the fourth derivative. An example of a CSRBF is the Wendland(3,2) RBF:

$$\varphi_{3,2}(r) = (1 - \epsilon r)_+^6 (35\epsilon^2 r^2 + 18\epsilon r + 3), \tag{7.2}$$

where $1/\epsilon$ represents the *support radius* and the subscript $+$ sign indicates that the associated bracketed term is a truncated power function, called the *cutoff function*, which is truncated at zero, that is,

$$(1 - \epsilon r)_+^6 = \begin{cases} (1 - \epsilon r)^6, & \epsilon r < 1, \\ 0, & \text{otherwise}. \end{cases} \tag{7.3}$$

Additional examples of Wendland CSRBFs are given in Appendix 7.C at the end of this chapter.

Another advantage of using CSRBFs is that for large problems the associated matrix operations can be more efficient. This is because the resulting A matrix and corresponding differentiation matrices $A_\mathcal{L}$ of Section 7.6 become sparse (usually banded), which means that sparse matrix techniques can be used to reduce the computational load. However, care has to be taken to set the correct support cutoff, as if support is too small, interpolation may be inaccurate, and if too large, the computational load may be too great. This situation is illustrated in Fig. 7.2. Also, some matrix operations destroy sparsity; for example, eqn. (7.52) would usually result in a dense matrix.

CSRBFs will not be discussed further here, and readers are referred to [Fas-07, Chapters 11 and 12] for additional information and example applications.

7.3 INTERPOLATION

Following Fasshauer [Fas-07], we assume that we can use a *radial basis function* expansion to interpolate a function $f(\mathbf{x})$ represented by scattered data as

$$\mathcal{P}_f(\mathbf{x}) = \sum_{k=1}^{N} c_k \varphi\left(\|\mathbf{x} - \mathbf{x}_N\|_2\right), \ \mathbf{x} \in \mathbb{R}^s. \tag{7.4}$$

Our discussions here apply to $s \in (1, 2, 3)$, see examples, although the ideas are readily extended to other dimensions. However, although meshless methods exhibit superior performance compared to other methods for two and higher dimensions, they rarely offer advantages over *Chebyshev polynomial* or *spline* interpolation of 1D data. In two dimensions we generally represent the components of $\mathbf{x}$ as (x, y) and in three dimensions as (x, y, z). The coefficients c_k are found by solving the following linear system:

$$\begin{bmatrix} \varphi(\|\mathbf{x}_1 - \mathbf{x}_1\|_2) & \varphi(\|\mathbf{x}_1 - \mathbf{x}_2\|_2) & \cdots & \varphi(\|\mathbf{x}_1 - \mathbf{x}_N\|_2) \\ \varphi(\|\mathbf{x}_2 - \mathbf{x}_1\|_2) & \varphi(\|\mathbf{x}_2 - \mathbf{x}_2\|_2) & \cdots & \varphi(\|\mathbf{x}_2 - \mathbf{x}_N\|_2) \\ \vdots & \vdots & \ddots & \vdots \\ \varphi(\|\mathbf{x}_N - \mathbf{x}_1\|_2) & \varphi(\|\mathbf{x}_N - \mathbf{x}_2\|_2) & \cdots & \varphi(\|\mathbf{x}_N - \mathbf{x}_N\|_2) \end{bmatrix} \begin{bmatrix} c_1 \\ c_2 \\ \vdots \\ c_N \end{bmatrix} = \begin{bmatrix} f(\mathbf{x}_1) \\ f(\mathbf{x}_2) \\ \vdots \\ f(\mathbf{x}_N) \end{bmatrix}, \tag{7.5}$$

which, as a matrix equation, we write $Ac = f$.

The terms $\varphi(\|\mathbf{x}_i - \mathbf{x}_j\|_2) = \varphi(r_{i,j})$ each represent a radial basis function, and r represents the Euclidean distance $r_{i,j} = \|\mathbf{x}_i - \mathbf{x}_j\|_2$.

Our objective is ultimately to solve

$$\begin{bmatrix} \varphi(\|\mathbf{x}_1^e - \mathbf{x}_1^c\|_2) & \varphi(\|\mathbf{x}_1^e - \mathbf{x}_2^c\|_2) & \cdots & \varphi(\|\mathbf{x}_1^e - \mathbf{x}_N^c\|_2) \\ \varphi(\|\mathbf{x}_2^e - \mathbf{x}_1^c\|_2) & \varphi(\|\mathbf{x}_2^e - \mathbf{x}_2^c\|_2) & \cdots & \varphi(\|\mathbf{x}_2^e - \mathbf{x}_N^c\|_2) \\ \vdots & \vdots & \ddots & \vdots \\ \varphi(\|\mathbf{x}_M^e - \mathbf{x}_1^c\|_2) & \varphi(\|\mathbf{x}_M^e - \mathbf{x}_2^c\|_2) & \cdots & \varphi(\|\mathbf{x}_M^e - \mathbf{x}_N^c\|_2) \end{bmatrix} \begin{bmatrix} c_1 \\ c_2 \\ \vdots \\ c_N \end{bmatrix} = \begin{bmatrix} \mathcal{P}_f(\mathbf{x}_1) \\ \mathcal{P}_f(\mathbf{x}_2) \\ \vdots \\ \mathcal{P}_f(\mathbf{x}_M) \end{bmatrix}, \tag{7.6}$$

which, as a matrix equation, we write $Bc = \mathcal{P}_f$. In this equation we have imposed the condition

$$\mathcal{P}_f(\mathbf{x}_i) = f(\mathbf{x}_i), \ i = 1, \ldots, M, \tag{7.7}$$

which represents the set of interpolation values that we wish to evaluate.

To determine $\mathcal{P}_f$ requires that we first solve the linear system $Ac = f$ of eqn. (7.5) for c, where the entries of A are given by

$$A_{i,j} = \varphi\left(\|\mathbf{x}_i - \mathbf{x}_j\|_2\right), \ i, j = 1, \ldots, N. \tag{7.8}$$

Here, $\mathbf{x}_i$ represent the *collocation points* and $\mathbf{x}_j$ represent the *centers* or *evaluation points*—these will generally be the same. Then we solve the linear system $Bc = \mathcal{P}_f$ for $\mathcal{P}_f$ using the vector c previously calculated. The entries of B are given by

$$B_{i,j} = \varphi\left(\left\|\mathbf{x}_i^e - \mathbf{x}_j^c\right\|\right), \ i = 1, \ldots, M, \ j = 1, \ldots, N. \tag{7.9}$$

Here, the collocation points $\mathbf{x}_i^e$ are the *estimation points* or *interpolation points* and $\mathbf{x}_j^c$ represent the *centers*. Note that the B matrix has N columns and M rows, that is, the same number of columns as the A matrix and one row for each estimation point. We now drop the use of bold characters to represent vectors, except where necessary to avoid confusion.

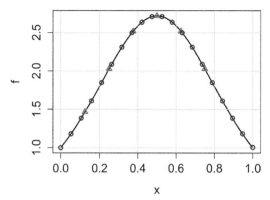

Figure 7.3. A 1D interpolation example showing the exact solution f(x) (line) with the estimated function points superimposed (circles) along with the Halton sequence centers (triangles).

The steps to obtaining an interpolated solution are as follows:

- Construct matrix A and vector f in eqn. (7.5) from centers x^c.
- Solve linear eqn. (7.5) for $c = [c_1, c_2, \ldots, c_N]^T$.
- Construct matrix $B_{i,j} = \varphi(\|x_i^e - x_j^c\|_2)$ in eqn. (7.6) from centers x^c and estimation points x^e.
- Finally, evaluate eqn. (7.6) to determine $\mathcal{P}_f$.

For further discussion on aspects relating to interpolation the reader is referred to [Fas-07]; and for discussion relating to the theoretical analysis of radial basis functions, the reader is referred to [Buh-03].

7.3.1 Interpolation Example: 1D

Consider a system of six scattered 1D data points represented by the following Halton sequence (discussed in Appendix 7.B at the end of this chapter) (see Fig. 7.3)

$$x = [0.500 \ 0.250 \ 0.750 \ 0.125 \ 0.625 \ 0.375]^T, \tag{7.10}$$

and the function to be interpolated is

$$f(x) = \exp(\sin(\pi x)). \tag{7.11}$$

Using the *multiquadratic* (MQ) RBF

$$\varphi(r) = \sqrt{(re)^2 + 1}, \tag{7.12}$$

eqn. (7.5) becomes

$$\begin{bmatrix} 1.000000 & 1.044031 & 1.044031 & 1.096586 & 1.011187 & 1.011187 \\ 1.044031 & 1.000000 & 1.166190 & 1.011187 & 1.096586 & 1.011187 \\ 1.044031 & 1.166190 & 1.000000 & 1.250000 & 1.011187 & 1.096586 \\ 1.096586 & 1.011187 & 1.250000 & 1.000000 & 1.166190 & 1.044031 \\ 1.011187 & 1.096586 & 1.011187 & 1.166190 & 1.000000 & 1.044031 \\ 1.011187 & 1.011187 & 1.096586 & 1.044031 & 1.044031 & 1.000000 \end{bmatrix} \begin{bmatrix} c_1 \\ c_2 \\ c_3 \\ c_4 \\ c_5 \\ c_6 \end{bmatrix} = \begin{bmatrix} 2.718282 \\ 2.028115 \\ 2.028115 \\ 1.466214 \\ 2.519044 \\ 2.519044 \end{bmatrix},$$

$$\tag{7.13}$$

where $r_{i,j} = x_i^c - x_j^c$, $\varphi_{i,j} = \phi(r_{i,j})$, $f_i = f(x_i^c)$, with $i, j = 1, \ldots, 4$. We have also used a shape parameter value of $e = 1.2$, which was arrived at by trial and error and is discussed subsequently.

Solving the relationship $Ac = f$ of eqn. (7.13), we obtain

$$c = [-4369.79 \; -737.372 \; -699.644 \; 11.2107 \; 2876.80 \; 2918.87]^T. \tag{7.14}$$

We are now in a position to interpolate the scattered data to estimate the value of a new set of data points. We choose the following 20 points at which to obtain estimated function values:

$$\begin{aligned}x^e = [\; & 0.00000000 \; 0.05263158 \; 0.10526316 \; 0.15789474 \; 0.21052632 \ldots \\ & 0.26315789 \; 0.31578947 \; 0.36842105 \; 0.42105263 \; 0.47368421 \ldots \\ & 0.52631579 \; 0.57894737 \; 0.63157895 \; 0.6842105 \; 0.73684211 \ldots \\ & 0.78947368 \; 0.84210526 \; 0.89473684 \; 0.94736842 \; 1.00000000 \;]^T\end{aligned} \tag{7.15}$$

(see Fig. 7.3), then the estimated function values are given by

$$\mathcal{P}_f(x_i^e) = Bc = \sum_{j=1}^{6} c_j \varphi(r_{i,j}^e), \quad i = [1, 2, \ldots, 20], \tag{7.16}$$

where $\varphi(r_{i,j}^e) = \varphi(x_i^e - x_j^c)$, c is the calculated column vector of eqn. (7.14) and B represents a 20×6 matrix from eqn (7.6). If more estimated data points are included in x_e, then the 20-row matrix would be extended by an extra row for each additional data point and $\mathcal{P}_f$ would become a column vector of length equal to the number of matrix rows.

Thus, calculating the final result from eqn. (7.16) (see Fig. 7.3), we obtain

$$\begin{aligned}\mathcal{P}_f = [\; & 1.000248 \; 1.179039 \; 1.383637 \; 1.609520 \; 1.848187 \ldots \\ & 2.086996 \; 2.309823 \; 2.498708 \; 2.636355 \; 2.709007 \ldots \\ & 2.709021 \; 2.636377 \; 2.498703 \; 2.309777 \; 2.086969 \ldots \\ & 1.848355 \; 1.610191 \; 1.385241 \; 1.182084 \; 1.005274 \;]^T.\end{aligned} \tag{7.17}$$

This compares with the exact values

$$\begin{aligned}f_{exact} = [\; & 1.000000 \; 1.178915 \; 1.383615 \; 1.609538 \; 1.848201 \ldots \\ & 2.086992 \; 2.309813 \; 2.498707 \; 2.636363 \; 2.709013 \ldots \\ & 2.709013 \; 2.636363 \; 2.498707 \; 2.309813 \; 2.086992 \ldots \\ & 1.848201 \; 1.609538 \; 1.383615 \; 1.178915 \; 1.000000 \;]^T,\end{aligned} \tag{7.18}$$

which corresponds to a maximum absolute error of $5.274235e-03$. This is a very good result achieved from six scattered data points, particularly as the values at the boundaries of the domain are extrapolations, where it is well known that meshless methods are least accurate. However, we must bear in mind that the function $f(x)$ used in this example is a simple well-behaved smooth function.

It is recommended that readers experiment with this example by obtaining estimates for $\mathcal{P}_f$ based on a variety of different functions and x^c, x^e values.

We now return to the value of $e = 1.2$ used for the shape parameter. Identifying the most suitable shape parameter value is a difficult area for meshless methods, particularly when applied to large real World problems, as there is currently no general method for obtaining the *best* value. In this example, where we are applying the method to a known function, we used a *trial-and-error* approach to find the best value for e. The results are

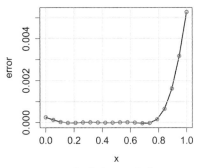 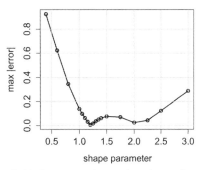

Figure 7.4. Results of 1D interpolation example. (left) Errors at the estimation points for the shape parameter value e = 1.2. (right) Maximum estimation errors for differently shape parameter values.

plotted in Fig. 7.4, from which it is clear that there are two minima—one at around $e = 1.2$ and one around $e = 2.0$. We chose to use the value of $e = 1.2$, which yields the smallest error. A more systematic approach to identifying the optimum shape parameter value is discussed in Section 7.5.2 for local RBFs.

The code for this example is included with downloads for this book but is omitted here as it is almost identical to the more complex example given in Section 7.3.3.

7.3.2 Interpolation Example: 2D

Consider a system of four scattered (x, y) data points represented by the following Halton sequence (discussed in Appendix 7.B at the end of this chapter) (see Fig. 7.5):

```
          x        y
[1,]  0.500  0.3333333
[2,]  0.250  0.6666667
[3,]  0.750  0.1111111
[4,]  0.125  0.4444444
```

and the function to be interpolated

$$f(x, y) = \sin(\pi x) \sin(\pi y). \tag{7.19}$$

Using the *multiquadratic* (MQ) RBF

$$\varphi(r) = \sqrt{(re)^2 + 1}, \tag{7.20}$$

eqn. (7.5) becomes

$$\begin{bmatrix} 1.000000 & 8.393119 & 6.764103 & 7.885954 \\ 8.393119 & 1.000000 & 14.981882 & 5.196449 \\ 6.764103 & 14.981882 & 1.000000 & 14.201917 \\ 7.885954 & 5.196449 & 14.201917 & 1.000000 \end{bmatrix} \begin{bmatrix} c_1 \\ c_2 \\ c_3 \\ c_4 \end{bmatrix} = \begin{bmatrix} 0.8660254 \\ 0.6123724 \\ 0.2418448 \\ 0.3768696 \end{bmatrix}, \tag{7.21}$$

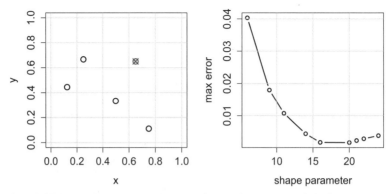

Figure 7.5. Small 2D interpolation example using four Halton points. (left) Computational domain showing the four data points (circles) and the interpolation point (cross in a circle). (right) Maximum estimation errors for different shape parameter values.

where $r_{i,j} = \sqrt{(x_i^c - x_j^c)^2 + (y_i^c - y_j^c)^2}$, $\varphi_{i,j} = \phi(r_{i,j})$, $f_{i,j} = f(x_i^c, y_j^c)$ with $i, j = 1, \ldots, 4$. We have also used a shape parameter value of $e = 20$ which was obtained by trial and error (see Fig. 7.5).

Solving the relationship $Ac = f$ of eqn. (7.21), we obtain

$$c = [-0.095946397,\ 0.001799701,\ 0.075244540\ 0.055529729]^T. \tag{7.22}$$

We are now in a position to interpolate the scattered data to estimate the value of a new data point. For example, if we choose $x^e = 0.65$, $y^e = 0.65$ (see Fig. 7.5), then the estimated function value is given by

$$\mathcal{P}_f(x^e, y^e) = Bc = \sum_{i=1}^{4} c_i \varphi(r_i^e), \tag{7.23}$$

where $r_i^e = \sqrt{(x^e - x_i^c)^2 + (y^e - y_i^c)^2}$, c is the calculated column vector of eqn. (7.22) and B represents a 1×4 matrix, that is, a single row vector with four elements from eqn. (7.6). Note that the term Bc in eqn. (7.23) represents an *inner product*. If more estimated data points are included, then the single row matrix would be extended by an extra row for each additional data point and $\mathcal{P}_f$ would become a column vector of length equal to the number of matrix rows.

Thus, calculating the final result from eqn. (7.23) (see Fig. 7.6), we obtain

$$\mathcal{P}_f(0.65, 0.65) = 0.792182. \tag{7.24}$$

This compares with the exact value $f_{exact} = 0.7938926$, which corresponds to an error of 0.00171058. This is a quite remarkable result considering it has been achieved from just four scattered data points. However, our admiration for this result must be tempered by the fact that $f(x, y)$ is a simple well-behaved smooth function and our choice for the new data point was fortuitous as it resulted in an accurate estimated value for $\mathcal{P}_f$. A different value may give a less accurate result, and it is recommended that readers experiment with this example by obtaining estimates for $\mathcal{P}_f$ based on a variety of different (x^e, y^e) values.

The code for this example is included with downloads for this book but is omitted here as it is almost identical to the more complex example given in Section 7.3.3.

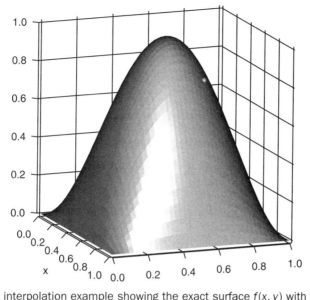

Figure 7.6. Small interpolation example showing the exact surface $f(x, y)$ with the single interpolation point superimposed (sphere).

7.3.3 Larger Interpolation Example: 2D

We now consider a more complex interpolation problem with 81 centers, x^c, generated from a Halton sequence and a requirement for 441 estimation points, x^e, evenly distributed over the 2D domain. A portion of the Halton sequence is given in Table 7.1 and both location sets are plotted in Fig. 7.7. The Halton sequence is used to generate data values from the so-called *Franke's* function, as detailed in Appendix 7.A at the end of this chapter, and the estimation points are located where we wish to calculate the interpolation values.

Table 7.1. *Halton sequence* of 81 centers for a larger interpolation problem on 2D domain

	x	y		x	y
[1,]	0.5	0.333333333	[71,]	0.8828125	0.950617284
[2,]	0.25	0.666666667	[72,]	0.0703125	0.098765432
[3,]	0.75	0.111111111	[73,]	0.5703125	0.432098765
[4,]	0.125	0.444444444	[74,]	0.3203125	0.765432099
[5,]	0.625	0.777777778	[75,]	0.8203125	0.209876543
[6,]	0.375	0.222222222	[76,]	0.1953125	0.543209877
[7,]	0.875	0.555555556	[77,]	0.6953125	0.87654321
[8,]	0.0625	0.888888889	[78,]	0.4453125	0.320987654
[9,]	0.5625	0.037037037	[79,]	0.9453125	0.654320988
[10,]	0.3125	0.37037037	[80,]	0.0390625	0.987654321
[11,]	0.8125	0.703703704	[81,]	0.5390625	0.004115226
(centers 12 to 70 omitted)					

Note: The full set of centers, x^c and the estimation points, x^e are plotted in Fig. 7.7.

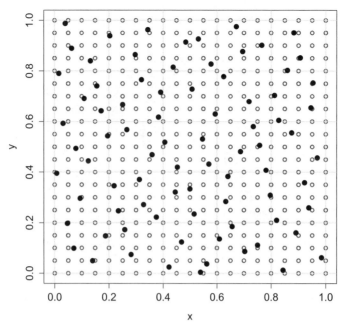

Figure 7.7. Larger interpolation example showing the 81 centers, x^c (points) and the 441 interpolation points, x^e (circles).

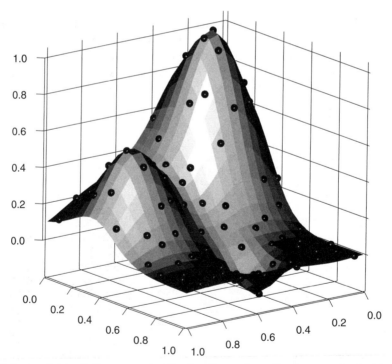

Figure 7.8. Larger interpolation example showing the surface $f(x, y)$ reconstructed from the 81 centers, x^c, superimposed (spheres). The reconstructed surface has a maximum error of −5.04e-02 when compared with the exact surface.

We follow the same procedure as in Section 7.3.2 for the small interpolation example, except that now the B matrix of eqn. (7.6) has 441 rows, one for each estimation point, x^e, and 81 columns, one for each center, x^c.

The resulting vector of estimation values, $\mathcal{P}_f$, has a maximum error equal to $-5.04\text{e}-02$, which occurs at location 1, that is, at $x_1^e(0,0)$. This is not unexpected, as the result at this point constitutes an extrapolation and meshless methods are generally least accurate at domain boundaries. This is a quite remarkable result of surface reconstruction which is unlikely to be achieved by polynomial type methods using such few data values. The estimated points, $\mathcal{P}_f$, are shown in Fig. 7.8, where they are overlaid on a plot of the reconstructed surface.

The R code that performs the interpolation calculations and which generates the plots of Fig. 7.7 and 7.8 is given in Listing 7.1.

```
# File: interp2D-Ex2.R
# 2D interpolation example
library(randtoolbox) # For Halton function
library(rgl)
source("mq.R")
source("circCtrs.R")
# Franke's test function, see: Fasshauer (2007)
testfunction <- function(x,y){0.75*exp(-((9*x-2)^2+(9*y-2)^2)/4)+
                0.75*exp(-((9*x+1)^2/49+(9*y+1)^2/10)) +
                0.5*exp(-((9*x-7)^2+(9*y-3)^2)/4) -
                0.2*exp(-((9*x-4)^2+(9*y-7)^2))}
# Multiquadratic RBF calculation
mq <- function(r, e){
  phi <- sqrt((r*e)^2 + 1)
  return((phi))
}
n <- 9 # 9
k <- n^2 # 81 #total number of interpolation points
ctrs<-halton(k,2) # generate k Halton sequence points (x,y)
Z<-testfunction(ctrs[,1],ctrs[,2]) # produces f(x,y)
#
epsilon <- 6 # shape parameter
# Create the interpolation matrix
A <- array(0,c(k,k));
for( i in 1:k){
  for( j in 1:k){
    r <- sqrt((ctrs[i,1]-ctrs[j,1])^2+(ctrs[i,2]-ctrs[j,2])^2)
    A[i,j] <- mq(r,epsilon) #r^5
  }
}
# find the coeficients c
c<-solve(A,Z)
#reconstruct of the surface on mxm point
```

```
m<-21
# Create mesh
N <- outer(seq(0,1,len=m),rep(1,m),FUN="*")
M <- outer(rep(1,m),seq(0,1,len=m),FUN="*")

E<-array(0,c(m,m))
for( i in 1:m){
  for( j in 1:m){
    r<-sqrt((M[i,j]-ctrs[1:k,1])^2+(N[i,j]-ctrs[1:k,2])^2)
    E[i,j] <- mq(r,epsilon)%*%c
  }
}
#
x <- N[,1]; y <- M[1,]
exact <- array(0,c(m,m))
for( i in 1:m){
  for( j in 1:m){
    exact[i,j] <- testfunction(x[i],y[j])
  }
}
#
err <- E-t(exact)
errMax <- max(abs(err))
errMaxLoc <- which.max(abs(err))
cat(sprintf("\nMax Error <- %e\n",err[errMaxLoc]))
#
# Plot results
plot(N,M,type="p",pch=1,col="red",lwd=2,cex=1,
     xlab="x",ylab="y",xlim=c(0,1),ylim=c(0,1))
grid(lty=1,lwd=1)
points(ctrs[,1],ctrs[,2],pch=16,cex=1.5,lwd=2)
#
open3d()
bg3d("white")
#Define color scheme
jet.colors = colorRampPalette(
  c("#00000F","#00007F", "blue", "#007FFF",
    "cyan", "#7FFF7F", "yellow", "#FF7F00", "red", "#7F0000"))
# Set palette
pal=jet.colors(100)
col.ind <- cut(E,100)
persp3d(M,N,E, aspect=c(1, 1, 1),
        color = "gray", #"cyan", #pal[col.ind],
        ylab = "", xlab = "", zlab = "",#zlim=c(-0.2,1),
        axes=FALSE, #main= "Reconstructed Surface",
        box=FALSE, smooth=FALSE)
```

```
um <- c( -0.6779972, 0.7350551, 0.0004425757, 0,
         -0.3594882, -0.3321092, 0.8720422983, 0,
          0.6411505, 0.5910874, 0.4894154668, 0,
          0.0000000, 0.0000000, 0.0000000000, 1)
UM <- matrix(data=um, byrow=TRUE,nrow=4,ncol=4)
rgl.viewpoint(fov=0) # set before userMatrix
par3d(userMatrix=UM)
par3d(windowRect=c(20,100,900,950), zoom=1.0)
decorate3d(type="n",axes=FALSE,
           col="black",aspect=TRUE,top=FALSE)
axes3d(c("x+", "y+", "z+"),col="black")
aspect3d(c(1, 1, 1)) # place after decorate
bg3d(color=c("white","white"))
grid3d(c("x-", "y-", "z"),col="black")
#
# Original data
xc <- array(ctrs[,1],c(n,n))
yc <- array(ctrs[,2],c(n,n))
zc <- array(Z,c(n,n))
plot3d(xc,yc,zc, type="s",size=1,col="red",add=TRUE)
# rgl.snapshot("reconstructedSurface_3D.png", fmt="png") # print to file
```

Listing 7.1. File: `interp2D-Ex2.R`—Code for Meshless RBF method 2D interpolation of the Halton sequence shown in Table 7.1 and that generates the plots of Figs. 7.7 and 7.8

7.3.4 Interpolation Example: 3D

We now consider the more more demanding 3D interpolation problem with $9^3 = 729$ centers, x^c, generated from a Halton sequence and a requirement for $20^3 = 8000$ estimation points, x^e, evenly distributed over the 3D domain. The Halton sequence is given in Table 7.2 and is used to generate data values from the function

$$f(x, y, z) = \sqrt{x^2 + y^2 + z^2}. \tag{7.25}$$

The estimation points are located where we wish to calculate the interpolation values.

We follow the same procedure as in previous interpolation examples, except that now the B matrix of eqn. (7.6) has 8,000 rows, one for each estimation point, x^e, and 729 columns, one for each center, x^c. For this example, $r_{i,j,k} = \sqrt{(x_i^c - x_j^c)^2 + (y_i^c - y_j^c)^2 + (z_i^c - z_j^c)^2}$, $\varphi_{i,j,k} = \phi(r_{i,j,k})$, $f_{i,j,k} = f(x_i^c, y_j^c, z_k^c)$ with $i, j, k = 1, \ldots, 20$. We have also used the multiquadratic RBF with a shape parameter value of $e = 1.25$, which was obtained by trial and error (see Fig. 7.9).

We use the `plot3D` package [Soe-14] to generate a slice plot of the estimated values, $\mathcal{P}_f$, which is shown in Fig. 7.10. This is indistinguishable from a similar plot of the exact function $f(x, y, z)$.

The resulting matrix of estimation error values has a maximum error equal to 2.97e-02, which occurs at the domain center, that is, at location (20,20,20) or $x^e(0, 0, 0)$ (see Fig. 7.10).

Table 7.2. *Halton sequence* of 729 centers for an interpolation problem on a 3D domain

	x	y	z		x	y	z
[1,]	0.000	−0.333333	−0.60	[720,]	−0.912109	−0.780521	−0.631360
[2,]	−0.500	0.333333	−0.20	[721,]	0.087891	−0.113855	−0.231360
[3,]	0.500	−0.777778	0.20	[722,]	−0.412109	0.552812	0.168640
[4,]	−0.750	−0.111111	0.60	[723,]	0.587891	−0.558299	0.568640
[5,]	0.250	0.555556	−0.92	[724,]	−0.662109	0.108368	0.968640
[6,]	−0.250	−0.555556	−0.52	[725,]	0.337891	0.775034	−0.935360
[7,]	0.750	0.111111	−0.12	[726,]	−0.162109	−0.336077	−0.535360
[8,]	−0.875	0.777778	0.28	[727,]	0.837891	0.330590	−0.135360
[9,]	0.125	−0.925926	0.68	[728,]	−0.787109	0.997257	0.264640
[10,]	−0.375	−0.259259	−0.84	[729,]	0.212891	−0.999086	0.664640
(centers 11 to 719 omitted)							

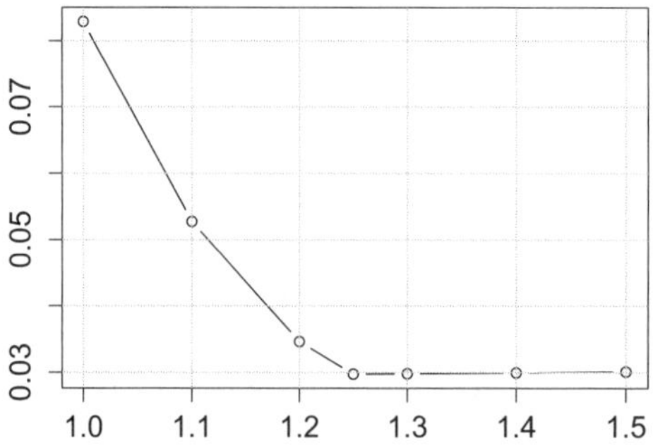

Figure 7.9. Maximum estimation errors for the 3D interpolation example for differently shaped parameter values.

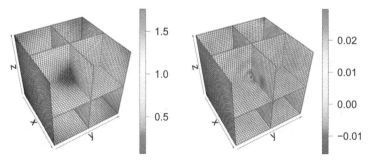

Figure 7.10. 3D interpolation example. (left) Slice through the 3D solution $u(x, y, z)$. (right) Slice through solution errors $err(x, y, z)$. The maximum error occurs at the center of the 3D domain.

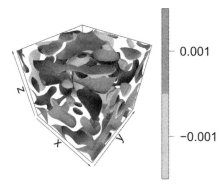

Figure 7.11. Iso-surface contours of the errors associated with the 3D interpolation problem. (See color plate 7.11)

It is clear from this and previous interpolation examples that problem complexity grows rapidly with the number of dimensions—*the curse of dimensionality*.[2]

As a further example of the versatility of the plot3D package, an iso-surface plot of errors equal to ±0.001 is shown in Fig. 7.11.

The R code that performs the interpolation calculations and which generates the plots of Figs. 7.10 and 7.11) is given in Listing 7.1.

```
# File: interp3D_main.R
# 3D interpolation example
library(randtoolbox)
library(plot3D) # for slice3D and isosurf3D plots
library(matlab) # for meshgrid
#
# testfunction <- function(x,y,z){ w <- sin(pi*x)*sin(pi*y)*sin(pi*z)}
testfunction <- function(x,y,z){ w <-sqrt(x^2+y^2+z^2)}
mq <- function(r, e){
  phi <- sqrt((r*e)^2 + 1)
  return((phi))
}
#
epsilon <- 1.25
n <- 9
q<-n^3 # Total number of interpolation points
ctrs <- halton(q,3)*2-1
dMin <- -1; dMax <- 1; dMid <- (dMax+dMin)/2
A <- array(0,c(q,q))
for( i in 1:q){
```

---

[2] The term *curse of dimensionality* was originally coined by Richard E. Bellman [Bel-57] when formulating his ideas in the area of dynamic optimization. His reasoning: in n-dimensional state space, for a fixed discretization scheme applied to each coordinate, the number of discrete states grows exponentially with n.

```r
    for( j in 1:q){
       r<-sqrt((ctrs[i,1]-ctrs[j,1])^2 +
               (ctrs[i,2]-ctrs[j,2])^2 +
               (ctrs[i,3]-ctrs[j,3])^2)
       A[i,j]<- mq(r,epsilon) # Multiquadratic RBF
    }
}
# find the coeficients c
Z<-testfunction(ctrs[,1],ctrs[,2],ctrs[,3]) # f(x,y,z)
#Z<-testfunction1(d,A)
#
c<-solve(A,Z)
m<-40
#
M <- meshgrid(seq(dMin,dMax,len=m),seq(dMin, dMax,len=m),
              seq(dMin, dMax,len=m),3)
E<-array(0,c(m,m,m))
#r <- with (M, sqrt(x^2 + y^2 + z^2))
for( i in 1:m){
  for( j in 1:m){
    for( k in 1:m){
      r<-sqrt((M$x[i,j,k]-ctrs[1:q,1])^2+(M$y[i,j,k]-ctrs[1:q,2])^2+
              ( M$z[i,j,k]-ctrs[1:q,3])^2)
      E[i,j,k] <- mq(r,epsilon)%*%c
    }
  }
}
exact <- testfunction(M$x,M$y,M$z)
err <- E-exact
errMax <- max(abs(err))
cat(sprintf("\nMax Error <- %e occurs at location: ",errMax))
cat(which( abs(err)==max(abs(err),na.rm=T) , arr.ind = T ))
# Plot shape parameter v error
par(cex=1,mar=c(5, 4, 4, 2) + 0.1)
s    <- c(1.0, 1.1, 1.2, 1.25, 1.3, 1.4, 1.5)
sErr <- c(8.289805e-02, 5.275579e-02, 3.463991e-02, 2.974280e-02,
          2.980937e-02, 2.995368e-02, 3.010911e-02)
plot(s,sErr,type="b",lwd=2,xlab="shape parameter",ylab="max error")
grid(lty=1,lwd=1)
#
x<-y<-z<-seq(dMin,dMax,len=m)
par(cex=2,mar=c(2.1,2.1,2.1,4.1))
#
slice3D(x, y, z, colvar = exact, d = 2, theta = 60, border = "black",
        xs = c(dMin, dMid), ys = c(dMin, dMid, dMax), zs = c(dMin, dMid))
#
```

```
slice3D(x, y, z, colvar = E, d = 2, theta = 60, border = "black",
    xs = c(dMin, dMid), ys = c(dMin, dMid, dMax), zs = c(dMin, dMid))
#
slice3D(x, y, z, colvar = err, d = 2, theta = 60, border = "black",
    xs = c(dMin, dMid), ys = c(dMin, dMid, dMax), zs = c(dMin, dMid))
#
levels<-seq(-0.001,0.001,len=2)
isosurf3D(x, y, z, colvar = err, level = levels,
    col = c("red", "blue"))
```

Listing 7.2. File: `interp3D_main.R`—Code for meshless RBF method 3D interpolation of the Halton sequence shown in Table 7.2

7.3.5 Interpolation with Polynomial Precision

It is difficult to achieve good accuracy using polynomials to interpolate *multivariate* (more than one variable) scattered data. Fortunately, the preceding method can be extended to include additional polynomial terms, which overcomes this difficulty. This is illustrated in what follows for a 2D system where the RBF system of eqn. (7.5) has been extended to provide degree 1 polynomial precision or linear precision:

$$\sum_{k=1}^{N} c_k \varphi (x_j - x_k) + d_1 + d_2 x_j + d_3 y_j = f_j, \quad j = 1, \ldots, N, \tag{7.26}$$

$$\sum_k c_k = 0, \quad \sum_k c_k y_k = 0,$$

which can be presented in matrix format as

$$\begin{bmatrix} A & P^T \\ P & 0 \end{bmatrix} \begin{bmatrix} c \\ d \end{bmatrix} = \begin{bmatrix} f \\ 0 \end{bmatrix}; \quad A = [\varphi(\|x_j - x_k\|)], \quad P^T = \begin{bmatrix} 1 & x_1 & y_1 \\ \vdots & \vdots & \vdots \\ 1 & x_N & y_N \end{bmatrix}. \tag{7.27}$$

Higher-order polynomial precision can be handled by extending the first-order method in an obvious way. For example, to achieve degree 2 or quadratic polynomial precision, the P matrix becomes

$$P^T = \begin{bmatrix} 1 & x_1 & y_1 & x_1^2 & y_1^2 & x_1 y_1 \\ \vdots & \vdots & \vdots & \vdots & \vdots & \vdots \\ 1 & x_N & y_N & x_N^2 & y_N^2 & x_N y_N \end{bmatrix}, \tag{7.28}$$

and the length of the d vector is extended to six elements. If the symmetric RBF interpolation matrix A is augmented by P and P^T of eqn. (7.28), similar to that shown in eqn. (7.27), then the scheme will interpolate exactly data to constant, degree 1 and degree 2 polynomial scattered data. If the scattered data are not precise, that is, they contain errors or noise, then the interpolation will not be exact. More details and examples of use can be found in [Fas-07].

7.4 DIFFERENTIATION

The RBFs $\varphi(r)$ can be differentiated by simple application of the chain rule for differentiation. Thus, for a two-dimensional domain in Cartesian coordinates we have

$$\varphi(r) = \varphi(\|\mathbf{x}\|) = \varphi(\sqrt{x^2 + y^2}), \tag{7.29}$$

when the first derivative with respect to x becomes

$$\frac{\partial}{\partial x}\varphi(\|\mathbf{x}\|) = \frac{d}{dr}\varphi(r)\frac{\partial}{\partial x}r(x,y) \tag{7.30}$$

$$= \frac{d}{dr}\varphi(r)\frac{x}{\sqrt{x^2+y^2}} \tag{7.31}$$

$$= \frac{x}{r}\frac{d}{dr}\varphi(r); \tag{7.32}$$

and similarly with respect to y we obtain

$$\frac{\partial}{\partial y}\varphi(\|\mathbf{x}\|) = \frac{y}{r}\frac{d}{dr}\varphi(r). \tag{7.33}$$

The second derivative with respect to x becomes

$$\frac{\partial^2}{\partial x^2}\varphi(\|\mathbf{x}\|) = \frac{x^2}{r^2}\frac{d^2}{dr^2}\varphi(r) + \frac{y^2}{r^3}\frac{d}{dr}\varphi(r); \tag{7.34}$$

and similarly with respect to y we obtain

$$\frac{\partial^2}{\partial y^2}\varphi(\|\mathbf{x}\|) = \frac{y^2}{r^2}\frac{d^2}{dr^2}\varphi(r) + \frac{x^2}{r^3}\frac{d}{dr}\varphi(r). \tag{7.35}$$

The *Laplacian* is simply given by

$$\left(\frac{\partial^2}{\partial x^2} + \frac{\partial^2}{\partial x^2}\right)\varphi(\|\mathbf{x}\|) = \frac{d^2}{dr^2}\varphi(r) + \frac{1}{r}\frac{d}{dr}\varphi(r); \tag{7.36}$$

and the second derivative with respect to *mixed partials* becomes

$$\frac{\partial^2}{\partial xy}\varphi(\|\mathbf{x}\|) = \frac{xy}{r^2}\frac{d^2}{dr^2}\varphi(r) - \frac{xy}{r^3}\frac{d}{dr}\varphi(r). \tag{7.37}$$

Now consider the linear system $Ac = u$, where A and c are defined by eqn. (7.5) and $u = [u_1, u_2, \ldots, u_N]$. Differentiating this system once with respect to x we obtain $A_x c = u'$ and twice with respect to x we obtain $A_{xx} c = u''$. This leads to $u' = D_x u$ and $u'' = D_{xx} u$, where $D_x = A_x A^{-1}$ and $D_{xx} = A_{xx} A^{-1}$ are known as *differentiation matrices*.

Table 7.3. Overall results from derivative test using a Halton sequence of six points

x	f	$\dfrac{df}{dx}$	Dx	Error
0.500	0.125000	−0.250000	−0.249983	1.7198E-05
0.250	0.140625	0.187500	0.187577	7.6611E-05
0.750	0.046875	−0.312500	−0.312541	−4.1053E-05
0.125	0.095703	0.546875	0.546405	−4.7007E-04
0.625	0.087891	−0.328125	−0.328137	−1.2391E-05
0.375	0.146484	−0.078125	−0.078153	−2.8171E-05

Corresponding results are obtained with respect to y and z for higher-dimensional systems and also for higher-order derivative matrices.

7.4.1 Derivative Example: 1D

We will now apply the preceding ideas for meshless differentiation using the multi-quadratic RBF. Consider the function $f(x)$ and corresponding first derivative

$$f(x) = x(x-1)^2, \quad \frac{df(x)}{dx} = (x-1)^2 + 2x(x-1). \quad (7.38)$$

We use the following Halton sequence of six data points as the meshless centers:

```
[1] 0.500
[2] 0.250
[3] 0.750
[4] 0.125
[5] 0.625
[6] 0.375
```

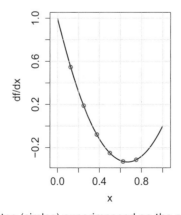

Figure 7.12. Derivative estimates (circles) superimposed on the analytical derivative function.

Table 7.4. Maximum first differentiation errors for different sets of *Halton* data points

No. of points, k	Shape parameter value, e	Maximum error
6	0.50	−4.70E-04
16	1.60	−2.65E-04
26	2.84	−1.47E-03
36	4.40	−5.12E-04
46	5.75	−6.77E-04

The associated A matrix for the MQ RBF $\varphi = \sqrt{(er_c)^2 + 1}$ using shape parameter $e = 0.5$ is calculated to be

$$A = \begin{bmatrix} 1.000000 & 1.007782 & 1.007782 & 1.017426 & 1.001951 & 1.001951 \\ 1.007782 & 1.000000 & 1.030776 & 1.001951 & 1.017426 & 1.001951 \\ 1.007782 & 1.030776 & 1.000000 & 1.047691 & 1.001951 & 1.017426 \\ 1.017426 & 1.001951 & 1.047691 & 1.000000 & 1.030776 & 1.007782 \\ 1.001951 & 1.017426 & 1.001951 & 1.030776 & 1.000000 & 1.007782 \\ 1.001951 & 1.001951 & 1.017426 & 1.007782 & 1.007782 & 1.000000 \end{bmatrix}, \quad (7.39)$$

where r_c represents the absolute distance between centers. The corresponding A_x matrix for the first derivative of the RBF, $\varphi_x = e^2 r_x / \sqrt{(er_c)^2 + 1}$, is

$$A_x = \begin{bmatrix} 0.00000 & 0.06202 & -0.06202 & 0.09214 & -0.03119 & 0.03119 \\ -0.06202 & 0.00000 & -0.12127 & 0.03119 & -0.09214 & -0.03119 \\ 0.06202 & 0.12127 & 0.00000 & 0.14914 & 0.03119 & 0.09214 \\ -0.09214 & -0.03119 & -0.14914 & 0.00000 & -0.12127 & -0.06202 \\ 0.03119 & 0.09214 & -0.03119 & 0.12127 & 0.00000 & 0.06202 \\ -0.03119 & 0.03119 & -0.09214 & 0.06202 & -0.06202 & 0.00000 \end{bmatrix}, \quad (7.40)$$

where r_x is the signed distance between centers.

It was shown that the first derivative matrix is given by $D_x = A_x A^{-1}$ and that the first derivative is given by $\frac{df}{dx} = D_x f$. After carrying out these calculations, the overall results are as given in Table 7.3 and plotted in Fig. 7.12.

The results shown in Fig. 7.12 are of high accuracy and clearly show the utility of the meshless methods using the MQ RBF. For this particular function, the accuracy of the first derivative does not change greatly if the Halton sequence is increased. Table 7.4 shows the results obtained for various Halton sequences, together with the associated shape parameter values, which were obtained by trial and error.

The R code that performed this calculation and generated the plot of Fig. 7.12 is included in Listing 7.3.

```
# File: derivTest1D_Ex1.R
library(randtoolbox) # For Halton function
library(pracma)      # For invers matrix
source("mq.R")
source("mqDerivatives.R")
```

```r
# Code for meshless derivative test
k <- 6
xc <- c(halton(k,1))#seq(0,1,len=k) #
# f <- function(x){exp(sin(pi*x))} # Analytical function
# d <- function(x){exp(sin(pi*x)*cos(pi*x)*pi)} # Analytical function
f <- function(x){x*(x-1)^2}       # Analytical function, f
d <- function(x){(x-1)^2+2*x*(x-1)} # Analytical function, fx
fc <- f(xc)
fd <- d(xc)
# Distance matrix - centers
Nc <- length(xc)
rx <- array(0,c(Nc,Nc))
for(i in 1:Nc){
  for(j in 1:Nc){
    rx[i,j] <- xc[i] - xc[j]
  }
}
rc   <- abs(rx)
e    <- 0.5 # optimum
A    <- mq(rc,e)
invA <- inv(A)
Ax   <- mqDerivatives(rc,rx,e,1)
Dx   <- (Ax%*%invA)%*%fc
#
error <- Dx-fd
errMax <- max(abs(error))
errMaxLoc <- which.max(abs(error))
cat(sprintf("\nMax Error <- %e\n",error[errMaxLoc]))
# Plot analytical curve plus calculated values
xi <- c(0:100)/100 # More points for plotting analytical curve
fdi <- d(xi)
# Plot results
plot(xi,fdi,type="l",lwd=2,xlab="x",ylab="df/dx",asp=1)
grid(lty=1)
points(xc,Dx,col="red",lwd=2) # Data points - Analytical values
```

Listing 7.3. File: derivTest1D_Ex1.R—Code to test derivative calculation using the meshless RBF method

7.5 LOCAL RBFS

The general meshless method using RBFs very quickly becomes impractical for many problems as the number collocation points increase. *Local RBFs* are a way of restricting the influence of RBFs to those located relatively close to the point being calculated. This can reduce the computational burden considerably. The reduced computation comes about because the system matrix in eqn. (7.5) becomes *sparse*, thus facilitating the use of sparse matrix techniques which greatly improve calculation efficiency. This approach is particularly effective where values change significantly over the solution domain.

For sparse matrix calculations we utilize the R package `Matrix`, which includes the S4 class `Matrix` along with many member functions. The S4 class is the latest object oriented programming class that provides a more secure environment that the S3 class system currently in wide use. Although we do not discuss S4 classes further here, nor do we use them, except via R packages, it important to be aware that the "@" symbol is used to access S4 member variables, rather than the "$" symbol, which is used for S3 classes. More information on S3 and S4 classes can be found in [Chap. 9][Mat-11].

We follow the general approach proposed by Sarra [Sar-12] where the main features are that it is *simple*, generally *applicable*, and *efficient* with no restrictions placed on the shape of the domain. The method can be employed to solve problems with many thousands of grid points spread over large domains.

The problem is discretized into a set of N distinct points as for a *global* RBF scheme. Then at each of the N centers, the local RBF method considers a local interpolant of the form of eqn. (7.4), but where the summations for a particular collocation point are restricted to the set of those RBFs associated with neighboring points. The neighboring point locations are selected according to some criterion, such as being the m closest neighbors. These points are collectively called a *stencil*. Therefore, associated with node n, there is a stencil S^n consisting of a set of m nodes, which may or may not be numbered consecutively. The numbers of these nodes, including the number of node n, are assigned to the vector $\mathcal{I}_n$.

Thus, for the nth collocation point, the interpolant becomes

$$f^n(x) = \sum_{k \in \mathcal{I}_n} c_k^n \varphi \left(\|x - x_k^n\|_2 \right), \quad n = [1, \ldots, N], \ x^n \in S^n, \ x \in \mathbb{R}. \tag{7.41}$$

Now, eqn. (7.41) represents N $m \times m$ linear systems

$$B^n \mathbf{c}^n = f^n, \tag{7.42}$$

to be solved for vector $\mathbf{c}^n$, where B^n is the interpolant matrix of the local system matrix. The elements of B^n are obtained from the chosen RBF, φ, that is,

$$B_{i,j}^n = \varphi \left(\|x_i - x_k\| \right), \quad k = \mathcal{I}_n(j), \ (i, j) = [1, \ldots, m]. \tag{7.43}$$

If we wish to assemble a global $N \times N$ system RBF matrix A from the local $m \times m$ RBF matrices $B^n, n = 1, \ldots, N$ calculated earlier, then the kth element of the nth row of A, $A_{n,k}$, is equal to

$$A_{n,k} = \begin{cases} B_{1,j}^n, & k = \mathcal{I}_n(j), \ j = [1, \ldots, m], \\ 0, & \text{otherwise}. \end{cases} \tag{7.44}$$

This assumes that the first row of B^n is associated with node n, which it will be if the neighboring nodes are ordered according to their distance from node n. Thus, it is clear that each row of A has m nonzero elements and $N - m$ zero elements, with the result that the completed matrix A is sparse. As the magnitude of $N - m$ increases, sparsity also increases, and sparse matrix calculations that involve A become more efficient.

If our intention is to solve PDEs, then we need to construct local *differentiation* matrices, and this is done in a similar way to constructing the local RBF matrix, but with some extra calculations. Recall from Section 7.4 that for global RBFs $u' = D_x u$ and $u'' = D_{xx} u$, where $D_x = A_x A^{-1}$ and $D_{xx} = A_{xx} A^{-1}$ are the global differentiation matrices. The same basic idea works for differentiation matrices derived from local RBFs.

To construct local differentiation matrices, we first construct the local RBF matrix associated with node n, B^n, and generate its inverse $(B^n)^{-1}$. Then, for the first derivative, we calculate just row one of the local RBF derivative matrix B_x^n associated with node n. The elements of this row are obtained from the chosen RBF, φ_x, that is,

$$B^n_{x_{1,j}} = \varphi_x\left(\|x_n - x_k\|\right), \quad k = \mathcal{I}_n(j), \quad j = [1, \ldots, m]. \tag{7.45}$$

Now, as for the global RBFs, the local differentiation matrix becomes

$$D_x^n = B_x^n (B^n)^{-1}, \tag{7.46}$$

which can be used to populate the global derivative RBF matrix D_x, as we did for the global RBF matrix A, that is,

$$D_{x_{n,k}} = \begin{cases} D^n_{x_{1,j}}, & k = \mathcal{I}_n(j), \; j = [1, \ldots, m], \\ 0, & \text{otherwise,} \end{cases} \tag{7.47}$$

where we have again assumed that the first row of D_x^n is associated with node n. The inverse matrix $(B^n)^{-1}$ of eqn. (7.46) is calculated using the results of a *singular value decomposition* (SVD). This is because the SVD calculation can also be used to estimate the *condition number* of the matrix B^n which is then used to help determine the best shape parameter e.

Higher-order derivative matrices D_{xx}, D_{xxx}, and so on, are derived in exactly the same way.

There are two outstanding practical aspects to consider before implementing this procedure: (1) assigning the stencil nodes for each collocation point and (2) determining the best shape parameter for calculating the RBFs. These are considered in the following sections.

The application of local RBFs is demonstrated in Sections 7.6.7.3 and 7.6.7.5, where the Fisher–Kolmogorov equation is solved in 1D and 2D, respectively.

7.5.1 Allocating Stencil Nodes

For each of the N centers we need to assign a stencil of neighboring grid points. Various possibilities exist for choosing the stencil nodes, but we will select these points as being the m closest neighbors, including itself. We calculate the distances from the nth node to all other nodes and then order them according to magnitude. The nodes associated with the m shortest distances are selected for the nth stencil. An $N \times m$ array is then populated with the node numbers, one row for each node stencil. The nth row contains the m nodes associated with node n, which we keep in descending ordered. We now have the necessary information in a suitable form to be able to implement the procedure. This has been done as part of the local meshless solution method used to solve the 1D and 2D problems of Sections 7.6.7.3 and 7.6.7.5 respectively.

Examples are shown in Fig. 7.13 of 21×21 grids superimposed with 9- and 21-point stencils.

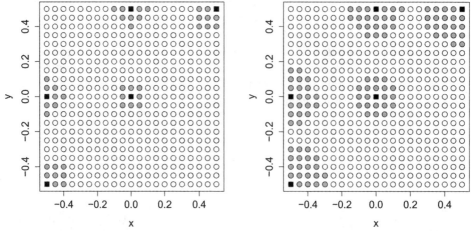

Figure 7.13. Example showing spatial domain with a 21 × 21 2D grid. (left) With five typical nine-point stencils superimposed. (right) With five typical 21-point stencils superimposed. Each example shows four stencils made up of boundary and interior nodes and one made up of entirely of interior nodes. The stencil centers are indicated by filled black squares and the supporting nodes by filled gray circles.

7.5.2 Choosing the Right Shape Parameter Value

As mentioned earlier, the inverse of matrix B^n is calculated using the results of a singular value decomposition. The SVD of the $m \times m$ matrix B^n is obtained by a call to the R function svd(), which returns three matrices of the same size as B^n, namely,

- Σ, a diagonal matrix of singular values, $\sigma_i, i = [1, \ldots, m]$, of B, in descending order by size
- U, a matrix whose columns contain the left singular vectors of B
- V, a matrix whose columns contain the right singular vectors of B

where U and V are orthogonal matrices. The reason these variables form a decomposition is because the SVD calculation *factorizes* B^n such that the following relationship holds:

$$B^n = U \Sigma V^T. \tag{7.48}$$

Because U and V are both orthogonal[3] and Σ is diagonal, it follows that the inverse of B^n is given by

$$(B^n)^{-1} = V \Sigma^{-1} U^T. \tag{7.49}$$

The condition number of matrix B^n is also obtained from the SVD calculation. It is given by the product of the Euclidean norms of B and B^{-1}, which is equal to the ratio of the largest to the smallest singular value, that is,

$$\kappa(B^n) = \|B\|_2 \|B^{-1}\|_2 = \frac{\sigma_{\max}}{\sigma_{\min}}. \tag{7.50}$$

It has been found heuristically that, for RBF methods, the error magnitude decreases as the condition number increases and that they are most accurate when their system matrix

---

[3] The inverse of an orthogonal matrix is simply equal to its transpose. Thus, $U^T U = I$ and $V^T V = I$.

is either ill-conditioned or close to being ill-conditioned. This observation leads to the idea of choosing the shape parameter value such that the condition number is acceptable. Sarra [Sar-12] proposes that when using double precision floating point arithmetic,[4] the shape parameter value should be chosen such that

$$10^{13} \leq \kappa(B^n) \leq 10^{15}. \tag{7.51}$$

If the error is considered as a function of the shape parameter, it has been found that, generally, this results in a shape parameter value that corresponds to a point on the error curve just before the curve begins to oscillate or rise again (see Figs. 7.4, 7.5, and 7.9).

We are now in a position to outline an algorithm for choosing the best shape value for calculating B^n and B_x^n. Algorithm 7.1 contains simplified pseudo code to calculate the *first derivative matrix* for a *single* particular node using the *optimum shape parameter* value, based on the above discussion. The algorithm would be called in a loop, once for each node. It has been implemented as part of the local meshless solution method used to solve the 1D and 2D problems of Sections 7.6.7.3 and 7.6.7.5, respectively. However, for these problems, the algorithm computer code was modified slightly to calculate a *second derivative* matrix.

Algorithm 7.1 To calculate *first derivative matrix* based on optimum local shape parameter value

1: **procedure** DERIVMATRIX($r, e_{\text{start}}, \kappa_{\min}, \kappa_{\max}, \delta$) ▷ Calculate local derivative matrix
2: $e \leftarrow e_{\text{start}}$ ▷ Set starting point for shape parameter value, e
3: $\kappa \leftarrow 1$ ▷ Set starting point for condition number, κ, of matrix B
4: **while** $\kappa < \kappa_{\min} \lor \kappa > \kappa_{\max}$ **do** ▷ Iterate while κ outside acceptable range
5:   $B \leftarrow \varphi(r, e)$ ▷ Calculate RBF from *radial distance* and *shape pameter*
6:   $tmp \leftarrow svd(B)$ ▷ Perform SVD. Object tmp contains u, s and v
7:   $U \leftarrow tmp\$u$ ▷ Orthogonal matrix containing left singular vectors of B
8:   $\Sigma \leftarrow tmp\$s$ ▷ diagonal matrix contains singular values, $\sigma_i, \ldots, \sigma_m$
9:   $V \leftarrow tmp\$v$ ▷ Orthogonal matrix containing right singular vectors of B
10: $\kappa \leftarrow \dfrac{\sigma_{\max}}{\sigma_{\min}}$ ▷ New κ, ratio of *max* to *min* singular values
11: **if** $\kappa < \kappa_{\min}$ **then**
12:   $e \leftarrow e - \delta$ ▷ Decrease shape parameter by δ
13: **else if** $\kappa > \kappa_{\max}$ **then**
14:   $e \leftarrow e + \delta$ ▷ Increase shape parameter by δ
15: **end if**
16: **end while** ▷ Optimum shape function, ϵ, found - gives κ within desired range.
17: $B^{-1} \leftarrow V\Sigma^{-1}U^T$ ▷ Calculate inverse of matix B
18: $B_x \leftarrow \varphi_x(r, e)$ ▷ Calculate first derivative RBF
19: $D_x \leftarrow B_x B^{-1}$ ▷ Calculate first derivative matrix
20: **return** D_x ▷ Return first derivative matrix
21: **end procedure**

---

[4] Refer to [Ove-01, Chapter 12] for a discussion on condition numbers in relation to floating point arithmetic.

7.6 APPLICATION TO PARTIAL DIFFERENTIAL EQUATIONS

We now outline a mesh-free method originally due to Kansa [Kan-90], who demonstrated how RBFs can be combined with classic numerical collocation methods to solve PDEs (see also [Pep-10]).

From the preceding we see that for a *linear differential operator* $\mathcal{L}$, with constant coefficients, we obtain

$$\mathcal{L}u = A_{\mathcal{L}}c = A_{\mathcal{L}}A^{-1}u$$
$$= D_{\mathcal{L}}u \tag{7.52}$$

where $\mathcal{L}$ consists of terms d/dx, $d^2/dx^2,\ldots$, and so on, and A and c are defined by eqn. (7.5).

Now consider the following PDE with Dirichlet boundary conditions:

$$\frac{\partial u}{\partial t} = \mathcal{L}u + f(x,t), \; u = u(x,t), \; t \geq 0, \; x \in \Omega, \; \Omega \subset \mathbb{R}^s, \; s \in (1,2,3),$$
$$u(x,t) = g(x,t), \qquad x \in \partial\Omega, \tag{7.53}$$
$$u(x,0) = u_0(x), \qquad t = 0.$$

We now apply the ideas of Section 7.4 to the system of eqn. (7.53) assuming $s = 1$. First we discretize the problem over the spatial domain and then apply meshless collocation, that is, $\mathcal{L}u = D_{\mathcal{L}}u$, to obtain

$$\therefore \frac{du_i(t)}{dt} = D_{\mathcal{L}}u_i(t) + f_i(t), \quad i = [1, \cdots, N],$$
$$u_b(t) = g_b(t), \tag{7.54}$$
$$u_i(0) = u_{i,0},$$

where $u_i(t) = u(x = x_i, t)$, $f_i(t) = f(x = x_i, t)$ and subscripts b and 0 indicate values of u_i on the boundary and at $t = 0$, respectively. Thus, we have now transformed a PDE with independent variable $u(x,t)$ into a set of ODEs with independent variables $u_i(t)$, $i = [1, 2, \ldots, N]$. Although we have assumed that $s = 1$, the analysis applies equally to a system of higher dimensions.

Equation (7.54) can be integrated numerically by any standard numerical integrator that will provide an accurate stable solution.

7.6.1 Explicit Euler Integration

A simple application of *explicit Euler* to the linear system of eqn. (7.53), ignoring boundary conditions, yields the following numerical scheme:

$$u^{n+1} = u^n + \Delta t \mathcal{L}u^n + \Delta t f^{n+1}, \tag{7.55}$$

where $u = [u_1, u_2, \ldots, u_N]$ and $f = [f_1, f_2, \ldots, _N]$ represents the discretization of continuous variables $u(x,t)$ and $f(x,t)$, respectively, Δt is the time step, and superscript n indicates a variable evaluated at the nth time step.

We are now in a position to solve the PDE iteratively by applying the RBF approximation $D_{\mathcal{L}} = A_{\mathcal{L}}A^{-1}$ of eqn. (7.52), when we obtain

$$u^{n+1} = (1 + \Delta t D_{\mathcal{L}})u^n + \Delta t f^{n+1}, \tag{7.56}$$

where $D_\mathcal{L}$ is a constant matrix, so has only to be evaluated once. This is an extremely simple method of solution and, for some problems, yields acceptable results. However, many problems are either unstable or require an impractically small time step.

Of course, boundary conditions have to be included, and these are imposed at the start of each time step calculation.

7.6.2 Weighted Average Integration

We now consider the *weighted average* or *θ-method* [Mor-94],

$$u^{n+1} - u^n = \Delta t \left[\theta \mathcal{L} u^{n+1} + (1-\theta) \mathcal{L} u^n \right], \tag{7.57}$$

which has superior stability characteristics to the explicit Euler method and has been used successfully to obtain meshless solutions to a number of PDE problems [Udd-11]. However, it does require the solution of a matrix equation.

Implementing this scheme on eqn. (7.53) yields

$$(1 - \Delta t \theta \mathcal{L}) u^{n+1} = (1 + \Delta t (1-\theta) \mathcal{L}) u^n + \Delta t f^{n+1}. \tag{7.58}$$

After taking into account boundary conditions and applying RBF approximations, we obtain

$$\begin{aligned} \left[A^I - \Delta t \theta A^I_\mathcal{L}(x) \right] c^{n+1} &= \left(A^I + \Delta t (1-\theta) A^I_\mathcal{L}(x) \right) c^n + \Delta t f^{n+1}(x), \quad x \in \Omega^I, \\ A^B c^{n+1} &= g^{n+1}(x), \quad x \in \partial\Omega, \end{aligned} \tag{7.59}$$

where superscript I denotes the interior nodes within Ω and superscript B denotes the boundary nodes on $\partial\Omega$. There are N^B boundary nodes and N^I interior nodes. Thus, the total number of nodes in Ω is equal to $N = N^B + N^I$.

Alternatively, we can represent the scheme in matrix form, that is,

$$\begin{aligned} G c^{n+1} &= H c^n + F^{n+1}, \\ \therefore c^{n+1} &= G^{-1} \left(H c^n + F^{n+1} \right), \end{aligned} \tag{7.60}$$

where the matrices are defined as

$$G = \begin{bmatrix} A^I - \Delta t \, \theta A^I_\mathcal{L} \\ A^B \end{bmatrix}, \quad H = \begin{bmatrix} \left(A^I + \Delta t (1-\theta) A^I_\mathcal{L} \right) \\ 0 \end{bmatrix}, \quad F^{n+1} = \begin{bmatrix} \Delta t f^{n+1} \\ g^{n+1} \end{bmatrix}. \tag{7.61}$$

Thus, we obtain the solution at t^{n+1} from

$$u^{n+1} = A c^{n+1}. \tag{7.62}$$

An equivalent but less efficient solution can be obtained from

$$\therefore u^{n+1} = A G^{-1} H A^{-1} u^n + A G^{-1} F^{n+1}. \tag{7.63}$$

It should be noted that for linear systems A^B, A^I, and $A^I_\mathcal{L}$ are constant matrices, and therefore G and H have only to be evaluated once. If f and g are non-time-varying, then F will also be a constant matrix. Matrix A^B has N columns and N^B rows, whereas matrices A^I and $A^I_\mathcal{L}$ each have N columns and N^I rows.

7.6.3 Method of Lines

The *method of lines*, as discussed in Chapter 3, can also be used to solve PDEs approximated by RBF collocation. When applied to the linear system $\partial u(x,t)/\partial t = \mathcal{L}u(x,t) + f(x,t)$, ignoring boundary conditions, we obtain the following numerical scheme:

$$u_t^n = \mathcal{L}u^n + f^n, \tag{7.64}$$

where $u = [u_1, u_2, \ldots, u_N]$ and $f = [f_1, f_2, \ldots, _N]$ represent the discretization of continuous variables $u(x,t)$ and $f(x,t)$, respectively, superscript n indicates variable evaluated at the nth time step, and subscript t indicates the differential taken with respect to t.

On applying the preceding RBF methods to eqn. (7.64) we obtain

$$u_t^n = D_{\mathcal{L}} u^n + f^n, \tag{7.65}$$

where we have used the relationship of eqn. (7.52). This equation is then integrated using one of the many standard ODE integrators. Of course, boundary conditions have to be included, and these are imposed at the start of each time step calculation. As for non-meshless methods, some numerical integrators appear to work better than others in particular situations, and some examples are discussed subsequently.

The MOL is simple to apply and yields good results. A number of examples are included in the following sections.

7.6.4 With Nonlinear Terms

For some nonlinear PDEs it is possible to modify the earlier solution method for linear PDEs, by treating the nonlinear term(s) separately. Consider the following equation, which is the same as eqn. (7.53), except that it has been modified to include the nonlinear operator $\mathcal{N}$:

$$\begin{aligned}\frac{\partial u}{\partial t} &= \mathcal{L}u + \mathcal{N}u + f(x,t), \; u = u(x,t), \; t \geq 0, \; x \in \Omega, \; \Omega \subset \mathbb{R}^s, \\ u(x,t) &= g(x,t), & x \in \partial\Omega, \\ u(x,0) &= u_0(x), & t = 0.\end{aligned} \tag{7.66}$$

The explicit Euler scheme numerical scheme of eqn. (7.55) therefore becomes

$$u^{n+1} = u^n + \Delta t \mathcal{L} u^n + \Delta t \mathcal{N} u^n + \Delta t f^{n+1}, \tag{7.67}$$

and the θ-method of eqn. (7.58) becomes

$$(1 - \Delta t \theta \mathcal{L})u^{n+1} = (1 + \Delta t (1-\theta)\mathcal{L})u^n + \Delta t \mathcal{N} u^n + \Delta t f^{n+1}. \tag{7.68}$$

In eqn. (7.68) the nonlinear term is handled explicitly and eqn. (7.60) becomes

$$\begin{aligned} Gc^{n+1} &= Hc^n + N^n + F^{n+1} \\ \therefore c^{n+1} &= G^{-1}\left(Hc^n + N^n + F^{n+1}\right),\end{aligned} \tag{7.69}$$

where matrix N represents the nonlinear term as

$$N = \begin{bmatrix} \Delta t \mathcal{N} u \\ 0 \end{bmatrix}. \tag{7.70}$$

The solution u^{n+1} at t^{n+1} is then obtained from eqn. (7.62).

7.6.4.1 Linearization

Some PDEs contain nonlinear terms of the form $u^2, u\frac{\partial u}{\partial x}, u^2\frac{\partial u}{\partial x}$, and so on. For such problems it may be appropriate when applying numerical methods to linearize these terms to facilitate obtaining a solution. We illustrate one way of achieving linearization by example.

Let n represent the nth iteration of the solution method and, assuming that

$$\left(U^{n+1} - U^n\right)\left(U^{n+1} - U^n\right) \approx 0, \tag{7.71}$$

then $\left(U^{n+1}\right)^2$ can be linearized as

$$(UU)^{n+1} \approx 2U^n U^{n+1} - (U^n)^2. \tag{7.72}$$

Similarly, $(UU_x)^{n+1}$ can be linearized as

$$(UU_x)^{n+1} \approx U^n U_x^{n+1} + U^{n+1} U_x^n - U^n U_x^n, \tag{7.73}$$

and $u^2\frac{\partial u}{\partial x}$ can be linearized as

$$(U^2 U_x)^{n+1} \approx (U^2)^n U_x^{n+1} + (2UU_x)^n U^{n+1} - 2(U^2)^n U_x^n, \tag{7.74}$$

and so on.

Thus, this approach can be employed to provide the RBF approximations

$$(UU)^{n+1} \approx 2U^n Ac^{n+1} - (U^2)^n, \tag{7.75}$$

$$(UU_x)^{n+1} \approx U^n A_x c^{n+1} + Ac^{n+1} A_x c^n - U^n A_x c^n, \tag{7.76}$$

and

$$(U^2 U_x)^{n+1} \approx (U^2)^n A_x c^{n+1} + 2U^n A_x c^n Ac^{n+1} - 2(U^2)^n A_x c^n, \tag{7.77}$$

and so on.

7.6.5 Initial Conditions (ICs) and Boundary Conditions (BCs)

To be able to simulate a partial differential equation, certain auxiliary conditions have to be specified in order that the problem is *well posed*. The number of required auxiliary conditions is determined by the *highest order derivative in each independent variable*. For example, for a PDE that is first order in t and second order in x, one auxiliary condition in t (an initial condition) is required together with two auxiliary conditions in x (boundary conditions). If the *value* at a boundary is specified, then this is known as a *Dirichlet* BC, whereas, if the *value of the first derivative* is specified at a boundary, then this is known as a *Neumann* BC. A mixture of the two is known as a *Robin* BC. BCs can be either fixed or time varying. Further discussion relating to the subject of boundary conditions can be found in Chapter 3, and their use in meshless methods is illustrated in the example solutions that follow.

7.6.6 Stability Considerations

Determining the stability of the basic numerical system is not always easy or straightforward to achieve. One key characteristic that is useful to know is whether the fundamental system is *stable* or *well posed*. This is particularly important because, if our numerical solution produces seemingly unstable results, we need to know if this is fundamental to the problem or whether it has been introduced by the solution method we have selected to implement. There are various tools at the analyst's disposal, for example, the stability of PDE meshless solutions can be analyzed in a similar way to other solution methods by investigating the *norm* of the *gain matrix* $G_{\text{gain}} = U^{n+1}/U^n$, as discussed in Chapters 2 and 4. Thus, for the explicit Euler scheme of eqn. (7.56), the stability condition is given by

$$\|G_{\text{gain}}\| = \|1 + \Delta t D_{\mathcal{L}}\| \leq 1, \tag{7.78}$$

and for the weighted average scheme, from eqns. (7.60) and (7.63), we have

$$\|G_{\text{gain}}\| = \|G^{-1}H\| \leq 1, \tag{7.79}$$

where we emphasize that G_{gain} and G are different matrices.

Stability can also be investigated by evaluating the eigenvalues of a discretized system to determine if any are located outside the stability region of the numerical integrator. Should this be the case, then the system would be unstable and an example is given in Section 7.6.7.4. See also [Pla-06] for an interesting discussion relating to stability of RBF schemes.

7.6.7 Time-Dependent PDEs

In this section we provide a number of evolutionary equation examples that bring out interesting aspects of the meshless solution of PDEs.

7.6.7.1 Korteweg–de Vries Equation (KdV): 1D

The KdV[5] equation was originally derived to describe shallow-water waves. Since then, it has found application in many other scientific and mathematical areas—refer to Chapter 3 for further discussion and analysis. It is defined in 1D as

$$\begin{aligned}\frac{\partial u}{\partial t} &= -\frac{\partial^3 u}{\partial x^3} - 6u\frac{\partial u}{\partial x}, \quad u = u(x,t), \; t_0 \leq t \leq t_f, \; x \in \Omega, \; \Omega \subset \mathbb{R}, \\ u(x,t) &= g(x,t), \qquad x \in \partial\Omega, \\ u(x,t=t_0) &= u_0(x), \qquad t = t_0.\end{aligned} \tag{7.80}$$

As an initial condition we use the known *two-soliton*[6] analytical solution described by

$$u_0 = \frac{12\left[3 + 4\cosh(2x - 8t) + \cosh(4x - 64t)\right]}{\left[3\cosh(x - 28t) + \cosh(3x - 36t)\right]^2 + \delta}, \tag{7.81}$$

---

[5] Named after D. J. Kortweg and G. de Vries, who published their analysis in 1895 [Kor-95].
[6] Solitons are solitary waves having the additional property that they can interact with other solitons such that they emerge following a collision without changing shape. However, collisions generally result in a small change in phase.

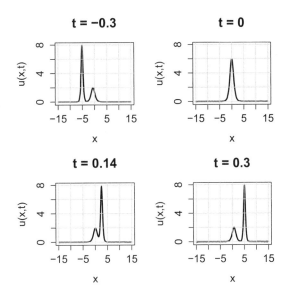

Figure 7.14. Results from a meshless method simulation of the KdV eqn. (7.80) from $t = -0.3$ to 0.3 on a grid of 301 grid points over the domain $x \in (-15, 15)$. The tall, narrow soliton moving from left to right catches up with the short, wider soliton when they coalesce. Both solitons then emerge, shapes unchanged, with the taller soliton having overtaken the shorter soliton.

where $\delta = 10^{-6}$ is a small number that has been included in the denominator to avoid division by zero problems.

After applying RBF collocation we have

$$\frac{\partial u}{\partial t} = -\left(D_{xxx}u + 3D_x(u^2)\right). \quad (7.82)$$

Note that we have represented the term $6u\frac{\partial u}{\partial x}$ in eqn. (7.80) by $3D_x(u^2)$.

The KdV equation is known as a difficult equation to simulate accurately, but the meshless method using multiquadratic RBF $\varphi(r, e) = \frac{1}{\sqrt{(er)^2+1}}$ performed well. A good shape parameter value of $e = 1.3$ was found by trial and error. The discretization consisted of 301 collocation points and numerical integration was performed by a fourth-order Runge–Kutta integrator, with a step size of $\Delta t = 0.0001$. The simulation ran from $t_0 = -0.3$ to $t_f = 0.3$ and executed in just over 15 s on a standard PC. The maximum error at the end of the run, when compared to the theoretical solution, was 4.9e-05, which corresponds to 0.0005%. A similar accuracy performance can be obtained using the lsodes integrator from the R package deSolve (code included with downloads), but the calculations run much more slowly.

This problem requires a *third derivative*, which for the MQ RBF is given by

$$\frac{d^3\varphi}{dx^3} = \frac{-3e^4 r}{\varphi^5}. \quad (7.83)$$

The main results are shown in Fig. 7.14, where characteristic soliton behavior is observed; that is, the tall, narrow soliton moving from left to right catches up with the short, wider soliton when they coalesce. Both solitons then emerge unchanged, with the taller soliton having overtaken the shorter soliton. The only effect is that they each experience a small phase shift, which can be seen in Figs. 7.15.

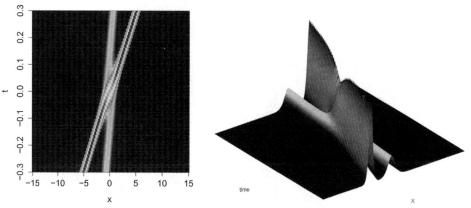

Figure 7.15. Solution $u(x, t)$ from a meshless method simulation of the 1D KdV eqn. (7.80) from $t = -0.3$ to 0.3. The calculation was carried out on a grid of 301 nodes over the domain $x \in (-15, 15)$. (left) x-t plot. (right) Surface plot.

The R code associated with this simulation is shown in Listing 7.4.

```
# File: KdVEqn1D_main.R
# 1D KdV Equation Model
# ut + (6*u*ux + uxxx) =0
#
cat("\014") # Clear console
rm(list = ls(all = TRUE)) # Delete workspace
#
require(compiler)
enableJIT(3)
#
library(rgl)
library("pracma") # For pinv()
#
ptm0 <- proc.time()
print(Sys.time())
#
# lineData_Sqr      # define nodes
source("lineDatCalc.R")
simTitle <- "KdV_1D"
#
# Derivative function
KdVEqn1D_deriv <- function(t, u, parms=NULL) {
  ncall <<- ncall + 1
  ut <- -(3*D1%*%(u^2) + D3%*%u)
  return((ut))
}
cat(sprintf("\nBoundary Nodes = %d, Inner Nodes = %d, Total Nodes = %d\n",
       Nb, Ni, N))
# Set weighted theta value for time integration
alpha <- 1
```

```r
x <- nodeDat[,2]
#
rbf1   <- function(e,r){ sqrt((e*r)^2+1)}        # MQ
dphi1  <- function(e,r,phi){ e^2*r/phi} # 1st derivative wrt r
d3phi1 <- function(e,r,phi){-3*e^4*r/phi^5} # 3rd derivative wrt r
#
r <- array(0,c(Nx,Nx))
for(i in 1:N){
  for(j in 1:N){
    r[i,j]=(xx[i]-xx[j])
  }
}
epsilon <- 0.65 #1.25# 1.54 #6
phi     <- rbf1(epsilon,r)
drphi   <- dphi1(epsilon,r,phi)
d3rphi  <- d3phi1(epsilon,r,phi)
#
t1 <- proc.time()-ptm0
cat(sprintf("\nTime to calculate RBFs, 1st and 2nd derivatives, t1 = %f", t1[3]))
A     <- phi
# Linear differential operators
invA = pinv(A)
D1 <- drphi %*% invA # solve(A)
D3 <- d3rphi %*% invA # solve(A)
#
Ub <- rep(0,Nb) # Dirichlet BC, fixed
#
t2 <- proc.time()-ptm0
cat(sprintf("\nTime to calculate L matrix, t2-t1 <- %f", (t2-t1)[3]) )
ua <- function(t,x){
  uaNum <- 12*(3+4*cosh(2*x-8*t)+cosh(4*x-64*t))
  uaDen <- (3*cosh(x-28*t)+cosh(3*x-36*t))^2 + 1e-06
  ua <- uaNum/uaDen
}
rk4 <- function(U0,t,h,f){
  k1 <- f(t,     U0)
  k2 <- f(t+h/2, U0 + h/2*k1)
  k3 <- f(t+h/2, U0 + h/2*k2)
  k4 <- f(t+h,   U0 + h*k3)
  u  <- U0 + h/6*(k1+2*k2+2*k3+k4)
}
#####################################################
# Set simulation times
#####################################################
nout=51
t0 <- -0.3 #0
```

```r
tf <- +0.3 #30
t <-rep(0,nout)
#####################################################
# ODE integration
#####################################################
ncall<<-0
h <- 0.0001
U0 <- ua(t0,x)
U1 <- U0
t[1] <- t0
tt <- t0
nsteps <- (tf-t0)/(h*(nout-1))
Uout <- array(0,c(nout,N))
Uout[1,] <- U1
for(iout in 2:nout){
  # Next output
  # Take nsteps in t to next output
  if(iout <= nout){
    # rk-4 integration
    for(it in 1:nsteps){
      U1 <- rk4(U1,tt,h,KdVEqn1D_deriv)
      tt <- tt+h
      U1[b_index] <- c(ua(tt,x[1]),ua(tt,x[Nx]))
      U1[(Nx-4):(Nx-1)] <- ua(tt,x[(Nx-4):(Nx-1)]) # needed for stability
    }
  }
  t[iout] <- tt
  cat(sprintf("\n t = %e",tt))
  Uout[iout,] <- U1
  # Next output
}
# Error calc
uanal <- ua(t[nout],x)
err <- Uout[nout,]-uanal
maxErrLoc <- which.max(abs(err))
cat(sprintf("\nMaximum error = %e\n",err[maxErrLoc]))
#####################################################
tFinish <- proc.time()-ptm0
cat(sprintf("\nCalculation time: %f\n",tFinish[3]))
source("KdVEqn1D_postSimCalcs.R")
```

Listing 7.4. File: KdVEqn1D_main.R—Code for the *main* program for meshless RBF method simulation of KdV eqn. (7.80). Calls code in lineDatCalc.R and KdVEqn1D_postSimCalcs.R, which are available in the downloads—these routines set up the computational domain and create associated plots, respectively

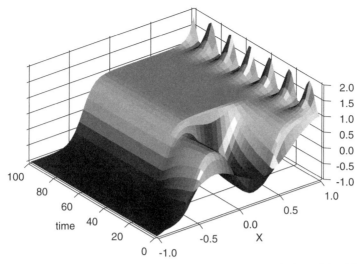

Figure 7.16. Solution $u(x, t)$ obtained from a meshless method simulation of the Allen–Cahn eqn. (7.84) from $t = 0$ to 100 on a grid of 41 nodes over the domain $x \in (-1, 1)$. (See color plate 7.16) `

7.6.7.2 Allen–Cahn Equation: 1D

The *Allen–Cahn* equation describes a *reaction–diffusion* process and is defined in 1D as

$$\frac{\partial u}{\partial t} = \mu \frac{\partial^2 u}{\partial x^2} + u - u^3, \; u = u(x, t), \; t \geq 0, \; x \in \Omega, \; \Omega \subset \mathbb{R},$$

$$u(x, t) = g(x, t), \qquad x \in \partial\Omega, \qquad (7.84)$$

$$u(x, 0) = u_0(x), \qquad t = 0,$$

where μ is a constant.

Consider the problem with $x \in (-1, 1)$ and $u(x, 0) = 0.53x + 0.47 \sin(-1.5\pi x)$ and Dirichlet boundary conditions $u(-1, t) = -1$ and $u(1, t) = 1 + \sin(0.2t)^2$.

After applying RBF collocation we have

$$\frac{\partial u}{\partial t} = \mu D_{xx} u + u - u^3. \qquad (7.85)$$

Figure 7.16 shows a surface plot of the results from a meshless method simulation of this system for $N = 41$ grid points and applying the MOL together with a fourth-order SHK integration scheme and step size $\Delta t = 0.01$ (see Chapter 1). The multiquadratic RBF was used with a shape parameter of $e = 2.5$, found by trial and error.

The simulation runs from 0 to $t = 100$ and executes in just over 3 s on a standard PC and provides good resolution of the solution, bringing out all the salient detail. Almost exactly the same result is obtained with $N = 20$ grid points. At this lower resolution of $N = 20$ the plot is the same as that obtained by Trefethen [Tre-00] and Fasshauer [Fas-07], who both used $N = 20$ for their analysis of the problem, using *spectral* and *meshless* methods, respectively.

The R code associated with this simulation is shown in Listing 7.5. A version using the RK-1 integrator is also provided with the downloads.

```r
# File: AllenCahn1D_SHK_main.R
# 1D Allen-Cahn Equation Model
# ut -mu*uxx -(u - u^3) = 0
#
library(rgl)
library(geometry)
library("pracma")
#
mq <- function(r, e){
  phi <- sqrt((r*e)^2 + 1)
  return((phi))
}
mqDerivatives <- function(r, rx, e, n){
  # r <- |xc-xi|, distance - Euclidian norm
  # rx <- (xc-xi), signed distance
  phi <- sqrt(1+(e*r)^2) # MQ rbf function
  if(n == 1){
    dphi <- e^2*rx/phi # MQ 1st derivative, dphdx
  }else if(n == 2){
    #dphi <- e^2/phi^3 # MQ 2nd derivative, d2phdr2
    dphi <- (e^2)/sqrt(1.0+(r*e)^2)-((rx^2)*(e^4))/(1.0+(r*e)^2)^(1.5) # d2phdx2
  }
  return((dphi))
}
AllenCahn1D_deriv <- function(t, u, parms) {
  ncall <<- ncall +1
  #
  ut <- mu*D2%*%u + u - u^3
  return((ut))
}
ptm0 <- proc.time()
print(Sys.time())
# define computational domain
source("lineDatCalc.R")

simTitle <- "AllenCahn_1D"
cat(sprintf("\nBoundary Nodes = %d, Inner Nodes = %d, Total Nodes = %d\n",
      Nb, Ni, N))
#
x <- nodeDat[,2]
epsilon <- 2.5
#
rs <- array(0,c(N,N)) # signed distance
for(i in 1:N){
  for(j in 1:N){
    rs[i,j]= xx[i]-xx[j]
```

```
  }
}
r1   <- abs(rs)
phi  <- mq(r1,epsilon)
d1phi <- mqDerivatives(r1,rs,epsilon,1)
d2phi <- mqDerivatives(r1,rs,epsilon,2)
#
A    <- phi
D20  <- d2phi
D20[b_index,] <- A[b_index,] # Dirichlet BCs
D2 = D20%*%solve(A)
mu <- 0.01
t1   <- proc.time()-ptm0
cat(sprintf("\nTime to calculate RBFs, 1st and 2nd derivatives, t1 = %f", t1))

Ub <- rep(0,Nb) # Dirichlet BC, fixed
#
t2   <- proc.time()-ptm0
cat(sprintf("\nTime to calculate L matrix, t2-t1 <- %f", t2-t1) )
# Initial condition
U0 <- 0.53*x + 0.47*sin(-1.5*pi*x) # Total nodes
#
#####################################################
# Set simulation times
#####################################################
nout=51
t0 <- 0
tf <- 100
t <-rep(0,nout)
#####################################################
# ODE integration
#####################################################
ncall<<-0
shkOrd <- 4
h <- 0.01
U1 <- U0
t[1] <- t0
tt <- t0
nsteps <- tf/(h*(nout-1))
Uout <- array(0,c(nout,N))
Uout[1,] <- U1
for(iout in 2:nout){
  # Take nsteps in t to next output
  if(iout <= nout){
    # SHK integration (equiv to RK, order=shkOrd)
    for(it in 1:nsteps){
```

```
          k <- 0
          Ub <- c(-1,1 + sin(tt/5)^2) # BC
          U1[b_index] <- Ub
          for(s in seq(shkOrd,1,-1)){
            f <- AllenCahn1D_deriv(tt,U1+k,NULL)
            k <- f*h/s
          }
        # Next output
        tt <- tt+h
        U1 <- U1 + k
        }
      }
      t[iout] <- tt
      cat(sprintf("\n t = %e\n",tt))
      Uout[iout,] <- U1
      # Next output
    }
    plot(Uout[11,])
    ####################################################
    tFinish <- proc.time()-ptm0
    cat(sprintf("\nCalculation time: %f\n",tFinish[3]))
    source("AllenCahn1D_postSimCalcs.R")
```

Listing 7.5. File: AllenCahn1D_SHK_main.R—Code for *main* program for meshless RBF method simulation of Allen–Cahn eqn. (7.84). Calls code in lineDatCalc.R and AllenCahn1D_postSimCalcs.R, which are available in the downloads. These routines set up the computational domain and create associated plots, respectively

7.6.7.3 Fisher–Kolmogorov Equation: 1D

The *Fisher–Kolmogorov* (KP) equation [Fis-37] [Kol-37], also known as the *Fisher–Kolmogorov–Petrovsky–Piscunov* (FKPP) equation, is written in 1D as

$$
\begin{aligned}
\frac{\partial u}{\partial t} &= D\frac{\partial^2 u}{\partial x^2} + ku(1-u), \quad u = u(x,t), \quad t \geq 0, \quad x \in \Omega, \quad \Omega \subset \mathbb{R}, \\
u(x,t) &= g(x,t), \qquad\qquad x \in \partial\Omega, \\
u(x,0) &= u_0(x), \qquad\qquad t = 0.
\end{aligned}
\tag{7.86}
$$

Fisher [Fis-37] first introduced this equation to investigate wave propagation of a gene in a population, and it is now used as a nonlinear model equation to study wave propagation in a large number of biological and chemical systems. It is also used to study logistic growth-diffusion phenomena.

By introducing the transformation

$$
t^* = kt, \quad x^* = x\left(\frac{k}{D}\right)^{\frac{1}{2}}, \tag{7.87}
$$

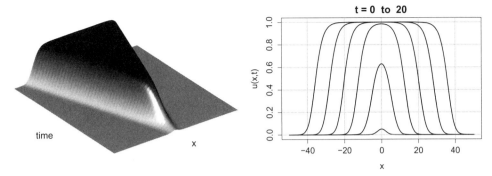

Figure 7.17. Results from a *local* meshless method simulation of the normalized 1D Fisher–Kolmogorov eqn. (7.88) showing how the solution $u(x, t)$ evolves smoothly over time from $t = 0$ to 20. The calculation was carried out on a grid of 301 nodes over the domain $x \in (-50, 50)$. The local stencil consisted of 17 nodes. (left) Surface plot. (right) Solutions at times $t = 0, 4, 8, 12, 16, 20$.

the PDE of eqn. (7.86) is transformed to the *normalized* version

$$\frac{\partial u}{\partial t} = \frac{\partial^2 u}{\partial x^2} + u(1 - u), \tag{7.88}$$

where we have dropped the asterisk. It is this equation that we have simulated using the *local* meshless method outlined in Section 7.5 when, after applying RBF collocation, we have

$$\frac{\partial u}{\partial t} = D_{xx}u + u(1 - u). \tag{7.89}$$

The results are shown in Fig. 7.17.

The simulation was performed over the 1D domain, $x \in (-50, 50)$, on a grid of $N = 301$ nodes and used the lsodes inegrator from the R package deSolve. The meshless RBF method was applied with a *local* stencil consisting of 17 nodes. The multiquadratic RBF was used for each stencil with a shape parameter value of $e = 0.268$, found by a direct search method (see Section 7.5). Dirichlet boundary conditions on $\partial \Omega$ of $u(x = \pm 50, t) = 0$ were imposed at each time step. The main differential matrix D_{xx} is sparse as a result of using local RBFs, and this facilitates the use of sparse matrix methods. Sparse matrix D2s was created using the function sparseMatrix() from the R package Matrix.

The simulation was performed on a standard PC with 8GB of RAM and took approximately 7.5 s to complete.

From the plots in Fig. 7.17, it apparent the solution evolves smoothly as expected and, when u reaches a value of unity, the front continues to propagate at a constant speed of 2. This is in accordance with other published results, for example, see the discussion by Murray [Mur-02, p. 439–449].

The R code associated with the main program for this simulation is shown in Listing 7.6. Functions associated with the local RBF calculations are given in Listings 7.7 and 7.8.

```
# File: Fisher1D_local.R
# 1D Fisher's Equation Model
# ut  <- alpha*uxx + beta*u*(1-u)
```

```
cat("\014") # Clear console
rm(list = ls(all = TRUE)) # Delete workspace
#
require(compiler)
enableJIT(3)
# Main program to simulate 1D Fisher-Kolmogorov Eqn.
library(rgl)
library(deSolve)
library(Matrix) # For sparse matrix functions
source("stencils1D.R")
source("localDiffMatrix1D.R")
#
ptm0 <- proc.time()
print(Sys.time())
#
mq <- function(r, e){
  phi <- sqrt((r*e)^2 + 1)
  return(phi)
}
mqDerivatives <- function(r, rx, e, n){
  # r <- |xc-xi|, distance - Euclidian norm
  # rx <- (xc-xi), signed distance
  phi <- sqrt(1+(e*r)^2) # MQ rbf function
  if(n == 1){
    dphi <- e^2*rx/phi # MQ 1st derivative, dphdx
  }else if(n == 2){
    dphi <- e^2/phi^3 # MQ 2nd derivative, d2phdr2
  }
}
FisherEqn1D_local_deriv <- function(t, u, parms=NULL) {
  ncall <<- ncall +1
  u[b_index] <- Ub # assign Dirichlet BCs
  # Note: The sparse matrix multiplication D2s%*%u is an S4 class
  # operation and the @ symbol is used to refer to a member variable
  ut <- alpha*(D2s%*%u)@x + beta*(1 - u)*u
  return(list(ut))
}
rk4 <- function(U0,t,h,f){
  k1 <- f(t,    U0)
  k2 <- f(t+h/2, U0 + h/2*k1)
  k3 <- f(t+h/2, U0 + h/2*k2)
  k4 <- f(t+h,  U0 + h*k3)
  u  <- U0 + h/6*(k1+2*k2+2*k3+k4)
}
# define nodes
source("lineDatCalc.R")
```

```
#
cat(sprintf("\nCalculating Stencils\n"))
ns <- 17 #17 # stencils size
st <- stencils1D(x ,ns)
# print(dim(st))
# scan(quiet=TRUE)
#
simTitle <- "FisherEqn_1D"
cat(sprintf("\nBoundary Nodes = %d, Inner Nodes = %d, Total Nodes = %d\n",
      Nb, Ni, N))
#
r <- array(0,c(Nx,Nx))
for(i in 1:N){
  for(j in 1:N){
    r[i,j]=abs(x[i]-x[j])
  }
}
########################################
epsilon <- 5.5 # 0.932
cat(sprintf("epsilon = %f",epsilon))
########################################
alpha <- 1; beta <- 1
t1 <- proc.time()-ptm0
cat(sprintf("\nTime to calculate RBFs, 1st and 2nd derivatives, t1 = %f",
    t1[3]))
##################################################
cat(sprintf("\nCalculating localDiffMatrix\n"))
shape <- epsilon #0.43 # initial shape parameter
dc    <- 0.001 # shape parameter increment
minK <- 1e15   # minimum condition number of the system matrix
maxK <- 1e16   # maximum condition number of the system matrix
D2    <- localDiffMatrix1D(st ,x ,shape ,minK, maxK, dc)
# Convert D2 to sparse matrix
nonZeroInds <- which(D2 !=0,arr.ind = TRUE)
ii <- nonZeroInds[,1]
jj <- nonZeroInds[,2]
dd <- D2[nonZeroInds]
rm(D2) # Dense matrix D2 not needed now - save memory!
D2s <- sparseMatrix(i=ii,j=jj,x=dd, dims=c(N,N)) # Create sparse matrix
t2  <- proc.time()-ptm0
cat(sprintf("\nTime to sparse matrix, t2 = %f", t2[3]))
##################################################
Ub <- rep(0,Nb) # Dirichlet BC, fixed
##################################################
# Set Initial Conditions
##################################################
```

```
case <- 0 # 0=Gaussian, 1=square
U0 <- rep(0,Nx)
lambda <- 0.1
A <- 0.05
Nc <- round(Nx/2)
offset_x <- (xl+xu)/2
for(i in 1:Nx){
  # IC starting point
  if(case == 0){
    U0[i] <- A*exp(-lambda*((offset_x-x[i])^2))
  }else {
    if (i>=(Nc-1) && i<=(Nc+1)){
      U0[i] = A
    }
  }
}
###################################################
# Set simulation times
###################################################
nout=51
t0 <- 0
tf <- 20
t <-seq(t0,tf,length=nout)
###################################################
# ODE integration- lsodes
###################################################
ncall <<- 0
out <- ode(method="lsodes",y=U0, times=t, func=FisherEqn1D_local_deriv,
           rtol=1e-12,atol=1e-12, parms=NULL,inz=nonZeroInds,nnz=6000,
           maxsteps=50000, maxord=5, sparsetype = "sparseuser")

Uout <- array(0,dim=c(nout,Nx))
Uout[1:nout,] <- out[1:nout,2:(Nx+1)]
###################################################
source("FisherEqn1DLocal_postSimCalcs.R")
tFinish <- proc.time()-ptm0
cat(sprintf("\nCalculation time: %f\n",tFinish[3]))
```

Listing 7.6. File: Fisher1D_local.R—Code for the *main* program for *local* meshless RBF method simulation of the 1D Fisher–Kolmogorov eqn. (7.88). This program calls code in lineDatCalc.R and FisherEqn1DLocal_postSimCalcs.R, which are available in the downloads. These routines set up the computational domain and create associated plots, respectively. In addition, calls are made to functions stencils1D() and localDiffMatrix1D(), for which the code is given in Listings 7.7 and 7.8

```
# R code to construct 1D stencil matrix
# Adapted from the Matlab function stencils()
# written by Scott A. Sarra of Marshall University
stencils1D <- function(xc, ns){
  N <- length(xc)
  st <- array(0,c(N,ns))
  for(i in 1:N){       # stencils for derivative approximation
    x0 <- xc[i]
    r  <- abs(xc - x0) # distance between center i and the rest of the centers
    ix <- order(r)     # list elements of r in order of magnitude
    st[i ,] <- ix[1:ns] # assign ns elements to ith node
  }
  return((st))
}
```

Listing 7.7. File: R code of function stencils1D() that constructs a 1D stencil matrix (refer to Section 7.5.1)

```
# Constructs a local differentiation matrix
# Adapted from the Matlab function weights()
# written by Scott A. Sarra of Marshall University
localDiffMatrix1D <- function(st,xc,shape,minK,maxK,dc){
  # INPUTS:
  # st indexes of the stencil centers
  # xc centers
  # shape initial shape parameter
  # minK min condition number of the RBF matrix (1 e+13)
  # maxK max condition number of the RBF matrix (1 e+15)
  # dc shape parameter increment
  # OUTPUTS:
  # D weights to discretize
  ndGrid <- function(x,y){
    Nx<-length(x); Ny = length(y)
    x1 <- outer(x,rep(1,Ny),"*") # Same as Matlab: [X Y]=ndgrid(x,y)
    y1 <- outer(rep(1,Nx),y,"*")
    return(list(X1=x1,Y1=y1))
  }
  # warning off
  N <- length(xc) # total centers
  n <- length(st[1, ])
  o <- rep(1,n)
#   D <- array(0, c(N, N)) # sparse???
  D <- matrix(0, nrow=N, ncol=N) # sparse???
  for(i in 1:N){ # interior centers
    pn <- st[i, ]
```

```
    RX <- ndGrid(xc[pn],xc[pn])
    rx <- RX$X1 - RX$Y1
    r <- abs(rx)
    K <- 1
    while(K<minK || K>maxK){
      B <- mq(r ,shape)
      # [U, S ,V] <- svd(B)
      tmp <- svd(B,nu=n,nv=n); U<-tmp$u; S<-diag(tmp$d); V<-tmp$v
      K <- S[1,1]/S[n,n]
      if(K<minK){
        shape <- shape - dc
      }else if(K>maxK){
        shape <- shape + dc
      }
    }
  print(shape)
    Bi <- V%*%(diag(1/tmp$d)%*%t(U))
    ri <- xc[i]-xc[pn] # signed distance
    h <- mqDerivatives( abs(ri), ri ,shape, 2)
    # D(i , pn) = h' * Bi;
    D[i,pn] <- t(h)%*%Bi
  }
  return(D)
}
```

Listing 7.8. File: R code of function `localDiffMatrix1D()` that constructs a 1D local differentiation matrix (refer to section 7.5.2)

7.6.7.4 Heat Equation: 2D

The well known 2 D *heat equation* is defined in Cartesian coordinates on a spatial domain Ω with boundary $\partial\Omega$ as

$$\frac{\partial u}{\partial t} = D\left(\frac{\partial^2 u}{\partial x^2} + \frac{\partial^2 u}{\partial y^2}\right), \quad u = u(x,y,t), \ (x,y) \in \Omega, \ \Omega \subset \mathbb{R}^2, \quad (7.90)$$
$$u(x,y,t) = u_b(x,y), \ (x,y) \in \partial\Omega,$$

where $D = k/(\rho c_p)$ represents *thermal diffusivity*, k *thermal conductivity*, ρ *density*, and c_p *specific heat*.

For this problem the x and y coordinates are each discretized into $N = 21$ grid points and the solution is required over the square domain $\Omega = [0, 1] \times [0, 1]$ with zero initial conditions. Dirichlet boundary conditions were imposed on $\partial\Omega = (x, y) \in (0, 1)$, where the boundary condition at $x = 1$ is given by $u_b(x, y) = \sin(\pi y)$ and $u_b(x, y) = 0$ elsewhere.

A fourth-order Runge–Kutta integrator was used with a fixed step size equal to $\Delta t = 0.0001$. The meshless method was applied with the multiquadratic RBF and a shape parameter value of $\epsilon = 3.2$ was found by trial and error. The result is shown in Fig. 7.18 in the form of a surface plot at the end of the simulation at time $t = 0.1$, for $D = 1$.

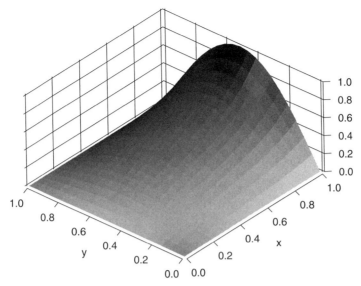

Figure 7.18. Surface plot of solution to heat eqn. (7.90) at time $t = 0.1$.

This is a fairly straightforward problem to solve using either classic finite differences, as in Chapter 3, or using the meshless method, as we have done here. However, because of its simplicity and its linearity, it also serves to demonstrate how stability can be inferred from the differential operator $\mathcal{L}$ of eqn. (7.52). For this problem we have $\mathcal{L} = D\left(\frac{\partial^2}{\partial x^2} + \frac{\partial^2}{\partial y^2}\right)$, which in terms of meshless variables becomes $\mathcal{L} = (A_{xx} + A_{yy})A^{-1}$ (see Section 7.4). Now, because we have

$$\frac{\partial u}{\partial t} = \mathcal{L}u, \tag{7.91}$$

we can estimate the stability of the system by determining if the eigenvalues of $\mathcal{L}$, scaled by Δt, lie within the stability contour of the numerical integrator to be used to obtain a solution, as mentioned in Section 7.6.6. To determine the eigenvalues, we need to modify the elements of the matrix $\mathcal{L}$ corresponding to boundary nodes to be equal to the corresponding elements of A. This accounts for the Dirichlet boundary conditions (see also comment later). The eigenvalues are calculated by a simple call to the R function eigen(). In this case we obtain 441 eigenvalues, the first 10 of which are

```
[1] -2.834044e+03+0.000000e+00i -2.831794e+03+0.000000e+00i
[3] -2.806850e+03+0.000000e+00i -2.806850e+03-0.000000e+00i
[5] -2.800830e+03+0.000000e+00i -2.799626e+03+0.000000e+00i
[7] -2.790135e+03+0.000000e+00i -2.699831e+03+1.470330e+00i
[9] -2.699831e+03-1.470330e+00i -2.699831e+03+1.470493e+00i
```

(remainder omitted to save space). The full set of eigenvalues, scaled by Δt, are plotted in Fig. 7.19, where it is seen that most eigenvalues lie on the negative real line, although

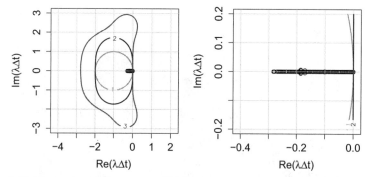

Figure 7.19. (left) Eigenvalues of heat eqn. (7.90) shown with stability contours overlaid for RK-1 (inner) RK-2 and RK-4 (outer). (right) Zoomed in to show more detail.

some do have imaginary parts. All the eigenvalues lie within the Runge–Kutta stability contours, and therefore the system is stable for our choice of RK-4.

The R code of the main simulation program for this problem is included in Listing 7.9, and the code for generating the eigenvalues and associated plots is included in Listing 7.10. Note that because the problem is linear, the differentiation matrix A_{xx} can be set equal to the RBF matrix A at the boundary, thus enforcing the fixed Dirichlet boundary conditions. This does not always work for nonlinear PDEs, where the boundaries may have to be imposed directly at each time step.

```
# File: heatEqn2D.R
# 2D heat Equation Model
# ut = mu*(uxx + uyy)
cat("\014") # Clear console
rm(list = ls(all = TRUE)) # Delete workspace
#
# Program to simulate the 2D heat equation
library(rgl)
library("pracma") # For function pinv()
source("rbfCalc_CS4.R")
#
ptm0 <- proc.time()
print(Sys.time())
#
rk4 <- function(U0,t,h,f){
  k1 <- f(t,     U0)
  k2 <- f(t+h/2, U0 + h/2*k1)
  k3 <- f(t+h/2, U0 + h/2*k2)
  k4 <- f(t+h,   U0 + h*k3)
  u  <- U0 + h/6*(k1+2*k2+2*k3+k4)
}
#
heatEqn2D_deriv <- function(t, u, parms=NULL) {
  ncall <<- ncall +1
```

```
    ut <- L%*%u
    return((ut))
}
source("sqrDatCalc.R")
simTitle <- "Heat"

cat(sprintf("\nBoundary Nodes = %d, Inner Nodes = %d, Total Nodes = %d\n",
         Nb, Ni, N))
#
D <- 1 # Thermal diffusivity
x <- nodeDat[,3]; y <- nodeDat[,4];
# Constant used in RBF
rbfType <- 5; epsilon <- 3.2 # Multiquadratic (MQ)  GSRBF
if(rbfType == 5){
    rbfText <- "Multiquadratic (MQ) Globally Supported RBF"
}
cat(sprintf("\nRBF Type = %s, epsilon = %5.2f\n",rbfText, epsilon))
# Perform RBF calculations
rbfDat <- rbfCalc_CS4(epsilon,x,y,rbfType)
#
r     <- rbfDat$r1;     phi <- rbfDat$phi1
X     <- rbfDat$X1;      Y  <- rbfDat$Y1
drphi <- rbfDat$drphi1; dxphi <- rbfDat$dxphi1; dyphi <- rbfDat$dyphi1
d2rphi <- rbfDat$d2rphi1; d2xphi <- rbfDat$d2xphi1; d2yphi <- rbfDat$d2yphi1
d2phi <- d2xphi+d2yphi # Lapacian
A   <- phi[1:N,1:N]
invA <- pinv(A)

t1 <- proc.time()-ptm0
cat(sprintf("\nTime to calculate RBFs, 1st and 2nd derivatives, t1 = %f", t1))
# Dirichlet BC, fixed
Ub      <- rep(0,Nb)              # Preallocate Ub
bInds   <- which(bNodes[,1]==yu)  # Nodes for U(x=1,y)
Ub[bInds] <- sin(pi*bNodes[bInds,2]/yu) # At BC x=1

D2<- d2phi                # Differentiation matrix
D2[b_index,] <- A[b_index,] # Dirichlet BCs
L <- D*D2%*%invA          # Differential operator

t2 <- proc.time()-ptm0
cat(sprintf("\nTime to calculate A, G and H matrices, t2-t1 <- %f", t2-t1) )

# Initial condition
U0 <- rep(0,N) # Total nodes
U0[b_index] <- Ub
#
```

```
U <- U0
nout <- 11
Uout <- array(0,c(nout,N))
Uout[1,] <- U
ncall <- 0
#
# Time integration
t0 <- 0
t  <- t0
nsteps <- 100          # Number of time steps
h <- 0.0001            # integration step size
beta <- 1/h
#
cat(sprintf("\n t = %7.4f\n",t))
for(iout in 2:nout){
  # Next output
  # Take nsteps in t to next output
  if(iout <= nout){    # Display solution at nout times in t
    for(it in 1:nsteps){
      U[b_index] <- Ub  # Dirichlet BCs
      Un <- U # Advance U
      t <- t+h # Advance time (here because F calculated at t_(n+1)
      U <- rk4(Un,tt,h,heatEqn2D_deriv)
    }
  }
  cat(sprintf("\n t = %7.4f\n",t))
  Uout[iout,] <- U
  # Next output
}
source("heatEqn2D_postSimCalcs.R")
source("stabilityPlot.R")
cat(sprintf("Calculation time: %f\n",(proc.time()-ptm0)[3]))
```

Listing 7.9. File: heatEqn2D.R—Code for *main* program for meshless RBF method simulation of the 2D heat eqn. (7.90). Calls code in sqrDatCalc.R and heatEqn2D_postSimCalcs.R, which are available as downloads. These routines set up the computational domain and create associated plots, respectively. In addition, this program also calls function rbfCalc_CS4() of Listing 7.13

```
# File: stabilityPlot.R
# Stability analysis
L2 <- L # Note: L has been modified for Dirichlet BCs in main program
#L2[b_index] <- (A%*%as.matrix(invA))[b_index] # Dirichlet BC at x[1]
#
lam <- eigen(L2)$values
lam_dt <- lam*h
specAbscissa <- max(Re(lam_dt))
```

```r
cat(sprintf("spectral abscissa, max{Re(lam*dt)} = %e",specAbscissa))
# Set plot dimensions
zoom <- 0
if(zoom ==0){
  xxl <- -4;   xxu <- 2
  yyl <- -3;   yyu <- 3
}else{
  xxl <- -0.4; xxu <- 0
  yyl <- -0.2; yyu <- 0.2
}
plot(Re(lam_dt),Im(lam_dt),xlim=c(xxl,xxu),ylim=c(yyl,yyu),
     xlab=expression(paste("Re(",lambda,Delta,t,")")),
     ylab=expression(paste("Im(",lambda,Delta,t,")")),
     main=expression(paste("Eigenvalues scaled by ",Delta,t)))
grid(lty=1)
# Overlay integrator stability contours
# Construct mesh
Nx2 <- 100; Ny2 <- 100
x2 <- seq (xxl ,xxu ,len= Nx2)
y2 <- seq (yyl ,yyu ,len= Ny2)
X2 <- matrix ( rep(x2 , each =Nx2),nrow =Nx2);
Y2 <- matrix ( rep(y2 , each =Ny2),nrow =Ny2)
# Calculate z
z <- t(X2) + 1i*Y2 # Forward Euler (Runge - Kutta 1)
G1 <- 1 + z;
# Runge - Kutta 2
G2 <- 1 + z + 0.5* z ^2
# Runge - Kutta 4
G4 <- 1 + z + 0.5* z^2 + (1/6) *z^3 + (1/24) *z ^4;
# Runge - Kutta RKF45
#G5 <- 1+z +(1/2) *z ^2+(1/6) *z ^3+(1/24) *z ^4+(1/120) *z ^5+(1/2080) *z^6
# Calculate magnitude of G
G_mag1 <- abs(G1);
G_mag2 <- abs(G2);
G_mag4 <- abs(G4);
# G_mag5 <- abs(G5)
# Plot contours of G_mag for RK -1 to RK -4
contour (x2 ,y2 , G_mag1, levels =1, cex.lab =2, labels = "1", col =" green ",
        lwd=2, add= TRUE )
contour (x2 ,y2 , G_mag2, levels =1, cex.lab =2, labels = "2", col =" blue ",
        lwd=2, add= TRUE )
contour (x2 ,y2 , G_mag4, levels =1, cex.lab =2, labels = "3", col =" red",
        lwd=2, add= TRUE )
```

Listing 7.10. File: stabilityPlot.R—Code for generating the eigenvalues of differential operator $\mathcal{L}$ of eqn. (7.91) and the plots in Fig. 7.19

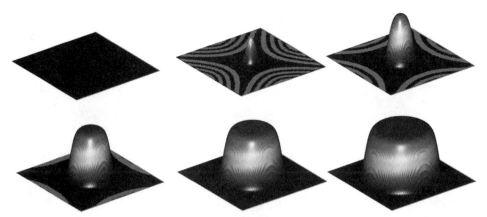

Figure 7.20. Sequence of *local* meshless method solutions $u(x, y, t)$ of the 2D Fisher–Kolmogorov equation (7.92). From left to right, top to bottom, results of simulation at times $t = 0, 0.0004, 0.0008, 0.0012, 0.0016,$ and 0.002. (See color plate 7.20)

7.6.7.5 Fisher–Kolmogorov Equation: 2D

This is the same problem as the example discussed in Section 7.6.7.3, except that it is calculated on a 2D domain and the PDE is not normalized. The 2D PDE that we simulate therefore becomes

$$\frac{\partial u}{\partial t} = D \left(\frac{\partial^2 u}{\partial x^2} + \frac{\partial^2 u}{\partial y^2} \right) + ku(1 - u), \tag{7.92}$$

and after applying RBF collocation, we have

$$\frac{\partial u}{\partial t} = D \left(D_{xx} + D_{yy} \right) u + ku(1 - u). \tag{7.93}$$

A 101×101 grid was employed over the 2D domain $(x, y) = [-0.5, 0.5] \times [-0.5, 0.5]$ using a 21-node local stencil. This is a more demanding problem than the 1D example from earlier, which required the use of the transformation of eqn. (7.87) to be able to obtain a stable solution. Sparse matrix methods were again employed by using sparse matrix D2s created by the function sparseMatrix() from the R package Matrix.

The parameter values were set to $D = 1$ and $k = 10,000$. Thus, the outward velocity for the 2D problem will be 100 times that of the speed of the normalized 1D problem. Therefore, to simulate the equivalent of the 1D problem, it follows from eqn. (7.87) that the spatial domain becomes $(x, y) \in [-0.5, 0.5] \times [-0.5, 0.5]$ with the simulation time being reduced to $t = 0.002$. As for the 1D example, zero Dirichlet boundary conditions were imposed on $\partial\Omega$, that is, $u(x = \pm 0.5, y, t) = 0$ and $u(x, y = \pm 0.5, t) = 0$.

The initial condition was set to zero, except for a small Gaussian perturbation from zero at the center of the domain, that is,

$$u(x, y, 0) = 0.05 \exp[-\lambda(x^2 + y^2)], \quad \lambda = 1000. \tag{7.94}$$

A fourth-order Runge–Kutta integrator was used with a fixed step size equal to $\Delta t = 0.00001$. The meshless method was applied with the multiquadratic RBF and a shape parameter value of $\epsilon = 7.8$ was used for each stencil, found by direct search method (see Section 7.5). The results are shown in Figs. 7.20 and 7.21. The simulated responses (speed and distance traveled) when transformed back to normalized units compare very well with the 1D results, as does the overall shape. A similar performance can be obtained

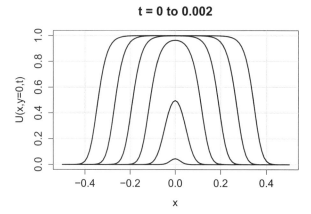

Figure 7.21. Sequence of 2D solutions through the center at $y = 0$ from a *local* meshless method simulation of the 2D Fisher–Kolmogorov eqn. (7.92). Plots are shown at times $t = 0, 0.0004, 0.0008, 0.0012, 0.0016,$ and 0.002. The small initial perturbation grows steadily until it reaches a value of 1, when the solution expands outward at a steady speed of 200. Note: This is equivalent to the *normalized* speed of 2 obtained in Section 7.6.7.3. Compare this plot to the 1D results shown in Fig. 7.17 at normalized times.

using the lsodes integrtor from the R package deSolve (code included with downloads), but the calculations run more slowly.

Decreasing the number of nodes from $N = 101$ to $N = 51$ did not significantly change the solution results. However, it did reduce the time to create the stencil data and to assemble the differentiation matrix from 23.6 to 2.8 s. The reason for this is that the size of the RBF matrix increases in proportion to the square of N. Thus the RBF matrix went from $10,201 \times 10,201$ to 2601×2601 (101^2 to 51^2). But, interestingly, the time to integrate the solution once the differentiation matrix was assembled only decreased from 2.6 to 1.2 s. This was because with local meshless methods, sparsity increases with N and sparse matrix methods become more efficient as sparsity increases. Unfortunately, it does not appear possible to make use of sparse matrix methods for creating the stencil data and assembling the differentiation matrix.

The use of sparse matrices also has the added advantage of using less memory. Using the R function object.size(), we find that the $10,201 \times 10,201$ dense matrix D2 uses 832 MB of RAM, whereas the sparse matrix D2s uses 2.5 MB—a reduction factor of 320. Therefore, to conserve memory, we remove the dense matrix by the command rm(D2) after D2s has been created, and D2 is no longer needed.

On trying to use the meshless method with dense matrices, the process became impracticable for $N > 81$ owing to memory allocation problems and increased execution times. However, more significantly, unlike the local method, this number of grid points was insufficient to obtain an accurate solution. Various compactly supported RBFs were also tried but although they gave better results, were too impractical for $N > 81$.

The R code associated with this simulation is shown in Listing 7.11.

```
# File: FisherEqn_2D_local.R
# Main program to simulate 2D Fisher-Kolmogorov Eqn.
# ut = alpha*(uxx + uyy) + beta*u(1-u)
cat("\014") # Clear console
```

```
rm(list = ls(all = TRUE)) # Delete workspace
#
library(rgl)
library("Matrix") # For sparse matrix functions
source("stencils2D.R")
source("localDiffMatrix2D.R")
source("mq.R")
source("mqDerivatives.R")
source("rk4.R")
source("FisherEqn2D_deriv.R")
#
ptm0 <- proc.time()
print(Sys.time())
#
source("sqrDatCalc.R")
x <- nodeDat[,3]; y <- nodeDat[,4];
#########################################
cat(sprintf("\nCalculating Stencils\n"))
ns <- 21 # 21 # stencils size (9 Or 21 to be symmetrical)
st <- stencils2D(x,y,ns)
#########################################
simTitle <- "Fishers_Eqn"

cat(sprintf("\nBoundary Nodes = %d, Inner Nodes = %d, Total Nodes = %d\n",
        Nb, Ni, N))
alpha <- 1; beta <- 10000
# Preallocate arrays for efficiency
FL <- rep(0,Ni); FB <- rep(0,Nb)
#
rbfText <-  "Multiquadratic (MQ) Globally Supported RBF"
cat(sprintf("\nRBF Type = %s\n",rbfText))
#
t1  <- proc.time()-ptm0
cat(sprintf("\nTime to calculate RBFs and 2nd derivatives, t1 = %f", t1))
# Dirichlet BC, fixed
Ub <- rep(0,Nb) # Dirichlet BC, fixed
# Initial condition
U0 <- rep(0,N) # Total nodes
U01 <- array(0,c(Nx,Ny))
#
len <- (xu-xl)/2
lambda <- 1000/(2*len)^2
A <- 0.05
offset <- 0
for(i in 1:Nx){
  for(j in 1:Ny){
```

```
      # IC starting point
      U01[i,j] <- A*exp(-lambda*((xx[i]-offset)^2+(yy[j]-offset)^2))
  }
}
U0 <- array(U01,c(N,1))

U0[b_index] <- Ub
###################################################
cat(sprintf("\nCalculating Weights\n"))
shape <- 300 #epsilon #0.43 # initial shape parameter
dc    <- 0.0001 # shape parameter increment
minK  <- 1e12   # minimum condition number of the system matrix
maxK  <- 1e15   # maximum condition number of the system matrix
D2    <- localDiffMatrix2D(st,x,y,shape ,minK, maxK, dc)
# Convert D2 to sparse matrix
nonZeroInds <- which(D2 !=0,arr.ind = TRUE)
ii <- nonZeroInds[,1]
jj <- nonZeroInds[,2]
dd <- D2[nonZeroInds]
rm(D2) # Dense matrix D2 not needed now - save memory!
D2s <- sparseMatrix(i=ii,j=jj,x=dd, dims=c(N,N))
t2  <- proc.time()-ptm0
cat(sprintf("\nTime to sparse matrix, t2 = %f", t2[3]))
###################################################
t2 <- proc.time()-ptm0
cat(sprintf("\nAbout to start numerical integration\n") )
# set up integration
###################################################
# Set time step parameters
nout <- 11 # Display solution at nout times in t
ncall <- 0
h <- 0.00001           # Time step
t0 <- 0; tf <- 0.002;
nsteps <- tf/(h*(nout-1)) #4      # Number of time steps
t   <- seq(from=t0,to=tf,length.out=nout)
tt <- t0
#
###################################################
# ODE integration
###################################################
#
Uout <- array(0,c(nout,N))
Uout[1,] <- U0
U <- U0
ncall <- 0
cat(sprintf("\n t[1] = %7.4f\n",t[1]))
```

```
for(iout in 2:nout){
  # Next output
  # Take nsteps in t to next output
  if(iout <= nout){
    for(it in 1:nsteps){
      U[b_index] <- Ub # Apply Dirichlet BCs
      Un <- U # Advance U
      tt <- tt+h # Advance time (here because F calculated at t_(n+1)
      U <- rk4(Un,tt,h,FisherEqn2D_deriv)
    }
  }
  t[iout] <- tt
  cat(sprintf("\n t[%d] = %7.4f\n",iout,t[iout]))
  Uout[iout,] <- U
  # Next output
}
source("FisherEqn_2D_postSimCalcs.R")
cat(sprintf("Calculation time: %f\n",(proc.time()-ptm0)[3]))
```

Listing 7.11. File: `FisherEqn_2D_local.R`—Code for *main* program for meshless RBF method simulation of the 2D Fisher–Kolmogorov eqn. (7.92). Calls code in `sqrDatCalc.R`, `stencils2D.R`, `localDiffMatrix1D.R`, `FisherEqn2D_deriv.R`, and `FisherEqn2DLocal_postSimCalcs.R`, which are available for download. These routines set up the computational domain, construct the stencil matrix, calculate the local 2D derivative matrix, calculate the ODE derivatives, and create associated plots, respectively.

7.6.8 Time-Independent PDEs

The solution of *time-independent* linear PDEs, such as the *Laplace* and *Poisson* equations, is well suited to meshless methods. Our discussion will be developed in terms of these two equations, but the ideas are readily extended to other linear, time-independent equations.

7.6.8.1 Laplace Equation

The Laplace equation is a *parabolic* PDE and is defined in *Cartesian* coordinates on a spatial domain Ω with boundary $\partial\Omega$ as

$$\frac{\partial^2 u}{\partial x^2} = 0, \quad u = u(x), \; x \in \Omega, \; \Omega \subset \mathbb{R}^s, \; s \in (1, 2, 3), \tag{7.95}$$
$$u(x) = g(x), \; x \in \partial\Omega.$$

After discretization and applying RBF approximations, we obtain the following matrix equation:

$$Gc = F$$
$$\therefore c = G^{-1}F, \tag{7.96}$$

and the solution is given by

$$u = Ac, \tag{7.97}$$

where A and c are defined by eqn. (7.5).

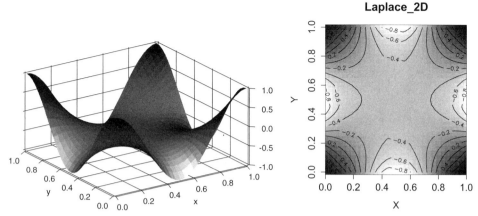

Figure 7.22. Laplace equation solution $u(x, y)$ for example 1, obtained by the meshless method using the Wendland(3,2) RBF with compact support. The x and y coordinates were discretized in to $N = 31$ grid points. (left) Surface plot. (right) A 2D plot with contours.

The matrices G and F are defined as

$$G = \begin{bmatrix} A_{xx}^I \\ A^B \end{bmatrix}, \quad F = \begin{bmatrix} 0^I \\ g^B \end{bmatrix}, \tag{7.98}$$

where we have used superscripts I and B to indicate *inner* and *boundary* nodes, respectively.

Thus we have reduced the solution of the Laplace eqn. (7.95) to that of solving a very simple matrix equation.

We now consider two Laplace equation examples.

Example 1

Consider a 2D problem described by eqn. (7.95) with coordinates x and y that are each discretized into $N = 31$ grid points. The solution is required over the square domain $\Omega = [0, 1] \times [0, 1]$ with Dirichlet boundary conditions on $\partial\Omega = (x, y) \in (0, 1)$. The problem therefore becomes

$$\begin{aligned} \frac{\partial^2 u}{\partial x^2} + \frac{\partial^2 u}{\partial y^2} &= 0, \quad u = u(x, y), \ (x, y) \in \Omega, \ \Omega \subset \mathbb{R}^2, \\ u(x, y) &= \cos[2\pi(x + y)], \quad (x, y) \in \partial\Omega. \end{aligned} \tag{7.99}$$

The problem of eqn. (7.99) is provided as an example of applying the meshless method using the Wendland(3,2) RBF with compact support. A shape parameter value of $e = 1.6$ was found by trial and error to give good results, which are shown in Fig. 7.22. The solution was obtained in 6.2 s. Simulations using the MQ and IMQ RBFs with global support also gave very similar results.

The R code for this problem is shown in Listings 7.12 and 7.13 with variable case=1.

```
# File: LaplaceEqn2d_main.R
# 2D Laplace Equation Model
# alpha*(uxx + uyy) = 0
cat("\014") # Clear console
```

```
rm(list = ls(all = TRUE)) # Delete workspace
#
library(rgl)
source("rbfCalc_CS4.R")
#
ptm0 <- proc.time()
print(Sys.time())

source("sqrDatCalc.R")
simTitle <- "Laplace_2D"

cat(sprintf("\nBoundary Nodes = %d, Inner Nodes = %d, Total Nodes = %d\n",
      Nb, Ni, N))
#
x <- nodeDat[,3]; y <- nodeDat[,4];
# Preallocate arrays
FL <- rep(0,Ni); FB <- rep(0,Nb)

# Define x and y domains
xmax <- max(x) # 1.5
ymax <- max(y) # 1
rmax <- sqrt(xmax^2+ymax^2)
# Set epsilon and RBF type by uncommenting appropriate line:
# ===========================================================
rbfType <- 1; epsilon <- 1.6 # WendLand phi(3,2)   CSRBF
# rbfType <- 5; epsilon <- 6 # Multiquadratic (MQ)   GSRBF
# rbfType <- 6; epsilon <- 4.0 # Inverse Multiquadric (IMQ) GSRBF
# RBF information for plot titles
if (rbfType == 1){
  rbfText <- "WendLand phi(3,2) Compactly Supported RBF"
}else if(rbfType == 5){
  rbfText <- "Multiquadratic (MQ) Globally Supported RBF"
}else if(rbfType == 6){
  rbfText <- "Inverse Multiquadric (IMQ) Globally Supported RBF"
}
cat(sprintf("\nRBF Type = %s, epsilon = %5.2f\n",rbfText, epsilon))
#
# Perform RBF calculations
rbfDat <- rbfCalc_CS4(epsilon,x,y,rbfType) # Call to RBF functions
r      <- rbfDat$r1;      phi   <- rbfDat$phi1
X      <- rbfDat$X1;      Y     <- rbfDat$Y1
drphi  <- rbfDat$drphi1; dxphi <- rbfDat$dxphi1; dyphi <- rbfDat$dyphi1
d2rphi <- rbfDat$d2rphi1; d2xphi <- rbfDat$d2xphi1; d2yphi <- rbfDat$d2yphi1
d2phi  <- d2xphi+d2yphi # Lapacian
A      <- phi[1:N,1:N]
```

```
t1 <- proc.time()-ptm0
cat(sprintf("\nTime to calculate RBFs, 1st and 2nd derivatives, t1 = %f",
    t1[3]))

f <- rep(0,Ni)
for(i in 1:Ni){
 f[i] <- 0
}
Ub <- rep(0,Nb)
# Set case
case <- 1
if(case == 1){
  Ub <- cos(2*pi*(bNodes[,1] + bNodes[,2]))
}else if(case ==2){
  bInds_yl <- which(bNodes[,2]==0, arr.ind = TRUE)
  Ub[bInds_yl] <- bNodes[bInds_yl,1]*(1-bNodes[bInds_yl,1]) # Dirichlet -
      overwrite
}
# Differential operator
L  <- d2phi[i_index,]
# Interior nodes
GL <- L
# Boundary Nodes (Ni+1 to N)
GB <- phi[b_index,]
#
G  <- rbind(GL, GB)
#
HB <- rep(0,Nb); HL <- rep(0,Ni)
#
t2 <- proc.time()-ptm0
cat(sprintf("\nTime to calculate A, G and H matrices, t2-t1 <- %f",
    (t2-t1)[3]) )
#
FL <- rep(0,Ni)
FB <- rep(0,Nb)
FL <- f
FB <- Ub
F  <- c(FL, FB)
# Nonlinear term handled explicitly!
NL <- 0
#
H  <-  c(HL+NL, HB)
# Calculate solution
c  <-  solve(G,F)
U  <-  A%*%c
#
```

```
source("LaplaceEqn2D_postSimCalcs.R")
cat(sprintf("Calculation time: %f\n",(proc.time()-ptm0)[3]))
```

Listing 7.12. File: LaplaceEqn2d_main.R—Code for meshless RBF method simulation of the Laplace eqn. (7.99). Calls code in sqrDatCalc.R and LaplaceEqn2D_postSimCalcs.R, which are available with the downloads. These routines set up the computational domain and create associated plots, respectively. This main program also calls the function rbfCalcCS4() of Listing 7.13, which performs various RBF calculations

```
# File: rbfCalc_CS4.R
# Calculates GSRBF and CSRBF values and their 1st and 2nd derivatives
rbfCalc_CS4 <- function(epsilon,x,y,rbfType){
# Usage ;
# ######
#   rbfDat <- rbfCalc_CS4(epsilon,x,y,rbfType)
# #
#   r        <- rbfDat$r1;     phi    <- rbfDat$phi1
#   X        <- rbfDat$X1;     Y      <- rbfDat$Y1
#   drphi    <- rbfDat$drphi1; dxphi  <- rbfDat$dxphi1; dyphi <- rbfDat$dyphi1
#   d2rphi   <- rbfDat$d2rphi1; d2xphi <- rbfDat$d2xphi1; d2yphi <- rbfDat$d2yphi1
#
  X <- outer(x,rep(1,N),"*") # Same as Matlab: [X Y]=ndgrid(x,y)
  Y <- outer(rep(1,N),y,"*")
  dX <- t(X)-X; dY <- t(Y)-Y
  r <- sqrt(dX^2+dY^2)
# N <- length(x)
  rm1 <- 1-epsilon*r       # Radial cut-off function
  rm1[rm1/epsilon<0] <- 0 # Set -ve elements of rm1 to zero

  rr <- r
# See: Fassheuar, 2007.
  if(rbfType == 1){          # Wendland phi(3,2) CSRBF
    phi     <- rm1^6*(3+18*epsilon*rr+35*epsilon^2*rr^2) # RBF
    drphi   <- -20*epsilon^2*rr*rm1^3                    # 1st derivative wrt r
    d2rphi  <- 20*epsilon^2*(4*epsilon*rr-1)*rm1^2       # 2nd derivative wrt r
    dxphi   <- -56*rm1^5*epsilon^2*dX*(5*epsilon*rr+1)   # 1st derivative wrt x
    dyphi   <- -56*rm1^5*epsilon^2*dY*(5*epsilon*rr+1)   # 1st derivative wrt y
    d2xphi  <- 56*rm1^4*epsilon^2*(-1-4*epsilon*rr+
               epsilon^2*(5*rr^2+30*dX^2))               # 2nd derivative wrt x
    d2yphi  <- 56*rm1^4*epsilon^2*(-1-4*epsilon*rr+
               epsilon^2*(5*rr^2+30*dY^2))               # 2nd derivative wrt y
  }else if(rbfType == 2){    # Wendland phi(3,3) CSRBF
    phi <- rm1^8*(32*epsilon^3*rr^3+        # RBF
           25*epsilon^2*rr^2+8*epsilon*rr+1)
    drphi   <- -22*rm1^7*epsilon^2*rr*(16*epsilon^2*
               rr^2+7*epsilon*rr+1)           # 1st derivative wrt r
```

```
            d2rphi  <- 22*rm1^6*epsilon^2*(-1-6*epsilon*rr+15*(epsilon*rr)^2+
                       160*(epsilon*rr)^3)            # 2nd derivative wrt r
            dxphi   <- -22*rm1^7*epsilon^2*dX*(16*epsilon^2*
                       rr^2+7*epsilon*rr+1)           # 1st derivative wrt x
            dyphi   <- -22*rm1^7*epsilon^2*dY*(16*epsilon^2*
                       rr^2+7*epsilon*rr+1)           # 1st derivative wrt y
            d2xphi  <- 352*rm1^6*(-1/16+rr*(rr^2+9*dX^2)*
                       epsilon^3+(-9*rr^2*(1/16)+3*dX^2*(1/2))*
                       epsilon^2-3*epsilon*rr*(1/8))*epsilon^2
            d2yphi  <- 352*rm1^6*(-1/16+rr*(rr^2+9*dY^2)*
                       epsilon^3+(-9*rr^2*(1/16)+3*dY^2*(1/2))*
                       epsilon^2-3*epsilon*rr*(1/8))*epsilon^2
    }else if(rbfType == 3){      # Wendand phi(6,4) CSRBF
            phi     <- rm1^10*(5+50*(epsilon*rr)+          # RBF
                       210*(epsilon*rr)^2+450*(epsilon*rr)^3+
                       +429*(epsilon*rr)^4)
            drphi   <- -5*rm1^9*epsilon^2*rr*              # 1st derivative wrt r
                       (1201.2*epsilon^3*rr^3+826.8*
                       epsilon^2*rr^2+234*epsilon*rr+26)
            d2rphi  <- 5*rm1^8*epsilon^2*                  # 2nd derivative wrt r
                       (15615.6*epsilon^4*rr^4+5116.8*
                       epsilon^3*rr^3+93.6*epsilon^2*rr^2-
                       208*epsilon*rr-26)
            dxphi   <- -26*rm1^9*epsilon^2*dX*             # 1st derivative wrt x
                       (231*epsilon^3*rr^3+45*
                       epsilon*rr+159*epsilon^2*rr^2+5)
            dyphi   <- -26*rm1^9*epsilon^2*dY*             # 1st derivative wrt y
                       (231*epsilon^3*rr^3+45*
                       epsilon*rr+159*epsilon^2*rr^2+5)
            d2xphi  <- 26*rm1^8*epsilon^2*                 # 2nd derivative wrt x
                       ((132*epsilon^2+1056*epsilon^3*rr+
                       2772*epsilon^4*rr^2)*dX^2-114*
                       epsilon^2*rr^2-40*epsilon*r-5-
                       72*epsilon^3*rr^3+231*epsilon^4*rr^4)
            d2yphi  <- 26*rm1^8*epsilon^2*                 # 2nd derivative wrt y
                       ((132*epsilon^2+1056*epsilon^3*rr+
                       2772*epsilon^4*rr^2)*dY^2-114*
                       epsilon^2*rr^2-40*epsilon*r-5-
                       72*epsilon^3*rr^3+231*epsilon^4*rr^4)
    }else if(rbfType == 4){      # Gaussian
            phi     <- exp(-(epsilon*r)^2)
            drphi   <- -2*epsilon^2*rr*exp(-(epsilon*rr)^2) # 1st derivative wrt r
            d2rphi  <- 2*epsilon^2*exp(-(epsilon*rr)^2)*    # 2nd derivative wrt r
                       (2*(epsilon*rr)^2-1)
            dxphi   <- -2*epsilon^2*dX*phi
            dyphi   <- -2*epsilon^2*dY*phi
```

```
      d2xphi <- (2*epsilon^2*dX)^2*phi+(-2*epsilon^2)*phi
      d2yphi <- (2*epsilon^2*dY)^2*phi+(-2*epsilon^2)*phi
    }else if(rbfType == 5){      # Multiquadric
      phi    <- sqrt((epsilon*r)^2+1)
      drphi  <- epsilon^2*rr/phi                     # 1st derivative wrt r
      d2rphi <- epsilon^2*rr/phi^3                   # 2nd derivative wrt r
      dxphi  <- epsilon^2*dX/phi
      dyphi  <- epsilon^2*dY/phi
      d2xphi <- epsilon^2*(1+epsilon^2*dY^2)/phi^3
      d2yphi <- epsilon^2*(1+epsilon^2*dX^2)/phi^3
    }else if(rbfType == 6){      # Inverse Multiquadric
      phi    <- 1/sqrt((epsilon*r)^2+1)
      drphi  <- -epsilon^2*rr/phi^3                  # 1st derivative wrt r
      d2rphi <- epsilon^2*((2*epsilon*rr)^2-1)/
                phi^5                                # 2nd derivative wrt r
      dxphi  <- -epsilon^2*dX*phi^3
      dyphi  <- -epsilon^2*dY*phi^3
      d2xphi <- epsilon^2*((2*dX^2-dY^2)*epsilon^2-1)*phi^5
      d2yphi <- epsilon^2*((2*dY^2-dX^2)*epsilon^2-1)*phi^5
    }else if(rbfType == 7){      # Inverse Quadratic (IQ)
      phi    <- 1/((epsilon*r)^2+1)
      drphi  <- -2*epsilon^2*rr/phi^2                # 1st derivative wrt r
      d2rphi <- 2*epsilon^2*(3*(epsilon*rr)^2-1)/
                phi^3                                # 2nd derivative wrt r
      dxphi  <- -epsilon^2*2*dX*phi^2
      dyphi  <- -epsilon^2*2*dY*phi^2
      d2xphi <- epsilon^2*((2*dX^2-dY^2)*epsilon^2-1)*phi^5
      d2yphi <- epsilon^2*((2*dY^2-dX^2)*epsilon^2-1)*phi^5
    }else if(rbfType == 8){      # Generalized Inverse Multiquadratic (GIMQ)
      phi    <- 1/((epsilon*r)^2+1)^2
      drphi  <- -4*epsilon^2*rr/phi^(3/2)            # 1st derivative wrt r
      d2rphi <- 4*epsilon^2*(5*(epsilon*rr)^2-1)/
                phi^(2)                              # 2nd derivative wrt r
      dxphi  <- -epsilon^2*4*dX*phi^3
      dyphi  <- -epsilon^2*4*dY*phi^3
      d2xphi <- epsilon^2*((2*dX^2-dY^2)*epsilon^2-1)*phi^5
      d2yphi <- epsilon^2*((2*dY^2-dX^2)*epsilon^2-1)*phi^5
    }
  return(list(r1=r,phi1=phi,X1=X,Y1=Y,drphi1=drphi,dxphi1=dxphi,
          dyphi1=dyphi,d2rphi1=d2rphi,d2xphi1=d2xphi,d2yphi1=d2yphi) )
}
```

Listing 7.13. File: rbfCalc_CS4.R—Code for function rbfCalc_CS4(), which performs calculations for a range of compact and globally supported RBFs

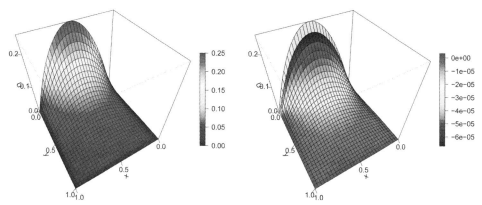

Figure 7.23. Meshless method solution u(x, y) of Laplace eqn. (7.100) discretized into a 31 × 31 grid. (left) Legend colors indicate solution values. (right) Rendering of the solution to indicate the corresponding error value at each location on the surface.

Example 2
This second Laplace equation example is also a 2D problem described by eqn. (7.95). The solution is required over the square domain $\Omega = [0, 1] \times [0, 1]$ with Dirichlet boundary conditions on $\partial\Omega = (x, y) \in (0, 1)$. The problem therefore becomes

$$\frac{\partial^2 u}{\partial x^2} + \frac{\partial^2 u}{\partial y^2} = 0, \quad u = u(x, y), \ (x, y) \in \Omega, \ \Omega \subset \mathbb{R}^2, \quad (7.100)$$

$$u(x, y = 0) = x(1 - x), \text{ zero elsewhere}, \ (x, y) \in \partial\Omega.$$

This problem has the following analytical *Fourier series* solution [Art-98, p. 255–257], which will enable us to determine the accuracy of the numerical solution:

$$u_a(x, y) = \sum_{n=0}^{\infty} \left(-4 \frac{\sin(n\pi x)((-1)^n - 1)\sinh(n\pi(1-y))}{\sinh(n\pi)n^3\pi^3}\right). \quad (7.101)$$

The coordinates x and y are again each discretized into $N = 31$ grid points, but this time the meshless method was applied with the inverse multiquadratic RBF. A shape parameter value of $e = 4$ was found by trial and error to give good results. The maximum numerical solution error when compared to the analytical solution u_a, was a very respectable −6.7e-05. The series solution of eqn. (7.101) was summed to $n = 50$, and larger values for n made insignificant difference to u_a. For example, the maximum difference between $u_{a,50}$ and $u_{a,150}$ was −5.4e-09, that is, when n increased from 50 to 150.

The results are shown in Fig. 7.23, where the plots were generated by the R function persp(). The legends were added by calls to the function image.plot(), which is provided by the R package fields.

Figure 7.23 shows a surface plot of the solution with colors indicating solution values for u, together with a plot with false coloring to indicate error values over the solution domain.

The R code for this problem is shown in Listings 7.12 and 7.13 with variable case=2.

7.6.8.2 Poisson Equation

The Poisson equation is also a *parabolic* PDE and has a similar form to that of the Laplace equation, except that instead of the Laplacian being equal to zero, it is equal to $f(x)$ and eqn. (7.95) becomes

$$\frac{\partial^2 u}{\partial x^2} = f(x), \; u = u(x), \; x \in \Omega, \; \Omega \subset \mathbb{R}^s, \; s \in (1, 2, 3), \tag{7.102}$$

$$u(x) = g(x), \; x \in \partial\Omega.$$

Now, after discretization and applying RBF approximations, as for the Laplace equation, we obtain the matrix equation

$$Gc = F,$$
$$\therefore c = G^{-1}F, \tag{7.103}$$

and the solution is given by

$$u = Ac. \tag{7.104}$$

Thus, the matrices for the Poisson problem are the same as for the Laplace equation, except that the upper subvector of F is now equal to the right-hand side of the PDE in eqn. (7.102). They are defined as

$$G = \begin{bmatrix} A^I_{xx} \\ A^B \end{bmatrix}, \quad F = \begin{bmatrix} f^I \\ g^B \end{bmatrix}, \tag{7.105}$$

where the superscripts I and B indicate *inner* and *boundary* nodes, respectively.

As for the Laplace equation, we have now reduced the solution of the Poisson equation (7.102) to solving a very simple matrix equation.

We will now use the preceding method to solve the following 2D electrostatic problems.

Example 1
A standard type of problem discussed in most courses in electrostatics concerns *electric potential*, defined as

$$V = \frac{q}{4\pi \epsilon_0 r}, \tag{7.106}$$

where q represents a *point-charge*, ϵ_0 *permitivity of free-space*, and r the *distance* from the charge to the point where V is measured. The discussion initially centers around *point-charges* located on an infinite domain. As the discussion becomes more advanced, the problem is extended to include the situation where part of the domain is grounded, then the solution is often obtained by using the *method of images*. A solution to this type of problem may also be found by solving *Poisson's equation* with suitable boundary conditions.

Consider a 2D problem described by eqn. (7.102) with coordinates x and y and where f represents charge density. Two stationary electrostatic charges, q_1 and q_2, are placed within a square domain $\Omega = [-1, 1] \times [-1, 1]$ [m×m] at locations $(x_1, y_1) = -0.5, -0.5$

[m] and $(x_2, y_2) = (0.5, 0.5)$ [m]. The charge densities associated with each of the charges are equal to $\rho_1 = -10^8$ and $\rho_2 = +10^8$ [Coulombs/m^2], respectively, and each charge occupies an area of one cell of the discretized domain, that is, $[2/(N-1)]^2$ [m^2], where $N = 40$ represents the number grid points used for each coordinate.

Constant *electric potential* Dirichlet boundary conditions on ∂x of $g(x, y) = 0$ [volts] are imposed. To apply the charges, values of $f(x, y)$ at the four nodes adjacent to the points, (x_1, y_1) and (x_2, y_2) are set to $-\rho_1/\epsilon_0$ and $-\rho_2/\epsilon_0$, respectively, with the charge densities for all other nodes being set to zero. The problem therefore becomes

$$\frac{\partial^2 u}{\partial x^2} + \frac{\partial^2 u}{\partial y^2} = -\frac{\rho(x,y)}{\epsilon_0}, \quad u = u(x,y), \ (x,y) \in \Omega, \ \Omega \subset \mathbb{R}^2, \quad (7.107)$$

$$u(x,y) = 0, \quad (x,y) \in \partial\Omega := (|x|=1) \cup (|y|=1),$$

where $\epsilon_0 = 8.854 \times 10^{-12}$ [Coulomb^2/N-m^2] and u represents volts [V].

Thus, it is clear that our problem is slightly more complicated than the point-charge problem discussed above. This is because we have distributed charges on a restricted domain with zero potential boundary conditions at the edges. Consequently, the problem is nontrivial to solve analytically. However, the meshless method readily finds a numerical solution. The method employed the *multiquadratic* (MQ) RBF using a value for the shape parameter of $e = 3.0$ that was found to be suitable by trial and error. Very similar results were obtained using the *inverse multiquadratic* (IMQ) RBF and shape parameter of $e = 6.0$. The computation process completed all calculations and plots in 3.1 s. This is quite impressive, considering that the calculations included the process of assembling the dense RBF matrix A consisting of $40^2 \times 40^2 = 2,560,000$ elements and then solving a matrix problem involving A.

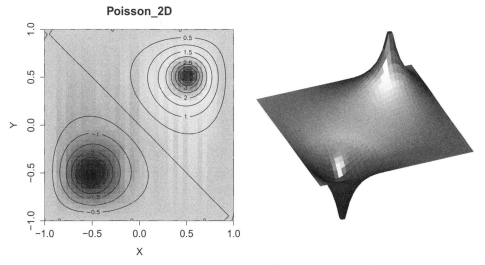

Figure 7.24. Meshless method solution $u(x, y)$ of the Poisson eqn. (7.107) where two opposite electrostatic charges are located within a square domain with Dirichlet boundary conditions. (left) 2D plot with contours. (right) Surface plot. (See color plate 7.24)

The results of the simulation are shown in Fig. 7.24 and the R code is included in Listing 7.14.

```r
# File: PoissonEqn2D_Ex1_main.R
# 2D Poisson Equation Model
# alpha*(uxx + uyy) = rho(x,y)
cat("\014") # Clear console
rm(list = ls(all = TRUE)) # Delete workspace
#
library(rgl)
#
ptm0 <- proc.time()
print(Sys.time())
#
source("sqrDatCalc.R") # set up square domain

simTitle <- "Poisson_2D"
cat(sprintf("\nBoundary Nodes = %d, Inner Nodes = %d, Total Nodes = %d\n",
      Nb, Ni, N))
#
x <- nodeDat[,3]; y <- nodeDat[,4];
# Preallocate arrays for efficiency
FL <- rep(0,Ni); FB <- rep(0,Nb)

# Define x and y domains
xmax <- max(x) # 1.5
ymax <- max(y) # 1
rmax <- sqrt(xmax^2+ymax^2)

# Constant used in RBF

# Set RBF type and epsilon by uncommenting appropriate line:
rbfType <- 1; epsilon <- 3.0 # Multiquadratic (MQ)     GSRBF
# rbfType <- 2; epsilon <- 6.0 # Inverse Multiquadric (IMQ) GSRBF

# Perform RBF calculations
X  <- outer(x,rep(1,N),"*") # Same as Matlab: [X Y]=ndgrid(x,y)
Y  <- outer(rep(1,N),y,"*")
dX <- t(X)-X; dY <- t(Y)-Y
r  <- sqrt(dX^2+dY^2)
if(rbfType == 1){
  rbfText <- "Multiquadratic (MQ) Globally Supported RBF"
  phi     <- sqrt((epsilon*r)^2+1)
  d2xphi <- epsilon^2*(1+epsilon^2*dY^2)/phi^3
  d2yphi <- epsilon^2*(1+epsilon^2*dX^2)/phi^3
}else if(rbfType == 2){
```

```
  rbfText <- "Inverse Multiquadric (IMQ) Globally Supported RBF"
  phi    <- 1/sqrt((epsilon*r)^2+1)
  d2xphi <- epsilon^2*((2*dX^2-dY^2)*epsilon^2-1)*phi^5
  d2yphi <- epsilon^2*((2*dY^2-dX^2)*epsilon^2-1)*phi^5
}
d2phi <- d2xphi+d2yphi # Laplacian
cat(sprintf("\nRBF Type = %s, epsilon = %5.2f\n",rbfText, epsilon))
####################################
A  <- phi

t1 <- proc.time()-ptm0
cat(sprintf("\nTime to calculate RBFs, 1st and 2nd derivatives, t1 = %f",
            t1[3]))

rho     <- rep(0,Ni)
#rInds <- c(97, 98, 99, 112, 113, 114, 127, 128, 129)
tol <- (xu-xl)/(Nx) # For adding charge - allows for differen grids

rInds1 <- which(abs(iNodes[,1]+0.25*(xu-xl))<tol &
                abs(iNodes[,2]+0.25*(yu-yl))<tol)
rInds2 <- which(abs(iNodes[,1]-0.25*(xu-xl))<tol &
                abs(iNodes[,2]-0.25*(yu-yl))<tol)
#
# Note: V = Q/(4*pi*epsilon*r) Electric Potential, [Volts]
permAir   <- 8.854*1e-12 # permitivity of free-space, [Coulomb^2/N-m]
rho[rInds1] <- -1e-8  # unit -ve charge, [Coulomb/m^2]
rho[rInds2] <- +1e-8  # unit +ve charge, [Coulomb/m^2]
f <- -rho/permAir  # Note the minus sign
#
Ub <- rep(0,Nb) # Dirichlet BC, fixed
# Differential operator
L  <- d2xphi[i_index,]+d2yphi[i_index,]#d2phi[i_index,]
# Interior nodes
GL <- L
# Boundary Nodes (Ni+1 to N)
GB  <- phi[b_index,]
#
G  <- rbind(GL, GB)
#
HB <- rep(0,Nb); HL <- rep(0,Ni)
#
t2 <- proc.time()-ptm0
cat(sprintf("\nTime to calculate A, G and H matrices, t2-t1 <- %f\n",
      (t2-t1)[3]) )
#
FL <- rep(0,Ni)
```

```
FB <- rep(0,Nb)
FL <- f
FB <- Ub
F  <- c(FL, FB)
# Nonlinear term handled explicitly!
NL <- 0
#
H  <- c(HL+NL, HB)
# Calculate solution
c  <- solve(G,F)
U  <- A%*%c # Electric potential
source("PoissonEqn2D_Ex1_postSimCalcs.R")
cat(sprintf("Calculation time: %f\n",(proc.time()-ptm0)[3]))
```

Listing 7.14. File: `PoissonEqn2D_Ex1_main.R`—Code for Meshless RBF method simulation of the Poisson eqn. (7.107). Calls code in `sqrDatCalc.R` and `PoissonEqn2D_Ex1_postSimCalcs.R`, which are available in the downloads. These routines set up the computational domain and create associated plots, respectively

Example 2

We again consider a 2D problem described by eqn. (7.102) with coordinates x and y and where f represents charge density. This is also an electrostatics problem, but on an *elliptical* domain Ω, with *semi-major* and *semi-minor* axes equal to 1 and 0.75, respectively. A uniform charge density is applied over the whole domain equal to $f(x, y) = -\rho/\epsilon_0 = -1$, and the boundary $\partial\Omega$, is grounded, that is, its electric potential is given by $u(x, y) = 0$ [V]. The problem therefore becomes

$$\frac{\partial^2 u}{\partial x^2} + \frac{\partial^2 u}{\partial y^2} = -1, \quad u = u(x, y), \ (x, y) \in \Omega, \ \Omega \subset \mathbb{R}^2, \tag{7.108}$$

$$u(x, y) = 0, \quad (x, y) \in \partial\Omega := \left(1 - \frac{x^2}{r_a^2} - \frac{y^2}{r_b^2}\right) = 0.$$

This problem has an analytical solution equal to

$$u(x, y) = \left(1 - \frac{x^2}{r_a^2} - \frac{y^2}{r_b^2}\right) \bigg/ \left(\frac{2}{r_a^2} + \frac{2}{r_b^2}\right), \tag{7.109}$$

which is easily demonstrated by plugging eqn. (7.109) into eqn. (7.108).

One of the really useful features of meshless methods is that they are able to readily solve problems on irregularly shaped domains. Methods that require regular grids often struggle to solve problems on irregular domains, or the calculations become tedious or have to be designed for a particular problem geometry. We use this problem to demonstrate the utility of meshless methods to find solutions on an irregular domain. Although this problem has an elliptical domain, and could be considered regular, the approach can be applied equally to any shaped domain.

A major decision to be made when solving problems on an irregularly shaped domain is where to locate the collocation points. For this problem it was decided to distribute

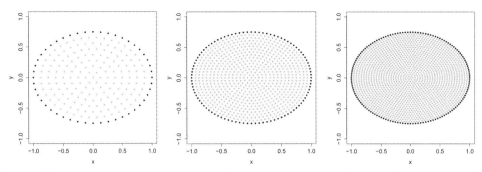

Figure 7.25. Examples from left to right of grid points generated by function `circCtrs(R, phi,offset,n)` for parameters $n = 8, 16$, and 24. The black outer nodes form the problem boundary. All calculations were performed with `R=c(1,0.75)` and `offset=c(0,0)`.

the grid points uniformly over the elliptical domain. This was achieved by use of the function `circCtrs()` that arranges the points in concentric ellipses, equally spaced from the center outward. The number of ellipses is set by parameter n, with the number of points starting at one in the center, six on the first ellipse, and then increasing by six for the next ellipse, and so on. The reason for choosing the increment of six is that it is close to 2π, the ratio of the circumference of a circle to its radius. The result is that the points are distributed more or less evenly over the elliptical domain, as shown in Fig. 7.25 for different values of n. The function `circCtrs(R=c(ra,rb),phi=2*pi,offset,n)` is quite general and can be used for any size elliptical or circular domain; `ra` and `rb` represent the semi-major and semi-minor axes of the ellipse, respectively, and, when equal, the result is a circular domain. The parameter `offset` allows the ellipse to be located away from the origin, and the parameter `phi` allows a grid points to be located on to a segment of an ellipse.

The associated R code is shown in Listing 7.15.

```
# File: circCtrs.R
circCtrs <- function(R=c(1,1),phi=2*pi,offset=c(0,0),n=4){
  # Generates evenly distributed grid points over an elliptical domain.
  # R = radii of ellipse, (ra,rb)
  # phi = angle (0-2*pi)
  # offSet = offset from domain center (0,0)
  # n = number of intervals along r
  # m = number of intervals along inner circle
  m <- 6 # Approx. ratio of circumferenc to radius
  ra <- c(0,(1:n)/n)*R[1] # radial grid spacing, for ra
  rb <- c(0,(1:n)/n)*R[2] # radial grid spacing, for rb
  p <- c(1,seq(m,n*m,by=m)) # grid points per circle
  p[2:(n+1)] <- ceiling(p[2:(n+1)]*phi/(2*pi))
  #print(p)
  pSum <- sum(p) # total number of points
  x <- rep(0,pSum);y<-x;h<-x
```

```
k <- 0
if((2*pi-phi) < 1e-3){tj0<-1}else{tj0<-0}
for(ri in 1:(n+1)){ # calculate x and y coords for each point
  dia <- phi*ra[(ri)]
  theta <- phi/p[(ri)]
  for(tj in tj0:p[(ri)]){
    k <- k+1
    x[k] <- ra[ri]*cos(tj*theta) + offset[1]
    y[k] <- rb[ri]*sin(tj*theta) + offset[2]
    if(ri==1)break # To avoid duplicate grid points at origin for phi != 2*pi
  }
}
return(list(x1=x,y1=y)) # return list of x-y cordinates for points
}
```

Listing 7.15. File: `circCtrs.R`—Code for function `circCtrs()`, which generates evenly distributed grid points on an elliptical domain. It is called by the function `circDatCalc()`, the code of which is contained in file `circDatCalc.R`, available with the downloads

If the domain were a different shape, we would need a different algorithm to assign grid point locations, such as included in [Per-04a, Per-04b] (see also http://persson.berkeley.edu/distmesh/). However, the current situation with R appears to be that, although it is straightforward to plot scattered data that generate a *convex hull*,[7] there are certain limitations in plotting scattered data that generate a *nonconvex hull*

Therefore it was decided to limit this demonstration to a problem that generates convex scattered data to show that irregular domains can be tackled using the meshless method outlined in this chapter. The scattered data are first assembled into a three-column matrix xyz, using the R function `cbind()` with rows equal to the number of scattered data points. The first two columns contain x and y coordinates of the points, and the third column contains the associated dependent variable values, z. At this point, the data are not in any particular order. The matrix xyz is then used by the R function `delaunayn()`, which interfaces to the *Qhull* library to generate a convex hull.[8] These data, together with xyz, are used by R function `surf.tri()` to generate a set of surface triangles for the convex hull. Finally, the triangles are displayed by a call to the R function `rgl.triangles()`. This process works well when the surface is convex; however, if the meshless method produces a solution with a nonconvex surface, the display is distorted as *concave* areas are shown filled. This is due to limitations in `delaunayn()`.

Some nonconvex surfaces can be plotted using the rgl package, but the x and y vectors must be provided in ascending order, which limits its use for some scattered data problems. Another R package, `alphashape3d`, does generate nonconvex hulls that can

---

[7] Computing a *convex hull* is a problem in computational geometry. A set S is defined as *convex* if, for any two points $p, q \in S$, a line segment is contained in S. In three dimensions, the *convex hull* of $P = \{p_1, p_2, \cdots, p_n\}$ is a triangular mesh with all its vertices in P.

[8] *Qhull* currently only produces a convex mesh, and full details are available online at http://www.qhull.org/.

Table 7.5. Poison equation simulation results for `ra=1`, `rb=0.6`

n	Nodes	Error	Time	ϵ
8	217	−1.57E-05	1.28	1.15
16	817	−5.80E-06	1.78	2.40
24	1801	−5.18E-06	4.91	4.00

Note: In the table, *n* refers to the number of radial grid points specified in the call to function `circCtrs()`, "nodes" to the total number of collocation nodes used in the simulation, "error" to the maximum error when compared to the analytical solution of eqn. (7.109), "time" the elapsed time in seconds to complete the calculations and render the plots, and ϵ to the shape parameter value.

be used to provide surface plots. However, it is not a general solution, as some parameter tuning is necessary to obtain acceptable results, and it does fail on some problems. It is hoped that the situation in R will improve in the future so that nonconvex hulls can be handled appropriately in a general way.

With regard to the specific problem in hand, it is defined on an elliptical domain with a convex solution (it could have been defined on an abrbitrary shaped domain, but the solution could then be nonconvex, which would present plotting difficulties). The solution was readily obtained using the earlier meshless method, and as for the previous Poisson problem, the method employed the *multiquadratic* (MQ) RBF using values for the shape parameter found by trial and error. The calculation runs quickly, and some solution statistics are provided in Table 7.5 for $n = 8$, 16, and 24. Similar results were also obtained using the *inverse multiquadratic* (IMQ) RBF. A surface plot of the solution for $n = 8$ (217 nodes) is shown in Fig. 7.26. The result compares very well with the analytical result, having a maximum error equal to −1.57e-05.

The R code for this problem is shown in Listing 7.16.

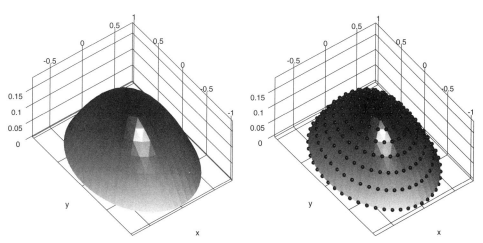

Figure 7.26. (left) Solution $u(x, y)$ of Poisson equation by meshless method, using 217 radially distributed collocation points on an elliptical domain. (right) With dots added to indicate the location of collocation points.

```r
# File: PoissonEqn2D_Ex2_main.R
# 2D Poisson Equation Model
# alpha*(uxx + uyy) = rho(x,y)
cat("\014") # Clear console
rm(list = ls(all = TRUE)) # Delete workspace
#
library(rgl)
library(geometry)
library(pracma)
source("circCtrs.R")
#
ptm0 <- proc.time()
print(Sys.time())
#
source("circDatCalc.R")

simTitle <- "Poisson_2D"
cat(sprintf("\nBoundary Nodes = %d, Inner Nodes = %d, Total Nodes = %d\n",
        Nb, Ni, N))
# Preallocate arrays for efficiency
FL <- rep(0,Ni); FB <- rep(0,Nb)

# Uncomment as required
rbfType <- 1; epsilon <- 1.15 #1.15, 2.4, 4.0 # Multiquadratic (MQ) GSRBF
# rbfType <- 2; epsilon <- 0.85 # Inverse Multiquadric (IMQ) GSRBF

# Perform RBF calculations
#########################################################
XX <- outer(xx,rep(1,N),FUN="*")
YY <- outer(rep(1,N),yy,FUN="*")
dX <- t(XX)-XX; dY <- t(YY)-YY
r  <- sqrt(dX^2+dY^2)
if(rbfType == 1){ # Multiquadratic
  rbfText <- "Multiquadratic (MQ) Globally Supported RBF"
  phi     <- sqrt((epsilon*r)^2+1)
  d2xphi  <- epsilon^2*(1+epsilon^2*dY^2)/phi^3
  d2yphi  <- epsilon^2*(1+epsilon^2*dX^2)/phi^3
}else{          # Inverse Multiquadric
  rbfText <- "Inverse Multiquadric (IMQ) Globally Supported RBF"
  phi     <- 1/sqrt((epsilon*r)^2+1)
  d2xphi  <- epsilon^2*((2*dX^2-dY^2)*epsilon^2-1)*phi^5
  d2yphi  <- epsilon^2*((2*dY^2-dX^2)*epsilon^2-1)*phi^5
}
d2phi <- d2xphi+d2yphi # Laplacian
cat(sprintf("\nRBF Type = %s, epsilon = %5.2f\n",rbfText, epsilon))
#########################################################
A <- phi
```

```
t1 <- proc.time()-ptm0
cat(sprintf("\nTime to calculate RBFs, 1st and 2nd derivatives, t1 = %f\n",
        t1[3]))

rho <- - rep(1,Ni) # set charge distribution
ua <- (1-(xx/ra)^2-(yy/rb)^2)/(2/ra^2+2/rb^2)
Ub  <- rep(0,Nb)  # Dirichlet BC, fixed at zero potential
#
# # Differential operator
L <- d2phi[i_index,] #d2phi[i_index,]
# # Interior nodes
GL <- L
# # Boundary Nodes (Ni+1 to N)
GB  <-  phi[b_index,]
# #
G  <- rbind(GL, GB)
# #
HB <- rep(0,Nb); HL <- rep(0,Ni)
# #
t2 <- proc.time()-ptm0
cat(sprintf("\nTime to calculate A, G and H matrices, t2-t1 <- %f",
    (t2-t1)[3]) )
#
FL <- rho
FB <- Ub
FF  <- c(FL, FB)
# # Nonlinear term handled explicitly!
NL <- 0
# #
H  <-  c(HL+NL, HB)
# # Calculate solution
c  <-  solve(G,FF)
# invG <- pinv(G)
# c <- invG%*%FF
U  <-  A%*%c
#
err <- U-ua
maxErrLoc <- which.max(abs(err))
cat(sprintf("\nMaximum error = %e\n",err[maxErrLoc]))
source("PoissonEqn2D_delaunayCalcs.R")
cat(sprintf("Calculation time: %f\n",(proc.time()-ptm0)[3]))
```

Listing 7.16. File: PoissonEqn2D_Ex2_main.R—Code for *main* program for meshless RBF method simulation of the Poisson eqn. (7.108). Calls code in circDatCalc.R and PoissonEqn2D_delaunayCalcs.R, which are available in the downloads. These routines set up the computational domain and create associated plots, respectively

Table 7.6. Halton sequence of 40 points for a 2D domain generated from Listing 7.17

	[,1]	[,2]		[,1]	[,2]
[1,]	0.50000	0.33333333	[21,]	0.656250	0.18518519
[2,]	0.25000	0.66666667	[22,]	0.406250	0.51851852
[3,]	0.75000	0.11111111	[23,]	0.906250	0.85185185
[4,]	0.12500	0.44444444	[24,]	0.093750	0.2962963
[5,]	0.62500	0.77777778	[25,]	0.593750	0.62962963
[6,]	0.37500	0.22222222	[26,]	0.343750	0.96296296
[7,]	0.87500	0.55555556	[27,]	0.843750	0.01234568
[8,]	0.06250	0.88888889	[28,]	0.218750	0.34567901
[9,]	0.56250	0.03703704	[29,]	0.718750	0.67901235
[10,]	0.31250	0.37037037	[30,]	0.468750	0.12345679
[11,]	0.81250	0.70370370	[31,]	0.968750	0.45679012
[12,]	0.18750	0.14814815	[32,]	0.015625	0.79012346
[13,]	0.68750	0.48148148	[33,]	0.515625	0.23456790
[14,]	0.43750	0.81481481	[34,]	0.265625	0.56790123
[15,]	0.93750	0.25925926	[35,]	0.765625	0.90123457
[16,]	0.03125	0.59259259	[36,]	0.140625	0.04938272
[17,]	0.53125	0.92592593	[37,]	0.640625	0.38271605
[18,]	0.28125	0.07407407	[38,]	0.390625	0.71604938
[19,]	0.78125	0.40740741	[39,]	0.890625	0.16049383
[20,]	0.15625	0.74074074	[40,]	0.078125	0.49382716

7.A FRANKE'S FUNCTION

Franke's function [Fra-79] is defined in eqn. (7.110). It is a reasonably complex smooth function that is used as a standard interpolation test. It provides a sufficiently demanding problem for performance comparison of interpolation codes:

$$\begin{aligned} f(x, y) = {} & \tfrac{3}{4} \exp\left[-\tfrac{1}{4}(9x - 2)^2 - \tfrac{1}{4}(9y - 2)^2\right] \\ & + \tfrac{3}{4} \exp\left[-\tfrac{1}{49}(9x + 1)^2 - \tfrac{1}{10}(9y + 1)^2\right] \\ & + \tfrac{1}{2} \exp\left[-\tfrac{1}{4}(9x - 7)^2 - \tfrac{1}{4}(9y - 3)^2)\right] \\ & - \tfrac{1}{5} \exp\left[-(9x - 4)^2 - (9y - 7)^2\right]. \end{aligned} \qquad (7.110)$$

7.B HALTON SEQUENCE

The *Halton sequence* [Hal-64] is a *deterministic* sequence in the interval [0,1] that can be used in lieu of a sequence of *random* data points produced by a random number generator. The call to function `halton()` in Listing 7.17, coded in R, produces *low discrepancy sequences*, that is, they are *quasi-random*, but are less random than can be obtained from a good random number generator. Nevertheless, they generally distribute the random points over the domain of interest more evenly than would pure random sequences, which can leave large holes in the data. Consequently, they are particularly useful for testing software designed to act on random data, for example, programs used for interpolation of scattered samples.

An example Halton sequence generated by a call to the built-in R function `halton()` is given, with the points plotted in Fig. 7.27. Note that a given Halton sequence is not ordered.

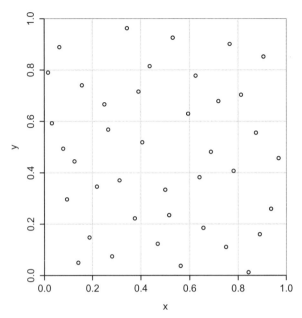

Figure 7.27. Example of a Halton sequence of 40 points on the domain $\Omega = [0, 1] \times [0, 1]$ generated from function haltonSeq() of Listing 7.17 called with $n = 40$ and $d = 2$.

```
library(randtoolbox)
H <- halton(n=40, dim=2)
print(H)
```

Listing 7.17. An R code example that can be used to generate a *Halton sequence* using the function halton() from the package randtoolbox

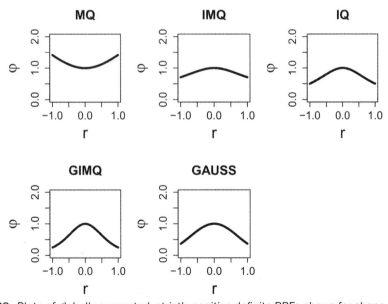

Figure 7.28. Plots of globally supported, strictly positive definite RBFs shown for shape parameter $\epsilon = 1$. Owing to radial symmetry, these plots apply to 1D and 2D implementations.

7.C RBF DEFINITIONS

Listing 7.13 contains R code for the function `rbfCalc_CS4()`, which calculates values for the RBFs defined in the following tables. Globally supported RBs are also plotted in Figure 7.28.

Globally supported, strictly positive definite RBFs

	MQ	IMQ	IQ
φ	$\sqrt{\epsilon^2 r^2 + 1}$	$\dfrac{1}{\sqrt{\epsilon^2 r^2 + 1}}$	$\dfrac{1}{(\epsilon^2 r^2 + 1)}$
$\dfrac{\partial \varphi}{\partial r}$	$\dfrac{\epsilon^2 r}{\sqrt{\epsilon^2 r^2 + 1}}$	$-\dfrac{\epsilon^2 r}{(\epsilon^2 r^2 + 1)^{3/2}}$	$\dfrac{-2\epsilon^2 r}{(\epsilon^2 r^2 + 1)^2}$
$\dfrac{\partial^2 \varphi}{\partial r^2}$	$\dfrac{\epsilon^2}{(\epsilon^2 r^2 + 1)^{3/2}}$	$\dfrac{2\epsilon^4 r^2 - \epsilon^2}{(\epsilon^2 r^2 + 1)^{5/2}}$	$\dfrac{6\epsilon^4 r^2 - 2\epsilon^2}{(\epsilon^2 r^2 + 1)^3}$
$\dfrac{\partial \varphi}{\partial x}$	$\dfrac{\epsilon^2 dX}{\sqrt{\epsilon^2 r^2 + 1}}$	$-\dfrac{\epsilon^2 dX}{(\epsilon^2 r^2 + 1)^{3/2}}$	$\dfrac{-2\epsilon^2 dX}{(\epsilon^2 r^2 + 1)^2}$
$\dfrac{\partial \varphi}{\partial y}$	$\dfrac{\epsilon^2 dY}{\sqrt{\epsilon^2 r^2 + 1}}$	$-\dfrac{\epsilon^2 dY}{(\epsilon^2 r^2 + 1)^{3/2}}$	$\dfrac{-2\epsilon^2 dY}{(\epsilon^2 r^2 + 1)^2}$
$\dfrac{\partial^2 \varphi}{\partial x^2}$	$\dfrac{\epsilon^2 \left(1 + dY^2 \epsilon^2\right)}{(\epsilon^2 r^2 + 1)^{3/2}}$	$-\dfrac{\epsilon^2 \left(3dY^2 \epsilon^2 - 2\epsilon^2 r^2 + 1\right)}{(\epsilon^2 r^2 + 1)^{5/2}}$	$\dfrac{-2\epsilon^2 \left(4dY^2 \epsilon^2 - 3\epsilon^2 r^2 + 1\right)}{(\epsilon^2 r^2 + 1)^3}$
$\dfrac{\partial^2 \varphi}{\partial y^2}$	$\dfrac{\epsilon^2 \left(1 + dX^2 \epsilon^2\right)}{(\epsilon^2 r^2 + 1)^{3/2}}$	$-\dfrac{\epsilon^2 \left(3dX^2 \epsilon^2 - 2\epsilon^2 r^2 + 1\right)}{(\epsilon^2 r^2 + 1)^{5/2}}$	$\dfrac{-2\epsilon^2 \left(4dX^2 \epsilon^2 - 3\epsilon^2 r^2 + 1\right)}{(\epsilon^2 r^2 + 1)^3}$
		GIMQ	Gaussian
φ		$\dfrac{1}{(\epsilon^2 r^2 + 1)^2}$	$e^{-\epsilon^2 r^2}$
$\dfrac{\partial \varphi}{\partial r}$		$\dfrac{-4\epsilon^2 r}{(\epsilon^2 r^2 + 1)^3}$	$-2\epsilon^2 r e^{-\epsilon^2 r^2}$
$\dfrac{\partial^2 \varphi}{\partial r^2}$		$\dfrac{20\epsilon^4 r^2 - 4\epsilon^2}{(\epsilon^2 r^2 + 1)^4}$	$4\left(\epsilon^2 r^2 - \dfrac{1}{2}\right)\epsilon^2 e^{-\epsilon^2 r^2}$
$\dfrac{\partial \varphi}{\partial x}$		$\dfrac{-4\epsilon^2 dX}{(\epsilon^2 r^2 + 1)^3}$	$-2\epsilon^2 e^{-\epsilon^2 r^2} dX$
$\dfrac{\partial \varphi}{\partial y}$		$\dfrac{-4\epsilon^2 dY}{(\epsilon^2 r^2 + 1)^3}$	$-2\epsilon^2 e^{-\epsilon^2 r^2} dY$
$\dfrac{\partial^2 \varphi}{\partial x^2}$		$\dfrac{-4\epsilon^2 \left(6dY^2 \epsilon^2 - 5\epsilon^2 r^2 + 1\right)}{(\epsilon^2 r^2 + 1)^4}$	$2\epsilon^2 e^{-\epsilon^2 r^2} \times \left(2\epsilon^2 r^2 - 2dY^2 \epsilon^2 - 1\right)$
$\dfrac{\partial^2 \varphi}{\partial y^2}$		$\dfrac{-4\epsilon^2 \left(6dX^2 \epsilon^2 - 5\epsilon^2 r^2 + 1\right)}{(\epsilon^2 r^2 + 1)^4}$	$2\epsilon^2 e^{-\epsilon^2 r^2} \times \left(2\epsilon^2 r^2 - 2dX^2 \epsilon^2 - 1\right)$

	Compactly supported, strictly positive definite RBFs		
	Wendland(3,1), C^4 at the origin	Wendland(3,2), C^4 at the origin	Wendland(3,3), C^6 at the origin
φ	$(1-\epsilon r)_+^4 (4\epsilon r + 1)$	$(1-\epsilon r)_+^6 (35\epsilon^2 r^2 + 18\epsilon r + 3)$	$(1-\epsilon r)_+^8$ $\times (32\epsilon^3 r^3 + 25\epsilon^2 r^2 + 8\epsilon r + 1)$
$\dfrac{\partial \varphi}{\partial r}$	$20(-1+\epsilon r)_+^3 \epsilon^2 r$	$280\epsilon^2 \left(\epsilon r + \dfrac{1}{5}\right) r(-1+\epsilon r)_+^5$	$22(-1+\epsilon r)_+^7$ $\times \epsilon^2 r (16\epsilon^2 r^2 + 7\epsilon r + 1)$
$\dfrac{\partial^2 \varphi}{\partial r^2}$	$80\epsilon^2 (-1+\epsilon r)_+^2 \left(\epsilon r - \dfrac{1}{4}\right)$	$56\epsilon^2 (-1+\epsilon r)_+^4$ $\times (-4\epsilon r + 35\epsilon^2 r^2 - 1)$	$22(-1+\epsilon r)_+^6 \epsilon^2$ $\times (160\epsilon^3 r^3 + 15\epsilon^2 r^2 - 6\epsilon r - 1)$
$\dfrac{\partial \varphi}{\partial x}$	$20(-1+\epsilon r)_+^3 \epsilon^2 dX$	$56\epsilon^2 (5\epsilon r + 1)(-1+\epsilon r)_+^5 dX$	$22(-1+\epsilon r)_+^7 \epsilon^2$ $\times (16\epsilon^2 r^2 + 7\epsilon r + 1) dX$
$\dfrac{\partial \varphi}{\partial y}$	$20(-1+\epsilon r)_+^3 \epsilon^2 dY$	$56\epsilon^2 (5\epsilon r + 1)(-1+\epsilon r)_+^5 dY$	$22(-1+\epsilon r)_+^7 \epsilon^2$ $\times (16\epsilon^2 r^2 + 7\epsilon r + 1) dY$
$\dfrac{\partial^2 \varphi}{\partial x^2}$	$-20(-1+\epsilon r)_+^2 \epsilon^2$ $\times (3\epsilon dY^2 - 4\epsilon r^2 + r)/r$	$-56(-1+\epsilon r)_+ \epsilon^2$ $\times (30 dY^2 \epsilon^2 - 35\epsilon^2 r^2$ $+ 4\epsilon r + 1)$	$-22(-1+\epsilon r)_+^6 \epsilon^2$ $\times (144\epsilon^3 r dY^2 - 160\epsilon^3 r^3$ $+ 24 dY^2 \epsilon^2 - 15\epsilon^2 r^2$ $+ 6\epsilon r + 1)$
$\dfrac{\partial^2 \varphi}{\partial y^2}$	$-20(-1+\epsilon r)_+^2 \epsilon^2$ $\times (3\epsilon dX^2 - 4\epsilon r^2 + r)/r$	$-56(-1+\epsilon r)_+ \epsilon^2$ $\times (30 dX^2 \epsilon^2 - 35\epsilon^2 r^2$ $+ 4\epsilon r + 1)$	$-22(-1+\epsilon r)_+^6 \epsilon^2$ $\times (144\epsilon^3 r dX^2 - 160\epsilon^3 r^3$ $+ 24 dX^2 \epsilon^2 - 15\epsilon^2 r^2$ $+ 6\epsilon r + 1)$

Note: The term $(1-\epsilon r)_+$ in the Wendland RBF definitions is to be interpreted as [Fas-07, p. 43]

$$(x)_+ = \begin{cases} x, & \text{for } x \geq 0 \\ 0, & \text{for } x < 0. \end{cases} \qquad (7.111)$$

REFERENCES

[Art-98] Articolo, G. A. (2003), *Partial Differential Equations and Boundary Valued Problems with Maple V*, Academic Press.
[Bel-57] Bellman, R. (1957), *Dynamic Programming*, Princeton University Press.
[Buh-03] Buhmann, M. D. (2003), *Radial Basis Functions*, Cambridge University Press.
[Fas-07] Fasshauer, G. E. (2007), *Meshfree Approximation Methods with Matlab*, World Scientific.
[Fis-37] Fisher, R. A. (1937), The Wave of Advance of Advantageous Genes, *Annals of Eugenics* **7**, 355–369.
[Fra-79] Franke, R. (1979), A Critical Comparison of Some Methods for Interpolation of Scattered Data, *Naval Postgraduate School Tech. Rep.* NPS-53-79-003.
[Fus-13] Fuselier, E. J. and G. B. Wright (2013), A High-Order Kernel Method for Diffusion and Reaction-Diffusion Equations on Surfaces, *Journal of Scientific Computing* **56**, 535–565.
[Gan-60] Gantmancher, F. R. (1960), *Matrix Theory*, vol. 2, Chelsea.
[Gan-77] Gantmancher, F. R. (1977), *Matrix Theory*, vol. 1, Chelsea.

[Gol-96] Golab, G. H. and C. F. van Loan (1996), *Matrix Computations*, The Johns Hopkins University Press.
[Hal-64] Halton, J. H. (1964), Algorithm 247: Radical-Inverse Quasi-Random Point Sequence, *ACM* **7**-12, 701–702.
[Har-71] Hardy R. L. (1971), Multiquadratic Equations of Topography and Other Irregular Surfaces, *Journal of Geophysical Research* **76**, 1905–15.
[Har-90] Hardy R. L. (1990), Theory and Applications of the Mutiquadratic-Biharmonic Method: 20 Years of Discovery, *Computational Mathematics Applications* **19**-8/9, 163–208.
[Kan-90] Kansa, E. J. (1990), Multiquadrics—A Scattered Data Approximation Scheme with Applications to Computational Fluid Dynamics I/II, *Computers and Mathematics with Applications* **19**-8/9, 147–161.
[Kol-37] Kolmogorov, A., I. Petrovsky and N. Piscounov (1937), Study of the Diffusion Equation with Growth of the Quantity of Matter and Its Applications to a Biological Problem, *Moscow University Mathematics Bulletin*, 1–25. (Trans. by Oliveira-Pinto, F. and B. W. Conolly (1982), *Applicable Mathematics of Non-Physical Phenomena*, Ellis Horwood, 169–184.)
[Kor-95] Korteweg, D. J. and G. de Vries (1895), On the Change of Form of Long Waves Advancing in a Rectangular Canal, and on a New Type of Long Stationary Waves, *Philosophical Magazine, Series 5* **39**, 422–443.
[Mat-11] Matloff, N. (2011), *The Art of R Programming*, No Starch Press.
[Mor-94] Morton, K. W. and D. F. Mayers. (1994), *Numerical Solution of Partial Differential Equations*, Cambridge University Press.
[Mur-02] Murray, J. D. (2002), *Mathematical Biology I: An Introduction*, 3rd ed., Springer.
[Ove-01] Overton, M. (2001), *Numerical Computing with IEEE Floating Point Arithmetic*, SIAM.
[Pep-10] Pepper, D. (2010), Meshless Methods for PDEs. *Scholarpedia* **5**-5, 9838. Available online at http://www.scholarpedia.org/article/Meshless_methods_for_PDEs.
[Per-04a] Persson, P.-O. and G. Strang (2004), A Simple Mesh Generator in MATLAB, *SIAM Review* **46**-2, 329–345.
[Per-04b] Persson, P.-O. (2004), *Mesh Generation for Implicit Geometries*. PhD thesis, Department of Mathematics, MIT.
[Pla-06] Platte, R. and T. Driscoll (2006), Eigenvalue Stability of Radial Basis Functions Discretizations for Time-Dependent Problems, *Computers and Mathematics with Applications* **51**, 1251–1268.
[Sar-08] Sarra, S. A. (2008), A Numerical Study of the Accuracy and Stability of Symmetric and Asymmetric RBF Collocation Methods for Hyperbolic PDEs, *Numerical Methods for Partial Differential Equations* **24**-2, 670–686.
[Sar-12] Sarra, S. A. (2012), A Local Radial Basis Function Method for Advection-Diffusion-Reaction Equations on Complexly Shaped Domains, *Applied Mathematics and Computation* **218**, 9853–9865.
[Soe-14] Soetaert, K. (2014), plot3D: *Tools for Plotting 3-D and 2-D Data*. R package version 1.0-1.
[Tre-00] Trefethen, L. N. (2000), *Spectral Methods in Matlab*, SIAM.
[Udd-11] Uddin, M. (2011), *RBF Approximation Method for Initial-Boundary Value Problems*, PhD thesis, Faculty of Engineering Sciences, Ghulam Ishaq Khan Institute of Engineering Sciences and Technology.

8
Conservation Laws

8.1 INTRODUCTION

The term *conservation laws* is applied throughout the physical sciences, where it is used to describe collectively that certain inherent physical properties of a system are conserved over time. For example, the so-called *mass* continuity equation for a fluid in motion is given by

$$\frac{\partial \rho}{\partial t} + \frac{\partial \rho u}{\partial x} = 0, \qquad (8.1)$$

where $u = u(x, t)$ (velocity), $\rho = \rho(x, t)$ (density), $t \in \mathbb{R}$ (time), $x \in \mathbb{R}^n$ (spatial vector), and, typically, $n = 1, 2,$ or 3. The conserved quantity is *mass*, m, which for the 3D case we have

$$\frac{\partial m}{\partial t} = \frac{\partial}{\partial t} \iiint_v \rho \, dv = \iint_S \rho u \, dS, \qquad (8.2)$$

where v represents a control volume having a corresponding surface S. Equation (8.1) is valid for any point within the fluid system, and eqn. (8.2) applies to any control volume, which could range from the infinitesimally small up to one that includes the entire boundary of the fluid system.

Equation (8.2) represents the *conservation of mass* law. It states mathematically that at any instant the rate of change of total mass, contained within a closed volume, is equal to the net flow of mass across its surface or boundary. This law, along with corresponding *conservation of momentum* and *conservation of energy* laws, applies generally to any system and is fundamental to classical physics. However, other conservation laws are more restrictive in their applicability and generally do not have a simple interpretation.

We now introduce the idea of a conservation law applied to the continuity eqn. (8.1) such that, by applying certain *decay conditions at infinity*, the *total mass* contained within a system is conserved, that is, it is constant. These conservation laws are also refered to as *local conservation laws*. From the preceding discussion, this implies that *equilibrium conditions must apply at the system boundary*; that is, the decay conditions at infinity are generally taken to mean that, for a PDE with dependent variable $u(x,t)$,

$$\left.\begin{array}{r} u \to \text{constant} \\ \dfrac{\partial^n u}{\partial x^n} \to 0 \end{array}\right\} \text{ as } |x| \to \infty. \qquad (8.3)$$

Following Drazin and Johnson [Dra-92], for eqn. (8.1) we make the assumption that the quantity $\rho u \to$ constant as $|x| \to \infty$ and that ρ and $(\rho u)_x$ are integrable over the entire spatial domain. Thus, for the case where the fluid velocity u, and density ρ, are constrained to vary in one direction only (say, $x \to x_1$), it follows that

$$\frac{d}{dt}\left(\int_{-\infty}^{\infty} \rho dx\right) = \int_{-\infty}^{\infty}\left(\frac{\partial \rho}{\partial t}\right) dx = \int_{-\infty}^{\infty}\left(-\frac{\partial \rho u}{\partial x}\right) dx$$
$$= [-\rho u]_{-\infty}^{+\infty} = 0, \qquad (8.4)$$

from which we see that

$$\int_{-\infty}^{\infty} \rho dx = \text{constant}. \qquad (8.5)$$

Equation (8.5) represents conservation of total mass in the system even though the density profile $\rho(x)$ may change over time.

We now generalize this idea to PDEs that can be transformed to the following conservative form:

$$\frac{\partial T}{\partial t} + \frac{\partial X}{\partial x} = 0, \qquad (8.6)$$

where $T = T(x, t, u, u_x, u_{xx}, \dots)$ (the *density*) and $X = X(x, t, u, u_x, u_{xx}, \dots)$ (the *flux*). Subscripts denote partial differentiation and will be used subsequently, where appropriate. Note that: it is not usually permitted for either T or X to be dependent upon derivatives of time.

Applying the same *decay conditions* on X as on ρu, that is

$$X \to \text{constant as } |x| \to \infty, \qquad (8.7)$$

yields

$$\frac{d}{dt}\left(\int_{-\infty}^{\infty} T dx\right) = 0 \qquad (8.8)$$

$$\therefore \int_{-\infty}^{\infty} T dx = \text{constant}. \qquad (8.9)$$

When applied to evolutionary equations for $u(x,t)$, the constant in eqn. (8.9) is a *conserved quantity* and is usually called a *constant of motion* or, an *invariant*, of eqn. (8.6). Evolutionary equations may have more than one conservation law and, hence, more than one constant of motion or invariant. These constants of motion are particularly useful when obtaining numerical solutions to evolutionary equations. The numerical results must represent accurately the solution fine structure as time advances. Thus, because the constants of motion should be invariant, they can be used to provide an indication of accuracy at each solution time [Sch-09, Chapter 7] [Zak-00]. Examples of how conservation laws or invariants are used to assess the accuracy of numerical solutions to PDEs are provided in Chapters 3 and 6.

We now discuss some additional examples related to particular evolutionary equations.

8.2 KORTEWEG–DE VRIES (KDV) EQUATION

The well-known *KdV equation*, which describes shallow-water waves, is defined as

$$\frac{\partial u}{\partial t} - 6u\frac{\partial u}{\partial x} + \frac{\partial u^3}{\partial x^3} = 0, \quad u = u(x,t), \ t \in \mathbb{R}, \ x \in \mathbb{R}. \tag{8.10}$$

8.2.1 The *First* Conservation Law, *u*

Equation (8.10) is readily presented in the conservative form of eqn. (8.6), where the density–flux pair becomes

$$T = u \quad \text{and} \quad X = -3u^2 + u_{xx}. \tag{8.11}$$

Using the same argument, it therefore follows that if T and X are integrable, we then have

$$\frac{d}{dt}\left(\int_{-\infty}^{\infty} u\, dx\right) = \int_{-\infty}^{\infty}\left(\frac{\partial u}{\partial t}\right) dx, = \int_{-\infty}^{\infty} \frac{\partial}{\partial x}\left(-3u^2 + u_{xx}\right) dx$$
$$= \left[-3u^2 + u_{xx}\right]_{-\infty}^{+\infty} = 0, \tag{8.12}$$

where we have applied the decay conditions that u is equal to a constant at $\pm\infty$ and that derivatives of $u \to 0$ as $|x| \to \infty$. It therefore follows that the relationship

$$\int_{-\infty}^{\infty} u\, dx = \text{constant} \tag{8.13}$$

applies for all solutions of eqn. (8.10). However, this result does not apply for periodic solutions, when the limits of the integral in eqn. (8.13) must be taken over one period of the wave.

The preceding constant of the motion described by eqn. (8.13), in which u evolves according to eqn. (8.10), can be interpretted as corresponding to *conservation of mass* or *conservation of horizontal momentum*.

8.2.2 The *Second* Conservation Law, u^2

The KdV equation also admits additional conservation laws. For example, multiplying eqn. (8.10) by u yields

$$\frac{\partial u^2/2}{\partial t} + \frac{\partial}{\partial x}\left(uu_{xx} - \frac{1}{2}u_x^2 - 2u^3\right) = 0, \tag{8.14}$$

from which, after applying the decay conditions, we obtain

$$\int_{-\infty}^{\infty} u^2\, dx = \text{constant} \tag{8.15}$$

and
$$T_2 = u^2. \tag{8.16}$$

The constant of the motion described by eqn. (8.15), in which u evolves according to eqn. (8.10), can be interpreted as corresponding to *conservation of horizontal momentum* or *conservation of energy*.

8.2.3 The *Third* Conservation Law, $u^3 + \frac{1}{2}u_x^2$

A third conservation law also applies to the KdV equation. This is obtained from

$$\left(3u^2 + u_x \frac{\partial}{\partial x}\right)\left(\frac{\partial u}{\partial t} - 6u\frac{\partial u}{\partial x} + \frac{\partial u^3}{\partial x^3}\right) = 0, \tag{8.17}$$

which yields

$$\frac{\partial}{\partial t}\left(u^3 + \frac{1}{2}u_x^2\right) + \frac{\partial}{\partial x}\left(-\frac{9}{2}u^4 + 3u^2 u_{xx} - 6uu_x^2 + u_x u_{xxx} - \frac{1}{2}u_{xx}^2\right) = 0. \tag{8.18}$$

Thus, similarly to the previous result, we obtain the third constant of motion

$$\int_{-\infty}^{\infty}\left(u^3 + \frac{1}{2}u_x^2\right) dx = \text{constant} \tag{8.19}$$

and

$$T_3 = u^3 + \frac{1}{2}u_x^2. \tag{8.20}$$

See also eqns. (8.47), (8.48), and (8.49).

The constant of the motion described by eqn. (8.19), in which u evolves according to eqn. (8.10), relates to the *Hamiltonian*.

The three conservation laws described are used in Chapter 3 to define invariants that are used to provide an estimate of the accuracy of the numerical scheme used to solve the KdV equation.

See Appendix 8.A at the end of this chapter for details relating to symbolic algebra source code to perform the preceding calculations.

8.2.4 Another Conservation Law

An interesting question posed by Drazin and Johnson [Dra-92, Q5.2], is to show that $T = xu + 3tu^2$ is a conserved density for the KdV eqn. (8.10). A clue to the solution is obtained by differentiating T with respect to t, when we obtain

$$xu_t + 6tuu_t + 3u^2. \tag{8.21}$$

This result suggests that we try to obtain the solution from $(x + 6tu) \times$KdV, that is,

$$(x + 6tu)\left(\frac{\partial u}{\partial t} - 6u\frac{\partial u}{\partial x} + \frac{\partial u^3}{\partial x^3}\right) = 0, \tag{8.22}$$

which, on expanding, yields

$$x\frac{\partial u}{\partial t} + 6tu\frac{\partial u}{\partial t} - 36tu^2\frac{\partial u}{\partial x} + 6tu\frac{\partial^3 u}{\partial x^3} - 6xu\frac{\partial u}{\partial x} + x\frac{\partial^3 u}{\partial x^3} = 0. \tag{8.23}$$

Putting this result into conservative form, we obtain

$$\frac{\partial}{\partial t}(xu + 3tu^2) - \frac{\partial}{\partial x}\left(12tu^3 - 6tu\frac{\partial^2 u}{\partial x^2} + 3t\left(\frac{\partial u}{\partial x}\right)^2 + 3xu^2 - x\frac{\partial^2 u}{\partial x^2} + \frac{\partial u}{\partial x}\right), \quad (8.24)$$

from which we obtain

$$\int_{-\infty}^{\infty} (xu + 3tu^2)\,dx = \text{constant} \quad (8.25)$$

and

$$T = xu + 3tu^2. \quad (8.26)$$

The constant of the motion described by eqn. (8.25), which evolves according to eqn. (8.10), represents an *undefined KdV conservation law*.

See Appendix 8.A at the end of this chapter for details relating to symbolic algebra source code to perform above calculations.

8.2.5 An *Infinity* of Conservation Laws

Now consider the Miura transformation [Miu-68a, Miu-68b]

$$u = v^2 + v_x, \quad v = v(x,t), \quad (8.27)$$

which transforms the KdV equation to

$$\left(2v + \frac{\partial}{\partial x}\right)\left(v_t - 6v^2 v_x + v_{xxx}\right) = 0, \quad (8.28)$$

where the second bracketed term must be equal to zero, that is,

$$v_t - 6v^2 v_x + v_{xxx} = 0. \quad (8.29)$$

Hence, if v is a solution to eqn. (8.29), the *modified KdV equation*, then u given by eqn. (8.27) is a solution to the KdV equation, but not vice versa, owing to the leading operator term in eqn. (8.28).

Further analysis using the Miura transformation demonstrates that the KdV equation admits an infinity of conservation laws. First we let

$$v = \frac{1}{2}\epsilon^{-1} + \epsilon w, \quad (8.30)$$

where $w = w(x,t)$ and ϵ is an arbitrary real parameter, which modifies the Miura transformation to

$$u = \frac{1}{4}\epsilon^2 + w + \epsilon w_x + \epsilon^2 w^2. \quad (8.31)$$

This equation can be simplified by letting $u = u + \frac{1}{4}\epsilon^2$ to

$$u = w + \epsilon w_x + \epsilon^2 w^2, \quad (8.32)$$

the so-called *Gardner transformation* [Gar-67, Gar-74]. Applying the Gardner transformation to eqn. (8.10) yields

$$\frac{\partial u}{\partial t} - 6u\frac{\partial u}{\partial x} + \frac{\partial u^3}{\partial x^3} = w_t + \epsilon w_{xt} + 2\epsilon^2 ww_t - 6\left(w + \epsilon w_x + \epsilon^2 w^2\right)\left(w_x + \epsilon w_{xx} + 2\epsilon^2 ww_x\right)$$

$$+ w_{xxx} + \epsilon w_{xxxx} + 2\epsilon^2 (ww_x)_{xx} = 0, \tag{8.33}$$

$$= \left(1 + \epsilon\frac{\partial}{\partial x} + 2\epsilon^2 w\right)\left[w_t - 6\left(w + \epsilon^2 w^2\right)w_x + w_{xxx}\right] = 0. \tag{8.34}$$

Hence, if w is a solution of

$$w_t - 6\left(w + \epsilon^2 w^2\right)w_x + w_{xxx} = 0, \tag{8.35}$$

then u given by eqn. (8.32) is a solution to the KdV equation, but not vice versa, owing to the leading operator term in eqn. (8.34). Clearly, for $\epsilon = 0$, eqn. (8.35) reduces to the KdV eqn. (8.10).

We now write eqn. (8.35) in a conservative form that applies to all values of ϵ,

$$\frac{\partial w}{\partial t} + \frac{\partial}{\partial x}\left(w_{xx} - 3w^2 - 2\epsilon^2 w^3\right) = 0, \tag{8.36}$$

from which we see that, as a consequence of the decay conditions, $u \to 0$ as $|x| \to \infty$,

$$\frac{\partial}{\partial t}\int_{-\infty}^{\infty} w\, dx = -\left[w_{xx} - 3w^2 - 2\epsilon^2 w^3\right]_{-\infty}^{\infty} = 0 \tag{8.37}$$

$$\therefore \int_{-\infty}^{\infty} w\, dx = \text{constant}. \tag{8.38}$$

We can now take advantage of this result to generate additional conservation laws for the KdV equation. Because $w \to u$ as $\epsilon \to 0$, w can be represented by the following asymptotic expansion in ϵ:

$$w(x, t; \epsilon) \approx \sum_{n=0}^{\infty} \epsilon^n w_n(x, t), \quad \text{as} \quad \epsilon \to 0. \tag{8.39}$$

It therefore follows that the expansion of eqn. (8.39) can also be also applied to the constant in eqn. (8.38), from which we obtain

$$\int_{-\infty}^{\infty} w_n\, dx = \text{constant}, \quad n = 0, 1, 2, \cdots \tag{8.40}$$

Clearly, the constants in eqns. (8.38) and (8.40) are not the same.

Substituting the expansion of eqn. (8.39) into the Gardner transformation, eqn. (8.32), yields

$$\sum_{n=0}^{\infty} \epsilon^n w_n \approx u - \epsilon \sum_{n=0}^{\infty} \epsilon^n (w_n)_x - \epsilon^2 \left(\sum_{n=0}^{\infty} \epsilon^n w_n\right)^2. \tag{8.41}$$

By equating coefficients of powers of ϵ, invoking the decay conditions, and solving recursively, we are able to obtain the following conservation laws for the KdV equation:

$$w_0 = u, \tag{8.42}$$

$$w_1 = -w_{0,x} = -u_x, \tag{8.43}$$

$$w_2 = -w_{1,x} - w_0^2 = (u_x)_x - u^2, \tag{8.44}$$

$$w_3 = -w_{2,x} - 2w_0 w_1 = -u_{xxx} + 4uu_x = \left(-u_{xx} + 2u^2\right)_x, \tag{8.45}$$

$$w_4 = -w_{3,x} - 2w_0 w_2 - w_1^2 = u_{xxxx} - 6uu_{xx} - 5u_x^2 + 2u^3$$
$$= 2u^3 + u_x^2 - (6uu_x - u_{xxx})_x, \tag{8.46}$$

$$\vdots$$

etc.

Integrals of differentiated quantities with respect to x evaluate to a constant. *It follows, therefore, as the solutions for w_1 and w_3 are seen to be exact differentials with respect to x, the corresponding integrals provide no useful information.* Also, as the bracketed terms in the solutions for w_2 and w_4 are exact differentials and therefore constant, they can be combined with the associated constants of motion. Finally, we see that the corresponding *useful* constant of motion equations therefore become

$$\int_{-\infty}^{\infty} T_0 \, dx = \int_{-\infty}^{\infty} u \, dx = \text{constant}, \tag{8.47}$$

$$\int_{-\infty}^{\infty} T_2 \, dx = \int_{-\infty}^{\infty} -u^2 \, dx = \text{constant}, \tag{8.48}$$

$$\int_{-\infty}^{\infty} T_4 \, dx = \int_{-\infty}^{\infty} 2u^3 + u_x^2 \, dx = \text{constant}. \tag{8.49}$$

Equations (8.47), (8.48), and (8.49) are equivalent to eqns. (8.13), (8.15), and (8.19), respectively, that were derived by different means. Thus, $T_0 = u$, $T_2 = -u^2$, and $T_4 = 2u^3 + u_x^2$ represent conserved quantities, or invariants, of the KdV eqn. (8.10).

See Appendix 8.A at the end of this chapter for details relating to symbolic algebra source code to perform the preceding calculations.

8.2.6 KdV Equation: 2D

The 2D KdV equation is defined as

$$\frac{\partial u}{\partial t} - 6u \frac{\partial u}{\partial x} + \frac{\partial u^3}{\partial x^3} + 3 \frac{\partial v}{\partial y} = 0, \quad \frac{\partial u}{\partial y} = \frac{\partial v}{\partial x}, \tag{8.50}$$

where $u = u(x, y, t)$, $v = v(x, y, t)$, $t \in \mathbb{R}$, and $x, y \in \mathbb{R}$. It describes shallow-water waves where the wavelengths are much greater than the corresponding amplitudes moving in the x-direction and which contain small variations in the y-direction.

Equation (8.50) is also known as the *Kadomtsev–Petviashvili* (KP) equation [Kad-70], when it is usually presented in the equivalent form

$$\left(\frac{\partial u}{\partial t} - 6u\frac{\partial u}{\partial x} + \frac{\partial u^3}{\partial x^3}\right)_x + 3\frac{\partial^2 u}{\partial y^2} = 0, \tag{8.51}$$

which in conservative form becomes

$$\frac{\partial}{\partial t}(u_x) + \frac{\partial}{\partial x}(-6uu_x + u_{xxx}) + \frac{\partial}{\partial y}(3u_y) = 0. \tag{8.52}$$

However, determining a conservation law for eqn. (8.50) (or eqn. (8.51)) is more complex than for the 1D case of eqn. (8.10). Here, following Johnson [Joh-97], the decay conditions are based on the assumption that the quantiies u and $v \to$ constant as $|x| \to \infty$ and, also, as $|y| \to \infty$. It is also assumed that u and v are integrable over the entire spatial domain.

On integrating the second of eqns. (8.50) with respect to x we obtain

$$\frac{\partial}{\partial y}\left(\int_{-\infty}^{\infty} u\, dx\right) = 0 \implies \int_{-\infty}^{\infty} u\, dx = f(t). \tag{8.53}$$

However, this equation applies for all y, so we further assume that the integral is evaluated far from any disturbances. Under this restricted situation the decay conditions mean that $f(t)$ must be a constant. Hence, we have

$$\int_{-\infty}^{\infty} u\, dx = \text{constant}. \tag{8.54}$$

Adopting a similar argument yields

$$\int_{-\infty}^{\infty} v\, dy = \text{constant}. \tag{8.55}$$

We now integrate the first of eqns. (8.50) to obtain

$$\frac{\partial}{\partial t}\left(\int_{-\infty}^{\infty} u\, dx\right) + \left[-3u^2 + \frac{\partial^2 u}{\partial x^2}\right]_{-\infty}^{\infty} + \frac{\partial}{\partial y}\left(\int_{-\infty}^{\infty} v\, dx\right) = 0. \tag{8.56}$$

From this result, eqns. (8.50), and (8.54), and on applying the decay conditions, we deduce that

$$\int_{-\infty}^{\infty} v\, dx = \text{constant}. \tag{8.57}$$

Thus, we can interpret from eqns. (8.55) and (8.57) that momentum is conserved in both the x direction and the y direction.

8.2.7 KdV Equation with Variable Coefficients (vcKdV)

The vcKdV equation

$$\frac{\partial u}{\partial t} + f(t)\frac{\partial u}{\partial x} + g(t)\frac{\partial u^3}{\partial x^3} = 0 \tag{8.58}$$

was first proposed by Grimshaw [Gri-79], where $u = u(x, t)$, $t \in \mathbb{R}$, and $x \in \mathbb{R}$. This equation was later reported by Joshi [Jos-87] that, to be integrable, it must satisfy the Painlevé integrable condition

$$g(t) = af(t)S(t), \quad S(t) = b + \int^t f(t)\,dt, \tag{8.59}$$

where $a \neq 0$ and b are arbitrary real constants.

Using the transformation due to Zhang [Zha-07]

$$u(x,t) = \frac{x}{S(t)} + \frac{6a}{S(t)} U(\mathcal{X}, \mathcal{T}), \tag{8.60}$$

$$\mathcal{X} = \frac{x}{S(t)}, \quad \mathcal{T} = \frac{a}{S(t)}, \tag{8.61}$$

eqn. (8.58) can be transformed to the standard KdV equation

$$\frac{\partial U}{\partial \mathcal{T}} - 6U \frac{\partial U}{\partial \mathcal{X}} + \frac{\partial U^3}{\partial \mathcal{X}^3} = 0, \quad U = U(\mathcal{X}, \mathcal{T}), \tag{8.62}$$

where we make use of the following relationships:

$$u_t = -\frac{xf(t)U_\mathcal{X}}{S(t)^2} - \frac{af(t)U_\mathcal{T}}{S(t)^2}, \tag{8.63}$$

$$u_x = \frac{U_\mathcal{X}}{S(t)}, \tag{8.64}$$

$$u_{xxx} = \frac{U_{\mathcal{XXX}}}{S(t)^3}, \tag{8.65}$$

$$\frac{dS(t)}{dt} = f(t), \tag{8.66}$$

which follow directly from eqns. (8.59), (8.60), and (8.61).

It therefore follows that eqn. (8.62) has an infinity of conservation laws, as demonstrated for eqn. (8.10) in Section 8.2.5. Thus, it further follows that applying the transformation eqn. (8.60) will yield an infinity of conservation laws for the vcKdV eqn. (8.58). Therefore, from eqns. (8.13), (8.15), and (8.19), we see that the first three conservation laws for the vcKdV equation become

$$\int_{-\infty}^{\infty} T_0 \, dx = \int_{-\infty}^{\infty} \left(\frac{x}{S(t)} + \frac{6a}{S(t)} u \right) dx = \text{constant}, \tag{8.67}$$

$$\int_{-\infty}^{\infty} T_1 \, dx = \int_{-\infty}^{\infty} \left(\frac{x}{S(t)} + \frac{6a}{S(t)} u \right)^2 dx = \text{constant}, \tag{8.68}$$

$$\int_{-\infty}^{\infty} T_3 \, dx = \int_{-\infty}^{\infty} \left[\left(\frac{x}{S(t)} + \frac{6a}{S(t)} u \right)^3 + \frac{1}{2} \left(\frac{1}{S(t)} u_x \right)^2 \right] dx = \text{constant}. \tag{8.69}$$

See Appendix 8.A at the end of this chapter for details relating to symbolic algebra source code to perform the preceding calculations.

8.3 CONSERVATION LAWS FOR OTHER EVOLUTIONARY EQUATIONS

8.3.1 Nonlinear Schrödinger Equation

The NLS equation is defined as

$$i\frac{\partial u}{\partial t} + \frac{\partial^2 u}{\partial x^2} + \epsilon u |u|^2 = 0, \quad i = \sqrt{-1}, \; \epsilon = \pm 1, \tag{8.70}$$

where $u = u(x, t), t \in \mathbb{R}$, and $x \in \mathbb{R}$; and the corresponding equation for the complex conjugate of u, that is, u^*, is given by

$$-i\frac{\partial u^*}{\partial t} + \frac{\partial^2 u^*}{\partial x^2} + \epsilon u^* |u|^2 = 0. \tag{8.71}$$

Now, following Johnson [Joh-97], from $u^* \times$ eqn. (8.70) $- u \times$ eqn. (8.71) we obtain

$$i\left(u^*\frac{\partial u}{\partial t} + u\frac{\partial u^*}{\partial t}\right) + u^*\frac{\partial^2 u}{\partial x^2} - u\frac{\partial^2 u^*}{\partial x^2} = 0, \tag{8.72}$$

which simplifies to

$$i\frac{\partial}{\partial t}(uu^*) + \frac{\partial}{\partial x}(u^* u_x - u u_x^*) = 0. \tag{8.73}$$

Integrating with respect to x over the whole domain and applying decay conditions whereby conditions at $\pm\infty$ are equal yields

$$i\frac{\partial}{\partial t}\int_{-\infty}^{\infty} |u|^2 \, dx = 0, \tag{8.74}$$

from which it follows directly that the *first* constant of motion is given by

$$\int_{-\infty}^{\infty} |u|^2 \, dx = \text{constant} \tag{8.75}$$

and

$$T_1 = |u|^2. \tag{8.76}$$

Recall that $uu^* = |u|^2$. Equation (8.75) is the *first* conservation law for both versions of the NLS equation, that is, eqns. (8.70) and (8.71), and can be interpreted as being associated with *conservation of mass*.

The second conservation law for the NLS equation is obtained by forming $u_x^* \times$ eqn. (8.70) $+ u_x \times$ eqn. (8.71), that is,

$$i(u_x^* u_t - u_x u_t^*) + u_x^* u_{xx} + u_x u_{xx}^* + \epsilon (u_x^* u + u_x u^*) |u|^2 = 0, \tag{8.77}$$

and $u^* \times \frac{\partial}{\partial x}$ eqn. (8.70) $+ u \times \frac{\partial}{\partial x}$ eqn. (8.71), that is,

$$i(u^* u_{xt} - u u_{xt}^*) + u^* u_{xxx} + u u_{xxx}^* + \epsilon \left[u^* \left(u |u|^2\right)_x + u \left(u^* |u|^2\right)_x\right] = 0; \tag{8.78}$$

and then subtracting the latter from the former gives

$$i\frac{\partial}{\partial t}(uu_x^* - u^* u_x) + \frac{\partial}{\partial x}(u_x u_x^*) - (u^* u_{xxx} + u u_{xxx}^*)$$
$$+ \epsilon |u|^2 (uu^*)_x - \epsilon \left[(uu^*)_x |u|^2 + 2uu^* \left(|u|^2\right)_x\right] = 0. \tag{8.79}$$

This equation can be expressed as

$$i\frac{\partial}{\partial t}(uu_x^* - u^*u_x) + \frac{\partial}{\partial x}\left[2u_xu_x^* - (u^*u_{xx} + uu_{xx}^*) - \epsilon|u|^4\right] = 0. \qquad (8.80)$$

As for the first conservation law, we integrate with respect to x over the whole domain and apply the same decay conditions. This yields

$$i\frac{\partial}{\partial t}\int_{-\infty}^{\infty}(uu_x^* - u^*u_x)\,dx = 0, \qquad (8.81)$$

from which it follows directly that the *second* constant of motion is given by

$$\int_{-\infty}^{\infty}(uu_x^* - u^*u_x)\,dx = \text{constant} \qquad (8.82)$$

and

$$T_2 = uu_x^* - u^*u_x. \qquad (8.83)$$

Equation (8.82) is the *second* conservation law for the NLS equation and can be interpreted as being associated with *conservation of momentum*.

The third conservation law for the NLS equation is generated from

$$\frac{\partial u^*}{\partial x} \times \frac{\partial}{\partial x}[\text{eqn. (8.70)}] + \frac{\partial u}{\partial x} \times \frac{\partial}{\partial x}[\text{eqn. (8.71)}]\, u_{xt}u_x^* + u_xu_{xt}^*, \text{which yields}$$

$$i\frac{\partial}{\partial t}\left(u_xu_x^* - \frac{1}{2}\epsilon(uu^*)^2\right) + \frac{\partial}{\partial x}\left[u_x^*u_{xx} - u_xu_{xx}^* - \epsilon\left(u^2u^*u_x^* + u^{*2}uu_x\right)\right] = 0, \qquad (8.84)$$

which can be expressed as

$$i\frac{\partial}{\partial t}\left(|u_x|^2 - \frac{1}{2}\epsilon|u|^4\right) + \frac{\partial}{\partial x}\left[u_x^*u_{xx} - u_xu_{xx}^* - \epsilon(uu_x^* + u^*u_x)|u|^2\right] = 0. \qquad (8.85)$$

As for the previous conservation laws, we integrate with respect to x over the whole domain and apply the same decay conditions. This yields

$$i\frac{\partial}{\partial t}\int_{-\infty}^{\infty}(uu_x^* - u^*u_x)\,dx = 0, \qquad (8.86)$$

from which it follows directly that the *third* constant of motion is given by

$$\int_{-\infty}^{\infty}\left(|u_x|^2 - \frac{1}{2}\epsilon|u|^4\right)dx = \text{constant} \qquad (8.87)$$

and

$$T_3 = |u_x|^2 - \frac{1}{2}\epsilon|u|^4. \qquad (8.88)$$

Equation (8.87) is the *third* conservation law for the NLS equation.

See Appendix 8.A at the end of this chapter for details relating to symbolic algebra source code to perform the preceding calculations.

The fourth conservation law for the NLS equation is generated from $\frac{\partial^3 u^*}{\partial x^3} \times$ eqn. (8.70) $+ u \times \frac{\partial^3}{\partial x^3}$ [eqn. (8.71)], which yields the following constant of motion

$$\int_{-\infty}^{\infty} \left(u u^*_{xxx} + \frac{3}{2} \epsilon |u|^2 u u^*_x \right) dx = \text{constant} \qquad (8.89)$$

and

$$T_4 = u u^*_{xxx} + \frac{3}{2} \epsilon |u|^2 u u^*_x. \qquad (8.90)$$

8.3.2 Boussinesq Equation

The following 1D *Boussinesq* equation describes weakly dispersive water waves propogating in both directions:

$$\frac{\partial^2 \eta}{\partial t^2} - \frac{\partial^2 \eta}{\partial x^2} + 3 \frac{\partial^2}{\partial x^2} (\eta^2) - \frac{\partial^4 \eta}{\partial x^4} = 0, \qquad (8.91)$$

where $\eta = \eta(x, t)$, $t \in \mathbb{R}$, and $x \in \mathbb{R}$. For our purposes, it is more convenient to express this single equation as a pair of coupled equations, that is,

$$\eta_t = -u_x, \quad u_t + \eta_x - 3(\eta^2)_x + \eta_{xxx} = 0, \qquad (8.92)$$

where we have reverted to subscript notation. The second of eqns. (8.92) is obtained by integrating eqn. (8.91) with respect to x and applying the usual decay conditions as $|x| \to \pm\infty$.

Integrating eqns. (8.92) with respect to x over the whole domain and applying decay conditions whereby conditions at $\pm\infty$ are equal yields

$$\int_{-\infty}^{\infty} \eta \, dx = -[u]_{-\infty}^{\infty} = 0, \quad \int_{-\infty}^{\infty} u_t \, dx = [3\eta^2 - \eta - \eta_{xx}]_{-\infty}^{\infty} = 0, \qquad (8.93)$$

from which it follows directly that the *first* and *second* constants of motion are given by

$$\int_{-\infty}^{\infty} \eta \, dx = \text{constant} \quad \text{and} \quad \int_{-\infty}^{\infty} u_t \, dx = \text{constant} \qquad (8.94)$$

and correspond to *conservation of mass* and *conservation of momentum*, respectively. Thus, we have

$$T_1 = \eta \quad \text{and} \quad T_2 = u_t. \qquad (8.95)$$

Further detailed discussion relating to conservation laws can be found in [Abl-91], [Dra-92, Chapter 5], [Deb-05, Chapter 9], and [Joh-97, Chapter 3].

8.A SYMBOLIC ALGEBRA COMPUTER SOURCE CODE

Many of the preceding conservation law calculations are quite tedious to perform by hand. So, as an extra resource that readers may find useful, we include with the downloads code written for symbolic algebra programs Maple and Maxima (open source) that derive conservation laws for

- KdV equation (main laws)
- KdV equation (undefined laws)
- KdV equation (an infinity of laws)
- variable coefficients KdV equation (main laws)
- nonlinear Schrödinger equation (main laws)

REFERENCES

[Abl-91] Ablowitz, M. J. and P. A. Clarkson (1991), *Solitons, Nonlinear Evolution Equations and Inverse Scattering*, London Mathematical Society Lecture Note Series 149, Cambridge University Press.

[Dra-92] Drazin, P. G. and R. S. Johnson (1992), *Solitons: An Introduction*, Cambridge University Press.

[Deb-05] Debnath, L. (2005), *Nonlinear Partial Differential Equations for Scientists and Engineers*, Birkhäuser.

[Gar-67] Gardener, C. S., J. Greene, M. Kruskal and R. M. Miura (1967), Method for Solving the Korteweg–de Vries Equation, *Physical Review Letters* **19**, 1095–1097.

[Gar-74] Gardener, C. S., J. Greene, M. Kruskal and R. M. Miura (1974), Korteweg–de Vries Equation and Generalizations. VI. Methods for Exact Solution, *Communications on Pure and Applied Mathematics* **27**, 97–133.

[Gri-79] Grimshaw R. (1979), Slowly Varying Solitary Waves. I. Korteweg–de Vries Equation, *Proceedings of the Royal Society, Series A* **368**, 359–375.

[Joh-97] Johnson, R. S. (1997), *A Modern Introduction to the Theory of Water Waves*, Cambridge University Press.

[Jos-87] Joshi, N. (1987), Painlevé Property of General Variable Coefficient Versions of the Korteweg–de Vries and Nonlinear Schrödinger Equations, *Physical Letters, Series A* **125**, 456–460.

[Kad-70] Kadomtsev, B. P. and V. I. Petviashvili (1970), On the Stability of Solitary Waves in Weakly Dispersing Media, *Soviet Physics–Doklady* **15**, 539–541.

[Miu-68a] Miura, R. A. (1968). Korteweg-de Vries Equation and Generalizations I. A remarkable explicit transformation. *J. Math. Phys.*, **9**, 1202-4.

[Miu-68b] Miura, R. A., C. S. Gardner and M. D. Kruskal (1968), Korteweg–de Vries Equation and Generalizations II. Existence of Conservation Laws and Constants of Motion, *Journal of Mathematical Physics* **9**, 1204–1209.

[Sch-09] Schiesser, W. E. and G. W. Griffiths (2009), *A Compendium of Partial Differential Equation Models: Method of Lines Analysis with Matlab*, Cambridge University Press.

[Zak-00] Zaki, S. I. (2000), Solitary Waves of the Korteweg–de Vries–Burgers' Equation, *Computer Physics Communications* **126**, 207–218.

[Zha-07] Zhang, D.-J. (2007), Conservation Laws and Lax Pair of Variable Coefficient KdV Equation, *Chinese Physics Letters* **24**-1, 3021–3023.

9

Case Study: Analysis of Golf Ball Flight

9.1 INTRODUCTION

The analysis of a golf ball in flight has been the subject of many papers, with some of the earliest being by famous scientists such as P. G. Tait [Tai-96] [Kno-11] and J. J. Thompson [Tho-10], who happened to be keen golfers. Tait's investigation into the flight of a golf ball was probably the earliest serious study and folklore has it that this was initiated by the feat of his son, F. G. Tait who, in 1895, managed to drive 295 yards on the *Old Course* at St. Andrews. Now Tait was one of the foremost mathematical physicists of the time, but was unable to explain this prodigious drive by the then current theory. This spurred him on to make a detailed analysis of golf ball dynamics, including the devising of some innovative experiments whereby he had golfers drive golf balls into wet clay so he could determine the initial flight and spin of the ball. He recorded his experiments thus:

> When we fastened one end of a long tape to the ball and the other to the ground, and induced a good player to drive the ball (perpendicularly to the tape) into a stiff clay face a yard or two off, we find that the tape is always twisted; no doubt to different amounts by different players â£" say from 40 to 120 or so turns per second. The fact is indisputable. [Kno-11]

There were no high speed cameras at that time!

Inevitably Tait's analysis entailed some simplifying assumptions in relation to *drag* and *lift* as very little data was available then. These assumptions enabled Tait to obtain a closed form solution to the golf ball trajectory problem, which he used to generate the interesting set of hand drawings included in Fig. 9.1. Parts of this figure were included in Tait's article, "Golf—Long Driving," published in the March 1896 edition of *Badminton Magazine*. Since this early research there has been a plethora of papers analyzing all aspects of the subject. In this chapter we shall present a more general analysis, illustrated by results from a numerical computer simulation, which demonstrate that Tait's approximate solutions, as represented by his excellent drawings, reproduced the main features of a golf ball trajectory.

The *total force* acting upon a golf ball whilst in flight is equal to the summation of three distinct separate forces: the *drag* force due to air resistance F_D, the *Magnus* force due to

Case Study: Analysis of Golf Ball Flight

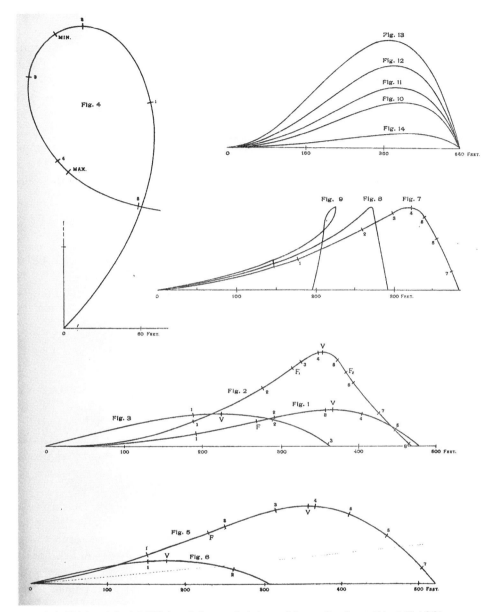

Figure 9.1. Tait's original 1896 hand drawn sketches of the path of a golf ball [Tai-96].

lift F_L, and the *gravitational* force F_g. The total force can be represented mathematically as

$$F_{total} = F_D + F_L + F_g, \quad (\text{N}). \tag{9.1}$$

Then, from Newton's second law (force = mass × acceleration) we can deduce an expression for acceleration, which takes the following form:

$$\frac{d\mathbf{v}}{dt} = -k_D |\mathbf{v}|\mathbf{v} + k_L |\mathbf{v}|(\hat{\boldsymbol{\omega}} \times \mathbf{v}) + \mathbf{g}, \quad (\text{m/s}^2), \tag{9.2}$$

where k_D and k_L are constants of proportionality for drag and lift forces, respectively.

Equation (9.2), together with an appropriate set of initial conditions, can be integrated once with respect to time to yield velocity and twice to yield distance traveled. We discuss below each of the terms in eqns. (9.1) and (9.2) in order that we may arrive at a set of equations that can be used to calculate a golf ball trajectory that is uniquely determined by the prevailing *ambient conditions* and its *initial velocity, angle of attack* and *spin*.

An additional force acting on a golf ball, not mentioned above, is *buoyancy*. Any body immersed in a fluid experiences an upward buoyancy force, also known as *Archimedes' principle*, that tends to oppose the downward gravitational force exerted on the object. This force is given by

$$F_B = \rho\, g\, vol, \quad (\text{N}), \tag{9.3}$$

where ρ represents density of fluid (kg/m^3), g gravitational acceleration (m/s^2), and vol volume of the object (m^3). However, in the case of a golf ball, the buoyancy force is equal to approximately 0.48×10^{-3} Newtons, whilst the gravitational force is approximately 1000 times greater. Therefore, buoyancy effects are generally ignored when calculating the trajectory of a golf ball.

9.2 DRAG FORCE

A sphere having a *very low* velocity moving through a viscous fluid experiences a drag force proportional to velocity. This force is defined by *Stokes Law*, $F_D = 3\pi \mu d v$, where μ represents *dynamic viscosity* [N.s/m^2], v velocity [m/s] and d diameter [m] of the sphere. Under low velocity conditions, the flow will be *laminar* and fluid will flow smoothly around the sphere. However, as the velocity increases, turbulence will be introduced and the Stokes force will eventually be dominated by the so called *drag force*, which is proportional to velocity squared.

The drag force acting on a body, such as a golf ball, is the result of two effects; *friction drag* due to *shear stress* at the surface, and *pressure drag* due to the *differential pressure* between the front and rear of the golf ball. The transition to turbulence occurs at the point where flow separates from the surface, which normally happens towards the rear of the ball (see Fig. 9.2).

At this point, the drag coefficient value falls off steeply allowing for a longer trajectory. It is this effect that golf ball manufacturers seek to take advantage of by designing surface dimple schemes to ensure that turbulence is induced at a low *Reynold's number*, *Re*. Reynold's number is a dimensionless quantity that provides a measure of the ratio of *inertial forces* to *viscous forces* for any object, and that can be used to estimate the onset of turbulence. It is defined as

$$\frac{\text{inertial acceleration}}{\text{viscous acceleration}} = \frac{v^2/L}{v v / L^2} = \frac{\rho v L}{\mu} = Re, \quad (-), \tag{9.4}$$

where ρ represents density of the flowing medium (air) (kg/m^3), v is the velocity (m/s) of the object relative to the medium, L is the characteristic length of the object which for a sphere is equal to its diameter, d (m), $v = \mu/\rho$ is the kinematic viscosity of the fluid (m^2/s)

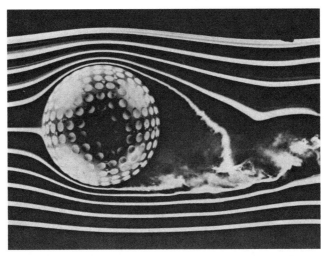

Figure 9.2. Photograph of golf ball undergoing a wind tunnel fluid dynamics test [Bro-71, p. 90]. In this test the ball is spinning clockwise about a fixed axis (normal to page) and air approaches rapidly from the left, simulating the flight of a golf ball moving from right to left. The effect is that the air is deflected downward, resulting in a net upward lift. The laminar flow stream lines can be seen breaking up at the rear of the ball, indicating turbulent conditions.

and μ is the dynamic viscosity of the fluid (kg/m/s). Figure 9.3 illustrates that, for a golf ball, the onset of turbulence occurs around $Re = 10^4$; and that for Reynold's numbers between 6×10^4 and 3×10^5, the drag coefficient of a dimpled golf ball is significantly lower than that for a smooth sphere. This counterintuitive effect has been verified experimentally numerous times over many years; for example, see the 1976 paper by Bearman and Harvey [Bea-76], or more recently the 2002 paper by Mizota and Simozono [Miz-02].

For a projectile in flight, there are generally three components to the drag force:

Form drag: is dependent upon the shape of a body. A smooth profile tends to maintain laminar flow at higher Reynolds numbers and minimize form drag. Larger cross-sectional areas normal to flow tend to increase form drag.

Frictional drag: is the result of viscous *shear stresses* between the object surface and fluid. As fluid flows along the surface of an object it sets up a *boundary layer* force which, when integrated, yields the amount of drag due to friction. Friction drag increases with the amount surface area in contact with the fluid, the *wetted area*, and also with viscosity.

Induced drag: also known as *lift induced drag*, is a component force parallel to the motion and results from the retarding effects of lift.

The drag force is a *vector quantity* and therefore has components in the three *Cartesian* co-ordinates x, y and z. We are able to ascertain the following expressions[1] for the

---

[1] The second of eqns. (9.5) is generally attributed to the English physicist Lord Rayleigh (John William Strutt, third Baron Rayleigh), 1842–1919, who discovered the relationship by employing the method of *dimensional analysis*.

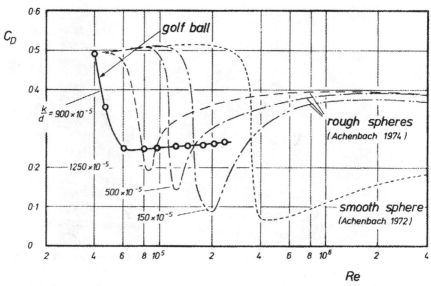

Figure 9.3. Plots of drag coefficients for various spheres (from [Bea-76]). The original data for rough and smooth spheres is from Achenbach [Ach-72] and [Ach-74]. For the range of Reynold's numbers between 6×10^4 and 3×10^5, the drag coefficient of a golf ball is clearly seen to be significantly lower than that for a smooth sphere. Assuming values for air density of $\rho = 1.225$ (kg m^{-3}) and for dynamic viscosity of $\mu = 1.8 \times 10^{-5}$ (kg m^{-1}s^{-1})), then for a golf ball having a diameter of $d = 0.04267$ (m), this corresponds to a golf ball velocity range of approximately 20–104 (m/s).

drag force as represented by the first term in eqn. (9.2),

$$\begin{aligned} F_D &= -k_D v^2 \frac{\mathbf{v}}{|\mathbf{v}|}, \quad (N), \\ &= -\tfrac{1}{2} C_D A \rho \left(v_x^2 \hat{\mathbf{x}} + v_y^2 \hat{\mathbf{y}} + v_z^2 \hat{\mathbf{z}} \right), \\ &= F_{Dx} \hat{\mathbf{x}} + F_{Dy} \hat{\mathbf{y}} + F_{Dz} \hat{\mathbf{z}}, \end{aligned} \quad (9.5)$$

where F_{Dx}, F_{Dy}, and F_{Dz} represent the component drag forces acting in the $\hat{\mathbf{x}}$, $\hat{\mathbf{y}}$, and $\hat{\mathbf{z}}$, directions respectively. The symbol $\hat{\mathbf{x}}$ represents a unit vector in the **x** direction; similarly for $\hat{\mathbf{y}}$ and $\hat{\mathbf{z}}$. C_D represents an experimentally determined drag coefficient $(-)$, A the area (m^2) normal to the direction of flight, $\mathbf{v}$ the velocity vector (m/s), $v = |\mathbf{v}|$ the absolute velocity value (m/s), $\frac{\mathbf{v}}{|\mathbf{v}|}$ a unit vector in the direction of $\mathbf{v}$ $(-)$, and ρ the ambient air density (kg/m^3). In general, subscripts x, y and z indicate component quantities acting in the directions of $\hat{\mathbf{x}}$, $\hat{\mathbf{y}}$, and $\hat{\mathbf{z}}$, respectively. The negative sign indicates that the force opposes motion.

Figure 9.4 contains a diagrammatic representation of the Cartesian co-ordinate system used in the golf ball trajectory calculations, and Figure 9.5 illustrates the instantaneous force vectors acting on a golf ball in flight.

The drag coefficient C_D depends on *spin-rate*, S $(-)$, defined as the ratio of *peripheral* velocity to *translational* velocity, that is,

$$C_D = f_D(S), \quad S = \frac{\omega r}{v}, \quad (9.6)$$

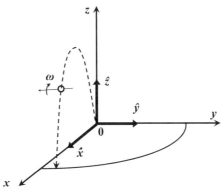

Figure 9.4. Diagram of the Cartesian co-ordinate system with unit vectors x̂, ŷ, and ẑ superimposed. The x vector is located along the intended direction of flight, while the y vector represents the lateral deviation from x and the z vector is directed vertically upward. The dotted line represents a typical trajectory where, in this case, the golf ball misses the center of the fairway to the left—indicating either a *pull* or *draw* shot. Variable ω represents the ball's angular velocity, for example, back spin.

where ω represents the magnitude of the ball's angular velocity vector (radians/s) and r the radius of the golf ball (m). It is assumed that the spin axis will tend to be in a constant relationship to the co-ordinate system during its flight due to the gyroscopic effect of rotation.

Varous values for the drag coefficients have been published in the literature, for example, those by Bearman and Harvey [Bea-76], Smits and Smith [Smi-94], and Mizota and Simozono [Miz-02] (see plots in Fig. 9.8). Simulation tests were performed using all of these coefficients and it was found that those reported by Mizota and Simozono gave

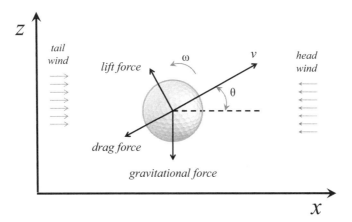

Figure 9.5. Vector diagram showing the instantaneous force vectors acting on a golf ball in flight, where v represents velocity along the direction of flight (m/s), ω represents the ball's angular velocity, for example, back spin (radians/s), θ the ball's angle of elevation (radians), and x, z Cartesian coordinates. In this diagram the trajectory is assumed to be parallel to the x axis, that is, it is contained within the x-z plane. The angular velocity is a vector quantity acting along the axis of rotation with its direction pointing at right angles out of the page.

results that fitted most closely the golf trajectory data given in Tables 9.2–9.4 (see Section 9.9).

9.3 MAGNUS FORCE

The *Magnus force* occurs when a spinning object moves in a velocity field and the spin is sufficient such that the boundary layer becomes turbulent (see Fig. 9.2). Consider an object rotating in the direction of motion where the air will tend to be dragged around the object and, consequently, the air passing over the top travels at a higher speed than the air at the bottom. This produces an unequal pressure across the object which results in a net downwards force or negative *lift*, that deflects, or tends to deflect, the object from its nominal path. It is named after the German scientist Heinrich Gustav Magnus, who published a theory describing the phenomenon in 1852 [Mag-52], which he developed whilst trying to improve the accuracy of German artillery.

This phenomena was previously reported by Isaac Newton in a letter to the Royal Society in 1671 in which he detailed results from his experiments into optics. He described the effect correctly and identified the cause after observing tennis players at Trinity College, Cambridge. Newton's description:

> Then I began to suspect whether the Rays, after their trajection through the Prisme, did not move in curve lines, and according to their more or less curvity tend to divers parts of the wall. And it increased my suspicion, when I remembered that I had often seen a Tennis ball, struck with an oblique Racket, describe such a curve line. For, a circular as well as a progressive motion being communicated to it by that stroak, its parts on that side, where the motions conspire, must press and beat the contiguous Air more violently than on the other, and there excite a reluctancy and reaction of the air proportionably greater. [New-71]

The Magnus effect is significant in many areas, for example, popular sports such as football, cricket and baseball, ballistic missiles, and wind turbines. There are ships that harness this effect as a propulsion mechanism, the most notable being the German vessel *The Backau* (see Fig. 9.6), which was equipped with *Flettner rotors*, named after their inventor, Anton Flettner.

Large modern cargo ships are still being built with Flettner rotors (see Fig. 9.7), which have become attractive due to the rising price of fossil fuels.

In the case of a golf ball in flight with negligible prevailing wind, the pressure below the ball is greater than the pressure above, due to back spin. This results in an upward lift.

Now, for a rotating object, the lift force in Newtons given by the *Kutta–Joukowski* theory is

$$F_L = \rho v \Gamma, \quad (\text{N}), \tag{9.7}$$

where ρ represents the density of ambient air, v the velocity relative to ambient air, and Γ *circulation*. Circulation is defined as the line integral about a closed path, at time t, of the tangential velocity component, $\mathbf{V}$, along the path of rotation [Sha-82, Chapter 13], that is,

$$\Gamma = \oint_C \mathbf{V} \cdot d\ell, \tag{9.8}$$

Figure 9.6. *The Backau*. The original Flettner rotor–powered ship built by Anton Flettner in 1924. The vessel was a refitted schooner which was equipped with two Flettner rotors approximately 15 m high and 3 m in diameter, driven by 37 KW electric motors. Source: US Library of Congress.

where C is the closed path and $d\ell$ a corresponding infinitesimal line segment. For a circulating cylinder of radius R, such as a Flettner rotor, the circulation is equal to $\Gamma = (\omega R)(2\pi R)$ per unit length (the tangential velocity multiplied by the circumference of the rotor). Therefore, approximating a sphere by a series of very short cylinders of decreasing diameter, we obtain the circulation of a rotating sphere by evaluating the following integral:

$$\Gamma = 2\int_0^R 2\pi\omega\left(R^2 - x^2\right)dx = \frac{8}{3}\pi\omega R^3. \tag{9.9}$$

Thus, from eqn. (9.7), the lift force is given by

$$F_L = \frac{16}{3}\pi^2 \rho N R^3 v, \quad (N), \tag{9.10}$$

where $N = \omega/(2\pi)$ represents sphere spin (revolutions per second).

Equation (9.10) represents an idealized situation and, in reality, the lift force would be reduced to something less than 40% of this value.

When it comes to the lift of a golf ball, eqn. (9.10) tends to overestimate the effect of spin, and wind tunnel tests have established a more accurate heuristic equation for

Figure 9.7. The E-Ship 1 is a *roll on roll off* cargo ship that made its first voyage with cargo in August 2010. The ship is owned by the third largest wind turbine manufacturer, Germany's Enercon GmbH, and is used to transport wind turbine components. The E-Ship 1 is equipped with four large Flettner rotor sails. Source: Wikipedia.

the lift force (see [Bea-76], [Smi-94], [Miz-02]), for which the second term in eqn. (9.2) becomes

$$F_L = k_L v^2 \frac{\boldsymbol{\omega} \times \mathbf{v}}{|\boldsymbol{\omega} \times \mathbf{v}|}, \quad (\text{N}),$$

$$= \tfrac{1}{2} C_L A \rho v^2 \frac{\boldsymbol{\omega} \times \mathbf{v}}{|\boldsymbol{\omega} \times \mathbf{v}|}, \qquad (9.11)$$

$$= F_{Lx}\hat{\mathbf{x}} + F_{Ly}\hat{\mathbf{y}} + F_{Lz}\hat{\mathbf{z}},$$

where F_{Lx}, F_{Ly}, and F_{Lz} represent component lift forces acting in the $\hat{\mathbf{x}}, \hat{\mathbf{y}}$, and $\hat{\mathbf{z}}$ directions, respectively; C_L represents the lift force coefficient $(-)$, $\boldsymbol{\omega}$ represents the *angular velocity* vector (1/s), and **v** represents the *velocity vector* (m/s) for the golf ball. The symbol × indicates a *cross product* (or *vector product*) and, hence, the term $\boldsymbol{\omega} \times \mathbf{v}$ can be represented by the determinant

$$\boldsymbol{\omega} \times \mathbf{v} = \begin{vmatrix} \hat{\mathbf{x}} & \hat{\mathbf{y}} & \hat{\mathbf{z}} \\ \omega_x & \omega_y & \omega_z \\ v_x & v_y & v_z \end{vmatrix}. \qquad (9.12)$$

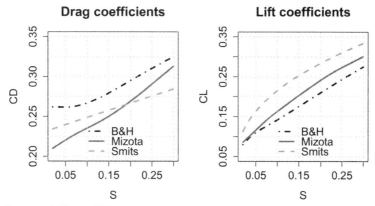

Figure 9.8. Drag and lift coefficients adapted from Bearman and Harvey [Bea-76], Smits and Smith [Smi-94], and Mizota and Simozono [Miz-02]. Note: The Smits and Smith drag coefficient correlation includes a dependence on Reynolds number, so only a representative curve, for $Re = 150,000$, has been included here.

Note that the *right-hand rule*[2] for a cross product indicates that $\boldsymbol{\omega}$ will have a negative y (i.e., right to left lateral) component. So, for a horizontal straight shot with back spin, we see from the coordinate system of Fig. 9.4 that $\mathbf{v}$ will point in the direction of $\hat{\mathbf{x}}$, $\boldsymbol{\omega}$ will point in the direction of $-\hat{\mathbf{y}}$, and $\boldsymbol{\omega} \times \mathbf{v}$ will point in the direction of $\hat{\mathbf{z}}$.

The lift force coefficient is a function of spin rate, defined previously, that is,

$$C_L = f_L(S), \quad S = \frac{\omega r}{|\mathbf{v}|}. \tag{9.13}$$

Various values for golf ball lift coefficients have been published in the literature, for example, those by Bearman and Harvey [Bea-76], Smits and Smith [Smi-94], and Mizota and Simozono [Miz-02] (see plots in Fig. 9.8). Simulation tests were carried on all these coefficients, and it was found that those reported by Mizota and Simozono gave results that fitted most closely the golf trajectory data given in Tables 9.2–9.4. Consequently, these coefficients were selected for use in the computer simulations discussed in Section 9.9.

9.4 GRAVITATIONAL FORCE

The gravitational force is simply the acceleration on the golf ball due to gravity, given by

$$F_g = -mg\hat{\mathbf{z}}, \quad (\text{N}), \tag{9.14}$$

where m represents mass (kg), g acceleration due to gravity (m/s^2), and $\hat{\mathbf{z}}$ is the unit vector co-directional with the *vertical* $\mathbf{z}$ axis, as discussed previously. The negative sign indicates that F_g is an attracting force, that is, directed toward the Earth.

---

[2] The direction of the vector $\boldsymbol{\omega}$ is predicted by wrapping the fingers of the right hand around the axis of rotation when the thumb, pointing along the axis, indicates the line along which the vector acts. Then, with the thumb, index, and middle fingers at right angles to each other and with the index finger pointed along the direction $\mathbf{v}$, the middle finger points in the direction of $\boldsymbol{\omega} \times \mathbf{v}$ and the thumb in the direction of $\boldsymbol{\omega}$.

9.5 GOLF BALL CONSTRUCTION

Modern golf balls come in a variety of different forms, from one- to five-piece construction. However, the golf ball types in common use are mainly of either a two- or three-piece construction. The two-piece ball usually has a solid central core with a high-durability *Surlyn*[3] cover. These balls are considered *hard* and are virtually *indestructible*. They have the advantage of giving the average player greater distance than other ball types. The three-piece construction usually consists of a either a solid rubber or liquid center covered by many turns of elastic windings, all covered by a high-durability *Balata* molded cover. Most players perceive Surlyn covered balls to feel harder when struck than Balata covered balls, and also to result in less back spin.

The distance that a golf ball flies depends not only upon the material it is made from, but also by the aerodynamic properties of its surface. Golf balls are designed for a specific region of air velocities by including dimples that reduce drag. Different manufactures go to great lengths to create dimple designs that maximize distance. Some designs suit the hard hitting professional or low handicap player, whilst other designs suit the average club golfer.

For our purposes we are mainly concerned with a golf ball's size and its weight. These characteristics are specified in Appendix III of *A Guide to the Rules on Clubs and Balls* [Roy-04, p. 56] of the Royal and Ancient Golf Club of St. Andrews (R & A), as follows:

Weight: A ball must not weigh more than 1.620 ounces *avoirdupois*[4] (45.93 g). There is no minimum weight.
Size: The ball must have a diameter of not less than 1.680 inches (42.67 mm).

A golf ball must also be spherically symmetric.

Similar rules are laid down for golf ball weight and size by the United States Golf Association (USGA).

9.6 AMBIENT CONDITIONS

The ambient conditions prevailing during the flight of a golf ball will vary according to altitude of location, temperature, relative humidity, wind speed, and direction of the air.

Ambient conditions can be calculated from the following set of correlation equations:

$$\begin{aligned}
\rho &= \frac{p}{1000\, ZRT_K} \left(\mathrm{MW}_{\mathrm{air}}(1-x) + \mathrm{MW}_{\mathrm{wat}} x \right), \quad (\mathrm{kg/m^3}), \\
p &= p_0 (1 - 2.25577 \times 10^{-5} h)^{5.25588}, \quad (\mathrm{Pa}), \\
\log_{10}(p_S) &= A - B/(C + T_C), \quad (\log_{10} \mathrm{Pa}), \\
p_V &= \mathrm{RH} \times p_S, \quad (\mathrm{Pa}), \\
x &= p_V / p, \quad (-),
\end{aligned} \qquad (9.15)$$

where ρ represents density of humid air (kg/m^3), p the humid air pressure (Pa), T_C/T_K the humid air temperature in degrees Centigrade/Kelvin, p_V the partial pressure of water

---

[3] A trade name for a group of proprietary thermoplastic resins developed by the Dupont Corporation.
[4] The avoirdupois system is the *common system of weight* in the United States and United Kingdom, although it is now largely superseded in the United Kingdom by the metric system. It is a system of weights based on units of pound (16 ounces). However, it is more correct to consider it a system of mass.

Table 9.1. A typical range of ambient conditions, illustrated for some famous golf club locations

Property	Summer high altitude	Mid-range conditions	Winter sea level
Location	Edgewood Tahoe, USA	Lausanne, Switzerland	Muirfield, Scotland
Altitude	1900m	500m	0m
Temperature	32°C	20°C	1°C
Humidity	5%	5%	35%
Air density	0.918 kg/m^3	1.134 kg/m^3	1.286 kg/m^3

Note: Values calculated from eqns. (9.15).

contained in the air (−), p_S the saturation pressure of water at temperature T_K, Z the equivalent compressibility factor for humid air, x the mole fraction of water with respect to the humid air, RH the relative humidity of humid air, and h height above sea level (m).

The following constants apply: the ambient air pressure at sea level $p_0 = 101,325$ (Pa), the gas constant $R = 8.31451$ (J mol^{-1}K^{-1}), the molecular weight of air MW$_{air}$ = 28.97 (g/mol), and the molecular weight of water MW$_{wat}$ = 18.02 (g/mol). Coefficient A, B, C for the *Antoine equation* for water vapor pressure are equal to 8.07131, 1730.63, 233.426, respectively. The equivalent compressibility factor is set to $Z = 1$ as variations are negligible at atmospheric pressures. The first of eqns. (9.15) includes 1000 in the denominator to convert density units from (g/cm^3) to (kg/m^3).

For summer at high altitude, winter at sea level, and mid-range conditions, reasonable figures for altitude, temperature, and humidity are given in Table 9.1, along with corresponding air densities. These figures have been used in our simulations in Section 9.9 to observe the effects of prevailing ambient conditions on golf ball flight. In addition, we have also simulated varying wind speed and direction to assess their effects on shot performance.

Note that the density of humid air is less than that of dry air. This follows from the first of eqns. (9.15) due to the lower molecular weight of water compared with that of air.

An R program that performs the above calculation and plots a 3D surface for ρ is included in Listing 9.1.

```
# File: File: airDensity.R
library("rgl")
#
airDensity <- function (TC=15, RH=0, Elev=0)
{
  # Function reurns the density in kg/m^3 of humid air
  # inputs: TC  - air temperature, [C]
  #         Elev - elevation, [m]
  #         RH  - relative humidity, [%]
  # constants
  # http://vle-calc.com/compound_properties.html?Comp1=04&numOfC=1&compnames=
          water_
  A <- 4.6543; B <- 1435.264; C <- 208.312 # Antoine coefficients
  MWa <- 28.9645  # Molecular weight - air, [g/mol]
```

```
  MWw <- 18.01528 # Molecular weight - water, [g/mol]
  R   <- 8.31451  # Ideal gas constant, [J/kmol/K]
  Z   <- 1        # Compressibility, [-]
  #
  TK <- TC+273.15 # Temperature, [K]
  p0 <- 101325    # Pressure at sea level, [Pa]
  # Location pressure, [Pa]
  p  <- p0*(1-2.25577*10^(-5)*Elev)^5.25588
  # Saturation pressure for water, [Pa]
  pS <- 10^(A-B/(C+TC))*10^5
  # Partial pressure of water contained in air
  pV <- RH*pS/100
  # Mole fraction of water in air
  x  <- pV/p
  # Humid air density, [kg/m^3]
  rho <- (p/(1000*Z*R*TK))*(MWa*(1-x)+MWw*x)
  return(c(rho))
}
###############################
# Air density test program
###############################
# Set data
N <- 20
RH <- seq(0,50,length.out=N)
temp <- seq(0,40,length.out=N)
rho <- matrix(0,N,N)
# Calculate air density
for(i in 1:N){
  rho[,i]<-airDensity(temp,RH[i],0)}
# Plot results
jet.colors <- colorRampPalette( c("red","blue") )
open3d()
persp3d(temp,RH,rho, aspect=TRUE, color =jet.colors(20),
        ylab = "RH", xlab = "Temperature", zlab="rho",
        smooth=FALSE,scales=TRUE);
um <- c( 0.6038209,  0.7971102, 0.001149754, 0,
        -0.2930455,  0.2206462, 0.930281579, 0,
         0.7412905, -0.5620638, 0.366821498, 0,
         0.0000000,  0.0000000, 0.000000000, 1)
UM <- matrix(data=um, byrow=TRUE,nrow=4,ncol=4)
rgl.viewpoint(fov=0) # set before userMatrix
par3d(userMatrix=UM)
```

Listing 9.1. File: airDensity.R—Code for function airDensity() that calculates the density of humid air followed by a test program that generates a surface plot of density as a function of RH and temperature

9.7 THE SHOT

9.7.1 Golf Ball Compression

When the club head strikes a golf ball, the ball rolls up the face of the club while at the same time the ball is being compressed. This process results in the ball achieving a high degree of *back spin* or a high *spin rate* as well as a high velocity. The launch speed of the golf ball is governed by the laws of *conservation of energy* and *conservation of momentum*.

The applicable equations for an ideal golf ball initially at rest are

$$\text{conservation of energy: } \tfrac{1}{2}m_C V_C^2 = \tfrac{1}{2}m_C v_C^2 + \tfrac{1}{2}m_B v_B^2,$$
$$\text{conservation of momentum: } m_C V_C = m_C v_C + m_B v_B, \quad (9.16)$$

where m represents mass (kg), V initial velocity (m/s), and v velocity (m/s) at launch. The subscripts C and B refer to club head and ball, respectively. Algebraic manipulation of eqns. (9.16) yields solutions for the postlaunch velocities, that is,

$$v_C = V_C \frac{m_C - m_B}{m_C + m_B}, \quad v_B = \frac{2V_C}{1 + m_B/m_C}. \quad (9.17)$$

This is effectively a statement of *Newton's law of restitution*.[5] However, as golf balls are not perfectly elastic, some kinetic energy is converted to heat, which reduces launch velocity. This loss of kinetic energy is accounted for by the ball's *coefficient of restitution*, C_R (–) (here we assume that the club head is perfectly rigid or elastic). This has a fractional value representing the ratio of speeds post- and preimpact, which, for a golf ball, the value is typically around 0.78 [Arn-10].[6] Thus, the final equation for launch speed becomes

$$v_B = \frac{(1 + C_R)V_C}{1 + m_B/m_C}. \quad (9.18)$$

Clearly, even for a highly elastic and very light golf ball, the launch speed cannot exceed twice the initial club head speed. Taking a typical driver with a head weighing 198 g, we find that $v_B = 1.78 v_C/(1 + 45.93/198) = 1.45 V_C$. With respect to golf club irons, there is generally a 7 g increase in mass between one club head and the next higher number [Jac-01, p. 86]. Hence, a typical 5-iron with a head weighing 257 g achieves a $v_B = 1.51 V_C$, while a typical 9-iron with a head weighing 285 g achieves a $v_B = 1.53 V_C$. Thus, the effect of compression is that ball launch speed is greater than club head speed by approximately 45% for a driver 51% for a 5-iron and and 53% for a 9-iron.

The R&A/USGA currently limits the coefficient of restitution to a maximum of 0.830. This means that following the club head-ball impact, no more than 83% of the club head energy can be transfered to the ball. In addition, they specify that the initial velocity of the ball shall not exceed 255 ft/s (76.5 m/s) maximum, when measured on apparatus approved by the R&A/USGA. A complementary overall distance standard states that the ball distance in carry and roll shall not exceed 296.8 yards (271.4 m) maximum using the same apparatus. These rules are updated every year.

---

[5] Most people will be aware of *Newton's cradle*, which illustrates this effect in an entertaining way.
[6] For cold temperatures around 0°C, C_R is reduced by around 10%, resulting in a correspondingly shorter carry. However, it should be noted that golf rules prohibit golf balls being purposely warmed during a stipulated round by use of a golf ball warmer or any other such device.

> An interesting slow-motion video of a golf ball impacting a heavy steel plate at 150 mph is available for viewing on YouTube [USG-10].

9.7.2 Spin

The amount of *spin* imparted to a golf ball when struck by a club has a considerable effect on distance and path the ball will travel. A ball with less spin cuts through the air better than a ball with more spin. However, too much back spin will cause the ball to rise quickly (ballooning) and lose its forward momentum. On the other hand, if there is too little back spin the ball will not stay airborne long enough to achieve maximum distance. Side spin in addition to back spin causes the ball to either *draw* or *fade*, and excessive side spin results in a *slice* or *hook*, usually with an attendant loss of distance.

A *draw* or *hook* is the name given to a shot when a right handed golfer strikes a ball in such a manner as to cause the ball to curve to the left. This is the result of the combined effects of back and left side spin, which causes the spinning axis to be inclined to the left. Similarly, a *fade* or *slice* is the name given to a shot when a right handed golfer strikes a ball in such a manner as to cause the ball to curve to the right. This is the result of the combined effects of back and right side spin, which causes the spinning axis to be inclined to the right.

9.7.2.1 Spin Rate Decay

The spin rate of a golf ball during a typical trajectory is slowed due to a retarding aerodynamic torque given by

$$T = -\frac{1}{2} C_T A \rho d v^2, \quad \text{(kg m)} \tag{9.19}$$

where C_T represents the coefficient of aerodynamic torque (–), d golf ball diameter (m), and other symbols are as defined previously. The coefficient C_T is a function of S, the ratio of *peripheral* velocity to *translational* velocity:

$$C_T = f_T(S), \quad S = \frac{\pi d N}{v}, \tag{9.20}$$

where N represents the spin rate (rps) and d the golf ball diameter (m). Very little in the way of reported data for the coefficient C_T is available in the literature, and those published by Mizota and Simozon are used in the computer simulation discussed in Section 9.9. These coefficients are plotted in Fig. 9.9.

The model for spin rate decay is discussed in more detail in Section 9.8.

9.7.3 Launch Angle

The initial angle at which a ball leaves the club face, the *launch angle*, is determined by a number of factors, including angle of attack, swing speed, club loft, and shaft flex. During the start of the downward part of the golf swing the club shaft flexes backwards but, as the club head approaches the ball it returns, and at impact it is actually ahead of where it would have been had the shaft not flexed. As a consequence, the loft at impact is, in general, increased and therefore it is properly referred to as *dynamic loft* or *effective loft*. High speed photographs have verified that dynamic loft is greater than club-face loft by

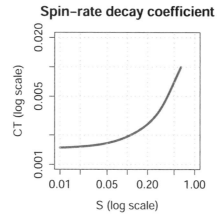

Figure 9.9. Plot of spin rate decay coefficient C_T, adapted from data by Mizota and Simozono [Miz-02].

several degrees [Pen-03]. However, after taking all factors into account, particularly the friction between club and ball, the launch angle is slightly lower than the dynamic loft angle of the club, generally by a factor of approximately 0.8 [Wes-09] (see Fig. 9.10).

A higher loft creates a higher launch angle, causing the ball to go higher. However, this means that more swing power is imparted to the vertical component of the strike, resulting in less distance. A lower loft results in the ball flying with a lower trajectory with more swing power being imparted to the horizontal component of the strike. This can, potentially, mean greater distances for a fast launch speed.

9.7.4 Bounce and Roll

When a golf shot falls to the ground, the ball ceases its descent and then, depending upon ground conditions, proceeds to bounce and/or roll in a variety of ways. For example, the ball can: (1) become plugged due to wet ground and stop completely; (2) hit wet ground with sufficient backspin to cause the ball to either bounce forward or

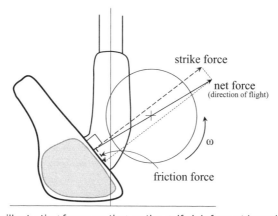

Figure 9.10. Diagram illustrating forces acting on the golf club face at launch. The launch angle is in the direction of the net force on the club face, being the result of the strike force normal to to the club face and friction force directed down the club face, the cause of back spin. The symbol $\omega = 2\pi N$ represents golf ball angular velocity (radians/s).

Table 9.2. Typical golf ball launch speeds, angles, and spin rates for different classes of player using a driver

Player class	Ball speed (mph)	Launch angle (deg)	Spin rate (rpm)
0–8	155	13	2500
9–16	145	12	2800
7–24	138	12	3000
24+	130	12	3100
Ave. PGA tour player	165	12	2400
Tiger Woods	180	11	2200
Ave. male long drive competitor	195	14	2000

Source: Anon.

cease forward motion and travel backwards; and (3) hit firm or hard ground and bounces forward followed by a roll.

All these different scenarios and others can be modeled mathematically, but the results can be so far ranging that here we adopt a simple heuristic model that seeks to provide a reasonable degree of perceived realism. The simulation results in Section 9.9 assume that following contact with the ground, the golf ball vertical velocity is reversed and reduced to 40% of its impact velocity. The forward and lateral velocities are simply reduced to 40% of their values at impact. The bounce calculation is performed in the main program at the end of each integration step (see Listing (9.2)).

Detailed discussion of the physics of golf ball bounce and roll can be found in [Wes-09].

9.7.5 Shot Statistics

Some *launch speed* and *carry distance* statistics are given in Tables 9.2–9.4.

Table 9.3. Carry distances in yards for some well-known top golfers [Win-06, p. 39]

	Adam Scott	Justin Leonard	Tiger Woods	Vijay Singh	Phil Michelson
Driver	285	270	285	275	300
3-wood	250	235	265	250	270
4-wood	–	225	–	–	255
2-iron	235	220	245	–	–
3-iron	220	215	230	220	230
4-iron	210	205	220	270	220
5-iron	200	190	208	195	205
6-iron	190	180	190	180	190
7-iron	175	165	172	155	175
8-iron	160	150	158	150	160
9-iron	150	140	142	140	150
PW	135	125	128	130	135
SW	115	105	106	115	110
LW	95	90	92	95	90

Table 9.4. Club strike statistics [Tra-10]

	Club speed (mph)	Attack angle (deg)	Ball speed (mph)	Smash factor	Vertical launch (deg)	Spin rate (rpm)	Apex (yds)	Land angle (deg)	Carry (yds)
Driver	112	−1.3	165	1.49	11.2	2685	31	39	269
3-wood	107	−2.9	158	1.48	9.2	3655	30	43	243
5-wood	103	−3.3	152	1.47	9.4	4350	31	47	230
Hybrid	100	−3.5	146	1.46	10.2	4437	29	47	225
3-iron	98	−3.1	142	1.45	10.4	4630	27	46	212
4-iron	96	−3.4	137	1.43	11.0	4836	28	48	203
5-iron	94	−3.7	132	1.41	12.1	5361	31	49	194
6-iron	92	−4.1	127	1.38	14.1	6231	30	50	183
7-iron	90	−4.3	120	1.33	16.3	7097	32	50	172
8-iron	87	−4.5	115	1.32	18.1	7998	31	50	160
9-iron	85	−4.7	109	1.28	20.4	8647	30	51	148
PW	83	−5.0	102	1.23	24.2	9304	29	52	136

Note: The location and weather conditions were not taken into consideration when gathering these statistics. Nevertheless, the data is based on a large number of shots and provides a good indication of shot statistics for tournament players.

Andrew Rice reports [Ric-12] that wearing a watch while playing golf can seriously affect the distance traveled by the golf ball. His tests indicate that wearing a 3 oz. watch can result in up to 10 yards lost when using a driver.

9.8 COMPLETING THE MATHEMATICAL DESCRIPTION

With the characteristics of the golf ball and ambient conditions specified, we can now assemble the necessary equations that will enable the flight of a golf ball to be determined, subject to the imposed initial conditions.

From Newton's second law, *force = mass × acceleration*, we obtain from eqns. (9.1), (9.5), (9.11) and (9.14) the following expressions for motion in the x, y, and z directions:

$$\begin{aligned} m\ddot{x} &= (F_{Dx} + F_{Lx}), \\ m\ddot{y} &= (F_{Dy} + F_{Ly}), \\ m\ddot{z} &= (F_{Dz} + F_{Lz} - mg). \end{aligned} \quad (9.21)$$

A single dot over a symbol represents differentiation with respect to time, and a double dot differentiation twice with respect to time. Thus, the symbol $\ddot{x}$ represents the second derivative of x with respect to time, d^2x/dt^2, that is, the acceleration component in the **x** direction; similarly for acceleration in the **y** and **z** directions. Other symbols have been defined previously.

To obtain the x, y, and z positions of the golf ball in flight, we need to integrate eqns. (9.21) twice with respect to time. This is carried out numerically and, in order to simplify the process, we first transform each of the second order differential equations into two first order differential equations. We achieve this for the x equation by letting

$$x_1 = x \quad \text{and} \quad x_2 = \dot{x}_1; \tag{9.22}$$

similarly for the y and z equations.

Finally, we include an equation for the retardation of the golf ball spin rate, for which we again invoke Newton's second law. In this case we have a rotating object, thus we need to calculate the rate-of-change of *angular momentum* $I\omega$, that is,

$$I\ddot{\theta} = I\dot{\omega} = 2\pi I \dot{N} = T, \tag{9.23}$$

where θ represents angular displacement about the spin axis, $\ddot{\theta} = \dot{\omega} = 2\pi \dot{N}$ angular acceleration, $\dot{N}$ rate of change of spin rate, I the golf ball *moment of inertia* [kg·m²] and T the spin-rate retarding torque (see eqn. (9.19)). The moment of inertia for a solid sphere rotating about its y axis is given by

$$\begin{aligned} I &= \tfrac{1}{2}\pi\rho \int_{-r}^{r} y^4 \, dz, \\ &= \tfrac{1}{2}\pi\rho \int_{-r}^{r} \left(r^2 - z^2\right)^2 dz, \\ &= \tfrac{2}{5}mr^2, \end{aligned} \tag{9.24}$$

where ρ [kg/m³] represents golf ball density, r radius [m], and $m = \tfrac{4}{3}\pi r^3 \rho$ [kg] mass.

This process yields seven coupled first-order differential equations, that is,

$$\begin{aligned} \dot{x}_2 &= x_1, \\ \dot{y}_2 &= y_1, \\ \dot{z}_2 &= z_1, \\ \dot{x}_1 &= (F_{Dx} + F_{Lx})/m, \\ \dot{y}_1 &= (F_{Dy} + F_{Ly})/m, \\ \dot{z}_1 &= (F_{Dz} + F_{Lz})/m - g, \\ \dot{N} &= T/(2\pi I), \\ x_i &= x_i(t), \; y_i = y_i(t), \; z_i = z_i(t), \; i \in \{1, 2\}, \; (x_i, y_i, z_i) \in \mathbb{R}, \; t \geq 0. \end{aligned} \tag{9.25}$$

To complete the model and enable us to solve the trajectory problem, we need a set of corresponding initial conditions for eqns. (9.25). These are

$$\begin{aligned} x_1(0) &= x_{10}, \; y_1(0) = y_{10}, \; z_1(0) = z_{10}, \\ x_2(0) &= x_{20}, \; y_2(0) = y_{20}, \; z_2(0) = z_{20}, \\ N(0) &= N_0. \end{aligned} \tag{9.26}$$

While applying the preceding transformations to the spatial variables, that is, eqn. (9.22), and so on may appear to complicate the situation, it does actually facilitate the task of integration as numerical integrators are generally designed to operate on first order differential equations.

9.8.1 The Effect of Wind

The effect of wind speed is additive to that of the effect of the golf ball velocity. Thus, wind having a positive speed W_x (assists motion in the **x** direction), will have the same effect on drag as if the ball's velocity v_x is decreased by W_x. On the other hand, if the wind

Table 9.5. Comparison of published golf shot carry data

Published carry data			Simulated	Diff. (%)
Player	Club	Yards	Yards	%
Tiger Woods	Driver	285	283	−0.7
Ave. tour pro.	Driver	269	263	−2.2
Ave. club plr.	Driver	230	226	−1.7
Ave. tour pro.	3-wood	243	251	3.3
Ave. tour pro.	5-wood	230	239	3.9
Ave. tour pro.	3-iron	212	218	2.8
Ave. tour pro.	5-iron	194	196	1.0
Ave. tour pro.	7-iron	172	163	−5.2
Ave. tour pro.	9-iron	148	138	−6.9
			Ave. diff. (%)	0.8

Note: See Tables 9.2–9.4, with simulated data.

opposes golf ball motion, it will have the same effect on drag as if the ball's velocity v_x is increased by W_x. Hence, we may write for the velocity components

$$\tilde{v}_x = v_x - W_x, \quad \tilde{v}_y = v_y - W_y, \quad \tilde{v}_z = v_z - W_z, \qquad (9.27)$$

where $\tilde{v}_x$ is the effective velocity in the **x** direction; similarly for $\tilde{v}_y$ and $\tilde{v}_z$.

9.9 COMPUTER SIMULATION

In this section the mathematical model described earlier is used to investigate various golf shot scenarios whereby different clubs impart differing velocities and spin on the ball. In addition, we look at shots made under different wind speeds and directions, and also under varying ambient conditions. Distances are generally given in yards, as this is the measure traditionally used by golfers. Calculations and terms generally relate to right-handed golfers, but those for left-handed golfers are given by the mirror image.

Parameters are set from Tables 9.1–9.4. However, although the data are from reputable sources, ball characteristics vary according to manufacturer and the full conditions (ambient, wind, elevation, etc.) under which the measurements took place are not known. In addition, some data have been averaged. This situation is also exacerbated by the trend over the last decade or so for manufactures to change their clubface loft angles. Until around the late 1990's most manufacures created sets of golf club irons with the same loft angles. It then appears that manufactures realized that they could reduce the loft angles and claim that their clubs went further, the result being that modern clubs have lofts around 4° less than the corresponding clubs of 20 or more years ago. Some additional details of these changes are included at http://en.wikipedia.org/wiki/Iron_(golf).

The drag and lift coefficients C_D, C_L, and C_T for the simulations discussed later have been taken from [Miz-02] (see Figs. 9.8 and 9.9). These coefficients give results comparable to data published for a wide range of golf shots made with a variety of clubs (see summary in Table 9.5).

Table 9.6. Launch data for driver shots simulation for various classes of player, taken from Table 9.2

Player	Speed (mph)	Angle (deg)	Back spin (rpm)
Tiger Woods	180	11	2200
Ave. tour pro.	165	11.2	2685
Ave. club player	145	12	2800

The lack of hard data is a real problem for investigating the flight of golf balls under playing conditions. Therefore, the right-hand column of Table 9.5 is labeled as *percentage differences* rather than *percentage errors* for the reasons stated earlier.

9.9.1 Driver Shots

In this section we simulate golf ball trajectories for driver shots made by various classes of player. We select the top professional Tiger Woods, an average tour professional and an average club player to simulate and use the launch data in Table 9.6.

The shots were assumed to be taken under ambient air conditions where the density is 1.225 kg/m³ (sea level, 15°C, 0% relative humidity) and the simulated trajectories are shown in Fig. 9.11.

The initial conditions are set by the line of code

```
source("init_drivers.R")
```

which should be selected by setting `testNo=1` in the main program of Listing 9.2.

The simulated carry distances are 283, 262, and 226 yards for Tiger Woods, the average tour professional, and the average club player, respectively. This compares with corresponding published data of 285, 269, and 230 yards, giving percentage differences of −0.7%, −2.2%, and −1.7%, respectively.

Figure 9.11. Simulated driver shots by different classes of player (see text).

Table 9.7. Launch data for wood shots simulation, taken from Table 9.4

Club	Speed (mph)	Angle (deg)	Back spin (rpm)
Driver	165	11.2	2685
3-wood	158	9.2	3655
5-wood	152	9.4	4350

9.9.2 Wood Shots

In this section we simulate golf ball trajectories for wood shots made by an average tour professional. We select the driver, 3-wood, and 5-wood, to investigate and use the launch data in Table 9.7.

The shots were assumed to be taken under ambient air conditions where the density is 1.225 kg/m³ (sea level, 15°C, 0% relative humidity), and the simulated trajectories are shown in Fig. 9.12.

The initial conditions are set by the line of code

```
source("init_woods.R")
```

which should be selected by setting testNo=2 in the main program of Listing 9.2.

The simulated carry distances are 262, 251, and 239 yards for the driver, 3-wood, and 5-wood, respectively. This compares with corresponding published data of 269, 243, and 230 yards, giving percentage differences of −2.2%, 3.3%, and 3.9%, respectively.

9.9.3 Iron Shots

In this section we simulate golf ball trajectories for iron shots made by an average tour professional. We select the 3-iron, 5-iron, 7-iron, and 9-iron to investigate and use the launch data in Table 9.8.

The shots were assumed to be taken under ambient air conditions where the density is 1.225 kg/m³ (sea level, 15°C, 0% relative humidity) and the simulated trajectories are shown in Fig. 9.13.

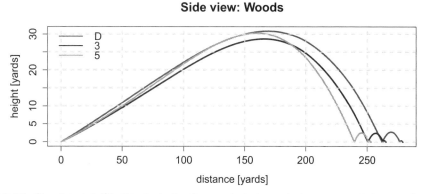

Figure 9.12. Simulated golf ball trajectories for wood shots made by an average tour professional (see text).

Table 9.8. Launch data for iron shots simulation, taken from Table 9.4.

Club	Speed (mph)	Angle (deg)	Back spin (rpm)
3-iron	142	10.4	4630
5-iron	132	12.1	5361
7-iron	120	16.3	7097
9-iron	109	20.4	8647

The initial conditions are set by the line of code

```
source("init_irons.R")
```

which should be selected by setting testNo=3 in the main program of Listing 9.2.

The simulated carry distances are 218, 196, 163, and 138 yards for the 3-iron, 5-iron, 7-iron, and 9-iron, respectively. This compares with corresponding published data of 212, 194, 172, and 148 yards, giving percentage differences of 2.8%, 1.0%, −5.2%, and −6.9%, respectively.

9.9.4 Effect of Wind

In Section 9.8.1, we discussed the effect of wind on a golf shot. Here we detail the results of simulated golf shots made by an average club golfer with a driver under head, tail, left, and right wind conditions. The four sets of simulated conditions are detailed in Table 9.9.

The shots were assumed to be taken under ambient air conditions where the density is 1.225 kg/m^3 (sea level, 15°C, 0% relative humidity) and the simulated trajectories are shown in Figs. 9.14 and 9.15.

The initial conditions are set by the line of code

```
source("init_windTest.R")
```

which should be selected by setting testNo=4 in the main program of Listing 9.2.

Figure 9.14 shows how the carry of a driver shot from an average club golfer is affected by a fairly moderate side wind. The 10 mph left to right wind moves the ball so it lands

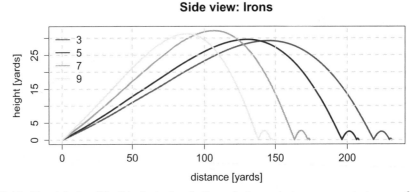

Figure 9.13. Simulated golf ball trajectories for iron shots made by an average tour professional (see text).

Table 9.9. Launch data for driver shots simulation subject to a 10 mph wind approaching from varying directions

| Wind | | Speed | Angle | Back spin |
Direction	deg	(mph)	(deg)	(rpm)
Head	−180	145	12	2800
Tail	0	145	12	2800
From left	−90	145	12	2800
From right	90	145	12	2800

12 yards to the right of the fairway center. Similarly, a 10 mph right to left wind moves the ball, so it lands 12 yards to the left of the fairway center. Both shots reach a height of 25 yards.

Figure 9.15 shows how the carry of a driver shot from an average club golfer is affected by fairly moderate head and tail winds. The 10 mph head wind increases drag so the ball only carries 214 yards to the middle of the fairway center and reaches a maximum height of 27 yards. On the other hand, a 10 mph tail wind reduces drag so that the ball carries 235 yards to the middle of the fairway center and reaches a maximum height of 23 yards. Thus, the performance difference between tail and head winds is a carry of 21 yards. This could mean at least another two or three club lengths for the player's second shot, which is very significant. This would, of course, depend upon other conditions, such as the amount of run achieved for each shot, which would be greater for the lower trajectory tail wind shot.

9.9.5 Effect of Differing Ambient Conditions

We discused the effect of varying altitude and temperature and relative humidity on the density of ambient air in Section 9.6. In this section we show the results of a computer simulation of golf ball carry and elevation for a driver under the three sets of ambient conditions detailed in Table 9.1.

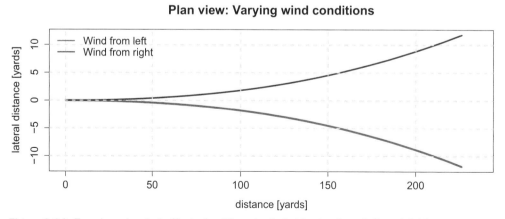

Figure 9.14. Top view showing effect of a 10 mph winds blowing from left and right.

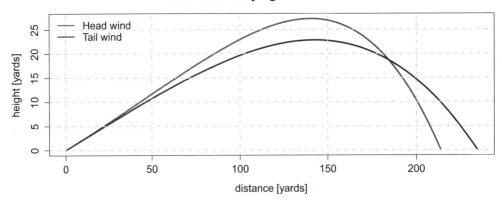

Figure 9.15. Side view showing effect of 10 mph head and tail winds.

The initial conditions are set by the line of code

```
source("init_ambientTest.R")
```

which should be selected by setting `testNo=5` in the main program of Listing 9.2.

The effects of different ambient conditions for a club golfer using a driver with launch speed, angle and spin of 150 mph, 12°, and 2800 rpm, respectively, are illustrated in Fig. 9.16. The maximum elevation and carry are both clearly seen to be affected. The maximum trajectory height and carry for each case were found to be, as follows:

High altitude: Maximum height = 23.2 yards (21.3 m), carry = 251 yards (230 m)
Mid-range: Maximum height = 26.0 yards (23.9 m), carry = 242 yards (221 m)
Sea level: Maximum height = 27.8 yards (25.4 m), carry = 233 yards (213 m)

The difference in maximum height of 4.6 yards (4.1 m) between a shot at high altitude and one at sea level is probably not that significant for most golf holes; however, the

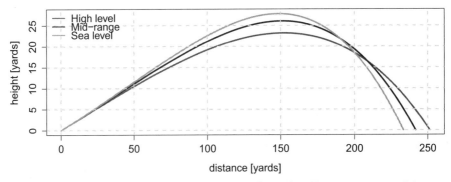

Figure 9.16. Trajectories for a golf ball showing the effect of varying ambient conditions on the carry and elevation of a golf ball (see text). The ambient conditions for altitude, temperature, relative humidity, and density were (a) 1900 m, 32°C, 0.918 kg/m^3, 5%; (b) 500 m, 20°C, 1.135 kg/m^3; (c) 1 m, 1°C, 1.286 kg/m^3, 35%.

Table 9.10. Launch data for simulated driver golf shots with push/pull and/or inclined golf ball spin axis.

Shot type	Speed (mph)	Angle (deg)	Back spin (rpm)	Axis (deg)	Push/pull (deg)
Straight	165	11.2	2685	0	0
Push	165	11.2	2685	0	−5
Pull	165	11.2	2685	0	5
Draw	165	11.2	2685	12	−5
Fade	165	11.2	2685	−12	5
Hook	160	11.2	2685	20	0
Slice	160	11.2	2685	−20	0
Duck hook	150	11.2	2685	45	5
Banana slice	150	11.2	2685	−45	−5

corresponding reduced carry of 18 yards (16.5 m) could mean a difference of two club lengths for a player's next shot, which is very significant.

Another effect, but not taken into account in this simulation, would be the difference in the golf ball coefficient of restitution which would further reduce the carry for low temperature situations, as discussed in Section 9.7.1.

9.9.6 Effect of Push/Pull and Inclined Golf Ball Spin Axis

In this section we illustrate a variety of golf shots that most golfers have experienced during a round of golf at one time or other. The simulation test conditions are shown in Table 9.10 for a driver. The terms fade, draw, hook, and slice were defined in Section 9.7.2, which leaves the terms pull and push to be explained. A *pull* shot is one where a right-handed player hits an otherwise straight shot, but pulled to the left by a few degrees. Similarly, a *push* shot is one where a right handed player hits an otherwise straight shot, but pushed to the right by a few degrees.

The shots were assumed to be taken under ambient air conditions where the density is 1.225 kg/m³ (sea level, 15°C, 0% relative humidity) and the simulated trajectories are shown in Fig. 9.17.

The initial conditions are set by the line of code

```
source("init_drawsAndSlices.R")
```

which should be selected by setting testNo=6 in the main program of Listing 9.2.

The effect of differing amounts of push/pull and axis angles on driver shots with the same launch angle and back spin can be clearly seen in Figs. 9.17 and 9.18. Some different launch speeds have been used to accentuate the various shot types. Unfortunately it has not been possible to compare the simulated results against actual trajectories, owing to lack of published data on this subject.

The salient features of the different simulated golf shots are as follows:

Straight shot: Maximum height = 30.8 yards (28.2 m), carry = 262 yards (240 m) and zero lateral deviation.

Push: Maximum height = 30.8 yards (28.2 m), carry = 262 yards (240 m) and final lateral deviation = −24.1 yards (−22 m).

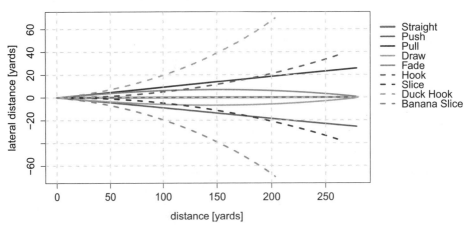

Figure 9.17. Top view showing a typical range of golf shots that most players have experienced, including combined pulled/fade and push/draw shots (see text). The driver trajectories include the golf ball bounce/run.

Pull: Maximum height = 30.8 yards (28.2 m), carry = 262 yards (240 m) and final lateral deviation = +24.1 yards (+22 m).

Push with draw: Maximum height = 30.0 yards (27.4 m), carry = 260 yards (238 m) and final lateral deviation = yards (0 m)

Pull with fade: Maximum height = 30.0 yards (27.4 m), carry = 260 yards (238 m) and final lateral deviation = 0 yards (0 m).

Hook: Maximum height = 26.5 yards (24.2 m), carry = 246 yards (225 m) and final lateral deviation = +31.9 yards (+29.2 m).

Slice: Maximum height = 26.5 yards (24.2 m), carry = 246 yards (225 m) and final lateral deviation = −31.9 yards (−29.2 m).

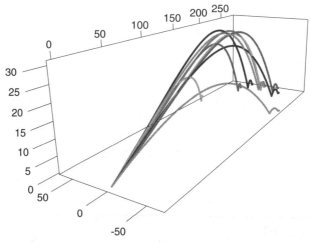

Figure 9.18. Perspective view showing the typical range of golf shots depicted in Fig. 9.17. Distances are shown in yards. (See color plate 9.18)

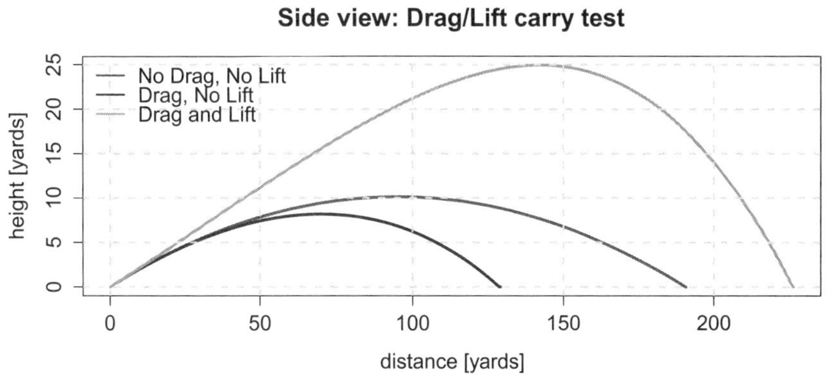

Side view: Drag/Lift carry test

Figure 9.19. The trajectories of golf shots under various conditions of *drag* and *lift* (see text).

Duck hook: Maximum height = 16.6 yards (15.2 m), carry = 199 yards (182 m) and final lateral deviation = +60.3 yards (+55.1 m).
Banna slice: Maximum height = 16.6 yards (15.2 m), carry = 199 yards (182 m) and final lateral deviation = −60.3 yards (−55.1 m).

9.9.7 Drag/Lift Carry Test

In this test we compare three versions of the same shot with (1) no drag and no lift; (2) drag, but no lift; (3) drag and lift. The initial conditions for this test are based on an average club player using a driver: launch speed = 145 mph, launch angle = 12°, back spin = 2800 rpm.

The shots were assumed to be taken under ambient air conditions where the density is 1.225 kg/m^3 (sea level, 15°C, 0% relative humidity). The simulated trajectories are shown in Fig. 9.19, from which the effects of drag and lift are clearly visible.

The initial conditions are set by the line of code

```
source("init_dragCarryTest.R")
```

which should be selected by setting testNo=7 in the main program of Listing 9.2.

The carry for the *drag only* case of 129 yards (118 m) is greatly reduced from the *no drag, no lift* case of 191 yards (175 m) (equivalent to *in vaccuo*) and would be further reduced were it not for the golf ball dimples, as discussed in Section 9.5. The greatest carry is achieved by the *drag and lift* case of 226 yards (207 m) with the increase over the drag only case being due to the spin induced lift, that is, the Magnus force, as discussed in Section (9.3). The trajectory height of 25 yards (22.8 m) is also greatest for this case, being 15 yards (13.7 m) higher than the no drag, no lift case and 17 yards (15.5 m) higher than the no lift case.

9.9.8 Drag Effect at Ground Level

We mentioned earlier that the drag effect of wind can affect a golf ball at ground level, and we now give a simple numerical example of how this effect can be evaluated.

Consider a standard golf ball resting on an upright tee, as shown in Fig. 9.20, with its center of mass at 5 cm above ground level. What velocity of wind would be required

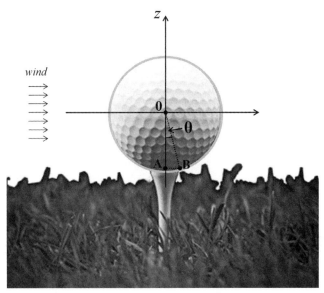

Figure 9.20. Force diagram for a golf ball on a tee, with wind blowing from the left (see text).

to blow the ball off the tee? Assume a standard golf ball which means that the ball area exposed to wind is $A = 14.3$ cm², its diameter $d = 4.267$ cm, $r = d/2$, and mass $m = 0.04593$ kg. Take air density $\rho = 1.225$ kg/m³, the drag coefficient for a rough sphere $C_D = 0.25$ and the width at the top of the tee $\ell = 1$ cm.

Now, moments need to be taken about the pivot point B with the drag force given approximately by the first of eqns. (9.5), that is,

$$F_D \simeq \frac{1}{2} C_D \rho A V_{wind}^2, \quad (\text{N}). \tag{9.28}$$

At the instant when the the drag force moment is just equal to the gravitational force moment, the ball will be about to move; and any increase in F_D will cause the ball to roll off the tee. Thus, we need to find the velocity for this situation, that is, when

$$F_g r \sin(\theta) = F_D r \cos(\theta) = \frac{1}{2} C_D \rho A V_{wind}^2 r \cos(\theta). \tag{9.29}$$

On rearranging eqn. (9.29) we obtain an expression for the required wind velocity:

$$V_{wind} = \sqrt{\frac{F_g \tan(\theta)}{\frac{1}{2} C_D \rho A}}. \tag{9.30}$$

Substituting in known values and noting that $\theta = \arcsin(\frac{1}{2}\ell/r) = 0.237$ radians (13.6°) and $F_g = mg$ where $g = 9.807$ m/s², we finally arrive at the following value for wind speed, above which the ball will be blown off the tee:

$$V_{wind} = 22.3 \text{ (m/s)}. \tag{9.31}$$

Now, wind velocity varies with height due to wind shear, and, for an open terrain, the *mean velocity distribution* of the *atmospheric boundary layer* can be represented by the following power law representation

$$\bar{u}(z) = az^{0.16} \tag{9.32}$$

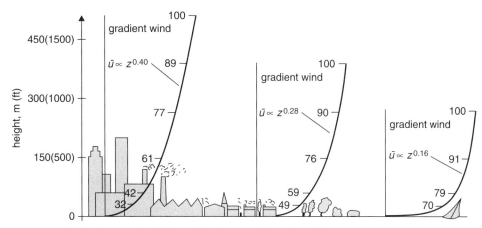

Figure 9.21. Typical atmospheric boundary layer profiles for different terrains: (left) city with high rise buildings; (center) urban area; (right) open terrain [Pla-71].

(see Fig. 9.21). From this relationship we can estimate the wind velocity at head height, say, 1.8 m. First we need to evaluate the coefficient a, which we do by utilizing the result of eqn. (9.31), that is,

$$\bar{u}_{z=0.05m} = 22.3 = a \times 0.05^{0.16}$$
$$\therefore a = 35.9$$
$$\Downarrow$$
$$\bar{u}_{z=1.8m} = 39.5 \text{ (m/s)}, \quad \text{or } 88.4 \text{ (mph)}. \tag{9.33}$$

Thus, we see that it is quite feasible for a golf ball to be blown off its tee by the force exerted by very strong gust of wind. However, it would be a brave or foolhardy golfer who attempted to tee his ball up in order to play in an almost 90 mph gale.

An excellent discussion relating to atmospheric boundary layer effects can be found in [Pla-71].

9.10 COMPUTER CODE

In this section we include the R computer code that was used to generate the plots in Section 9.9. The code for only one set of initial conditions has been included to save space. However, all the initial condition files are included with the downloads available for this book.

9.10.1 Main Program

The R computer code for the main golf ball trajectory program is provided in Listing 9.2. We note the following points about this program:

Libraries: The library packages needed for this program are `rgl`, which provides graphical routines necessary for displaying and saving plots images, and `RSEIS` which provides the vector product routine used in the derivative function.

Initial conditions: Seven sets of initial conditions (described earlier) are available for use with this program. A selection is made by setting the variable `testNo` to the appropriate value at the beginning of the program.

Integration: The numerical integration algorithm selected for this application is the fixed step size *the SHK method*. Its order is set to 4 in the code, but can be changed by modifying the variable s. A step size of 0.01s was found to give sufficient accuracy, but this can be changed by modifying variable h in the code. It provides the same levels of accuracy as the Runge–Kutta methods and is discussed in detail in Chapters 1 and 2.

Plotting: The program also calls the code in file golfballFlight_postSimCalcs.R which generates a variety of plots. For each case the side view is plotted together with optional plots that are controlled by the flags, PlotY (top view), PlotV (velocities), PlotN (spin rates), and Plot3D (3D), the values of which are to be set by the user as required: 1 for *plot* and 0 for *no plot*. Plotting in 3D is useful for illustrating pull/push or slice/hook shots. The call to rgl.snapshot() saves a copy of the 3D image for later viewing.

```
# File: golfballFlight_main.R
rm(list = ls(all = TRUE)) # Delete workspace
library("rgl")
library("RSEIS")
################################################################
# Initial condition options - set variable testNo as required
################################################################
plotY <- 0
testNo <- 1
if(testNo == 1){
  source("init_drivers.R")
}else if(testNo == 2){
  source("init_woods.R")
}else if(testNo == 3){
  source("init_irons.R")
}else if(testNo == 4){
  source("init_windTest.R")
}else if(testNo == 5){
  source("init_ambientTest.R")
}else if(testNo == 6){
  source("init_drawsAndSlices.R")
  plotY <- 1 # 0=no plot, 1=plot
}else if(testNo == 7){
  source("init_dragCarryTest.R")
}else if(testNo == 8){
  source("init_angleTest.R") # Erlichson launch angle test
}
################################################################
# Derivative routine
source("golfballFlight_deriv.R")
#
ptm = proc.time()
```

```
#
# Additional ambient conditions
g       <- 9.8066               # gravitational acceleration [m/s2]
kvisc   <- 15.68*10^(-6)        # Kinematic viscosity [m^2/s]
# Ball parameters
d       <- 42.67/1000           # ball diameter [m] (1.68 in)
R       <- d/2                  # radius of ball [m]
A       <- pi*R^2               # area of ball normal to flight
Wt      <- 45.93/1000           # weight of [kg] (1.62 oz)
m       <- Wt                   # mass of ball [kg]
MI      <- (2/5)*m*R^2          # Moment of inertia [kg.m2]
# Other parameters
t0      <- 0                    # start time [s]
tf      <- 10.5                 # max finish time [s]
#
h <- 0.01
tn <- tf/h
times <- rep(0,tn)
# Allocate arrays
x_mtrs <- array(rep(0,(nCase*tn*6)),c(nCase,tn,6))
x_yrds <- x_mtrs
N      <- array(rep(0,(nCase*tn)),c(nCase,tn))
V      <- array(rep(0,(nCase*tn)),c(nCase,tn))
#
# Iterate through golf club cases as set during initilization
for(case in 1:nCase){
  # Case parameters (from init file)
  cat(sprintf("\n%s\n",titleTxt[case]))
  ncall <<- 0                               # dervative counter
  nu    <- 1.5*10^(-5)                      # kinematic viscosity [m2/s]
  theta <- LA[case]*pi/180                  # Launch angle [radians]
  phi   <- LB[case]*pi/180                  # Push/pull angle [radians]
  psi   <- LC[case]*pi/180                  # Slice/draw angle [radians]
  vx    <- LS[case]*1.609*1000/3600*cos(theta)*cos(phi)
  vy    <- LS[case]*1.609*1000/3600*cos(theta)*sin(phi)
  vz    <- LS[case]*1.609*1000/3600*sin(theta)
  vtot  <- sqrt(vx^2+vy^2+vz^2)
  eta   <- WA[case]*pi/180                  # [radians]
  windx <- WS[case]*1.609*1000/3600*cos(eta)
  windy <- WS[case]*1.609*1000/3600*sin(eta)
  bSpin <- revs[case]
  N0    <- abs(bSpin)/60                    # Back spin, [rps]
  rho   <- den[case]                        # air density
  Aconst <- (1/2)*rho*A/m                   # Aerodynamic constant [1/m]
  CD_Off <- 0; CL_Off <<- 0
  if(exists(as.character(substitute(CD_CHK))) ){
```

```
    CD_Off <- CD_CHK[case]}                    # Aerodynamic constant [1/m]
if(exists(as.character(substitute(CL_CHK))) ){
  CL_Off <- CL_CHK[case]}
#              x,  vx, y,  vy, z,  vz, Nb
y0       <- c( 0, vx, 0,  vy, 0,  vz, N0)      # initialize states
#
grdFlg <- 0 # grdFlg set to 1 when golf ball hits the ground
# calculate solution
out <- matrix(0,ncol=7,nrow=tn)
out[1,] <- y0
carryPrt <- 0
k <- rep(0,7)
for(i in 2:tn){
  if((out[i-1,5]+h*out[i-1,6]) >= -0.01){
    dt <- h
  }else{
    dt <- out[i-1,5]/out[i-1,6] # Don't let height go negative
  }
  times[i] <- times[i-1]+dt
  s <- 4 # order (SHK itegration method)
  for(j in seq(s,1,by=-1)){
    k[1:7] <- (dt/j)*golfballFlight_deriv(times[i],
              (out[(i-1),1:7] + k[1:7]), parms=NULL)
  }
  out[i,1:7] <- out[(i-1),1:7] + k[1:7]
  # Check if ball has reached ground level
  if(out[i,5] <= 0){
    if (carryPrt == 0){ # Print carry stats
      carryDist  <- 1.0936*sqrt(out[i,1]^2+out[i,3]^2) # yards
      htMax      <- 1.0936*max(out[,5])                # yards
      lateralDev <- 1.0936*max(abs(out[,3]))           # yards
      cat(sprintf(" case=%d, t=%f s, carry=%f yards\n",
              case,times[i],carryDist))
      cat(sprintf("      max height=%f yards\n",htMax))
      cat(sprintf("      max lateral deviation,=%f yards\n",
              lateralDev))
      carryPrt <- 1     # turn off 'print carry stats'
    }
    # Bounce calculation
    if(bounce==0){
      Kbounce <- 0
    }else{
      Kbounce <- 0.4
    }
    out[i,5] <- 0.0                  # set height to zero when ball hits ground
    out[i,2] <- Kbounce*out[i,2] # reduce Vx
```

```
      out[i,4] <- Kbounce*out[i,4]    # reduce Vy
      out[i,6] <- -Kbounce*out[i,6]   # reduce Vz
    }
  }
  # Plotting Variables
  t <- times                                    # time [s]
  x_mtrs[case,,1:6] <- out[,1:6]                # distance [m]
  x_yrds[case,,1:6] <- 1.0936*x_mtrs[case,,1:6] # distance [yds]
  N[case,] <- out[,7]                           # spin [rps]
  V[case,] <- sqrt(x_mtrs[case,,2]^2+           # velocity, [m/s]
                   x_mtrs[case,,4]^2+
                   x_mtrs[case,,6]^2)
}
# Plot results
source("golfballFlight_postSimCalcs.R")
cat(sprintf("\nCalculation time = %f\n",(proc.time() - ptm)[3]))
```

Listing 9.2. File: golfballFlight_main.R—Code for *main* simulation program that calculates a golf ball trajectory. It calls the derivative function golfballFlight_deriv() given in Listing 9.3, and also calls the code in file golfballFlight_postSimCalcs.R, which plots the results

9.10.2 Derivative Function

The R computer code for the derivative function is given in Listing 9.3. We note the following points about this code:

Correlations: The correlations used for drag, lift and spin rate decay coefficients CD, CL and CM are each stated as pairs of arrays that are then interpolated using the R function splinefun() with method=natural. This provides a very simple and effective way of using data provided in tabulated form.

Angular momentum: The variable spiny has a negative sign because ω has a negative y component, as indicated by the right-hand rule (see Section 9.3).

Cross product: The cross product $\omega \times \mathbf{v}$ described by eqn. (9.12) is evaluated by using the R function xprod(). The result is stored in the three-element vector variable wxU, with the corresponding direction vector $\frac{\omega \times \mathbf{v}}{|\omega \times \mathbf{v}|}$ stored in variable dir_wxU.

```
# File: golfballFlight_deriv.R
golfballFlight_deriv <- function (t,y, parms=NULL)
{
  X <-y[1]; Y<-y[3]; Z<-y[5]
  Vx<-y[2]; Vy<-y[4]; Vz<-y[6]
  Nb<-y[7];
  ncall <<- ncall + 1
  # Magnus
  spinx <- Nb*cos(psi)*sin(phi)
```

```
spiny <- -Nb*cos(psi)*cos(phi)   # Note -ve sign (due to right hand rule)
spinz <- Nb*sin(psi)
w_x   <- spinx                   # [rad/s]
w_y   <- spiny*(2*pi)            # [rad/s]
w_z   <- spinz*(2*pi)            # [rad/s]
# spin
omega     <- c(w_x,w_y,w_z)      # spin vector, [1/s]
abs_omega <- sum(omega^2)^(1/2)  # total spin, [1/s]
# velocity
U     <- c(Vx-windx,Vy-windy,Vz) # velocity vector, [m/s]
abs_U <- sum(U^2)^(1/2)          # absolute velocity, [m/s]
# cross product, omeg x U
wxU       <- xprod(omega,U)      # omega X U, [m/s2] (backspin -ve)
abs_wxU   <- sum(wxU^2)^(1/2)    # absolute of cross product, [N]
dir_wxU   <- wxU/(abs_wxU+0.01)  # unit vector in direction of cross product, [-]
# spin-rate parameter
Sp        <- abs_omega*R/(abs_U+0.01) # Mizota's Sp=pi*d*N/U, [-]
# Mizota correlations
SpmCD <- c(0.0335,0.0734,0.1355,0.2215,0.3050,0.4203,  # Sp values
           0.5380,0.7411,1.0462,1.3382)
mCD   <- c(0.2146,0.2277,0.2453,0.2781,0.3153,0.3657,  # CD
           0.4161,0.4599,0.5058,0.5365)
SpmCL <- c(0.0314,0.04878,0.0741,0.1102,0.1532,0.2130, # Sp values
           0.2711,0.3451,0.4298,0.5846,0.8130,1.2346)
mCL   <- c(0.0971,0.1146,0.1405,0.1708,0.2053,0.2505,  # CL
           0.2849,0.3236,0.3602,0.3947,0.4292,0.4701)
# Sp values for CM correlation - Mizota
spmCM <- c(0.0113,0.0206,0.0404,0.0737,0.1476,0.2690,  # Sp values
           0.3979,0.5242,0.6440)
# CM correlation values
fCM   <- c(0.00142,0.00146,0.00151,0.00167,0.0019,     # CM
           0.00296,0.00437,0.00627,0.00948)
#
if (Z <= -0.01){
  grdFlg <<- 1 # golf ball has hit the ground
}
if(grdFlg == 0){
  CD <- splinefun(SpmCD, mCD, method = "natural")(Sp) # Mizota
}else{
  CD <- 0.33
}
if(Nb > 10){
  CM <- splinefun(spmCM, fCM, method = "natural")(Sp) # Mizota
}else{
  CM <- 0.0004
}
```

```
    CL <- splinefun(SpmCL, mCL, method = "natural")(Sp) # Mizota
    if(CD_Off == 1){
       CD <- 0
    }
    if(CL_Off == 1){
       CL <- 0
    }
    aD   <- Aconst*CD*abs_U*U        # Drag acceleration vector
    aL   <- Aconst*CL*abs_U^2*dir_wxU # Lift acceleration vector
    TM   <- (1/2)*CM*rho*A*d*abs_U^2  # spin-rate decay torque
    aM   <- TM/(2*pi*MI)              # spin-rate derivative
    #
    xdot <- rep(0,7); # allocate array
    xdot[1] <- Vx;              # x
    xdot[2] <- -aD[1] + aL[1]   # dx/dt
    xdot[3] <- Vy;              # y
    xdot[4] <- -aD[2] + aL[2]   # dydt
    xdot[5] <- Vz;              # z
    xdot[6] <- -aD[3] + aL[3] - g # dzdt
    xdot[7] <- -aM              # dNdt

    return(xdot)     # return time derivatives
}
```

Listing 9.3. File: golfballFlight_deriv.R—Code for function golfballFlight_deriv() called from the *main* program described in Section 9.10.1. It calculates time derivatives for golf ball trajectory dynamics

9.10.3 Initial Conditions

In Listing 9.4 we include example R computer code to set simulation initial conditions. Only one set (for the driver test) has been included to save space; however, initial condition files for all tests described in Section 9.9 are included with the downloads.

```
# File: init_drivers.R
testTxt <- "Drivers"
testLgd <- c("TW","ATP","ACP") # legend text
testCol <- c("red","blue","green")
titleTxt <- c("Tiger Woods, driver=285yds",
              "Ave. Tour Pro, driver=269yds",
              "Ave. Club Player, driver=230yds")
LS      <- c(180,165,145)      # Launch speed [mph]
LA      <- c(11,11.2,12)       # Launch angle-elevation [deg]
LB      <- c(0,0,0)            # Launch angle-push/pull [deg]
LC      <- c(0,0,0)            # Launch angle-slice/draw [deg]
revs    <- c(2200,2685,2800)   # Ball spin [rpm]
```

```
WS       <- c(0,0,0)              # wind speed [mph]
WA       <- c(0,0,0)              # Wind angle [deg]
den      <- c(rep(1.225,3))       # density [kg/m3]
nCase    <- length(LS)            # Number of cases [-]
bounce   <- 1                     # Bounce flag, 0=off, 1=on
```

Listing 9.4. File: init_drivers.R—Code that sets all the initial conditions for the *driver test* required for the *main* program described in Section 9.10.1. Initial conditions for other tests follow a similar pattern

REFERENCES

[Ach-72] Achenbach, E. (1972), Experiments on the Flow Past Spheres at Very High Reynolds Numbers, *Journal of Fluid Mechanics* **54**, 565–575.

[Ach-74] Achenbach, E. (1974), The Effects of Surface Roughness and Tunnel Blockage on the Flow Past Spheres, *Journal of Fluid Mechanics* **65**, 113–115.

[Arn-10] Arnold, D. N. (2010), The Science of a Drive, *Notices of the American Mathematical Society* **57**-4, 498–501.

[Bea-76] Bearman, P. W. and J. K. Harvey (1976), Golf Ball Aerodynamics, *Aeronautical Quarterly*, January, 112–122.

[Bro-71] Brown, F. N. M. (1971), *See the Wind Blow*, Dept. Aerospace. Mech. Eng. Report, University of Notre Dame.

[Jac-01] Jackson, J. (2001), *The Modern Guide to Golf Clubmaking: The Principles and Techniques of Component Golf Club Assembly and Alteration*, 4th ed., Dynacraft Golf Products.

[Kno-11] Knott, C. G. (1911), *Life and Scientific Work of Peter Guthrie Tait*, Cambridge University press.

[Mag-52] Magnus, H. G. (1852), Über die Abweichung der Geschosse, *Abhandlungen der Königlichen Akademie der Wissenschaften zu Berlin*, 1–234.

[Miz-02] Mizota, T. and H. Simozono (2002), 3-Dimensional Trajectory Analysis of Golf Balls, in *World Scientific Congress of Golf IV*, Eric Thain, editor. Routledge, pp. 349–358.

[New-71] Newton, I. (1671), New Theory about Light and Colors, *Philosophical Transactions of the Royal Society of London* **6**, 3075–3087.

[Pen-03] Penner, A. R. (2003), The Physics of Golf, Reports on Progress, *Physics* **66**, 131–171.

[Pla-71] Plate, E. J. (1971), *Aerodynamic Characteristics of Atmospheric Boundary Layers*, U.S. Atomic Energy Commission, Division of Technical Information, Springfield, Va., USA.

[Ric-12] Rice, A. (2012), *When You Play: Watch or No Watch?*, Andrew Rice website, available online at http://www.andrewricegolf.com/tag/carry-distance/.

[Roy-04] Royal and Ancient Golf Club (2004), *A Guide to the Rules on Clubs and Balls*, R & A Rules Limited.

[Sha-82] Shames, I. H. (1982), *Fluid Mechanics*, 2nd ed., McGraw-Hill.

[Smi-94] Smits, A. J. and D. R. Smith (1994), A New Aerodynamic Model of a Golf Ball in Flight, in *World Scientific Congress of Golf II*, A. J. Cochran and M. R. Farrally, editors. E & F N Spon, pp. 411–420.

[Tai-96] Tait, P. G. (1896), On the Path of a Rotating Spherical Projectile, *Transactions of the Royal Society, Edinburgh* **16** (part 2), 491–506.

[Tho-10] Thompson, J. J. (1910), The Dynamics of a Golf Ball, *Nature* **85**, 2151–2157.

[Tra-10] Trackman A/S (2010), Analysis: Long Drivers of America, *Trackman News Letter* **6**, 13. Available online at: http://www.trackman.dk/download/newsletter/newsletter6.pdf.

[USG-10] USGA (2010), Video of an Actual Golf Ball Hitting a Heavy Steel Plate at 150 mph Taken at 40,000 Frames per Second, USGA Research and Test Center. Available online at http://www.youtube.com/watch?v=00I2uXDxbaE.
[Wes-09] Wesson, J. (2009), *The Science of Golf*, Oxford University Press.
[Win-06] Winther, S. (2006), *The Scientific Truth of the Golf Swing*, Lulu.com.

10

Case Study: Taylor–Sedov Blast Wave

10.1 BRIEF BACKGROUND TO THE PROBLEM

Back in 1945 Sir Geoffrey Ingram Taylor was asked by the British MAUD (Military Application of Uranium Detonation) Committee to deduce information regarding the power of the first atomic explosion at the Trinity site in the New Mexico desert. He derived some remarkable results, which were based on his earlier classified work [Tay-41], and was able to estimate, using only public domain photographs of the blast, that the yield of the bomb was equivalent to between 16.8 and 23.7 kilotons of TNT, depending upon which value for the isentropic exponent for air was assumed. Each of these photographs, declassified in 1947, crucially, contained a distance scale and precise time (see, e.g., Figs. 10.1 and 10.3). Ingram's calculation was classified secret but, five years later he published the details [Tay-50a, Tay-50b], much to the consternation of the British government. John von Neumann and Leonid Ivanovitch Sedov published similar independently derived results [Bet-47, Sed-46]. Later Sedov also published a full analytical analysis [Sed-59] (see Appendix 10.B at the end of this chapter). For further interesting discussion relating to the theory, refer to [Deb-58, Kam-00, Pet-08].

The *Manhattan project*, also known as the *Trinity project*, was the code name for the secret U.S. project set up in 1942 to develop an atomic bomb.

10.2 SYSTEM ANALYSIS

The Taylor–Sedov blast is generally defined to be a spherical explosion caused by the point injection of energy. In general physics this energy could derive from the detonation of an explosive device and in astrophysics it could derive from a supernova. Here, we are primarily concerned with the detonation of a powerful explosive device.

Taylor analyzed the blast wave using the equivalent of the following equations:

$$\frac{\partial \rho}{\partial t} + \frac{\partial \rho u}{\partial r} + (\nu - 1)\frac{\rho u}{r} = 0,$$
$$\frac{\partial u}{\partial t} + u\frac{\partial u}{\partial r} + \frac{1}{\rho}\frac{\partial p}{\partial r} = 0, \quad (10.1)$$
$$\frac{\partial (p/\rho^\gamma)}{\partial t} + u\frac{\partial (p/\rho^\gamma)}{\partial r} = 0,$$

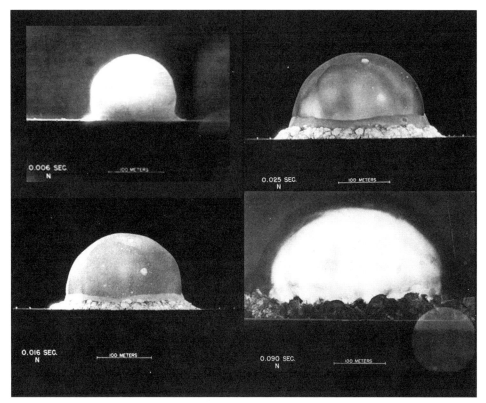

Figure 10.1. Time-lapse photographs with distance scales (100 m) of the first atomic bomb explosion in the New Mexico desert at 5:29 a.m. on July 16, 1945 [Mac-49]. Times from instant of detonation are indicated in bottom left corner of each photograph. See also Fig. 10.3.

where t represents *time* and ρ, u, p, r, and γ represent *density, velocity, pressure, radial coordinate*, and *isentropic exponent* (ratio of specific heats) of the medium, respectively. In addition, ν is a constant dependent upon the geometry of the problem: $\nu = 1$ for *planar flow*, $\nu = 2$ for *cylindrical symmetry*, and $\nu = 3$ for *spherical symmetry*. Also, like Taylor, we will be dealing with spherical symmetry where gravitation effects are ignored.

The effect of the explosion is to force most of the air within the shock wave into a thin shell just inside blast front. As the front expands, the maximum pressure decreases until, at about 10 atm, the analysis ceases to be accurate.

Now, if we assume we are dealing with an ideal problem such that

- the blast can be considered to result from a point source of energy
- the process is *isentropic* and the medium can be represented by the *equation of state* $(\gamma - 1)e = p/\rho$, where e represents *internal energy per unit mass*
- there is spherical symmetry

then, similarity considerations (see Appendix 10.A at the end of this chapter) lead to the following equation [Tay-50b]:

$$R = S(\gamma) t^{2/5} E^{1/5} \rho_0^{-1/5}, \tag{10.2}$$

where, for a consistent set of units, $S(\gamma)$ is a *constant* that depends solely on γ, R is the *radius of the wave front*, ρ_0 is initial density of the medium, and E is the *total energy* released by the explosion. Equation (10.2) together with the *gas laws* enables a solution to eqns. (10.1) to be found.

Sir Geoffrey Ingram solved the PDEs (10.1) by first transforming them to a set of ordinary differential equations (ODEs), and, then applying numerical integration. By this method he arrived at a value of $K = 0.856 = S^{-5}$ [Tay-50b, Table 3], for an explosion in air, which compares well with the analytical value of $K = 0.851$ derived subsequently by Sedov via a lengthy complex calculation [Sed-59, p231]. The fact that Taylor could obtain such a close value is remarkable considering the lengthy serial calculations involved and that he would have had to use pencil and paper techniques, for example, *log tables*. Computer calculations using modern numerical integrators show that Taylor's method converges to the correct analytical value (discussed subsequently).

Following the general approach and nomenclature of Taylor [Tay-50a], we adopt the same similarity assumptions[1] for an expanding, spherically symmetric blast wave of constant total energy:

$$\text{Pressure: } p/p_0 = y = R^{-3} f_1,$$
$$\text{Density: } \rho/\rho_0 = \psi, \qquad (10.3)$$
$$\text{Radial Velocity: } u = R^{-\frac{3}{2}} \phi_1,$$

where R is the radius of the shock wave forming the outer edge of the disturbance, with p_0 and ρ_0 being the pressure and density of the undisturbed atmosphere. If we define r as being the radial co-ordinate, then $\eta = r/R$ and f_1, ϕ_1, and ψ are functions of η. It will be shown that these assumptions are consistent with the equations of motion and also with the equation of state of a perfect gas.

Substituting eqns. (10.3) into the second of eqns. (10.1), the *momentum equation*, we obtain

$$-\left(\frac{3}{2}\phi_1 + \eta\phi_1'\right) R^{-\frac{5}{2}} \frac{dR}{dt} + R^{-4}\left(\phi_1 \phi_1' + \frac{p_0}{\rho_0}\frac{f_1'}{\psi}\right) = 0, \qquad (10.4)$$

where we use a *prime* to denote differentiation with respect to η.

Now it is clear that for eqn. (10.4) to be correct and consistent with eqn. (10.2), it follows that we must have

$$\frac{dR}{dt} = AR^{-\frac{3}{2}}, \qquad (10.5)$$

where

$$A = 2\frac{\phi_1 \phi_1' \rho_0 \psi + p_0 f_1'}{\rho_0 \psi (2\phi_1' \eta + 3\phi_1)} = \text{constant}. \qquad (10.6)$$

This is because p_0 and ρ_0 are constants and ϕ, ψ, and f are only functions of η. Thus, if A was not a constant, then the blast speed dR/dt would be a function of η which is not physically realistic. The value $A = 3.267 \times 10^{11}$ is derived in what follows, where it is assumed that the units for R and t are centimeters and seconds respectively (see eqn. (10.49)).

---

[1] These assumptions facilitate the calculations that follow and demonstrate Taylor's deep understanding of the problem. They are not unique, and Sedov uses a slightly different set [Sed-59, p. 217].

Recall from the differential calculus that

$$\frac{\partial (\cdot)}{\partial t} = \frac{\partial (\cdot)}{\partial \eta} \frac{\partial \eta}{\partial R} \frac{\partial R}{\partial t}. \tag{10.7}$$

Substituting eqns. (10.3) and (10.5) into the first of eqns. (10.1), the *continuity equation*, and making use of eqn. (10.7), we obtain

$$-A\eta\psi' + \psi'\phi_1 + \psi\left(\phi_1' + \frac{2}{\eta}\phi_1\right) = 0. \tag{10.8}$$

Substituting eqns. (10.3) and (10.5) into the third of eqns. (10.1), the *equation of state*, and again making use of eqn. (10.7), we obtain

$$-A(3f_1 + \eta f_1') + \frac{\gamma f_1}{\psi}\psi'(-A\eta + \phi_1) - \phi_1 f_1' = 0. \tag{10.9}$$

We can now simplify the above equations by employing the following transformations

$$\begin{aligned} f &= f_1 a^2/A^2, \\ \phi &= \phi_1/A, \end{aligned} \tag{10.10}$$

where $a^2 = \gamma p_0/\rho_0$ is the velocity of sound in air, when we obtain

$$\begin{aligned} \phi'(\eta - \phi) &= \frac{1}{\gamma}\frac{f'}{\psi} - \frac{3}{2}\phi, \\ \frac{\psi'}{\psi} &= \frac{\phi' + 2\phi/\eta}{\eta - \phi}, \\ 3f + \eta f' + \frac{\gamma\psi'}{\psi} f(-\eta + \phi) - \phi f' &= 0. \end{aligned} \tag{10.11}$$

On solving for the primed variables in eqns. (10.11), we obtain the following three coupled ODEs:

$$\begin{aligned} f' &= \frac{6f\psi(-\eta + \phi)\eta + f\eta\phi\gamma\psi - 4f\phi^2\gamma\psi}{2\eta^3\psi - 4\eta^2\psi\phi - 2f\eta + 2\phi^2\psi\eta}, \\ \phi' &= \frac{1}{2}\frac{3\eta\gamma\psi\phi^2 - 3\phi\eta^2\gamma\psi + 4\phi f\gamma - 6f\eta}{\eta(\eta^2\psi - 2\eta\psi\phi - f + \phi^2\psi)\gamma}, \\ \psi' &= -\frac{1}{2}\frac{(4\gamma\psi\phi^3 - 5\eta\gamma\psi\phi^2 + \phi\eta^2\gamma\psi - 6f\eta)\psi}{\eta(\eta^2\psi - 2\eta\psi\phi - f + \phi^2\psi)\gamma(-\eta + \phi)}, \end{aligned} \tag{10.12}$$

which is a quite extraordinary result.

We have now reached a situation where we have transformed the problem represented by the three partial differential equations of eqn, (10.1) to a simpler problem represented by the three ordinary differential equations of eqn. (10.12). Thus, if values for f, ϕ, and ψ are known for a given value of η, then additional values for f, ϕ, and ψ can be obtained by means of numerical integration.

The preceding calculations are a testament to Taylor's ability and insight. However, they are tedious to perform by hand. So, as an extra resource that readers may find useful, we include with the downloads code written for symbolic algebra programs Maple and Maxima (open source) that derive eqns. (10.12) from Taylor's orginal PDEs.

10.3 SOME USEFUL GAS LAW RELATIONS

We assume that we are dealing with a gas that obeys the *ideal gas law*,

$$pv = \mathcal{R}T, \tag{10.13}$$

where p represents *pressure*, $v = 1/\rho$ *specific volume*, ρ *density*, T *temperature*, and $\mathcal{R}$ the *gas constant*.

We define the following simple relationships:

$$\begin{aligned} de &= C_v dT, \\ dh &= C_p dT, \\ h &= e + p/\rho, \\ a^2 &= \gamma p/\rho = \gamma \mathcal{R}T, \end{aligned} \tag{10.14}$$

where e represents *internal energy* h *enthalpy*, C_v *specific heat at constant volume*, C_p *specific heat at constant pressure*, a *speed of sound* in the gas, and $\gamma = C_p/C_v$ *ratio of specific heats*.

Additionally, we assume that C_p and C_v are constants and independent of temperature. Such a gas is called a *calorically perfect gas* and, under these conditions, the internal energy and enthalpy relationships of eqns. (10.14) become

$$\begin{aligned} e &= C_v T \\ h &= C_p T. \end{aligned} \tag{10.15}$$

The differential forms of eqn. (10.13) and h are

$$\begin{aligned} p\,dv + v\,dp &= \mathcal{R}\,dT \\ dh &= de + p\,dv + v\,dp. \end{aligned} \tag{10.16}$$

Therefore, it follows from eqns. (10.14) and (10.16) that

$$\mathcal{R} = C_p - C_v. \tag{10.17}$$

The following represent the *Rankine–Hugonoit* conditions, which relate thermodynamic properties across a shock. They always hold for flow velocities under adiabatic conditions, regardless of whether the shock is stationary or moving:

$$\begin{aligned} \text{Continuity:} \quad & \rho_1 u_1 = \rho_0 u_0, \\ \text{Momentum:} \quad & p_1 + \rho_1 u_1^2 = p_0 + \rho_0 u_0^2, \\ \text{Energy:} \quad & \rho_1 u_1 \left(e_1 + \tfrac{1}{2} u_1^2 + p_1/\rho_1 \right) = \rho_0 u_0 \left(e_0 + \tfrac{1}{2} u_0^2 + p_0/\rho_0 \right). \end{aligned} \tag{10.18}$$

Subscripts 1 and 0 indicate conditions immediately upstream and downstream of the shock, respectively.

Now consider the situation shown in Fig. 10.2, where a moving piston introduces a shock wave into the system. The piston creates a shock that proceeds at a velocity greater than that of the piston itself. In this situation, the velocity immediately behind the shock wave is equal to $u_s - u_p$ and eqns. (10.18) become

$$\begin{aligned} \text{Continuity:} \quad & \rho_1(u_s - u_p) = \rho_0 u_s, \\ \text{Momentum:} \quad & p_1 + \rho_1(u_s - u_p)^2 = p_0 + \rho_0 u_s^2, \\ \text{Energy:} \quad & e_1 + \tfrac{1}{2}(u_s - u_p)^2 + p_1/\rho_1 = e_0 + \tfrac{1}{2} u_s^2 + p_0/\rho_0, \end{aligned} \tag{10.19}$$

where the continuity equation has been used to simplify the energy equation.

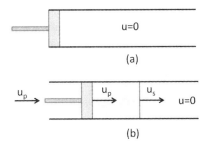

Figure 10.2. Shock tube containing a piston. (a) With piston stationary. (b) With piston moving at constant velocity, u_p, and the shock wave moving with velocity, u_s, into a stationary gas.

On substituting the continuity equation into the momentum equation and rearranging, we obtain

$$(u_s - u_p)^2 = \frac{p_0 - p_1}{\rho_0 - \rho_1}\left(\frac{\rho_0}{\rho_1}\right). \tag{10.20}$$

Substituting this result and the continuity equation into the energy equation yields

$$e_0 - e_1 = \frac{p_1 + p_0}{2}\left(\frac{1}{\rho_1} - \frac{1}{\rho_0}\right). \tag{10.21}$$

Eliminating internal energy by use of the equation of state, $(\gamma - 1)e = p/\rho$, and rearranging leads to the following important result [Tay-50a, eqn. (15)]:

$$\frac{\rho_1}{\rho_0} = \frac{1 + \dfrac{\gamma+1}{\gamma-1}\dfrac{p_1}{p_0}}{\dfrac{\gamma+1}{\gamma-1} + \dfrac{p_1}{p_0}}. \tag{10.22}$$

Alternatively, we can rearrange eqn. (10.21) by using the relationships $e = C_v T$ and $\rho = p/RT$ to obtain

$$\frac{T_1}{T_0} = \frac{p_1}{p_0}\left(\frac{\dfrac{\gamma+1}{\gamma-1} + \dfrac{p_1}{p_0}}{1 + \dfrac{\gamma+1}{\gamma-1}\dfrac{p_1}{p_0}}\right) = \frac{p_1}{p_0}\frac{\rho_0}{\rho_1}. \tag{10.23}$$

Rearranging the continuity equation of eqns. (10.19), we obtain

$$\frac{u_p}{u_s} = 1 - \frac{\rho_1}{\rho_0}, \tag{10.24}$$

which, on using eqn. (10.22), yields a second important result [Tay-50a, eqn. (17)]:

$$\frac{u_p}{u_s} = \frac{2(p_0/p_1 - 1)}{\gamma + 1 + (\gamma - 1)(p_0/p_1)}. \tag{10.25}$$

Now, a basic relationship that links the pressure ratio, $y_1 = p_1/p_0$, and the *Mach number*, $M_s = u_s/a_0$, is

$$\frac{p_1}{p_0} = 1 + \frac{2\gamma}{\gamma + 1}(M_s^2 - 1), \tag{10.26}$$

which, on rearranging, yields a third important result [Tay-50a, eqn. (16)]:

$$\frac{u_s^2}{a_0^2} = \frac{1}{2\gamma}\left\{\gamma - 1 + (\gamma + 1)\frac{p_1}{p_0}\right\}. \qquad (10.27)$$

The important results derived here facilitate solution of the blast wave eqns. (10.1).

The downloads include code written for symbolic algebra programs Maple and Maxima (open source) that derive eqns. (10.22), (10.25), and (10.27) from the Rankine–Hugonoit conditions.

10.4 SHOCK WAVE CONDITIONS

The following equations were derived earlier from the Rankine–Hugonoit eqns. (10.19):

$$\frac{\rho_1}{\rho_0} = \frac{(\gamma - 1) + (\gamma + 1)y_1}{(\gamma + 1) + (\gamma - 1)y_1},$$

$$\frac{U^2}{a^2} = \frac{1}{2\gamma}\{\gamma - 1 + (\gamma + 1)y_1\}, \qquad (10.28)$$

$$\frac{u_1}{U} = \frac{2(y_1 - 1)}{\gamma - 1 + (\gamma + 1)y_1},$$

where ρ_1, u_1, and y_1 represent values of ρ, u, and p, respectively, immediately behind the shock wave and $U = dR/dt$ ($U = u_s$ and $u_1 = u_p$ in Section 10.3) is the radial velocity of the shock wave. Ambient conditions are indicated by the subscript 0.

It follows from eqns. (10.28) that the *strong shock* conditions, where $y_1 = p_1/p_0 \gg 1$ at the shock boundary, are

$$\frac{\rho_1}{\rho_0} = \frac{\gamma + 1}{\gamma - 1},$$

$$\frac{U^2}{a^2} = \frac{\gamma + 1}{2\gamma}y_1, \qquad (10.29)$$

$$\frac{u_1}{U} = \frac{2}{\gamma + 1},$$

Note that the second of eqns. (10.29) is a corrected version of Taylor's eqn. (16a) [Tay-50a].

Now, from eqns. (10.3), (10.5), and (10.10), we have

$$\psi = \rho_1/\rho_0,$$

$$\phi = \phi_1/A = u_1 R^{3/2}/A = u_1/U, \qquad (10.30)$$

$$f = f_1 a^2/A^2 = y_1 a^2 R^3/A^2 = y_1 a^2/U^2 = 2\gamma/(\gamma + 1).$$

Thus, it follows that at the shock boundary, that is, at $\eta = 1$, we have

$$\psi(1) = \frac{\gamma + 1}{\gamma - 1},$$

$$f(1) = \frac{2\gamma}{\gamma + 1}, \qquad (10.31)$$

$$\phi(1) = \frac{2}{\gamma + 1}.$$

These values can be used as the starting point to obtain a (step by step) solution to eqns. (10.12) by means of numerical integration.

10.5 ENERGY

The total energy E of the disturbance may be regarded as consisting of two parts, the kinetic energy

$$\text{K.E.} = 4\pi \int_0^R \frac{1}{2} \rho u^2 r^2 \, dr \tag{10.32}$$

and the heat energy

$$\text{H.E.} = 4\pi \int_0^R \frac{p r^2}{\gamma - 1} \, dr. \tag{10.33}$$

In terms of the variables f, ϕ, ψ, and η, we have

$$E = \text{K.E.} + \text{H.E.} = 4\pi A^2 \left\{ \frac{1}{2} \rho_0 \int_0^1 \psi \phi^2 \eta^2 \, d\eta + \frac{p_0}{a_0^2(\gamma - 1)} \int_0^1 f \eta^2 \, d\eta \right\}, \tag{10.34}$$

or, alternatively, because $p_0 = a_0^2 \rho_0 / \gamma$,

$$E = B \rho_0 A^2, \tag{10.35}$$

where B is a function of γ only, and whose value is

$$B = 2\pi \int_0^1 \psi \phi^2 \eta^2 \, d\eta + \frac{4\pi}{\gamma(\gamma - 1)} \int_0^1 f \eta^2 \, d\eta. \tag{10.36}$$

Because the integrals in (10.36) are both functions of γ only, it appears that for a given value of γ, A^2 is simply proportional to E/ρ_0.

Equating eqn. (10.5) and (10.70) from Appendix 10.A at the end of this chapter yields

$$A = \frac{2}{5} R^{\frac{5}{2}} t^{-1}. \tag{10.37}$$

Recall from eqn. (10.2) that

$$R = S(\gamma) t^{2/5} E^{1/5} \rho_0^{-1/5}. \tag{10.38}$$

It therefore follows from eqns. (10.35) and (10.37) that

$$S(\gamma) = \left(\frac{25}{4B} \right)^{1/5}. \tag{10.39}$$

Thus, once B is found by numerical integration, $S(\gamma)$ is known and, hence, the total energy of the explosion can be calculated. Taylor actually derives the variable $K = S^{-5}$ [Tay-50b, eqn. (8)]. Note that the constant $S(\gamma)$ in eqn. (10.39) is the same S as derived in eqn. (10.67) and that K is equal to the constant α used by Sedov [Sed-59, p. 213].

10.6 PHOTOGRAPHIC EVIDENCE

The photograph strips shown in Fig. 10.3 provided Taylor with crucial evidence upon which he could finally narrow down the physical solution of the Trinity blast wave problem. By carefully measuring the width of the blast wave for each photograph and plotting these values against the corresponding time it was taken following the blast, the relationship between blast radius R and elapsed time t, was determined. The figures published by Taylor are shown in Table 10.1 and have also been plotted in Fig. 10.4. These data enabled Taylor to determine the constant in eqn. (10.69) and establish the following relationship between R and t:

$$\frac{5}{2} \log_{10} R - \log_{10} t = 11.915. \quad (10.40)$$

Equation (10.40) is equivalent to

$$R^5 t^{-2} = 6.67 \times 10^{23}, \quad (10.41)$$

where the units for R are centimeters and the units for t are seconds.

Table 10.1. Radius R of blast wave at time t after the explosion. The data are based on Table 1 from [Tay-50b], which tabulated measurements from high-speed photographs that were taken of the Trinity explosion. The units for time in column 1 are milliseconds, and for column 3 they are seconds. The units for R in column 2 are meters and in columns 4 and 5 they are centimeters

t	R	$\log_{10}(t)$	$\log_{10}(R)$	$\frac{5}{2}\log_{10}(R)$
0.1	11.1	$\overline{4}.0$	3.045	7.613
0.24	19.9	$\overline{4}.38$	3.298	8.244
0.38	25.4	$\overline{4}.58$	3.405	8.512
0.52	28.8	$\overline{4}.716$	3.458	8.646
0.66	31.9	$\overline{4}.82$	3.504	8.759
0.8	34.2	$\overline{4}.903$	3.535	8.836
0.94	36.3	$\overline{4}.973$	3.56	8.901
1.08	38.9	$\overline{3}.033$	3.59	8.976
1.22	41.0	$\overline{3}.086$	3.613	9.032
1.36	42.8	$\overline{3}.134$	3.631	9.079
1.5	44.4	$\overline{3}.176$	3.647	9.119
1.65	46.0	$\overline{3}.217$	3.663	9.157
1.79	46.9	$\overline{3}.257$	3.672	9.179
1.93	48.7	$\overline{3}.286$	3.688	9.22
3.26	59.0	$\overline{3}.513$	3.771	9.427
3.53	61.1	$\overline{3}.548$	3.786	9.466
3.8	62.9	$\overline{3}.58$	3.798	9.496
4.07	64.3	$\overline{3}.61$	3.809	9.521
4.34	65.6	$\overline{3}.637$	3.817	9.543
4.61	67.3	$\overline{3}.688$	3.828	9.57
15.0	106.5	$\overline{2}.176$	4.027	10.068
25.0	130.0	$\overline{2}.398$	4.114	10.285
34.0	145.0	$\overline{2}.531$	4.161	10.403
53.0	175.0	$\overline{2}.724$	4.243	10.607
62.0	185.0	$\overline{2}.792$	4.267	10.668

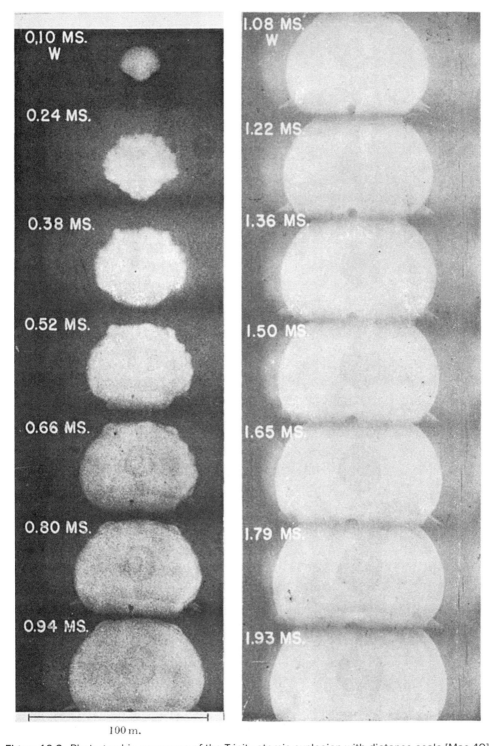

Figure 10.3. Photographic sequence of the Trinity atomic explosion with distance scale [Mac-49]. See also Fig. 10.1.

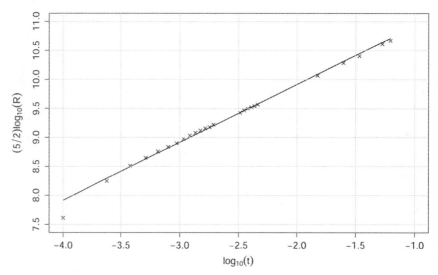

Figure 10.4. Data from Table 10.1 superimposed on a plot of the relation $\frac{5}{2}\log_{10} R - \log_{10} t = 11.915$. The units of R are centimeters.

Thus, Taylor showed that the explosion fireball expanded very closely in accordance with the theoretical prediction [Tay-46] he made more than four years previously.

Note that the bar over the leading figure in the third column of Table 10.1 is how logarithms used to be presented for fractional numbers. To preserve precision, fractional quantities were shifted to the left until the leading figure immediately preceded the decimal point. Then this number was divided by a power of 10 to restore the number back to its correct value. For example, the number 0.00456 becomes 4.56/1000 and the logarithm was written as $\bar{3}.659$. This was understood to be equivalent to $-3 + 0.659$ or -2.341. This may seem cumbersome to the modern reader who has access to computers or calculators that are accurate to many significant figures. However, before these were available, hand calculations employed *logarithm* and *antilogarithm* tables with a limited number of significant figures (commonly five), and this method ensured that maximum precision was maintained.

The data for Table 10.1, in the form of a csv file, along with a R program to generate Fig. 10.4, have been included with the downloads for this book.

10.7 TRINITY SITE CONDITIONS

Operation TRINITY, conducted by the Manhattan Engineer District (MED), was designed to test and assess the effects of a nuclear weapon. The TRINITY device was detonated on a 100-foot tower at the Alamogordo Bombing Range in south central New Mexico at 5:30 A.M. on July 16, 1945. The yield of the detonation was equivalent to the energy released by detonating 21 kilotons of TNT. At shot time, the temperature was 21.8°C, and surface air pressure was 850 millibars. The winds were nearly calm at the surface; at 10,300 feet above mean sea level, they were from the southwest at 10 knots. The winds blew the cloud resulting from the detonation to the northeast. From July 1945

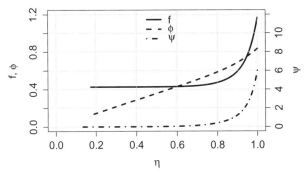

Figure 10.5. Plots of f, ϕ, and ψ resulting from numerical integration of eqns. (10.12).

through 1946, about 1,000 military and civilian personnel took part in Project TRINITY or visited the test site [Dtr-07].

10.8 NUMERICAL SOLUTION

Using the values given in eqns. (10.29) for f, ϕ, and ψ at the blast wave boundary, we are now in a position to integrate eqns. (10.12) numerically using the integrator lsodes from the package deSolve. The solutions are plotted in Fig. 10.5 using the ratio of specific heats for air, that is, $\gamma = 1.4$.

Now that f, ϕ, and ψ are known and so we can evaluate the integrals in eqn. (10.36), that is,

$$I_1 = \int_0^1 \psi \phi^2 \eta^2 d\eta \qquad (10.42)$$

$$I_2 = \int_0^1 f\eta^2 d\eta. \qquad (10.43)$$

However, we only have numerical values for f, ϕ, and ψ, and therefore a quadrature method is required. We choose to employ the general-purpose *composite Simpson's rule*, that is,

$$\int_a^b f(x)\,dx \approx \frac{h}{3}\left[f(x_0) + 2\sum_{j=1}^{n/2-1} f(x_{2j}) + 4\sum_{j=1}^{n/2} f(x_{2j-1}) + f(x_n)\right], \qquad (10.44)$$

which has a fourth-order error equal to

$$-\frac{h^4}{180} h \max_{\xi \in [a,b]} |f^{(4)}(\xi)|. \qquad (10.45)$$

The η-domain was represented by $N = 10{,}001$ points with an interval $h = (b-a)/(N-1)$, and application of Simpson's rule resulted in the following values for I_1 and I_2:

$$I_1 = 0.1851699, \quad I_2 = 0.1851963. \qquad (10.46)$$

The limits of integration were chosen as $a : \eta = 10^{-8}$ and $b : \eta = 1$. The lower limit can not be set exactly to *zero* as this causes the integral to become *singular*. However, this is not important as we can get sufficiently close to zero so that any error is insignificant. The value of B can now be calculated as

$$B = 2\pi I_1 + \frac{4\pi}{\gamma(\gamma-1)} I_2 = 5.319252, \tag{10.47}$$

which, in turn, enables S and K to be evaluated:

$$\begin{aligned} S &= \left(\tfrac{25}{4B}\right)^{1/5} = 1.0328 \\ K &= S^{-5} = 0.8510804. \end{aligned} \tag{10.48}$$

In addition the constant A can be calculated from eqns. (10.37) and (10.41):

$$A = \frac{2}{5} \frac{R^{5/2}}{t} = 3.267 \times 10^{11}, \tag{10.49}$$

where the units of R are centimeters.

The total energy of the blast can now be determined from eqn. (10.35) using $\rho_0 = 0.00125 (\text{g/cm}^3)$ to give

$$\boxed{E_{tot} = B\rho_0 A^2 = 7.096 \times 10^{20} \text{ (ergs)}. \; [7.096 \times 10^4 \text{ (GJ)}]} \tag{10.50}$$

The fact that the total energy of the blast can be predicted with good accuracy from, essentially, a set of time-lapsed photographs, is nothing less than a spectacular result.

Taylor defines the amount of energy liberated by one long ton[2] of TNT,[3] when exploded, to be 4.25×10^{16} erg (4.25 GJ). Therefore, the amount of TNT, M_{TNT}, that the blast can be considered equivalent to is

$$M_{TNT} = E_{tot}/4.25 \times 10^{16} = 16{,}669 \text{ (ton)}. \tag{10.51}$$

This is also referred to as the *yield* of the explosive device.

Values for p and T at any value of η and R can now be determined from eqns. (10.3), (10.10), (10.23), and (10.35) to give, for $\gamma = 1.4$,

$$p(\eta, R) = p_0 R^{-3} f_1(\eta) = R^{-3} f(\eta) \frac{\rho_0 A^2}{\gamma} = R^{-3} f(\eta) \frac{E_{tot}}{B\gamma}, \tag{10.52}$$
$$\therefore p_s = p(1, R) = 0.1343 R^{-3} f(1) E_{tot},$$

and

$$T(\eta, R) = T_0 \frac{p(\eta, R)}{p_0} \frac{1}{\psi(\eta)}, \tag{10.53}$$
$$\therefore T_s = T(1, R) = 0.1343 R^{-3} f(1) E_{tot} T_0 / (p_0 \psi(1)).$$

The shock pressure and temperature will occur at the blast wave position, which corresponds to $\eta = 1$, when $f(1) = 1.1667$ and $\psi(1) = 6$. Figure 10.6 shows a plot of p_s and T_s with respect to the distance from the center of the explosion, where p_s has been divided by 10^6 to convert from dynes/cm^2 to atm.

---

[2] One long ton is equal to 2240 lbs or 1016 kg.
[3] TNT is the acronym for the explosive substance *trinitrotoluene*. The explosive yield of TNT is considered to be the standard measure of strength of bombs and other explosives.

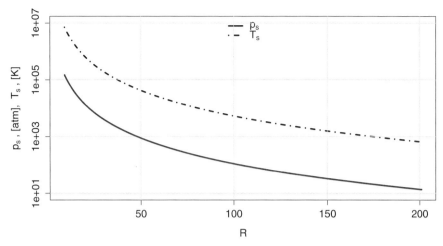

Figure 10.6. Plots of p_{max} and T_{max}.

It also follows from eqns. (10.52) and (10.53) that

$$\frac{p(\eta)}{p_s} = \frac{p(\eta, R)}{p(1, R)} = \frac{f(\eta)}{f(1)} = \frac{1}{1.1667} f(\eta) \qquad (10.54)$$

and

$$\frac{T(\eta)}{T_s} = \frac{T(\eta, R)}{T(1, R)} = \frac{p(\eta)}{p(1)} \frac{\psi(1)}{\psi(\eta)} = 5.143 \frac{f(\eta)}{\psi(\eta)}. \qquad (10.55)$$

Figure 10.7 shows a plot of $p(\eta)/p_s$ and $T(\eta)/T_s$ with respect to η. Note that there is no dependency on R in eqns. (10.54) and (10.55), as the R terms cancel.

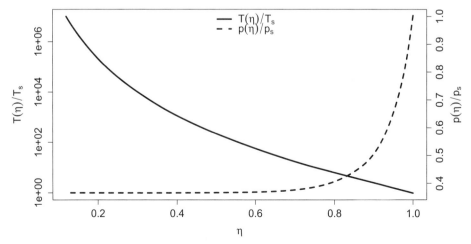

Figure 10.7. Plots of $p(\eta)/p_s$ and $T(\eta)/T_s$.

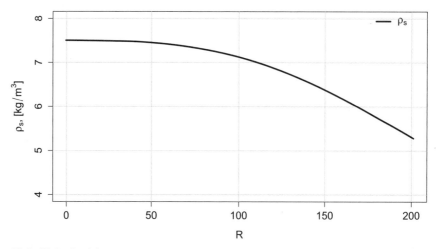

Figure 10.8. Plot of $\rho_s(\eta)$.

From eqn. (10.22) we see that

$$\rho_s = \rho(1, R) = \rho_0 \left(\frac{1 + \dfrac{\gamma+1}{\gamma-1} \dfrac{p_s}{p_0}}{\dfrac{\gamma+1}{\gamma-1} + \dfrac{p_s}{p_0}} \right). \tag{10.56}$$

Thus, using eqn. (10.52), we are able to plot ρ_s as it evolves with the blast wave front (see Fig. 10.8). From eqns. (10.41) and (10.70) we see that the shock velocity is

$$u_s(t) = \frac{2}{5}\frac{R}{t} = \frac{2}{5}\left(\frac{6.67 \times 10^{23}}{t^3}\right)^{\frac{1}{5}}. \tag{10.57}$$

Therefore we can plot u_s as it evolves with time. Similarly, from eqn. (10.38), we can plot the evolution of the blast front R with time (see Fig. 10.9).

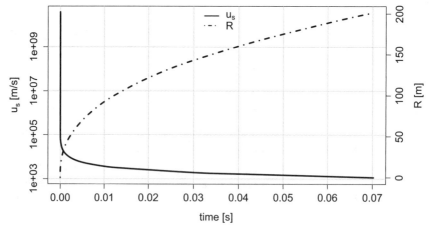

Figure 10.9. Plots of $u_s(t)$ and $R(t)$.

Table 10.2. Comparison of primary variable values from Taylor with those generated by computer

Primary variable	Taylor result	Computer result
I_1	0.185	0.185
I_2	0.187	0.185
K	0.856	0.851
S	1.0316	1.0328
E	7.140×10^{20} (erg)	7.096×10^{20} (erg)
M_{TNT}	16,800 (long-ton)	16,696 (long-ton)

Note: 1 erg = 10^{-7} J and 1 long ton = 2240 lbs/1016 kg.

From eqns. (10.3) and (10.10) we see that $u(\eta, R) = \phi(\eta) A R^{-3/2}$. But the shock velocity will occur at the blast wave position, which corresponds to $\eta = 1$. Therefore, it follows that

$$u(\eta)/u_s = u(\eta, R)/u(1, R) = \phi(\eta)/\phi(1), \quad (10.58)$$

where for $\gamma = 1.4$, we see from eqn. (10.31) that $\phi(1) = 2/2.4 = 0.833$.

We are now able to plot $u(\eta)/u_s$ against η (see Fig. 10.10). Note that there is no dependency on R in eqn. (10.58), as the R terms cancel.

It is not certain exactly what numerical methods Taylor used in his step-by-step approach. However, he does indicate that steps of $\Delta\eta = 0.02$ were used together with a form of predictor corrector algorithm [Tay-50a, p. 163]. Nevertheless, whatever approach he did employ, it is clear that the method was extremely effective and produced very good results.

For comparison purposes, Table 10.2 lists Taylor's primary results alongside computer generated results obtained by the same method, but which employed a modern adaptive numerical integrator with the *absolute* and *relative errors* specified as 10^{-14}.

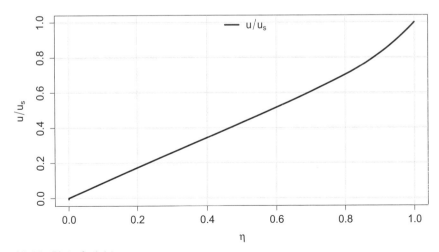

Figure 10.10. Plot of $u(\eta)/u_s$.

R programs that perform the preceding calculations are included in Listings 10.1–10.3.

```r
# File: TaylorSedov_main.R
# ---------------------------------------------------------------------------
# Taylor-Sedov explosion example
# ---------------------------------------------------------------------------
#
rm(list = ls(all = TRUE)) # Delete workspace
#
# Load libraries
library("deSolve")
library("rgl")
#
# Load derivative function
source("TaylorSedov_deriv.R")
#
ptm = proc.time()
# Declare array sizes
N=100001; # Make an odd number for Simpson's Rule calc.
#
# Set parameters
h<-1/(N-1)
gamma <- 7/5
Rgas <- 287.058
# Site conditions
T0_site <- 21.8 +273 # [K], 21.8 [C]
p0_site <- 85000 # Pascals, 850000 dynes/cm2 (850mBar, 0.83888 Std Atmosphere)
rho0_site <- p0_site/(Rgas*T0_site) # kg/m3
# Initialize state variables - see book text
f0   <- 2*gamma/(gamma+1)
phi0 <- 2/(gamma+1)
psi0 <- (gamma+1)/(gamma-1)
etaf <- 1; eta0 <- 0.00000001
eta=seq(etaf, eta0, by=-1/N)
nout=length(eta)
#
# Initial condition
uini = c(f0,phi0,psi0)
#
# Initialze other variables
ncall=0
#
# ODE integration
parms<-c(N,gamma)
```

```
out<-lsodes(y=c(uini), times=eta, func=TaylorSedov_deriv, #verbose = TRUE,
            sparsetype = "sparseint", ynames = FALSE,
            rtol=1e-12,atol=1e-12,maxord=5,parms=parms)
# Extract solution data from integrator output
f   <- out[1:(N),2]
phi <- out[1:(N),3]
psi <- out[1:(N),4]
# Plot results
source("TaylorSedov_postSimCalcs.R")
cat(sprintf("\nCalculation time: %f\n",(proc.time()-ptm)[3]))
```

Listing 10.1. File: TaylorSedov_main.R—Code for *main* program that performs the numerical analysis of the Taylor–Sedov blast wave problem. It calls the derivative function TaylorSedov_deriv() and code to perform postsimulation calculations and generate plots given in Listings 10.2 and 10.3, respectively

```
# File: TaylorSedov_deriv.R
TaylorSedov_deriv = function(eta, u, parms) {
    #
    # Reshape vector u to f, phi and psi vectors
    f   <- u[1]
    phi <- u[2]
    psi <- u[3]
    #
    dfdeta <- (6*f*psi*(-eta+phi)*eta+f*eta*phi*gamma*psi-4*f*phi^2*gamma*psi)/
              (2*eta^3*psi-4*eta^2*psi*phi-2*f*eta+2*phi^2*psi*eta)
    #
    dphideta <- (1/2)*(3*eta*gamma*psi*phi^2-3*phi*eta^2*gamma*psi+4*phi*f*gamma
        -6*f*eta)/
                  (eta*(eta^2*psi-2*eta*psi*phi-f+phi^2*psi)*gamma)
    #
    dpsideta <- -(1/2)*(4*gamma*psi*phi^3-5*eta*gamma*psi*phi^2+phi*eta^2*gamma
        *psi-6*f*eta)*psi/
                  (eta*(eta^2*psi-2*eta*psi*phi-f+phi^2*psi)*gamma*(-eta+phi))
    # Calculate time derivative
    dudeta = c(dfdeta,dphideta,dpsideta)
    #
    # update number of derivative calls
    ncall <<- ncall+1
    #
    # Return solution
    return(list(dudeta))
}
```

Listing 10.2. File: TaylorSedov_deriv.R—Code for derivative function TaylorSedov_deriv(), called from the main program of Listing 10.1

```r
# File: TaylorSedov_postSimCalcs.R
#
# Energy integrals
dI1de <- psi*phi^2*eta^2
dI2de <- f*eta^2
#
# Integrate by Simpson's rule
I1 <- h*(1/3*(dI1de[1] + dI1de[N]) + 4/3*sum(dI1de[seq(2,(N-1),2)])
        + 2/3*sum(dI1de[seq(3,(N-2),2)]))
I2 <- h*(1/3*(dI2de[1] + dI2de[N]) + 4/3*sum(dI2de[seq(2,(N-1),2)])
        + 2/3*sum(dI2de[seq(3,(N-2),2)]))
#
B <- (2*pi*I1+(4*pi)/(gamma*(gamma-1))*I2)
S <- (25/(4*B))^{1/5}
K <- S^(-5)
c <- 1/K
#
TaylorData <- 1
###########################
# Set simulation conditions
if (TaylorData == 1) {
  # Taylor paper
  p_0 <- 1 #1.013250 # [atm] (1013250 dynes/cm2) # CHECK !!!
  T_0 <- 288       # [K] (15 [C])
  rho_0 <- 1.25e-3 # [g/cm^3] {p178, Taylor II}
}else {
  # DTRA data for New mexico desert at time of blast
  p_0 <- p0_site/101325 # [atm]
  T_0 <- T0_site        # [K]
  rho_0 <- rho0_site/1000 # [g/cm3]
}
###########################
# Similarity Eqn
R5tm2 <- 6.67e23 # (R*100)^5/t^2 [cm^5/s^2]
E_tot <- K*rho_0*R5tm2 # ergs
E_TNT <- 4.25e16       # ergs
M_eqivTNT <- E_tot/E_TNT
R_max <- 201
R <- seq(1,N,1)*R_max/N
#R2 <- seq(1,R_max,8)
t <- sqrt(K*(R*100)^5*rho_0/E_tot)
us <- (2/5)*R/t # [m/s]
A <- (2/5)*(R[1]*100)^(5/2)/t[1] # = sqrt(E_tot/rho_0/B)
u_us <- phi/phi[1]
# Max pressure at any point from explosion
ps <- (1/(B*gamma))*E_tot*f[1]/(R*100)^3/(p_0*1013250) # Atm=(dynes/cm2)/1013250
```

```r
# p_max2 <- 0.155*E_tot/(R*100)^3/10^6 eqn(35), Taylor Part 1
#
eta2 <- seq(eta0, etaf, by=1/(R_max))
Ts <- T_0*ps/p_0/psi[1] # [K]
#Ts_T0 <- (ps/p_0)^(1/gamma)/psi[1]
T_Ts <- (f/f[1])*(psi[1]/psi)
#
p_ps <- f/f[1]
#
rhos <- 1000*rho_0*(1+(ps/p_0)*(gamma+1)/(gamma-1))/
    ((gamma+1)/(gamma-1)+ps/p_0) # [kg/m3]
#
# rho_max <- rho_0*((gamma+1)/(gamma-1)+p_0/p_max)/(1+(p_0/p_max)*(gamma+1)/
    (gamma-1))
rho_rhos <- psi/psi[1]
#
# psi_max=psi[1]=max{rho/rho_0}=6; {Eqn(9), Taylor, Part II}
##############################
# Plot f, phi and psi againts eta
# Set margin to allow for multiple axes
par(mar = c(5,5,2,5))
plot(eta, f, ylab = expression(paste(f,", ",phi)),
    xlab=expression(eta), ylim=c(0,1.2),type="l",
    col="black", lty=1, lwd=3)
grid(nx = NULL, ny = NULL, col="lightgray", lwd = 1, lty=1)
#
lines(eta, phi, ylim=c(0,max(f)), xlab="", ylab="",
      type="l",lty=2,col="black", main="",
      xlim=c(etaf,eta0),lwd=3)
#
par(new = T)
plot(eta, psi, ylim=c(0,12),type="l", lwd=3,
    lty=4, col = "black",
    axes = F, xlab = NA, ylab = NA)
axis(side = 4)
mtext(text=expression(psi), side = 4, line = 3 )
#
legend("top",legend=c(expression(f),expression(phi),expression(psi)),
      lwd=c(3,3,3),lty=c(1,2,4),bty="n", col=c("black","black","black"))
#
# Plot V_blast and R againts time
# Set margin to allow for multiple axes
par(mar = c(5,5,2,5))
plot(t, us, log="y", ylab = expression(paste(u[s], " , [m/s]")),
    xlab="time , [s]", type="l",
    lty=1, lwd=3)
```

```
grid(nx = NULL, ny = NULL, col="lightgray", lwd = 1, lty=1)
#
par(new = T)
plot(t, R, type="l", lwd=3,
    lty=4, col = "black",
    axes = F, xlab = NA, ylab = NA)
axis(side = 4)
mtext(text="R , [m]", side = 4, line = 3 )
#
legend("top",legend=c(expression(paste(u[s])) , "R"),
     lwd=c(2,2),lty=c(1,4),bty="n", col=c("black","black"))
#####################
N2<-round(0.88*N) # reduced eta domain - central area not acurate
par(mar = c(5,5,2,5))
plot(eta[1:N2], T_Ts[1:N2],
    ylab = expression(paste(T(eta)/T[s])),
    xlab=expression(eta), log="y", ylim=c(1,10^7)
    ,type="l", col="black", lty=1, lwd=3)
#
par(new = T)
plot(eta[1:N2], p_ps[1:N2], ,type="l",
    lwd=3, lty=2, col = "black",
    axes = F, xlab = NA, ylab = NA)
axis(side = 4)
mtext(text=expression(paste(p(eta)/p[s])),
    side = 4, line = 3 )
legend("top",lwd=c(3,3),lty=c(1,2),bty="n",
     legend=c(expression(paste(T(eta)/T[s], " ")),
             expression(paste(p(eta)/p[s], " "))),
     col=c("black","black"))
#####################
# Plot p_max and T againts R
# Set margin to allow for multiple axes
par(mar = c(5,5,2,5))
plot(R[4500:N], ps[4500:N],
    ylab = expression(paste(p[s], " , [atm], ", T[s], " , [K]")),
    xlab=expression(R), log="y", ylim=c(10,10^7)
    ,type="l", col="black", lty=1, lwd=3)
#
lines(R[4500:N], Ts[4500:N], xlab="", ylab="",
     type="l",lty=2,col="black", main="",
     xlim=c(0,R_max),lwd=3)
#
legend("top",lwd=c(3,3),lty=c(1,2),bty="n",
     legend=c(expression(paste(p[s], " ")),
             expression(paste(T[s], " "))),
```

```
        col=c("black","black"))
grid(nx = NULL, ny = NULL, col="lightgray", lwd = 1, lty=1)
#
#######################
plot(R, rhos, ylim=c(4,8),type="l",
    lwd=3, lty=1, col = "black",
    xlab = expression(paste(R)),
    ylab = expression(paste(rho[s], ", ", ["", kg/m^3,"]")))
legend("topright",lwd=c(3),lty=1,bty="n",
       legend=expression(paste(rho[s], "         ")),
       col=c("black"))
grid(nx = NULL, ny = NULL, col="lightgray", lwd = 1, lty=1)
#######################
plot(eta, u_us, type="l",
    lwd=3, lty=1, col = "black",
    xlab = expression(paste(eta)),
    ylab = expression(paste(u/u[s])))
legend("top",lwd=c(3),lty=1,bty="n",
       legend=expression(paste(u/u[s])),
       col=c("black"))
grid(nx = NULL, ny = NULL, col="lightgray", lwd = 1, lty=1)
#########################
```

Listing 10.3. File: TaylorSedov_postSimCalcs.R—Code to perform post simulation calculations and generate plots, called from the main program of Listing 10.1

10.9 INTEGRATION OF PDES

As an alternative to eqns. (10.1), we can choose to solve a set of *conservative equations* with *source terms*, that is, in the form

$$\frac{\partial \mathbf{U}}{\partial t} + \frac{\partial \mathbf{F}(\mathbf{U})}{\partial r} = -\mathbf{S}(\mathbf{U}), \tag{10.59}$$

where $\mathbf{U}$ represents the state variable vector, $\mathbf{F}$ the flux vector, r the radial direction, and $\mathbf{S}$ vector of source terms.

We define the problem in terms of the following *Euler equations* with $\mathbf{U} = [\rho, \rho u, E]^T$, $\mathbf{F} = [\rho u, \rho u^2 + p, (E+p)u]^T$, and $\mathbf{S} = \frac{(\nu-1)}{r}[\rho u, \rho u^2, (E+p)u]^T$:

$$\begin{aligned}
\frac{\partial \rho}{\partial t} + \frac{\partial \rho u}{\partial r} &= -\frac{(\nu-1)}{r}\rho u, \\
\frac{\partial \rho u}{\partial t} + \frac{\partial (\rho u^2 + p)}{\partial r} &= -\frac{(\nu-1)}{r}\rho u^2, \\
\frac{\partial E}{\partial t} + \frac{\partial ((E+p)u)}{\partial r} &= -\frac{(\nu-1)}{r}(E+p)u,
\end{aligned} \tag{10.60}$$

where, again, $\nu = 1$ for the planar case, $\nu = 2$ for the cylindrical case, and $\nu = 3$ for the spherically symmetric case. The variables are as previously defined, with $E = \rho(e + \frac{1}{2}u^2)$ representing *total energy*. Equations (10.60) are presented in *conservative form with source* as this facilitates solution by the finite volume method. Some 1D numerical solution examples are included in Chapter 6.

10.A APPENDIX: SIMILARITY ANALYSIS

Similarity analysis, also known as *dimensional analysis*, indicates that the fireball expands at a rate that depends upon the follow relevant physical variables: explosion energy E_0, ambient density ρ_0, instantaneous blast wave radius R, and time t. We employ the ansatz that these can be combined into an algebraic function, f, to form a dimensionless constant c, that is, $c = f(R, E_0, \rho_0, t)$.

Now, for a geometry-dependent system, the dimensions of each variable are

$$E_0 = [ML^{\nu-1}T^{-2}], \rho_0 = [ML^{-3}], R = [L], t = [T],$$

where ν is a constant dependent upon the geometry of the problem: $\nu = 1$ for *planar flow*, $\nu = 2$ for *cylindrical symmetry*, and $\nu = 3$ for *spherical symmetry*. On the basis of these assumptions, R can be represented, dimensionally, by

$$[R] = [E_0]^x [\rho_0]^y [t]^z. \tag{10.61}$$

If we substitute in the appropriate dimensions for each variable, we obtain

$$[L] = [ML^{\nu-1}T^{-2}]^x [ML^{-3}]^y [T]^z. \tag{10.62}$$

Then, on expanding exponents, we obtain the following set of linear equations:

$$[L^1] = [M^{x+y} L^{(\nu-1)x-3y} T^{z-2x}]. \tag{10.63}$$

But the variable exponents for L, M, and T must be equal on both sides of eqn. (10.63). This yields the following set of linear equations:

$$\begin{aligned} 1 &= (\nu - 1)x - 3y, \\ 0 &= x + y, \\ 0 &= -2x + z, \end{aligned} \tag{10.64}$$

which has the solution

$$x = 1/(\nu + 2), \quad y = -1/(\nu + 2), \quad z = 2/(\nu + 2). \tag{10.65}$$

Thus, we arrive at the dimensionally consistent equation

$$[R] = [E_0]^{1/(\nu+2)} [\rho_0]^{-1/(\nu+2)} [t]^{2/(\nu+2)}. \tag{10.66}$$

Hence, it follows that

$$R = S \left(\frac{E_0 t^2}{\rho_0} \right)^{\frac{1}{(\nu+2)}} = \left(\frac{E_0 t^2}{\alpha \rho_0} \right)^{\frac{1}{(\nu+2)}}, \tag{10.67}$$

where, for the spherically symmetric case, $\nu = 3$, and we arrive at eqn. (10.2). For $\gamma = 1.4$ and using *constant of proportionality* values for α and S from Table 10.4, the problem

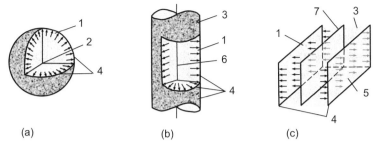

Figure 10.11. Blast waves in a compressible medium for (a) spherical, (b) cylindrical, and (c) planar geometry cases. Key: (1) moving medium, (2) symmetry center (of blast), (3) medium at rest, (4) shock wave front, (5) moving medium, (6) symmetry axis (of blast), (7) symmetry plane (of blast). Adapted from [Kor-91].

is fully defined.[4] The forms that a blast wave would take for the planar, cylindrical, and spherical cases are shown in Fig. 10.11.

For a point explosion in uniform ambient conditions, E_0 and ρ_0 are also constant. Therefore, for the spherically symmetric case, we have

$$R^5 t^{-2} = \text{constant}, \tag{10.68}$$

$$\Downarrow$$

$$\frac{5}{2}\log_{10}(R) - \log_{10}(t) = \text{constant}, \tag{10.69}$$

which is consistent with data from the time-lapsed photographs (see eqn. (10.40)). Also, on differentiating R in eqn. (10.67) with respect to t, we obtain the velocity of the moving blast front, that is,

$$\frac{dR}{dt} = U = \frac{2}{5}\frac{R}{t}. \tag{10.70}$$

The downloads include code written for symbolic algebra programs Maple and Maxima (open source) that perform the preceding blast wave similarity analysis.

10.B APPENDIX: ANALYTICAL SOLUTION

Shortly after the numerical solutions published by Taylor, von Neuman, and Sedov, Sedov published a full analytical solution [Sed-59]. The analysis is rather lengthy, so we will just include some aspects that enable *point detonation* explosions in uniform ambient initial conditions to be tackled.

Using eqn. (10.67), which we repeat for convenience,

$$R = S\left(\frac{E_0}{\rho_0}t^2\right)^{\frac{1}{\nu+2}}, \tag{10.71}$$

---

[4] From Table 10.4 for $\gamma = 1.4$, we have the analytical value $\alpha_{sp} = E_0/E = 0.851$ for spherical geometry, $\alpha_{cyl} = 0.984$ for cylindrical geometry, and $\alpha_{pl} = 1.077$ for planar geometry. But $S = \alpha^{-1/(\nu+2)}$; therefore $S_{sp} = 1.0328, S_{cyl} = 1.0040$, and $S_{pl} = 0.9755$. See also [Feo-96] and [Kor-91, p74] for a discussion on approximations to α.

the analytical solution can be derived from eqns. (10.1), the shock conditions

$$v_1 = \frac{2}{\gamma+1}u_s, \quad \rho_1 = \frac{\gamma+1}{\gamma-1}\rho_0, \quad p_1 = \frac{2}{\gamma+1}\rho_0 u_s^2, \quad u_s = \frac{2}{\nu+2}\frac{r_1}{t}, \quad (10.72)$$

where subscripts 0 and 1 refer to undisturbed ambient conditions and conditions immediately behind the shock wave respectively, together with the following set of relationships due to Sedov [Sed-59, p. 211, 217, 219]:

$$\frac{r}{r_1} = \lambda = \left[\frac{(\nu+2)(\gamma+1)}{4}V\right]^{-2/(2+\nu)} \left[\frac{\gamma+1}{\gamma-1}\left(\frac{(\nu+2)\gamma}{2}V-1\right)\right]^{-\beta_2}$$
$$\times \left[\frac{(\nu+2)(\gamma+1)}{(\nu+2)(\gamma+1)-2[2+\nu(\gamma-1)]}\left(1-\frac{2+\nu(\gamma-1)}{2}V\right)\right]^{-\beta_1},$$

$$\frac{v}{v_1} = f = \frac{(\nu+2)(\gamma+1)}{4}V\frac{r}{r_1},$$

$$\frac{\rho}{\rho_1} = g = \left[\frac{(\gamma+1)}{(\gamma-1)}\left(\frac{(\nu+2)\gamma}{2}V-1\right)\right]^{\beta_3} \left[\frac{\gamma+1}{\gamma-1}\left(1-\frac{\nu+2}{2}V\right)\right]^{-\beta_5} \quad (10.73)$$
$$\times \left[\frac{(\nu+2)(\gamma+1)}{(\nu+2)(\gamma+1)-2[2+\nu(\gamma-1)]}\left(1-\frac{2+\nu(\gamma-1)}{2}V\right)\right]^{-\beta_4},$$

$$\frac{p}{p_1} = h = \left[\frac{(\nu+2)(\gamma+1)}{4}V\right]^{2\nu/(2+\nu)} \left[\frac{\gamma+1}{\gamma-1}\left(1-\frac{(\nu+2)}{2}V\right)\right]^{\beta_5+1}$$
$$\times \left[\frac{(\nu+2)(\gamma+1)}{(\nu+2)(\gamma+1)-2[2+\nu(\gamma-1)]}\left(1-\frac{2+\nu(\gamma-1)}{2}V\right)\right]^{\beta_4-2\beta_1},$$

$$\frac{T}{T_1} = \frac{p}{p_1}\frac{\rho_1}{\rho}, \quad T_1 = \frac{p_1}{\mathcal{R}\rho_1},$$

where

$$\beta_1 = \frac{(\nu+2)\gamma}{2+\nu(\gamma-1)}\left[\frac{2\nu(2-\gamma)}{\gamma(\nu+2)^2}-\beta_2\right], \quad \beta_2 = \frac{1-\gamma}{2(\gamma-1)+\nu},$$

$$\beta_3 = \frac{\nu}{2(\gamma-1)+\nu}, \quad \beta_4 = \frac{\beta_1(\nu+2)}{2-\gamma}, \quad (10.74)$$

$$\beta_5 = \frac{2}{\gamma-1}, \quad \beta_6 = \frac{\gamma}{2(\gamma-1)+\nu},$$

$$\beta_7 = \frac{[2+\nu(\gamma-1)]\beta_1}{\nu(2-\gamma)}.$$

T represents temperature, $\mathcal{R}$ the ideal gas constant, and u_s shock wave speed. Other symbols have been defined previously. Note that subscript 1 is used to indicate conditions immediately behind the shock, whereas in [Sed-59], the subscript 2 is used. Also, Sedov uses the symbol α_i for the indices in eqns. (10.73), rather than β_i.

With $\gamma > 1$ and $\nu \in (1,2)$, values for $r = r_1\lambda, v = v_1 f, \rho = \rho_1 g,$ and $p = p_1 h$ can be calculated by varying the nondimensional parametric variable V in eqns. (10.73) and (10.74)

Table 10.3. Analytical data for the planar case $\nu = 1$ and $\gamma = 1.4$. Note the variable f used by Sedov is not the same variable f used by Tayor. Also, Sedov uses subscript 2 rather than 1 to indicate conditions appertaining immediately behind the shock wave position

$\lambda = \dfrac{r}{r_1}$	$\dfrac{r_0}{r_1}$	$\dfrac{v}{v_1} = f$	$\dfrac{\rho}{\rho_1} = g$	$\dfrac{p}{p_1} = h$	f'	g'	h'
1	1	1	1	1	1.5	7.5	4.5
0.9797	0.8873	0.9699	0.8625	0.9162	1.4688	6.1575	3.8131
0.9420	0.7151	0.9156	0.6659	0.7915	1.4067	4.3615	2.8352
0.9013	0.5722	0.8599	0.5160	0.6923	1.3367	3.1109	2.0959
0.8565	0.4501	0.8017	0.3982	0.6120	1.2597	2.2183	1.5250
0.8050	0.3427	0.7390	0.3019	0.5457	1.1758	1.5638	1.0723
0.7419	0.2448	0.6678	0.2200	0.4904	1.0802	1.0738	0.7067
0.7029	0.1980	0.6263	0.1823	0.4661	1.0390	0.8726	0.5483
0.6553	0.1514	0.5780	0.1453	0.4437	0.9921	0.6919	0.4022
0.5925	0.1040	0.5172	0.1074	0.4229	0.9438	0.5232	0.2648
0.5396	0.0741	0.4682	0.0826	0.4116	0.9144	0.4213	0.1838
0.4912	0.0529	0.4244	0.0641	0.4038	0.8947	0.3473	0.1289
0.4589	0.0415	0.3957	0.0536	0.4001	0.8849	0.3020	0.1003
0.4161	0.0293	0.3580	0.0415	0.3964	0.8750	0.2551	0.0453
0.3480	0.0156	0.2988	0.0263	0.3929	0.8651	0.1905	0.0237
0.2810	0.0074	0.2410	0.0153	0.3911	0.8602	0.1370	0.0111
0.2320	0.0038	0.1989	0.0095	0.3905	0.8584	0.1025	0.0089
0.1680	0.0012	0.1441	0.0042	0.3901	0.8574	0.0630	0.0029
0.1040	0.0002	0.0891	0.0013	0.3900	0.8572	0.0307	0.0005
0	0	0	0	0.3900	0.8571	0	0

subject to the constraints

$$\frac{2}{(\nu+2)\gamma} \leq V \leq \frac{4}{(\nu+2)(\gamma+1)}. \tag{10.75}$$

If $\nu = 3$ and $\gamma < 7$, then eqn. (10.75) also applies. For $\nu = 3$ and $\gamma > 7$, V is subject to the following alternative constraints:

$$\frac{4}{5(\gamma+1)} \leq V \leq \frac{2}{5}. \tag{10.76}$$

Table 10.3 has been taken from [Sed-59] and contains solution data for $\nu = 1$ and $\gamma = 1.4$. It was generated using the preceding method and has been used in a numerical example included in Chapter 6.

Note that the Sedov variables $\lambda, f, g,$ and h correspond to the Taylor variables $\eta, \frac{\gamma+1}{2}\phi,$ $\frac{\gamma-1}{\gamma+1}\psi,$ and $\frac{\gamma+1}{2}f$, respectively (see eqns. (10.3), (10.29), (10.31), and (10.54)).

R code that can be used to generate similar tables to Table 10.3 for $\nu = 2, 3$ and plot the results is included in Listing 10.4.

10.B.1 Closed-Form Solution

If we neglect energy loss due to radiation and consider only very short periods following detonation, the total energy contained within the region from the point of detonation to the edge of the shock wave is constant throughout the period of blast expansion. It is equal to the sum of kinetic and internal energies, which can be expressed by the following integral:

$$E_0 = \mathrm{Co} \int_0^{r_2} \left(\tfrac{1}{2}\rho v^2 + \rho e\right) r^{\nu-1} dr,$$

$$\mathrm{Co} = \begin{cases} 4\pi, & \nu = 3, \text{: spherical case,} \\ 2\pi, & \nu = 2, \text{: cylindrical case,} \\ 2, & \nu = 1, \text{: planar case,} \end{cases} \tag{10.77}$$

which, for the spherical case, is equivalent to eqns. (10.32) and (10.33) used by Taylor.

If we substitute $r = \lambda r_1$, $v = v_1 f$, $r = \lambda r_1$, and $\rho e = p/(\gamma - 1)$ into eqns. (10.77), then on using the shock conditions of eqns. (10.72), we arrive at

$$E_0 = \frac{8 \,\mathrm{Co}\, \rho_1 r_1^{\nu+2}}{(\nu+2)^2(\gamma^2-1)t^2} \int_0^1 (gf^2 + h)\lambda^{\nu-1} d\lambda. \tag{10.78}$$

But from eqn. (10.67) we see that $\rho_1 r_1^{\nu+2}/t^2 = E_0/\alpha$. Therefore, on rearranging and splitting the integral into two parts, we obtain

$$J_1 = \int_{V_0}^{V_1} \frac{1}{2}\frac{\gamma+1}{\gamma-1} g V^2 \lambda^{\nu+1} \left(\frac{d\lambda}{dV}\right) dV,$$

$$J_2 = \int_{V_0}^{V_1} \frac{8h\lambda^{\nu-1}}{(\nu+2)^2(\gamma^2-1)} \left(\frac{d\lambda}{dV}\right) dV, \tag{10.79}$$

$$\alpha = \mathrm{Co}(J_1 + J_2),$$

where we have used the relationship $f = \tfrac{1}{4}(\nu+2)(\gamma+1)V\lambda$ from eqns. (10.73). We have also changed the integration variable from λ to V, with a corresponding change in integration limits from eqn. (10.75). Also, the following expression for the analytical derivative of λ with respect to V is obtained from eqns. (10.73):

$$\frac{d\lambda}{dV} = -\frac{\lambda(8 + \gamma(\nu+2)^2(\gamma+1)V^2 - 4(\nu+2)(\gamma+1)V)}{(-2+(2+\nu\gamma-\nu)V)V(-2+\gamma(\nu+2)V)(\nu+2)}. \tag{10.80}$$

Finally, for a given ν and γ, eqns. (10.79) can be integrated (numerically), using the relationships for g, h, and λ from eqns. (10.73), to give a value for alpha. Thus, for known ambient gas conditions and geometry, the complete solution has been established.

The downloads include code written for symbolic algebra programs Maple and Maxima (open source) that derive the Sedov integrals of eqns. (10.79).

Some α and S values for different cases are given in Table 10.4—recall from eqn. (10.67) that $S = \alpha^{-1/(\nu+2)}$.

The R code used to generate the values in Table 10.4 is included in Listing 10.4. It also generates Figs. 64–66 from [Sed-59] for given values of ν and γ. The code should be self-explanatory, but it is worth mentioning that the program uses the excellent function quad(), from the R package pracma, that performs the numerical integration of the two parts of the energy integral, J_1 and J_2. A theoretical discussion relating to the algorithm used in function quad() is given in [Gan-00].

Table 10.4. Values for constants α (used by Sedov) and S (used by Taylor) calculated for planar, cylindrical, and spherical cases with $\gamma = 7/5$ and $\gamma = 5/3$

Geometry	α		S	
	$\gamma=7/5$	$\gamma=5/3$	$\gamma=7/5$	$\gamma=5/3$
Pl, $\nu = 1$	1.077	0.603	0.9755	1.1837
Cyl, $\nu = 2$	0.984	0.564	1.0040	1.1538
Sp, $\nu = 3$	0.851	0.494	1.0328	1.1517

```
# File: SedovCalc.R
require(compiler)
enableJIT(3)
rm(list = ls(all = TRUE)) # Delete workspace
library("pracma")

f <- function(V,gamma,nu){
  f <- (nu+2)*(gamma+1)/4 * V * lambda(V,gamma,nu)
}
g <- function(V,gamma,nu){
  g <- ((gamma+1)/(gamma-1)*
        ( (nu+2)*gamma/2 * V-1))^(beta[3]) *
    ((gamma+1)/(gamma-1)*(1-(nu+2)/2 * V))^(beta[5])*
    ((nu+2)*(gamma+1)/((nu+2)*(gamma+1)-2*(2+nu*(gamma-1)))*
       (1-(2+nu*(gamma-1))/2 *V ) )^(beta[4])
}
h <- function(V,gamma,nu){
  h <- ((nu+2)*(gamma+1)/4*V)^(2*nu/(2+nu)) *
    ((gamma+1)/(gamma-1)*(1-(nu+2)/2 * V))^(beta[5]+1)*
    ((nu+2)*(gamma+1)/((nu+2)*(gamma+1)-2*(2+nu*(gamma-1)))*
       (1-(2+nu*(gamma-1))/2 *V ) )^(beta[4]-2*beta[1])
}
lambda <- function(V,gamma,nu){
  lambda <- ((nu+2)*(gamma+1)*V/4)^(-2/(2+nu)) *
    ((gamma+1)/(gamma-1)*((nu+2)*(gamma/2)*V - 1))^(-beta[2])*
    ((nu+2)*(gamma+1)/((nu+2)*(gamma+1)-2*(2+nu*(gamma-1)))*
       (1-(2+nu*(gamma-1))*V/2))^(-beta[1])
}
dlambdadV <- function(V,gamma,nu){
  dlambdadV <- -lambda(V,gamma,nu)*(8+gamma*(nu+2)^2*(gamma+1)*V^2-
                (4*(nu+2))*(gamma+1)*V)/
                (((-2+(2+nu*gamma-nu)*V)*V*(-2+gamma*(nu+2)*V)*(nu+2)+eps)
}
betaSet <- function(gamma,nu){
  beta <- rep(0,7)
```

```
  beta[2] <- (1-gamma)/(2*(gamma-1)+nu)
  beta[1] <-(nu+2)*gamma/(2+nu*(gamma-1)) *
           ((2*nu*(2-gamma))/(gamma*(nu+2)^2) - beta[2])
  beta[3] <- nu/(2*(gamma-1)+nu)
  beta[4] <- beta[1]*(nu+2)/(2-gamma)
  beta[5] <- 2/(gamma-2)
return(beta)
}
dJ1 <- function(V){
  dJ1dV <- (gamma+1)/(2*(gamma-1))*lambda(V,gamma,nu)^(nu+1)*
          g(V,gamma,nu)*V^2*dlambdadV(V,gamma,nu)
  return(dJ1dV)
}
dJ2 <- function(V){
  dJ2dV <- 8/((gamma^2-1)*(nu+2)^2)*lambda(V,gamma,nu)^(nu-1)*
          h(V,gamma,nu)*dlambdadV(V,gamma,nu)
  return(dJ2dV)
}
aS <- function(V){
#    print(dJ1())
#    scan(quiet=TRUE)
  # Energy integral and beta calc - eqn. (54), Kamm (2007).
  J1 <- quad(dJ1, V0, V1, tol=10^-6)
  J2 <- quad(dJ2, V0, V1, tol=10^-6)
  #
  if(nu == 1){
    C0 <- 2
  }else{
    C0 <- 2*(nu-1)*pi
  }
  alpha <- C0 * (J1 + J2)
  S <- alpha^(-1/(nu+2))
  return(c(alpha,S))
}
# Eqns. (11.15) and (11.16) from Sedov (1959).
N <- 100
eps <- 1e-60
nu <- 3 # 1,2,3
gamma <- 7/5 # 5/3
V0 <- 2/((nu+2)*gamma)
V1 <- 4/((nu+2)*(gamma+1)) # eqn. (11.14a)
V <- seq(V0, V1, length.out=N)
beta <- betaSet(gamma,nu)
# Plot results for given nu and gamma - Sedov Figs (64,65,66)
plot(lambda(V,gamma,nu),f(V,gamma,nu),type="l",lwd=3,col="red",
     xlab=expression(paste(lambda)),ylab="f, g, h")
```

```
lines(lambda(V,gamma,nu),g(V,gamma,nu),lwd=3,col="blue")
lines(lambda(V,gamma,nu),h(V,gamma,nu),lwd=3,col="green")
legend("top",lwd=c(3),lty=1,bty="n",
       legend=c("f","g","h"),
       col=c("red","blue","green"))
grid(nx = NULL, ny = NULL, col="lightgray", lwd = 1, lty=1)
#
aSvec <- aS(V)
alpha <- aSvec[1]
S <- aSvec[2]
```

Listing 10.4. File: `SedovCalc.R`—Code of program that performs the Sedov analytical calculations of eqns. (10.73) and (10.74) and plots results

10.B.2 Additional Complexity

The problem considered previously with constant ambient conditions can be readily extended to situations where the blast takes place in an incompressible medium or in a gas with variable density. For discussion on these and other cases, the reader is referred to [Sed-59], [Kor-91], [Boo-94], [Kam-00], and [Kam-07].

10.B.3 The Los Alamos Primer

As additional background to the preceding analysis, it is interesting to read the technical briefing that new recruits were given as an introduction to the Manhatton Project. The briefing was given as a course of five lectures by American phyisicist Dr. Robert Serber as an *indoctrination course* during the first two weeks of April 1943. A set of classified notes, written up by nuclear physicist Edward U. Condon, was printed at Los Alamos under the title *The Los Alamos Primer*. It was declassified in 1963 and is now in the public domain and available online at https://commons.wikimedia.org/wiki/File:Los_Alamos_Primer.pdf. This document summarizes what was known regarding nuclear explosions around 1943. It was subsequently published as a book by the University of California Press [Ser-92].

At the end of World War II, Serber joined the faculty at Columbia University and later became professor and chair of its physics department.

REFERENCES

[Bet-47] Bethe, H. A., K. Fuchs, J. O. Hirschfelder, J. L. Magee, R. E. Peierls J. von Neumann (1947), Blast Wave, *Los Alamos Scientific Laboratory Report LA-2000*.

[Boo-94] Book, D. L. (1994), The Sedov Self-Similar Point Blast Solutions in Nonuniform Media, *Shock Waves* **4**-1, 1–10. doi:10.1007/BF01414626.

[Deb-58] Deb Ray, G. (1958), An Exact Solution of a Spherical Blast Under Terrestrial Conditions, *Proc. Natn. Inst. Sci. India* A **24**, 106–112.

[Dtr-07] DTRA (2007), Operation Trinity, Public Affairs Fact Sheet, *Defense Threat Reduction Agency (DTRA)*, available online at http://www.dtra.mil/documents/ntpr/factsheets/Trinity.pdf.

[Feo-96]   Feoktistov, L. P. (1996), Self-Similarity-Particular Solution or Asymptote?, *JETP* **83**, 996–999.
[Gan-00]   Gander, W. and W. Gautschi (2000), Adaptive Quadrature-Revisited, *BIT Numerical Mathematics* **1**, 84–101.
[Kam-00]   Kamm, J. R. (2000), Evaluation of the Sedov–von Neumann–Taylor Blast Wave Solution, *Los Alamos National Laboratory Report LA-UR-00-6055*.
[Kam-07]   Kamm, J. R. and F. X. Timmes (2007), On Efficient Generation of Numerically Robust Sedov Solutions, *Los Alamos National Laboratory Report LA-UR-07-2849*.
[Kor-91]   Korobeinikov, V. P. (1991), *Problems of Point-Blast Theory*, American Institute of Physics.
[Mac-49]   Mack, J. E. (1947), Semi-popular motion picture record of the Trinity explosion, MDDC221, U.S. Atomic Energy Commission.
[Pet-08]   Petruk, O. (2008). Approximations of the Self-Similar Solution for Blastwave in a Medium with Power-Law Density Variation, *arXiv Preprint Astro-ph/0002112v1*, 1–17, available online at http://arxiv.org/abs/astro-ph/0002112.
[Sed-46]   Sedov, L. I. (1946), Propagation of Strong Shock Waves, *Journal of Applied Mathematics and Mechanics* **10**, 241–250.
[Sed-59]   Sedov, L. I. (1959), *Similarity and Dimensional Methods in Mechanics*, Academic Press.
[Ser-92]   Serber, R. (1992), *The Los Alamos Primer, the First Lectures on How to Build an Atomic Bomb*. Annotated by Robert Serber. Edited with an introduction by Richard Rhodes. University of California Press.
[Tay-41]   Taylor, G. I. (1941), The Formation of a Blast Wave by a Very Intense Explosion, *British Civil Defence Research Committee*, Report RC-210.
[Tay-46]   Taylor, G. I. (1946), The Air Wave Surrounding an Expanding Sphere, *Proceedings of the Royal Society of London, Series A* **186**, 273–292.
[Tay-50a] Taylor, G. I. (1950), The Formation of a Blast Wave by a Very Intense Explosion. I. Theoretical Discussion, *Proceedings of the Royal Society of London, Series A* **201**, 159–174.
[Tay-50b] Taylor, G. I. (1950), The Formation of a Blast Wave by a Very Intense Explosion. II. The Atomic Explosion of 1945, *Proceedings of the Royal Society of London, Series A* **201**, 175–186.

11

Case Study: The Carbon Cycle

11.1 INTRODUCTION

One of the great scientific endeavors of the twenty-first century is the concerted effort by international research groups to understand how the global climate is changing. There is general agreement that sustained emissions of carbon dioxide into the atmosphere as a result of anthropological activities is the main cause of rising carbon dioxide (CO_2) in the atmosphere. Increased levels of CO_2 results in more heat, in the form of infra-red radiation, being trapped in the atmosphere—the so-called *greenhouse effect*. This additional heat tends, over time, to increase the global temperature. Thus, the rising concentrations of CO_2 in the atmosphere is a prime candidate for the cause of *climate change*, which is predicted to increase weather extremes, melting of glaciers and the rise in global sea levels [IPCC-07].

This case study is an attempt to illustrate some of the methods used to study climate change using a relatively simple model consisting of a system of algebraic and differential equations. We restrict our discussion to the major physical processes in the *carbon cycle*, with the associated equations and solution methods being explained in some detail.

11.2 THE MODEL

The mathematical model used here is based on that developed by James Walker [Walk-91], which has also been featured elsewhere to demonstrate numerical procedures [Mol-05]. However, the scope has been extended to include detailed representation of air-gas exchange between atmosphere and ocean, improved seawater chemistry calculations, including pH levels, and the inclusion of basic radiation calculations for the atmosphere.

We will use the model to illustrate the main processes involved when large quantities of CO_2 are emitted into the atmosphere. The model consists of three main reservoirs: the atmosphere, surface water of the oceans and ocean deep water. The calculations attempt to predict how fossil fuel emissions result in higher concentrations of CO_2 in the atmosphere and how a large fraction of these emissions is absorbed into the oceans. The effect of increased levels of CO_2 is shown to increase absorption of terrestrial infra-red radiation by the atmosphere and, hence, a corresponding rise in the effective global temperature of the Earth. The main model parameters are detailed in Table 11.1.

Table 11.1. Model parameters set by R computer code in file carbonInit.R, which also performs some other initialization tasks

Symbol	Description	Value	Units
A_s	surface area of oceans	3.58E+14	m^2
d	depth of surface water (nominal)	250	m
M_{atm}	mass of atmosphere	5.13E+21	g
m_{atm}	mass of atmosphere (M_{atm}/MW_{air})	1.77E+20	mol
MW_{air}	molecular weight of air	29	g/mol
MW_C	molecular weight of carbon	12	g/mol
MW_{CO_2}	molecular weight of carbon dioxide	44	g/mol
PAL	atmos. partial pressure of CO_2—PAL*	2.8E-04	ppmv
r_{bio}	settling rate of organic matter	1.75E+14	mol/y
R_{oc}	ratio of carbonate to organic carbon	0.25	–
S	salinity	35	‰
T_C	temperature, ambient	15	C
T_K	temperature, ambient	288	K
v_d	volume, deep seawater ($v_t - v_s$)	1.25E+18	m^3
v_s	volume, surface seawater ($d \times A_s$)	8.95E+16	m^3
v_t	volume, total ocean	1.34E+18	m^3
w_{exch}	upwelling/downwelling	1.00E+15	m^3/y

Note: carbonInit.R is called from the main program at the start of the simulation.
*PAL = preindustrial activity level.

Although this model is extremely simplified, it does bring out the salient features of the effect of increasing levels of carbon dioxide emissions. The simulation results are generally consistent with those from more detailed models published in the scientific literature, and references are provided to data used for comparison. Also, including additional complexity and/or new features would require a more advanced treatment of the technical aspects and would significantly increase the space needed to adequately describe this case study.

The R computer code for the main program that controls this simulation is given in Listing 11.1.

```
# File: carbonCycle_main.R
cat("\014") # Clear console
rm(list = ls(all = TRUE)) # Delete workspace
#
require(compiler)
enableJIT(3)
#
library(deSolve)
library(pracma)
source("carbonCycle_deriv.R")
source("carbonCycle_ratesOfChange.R")
source("carbEq.R")
source("carbonCycle_fossilFuel.R")
```

```r
# The model simulates the interaction of the various forms of
# carbon that are stored in three regimes: the atmosphere, the shallow
# ocean, and the deep ocean. The array y[t,1:5], of five principal variables
# in the model are all functions of time:

# y[,1] <- partial pressure of co2 in the atmosphere (normalized to 1 at 1750)
# y[,2] <- total dissolved carbon concentration in the shallow ocean (mol/m3)
# y[,3] <- total dissolved carbon concentration in the deep ocean (mol/m3)
# y[,4] <- alkalinity in the shallow ocean (mol/m3)
# y[,5] <- alkalinity in the deep ocean (mol/m3)
# The rate of change of these five principal variables is described by five
# ordinary differential equations.
# Other variables are described by comments in the code

# Initialise simulation parameters
source("carbonCycle_init.R")
#
ncall <- 0  # Initialize global variable - No of calls to deriv. function
nout <- 100 # Number of output steps
t0   <- 1750 # simulation start time
tf   <- 3000 # simulation end time
# tspan <- seq(tStart,tf,by=1000) # linspace(tStart, tf, 1001)
tspan <- seq(t0, tf,len=nout)
y0 <- c(y=y)

cat(sprintf('Carbon calc problem!'))
cat(sprintf('\n===================\n'))
cat(sprintf('\nStart time ....................: %g',t0))
cat(sprintf('\nEnd time ......................: %g\n',tf))
#
cases <- 1:4
yout <- array(0,c(nout,5,4))
# Run simulation for each emissions/reserves case
for(caseNo in cases){
  out  <- ode(y=y0, times=tspan, func=carbonCycle_deriv, parms=NULL,
              method="lsodes", rtol=1e-6, atol=1e-6)
  yout[1:nout,1:5,caseNo] <- out[,2:6]
}
tout <- out[,"time"]
nout <- length(tout)

cat(sprintf('\nNumber of derivative calls ....: %d\n',ncall))
#
# Preallocate arrays for carbonate variables
pH_s   <- array(0,c(nout,4)) # acidity - shallow water (-)
pco2_s <- array(0,c(nout,4)) # CO2 partial press. - shallow water (normal)
```

```
co2_s  <- array(0,c(nout,4)) # dissolved CO2 - shallow water (mol/kg)
co3_s  <- array(0,c(nout,4)) # bi-carbonate - shallow water (mol/kg)
hco3_s <- array(0,c(nout,4)) # carbonate - shallow water (mol/kg)
R      <- array(0,c(nout,4)) # Revelle factor (-)
# Carbonate equilibrium calculations for each case
for(i in cases){
  for(j in 1:nout){
    carbSys_s<- carbEq(288,salinity,0,yout[j,4,i],yout[j,2,i])
    pH_s[j,i]   <- carbSys_s$pH
    pco2_s[j,i] <- carbSys_s$fco2/pco2_PAL
    co2_s[j,i]  <- carbSys_s$co2
    co3_s[j,i]  <- carbSys_s$co3
    hco3_s[j,i] <- carbSys_s$hco3
    R[j,i]      <- carbSys_s$R
    #
  }
}
dic_s <- co2_s+hco3_s+co3_s # dic - surface water (mol/kg)
dic_d <- yout[,3,]      # dic - deep ocean (mol/kg)
#
source("carbonCycle_postSimCalcs.R")
```

Listing 11.1. File: carbonCycle_main.R—Code for *main* program that controls the simulation. It calls the lsode numerical integrator that solves the five ODEs, followed by calls to the carbonate chemistry function carbEq and the plotting code in postSimCalcs.R.

11.2.1 Atmosphere

We treat the atmosphere as a single well-mixed homogeneous mass, the CO_2 composition of which is modeled by the following ordinary differential equation:

$$\frac{d\,pCO_2}{dt} = \left(\delta(pCO_2^{aq} - pCO_2^{atm}) + \frac{f}{m_{atm} MW_C} \right) \frac{1}{PAL}, \quad (11.1)$$

where pCO_2^{aq} and pCO_2^{atm} represent the partial pressure of CO_2 in the surface water of the oceans and the atmosphere, respectively, f fossil fuel emissions (GtC/y) (see Section 11.2.7), MW_C molecular weight of carbon, m_{atm} the mass of the atmosphere (mol-air), and t time (yr). The parameter $\delta = k_p A K_H / m_{atm}$ (1/y) is a gas exchange constant where k_p is the piston velocity (m/yr), A surface area of air–sea exchange (m²), and K_H is Henry's constant for solubility of gas in seawater (mol/m³/atm) (see Section 11.2.3 for detailed discussion). The term f/MW_C converts mass of carbon per year to moles of carbon per year (also equal to moles of carbon dioxide per year). On further dividing by m_{atm}, this result is converted to change in carbon dioxide concentration per year. The constant PAL $= 280 \times 10^{-6}$ represents atmospheric CO_2 concentration at the *preindustrial activity level* of 280 (ppm). It is included to normalizes pCO_2 so that at the start of the simulation it is equal to unity. Other sources and sinks of CO_2, such as land uptake, are not sufficiently understood to have reached a consensus as to their magnitudes [Sar-02].

Therefore, we considered them to be either more or less in balance, or their net effect to be insignificant compared to the uncertainties in fossil fuel emission rates. See Appendix 11.A.1 at the end of this chapter for additional detail. A model that does include terms for land *biota* and the marine *biosphere* is described in [McH-15].

The atmospheric concentration of CO_2 tends to increase at present due to a constant supply of fossil fuel emissions; however, this is partially mitigated as a result of absorption of CO_2 by the oceans. This gas–sea exchange flux is driven by the CO_2 partial pressure difference between atmosphere and ocean, that is, $\Delta pCO_2 = (pCO_2^{aq} - pCO_2^{atm})$. Once in the ocean, CO_2 rapidly dissociates into bicarbonate and carbonate ions (see Section 11.2.4).

Equation (11.1) is a more complex ODE than we have discussed in previous chapters because the term pCO_2^{aq} has to be obtained by solving the carbonate chemistry equations of Section 11.2.4 at each step, which adds a significant computational overhead. The ODE is then integrated numerically, along with eqns. (11.2) to (11.5), to obtain the atmospheric concentration of CO_2 over time.

11.2.2 Oceans

The oceans are modeled as two well-mixed homogeneous masses, the dynamics of which are described by the following ordinary differential equations:

$$\frac{dc_{dic}^s}{dt} = \left(-\delta m_{atm} \text{PAL}(pCO_2^{aq} - pCO_2^{atm}) - (1 + R_{oc})r_{bio} + (c_{dic}^d - c_{dic}^s)w_{exch}\right)/vol_s, \tag{11.2}$$

$$\frac{dc_{dic}^d}{dt} = \left((1 + R_{oc})r_{bio} - (c_{dic}^d - c_{dic}^s)w_{exch}\right)/vol_d, \tag{11.3}$$

$$\frac{dc_{alk}^s}{dt} = \left((c_{alk}^d - c_{alk}^s)w_{exch} - (2R_{oc} - 0.15)r_{bio}\right)/vol_s, \tag{11.4}$$

$$\frac{dc_{alk}^d}{dt} = \left((2R_{oc} - 0.15)r_{bio} - (c_{alk}^d - c_{alk}^s)w_{exch}\right)/vol_d, \tag{11.5}$$

where c_{dic} and c_{alk} represent concentrations of dissolved inorganic carbon (DIC) and alkalinity (Alk or TA), respectively, with superscripts s and d indicating surface and deep water, respectively. Variables vol_s and vol_d are constants representing the ocean surface and deep water volumes (m³). The constant w_{exch} represents exchange of seawater (m³/y) between the surface and deep ocean waters, known as *upwelling* and *downwelling*. We use this simple approach because it appropriate for the reduced-dimensional model being considered. In reality, ocean mixing is complex, with near-surface circulation being wind-driven, whereas in deeper waters, circulation is predominately density driven. The variable R_{oc} represents the ratio of carbonate to organic carbon, and r_{bio} represents the "rain" of particulate organic carbon (POC) (mol/y).

Equations (11.2) to (11.5) are integrated numerically to obtain the four concentrations over time. Each term in these equations represents an incremental change per year in carbon or alkalinity concentration of the associated reservoir. See Appendix 11.A.1 at the end of this chapter for additional detail.

Note that the atmosphere and ocean masses are very large compared with the changes due to emissions and gas–ocean exchange. Therefore, we consider their overall masses to be constant in the mass balance calculations, with only concentrations varying over time.

The function that calculates the preceding time derivatives is included in Listing 11.2.

```
# File: carbonCycle_deriv.R
carbonCycle_deriv <- function(t, y, parms=NULL){
# carbonDeriv <- function(y){
# Global variable set during initialization
  ncall<<-ncall+1
  carbSys <- carbEq(288,salinity,0,y[4],y[2])
  pco2_s <- carbSys$fco2/pco2_PAL
#
  dydt <- rep(0,5)
# fossil fuel consumption rate (convert from GtC/y to gC/y)
  f <- carbonCycle_fossilFuel(t,caseNo)*10^15 # [gC/year]
#
  dydt[1] <- delta*(pco2_s - y[1]) + (f/MW_C)/m_atm_PAL # pCO2 (normalized)
# ODEs describing the exchange between the shallow and deep oceans
# involve vol_s and vol_d, the volumes of the two regimes:

  dydt[2] <- ( - delta*m_atm_PAL*(pco2_s - y[1]) - K_1 + # dic_s
                      (y[3] - y[2])*w_exch)/vol_s
  dydt[3] <- ( K_1 - (y[3] - y[2])*w_exch )/vol_d       # dic_d
  dydt[4] <- ((y[5] - y[4])*w_exch - K_2)/vol_s          # alk_s
  dydt[5] <- (K_2 - (y[5] - y[4])*w_exch)/vol_d          # alk_d

  return(list(dydt))
}
```

Listing 11.2. File: `carbonCycle_deriv.R`—Code of function `carbonCycle_deriv()` that calculates time derivatives and that is called from the `lsode` numerical integrator

11.2.3 Air–Ocean Exchange

The mass transfer flux between the atmosphere and ocean is governed by many mechanisms, including diffusion, wind speed, turbulence/bubbles, or churning of water waves. The process is extremely complex, and a global calculation, almost by definition, has to be an approximation, as weather conditions are continually changing throughout the globe. A method that has become very popular with climate modelers is the following *wind-driven* gas–sea exchange relationship due to Wanninkhof [Wan-92]:

$$F = zk_p A K_H \left(pCO_2^{atm} - pCO_2^{aq} \right), \tag{11.6}$$

Table 11.2. Piston velocity constants used in eqn. (11.7)

Var	Warm sea	Cold sea
D_c	1.680 × 1e-5	1.090 × 1e-5
ν	1.004 × 1e-2	1.562 × 1e-2
R_{ice}	0	0
$\bar{u}^2$	7.5	7.5
$\overline{u^2}$	9	9
S_c	597.6	921.1
k_p	19.25	23.9

Note: Values for D_c and ν are from [Eme-08, pp. 344–345], and those for $\bar{u}^2$ and $\overline{u^2}$ have been adapted from data in [Mil-09]. Sea ice is assumed to be negligible, hence R_{ice} is set to zero. The value of k_p shown has units of (cm/h) and has to be converted to (m/y) for use in eqn. (11.6). See text for other units.

where F represents gas exchange flux from atmosphere to ocean (mol/y), k_p piston velocity (m/yr), A surface area of air–sea exchange, and K_H Henry's constant for solubility of gas in seawater (mol m^{-3} atm^{-1}); $p\text{CO}_2^{atm}$ represents partial pressure of atmospheric CO_2, whereas $p\text{CO}_2^{aq}$ represents partial pressure of CO_2 in equilibrium with seawater; $z = 0.7$ represents a tuning constant that has been chosen so that the net CO_2 exchange rate matches net fluxes given in [IPCC-13, Fig. 6.1].

It should be appreciated that gas is continuously released from some parts of the ocean whereas at the same time being dissolved in other parts, depending upon the prevailing conditions at the time. The resulting changes in atmosphere/ocean concentration are therefore dependent upon the net exchange. According to the IPCC, at the present time, the net exchange of CO_2 from atmosphere to oceans is around 2 (GtC/y).

Equation (11.6) divided by the mass of the atmosphere in moles forms part of the dynamic mass balance ODE (11.1) associated with the atmosphere and surface ocean interface.

The *piston velocity* k_p (m/y), used in eqn. (11.6), is a parameter that represents the rate at which gas diffuses into a liquid, hence the name. The value of k_p (cm/h) is obtained from the empirical relationship [Wan-92]

$$k_p = k_o(1 - R_{ice})\left(\bar{u}^2 + \overline{u^2}\right)\sqrt{S_c/660}, \qquad (11.7)$$

where $k_o = 0.31$ (cm h^{-1} s^2 m^{-2}) represents an empirically derived constant, $\bar{u}^2$ the mean wind velocity squared (m/s)^2, $\overline{u^2}$ the wind velocity varience (m/s)^2, R_{ice} the sea-ice ratio, $S_c = \nu/D_c$ the *Schimdt number*, ν kinematic viscosity (cm^2/s), and D_c an appropriate molecular diffusion coefficient (cm^2/s).

Constants for eqn. (11.7) that we use in the simulation, along with the corresponding calculated values for k_p, are included in Table 11.2.

Computer code that performs the preceding calculations, and which is called during initialization, is shown in Listing 11.3.

```
# File: carbonCycle_pistonCalc.R
# Ref: 1) Wanninkhof, R.H. (1992). Relationship between gas exchange and wind
#         speed over the ocean. Journal of Geophysical Research, 97(5),7373-7382.
#      2) Emerson (2008), p344-5
#      3) Sarmiento (2002), p80-95
#
ko      <- 0.31        # constant [cm/h s^2 /m2] # 0.24
nu      <- 1.004*1e-2  # [cm^2/s] at 20 deg C
DC      <- 1.68*1e-5   # [cm^3/s] at 20 deg C
Rsi     <- 0           # Sea ice ratio, [-]
v_ave   <- 7.5         # monthly wind speed mean at 10m, [m/s]
v_var   <- 9.0         # monthly wind speed varience at 10m, [m/s]
#
Sc_chk  <- 2073.1-125.62*TC+3.6276*TC^2-0.043219*TC^3 # Sarmiento, p85
           (665.6 at 20 deg C)
Sc      <- Sc_chk #nu/DC # Schmidt No [-]
# Wind-based model due to Wanninkhof (1994)
kpiston <- ko*(1-Rsi)*(v_ave^2+v_var)*sqrt(Sc/660) # [cm/h]
# Below, the 0.7 tunes Net CO2 exchange rate for IPCC net fluxes
kpd     <- 0.7*kpiston*24/100 # [m/d]
kpy     <- kpd*365
kex     <- kpy*(KH*1000)*surface
delta   <- kex/m_atmos # [1/y] *1e14
```

Listing 11.3. File: `carbonCycle_pistonCalc.R`—Code to calculate *piston velocity* associated with CO_2 gas–ocean exchange

11.2.4 Carbonate Chemistry

The world's oceans are a primary sink for atmospheric CO_2, currently absorbing 25%–30% of the fossil fuel emissions. Since the Industrial Revolution, they have taken up almost half of accumulated emissions [Sab-04]. Present estimates [Fal-00, Stew-07] indicate that the oceans contain around 40,000 Gt (Giga metric tonnes) of carbon in the form of dissolved inorganic carbon (DIC), dissolved organic carbon (DOC), and particulate organic carbon (POC) (living and dead). Measurements reported in 2000 indicated that they exist in the approximate ratio DIC:DOC:POC = 2000:38:1 [IPCC-07, Chapter 7]. The corresponding figure for the atmosphere is 750 Gt of carbon in the form of CO_2.

At equilibrium, the partial pressure of gaseous carbon dioxide pCO_2^{aq} is in equilibrium with dissolved carbon dioxide and is proportional to the concentration of carbon dioxide dissolved in seawater [CO_2(aq)] (Henry's law), that is,

$$pCO_2^{aq} = \frac{[CO_2(aq)]}{K_H}, \tag{11.8}$$

where K_H (mol/kg/atm) is Henry's constant. Square brackets indicate, here and subsequently, that the enclosed variable has units of concentration. So [CO_2(aq)] refers to the concentration of CO_2 in solution. Note that eqn. (11.8) should, theoretically, use *fugacity* $f_{CO_2^{aq}}$ rather than partial pressure pCO_2^{aq}. This is because partial pressure is equal to fugacity only for ideal gases. However, for the temperature range 0°C to 30°C, it has been shown [Wei-74] that the ratio of fugacity to partial pressure is between 0.995 and 0.997. Thus, for our purposes, we choose to use partial pressure because the nonideal behavior of CO_2 in seawater is negligibly small.

The dissolved carbon dioxide in seawater rapidly takes other forms as a result of fast chemical reactions involving the species H_2O *water*, CO_2(aq) *dissolved carbon dioxide*, H_2CO_3 *carbonic acid*, HCO_3^- *bicarbonate ions*, CO_3^{2-} *carbonate ions*, and H^+ *positive hydrogen ions*. The reactions are

$$CO_2(aq) + H_2O \rightleftharpoons H_2CO_3, \tag{11.9}$$

$$H_2CO_3 \rightleftharpoons HCO_3^- + H^+, \tag{11.10}$$

$$HCO_3^- \rightleftharpoons CO_3^{2-} + H^+. \tag{11.11}$$

Not all the CO_2 dissolved in seawater reacts to make carbonic acid, and therefore seawater also contains dissolved gaseous carbon dioxide, CO_2(aq). A major effect of dissolving CO_2 in seawater is to increase the hydrogen ion concentration in the ocean, and thus decrease ocean pH (discussed subsequently).

As the forward and reverse reaction rates are fast and we are considering time over decades and centuries, we may consider the system to be at equilibrium. At equilibrium, the concentrations are governed by the *mass action law* (see Appendix 11.A.5 at the end of this chapter), and eqns. (11.9) and (11.10) yield the following relationships:

$$K_{co2(aq)} = [CO_2(aq)]/[H_2CO_3]$$
$$K_{h2co3} = [H^+][HCO_3^-]/[H_2CO_3], \tag{11.12}$$

where $K_{co2(aq)}$ and K_{h2co3} are equilibrium constants.

Eliminating H_2CO_3 from eqns. (11.12), we obtain

$$K_{a1} = [H^+][HCO_3^-]/[CO_2(aq)], \tag{11.13}$$

where $K_{a1} = K_{h2co3}/K_{co2(aq)}$ is known as the *first dissociation constant* for carbonic acid.

The equilibrium equation for the reaction of eqn. (11.11) is

$$K_{a2} = [H^+][CO_3^{2-}]/[HCO_3^-], \tag{11.14}$$

where K_{a2} is known as the *second dissociation constant* for carbonic acid. The variables [CO_2(aq)], [CO_3^{2-}], [H^+], [HCO_3], [H_2CO_3], and so on, all represent species mole concentrations and, unless stated to the contrary, will have units of micromoles per kg (μmol/kg). It therefore follows that K_{a1} and K_{a2} will also have units of μmol/kg.

Dissolved carbon dioxide in seawater exists mainly in three inorganic forms: free aqueous carbon dioxide CO_2(aq), bicarbonate ions HCO_3^-, and carbonate ions CO_3^{2-} (a carbonate is a salt or ester of carbonic acid). A minor form is true carbonic acid H_2CO_3 whose concentration is less than 0.3% of CO_2(aq). At typical surface seawater pH of 8.2, the *speciation* (quantitative distribution) between CO_2(aq), HCO_3^-, and CO_3^{2-} is 0.5%, 89%, and 10.5%, showing that most of the dissolved carbon dioxide is in the form of

HCO_3^- and not CO_2 [Zee-06]. We will therefore follow usual practice and now denote the sum of $[CO_2(aq)]$ and $[H_2CO_3]$ as $[CO_2(aq)]$.

The concentrations of different carbon species depend upon total dissolved carbon along with the requirement that the solution be *electrically neutral*, that is, the positive and negative charges per unit volume must be equal. The total *dissolved inorganic carbon* (DIC) concentration is given by

$$DIC = [HCO_3^-] + [CO_3^{2-}] + [CO_2(aq)]. \tag{11.15}$$

We can now combine eqns. (11.13), (11.14), and (11.15) to establish relationships for $[CO_2(aq)]$, $[CO_3^{2-}]$, and $[HCO_3]$, as follows:

$$[CO_2(aq)] = \frac{DIC}{1 + \frac{K_{a1}}{[H^+]} + \frac{K_{a1}K_{a2}}{[H^+]^2}}, \tag{11.16}$$

$$[HCO_3^-] = \frac{DIC}{\frac{[H^+]}{[K_{a1}]} + 1 + \frac{K_{a2}}{[H^+]}}, \tag{11.17}$$

$$[CO_3^{2-}] = \frac{DIC}{1 + \frac{[H^+]^2}{K_{a1}K_{a2}} + \frac{[H^+]}{K_{a2}}}. \tag{11.18}$$

Another acid–base pair important with respect to seawater chemistry is *boric acid*. The associated reaction is

$$B(OH)_3 + H_2O \rightleftharpoons B(OH)_4^- + H^+, \tag{11.19}$$

where $B(OH)_3$ and $B(OH)_4^-$ represent *boric acid* and its base, *tetrahydroxyborate*, respectively. At equilibrium, the concentrations are governed by

$$K_B = [B(OH)_4^-][H^+]/[B(OH)_3], \tag{11.20}$$

where K_B is the dissociation constant for boric acid.

Total boron B_T is a conserved quantity, and its concentration in mol/kg is in constant ratio to the *salinity* of seawater, that is,

$$B_T = 0.000416 \times s/35.0, \tag{11.21}$$

where s represents salinity in parts per thousand. It is equal to

$$B_T = [B(OH)_3] + [B(OH)_4^-]. \tag{11.22}$$

Now the total concentration of charge (recall that charge is a negative quantity) carried by species in seawater is called *total alkalinity* (TA) or just *alkalinity*. The major concentrations of species that contribute to total alkalinity are included in the following equation:

$$TA = [HCO_3^-] + 2[CO_3^{2-}] + [B(OH)_4^-] + [OH^-] - [H^+] + \epsilon, \tag{11.23}$$

where ϵ represents minor species[1] that we shall ignore. Equation (11.23) follows because the bicarbonate ion carries one unit of charge and the carbonate ion carries two.

The final equation we need to complete our definition of the problem is the self-ionization water reaction

$$H^+ + OH^- = H_2O, \qquad (11.24)$$

for which the dissociation constant is described by

$$K_W = [OH^-][H^+]. \qquad (11.25)$$

The species OH^- represents hydroxide ions, which are a naturally occurring constituent of water.

We are now in a position to start solving the preceding equations. Following Bacastow and Keeling [Bac-73], we substitute eqns. (11.17), (11.18), (11.20), (11.22), and (11.25) into eqn. (11.23) and rearrange to yield the following fifth-degree polynomial in H^+:

$$0 = \sum_{i=0}^{5} a_i [H^+]^i, \qquad (11.26)$$

where

$$\begin{aligned}
a_0 &= K_W K_{a1} K_{a2} K_B \\
a_1 &= -2K_{a2} K_{a1} \text{DIC} K_B - K_B B_T K_{a1} K_{a2} - K_W K_{a1} K_B + \text{TA} K_{a1} K_{a2} K_B \\
&\quad - \text{TA} K_{a1} K_{a2} K_B - K_W K_{a1} K_{a2} \\
a_2 &= -K_{a1} \text{DIC} K_B - 2K_{a2} K_{a1} \text{DIC} - K_B B_T K_{a1} - K_W K_{a1} + \text{TA} K_{a1} K_{a2} \\
&\quad + K_{a1} K_{a2} K_B + \text{TA} K_{a1} K_B - K_W K_B \\
a_3 &= -K_{a1} \text{DIC} K_B - 2K_{a2} K_{a1} \text{DIC} - K_B B_T K_{a1} - K_W K_{a1} + \text{TA} K_{a1} K_{a2} \\
&\quad + K_{a1} K_{a2} K_B + \text{TA} K_{a1} K_B - K_W K_B \\
a_4 &= K_{a1} + \text{TA} + K_B \\
a_5 &= 1
\end{aligned} \qquad (11.27)$$

Total boron B_T is fixed by salinity, and TA and DIC are integrated variables that are calculated at each time step. Therefore, assuming that we know B_T, TA, and DIC, along with constants K_W, K_B, K_H, K_{a1}, and K_{a2}, we are able to obtain the corresponding value of $[H^+]$ by solving eqns. (11.27) using the R function `roots` (see Listing 11.4).

Knowing $[H^+]$, we can now use eqns. (11.16), (11.17), and (11.18) to obtain $[CO_2(aq)]$, $[HCO_3^-]$, and $[CO_3^{2-}]$. Using values for B_T and $[H^+]$, we obtain concentrations $[B(OH)_3]$ and $[B(OH)_4^-]$ from eqns. (11.20) and (11.22). From eqn. (11.25) we obtain $[OH^-]$ and from eqn. (11.8) pCO_2^{aq} the partial pressure of CO_2 in equilibrium with $CO_2(aq)$. This means we now have calculated values for all the unknowns.

As carbonate concentration falls, it has serious consequences for various marine life that form shells. This is because of the following reaction, where calcium carbonate ($CaCO_3$) can react in seawater to form calcium ions (Ca^{2+}) and carbonate ions (CO_3^{2-}).

---

[1] Minor species $[H_3SiO_4^-]$, $[HPO_4^{2-}]$, $2[PO_4^{3-}]$, $[HSO_4^-]$, $[HF]$, and $[H_3PO_4]$ occur in such low concentration at the pH levels we are concerned with that they are considered to be insignificant. The two uncharged species [HF] and $[H_3PO_4]$ are included to conform with accepted practice [Eme-08].

The equation for dissolution/formation of pure calcium carbonate is

$$\text{CaCO}_3 \underset{\text{dissolution} \rightarrow}{\overset{\leftarrow \text{ mineral formation}}{\rightleftharpoons}} \text{Ca}^{2+} + \text{CO}_3^{2-}, \quad (11.28)$$

which has the simple *apparent solubility*[2] product in seawater

$$K'_{sp} = [\text{Ca}^{2+}][\text{CO}_3^{2-}]. \quad (11.29)$$

Carbonate minerals are used by crustaceans and coral for their formation and the increasing levels of CO_3^{2-} that accompany falling pH (defined below) result in the reaction of eqn. (11.28) moving to the right, which causes dissolution of the minerals. Thus, as pH falls, existing exoskeletons of certain species and coral reefs become vulnerable to dissolution. Also, the formation of new exoskeletons and reefs is more difficult due to inhibited growth. This situation occurs mainly in surface waters, because at deep levels in the ocean, there is an abundance of CaCO_3 sediments that act as a buffer and tend to keep pH levels at around 8.0. For detailed discussion on the effects of acidification on crustaceans refer to [Whi-11], and for the effects on coral refer to [Kle-09].

Correlations for the constants used in the preceding calculations are included in Appendix 11.A.2 at the end of this chapter, and R computer code that solves the equations is given in Listing 11.4.

```
# File: carbEq.R
carbEq <- function (TK, s, z, alk, dic){
  # Ref: Emerson and Hedges (2008) p129/130. Adapted and extended to add
  #      calculations for Revelle factor and pH.
  # Function to calculate fCO2, HCO3, and CO3 from ALK and DIC
  # Example ...
  # Input:     res <- carbEq (293.15, 35, 0, 2.300, 1.970)
  # Output: res$fco2 <- 0.00028007
  #         res$pH <- 8.1715
  #         res$co2 <- 9.0764e-006
  #         res$hco3 <- 0.0017
  #         res$co3 <- 0.00020373
  ##########################################
  alk<-alk/1000 # alk now in (mol/m3)
  dic<-dic/1000 # dic now in (mol/m3)
  ##########################################
  R<-83.131
  # Calculate total borate (Tbor) from chlorinity
  Tbor<-.000416 * s / 35.0;
  # Calculate Henry's Law coeff, KH (Weiss, 1974)
  U1<- -60.2409+93.4517 * (100/TK)+23.3585 * log(TK/100)
```

---

[2] Concentration of material at apparent equilibrium (supersaturation) conditions. That is not to be confused with thermodynamic solubility, which reaches equilibrium at infinite time.

```
U2<- s * (.023517 - .023656 * (TK/100)+.0047036 * (TK/100) ^ 2)
KH<- exp(U1+U2)
# Calculate KB from TK & sal (Dickson, 1990)
KB<-exp((-8966.9-2890.53 * s ^0.5-77.942 * s+1.728 * s^1.5 -0.0996*s^2)/TK
     +148.0248+137.1942 * s^0.5+1.62142 * s -(24.4344+25.085 * s^0.5
     +0.2474 * s) * log(TK)+0.053105 * s^0.5 * TK)
# Calculate K1 and K2 (Luecker et al., 2000)
K1<-10^(-(3633.86/TK-61.2172+9.67770 * log(TK)-0.011555*s+0.0001152 * s^2))
K2<-10^(-(471.78/TK+25.9290-3.16967 * log(TK)-0.01781*s+0.0001122 * s^2))
##############################################
# Temperature dependence of Kw (DoE, 1994)
KW1<-148.96502-13847.26/TK-23.65218*log(TK)
KW2<-(118.67/TK-5.977+1.0495*log(TK))*s ^0.5-0.01615*s
KW<-exp(KW1+KW2)
##############################################
# solve for H ion (Zeebe and Wolf-Gladrow, 2000)
a1 <-1
a2 <-(alk+KB+K1);
a3 <-(alk*KB-KB*Tbor-KW+alk*K1+K1*KB+K1*K2-dic*K1)
a4 <-(-KW*KB+alk*KB*K1-KB*Tbor*K1-KW*K1+alk*K1*K2
    +KB*K1*K2-dic*KB*K1-2*dic*K1*K2)
a5 <- (-KW*KB*K1+alk*KB*K1*K2-KW*K1*K2-KB*Tbor*K1*K2-2*dic*KB*K1*K2)
a6 <- -KB*KW*K1*K2
p <- c(a1, a2, a3, a4, a5, a6)
#p <- c(a6, a5, a4, a3, a2, a1)
r <- roots(p) # Note: polyroot(p) requires coeff's in reverse order
h <- max(Real(r)) # Need the largest real root!
#
# calculate the HCO3, CO3 and CO2aq using DIC, AlK and H+
hco3<-dic/(1+h/K1+K2/h)
co3<-dic/(1+h/K2+h*h/(K1*K2))
co2<-dic/(1+K1/h+K1*K2/(h*h))
fco2<-co2 / KH
pH<--log10(h)
# calculate B(OH)4 and OH
BOH4<-KB*Tbor/(h+KB)
oh<-KW/h
BOH3<-Tbor-BOH4
# Calculate the Revelle factor
# Ref: Egleston, et al, (2010). Revelle revisited: Buffer factors that
#      quantify the response of ocean chemistry to changes in DIC and
#      alkalinity, 'Global Biogeochemical Cycles, vol 24, 1-9.
ATc  <- (2 * co3 + hco3) # alkalinity due to carbonate species
Gdic <- (hco3 + 4*co3 + h*BOH4/(KB+h) + h + oh)
R    <- dic/(dic - ATc^2/Gdic)
#
```

```
       return(list(fco2=fco2, pH=pH, co2=co2, hco3=hco3, co3=co3,
               h=h, oh=oh, BOH3=BOH3, BOH4=BOH4,R=R))
}
```

Listing 11.4. File: `carbEq.R`—Code of function `carbEq` to solve the carbonate system equations

11.2.5 Acidity of Surface Seawater

The effect on increased CO_2 concentrations in the atmosphere is to increase the level of dissolved CO_2 in the surface seawater. This increase in dissolved CO_2 tends to reduce the pH value and make the seawater more acid. The symbol "pH" stands for *potential of hydrogen* and is a measure of the acidity or alkalinity of a solution. Aqueous solutions at 25°C with a pH less than 7 are considered acidic, whereas those with a pH greater than 7 are considered basic (alkaline). A pH value of 7.0 is considered neutral at 25°C, because at this pH the concentration of H_3O^+ approximately equals the concentration of OH^- in pure water. pH can be calculated from the following simple equation:

$$\text{pH} = -\log_{10}[H^+], \tag{11.30}$$

where the concentration of positive hydrogen ions $[H^+]$ has units of (mol/liter).

11.2.5.1 Organic Carbon Pump

The so-called *organic carbon pump* results in changes in carbon fixation to particulate organic carbon (POC) in surface waters by *photosynthesis* and export of this carbon through *sinking of organic particles* out of the surface layer—this process is limited to first order by availability of light and nutrients (phosphate, nitrate, silicic acid, and micronutrients such as iron). Organic particles are remineralized (oxidized to dissolved inorganic carbon (DIC) and other inorganic compounds through the action of bacteria, primarily in the upper 1000 m of the ocean, with an accompanying decrease in dissolved oxygen. Upwelling of carbon from the deep ocean results an increase in CO_2 in surface seawater, which is subsequently released back into the atmosphere. Although the POC reservoir is small, it plays an important role in keeping DIC concentrations low in surface waters and high in deep waters (see [IPCC-07, Chapter 7] for more details). For our model this process is considered to be more or less in balance with the photosynthesis of land based vegetation such that the net exchange will be small compared with the uncertainty of fossil fuel emission predictions.

11.2.5.2 $CaCO_3$ Counterpump

The *calcium carbonate*, $CaCO_3$, *counterpump*—so-called because it exerts an opposite effect on atmospheric CO_2 to the organic carbon pump (also known as the *alkalinity, hard tissue*, or *carbonate pump*)—results in the release of CO_2 in surface waters during formation of calcium carbonate shell material by plankton. In general, $CaCO_3$ particles sink deeper than POC before they undergo dissolution, as deep waters are undersaturated with respect to $CaCO_3$. The remainder of the particle flux enters marine sediments and is subject to either redissolution or accumulation within the sediments. Upwelling from the deep ocean results in significant dissolution of $CaCO_3$ occurring in the upper

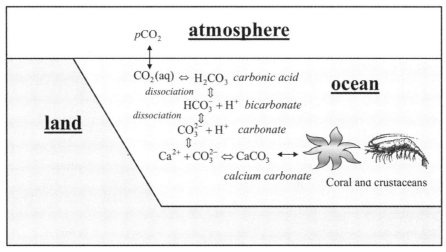

Figure 11.1. Diagram showing chemical route from atmospheric CO₂ to CaCO₃ in the ocean.

1000 m of the ocean under conditions oversaturated with respect to *calcite* or *aragonite*, the two main forms of $CaCO_3$.

Saturation applies to a solution containing just sufficient of a substance to achieve equilibrium. The saturation state Ω of seawater with respect to $CaCO_3$ is determined from

$$\Omega = [Ca^{2+}][CO_3^{2-}]/K'_{sp}. \tag{11.31}$$

A solution is in *equilibrium* when $\Omega = 1$ and is *supersaturated* when $\Omega > 1$. If Ω falls below 1 the solution is *undersaturated*, and this can be caused by falling pH values. The effect of decreasing pH in the oceans is therefore to move coral and the shells of crustacean animals closer to dissolution. Supersaturated waters should, theoretically, precipitate minerals; however, in practice, this has been found to be a rare occurrence in seawater (see [Eme-08, pp. 420–421] for further discussion). Of the two forms of $CaCO_3$, aragonite is more easily dissolved due to the higher value of its apparent solubility constant K'_{sp}. Consequently, for a given concentration of CO_3^{2-}, we have $\Omega_{ar} < \Omega_{ca}$.

This process increases the ability of surface seawater to absorb atmospheric CO_2 (see [IPCC-07, Chapter 7] for more details). For our model this process is considered to be more or less in balance with the net exchange being small compared with the uncertainty of fossil fuel emission predictions.

11.2.6 Ocean Circulation

The ocean is stratified into progressively dense layers as one descends from the surface to deep water, and circulation results from thermal and density gradients, *Ekman transport* (which causes upwelling along sloping density layers from the deep ocean to the surface), and westerly winds combined with the Coriolis force.[3] Ocean circulation is a complex and

---

[3] A force describing the acceleration acting on particles on a rotating sphere.

intermittent process that is not entirely understood at present. A good discussion can be found in [Mor-15].

11.2.6.1 Ocean Layers
Oceans are often described by reference to surface, intermediate, and deep-water layers, where

- *surface* water or the *mixed* layer has a more or less uniform temperature and is the warmest layer.
- *intermediate* water or the *thermocline* layer has a temperature gradient from the warm surface to the cold deep water
- *deep* water layer has a more or less uniform temperature and is the coldest layer

For our purposes, the model divides the ocean into two layers, surface water, and deep water, with circulation between the layers. The discussion assumes that each of these layers can be represented by a single reservoir with an average temperature. Although this is a gross simplification, the simulation does produce surprisingly good representative values (see results in Section 11.3).

11.2.6.2 Upwelling and Downwelling
Upwelling is the upward flow of subsurface waters to the surface, whereas *downwelling* is the movement of surface waters to the subsurface. Most of the upwelling in the ocean occurs in two forms: coastal upwelling along the shores subject to boundary currents and equatorial upwelling along the equator. Downwelling is the opposite of upwelling and is a direct consequence of *conservation of mass* and *convection* phenomena. The upwelling and downwelling processes create the so-called *thermohaline* circulation or *global oceanic conveyor belt*, which distributes heat and salt throughout the Earth's oceans.

11.2.7 Emission Profiles

Realistic *carbon dioxide emission profile* scenarios are important for establishing bounds on the rise in atmospheric CO_2 concentrations. This aspect is discussed briefly subsequently, along with various emission scenarios based on outstanding fossil fuel reserves.

11.2.7.1 Profile Calculations
The following *bell-shaped* model for defining the *buildup* and *fall-off* of carbon emission flux was proposed by Caldeira and Wickett [Cal-05]:

$$\frac{dF(t)}{dt} = aF(t)(F_0 - F(t)), \quad t \geq t_1, \tag{11.32}$$

where $F(t)$ represents the *fossil fuel reserves* at time t and F_0 is the *estimated fossil fuel reserves* at some time t_0, in the past. The constant a is chosen so that the rate of fossil fuel emissions $\frac{dF(t)}{dt}$ is equal to the known emission flux value at the start of the calculation sequence.

Table 11.3. Reserves for years 1750 and 2010 and emissions values for year 2010, together with constants for eqns. (11.32) and (11.34)

Variable	Reserves and emission scenario values			
$F(1750)$	1365	2365	3365	4365
$F(2010)$	1000	2000	3000	4000
$dF(2010)/dt$	9.167	9.167	9.167	9.167
a	2.51E-05	1.26E-05	8.38E-06	6.28E-06
b	3.05E-34	8.83E-31	8.92E-30	2.39E-29

Note: Symbols are defined in the text. Reserves are given in GtC and emissions in GtC/y. To convert between mass of carbon and corresponding mass of carbon dioxide, multiply by the ratio of molecular weights, that is, $MWCO_2/MWC = 44/12 = 3.67$.

The cumulative total carbon emission can be obtained by integrating eqn. (11.32) by the *variable separable* method, that is,

$$\int \frac{dF}{aF(F_0 - F)} = \int dt, \qquad (11.33)$$

to give

$$F(t) = \frac{F_0}{1 + bF_0 \exp(aF_0 t)}, \quad t \geq t_1, \qquad (11.34)$$

where the constant of integration b is chosen so that the reserves at year t_1 are equal to $F(t_1)$.

Notes: 1. In eqn. (11.32) the term $F_0 - F(t)$ represents cumulative emissions from year F_0 to year $F(t)$.
2. Reserves and emission values prior to t_1 are based on historic data.

11.2.7.2 Year 2010 Fossil Fuel Emission Scenarios

The preceding method will now be used to estimate emission profiles for years 2010 onward. Historic data for the years 1751 to 2010, when integrated, give cumulative emissions of 364 GtC for this period [Bod-13]. We will use four emission scenarios for our model and take the fossil fuel reserves at year 2010 to be, respectively, 1000, 2000, 3000, and 4000 GtC. As an example, we will calculate the profile for the 1000 GTC reserve to illustrate the method and use the following data:

$$t_0 = 1750, \; t_1 = 2010,$$
$$F_0 \simeq 1364 \text{ GtC}, \; F(t_1) = 1000 \text{ GtC}, \; \frac{dF(t_1)}{dt} = 9.167 \text{ GtC/y}. \qquad (11.35)$$

Thus, from eqn. (11.32), we get $a = 2.51 \times 10^{-5}$. Similarly, using the same data along with the preceding value for a, we find that constant $b = 3.05 \times 10^{-34}$. A similar approach can be used to arrive at a variety of emission profiles. Thus, adopting the same procedure for other assumed reserves at year 2010, we arrive at appropriate values for the corresponding constants a and b (see Table 11.3).

Flux and fossil fuel reserve values based on the above constants have been plotted in Fig. 11.2 from year 1750 to 2500.

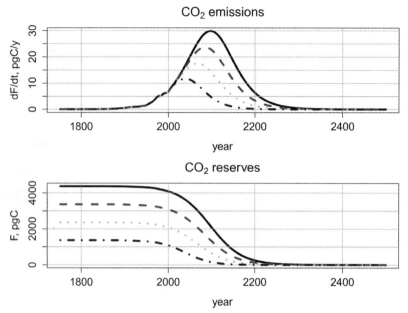

Figure 11.2. Emission scenario plots showing (top) fossil fuel *emission rate* rising to a peak and then falling off to zero and (bottom) *reserve* profiles initially at maximum value at year 1750 and then noticeably starting to fall off around year 1900 as reserves become depleted. The curves represent reserves in Gt of carbon: 4000 (solid), 3000 (dashed), 2000 (dots), and 1000 (dash-dotted) at year 2010. Values from 1750 to year 2010 are based on historical data.

A full set of tabulated emissions data covering historical values from 1751 to year 2010 is provided as a .csv file with the downloads. These data are interpolated using the R function interp1 to drive the carbon model up to year 2010, after which eqn. (11.34) with appropriate parameters from Table 11.3 is used for forward projections.

The preceding calculations for determining fossil fuuel emission for a particular year are performed by the R computer code given in Listing 11.5.

```
# File: carbonCycle_fossilFuel.R
carbonCycle_fossilFuel<-function(t,case){
  #a1 <- 263.889; b1 <- -0.576518; c1 <- 0.000368699; d1 <- -7.29599e-08
  a <- c(6.283501e-06, 8.378002e-06, 1.256700e-05, 2.513401e-05)
  c <- c(2.394043e-29, 8.917246e-30, 8.826662e-31, 3.051255e-34)
  F_2010 <- c(4000,3000,2000,1000)
  F_1750 <- c(4364.725, 3364.725, 2364.725, 1364.725)
  #
  if(t<2010){
    if(t<ff1[1,1]){
      ef <- ff1[1,2]/1000 # (GtC/y)
    }else{
      ef <- interp1(x=ff1[,1],y=ff1[,2], xi=t,method="linear")/1000 # (GtC/y)
    }
  }else{
```

```
    ff <- F_1750[case]/(1+c[case]*F_1750[case]*exp(a[case]*F_1750[case]*t))
    #print(ff)
    ef <- a[case]*ff*(F_1750[case]-ff)
    #print(ef)
  }
  return(ef)
}
```

Listing 11.5. File: `carbonCycle_fossilFuel.R`—Code of function `fossilFuel` that returns the fossil fuel emissions at a given year for a particular emissions/reserves scenario. This function is called by function `carbonCycle_deriv()` and the code in file `carbonCycle_postSimCalcs.R`.

11.2.8 Earth's Radiant Energy Balance

The overall energy balance of the Earth is illustrated in Fig. 11.3, where all the major radiation energy flow paths are detailed.

11.2.8.1 Radiation from the Sun

Max Planck discovered the following *radiation energy spectrum density* relationships for a *black body* at a temperature T (K), which are given in terms of *wavelength* and *frequency*,

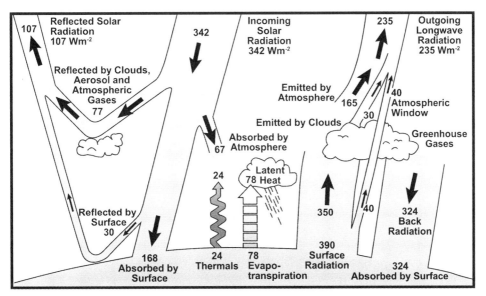

Figure 11.3. Estimated annual and global mean energy balance for the Earth. Over the long term, the amount of incoming solar radiation absorbed by the Earth and atmosphere is balanced by the Earth and atmosphere releasing the same amount of outgoing long wave radiation. About half of the incoming solar radiation is absorbed by the Earth's surface. This energy is transferred to the atmosphere by warming the air in contact with the surface (thermals), by evapotranspiration, and by long wave radiation that is absorbed by clouds and greenhouse gases. The atmosphere in turn radiates long wave energy back to Earth as well as out to space. Adapted from *Climate Change 2007* [IPCC-07].

respectively:

$$u_\lambda(T) = \frac{8\pi hc}{\lambda^5} \frac{1}{e^{hc/\lambda kT} - 1}, \quad \left(\text{J m}^{-3}/\text{m}\right) \quad \text{or}$$

$$u_\nu(T) = \frac{8\pi h\nu^3}{c^3} \frac{1}{e^{h\nu/kT} - 1}, \quad \left(\text{J m}^{-3}/\text{Hz}\right).$$

(11.36)

Planck's radiation equations (11.36) are in the original form of his seminal paper [Pla-00], where *Planck's constant* $h = 6.626 \times 10^{-34}$ (J s) and *Boltzmann's constant* $k = 1.38066 \times 10^{-23}$ (J/K). Using the relationships $\nu = c/\lambda$, $u_\lambda d\lambda = -\nu d\nu$ and $d\lambda = -\left(\frac{2}{\lambda^2}\right) d\nu$ the second of eqns. (11.36) follows from the first. For our purposes it is convenient to convert from energy density per unit volume u, to power intensity per unit surface area, $B(T)$; since $u = 4B(T)/c$ [Hot-67], we obtain[4]

$$B_\lambda(T) = \frac{2\pi hc^2}{\lambda^5} \frac{1}{e^{hc/\lambda kT} - 1}, \quad \left(\text{W m}^{-2}/\text{m}\right) \quad \text{or}$$

$$B_\nu(T) = \frac{2\pi h\nu^3}{c^2} \frac{1}{e^{h\nu/kT} - 1}, \quad \left(\text{W m}^{-2}/\text{Hz}\right).$$

(11.37)

If we integrate the first of eqns. (11.37) over the entire spectrum, we obtain the *Stefan–Boltzmann law* [Ste-79, Bol-84]

$$F(T) = \int_0^\infty B_\lambda(T) d\lambda = \sigma T^4, \quad \left(\text{W m}^{-2}\right)$$

(11.38)

and find that $\sigma = \frac{2\pi^5 k^4}{15 c^2 h^3} = 5.6704 \times 10^{-8}$, which is known as the Stefan–Boltzmann constant. Similarly, if we differentiate eqn. (11.37) with respect to λ, then on setting $dB_\lambda(T)/d\lambda$ equal to zero and solving for λ, we obtain *Wien's displacement law*[5] [Wie-93]. This law states that, for a blackbody, the wavelength at which peak radiant emission occurs is inversely proportional to temperature, that is,

$$\lambda^p = a/T, \quad (\text{m}),$$

(11.39)

where $a = 2.8977685 \times 10^{-3}$. A summary of electromagnetic spectrum data is given in Section 11.9.

The total rate of energy radiated per unit area from the surface of a blackbody is governed by the Stefan–Boltzmann law which, for a *graybody* that does not emit the full amount of radiant power, becomes

$$F = \varepsilon \sigma T^4, \quad \left(\text{W m}^{-2}\right),$$

(11.40)

where ε is *surface emissivity* ($\varepsilon = 1$ for an ideal radiator, i.e., a *blackbody*). The *total radiant power*, also known as *radiant emittance, radiant exitance*, or *radiant flux*, is given by

$$\Phi = \varepsilon \sigma A T^4, \quad (\text{W})$$

(11.41)

where A represents the surface area of the radiant body.

---

[4] The spectral power intensity per unit surface area is often given in units per solid angle when eqn. (11.37) becomes $B_\lambda(T) = \frac{2hc^2}{\lambda^5} \frac{1}{e^{hc/\lambda kT}-1}$ (W m^{-2}sr^{-1}/m), or $B_\nu(T) = \frac{2h\nu^3}{c^2} \frac{1}{e^{h\nu/kT}-1}$ (W m^{-2}/Hz), where sr = steradian.

[5] *Wien* is pronounced "veen."

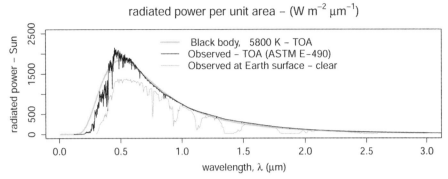

Figure 11.4. Idealized radiant power frequency spectrum of the Sun at TOA with ASTM observed TOA data and observed radiation reaching the Earth's surface (clear day) superimposed.

Now we can use the preceding equations to model the heat from the Sun. A reasonable assumption is that its radiation results from a surface temperature of 5800 K[6] (see Fig. 11.4), and that it has a high emissivity of, say, $\varepsilon_{Sun} = 0.986$ and a mean diameter of $r_{Sun} = 1,392 \times 10^6$ m. From these parameters we calculate that the Sun radiates energy at the rate of

$$\Phi_{Sun} \simeq 3.85 \times 10^{26}, \quad (W). \tag{11.42}$$

From Wien's displacement law we calculate that *peak radiant emission* occurs at the following wavelength and frequency

$$\lambda^p_{Sun} = 0.499, \quad (\mu m), \quad \nu^p_{Sun} = 6.0 \times 10^{14}, \quad (Hz). \tag{11.43}$$

See Appendix 11.A.6 at the end of this chapter for spectrum wavelength/frequency ranges.

The radiative power intensity reduces over distance due to the *inverse-square-law* effect, so we also need to determine the reduced power of the Sun's radiation when it actually strikes *normal* to the Earth. To obtain this value,[7] we multiply by $\left(\frac{r_{Sun}}{R}\right)^2$, which represents the radiation *view factor* or *exchange area* [Hot-67], and where $R = 149 \times 10^6$ km is the mean distance from Sun to Earth. A plot of the (idealized) blackbody radiation power spectrum for the Sun (per unit area) is given in Fig. 11.4 along with ASTM E-490 observed data [ASTM-05] and the radiation that reaches the Earth's surface on a clear day.

The difference between the *top of atmosphere* (TOA) and surface plots is equal to the solar irradiance absorption by the atmosphere and represents approximately 67 W/m² (see Fig. 11.3).

---

[6] However, it should be noted that as a result of immense pressure due to gravity, *nuclear reactions* take place within the Sun's interior. These reactions generate energy through the *fusion of hydrogen atoms into helium*, which raises the temperature at the Sun's core to an estimated 15.5 million K.

[7] Usually taken to mean the value at the *top of the atmosphere* (TOA).

To arrive at the total rate of energy from the Sun that irradiates the Earth (per unit area), the so-called *solar constant* S_O, or *solar irradiance*,[8] we employ the *Stefan–Boltzmann* eqn. (11.38) and calculate that

$$S_O = \varepsilon_{\text{Sun}} \sigma \left(\frac{r_{\text{Sun}}}{R}\right)^2 T_{\text{Sun}}^4 \simeq 1368, \quad (\text{W m}^2). \tag{11.44}$$

Because the radiation power of the Sun falls off according to the inverse square of distance, we have included the term $\left(\frac{r_{\text{Sun}}}{R}\right)^2$ when calculating S_o, as discussed earlier. Also, because this figure applies to radiation striking an equivalent *disk* of diameter equal to that of the Earth, the average radiant energy striking the entire Earth's *spherical* surface, *at top of atmosphere*, is given by

$$F_{\text{TOA}} = S_O \times \frac{\pi r_{\text{Earth}}^2}{4\pi r_{\text{Earth}}^2} = 342, \quad (\text{W m}^{-2}). \tag{11.45}$$

If we integrate the idealized Sun's blackbody radiation power at the TOA over all wavelengths, we obtain the same figure of 1368 W/m² that we obtained from the *Stefan–Boltzmann* equation, as expected.

11.2.8.2 Radiation from the Earth

The Earth has an *effective albedo* of $a \simeq 0.3$ ($a = \frac{\text{reflected solar radiation}}{\text{incident solar radiation}}$), which means that it absorbs approximately 70% of the radiation energy that it receives, the other 30% being reflected back into space. To maintain an equilibrium, the Earth must re-radiate the same amount of energy that it absorbs averaged over time. Thus, we can now use the Stefan–Boltzmann equation to estimate the *effective* blackbody temperature of the Earth at TOA. Note that we use the term *effective temperature* rather than *average temperature* when calculating radiant energy due to the nonlinear relationship between temperature and radiated energy, as defined by eqn. (11.40). The effective temperature of a body with surface area A is given by

$$T_{\text{eff}} = \frac{1}{A} \left(\oiint_A T^4 dA \right)^{\frac{1}{4}}, \tag{11.46}$$

whereas the average temperature is given by

$$T_{\text{ave}} = \frac{1}{A} \left(\oiint_A T dA \right). \tag{11.47}$$

However, if the radiant energy is also known over the surface, then a simple average can be calculated and used together with T_{eff} to obtain the *effective* emittance of the surface.

---

[8] This figure for the *solar constant* is consistent with the value reported by the IPCC [IPCC-07, Chapter 2] (±3.4% due to the eccentricity of Earth's orbit around Sun and changes in the tilt of the Earth's axis), also confirmed by NASA from satellite measured yearly average data. The NASA SOlar Radiation & Climate Experiment (SORCE) uses the Total Irradiance Monitor (TIM), launched in 2003, as part of its satellite monitoring of Solar radiation [Kop-04, TIM-07].

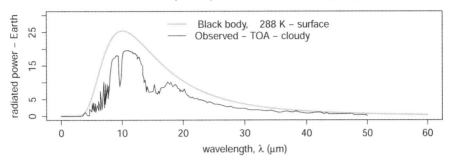

Figure 11.5. Idealized frequency spectrum of the Earth's surface radiant emittance, assuming blackbody radiation at 288 K, with average observed data at TOA for cloudy conditions superimposed. The dips in the observed spectrum represent absorption bands corresponding to various chemicals that make up the atmosphere. The wide dip centered at around 15 μm corresponds to CO_2 and the narrow one at 9 μm to O_3 (ozone). Adapted from [Kie-97].

The radiant flux and effective temperature of the Earth is calculated as follows:

$$\Phi_{\text{Earth}} = \varepsilon_{\text{TOA}} \sigma A_{\text{Earth}} T^4_{\text{Earth,TOA}} = (1-a) A_n S_O,$$
$$\Downarrow$$
$$T_{\text{Earth,TOA}} = \sqrt[4]{\frac{1}{4} \frac{(1-a) S_O}{\varepsilon_{\text{TOA}} \sigma}} \simeq 255 \text{ K,} \qquad (11.48)$$

where A_n represents the area of the Earth normal to the radiation. Thus, it is assumed incident radiation is equivalent to that striking a disc of diameter equal to that of the Earth and that radiation from the Earth is emitted uniformly at TOA. We have also assumed an effective emissivity $\varepsilon_{\text{TOA}} = 1.0$ over the Earth's surface, which gives an effective surface temperature of 255 K. The Earth's atmosphere therefore radiates at its effective blackbody temperature of 255 K, and this equates to an average radiation emittance of 239 W/m² compared with the IPCC figure of 235 W/m² (see Fig. 11.3). From eqn. (11.39) we calculate that its peak radiation occurs in the infrared spectrum at a wavelength $\lambda^p_{\text{Earth,TOA}} = 11.4$ μm with corresponding frequency $\nu^p_{\text{Earth,TOA}} = 0.264 \times 10^{14}$ Hz.

Now the effective temperature at the Earth's surface is $\sim$288 K, and from eqn. (11.39) we calculate that its peak radiation occurs in the infrared spectrum at a wavelength $\lambda^p_{\text{Earth,Surf}} = 10.1$ μm with corresponding frequency $\nu^p_{\text{Earth,Surf}} = 0.298 \times 10^{14}$ Hz.

A plot of the Earth's (idealized) blackbody radiation spectrum at the surface is given in Fig. 11.5 with corresponding global cloudy data, as observed at TOA, superimposed [Kie-97]. The curve for a noncloudy day would be closer to the blackbody curve. The difference between the surface and TOA plots represents infrared absorption by the atmosphere.

Integrating the idealized blackbody radiation power at the Earth's surface over all wavelengths gives a figure of 389.27 W/m². This figure is in agreement with the value 390 W/m² reported by the IPCC [IPCC-07, Chapter 2].

For comparison purposes, *radiation emittance* plots at the surface of the Sun and Earth are shown superimposed in Fig. 11.6, where the terrestrial radiation has been scaled up by a factor of 10^6.

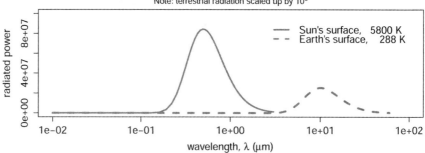

Figure 11.6. Surface solar and terrestrial radiation emittance spectra.

11.2.9 How the Atmosphere is Affected by Radiation

To appreciate how the Sun's radiation warms the Earth, we need to understand some of the underlying processes at work. We will not attempt a rigorous detailed analysis here but will introduce the following primary concepts: *Kirchoff's law*, *Beer's law*, and *air mass*.

11.2.9.1 Kirchoff's Law

In general, a body will not only *absorb* but will also *reflect* part of the incident radiation and *transmit* the remainder [Pei-92]. Thus, at wavelength λ, for a layer of gas or atmosphere, we have the following:

$$\alpha_\lambda + r_\lambda + \tau_\lambda = 1, \tag{11.49}$$

where $\alpha_\lambda = I_{\lambda a}/I_\lambda$ represents *absorptivity*, $r_\lambda = I_{\lambda r}/I_\lambda$ represents *reflectivity* (or *albedo*), and $\tau_\lambda = I_{\lambda \tau}/I_\lambda$ represents *transmitivity* of the layer.

The emissivity ϵ_λ of a body at a particular wavelength is defined as the ratio of the emitted intensity to the Planck function. *Kirchoff's law* states that

> in thermodynamic equilibrium and at a given wavelength the ratio of the intensity of emission I_λ to the absorptivity α_λ of any substance does not depend upon the nature of the substance. It is a function of temperature and wavelength only, i.e.

$$I_\lambda/\alpha_\lambda = f(\lambda, T). \tag{11.50}$$

Kirchoff's law can also expressed as

$$\alpha_\lambda = \epsilon_\lambda. \tag{11.51}$$

For situations where it can be assumed that α_λ is the same for all wavelengths, then the body is defined as being a *graybody*.

11.2.9.2 Beer's Law

The rate of scattering or absorption of incident radiation I_λ (W/m^2) in a gaseous medium can be expressed as

$$dI_\lambda = -I_\lambda \rho c k_\lambda m \, dz, \tag{11.52}$$

where

- k_λ = mass absorption coefficient for species (m²/kg)
- m = air mass factor (−) (see Section 11.2.9.3)
- ρ = density of medium (kg/m³)
- c = mass concentration of absorbing species (−)
- z = thickness of medium (m)

For an atmosphere consisting of a number of chemical species, the individual contributions are additive, that is,

$$ck_\lambda = \sum c_i k_{\lambda_i}. \tag{11.53}$$

On integration through the atmosphere, we obtain

$$-\int \frac{dI_\lambda}{I_\lambda} = \ln I_{\lambda\infty} - \ln I_\lambda = m \int_0^b \rho c k_\lambda dz$$
$$\Downarrow \tag{11.54}$$
$$I_\lambda = I_{\lambda\infty} e^{-m u_\lambda} = I_{\lambda\infty} \tau_\lambda,$$

where $I_{\lambda\infty}$ represents intensity of incident radiation *prior to absorption* and the *slant path optical thickness* u_λ is related to the *thickness of the atmosphere b* by

$$u_\lambda = \int_0^b \rho c k_\lambda dz. \tag{11.55}$$

The *transmissivity* of the layer is defined by

$$\tau_\lambda = \frac{I_\lambda}{I_{\lambda\infty}} = e^{-m u_\lambda}, \tag{11.56}$$

and the corresponding *absorptivity* is given by

$$\alpha_\lambda = 1 - \tau_\lambda = 1 - e^{-m u_\lambda}. \tag{11.57}$$

Beer's law, also known as the *Beer–Lambert law* or *Beer–Bouguer–Lambert law*, can be applied to describe the *attenuation* of solar (or stellar) radiation as it travels through the Earth's atmosphere. However, in this case, there is *scattering* as well as *absorption*, and it can be written as

$$I_\lambda = I_{\lambda\infty} e^{-m(u_a + u_g + u_{NO_2} + u_w + u_{O_3} + u_r)}, \tag{11.58}$$

where the various optical thicknesses (or depths) are defined in Table 11.4.

The concentration of greenhouse gases can make the atmosphere largely opaque in a particular band. If the atmosphere absorbs 100% of the radiation in a band, the absorption will not be increased when additional greenhouse gases are added. The atmosphere would then be said to be saturated in that particular frequency band.

Although the preceding deals properly with atmospheric absorption of radiation by greenhouse gases, the effect of changes in cloudiness on global temperatures are more important than the changes in the atmospheric greenhouse gases. For example, when we have a clear night sky, the atmospheric temperature falls significantly compared to a situation of dense cloud cover. As the greenhouse gas composition in the atmosphere is the same regardless of cloud cover, the difference results from the degree of cloudiness.

Table 11.4. Definition of symbols associated with eqn. (11.58)

Symbol	Description
u_a	aerosols (absorbtion and scattering)
u_g	uniformly mixed gases (mainly carbon dioxide CO_2 and molecular oxygen O_2 absorbtion)
u_{NO_2}	nitrogen dioxide, mainly due to urban pollution (absorption)
u_w	water vapor (absorption)
u_{O_3}	ozone (absorption)
u_r	Rayleigh scattering from molecular oxygen O_2 and nitrogen N_2 (responsible for the blue color of the sky)

Also, overlapping of spectral bands complicates the process; however, this aspect will not be discussed here, but further detailed discussion on this and other aspects relating to the transmission of radiation in the atmosphere can be found in [Lio-02] and [Jac-05, Chapter 9].

11.2.9.3 Air Mass

The term *airmass* is used to represent the relative path length of the direct solar beam radiance through the atmosphere. When the sun is directly above a sea-level location the path length is defined as airmass 1 (AM 1.0). AM 1.0 is not synonymous with solar noon because the sun is usually not directly overhead at solar noon in most seasons and locations. When the angle of the sun from zenith (directly overhead) increases, the airmass increases approximately by the secant of the zenith angle (see Fig. 11.7).

A reasonably accurate calculation from Kasten and Young [Kas-89] for a zenith angle up to 90° is given by

$$AM = \frac{1}{\cos(z) + 0.50572 \times \left(96.07995 - \frac{180}{\pi}z\right)^{-1.6364}}, \quad (11.59)$$

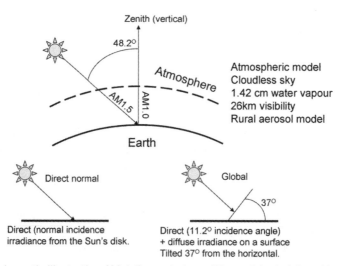

Figure 11.7. Schematic illustrating AM 1.5 spectral conditions. Adapted from Kasten and Young [Kas-89].

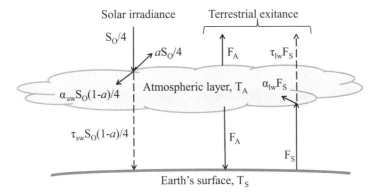

Key: a = albedo, α = absorptivity, τ = transmissivity
S_O = solar constant (W/m²), T = temperature, (K)
lw = long wave, sw = short wave, S = surface, A = atmosphere

Figure 11.8. Simplified single-layer radiation energy balance for the Earth.

where z is the *solar zenith angle*. A high-accuracy calculation by Young [You-94] is

$$AM = \frac{1.002432\cos^2 z + 0.148386 \cos z + 0.0096467}{\cos^3 z + 0.149864 \cos^2 + 0.0102963 \cos z + 0.000303978}. \tag{11.60}$$

11.2.9.4 Single-Layer Radiation Model

We will now provide a simple illustration of a single-layer model of the radiation balance for the Earth based on Fig. 11.8. The value of the solar irradiance incidence on the Earth was determined above as $S_O/4$. The albedo effect causes $aS_O/4$ to be reflected back into space, with the remaining, $S_O(1-a)/4$, actually left to warm the Earth.

If we now perform an overall radiation energy balance on the Earth surface, we have

$$\tau_{sw}(1-a)S_O/4 + F_A = F_S. \tag{11.61}$$

Similarly, an energy balance over the atmospheric layer gives

$$\alpha_{sw}(1-a)S_O/4 + \alpha_{lw}F_G = 2F_A, \tag{11.62}$$

which, on substituting into eqn. (11.61), yields

$$F_A = \frac{(1-a)S_O(1-\tau_{sw}\tau_{lw})}{4(1+\tau_{lw})}$$
$$F_S = \frac{(1-a)S_O(1+\tau_{sw})}{4(1+\tau_{lw})}. \tag{11.63}$$

To arrive at eqn. (11.63), we have used the following Kirchoff relationship, assuming that, within the atmosphere, reflectivity (albedo) $r = 0$:

$$\alpha_\lambda + r_\lambda + \tau_\lambda = 1, \tag{11.64}$$

where $\alpha_\lambda = I_{\lambda a}/I_\lambda$ represents *absorptivity*, $r_\lambda = I_{\lambda r}/I_\lambda$ represents *reflectivity* (or albedo), and $\tau_\lambda = I_{\lambda \tau}/I_\lambda$ represents *transmitivity* of the layer.

Example: Given the effective global values for the single-layer model constants $S_O = 1370$ W/m², $a = 0.3$, and $\sigma = 5.6704 \times 10^{-8}$ and the reasonable values for transmissivity $\tau_{lw} = 0.2$ and $\tau_{sw} = 0.9$, calculate the corresponding radiant flux and associated temperature values, and also the unreflected solar radiance.

Using the preceding equations along with the Stephan–Boltzmann relationship of eqn. (11.40), we find the radiant fluxes to be

$$F_A = 163.8 \text{ W m}^{-2}, \quad F_S = 379.6 \text{ W m}^{-2},$$

where the associated temperatures are

$$T_A = 245.1 \text{ K}, \quad T_S = 286.0 \text{ K}.$$

The mean unreflected solar irradiance is given by

$$F_O = \frac{1}{4}(1-a)S_O = F_A + \tau_{lw}F_S = 239.8 \text{ W m}^{-2}. \tag{11.65}$$

If the atmosphere were entirely opaque to infrared and solar radiation (zero transmittance), we would have $\tau_{lw} = \tau_{sw} = 0$. Then, assuming no change in albedo, this simplified model gives $T_A = 255$ K and $T_S = 255$ K, which is what we would expect (see eqn. 11.48).

It should be borne in mind that these figures are for the effective global situation. In reality the situation is quite diverse. Consider the polar ice caps with surfaces that can fall to $-70°$C, and desert sand that can reach $65°$C; surfaces at these locations would radiate approximately 94 and 564 W/m², respectively—quite a range. These figures are based upon emissivities of $\epsilon = 0.97$ for ice and $\epsilon = 0.76$ for sand.

Note that the preceding simplified analysis ignores energy transfer mechanisms, other than radiation, that should be included for a complete description of the problem, for example, convection, conduction, and phase change.

11.2.9.5 Climate Sensitivity

The sensitivity of the Earth's effective global surface temperature, T_s, to the radiative forcing, F, is defined as

$$\lambda = \frac{dT_s}{dF}, \quad (\text{K W}^{-1}\text{m}^2). \tag{11.66}$$

The quantity λ is known as climate sensitivity [Wall-06, pp. 444–445] and includes various climate feedback effects, such as atmospheric water vapor, fraction of snow/ice cover, and cloud cover, that affect the surface temperature, T_s (see Fig. 11.9).

These feedback effects can each be represented by a variable y_i, such that, on expanding the total derivative of eqn. (11.66), we obtain

$$\lambda = \frac{dT_s}{dF} = \frac{\partial T_s}{\partial F} + \sum_i \frac{\partial T_s}{\partial y_i}\frac{dy_i}{dF}, \tag{11.67}$$

where the summation term represents all the feedback contributions.

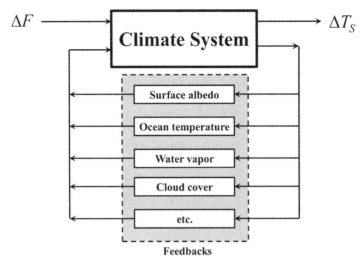

Figure 11.9. Schematic showing climate feedbacks, as represented by the summation term of eqn. (11.67), where ΔT_S represents the overall change in surface temperature due to ΔF radiative forcing brought about by a change in climatic conditions.

The partial derivative term $\partial T_s/\partial F$ represents the climate sensitivity λ_0 under conditions of zero feedback terms. Thus, we have

$$\lambda_0 = \frac{\partial T_s}{\partial F} \approx \frac{\partial T_E}{\partial F}, \qquad (11.68)$$

where T_E is the equivalent blackbody temperature (refer to Section 11.2.8.1).

But from basic calculus we have $\frac{dy_i}{dF} = \frac{dy_i}{dT_s}\frac{dT_s}{dF}$. Therefore, eqn. (11.67) becomes

$$\lambda = \frac{dT_s}{dF} = \frac{\partial T_s}{\partial F} + \frac{dT_s}{dF}\sum_i f_i, \qquad (11.69)$$

where the *i*th *feedback factor* f_i is defined as

$$f_i = \frac{\partial T_s}{\partial y_i}\frac{dy_i}{dT_s}. \qquad (11.70)$$

Solving for the sensitivity of T_s with respect to the forcing F, we obtain

$$\frac{dT_s}{dF} = \frac{\partial T_s/\partial F}{1-f}, \qquad (11.71)$$

where $f = \sum_i f_i$.

It therefore follows that the system *gain* due to the cumulative effects of *all* climate feedbacks is given by

$$g = \frac{\lambda}{\lambda_0} = \frac{1}{1-f}, \quad f < 1. \qquad (11.72)$$

If f is positive due to, say, an increase in greenhouse gases, then $g > 1$ (warming); and if f is negative due to, say, increased albedo, then $g < 1$ (cooling).

Now, from the Stefan–Boltzmann law, we have

$$T_E = \left(\frac{F_s}{\sigma}\right)^{\frac{1}{4}}, \tag{11.73}$$

and, after taking logarithms, differentiating, and rearranging, we obtain

$$\frac{\partial T_E}{\partial F_s} = \frac{1}{4}\frac{T_E}{F_s}. \tag{11.74}$$

Substituting in values of $F_s = 240$ W/m^2 and $T_E = 255$ K, we obtain the following value for the base sensitivity:

$$\lambda_0 = \frac{\partial T_s}{\partial F_s} = 0.266 \text{ K/(W/m}^2). \tag{11.75}$$

This means that the Earth's blackbody temperature rises 1°K for each 3.76 W/m^2 of radiant forcing, which is the figure quoted in [IPCC-01].

The IPCC report [IPCC-01, Chapter 6.35] also lists various empirically derived equations that have been proposed for calculating overall radiative forcing, as a result of changing levels of CO_2 in the atmosphere. We choose to use the following IPCC expression (although we will use a different value for the constant α (see later discussion)):

$$\Delta F = \alpha \ln(c/c_0), \quad \alpha = 5.35, \tag{11.76}$$

where c represents atmospheric carbon concentration (ppm) and c_0 the unperturbed concentration. Thus, for $c = 2c_0$ we obtain $\Delta F = 3.708$ W/m^2, and from eqn. (11.75) the corresponding temperature change *with no feedbacks* would be

$$\Delta T = 0.266 \times 3.708 = 0.986 \text{ K}. \tag{11.77}$$

However, a change in global temperature would also result in various feedback processes (represented by the summation term of eqn. (11.67)) becoming active. This would, in turn, modify the value of ΔT. For example, an increase in albedo would tend to reduce ΔT due to more solar radiation being reflected away from the Earth. On the other hand, a rise in water vapor would tend to increase ΔT due to increased absorption by the atmosphere of terrestrial infrared radiation. Another important feedback effect is the lower uptake of CO_2 as ocean temperatures increase—this is due to its reduced solubility as the value of K_H in eqn. (11.8) falls. The net result of all feedbacks for increasing levels of CO_2 is to increase radiant forcing, consequently, the above calculations tend to underestimate ΔT.

Including individual feedback effects is beyond the scope of this model and will not be pursued further here. However, we are still able to improve on the estimate given by eqn. (11.77) for a doubling of atmospheric CO_2. Consider Figures 11.10 and 11.11, which show *radiant forcing* and *temperature anomaly* data, respectively. The term *anomaly* is used to refer to a departure from a reference value or long-term average. A positive anomaly indicates that the value is greater than the reference figure, whilst a negative anomaly indicates that the value is less than the reference figure. The reason that temperature anomalies rather than absolute temperatures are used for comparison purposes is that they can be used with an *arbitrary datum*. This is important as there is no generally agreed way to establish absolute globally averaged temperatures from historical times to the present (let alone effective global temperatures). However, as long as the methodology

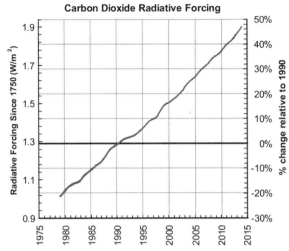

Figure 11.10. NOAA Annual Greenhouse Gas Index (AGGI). Radiative forcing, relative to 1750, due to carbon dioxide alone since 1979. The percentage change from January 1, 1990, is shown on the right axis. Source: [NOAA-14].

for determining the anomaly is consistent then the temperature anomaly can be used to establish long-term trends.

Because we are looking at very long term changes in temperature, it is appropriate to smooth out the time series data in Figs. 11.10 and 11.11 to determine the underlying trends. On fitting first-order relationships to these data between years 1975 and 2010, we

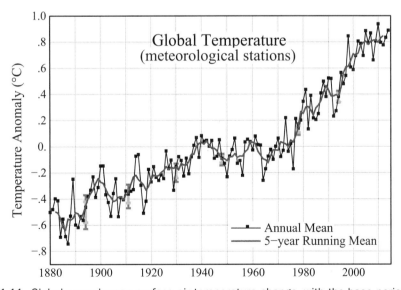

Figure 11.11. Global annual-mean surface air temperature change, with the base period 1951–1980, derived from the meteorological station network (this is an update of Plate 6(b) in [Han-01]). Uncertainty bars (95% confidence limits) are shown for both the annual and five-year means. They account only for incomplete spatial sampling of data. Source: NASA, Goddard Space Flight Center, available online at http://data.giss.nasa.gov/gistemp/graphs_v3/. (See color plate 11.11)

obtain for radiant forcing

$$dF = 0.025t - 49.375 \; (\text{W/m}^2), \qquad (11.78)$$

where t represents time (y); and, similarly for temperature anomaly,

$$dT = 0.0187t - 36.933 \; (\text{K}). \qquad (11.79)$$

Neither eqn. (11.78) nor eqn. (11.79) can be used independently to estimate dF or dT outside of the time range for which they are valid. However, we can use them together to eliminate t and establish the relationship

$$dT = 0.748 \, dF. \qquad (11.80)$$

Substituting this result into eqn. (11.76), but using a value of $\alpha = 4.8$, we obtain the following improved relationship (from that of eqns. (11.75) and (11.76)) between the change in temperature anomaly due to a change in atmospheric concentration of CO_2:

$$dT = 3.59 \ln(c/c_0) - 0.5 \; (\text{K}). \qquad (11.81)$$

The constant of -0.5 in eqn. (11.81) establishes the zero datum at around year 1975,[9] and the amended value for α was chosen to better fit the data of Fig. 11.11.

For a doubling of atmospheric CO_2 concentration from preindustrial levels of 280 ppm to 560 ppm, eqn. (11.81) predicts a temperature rise of $dT \simeq 2°C$. This is within the range of current estimates [Han-11], [IPCC-13, Table 11.3]. Also, this equation gives an estimated value of $dT \simeq 1.3°C$ for the temperature rise from preindustrial CO_2 levels of 280 ppm to 400 ppm, which again is in line with current estimates. However, it must be emphasized that this relationship is only valid strictly between years 1975 and 2010, and outside of this range the relationship is likely to be unreliable due to various factors.[10] Nevertheless, in the absence of a better simple estimate, we will use it in our model to provide an estimate of the globally averaged temperature rise over time as atmospheric CO_2 levels change. The simulated temperature anomaly results for each emissions scenario calculated using eqn. (11.81) are plotted in Fig. 11.17. The estimated years when pCO_2 reaches $2 \times \text{PAL} = 560$ ppm and associated temperature anomaly rises are listed for each emissions scenario in Table 11.5.

The simple temperature anomaly calculation of eqn. (11.81) is performed by R computer code in `tempChange.R`, which also generates appropriate plots. The computer code of file `tempChange.R` is called from `postSimCalcs`. All the computer code is available for download.

For more detailed discussions relating to the Earth's radiation energy balance, interested readers are referred to [And-10], [Pie-10], and [Zdu-07].

11.2.9.6 Sea Level Rise

One of the consequences of an increasing global temperature, resulting from increasing levels of atmospheric CO_2, is that sea levels also rise. This is due to a number of factors, including melting of glaciers, reduced polar ice, and ocean thermal expansion. Although

---

[9] The choice of datum year is arbtrary, but year 1975 was chosen to make the temperature anomaly consistent with Fig. 11.11.
[10] For example, the time lag from a change in pCO_2 to reach a new temperature equilibrium, feedback effects illustrated in Fig. 11.9, nonlinearities due to radiated energy transmission.

Table 11.5. Estimated years for each emissions scenario when pCO_2 reaches $2 \times PAL = 560$ ppm along with the corresponding temperature anomalies

Scenario (GtC)	$pCO_2 = 2 \times PAL$ yr	Temp. anomaly (°K)
1000	2050	1.98
2000	2045	1.99
3000	2043	1.99
4000	2043	1.99

Note: The pCO_2 figure is that which the UN, IPCC, and most research institutions recommend should not be exceeded, as this will lead to a temperature anomaly rise of 2°K—a possible tipping point for climate change due to amplifying feedback effects.

we will not attempt to quantify sea level changes here, it is appropriate to mention that this phenomenon has been observed for some considerable time in data obtained from tide gauges at coastal stations around the world.

Some good news on the subject comes from a recent report [Hay-15], which has concluded that previous global mean sea level (GMSL) rise estimates for the period 1901 to 1990 should be revised down from 1.6–1.9 mm/y, to 1.2 ± 0.2 mm/y, with a 90% confidence interval. However, worryingly, the same study also reports that data now indicates that the GMSL rose at a rate of 3.0 ± 0.7 mm/yr between 1993 and 2010. If this trend continues, many shoreline communities could be devastated in the not-too-distant future.

11.3 SIMULATION RESULTS

The carbon cycle model described in Section 11.2 was simulated using the R computer language, and the results are discussed in the following. A set of minor housekeeping calculations is performed by R computer code in postSimCalcs.R, which also controls plotting of most results and is called from the main program.

A summary of the main data values for each emissions scenario is listed in Table 11.6. Discussion and plots relating to specific topics are provided in subsequent sections.

11.3.1 Carbon Buildup in the Atmosphere

The emissions were calculated for four fossil fuel emission scenarios as outlined in Section 11.2.7, and Fig. 11.12 shows how each scenario results in a corresponding buildup of atmospheric CO_2. The emissions used to drive the model for years 1751 to 2010 are based on the actual yearly average estimates published by [Bod-13] and thereafter are based on projections arrived at using the methods proposed in [Cal-05], as discussed in Section 11.2.7. It is seen that for each scenario the level of atmospheric CO_2 rises to reach a corresponding peak level and then falls off. Measured and estimated atmospheric CO_2 levels to year 2008, represented by circles, are superimposed on the pCO_2 plot, which

Table 11.6. Main simulation data values for each emissions scenario

Scenario (GtC)	Max Emissions (GtC/yr)	Max pCO_2 (ppm)	Max DIC (mol/m^3)	Max net flux (GtC/yr)	Max Revelle	Min uptake factor (μmol/kg/atm)	Min pH	Max temp. rise (°C)
1000	11.7	715	2.18	2.74	14.80	0.209	7.83	3.34
2000	17.6	1077	2.23	3.35	17.14	0.124	7.66	4.82
3000	23.5	1465	2.27	3.79	17.90	0.088	7.54	5.93
4000	29.7	1869	2.30	4.10	18.00	0.070	7.44	6.80

Note: The scenarios: 1000, 2000, 3000, and 4000 relate to recoverable fossil fuel reserves in GtC at year 2010, as discussed in Section 11.2.7, and the maximum temperature rises are from year 1750. The data clearly predict worsening situations as total emissions increases.

shows good agreement. These values consist of *ice core* data [Nef-94] merged with *Mauna Loa Observatory* data [Kee-09].

All four scenarios exceed the $2 \times pCO_2$ preindustrial level recommended by IPCC to avoid an excessive rise in global temperature. Also, what is quite clear from Fig. 11.12 is the time frame over which the changes take place and how long the atmosphere will take to recover to reasonable levels of, say, 2 times preindustrial levels. Clearly, for the more extreme emission scenarios, this will take many centuries—not an attractive prospect.

11.3.2 Carbon Buildup in Surface Seawater and Accompanying Acidification

11.3.2.1 Atmosphere–Seawater Fluxes

Fluxes to and from atmosphere and oceans are calculated using eqn. (11.6), as described in Section 11.2.3. The straightforward calculations are performed in R function ratesOfChange.

As atmospheric levels of CO_2 rise, the net amount of carbon dioxide absorbed by the oceans increases, as does the net exchange flux between atmosphere and ocean. This is shown by the plots in Fig. 11.13, where measured values, averaged over years 2000 to 2009 [IPCC-13, Fig. 6.1], have been superimposed (circular dots). Model predictions are in good agreement with these measured data.

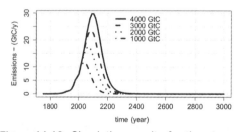
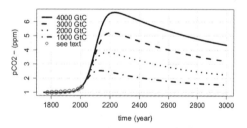

Figure 11.12. Simulation results for the atmosphere due to the year 2010 reserves scenarios. (left) The four scenarios discussed in Section 11.2.7. (right) Corresponding changes in partial pressure of atmospheric CO_2 from preindustrial levels, with estimated (ice core samples) and measured (Mauna Loa) values represented by circles (see Section 11.3.1).

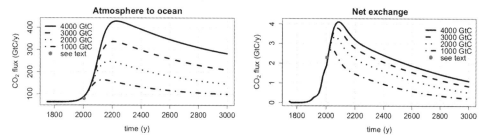

Figure 11.13. Simulation results for CO_2 exchange between atmosphere and surface seawater. (left) Atmospheric CO_2 flux absorbed by the oceans for each scenario. (right) The net CO_2 flux entering the oceans due to absorption of CO_2 by the oceans from the atmosphere, and also outgassing from the oceans to the atmosphere. The dots indicate values reported by [IPCC-13]) (see text).

11.3.2.2 Surface Seawater Dissolved Inorganic Carbon Concentration and pH

The effect of changes in atmospheric CO_2 is shown in Fig. 11.14, where for each of the four scenarios the buildup of DIC and reduction in pH in surface seawater are clearly visible. For the 4000 GtC reserve scenario the maximum DIC value is predicted to reach a value of 2.30 (mol/m^3) and the minimum pH value is predicted to fall to a value of 7.44, both in year 2230. In contrast, for the 1000 GtC reserve scenario, the maximum DIC value is predicted to reach a value of 2.18 (mol/m^3) and the minimum pH value is predicted to fall to a value of 7.83, both in year 2129.

11.3.2.3 Deep Ocean Dissolved Inorganic Carbon Concentration and pH

The changes in the deep ocean are very much more gradual than occurs in surface seawater. This is shown in Fig. 11.15, where deep ocean and surface seawater DIC values are plotted over time for each of the four year 2010 reserves scenarios.

The simulation does *not* calculate deep ocean pH values, as the carbonic acid dissociation correlations do not include for pressure variations. Pressures can vary from 25 bar at 250 m depth to 400 bar at 4000 m (assumes 1 bar increase per 10 m depth). Readers wishing to pursue this aspect further can find more information on pressure correction calculations in [Eme-08] and [Mil-06].

11.3.2.4 Seawater Dissolved Inorganic Carbon Distribution

It was shown in Section 11.2.4 that dissolved CO_2 in seawater ends up distributed into predominately three species: $CO_2(aq)$, HCO_3^- and CO_3^{2-}, the sum of which is equal to

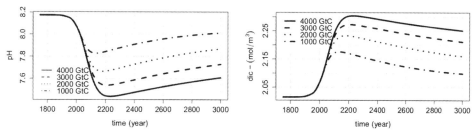

Figure 11.14. Surface seawater pH and DIC simulation results for each of the four year 2010 reserves scenarios.

Table 11.7. Distribution of carbon species that make up surface seawater DIC

	Species concentration (%)			
year	[CO_2(aq)]	[HCO_3^-]	[CO_3^{2-}]	DIC
1750	0.523	89.576	9.900	100
2010	0.693	91.507	7.799	100

Note: These values are the same for all emissions/reserves scenarios as changes occur only after year 2010. Peak values of DIC for each scenario are given in Table 11.6.

DIC. The percentages depend upon the total amount of DIC, and Table 11.7 shows the distributions for years 1750 and 2010. From this table it is seen that most of the DIC ends up as HCO_3^-. The percentage of HCO_3^- has increased from 89.6% in 1750 to 91.5% in year 2010, and this is due to the increased concentration of DIC over that period (see Fig. 11.15).

11.3.2.5 Seawater Buffering

As the level of DIC concentration in surface seawater increases, the ability of the oceans to absorb further atmospheric CO_2 (buffering) decreases (refer to discussion on important concepts involving Revelle and uptake factors in Appendix 11.A.3 at the end of this chapter).

Figure 11.16 shows how the Revelle factor increases (less buffering) over time until the level of pCO_2 reaches a peak (see Fig. 11.12), when it starts to decrease (more buffering). Similarly, it is seen that the uptake factor decreases (less buffering) until the peak level of pCO_2 is reached, when it starts to increase (more buffering). What is noticeable here is that the recovery times are much longer than the time it takes buffering to reduce.

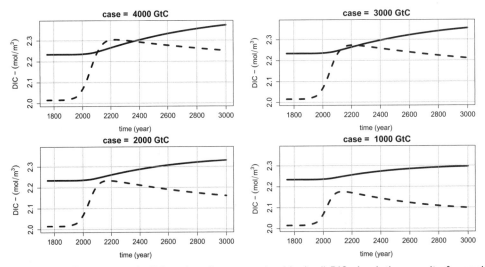

Figure 11.15. Deep ocean (solid) and surface seawater (dashed) DIC simulation results for each of the four year 2010 reserves scenarios.

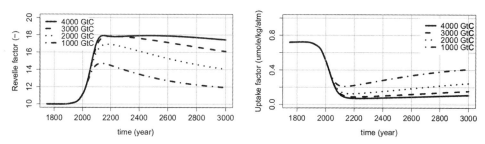

Figure 11.16. (left) *Revelle factor* plots and (right) *update factor* plots for each of the four year 2010 reserve scenarios.

11.3.3 Surface Temperature Changes

We discussed in Section 11.2.9.5 a method of calculating the globally averaged temperature rise corresponding to an increase in the level of carbon dioxide concentration in the atmosphere. Now, although we did not assert that this method was rigorous, it does give values that are within the expected range, according to the literature. Therefore, with the caveat that the predictions are likely to be unreliable in the long term (along with most current estimates in the literature), we have included this calculation in the simulation in order to provide some idea of possible future globally averaged temperature anomalies.

Figure 11.17 provides simulated temperature anomaly predictions for each of the four emissions scenarios. The scenario having reserves of 1000 GtC in year 2010 gives the lowest maximum predicted temperature anomaly of 2.93C at year 2129; and the scenario having reserves of 4000 GtC in year 2010 gives the highest maximum predicted temperature anomaly of 6.78C at year 2230. For our calculations we assume a zero anomaly at 1975. From 1750 to year 1975 the simulation estimates the anomaly to be 0.53°C, and from

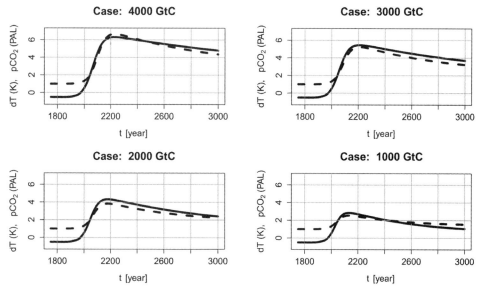

Figure 11.17. Temperature anomalies dT (solid), shown along with corresponding atmospheric carbon dioxide concentrations pCO_2 (dashed), for the four emission scenario cases.

Table 11.8. Definitions for computer state variables along with initial values, consistent with [Eme-08, p. 113]

Computer variable, y_n	Symbol	Initial value	Units
y_1	pCO_2	1	–
y_2	c_{dic}^s	2.015	mol m^{-3}
y_3	c_{dic}^d	2.234	mol m^{-3}
y_4	c_{alk}^s	2.300	mol m^{-3}
y_5	c_{alk}^d	2.365	mol m^{-3}

Note: Computer variable y_1 represents the normalized state variable pCO_2 at a preindustrial activity level for atmospheric CO_2, i.e., a concentration of 280 ppm.

1975 to year 2010, 0.77°C. This makes a total estimated anomaly from 1750 to year 2010 of 1.3°C. But remember that this value is based on many simplifying assumptions.

All scenarios indicate temperature rises in excess of the 2°C maximum adopted/recommended by the United Nations and G-20. This supports the case for further research into the important area of temperature anomaly prediction and whether a 2°C ceiling is feasible.

11.A APPENDICES

11.A.1 Model Differential Equations

The *ordinary differential equations* (ODEs) that describe the dynamic behavior of the model are given by

$$\frac{d\mathbf{y}(t)}{dt} = g(\mathbf{y}, t) + f(t), \quad \mathbf{y}(t=0) = \mathbf{y}_0, \quad t \geq 0, \tag{11.82}$$

where $\mathbf{y} = [y_1, y_2, \ldots, y_5]^T$ represents a vector of the five state variables that describe the main dynamic variables of the system, t represents time, $g(t)$ is a nonlinear function representing interactions between state variables and time, and $f(t)$ is a function of time representing fossil fuel emissions. The state variables are defined in Table 11.8, along with values that make up the initial condition vector $\mathbf{y}_0$.

Equations (11.82), which should be read in conjunction with Sections 11.2.1 and 11.2.2, are integrated numerically to obtain the model solution over time.

11.A.2 Correlations for Chemical Equilibrium and Dissociation Constants

The boron dissociation constant K_B [Dic-90] used in eqn. (11.20) is calculated from

$$K_B = \exp(A + B + C\log(T_K) + 0.053105 S^{0.5} T_K), \tag{11.83}$$

where

$$\begin{aligned}
A &= -8966.9 - 2890.53 S^{0.5} - 77.942 S + 1.728 S^{1.5} - 0.0996 s^2)/T_K, \\
B &= 148.0248 + 137.1942 S^{0.5} + 1.62142 S, \\
C &= -(24.4344 + 25.085 S^{0.5} + 0.2474 S).
\end{aligned} \tag{11.84}$$

The variables S and T_K represent salinity in parts per thousand and temperature in degrees Kelvin, respectively.

The carbonic acid dissociation constants [Lue-00] used in eqns. (11.13) and (11.14) are calculated from

$$\log_{10}(K_{a1}) = -(3633.86/T_K - 61.2172 + 9.67770\ln(T_K) \tag{11.85}$$
$$-0.011555S + 0.0001152S^2) \tag{11.86}$$
$$\log_{10}(K_{a2}) = -(471.78/t + 25.9290 - 3.16967\ln(T_K) \tag{11.87}$$
$$-0.01781S + 0.0001122S^2). \tag{11.88}$$

The hydroxyl dissociation constant [DoE-94] used in eqn. (11.25) is calculated from

$$K_W = \exp(A+B), \tag{11.89}$$

where

$$A = 148.96502 - 13847.26/T_K - 23.65218\ln(T_K) \tag{11.90}$$
$$B = (118.67/T_K - 5.977 + 1.0495\ln(t))S^{0.5} - 0.01615S. \tag{11.91}$$

Henry's constant [Wei-74] used in eqn. (11.8) is calculated from

$$K_H = \exp(A+B), \tag{11.92}$$

where

$$A = -60.2409 + 93.4517(100/T_K) + 23.3585\ln(T_K/100) \tag{11.93}$$
$$B = S(0.023517 - .023656(T_K/100) + .0047036(T_K/100)^2). \tag{11.94}$$

Values from [Mil-06, Table 7.4] for the solubility product in the calcium carbonate reactions of eqns. (11.29) and (11.31) at 15°C are

$$\begin{aligned} \text{Calcite: } K'_{sp} &= 4.37\text{E-}07 \\ \text{Aragonite: } K'_{sp} &= 6.76\text{E-}07. \end{aligned} \tag{11.95}$$

11.A.3 Revelle and Uptake Factors

We know from *carbonate chemistry*, outlined in Section 11.2.4, that under equilibrium conditions, there is a causal link between concentration of *carbon dioxide* in the atmosphere pCO_2 and concentration of *dissolved inorganic carbon* DIC=$[CO_2(aq)]$ + $[H_2CO_3^-]$ + $[CO_3^{2-}]$ in the upper layer of the oceans. Here the term pCO_2 represents the partial pressure of atmospheric carbon dioxide, which is very nearly equal to concentration in parts per million by volume (ppmv).

We now make the reasonable assumption that there is a power law relationship between partial pressure of carbon dioxide in the atmosphere and concentration of dissolved inorganic carbon. This means that it can be represented on a log-log plot as a straight line, which implies a relationship of the form

$$\ln \text{DIC} = \text{constant} + \frac{1}{R}\ln pCO_2. \tag{11.96}$$

Rearranging and differentiating with respect to pCO_2, we obtain

$$R = \frac{\partial \ln (pCO_2)}{\partial \ln (\text{DIC})}. \tag{11.97}$$

Expanding and representing the derivative term by the ratio of small increments, we obtain an expression for the relative fractional change of pCO_2 with respect to the fractional change of DIC:

$$R = \left(\frac{\delta pCO_2}{pCO_2}\right) \Big/ \left(\frac{\delta \text{DIC}}{\text{DIC}}\right). \tag{11.98}$$

The ratio R is the so-called *Revelle factor*, also known as the *buffer factor* or *evasion factor*, and is calculated after reequilibration, that is, when a new state of equilibrium has been reached. The term is similar to that used by Revelle and Suesse [Rev-57], later called the *Revelle factor* by Broecker and Peng [Bro-82]. The *Revelle factor* provides a simple way of determining the expected concentration change in DIC due to changes in pCO_2.

To put this into context using our R function carbEq, we see that at preindustrial levels of atmospheric carbon dioxide, say, $pCO_2 = 280$ ppmv, DIC $= 2000$ (μmol/kg), CO_2(aq)$=9.8$ (μmol/kg), and the corresponding Revelle factor is $R = 9.73$. This indicates that the fractional change in pCO_2 at equilibrium is approximately 9.73 times that of the fractional change in DIC. However, because DIC contains around $2000/9.8 = 204$ times more carbon moles per liter than the corresponding CO_2(aq), this means that carbon uptake by seawater is around $204/9.73 = 21$ times greater than if there were no chemical reactions. This is known as seawater *buffering*. These figures change quite rapidly as the level of CO_2 in the atmosphere rises. For example, at $pCO_2 = 400$ ppmv the figures become DIC $= 2040$ (μmol/kg) and CO_2(aq)$=12.0$ (μmol/kg), and the corresponding Revelle factor is $R = 10.5$. Now seawater buffering is around $2040/12.0/10.5 = 16.2$. Thus, as the concentration of atmospheric CO_2 increases, the ability of seawater to absorb CO_2 decreases. If pCO_2 should reach 1000 ppmv, then the carbon uptake by seawater will fall to around 3.5 times greater than if there were no chemical reactions. So, to summarize, seawater buffering capability will fall from 21 to 3.5 as pCO_2 rises from 280 to 1000 ppmv.

The following analytical expression for the Revelle factor can be obtained from the carbonate chemistry relationships given in Section 11.2.4:

$$R = \frac{\text{DIC}}{\text{DIC} - \dfrac{\left(2[CO_3^{2-}] + [HCO_3^-]\right)^2}{[HCO_3^-] + 4[CO_3^{2-}] + \dfrac{[H^+][B(OH)_4^-]}{K_B + [H^+]} + [H^+] + [OH^-]}}. \tag{11.99}$$

Further detailed analysis and discussion relating to the Revelle factor and derivation of eqn. (11.99) can be found in [Egl-10].

Another important related concept is the *uptake factor*, due to Pilson [Pil-88], which is defined as the ratio of incremental change in DIC in (mol kg^{-1}) due to the incremental change in pCO_2 (atm), that is,

$$\text{UF} = \frac{\delta \text{DIC}}{\delta pCO_2} = \frac{1}{R} \frac{\text{DIC}}{pCO_2}, \quad (\text{mol kg}^{-1} \text{ atm}^{-1}). \tag{11.100}$$

Using the same numbers as earlier gives UF=0.49. Thus, if pCO_2 increases from 400 to 410 [ppm], we can expect that

$$\delta \text{DIC} = 0.49 \times \frac{10}{10^6} = 4.9 \times 10^{-6}, \quad (\text{mol kg}^{-1}). \tag{11.101}$$

In other words, on rounding up to 0.5, we would expect the concentration of DIC to increase from 2040 to 2045 (μmol/kg).

The Revelle factor is calculated in R function `carbEq` (see Listing 11.4 in Section 11.2.4). The uptake factor is calculated in R function `Revelle`, which also generates appropriate plots for both factors.

11.A.4 Residence Time

Consider a well-mixed vessel containing a volume V of a particular fluid and subject to volumetric flows of fluid in and out. Assuming constant fluid density, a mass balance around this volume yields

$$V \frac{dx_o}{dt} = x_i Q_i - x_o Q_o, \tag{11.102}$$

where x represents the volume fraction of a chemical species and Q a constant volumetric flow. Subscripts i and o denote quantities flowing into and out of the vessel.

Without loss of generality, assume that $Q_i = Q_o = Q$; then, on dividing through by Q, we obtain

$$\tau \frac{dx_o}{dt} = (x_i - x_o), \tag{11.103}$$

where $\tau = V/Q$ is the so-called *residence time* or *time constant*.

Equation (11.103) is a first-order ordinary differential equation (ODE) to which we can readily find an analytical solution, as it is *separable*. We proceed by integration as follows:

$$\int \frac{\tau \, dx_o}{x_i - x_o} = dt \quad \Rightarrow \quad \tau \ln(x_i - x_o) = t + K. \tag{11.104}$$

On rearranging eqn. (11.104) and taking the initial condition at $t = 0$ as $x_o = x_o(0)$, we obtain the general solution

$$x_o = x_i - (x_i - x_o(0)) \exp(-t/\tau). \tag{11.105}$$

Thus, we see that the residence time τ is also equal to the inverse of the *eigenvalue* of eqn. (11.103), that is, $\tau = 1/\lambda$. From eqn. (11.105) we see that, for constant x_i, the residence time can also be thought of as the time it takes for x_o to reach 63% of x_i.

We now consider the residence time of the atmosphere in relation to CO_2 absorption by the oceans and apply the ideal gas law $PV = nRT$, where P represents pressure (Pascals), R gas constant 8.314 (J/mol/K), V volume (m^3), and n number of moles (mol). The mass of the atmospheric is equal to 5.13×10^{21} (g), and the mass of carbon absorbed by the oceans from the atmosphere each year is equal 80.0×10^{15} (g/yr) (average for years 2000 to 2009 [IPCC-13, Fig. 6.1]). On converting to moles using $MW_{air} = 29$, $MW_C = 12$, along with the ideal gas law, we obtain the residence time for atmosphere/ocean gas exchange:

$$\tau = \frac{PV_{CO_2}^{atm}}{PQ_{CO_2}^{ocean}} = \frac{5.13 \times 10^{21} \times 380 \times 10^{-6}/29\ RT}{80.0 \times 10^{15}/12\ RT},$$

(11.106)

$$\therefore \tau = \frac{V_{CO_2}^{atm}}{Q_{CO_2}^{ocean}} \simeq 10\ (\text{yr}).$$

Recall that the number of carbon dioxide moles exchanged is the same as the number of carbon moles exchanged. In eqn. (11.106) we have used the average value for atmospheric concentration of carbon dioxide during years 2000 to 2009 of $pCO_2 = 380 \times 10^{-6}$ [Kee-09].

11.A.5 Mass Action

Given the reaction

$$a\text{A} + b\text{B} \rightleftharpoons c\text{C} + d\text{D},$$

where a, b, c, d are the coefficients for a balanced chemical equation, the *mass action law* states that if the system is at equilibrium at a given temperature, then the following ratio is a constant:

$$\frac{[\text{C}]^c[\text{D}]^d}{[\text{A}]^a[\text{B}]^b} = K_{eq}.$$

This is the *ideal law of chemical equilibrium* or *law of mass action*. K_{eq} is the so-called *equilibrium constant* for the particular reaction. Generally, in the literature, square brackets around a chemical species indicate a concentration quantity, also known as *molarity*.

The units for K_{eq} depend upon the units used for concentration. If M is used for all concentrations, then K_{eq} will have units $M^{c+d-(a+b)}$.

11.A.6 The Electromagnetic Spectrum

Table 11.9. Radiation spectrum range data

Spectrum name	Wavelength, λ	Frequency, ν	Energy
Radio frequency (RF)	> 1 m	< 300×10^6 Hz	< 0.12×10^{-5} eV
Microwave	1 m → 1 mm	300×10^6 → 300×10^9 Hz	0.12×10^{-5} → 0.0012 eV
Infra red (IR)	1 mm → 700 nm	300×10^9 → 4.3×10^{14} Hz	0.0012 → 1.65 eV
Visible	700 → 400 nm	4.3×10^{14} → 7.5×10^{14} Hz	1.65 → 3.1 eV
Ultra violet (UV)	400 → 10 nm	7.5×10^{14} → 3×10^{16} Hz	3.1 → 120 eV
X-ray	< 10 nm	> 3×10^{16} Hz	> 120 eV

Note: Refer to Section 11.2.8.

REFERENCES

[And-10] Andrews, D. (2010), *An Introduction to Atmospheric Physics*, Cambridge University Press.

[ASTM-05] ASTM (2005), ASTM G173-03e1 Standard Tables for Reference Solar Spectral Irradiances: Direct Normal and Hemispherical on 37° Tilted Surface, January. Year 2000 ASTM Standard Extraterrestrial Spectrum Reference E-490-00 data are available for download at Solar Spectra: Air Mass Zero: http://rredc.nrel.gov/solar/spectra/am0/.

[Bac-73] Bacastow, R. and C. D. Keeling (1973), *Atmospheric Carbon Dioxide and Radiocarbon in the Natural Carbon Cycle: II. Changes from A.D. 1700 to 2070 as Deduced from a Geochemical Model*, in Carbon and the Biosphere, G. M. Woodwell and E. V. Pecan, editors, Atomic Energy Commission, Technical Information Service, pp. 86–135.

[Bee-11] Wikipedia (2011), The Beer–Lambert Law, available online at http://http://en.wikipedia.org/wiki/Beer-Lambert_law.

[Bod-13] Boden, T. A., G. Marland and R. J. Andres (2013), *Global, Regional, and National Fossil-Fuel CO2 Emissions*, Carbon Dioxide Information Analysis Center, Oak Ridge National Laboratory, U.S. Department of Energy, doi: 10.3334/CDIAC/00001_V2013, available online at http://cdiac.ornl.gov/ftp/ndp030/global.1751_2010.ems

[Bol-84] Boltzmann, L. (1884), Ableitung des Stefan'schen Gesetzes, betreffend die Abhängigkeit der Wärmestrahlung von der Temperatur aus der electromagnetischen Lichttheorie, *Annalen der Physik und Chemie* **22**, 291–294.

[Bro-82] Broecker, W. S. and T.-H. Peng (1982), *Tracers in the Sea*, Lamont-Doherty Earth Observatory.

[Cal-05] Caldeira, K. and M. E. Wickett (2005), Ocean Model Predictions of Chemistry Changes from Carbon Dioxide Emissions to the Atmosphere and Ocean, *Journal of Geophysical Research* **110**, 1–12.

[Dic-90] Dickson, A. G. (1990), Thermodynamics of the Dissociation of Boric Acid in Synthetic Seawater from 273.15 to 298.15 K, *Deep-Sea Research* **37**, 755–766.

[DoE-94] Dickson, A. G. and C. Goyet, eds. (1994), *Handbook of Methods for the Analysis of the Various Parameters of the Carbon Dioxide System in Seawater*, version 2, ORNL/CDIAC-74.

[Egl-10] Egleston, E. S., C. L. Sabine and F. M. M. Morel (2010), Revelle Revisited: Buffer Factors that Quantify the Response of Ocean Chemistry to Changes in DIC and Alkalinity, *Global Biogeochemical Cycles* **24**-1, 1–9.

[Eme-08] Emerson, S. and J. Hedges (2008), *Chemical Oceanography and the Marine Carbon Cycle*, Cambridge University Press.

[Fal-00] Falkowski, P., et al. (2000), The Global Carbon Cycle: A Test of Our Knowledge of Earth as a System, *Science*, October 13.

[Han-01] Hansen, J. E., R. Ruedy, M. Sato, M. Imhoff, W. Lawrence, D. Easterling, T. Peterson, and T. Karl (2001), A Closer Look at United States and Global Surface Temperature Change, *Journal of Geophysical Research* **106**, 23947–23963.

[Han-11] Hansen, J., M. Sato, P. Kharecha, and K. von Schuckmann (2011), Earth's Energy Imbalance and Implications, *Atmospheric Chemistry and Physics* **11**, 13421–13449.

[Hay-15] Hay, C. C., E. Morrow, R. E. Kopp and J. X. Mitrovica (2015), Probabilistic Reanalysis of Twentieth-Century Sea-Level Rise, *Nature*, **517**-7535, 481–484.

[Hot-67] Hottel, H. C. and A. F. Sarofim, (1967), *Radiative Transfer*, McGraw-Hill.

[IPCC-01] IPCC (2001), Radiative Forcing of Climate Change, in *Climate Change 2001: The Scientific Basis: Contribution of Working Group I to the Third Assessment Report of the Intergovernmental Panel on Climate Change*, [J. T. Houghton, Y. Ding, D. J. Griggs, M. Noguer, P. J. van der Linden, X. Dai, K. Maskell and C. A. Johnson, editors, Cambridge University Press, pp. 349–416.

[IPCC-07] IPCC (2007), *Climate Change 2007: The Physical Science Basis. Contribution of Working Group I to the Fourth Assessment Report of the Intergovernmental Panel on Climate Change*, S. Solomon, D. Qin, M. Manning, M. Marquis, K. Averyt, M. M.B. Tignor, H. L. Miller Jr. and Z. Chen, editors, Cambridge University Press.

[IPCC-13] IPCC (2013), *Climate Change 2013: The Physical Science Basis. Contribution of Working Group I to the Fifth Assessment Report of the Intergovernmental Panel on Climate Change*, T. F. Stocker, D. Qin, G.-K. Plattner, M. Tignor, S. K. Allen, J. Boschung, A. Nauels, Y. Xia, V. Bex and P. M. Midgley, editors, Cambridge University Press.

[Jac-05] Jacobson, M. Z. (2005), *Fundamentals of Atmospheric Modeling*, 2nd ed., Cambridge University Press.

[Kas-89] Kasten, F. and A. T. Young (1989), Revised Optical Air Mass Tables and Approximation Formula, *Applied Optics* **28**-22, 4735–4738.

[Kee-09] Keeling, R. F., S. C. Piper, A. F. Bollenbacher and S. J. Walker (2009), *Mauna Loa Observatory Data*, Scripps Institution of Oceanography, University of California, available online at http://cdiac.ornl.gov/ftp/trends/co2/maunaloa.co2.

[Kie-97] Kiehl, J. T. and K. E. Trenberth (1997), Earth's Annual Global Mean Energy Budget, *Bulletin of the American Meteorological Association* **78**, 197–206.

[Kle-09] Kleypas, J. A. and K. K. Yates (2009), Coral Reefs and Ocean Acidification, *Oceanography* **22**-4, 108–117.

[Kop-04] Kopp, G., G. Lawrence and G. Rottman (2004), Total Irradiance Monitor Design and On-Orbit Functionality, *SPIE Proceedings* **5171**, 14–25.

[Lio-02] Liou, K. N. (2002), *An Introduction to Atmospheric Radiation*, 2nd ed., Academic Press.

[Lue-00] Luecker, T. J., A. G. Dickson and C. D. Keeling (2000), Ocean CO_2 Calculated from Dissolved Inorganic Carbon, Alkalinity, and the Equation for K_1 and K_2: Validation Based on Laboratory Measurements of CO_2 in Gas and Seawater at Equilibrium, *Marine Chemistry* **70**, 105–119.

[McH-15] McHugh, A. J., G. W. Griffiths and W. E. Schiesser (2015), *An Introductory CO_2 Model*, World Scientific Press.

[Mar-13] Marland, G., T. A. Boden and R. J. Andres (2013), Global, Regional, and National CO2 Emissions, in *Trends: A Compendium of Data on Global Change*, Carbon Dioxide Information Analysis Center, Oak Ridge National Laboratory, U.S. Department of Energy.

[Mil-09] Miller, S., C. Marandino, W. de Bruyn, W. and E. S. Saltzman (2009), Air Sea Gas Exchange of CO2 and DMS in the North Atlantic by Eddy Covariance, *Geophysical Research Letters* **36**, L15816.

[Mil-06] Millero, F. (2006), *Chemical Oceanography*, 3rd ed., CRC Press.

[Mol-05] Moler, C. (2005), *Numerical Computing with MATLAB*, Mathworks, available online at http://www.mathworks.com/moler/odes.pdf.

[Mor-15] Morrison, A. K., T. L. Frölicher and J. L. Sarmiento (2015), Upwelling in the Southern Ocean, *Physics Today* **68**-1, 27–32.

[Nef-94] Neftel, H., H. Friedli, E. Moor, H. Lotscher, H. Oeschger, U. Siegenthaler and B. Stauffer (1994), *Historical CO2 Record from the Siple Station Ice Core*, Physics Institute, University of Bern, available online at http://cdiac.ornl.gov/ftp/trends/co2/siple2.013.

[NOAA-14] NOAA (2014), *Annual Greenhouse Gas Index (AGGI)*, National Oceanic and Atmospheric Administration, Earth System Research Laboratory, Global Monitoring Division, available online at esrl.noaa.gov/gmd/aggi/aggi.html.

[Pei-92] Peixoto, J. P. and A. H. Oort (1992), *Physics of Climate*, Springer.

[Pie-10] Pierrehumbert, R. T. (2010), *Principles of Planetary Climate*, Cambridge University Press.

[Pil-88] Pilson, M. E. Q. (1998), *An Introduction to the Chemistry of the Sea*, Prentice Hall.

[Pla-00] Planck, M. (1900), Ueber das Gesetz der Energieverteilung im Normalspectrum, *Verh. D. Physik* **2**, 237–239. [Subsequently in English: Planck, M. (1901), On the Law of Distribution of Energy in the Normal Spectrum, *Annals of Physics* **4**, 553.]

[Rev-57] Revelle, R. and H. E. Suess (1957), Carbon Dioxide Exchange between Atmosphere and Ocean and the Question of an Increase of Atmospheric CO2 during Past Decades, *Tellus* **9**, 18–27.

[Sab-04] Sabine, C. L. et al. (2004), The Oceanic Sink for Anthropogenic CO2, *Science*, **305**-5682, 367–371.

[Sar-02] Sarmiento, J. L. and N. Grubar (2002), Sinks for Anthropogenic Carbon, *Physics Today*, August, 30–36.

[Ste-79] Stefan, J. (1879), Über die Beziehung zwischen der Wärmestrahlung und der Temperatur [On the relation between heat radiation and temperature], *Sitzungsberichte der mathematisch-naturwissenschaftlichen Classe der kaiserlichen Akademie der Wissenschaften* **79**, 391–428.

[Stew-07] Stewart, R. H. (2007), *Introduction to Physical Oceanography*, Texas A&M University, available online at http://oceanworld.tamu.edu/resources/ocng_textbook/PDF_files/book_pdf_files.html.

[TIM-07] TIM (2007), Total Solar Irradiance (TSI) Data. TIM data are regularly updated and available for download at the Solar Radiation and Climate Experiment (SORCE), available online at http://lasp.colorado.edu/sorce/data/tsi_data.htm

[Wall-06] Wallace, J. M. and P. V. Hobbs (2006), *Atmospheric Science: An Introductory Survey*, 2nd ed., Academic Press.

[Walk-91] Walker, J. C. G. (1991), *Numerical Adventures with Geochemical Cycles*, Oxford University Press.

[Wan-92] Wanninkhof, R. (1992), Relationship between Wind Speed and Gas Exchange over the Ocean, *Journal of Geophysical Research* **97**-C5, 7373–7382.

[Wie-93] Wien, W. (1893), Eine neue Beziehung der Strahlung schwarzer Körper zum zweiten Hauptsatz der Wärmetheorie [A new relationship between black body radiation and the second law of thermodynamics], *Sitzungsberichte der Königlichen Preußischen Akademie der Wissenschaften zu Berlin*, 55–62

[Wei-74] Weiss, R. F. (1974), Carbon Dioxide in Water and Seawater: The Solubility of a Non-ideal Gas, *Marine Chemistry* **2**, 203–215.

[Whi-11] Whiteley, N. (2011), Physiological and Ecological Responses of Crustaceans to Ocean Acidification, *Marine Ecology Progress Series* **430**, 257–271.

[You-94] Young, A. T. (1994), Air Mass and Refraction, *Applied Optics* **33**, 1108–1110.

[Zdu-07] Zdunkowski, W., T. Trautman and A. Bott (2007), *Radiation in the Atmosphere: A Course in Theoretical Meteorology*, Cambridge University Press.

[Zee-06] Zeebe, R. (Lead Author) and J.-P. Gattuso (Topic Editor) (2006), *Marine Carbonate Chemistry*, in *Encyclopedia of Earth*, C. J. Cleveland, editor, Environmental Information Coalition, National Council for Science and the Environment, available online at http://www.eoearth.org/article/Marine_carbonate_chemistry.

Appendix: A Mathematical Aide-Mémoire

A.1 NUMBER SYSTEMS

Natural numbers: $\mathbb{N} = \{1, 2, 3, \ldots\}$.
Integers: $\mathbb{Z} = \{0, \pm 1, \pm 2, \pm 3, \ldots\}$.
  Positive/negative integers are sometimes denoted $\mathbb{Z}^+/\mathbb{Z}^-$.
Rational numbers: $\mathbb{Q} = \{p/q : p, q \in \mathbb{Z}, q \neq 0\}$.
  Numbers that can be represented by a finite or periodic decimal expansion.
Real numbers: $\mathbb{R} =$ numbers that fall into the following categories,
  Rational: numbers that can be expressed as the ratio of two integers (see preceding).
  Irrational: numbers that are *not* rational
  Algebraic: nonzero numbers that can be either real or complex and that are a solution to a polynomial $p(x)$, with rational coefficients (or equivalently with cleared denominators, integer coefficients), that is,

$$p(x) := a_n x^n + a_{n-1} x^{n-1} + \cdots, a_0 = 0, \; a_j \in \mathbb{Q} \qquad (A.1)$$

  Transcendental: numbers that are *not* algebraic
Complex numbers: $\mathbb{C} = \{x + iy : x, y \in \mathbb{R}, i = \sqrt{-1}\,\}$.

A.2 INTERVALS

When defining the *interval* over which a variable is *valid*, we use the following notation:

$$
\begin{aligned}
x \in (a, b) &= \{x \in \mathbb{R} : a < x < b\} \text{ open;} \\
x \in [a, b] &= \{x \in \mathbb{R} : a \leq x \leq b\} \text{ closed;} \\
x \in (a, b] &= \{x \in \mathbb{R} : a < x \leq b\} \text{ left-open, right-closed;} \\
x \in [a, b) &= \{x \in \mathbb{R} : a \leq x < b\} \text{ left-closed, right-open.}
\end{aligned}
$$

A.3 VECTOR AND MATRIX DEFINITIONS

A.3.1 Vectors

A *vector* is defined as a 1D array whose elements can be real or complex. As an example, we write $a \in \mathbb{C}^n$ to specify a vector with n complex elements. Similarly, we write $a \in \mathbb{R}^n$ to specify a vector with n real elements. In general, a vector is considered to be a column vector, which we represent as

$$a = \begin{bmatrix} a_1 \\ a_2 \\ \vdots \\ a_n \end{bmatrix}. \tag{A.2}$$

A row vector is simply the transpose of a column vector, which we write as

$$a^T = \begin{bmatrix} a_1 \\ a_2 \\ \vdots \\ a_n \end{bmatrix}^T = \begin{bmatrix} a_1 & x_2 & \cdots & a_n \end{bmatrix}. \tag{A.3}$$

The addition of two vectors $a, b \in \mathbb{C}^n$ is a vector $c \in \mathbb{C}^n$, defined as

$$c = a + b = \begin{bmatrix} a_1 \\ a_2 \\ \vdots \\ a_n \end{bmatrix} + \begin{bmatrix} b_1 \\ b_2 \\ \vdots \\ b_n \end{bmatrix}$$

$$= \begin{bmatrix} a_1 + b_1 \\ a_2 + b_2 \\ \vdots \\ a_n + b_n \end{bmatrix}.$$

A.3.1.1 Dot Product

The *dot product* (also known as *scalar product* or *inner product*) of two vectors $a, b \in \mathbb{R}^n$ is a scalar c, defined as

$$c = a^T \cdot b = [a_1 a_2 \cdots a_n] \cdot \begin{bmatrix} b_1 \\ b_2 \\ \vdots \\ b_n \end{bmatrix}$$

$$= a_1 b_1 + a_2 b_2 \cdots + a_n b_n$$

$$= \sum_{i=1}^{n} a_i b_i.$$

Numerically, this is equal to $a = |x| \, |y| \cos(\theta)$, where θ is the angle between x and y.

A.3.1.2 Cross Product

The *cross product* (also known as *vector product*) of two vectors $a, b \in \mathbb{R}^n$ is vector c. It is defined as

$$c = a \times b = \begin{vmatrix} e_x & e_y & e_z \\ a_x & a_y & a_z \\ b_x & b_y & b_z \end{vmatrix},$$

where the third term represents a *determinant* (see later) and e_x, e_y, e_z are unit vectors in the direction of orthogonal vectors x, y, z. The vector $c \in \mathbb{R}^n$ is *orthogonal* to the plane formed by a and b, with magnitude $|c| = |a| |b| \sin(\theta)$.

A.3.1.3 Outer Product

The *outer product* (also known as a *dyadic product* or *tensor product*) of two vectors $a \in \mathbb{R}^m, b \in \mathbb{R}^n$ is matrix $c \in \mathbb{R}^{m \times n}$. It is defined as

$$c = a \otimes b = \begin{bmatrix} b_1 a_1 & b_2 a_1 & \cdots & b_n a_1 \\ b_1 a_2 & b_2 a_2 & \cdots & b_n a_2 \\ \vdots & \vdots & \ddots & \vdots \\ b_1 a_m & b_2 a_m & \cdots & b_n a_m \end{bmatrix},$$

where this time the third term represents a $m \times n$ *matrix*.

A.3.2 Matrices

A *matrix* is defined as a 2D array having elements that can be real or complex. As an example, we write $A \in \mathbb{C}^{m \times n}$ to specify a matrix having m rows and n columns whose elements are complex. Similarly, we write $A \in \mathbb{R}^{m \times n}$ to specify a matrix having m rows and n columns whose elements are real. We represent such matrices as

$$A = \begin{bmatrix} a_{11} & a_{12} & \cdots & a_{1n} \\ a_{21} & a_{22} & \cdots & a_{2n} \\ \vdots & \vdots & \ddots & \vdots \\ a_{m1} & a_{m2} & \cdots & a_{mn} \end{bmatrix},$$

with elements a_{ij} being real or complex as appropriate. If $n = m$, then A is a *square matrix*.

We write $\mathbb{R}^{m \times n} \times \mathbb{R}^n \to \mathbb{R}^m$ to indicate the linear operation of matrix $A \in \mathbb{R}^{m \times n}$, multiplied by vector $x \in \mathbb{R}^n$, which is defined as

$$b = Ax = \begin{bmatrix} a_{11}x_1 + a_{12}x_2 + \cdots + a_{1n}x_n \\ a_{21}x_1 + a_{22}x_2 + \cdots + a_{2n}x_n \\ \vdots \\ a_{m1}x_1 + a_{m2}x_2 + \cdots + a_{mn}x_n \end{bmatrix},$$

where $b \in \mathbb{R}^m$ whose elements are given by

$$b_i = \sum_{j=1}^{n} a_{ij} x_j.$$

Note that vector x has the same number of rows as the columns in matrix A, while the resulting vector b has the same number of rows as matrix A.

We write $\mathbb{R}^{m \times n} \times \mathbb{R}^{m \times n} \to \mathbb{R}^{m \times n}$ to indicate the linear operation of adding matrix $A \in \mathbb{R}^{m \times n}$ to matrix $B \in \mathbb{R}^{m \times n}$, which is defined as

$$C = \begin{bmatrix} a_{11} & a_{12} & \cdots & a_{1n} \\ a_{21} & a_{22} & \cdots & a_{2n} \\ \vdots & \vdots & \ddots & \vdots \\ a_{m1} & a_{m2} & \cdots & a_{mn} \end{bmatrix} + \begin{bmatrix} b_{11} & b_{12} & \cdots & b_{1n} \\ b_{21} & b_{22} & \cdots & b_{2n} \\ \vdots & \vdots & \ddots & \vdots \\ b_{m1} & b_{m2} & \cdots & b_{mn} \end{bmatrix}$$

$$= \begin{bmatrix} a_{11}+b_{11} & a_{12}+b_{12} & \cdots & a_{1n}+b_{1n} \\ a_{21}+b_{21} & a_{22}+b_{22} & \cdots & a_{2n}+b_{2n} \\ \vdots & \vdots & \ddots & \vdots \\ a_{m1}+b_{m1} & a_{m2}+b_{m2} & \cdots & a_{mn}+b_{mn} \end{bmatrix},$$

where $C \in \mathbb{R}^{m \times n}$ whose elements are given by

$$c_{ij} = a_{ij} + b_{ij}.$$

Note the number of rows and columns in B must equal the number of rows and columns in A.

We write $\mathbb{R}^{m \times n} \times \mathbb{R}^{n \times \ell} \to \mathbb{R}^{m \times \ell}$ to indicate the linear operation of matrix $A \in \mathbb{R}^{m \times n}$ multiplied by matrix $B \in \mathbb{R}^{n \times \ell}$, which is defined as

$$C = \begin{bmatrix} a_{11} & a_{12} & \cdots & a_{1n} \\ a_{21} & a_{22} & \cdots & a_{2n} \\ \vdots & \vdots & \ddots & \vdots \\ a_{m1} & a_{m2} & \cdots & a_{mn} \end{bmatrix} \begin{bmatrix} b_{11} & b_{12} & \cdots & b_{1\ell} \\ b_{21} & b_{22} & \cdots & b_{2\ell} \\ \vdots & \vdots & \ddots & \vdots \\ b_{n1} & b_{n2} & \cdots & b_{n\ell} \end{bmatrix},$$

where $C \in \mathbb{R}^{m \times \ell}$, whose elements are given by

$$c_{ij} = \sum_{k=1}^{n} a_{ik} b_{kj}.$$

Note the number of rows in B must equal the number columns in A.

We write $\mathbb{R}^{m \times n} \to \mathbb{R}^{n \times m}$ to indicate a basic matrix operation, for example, to indicate a *matrix transpose* operation.

A.3.2.1 Trace of a Matrix

The *trace* of a *square* matrix is defined as [Gol-96]

$$tr(A) = \sum_{i=1}^{m} a_{ii},$$

from which it follows, because A is always similar to its *Jordan form*, that

$$tr(A) = \sum_{i=1}^{m} \lambda_i.$$

A.3.2.2 Symmetric Matrix
A *symmetic* matrix, $A \in \mathbb{R}^{m \times m}$, is *symmetric* about its *northwest–southeast diagonal*.

A.3.2.3 Persymmetric Matrix
A *persymmetic* matrix, $A \in \mathbb{R}^{m \times m}$, is defined whereby

- it is *symmetric* about its *northeast-southwest diagonal* or
- the values for each perpendicular to the main diagonal (i.e., each antidiagonal) are the same for a particular line—often referred to as a *Hankel* matrix

A.3.2.4 Toeplitz Matrix
A *Toeplitz* matrix, $T \in \mathbb{R}^{m \times m}$, belongs to the class of *persymmetic* matrices, where additionally, the elements of each diagonal are equal, that is, Toeplitz matrices are *banded*. For example, the matrix

$$T = \begin{bmatrix} t_0 & t_1 & & & & & \\ t_{-1} & t_0 & t_1 & & & & \\ & t_{-1} & t_0 & & & & \\ & & & \ddots & & & \\ & & & & t_0 & t_1 & \\ & & & & t_{-1} & t_0 & t_1 \\ & & & & & t_{-1} & t_0 \end{bmatrix}$$

is *Toeplitz*. Toeplitz matrices can occur naturally when dealing with finite difference equations with periodic boundary conditions. See also *circulant* matrix.

A.3.2.5 Transpose
The *transpose* of matrix $A \in \mathbb{R}^{m \times n}$ is defined as

$$B = A^T = \begin{bmatrix} a_{11} & a_{21} & \cdots & a_{m1} \\ a_{12} & a_{22} & \cdots & a_{m2} \\ \vdots & \vdots & \ddots & \vdots \\ a_{1n} & a_{2n} & \cdots & a_{mn} \end{bmatrix},$$

where $B \in \mathbb{R}^{n \times m}$, and the elements of B are given by

$$b_{ij} = a_{ji}.$$

Some properties of the transpose operation include the following:

Transpose of the Transpose: $(A^T)^T = A$
Transpose of a Matrix Sum: $(A + B)^T = A^T + B^T$
Transpose of a Matrix Product: $(AB)^T = B^T A^T$

If $A = A^T$, then A is a *symmetric matrix*.

We define a matrix B to be the *conjugate transpose* of $A \in \mathbb{C}^{m \times n}$ if

$$B = A^* = \begin{bmatrix} a_{11}^* & a_{21}^* & \cdots & a_{m1}^* \\ a_{12}^* & a_{22}^* & \cdots & a_{m2}^* \\ \vdots & \vdots & \ddots & \vdots \\ a_{1n}^* & a_{2n}^* & \cdots & a_{mn}^* \end{bmatrix},$$

where $B \in \mathbb{C}^{n \times m}$, and the elements of B are given by

$$b_{ij} = a_{ji}^*,$$

and a_{ji}^* represents the complex conjugate of a_{ji}.

A.3.2.6 Inverse

Matrix $A \in \mathbb{R}^{m \times m}$ has an *inverse matrix* A^{-1}, if it exists, such that

$$AA^{-1} = I,$$

where I is the identity matrix.

A square matrix for which an inverse does not exist (*noninvertible*) is called *singular*. A square matrix A is singular iff its determinant is zero, that is, $\det(A) = 0$.

See also *pseudoinverse* discussed in Section A.3.7.

A.3.2.7 Circulant Matrix

A *circulant* matrix, $C \in \mathbb{R}^{m \times m}$, belongs to the class of Toeplitz matrices. It is fully specified by its first column (row) vector, with each column (row) being a down (right) *cyclic shifted* version of its predecessor. Thus, the following matrix is circulant:

$$C = \begin{bmatrix} c_0 & c_1 & c_2 & c_3 & \cdots & c_{m-1} \\ c_{m-1} & c_0 & c_1 & c_2 & \cdots & \vdots \\ c_{m-2} & c_{m-1} & c_0 & c_1 & \cdots & \vdots \\ \vdots & \ddots & \ddots & \ddots & & \vdots \\ & & & & & c_1 \\ c_1 & c_2 & c_3 & c_4 & \cdots & c_0 \end{bmatrix}.$$

Every circulant matrix, $C \in \mathbb{R}^{m \times m}$, has a set of eigenvectors

$$\mathbf{r}_k = \frac{1}{\sqrt{m}} \begin{bmatrix} 1 \\ \rho_k \\ \rho_k^2 \\ \vdots \\ \rho_k^{m-1} \end{bmatrix}, \quad \rho_k = e^{-2\pi i (k/m)}, \quad k \in \{0, \ldots, m-1\}, \; i = \sqrt{-1},$$

with corresponding eigenvalues

$$\lambda_k = \sum_{j=0}^{m-1} c_j \rho_k^j.$$

Thus, all circulant matrices have the same eigenvectors, and they can be expressed in the form $C = R \Lambda R^*$, where R is a matrix formed from the eigenvectors, $\mathbf{r}_j$, in order, R^* is the conjugate transpose of R and $\Lambda = \text{diag}(\lambda_j)$ [Gra-96]. All circulant matrices *commute* and are *normal*. If C is symmetric, then the eigenvalues will all be real.

Circulant and Toeplitz matrices can also be complex. In addition, they have properties that enable vector and matrix operations to be performed extremely efficiently. For more information relating to operations on Topelitz and circulant matrices refer to [Gra-96, Gol-96, Tee-05].

A.3.2.8 Useful Matrix Properties

Complex matrices

Consider $A \in \mathbb{C}^{m \times m}$ with inverse A^{-1}, transpose A^T, and conjugate transpose A^*.

- A is *Hermitian* if $A^* = A$.
- A is *unitary* if $A^* = A^{-1}$.
- A is *normal* if $AA^* = A^*A$ (A will also be Hermitian and unitary).

Real matrices

- If A is real, then $A^* = A^T$.
- A is *Hermitian* iff it is *symmetric*.
- A is *orthogonal* if $AA^T = I$.
- A is *unitary* iff it is *orthogonal*.

A.3.3 Determinant

The *determinant* of matrix $A \in \mathbb{R}^{m \times m}$ is denoted $|A|$ or $\det(A)$.

If $A = a$, $a \in \mathbb{R}^{1 \times 1}$, then its determinant [Gol-96] is given by $\det(A) = a$, also written as $|A|$. The determinant of $A \in \mathbb{R}^{m \times m}$ is defined in terms of order $n-1$ determinants, that is,

$$\det(A) = \sum (-1)^{j+1} a_{1j} \det(A_{1j}),$$

where A_{1j} is a $(n-1) \times (n-1)$ matrix obtained by deleting the first row and jth column of A. For example, if $A \in \mathbb{R}^{2 \times 2}$, the determinant is given by

$$\begin{vmatrix} a_{11} & a_{12} \\ a_{12} & a_{22} \end{vmatrix} = a_{11}a_{22} - a_{12}a_{21}.$$

Similarly, if $A \in \mathbb{R}^{3 \times 3}$, the determinant is given by

$$\begin{vmatrix} a_{11} & a_{12} & a_{13} \\ a_{12} & a_{22} & a_{23} \\ a_{13} & a_{23} & a_{33} \end{vmatrix} = a_{11}(a_{22}a_{33} - a_{23}a_{32}) - a_{12}(a_{21}a_{33} - a_{23}a_{31}) \\ + a_{13}(a_{21}a_{32} - a_{22}a_{31}).$$

Some useful properties of the determinant include the following:

$$\begin{aligned} \det(AB) &= \det(A)\det(B), & A, B \in \mathbb{R}^{m \times m}; \\ \det(A^T) &= \det(A), & A \in \mathbb{R}^{m \times m}; \\ \det(cA) &= c^m \det(A), & c \in \mathbb{R}, A \in \mathbb{R}^{m \times m}; \\ \det(A) &\neq 0 \iff A \text{ is nonsingular}, & A \in \mathbb{R}^{m \times m}. \end{aligned}$$

A.3.4 Eigenvalues and Eigenvectors

Given $A \in \mathbb{C}^{m \times m}$, then A has a set of scalar *eigenvalues* (characteristic values) λ_i, generally complex, and a set of nonzero *eigenvectors* (characteristic vectors) $x_i = [x_1, x_2, \ldots, x_m]^T$, which satisfies the following equation:

$$(A - \lambda I) x = 0.$$

The eigenvalues of a A are equal to the roots of its *characteristic polynomial* $p(\lambda) = 0$ of degree n obtained from the following determinant equation:

$$p(\lambda) = \det(A - I\lambda) = \begin{vmatrix} a_{11} - \lambda & a_{12} & \cdots & a_{1m} \\ a_{21} & a_{22} - \lambda & \cdots & a_{2m} \\ \vdots & \vdots & \ddots & \vdots \\ a_{m1} & a_{m2} & \cdots & a_{mm} - \lambda \end{vmatrix} = 0.$$

Now, we know that the roots of the characteristic polynomial $p(\lambda) = 0$ are the eigenvalues of A. Thus, it follows that

$$p(\lambda) = (\lambda_1 - \lambda)(\lambda_2 - \lambda) \cdots (\lambda_m - \lambda) = 0.$$

But on evaluating $p(\lambda)$ at $\lambda = 0$, we obtain

$$p(\lambda)\big|_{\lambda=0} = \det(A - I\lambda)\big|_{\lambda=0}$$
$$\therefore p(0) = \det(A),$$

from which it follows that

$$\det(A) = \prod_{i=1}^{m} \lambda_i.$$

A.3.4.1 Eigenvalue Properties

1. λ_i need not be *distinct*.
2. λ is real $\forall i$ if A is Hermitian.
3. A is positive definite if $\lambda_i > 0$ $\forall i$ and A is Hermitian.
4. The set of all eigenvalues of A is known as the *spectrum* of A.

A.3.5 Singular Values

The *singular values* of a $n \times m$ matrix A are equal to the positive square roots of the eigenvalues of A^*A.

- If $n = m$ and A is *normal*, its singular values are equal to the absolute values of its eigenvalues, that is, $\sigma_i = |\lambda_i|$, $i = (1, \ldots, m)$.
- If $n = m$, A is *nonsingular* iff $\sigma_i > 0$, $i = (1, \ldots, m)$.
- The *condition number* of a matrix is equal to $\kappa = \frac{\sigma_{max}}{\sigma_{min}}$ (see Section A.4.3).

A.3.6 Singular Value Decomposition (SVD)

Given a a matrix $A \in \mathbb{R}^{m \times n}$, then there exists a factorization

$$A = U\Sigma V^T,$$

where U is an $m \times n$ matrix, Σ is an $m \times n$ *diagonal* matrix with *nonnegative* real numbers on the diagonal (usually listed in decreasing order), and V is an $n \times n$ matrix with V^T representing its *transpose*. Matrices U and V are *orthogonal*, that is, having a transpose is equal to its inverse.

The diagonal entries σ_i of Σ are the *singular values* of A, and the preceding factorization is the so-called *singular value decomposition* of A.

If A is complex, then we have

$$A = U\Sigma V^*,$$

where U and V are *unitary* matrices and V^* represents the *conjugate transpose* of V.

A.3.7 Pseudoinverse

The *pseudoinverse* of a matrix $A \in \mathbb{C}^{m \times n}$ is defined as

$$A^+ = (A^*A)^{-1}A^* \in \mathbb{C}^{n \times m},$$

where the symbol $^+$ is used to denote a pseudoinverse. It can be obtained from its singular value decomposition $A = U\Sigma V^*$ (see earlier) and is defined as

$$A^+ = V\Sigma^+ U^*,$$

where Σ^+ is obtained by inverting nonzero diagonal entries of Σ and transposing the result. The pseudoinverse is applied in many situations to solve practical problems, for example, *least squares* minimization. If A is invertable, then $A^+ = A^{-1}$.

A.3.8 Spectral Abscissa

The *spectral abscissa* [Tre-97] of $A \in \mathbb{C}^{m \times m}$ with real or complex *eigenvalues* λ_i, $i = (1, \ldots, m)$, is defined as

$$\phi(A) = \max_{1 \le i \le m} \text{Re} \{\lambda_i\}.$$

A.3.9 Spectral Radius

The *spectral radius* of $A \in \mathbb{C}^{m \times m}$ with real or complex *eigenvalues* λ_i, $i \in \{1, 2 \cdots, m\}$, is defined as

$$\rho(A) = \max_{1 \le i \le m} |\lambda_i|.$$

Generally, for any square matrix [Tre-97],

$$\rho(A) \le \|A\|_2.$$

But, if A is *normal* (see Section A.3.2.8), then

$$\rho(A) = \|A\|_2.$$

The spectral radius of A is also equal to its *dominant eigenvalue*, which is defined such that if

$$|\lambda_1| > |\lambda_2| \geq \cdots \geq |\lambda_m|,$$

then $|\lambda_1|$ is the dominant eigenvalue. The eigenvector associated with the dominant eigenvalue is known as the *dominant eigenvector*.

Also, $\rho(A)$ can be thought of as being equal to the radius of the smallest circle, centered on the origin in the complex plane, whereby all the eigenvalues of A are either inside or on the circle.

A.3.10 Gershgorin's Theorem

A bound on the *maximum eigenvalues* of a matrix can be established by use of the *Gershgorin theorem*, which states the following:

All the eigenvalues λ, of an arbitrary matrix $A \in \mathbb{C}^{m \times m} = (a_{ij})$ over the complex field are located in the union of the set of circular discs defined by

$$|\lambda - a_{ii}| \leq \sum_{\substack{i=1 \\ i \neq j}}^{m} |a_{ij}| \quad \text{OR} \quad |\lambda - a_{ii}| \leq \sum_{\substack{j=1 \\ j \neq i}}^{m} |a_{ij}|.$$

Furthermore, if n of these disks form a connected domain that is disjoint from the remaining m-n disks, there are precisely n eigenvalues of A located within this domain.

For a proof, see [Bur-93].

A.3.11 Diagonal Dominance

A matrix $A \in \mathbb{C}^{m \times m}$ is said to be

- *diagonally dominant*
  by *row* if $|a_{ii}| \geq \sum_{\substack{i=1 \\ i \neq j}} |a_{ij}|$, or
  by *column* if $|a_{ii}| \geq \sum_{\substack{j=1 \\ j \neq i}} |a_{ij}|$.
- *strictly diagonally dominant*
  by *row* if $|a_{ii}| > \sum_{\substack{i=1 \\ i \neq j}} |a_{ij}|$, or
  by *column* if $|a_{ii}| > \sum_{\substack{j=1 \\ j \neq i}} |a_{ij}|$.

As a consequence, if A is strictly diagonally dominant, then it has an inverse.

A.3.12 Matrix Diagonalization

If $A \in \mathbb{R}^{m \times m}$ has m linearly independent eigenvectors r^i, $i \in \{1, 2, \ldots m\}$, then it is nonsingular and can be diagonalized. With A as described, we have

$$R^{-1}AR = \Lambda \text{ and } A = R\Lambda R^{-1},$$

where $\Lambda = \text{diag}\{\lambda_i\}$ and

$$\Lambda = \begin{bmatrix} \lambda_1 & & & \\ & \lambda_2 & & \\ & & \ddots & \\ & & & \lambda_m \end{bmatrix}$$

and R is a square matrix formed from the m linearly independent eigenvectors, that is,

$$R = \begin{bmatrix} r^1, & r^2, & \cdots, & r^m \end{bmatrix}.$$

As such, R will be nonsingular and have an inverse. Matrices A and R are called similar, as they have the same characteristic polynomial. This is important in linear algebra and many applications of numerical analysis as it allows certain coupled problems to be decoupled. For instance, consider the following ODE:

$$\frac{dy(t)}{dt} = Ay(t), \quad y(x,t) = [y_1, y_2, \ldots, y_m], \quad y(0) = y_0.$$

If we assume that matrix A of the preceding equation has distinct eigenvectors, this means it can be diagonalized and we can transform the ODE to

$$R^{-1}\frac{dy}{dx} = R^{-1}ARR^{-1}y.$$

Now, if we define $q(t) \equiv R^{-1}y$, we obtain

$$\frac{dq(t)}{dx} = \Lambda q(t), \quad q(t) = [q_1, q_2, \ldots, q_m], \quad q(0) = q_0.$$

Because Λ is diagonal, the problem has been transformed into a set of m decoupled ODE equations, where $q(t)$ are called the characteristic variables. Therefore,

$$\frac{dq_i}{dt} = \lambda_i q_i, \quad i \in \{1, 2, \ldots m\}.$$

This is an important result as it means that each of the decoupled equations can be integrated separately at each time step, and the original variables recovered from $y = Rq$. A similar approach can be taken with PDEs.

A.3.13 Matrix Powers

If $A \in \mathbb{R}^{m \times m}$ can be diagonalized, such that $A = R\Lambda R^{-1}$, with Λ and R being $m \times m$ matrices and $\Lambda = \text{diag}\{\lambda_i\}$, $i = (1, \ldots, m)$, then we can decompose the *nth power* of A,

A^n, as follows:

$$A^2 = (R\Lambda R^{-1})(R\Lambda R^{-1}) = (R\Lambda^2 R^{-1}),$$
$$A^3 = (R\Lambda R^{-1})(R\Lambda R^{-1})(R\Lambda R^{-1}) = (R\Lambda^3 R^{-1}),$$
$$\vdots$$
$$A^n = (R\Lambda R^{-1})(R\Lambda R^{-1})\cdots(R\Lambda R^{-1}) = (R\Lambda^n R^{-1}).$$

For a diagonalizable matrix A, and consistent matrix norm, we can put a condition on A such that $\|A^n\|$ will remain bounded as $n \to \infty$. Recalling the definition of spectral radius ρ, and that $\rho(\Lambda) = \rho(A)$, it follows that

$$A^n = R\Lambda^n R^{-1}$$
$$\therefore \|A^n\| \leq \|R\| \|\Lambda^n\| \|R^{-1}\| = \|\Lambda\|^m \operatorname{cond}(R)$$
$$\leq [\rho(\Lambda)]^n \operatorname{cond}(R)$$
$$\leq \left[\max_i |\lambda_i|\right]^n \operatorname{cond}(R),$$

where $\operatorname{cond}(R) = \|R\|\|R^{-1}\|$ is the condition number of the transformation. Consequently, if $\|A^n\|$ is to remain bounded as $n \to \infty$, then we must have the condition that $|\lambda_i| \leq 1$ for all i.

A.3.14 Matrix Exponential

If $A \in \mathbb{R}^{m \times m}$ can be diagonalized, then we define an *exponential matrix* in terms of a matrix power series, as follows:

$$e^{At} = \sum_{n=0}^{\infty} \frac{t^n A^n}{n!} = I + tA + \frac{t^2 A^2}{2} + \frac{t^3 A^3}{6} + \cdots$$

It can be shown that the preceding matrix series converges absolutely and uniformly for finite t. Furthermore, it follows from application of the power matrix definition that

$$e^{At} = \sum_{n=0}^{\infty} \frac{t^n R\Lambda^n R^{-1}}{n!}$$
$$= I + tR\Lambda R^{-1} + \frac{t^2 R\Lambda^2 R^{-1}}{2} + \frac{t^3 R\Lambda^3 R^{-1}}{6} + \cdots$$
$$= R\left(I + t\Lambda + \frac{t^2}{2}\Lambda^2 + \frac{t^3}{6}\Lambda^3 + \cdots\right)R^{-1}$$
$$= Re^{\Lambda t}R^{-1},$$

where $\Lambda = \operatorname{diag}\{\lambda_i\}$ and

$$e^{\Lambda t} = \begin{bmatrix} e^{\lambda_1 t} & & & \\ & e^{\lambda_2 t} & & \\ & & \ddots & \\ & & & e^{\lambda_m t} \end{bmatrix}.$$

From the preceding definition of the exponential matrix, it follows that we can apply differentiation as follows:

$$\frac{d}{dt}\left[e^{At}\right] = A + \frac{A^2 t}{2} + \frac{A^3 t^2}{6} + \cdots = Ae^{At} = e^{At}A,$$

and in general,

$$\frac{d^k}{dt^k}\left[e^{At}\right] = e^{At}A^k.$$

Taking a similar approach, we can apply integration as follows:

$$\int_0^t e^{A\tau} d\tau = It + \frac{At^2}{2!} + \frac{A^2 t^3}{3!} + \frac{A^3 t^4}{4!} + \cdots,$$

$$\therefore A \int_0^t e^{A\tau} d\tau = e^{At} - I,$$

$$\Downarrow$$

$$\int_0^t e^{A\tau} d\tau = \left(e^{At} - I\right) A^{-1}.$$

Evaluation of e^{At} for known A can be accomplished to within a desired tolerance by an iterative procedure [Shi-75, pp. 81–83]. A comprehensive review of calculating the exponential matrix is given by Moler and Loan [Mol-03].

Other interesting properties of the matrix exponential [Elg-67] are

- $e^{At} e^{-At} = I$. Thus, the inverse of e^{At} is computed simply by replacing t with $-t$.
- $e^{A(t_1+t_2)} = e^{At_1} e^{At_2} = e^{At_2} e^{At_1}$. Thus, e^{At_1} and e^{At_2} commute.
- $\frac{d}{dt}(A) = Ae^{At} = e^{At}A$. Thus, e^{At_1} and A commute.

However, it should be cautioned that $e^{At} e^{Bt} = e^{(A+B)t}$ iff A and B commute, that is, $AB = BA$.

It follows that for a diagonalizable matrix A, and consistent matrix norm,

$$e^{At} = e^{R \Lambda R^{-1} t} = R e^{\Lambda t} R^{-1},$$

$$\therefore \left\| e^{At} \right\| \leq \|R\| \left\| e^{\Lambda t} \right\| \|R^{-1}\|,$$

$$\leq \left\| e^{\Lambda t} \right\| \operatorname{cond}(R),$$

where $\operatorname{cond}(R) = \|R\| \|R^{-1}\|$ is the condition number of the transformation. Consequently, if $\|e^{At}\|$ is to remain bounded as $t \to \infty$, then we must have the condition that the *spectral abscissa* $\phi(\Lambda) \leq 0$, or $Re\{\lambda_i\} \leq 0$, $\forall i$. Thus, the condition for boundedness becomes

$$\left\| e^{At} \right\| \leq \left\| e^{\Lambda t} \right\| \operatorname{cond}(R), \quad \phi(\Lambda) \leq 0.$$

A.3.15 Jordan Blocks

A matrix $J_B \in \mathbb{C}^{m \times m}$ is a Jordan block if it is constructed such that

1. all diagonal elements are equal, that is, $j_{ii} = j_{kk}, i, k = (1, \ldots, m)$
2. each element directly above a diagonal element is equal to 1, that is, $j_{i-1\,i} = 1, i = (2, \ldots, m)$

3. all other entries are zero

Thus, the matrix $\begin{bmatrix} 3 & 1 & 0 \\ 0 & 3 & 1 \\ 0 & 0 & 3 \end{bmatrix}$ is a Jordan block, while matrix $\begin{bmatrix} 3 & 1 & 0 \\ 0 & 3 & 0 \\ 0 & 0 & 3 \end{bmatrix}$ is not. The eigenvalues of a $m \times m$ Jordan block matrix are equal to its diagonal elements; that is, it has a single distinct eigenvalue of multiplicity m.

A.3.16 Jordan Canonical Form

A matrix $J \in \mathbb{C}^{m \times m}$ is in *Jordan canonical form* if it consists of Jordan blocks, placed corner to corner along the main diagonal, with only zero elements outside these blocks. Thus, by definition, all diagonal matrices are in Jordan canonical form. An extremely useful consequence of a matrix being in Jordan canonical form is that its *eigenvalues are equal to the principle diagonal elements*. Thus, the matrix has a set of distinct eigenvalues *equal to the number of Jordan blocks*, each of which has *multiplicity equal to the size of its associated Jordan block*.

The following is an example of a 6×6 matrix in Jordan canonical form:

$$J = \begin{bmatrix} \lambda_1 & 1 & 0 & 0 & 0 & 0 \\ 0 & \lambda_1 & 1 & 0 & 0 & 0 \\ 0 & 0 & \lambda_1 & 0 & 0 & 0 \\ 0 & 0 & 0 & \lambda_2 & 0 & 0 \\ 0 & 0 & 0 & 0 & \lambda_3 & 1 \\ 0 & 0 & 0 & 0 & 0 & \lambda_3 \end{bmatrix},$$

where the individual Jordan blocks are

$$J_{B1} = \begin{bmatrix} \lambda_1 & 1 & 0 \\ 0 & \lambda_1 & 1 \\ 0 & 0 & \lambda_1 \end{bmatrix}, \quad J_{B2} = [\lambda_2], \quad J_{B3} = \begin{bmatrix} \lambda_3 & 1 \\ 0 & \lambda_3 \end{bmatrix},$$

and the eigenvalues are equal to λ_1 with multiplicity of 3; λ_2 with multiplicity of 1; and λ_3 with multiplicity of 2.

A.3.17 Cayley–Hamilton Theorem

The Cayley–Hamilton theorem states that *every square matrix A satisfies its own characteristic equation*. Thus, for $A \in \mathbb{R}^{n \times n}$, if the characteristic equation is represented by

$$a_n \lambda^n + a_{n-1} \lambda^{n-1} + \cdots + a_1 \lambda^1 + a_0 = 0,$$

where λ is an eigenvalue of A, then the following matrix equation, in powers of A, is also valid:

$$a_n A^n + a_{n-1} A^{n-1} + \cdots + a_1 A^1 + A_0 = 0.$$

Example

Find the inverse of matrix $A = \begin{bmatrix} 1 & 1 & 1 \\ 0 & 1 & 2 \\ 1 & 2 & 4 \end{bmatrix}$ using the Cayley–Hamilton theorem.

The characteristic equation is found from $|A - I\lambda| = 0$ to be

$$\lambda^3 - 6\lambda^2 + 4\lambda - 1 = 0.$$

Thus, by definition,

$$A^3 - 6A^2 + 4A - I = 0,$$

from which it follows that

$$A^{-1} = A^2 - 6A + 4I = \begin{bmatrix} 0 & -2 & 1 \\ 2 & 3 & -2 \\ -1 & -1 & 1 \end{bmatrix}.$$

A.4 NORMS

A.4.1 Vector Norms

A *norm* is a function that assigns a real-valued length to a vector. There are different notions of length, and to conform to this definition, norms must satisfy the following three criteria for all vectors $x, y \in \mathbb{C}^n$ and all scalars $\alpha \in \mathbb{C}$:

1. $\|x\| \geq 0$, and $\|x\| = 0$ iff $x = 0$
2. $\|x + y\| \leq \|x\| + \|y\|$
3. $\|\alpha x\| = |\alpha| \|x\|$

Various vector norm definitions

- ℓ_∞ norm (infinity norm)
  $$\|x\|_\infty = \max_{1 \leq i \leq n} |x_i|$$
- ℓ_1 norm
  $$\|x\|_1 = \sum_{i=1}^n |x_i|$$
- ℓ_2 norm
  $$\|x\|_2 = \left(\sum_{i=1}^n |x_i|^2 \right)^{1/2}$$
- ℓ_p norm
  $$\|x\|_p = \left(\sum_{i=1}^n |x_i|^p \right)^{1/p}$$
- Parseval's equality
  $$\|v\|_2 = \|V\|_2,$$
  where V represents the Fourier transform of v.

A.4.2 Matrix Norms Induced by Vector Norms

Given $A \in \mathbb{C}^{m \times n}$, $x \in \mathbb{C}^n$, and a *vector norm* $\| \cdot \|$, the corresponding *natural* or *induced matrix norm* is defined by

$$\|A\| = \max_{x \neq 0} \frac{\|Ax\|}{\|x\|}.$$

Matrix norms need not be induced by vector norms. However, they must satisfy the following four criteria for $A, B \in \mathbb{C}^{m \times n}$, and all scalars $\alpha \in \mathbb{C}$:

1. $\|A\| \geq 0$, where $\|A\| = 0$ iif $A = 0$ (matrix with all zero entries)
2. $\|\alpha A\| = |\alpha| \|A\|$
3. $\|A + B\| \leq \|A\| + \|B\|$
4. $\|AB\| \leq \|A\| \|B\|$ (Cauchy–Schwarz inequality)

Various matrix norm definitions

- ∞ norm (infinity norm)
$$\|A\|_\infty = \max_{1 < i < m} \sum_{j=1}^{n} |a_{ij}|$$
- 1-norm
$$\|A\|_1 = \max_{1 < j < n} \sum_{i=1}^{m} |a_{ij}|$$
- 2-norm
$$\|A\|_2 = \sqrt{\rho(A^*A)} \geq \rho(A),$$
where $\rho(\cdot)$ is the *spectral radius* (see section A.3.9)
- Frobenius norm
$$\|A\|_F = \sqrt{\sum_i \sum j (a_{ij})^2} = \sqrt{Tr(AA^*)}$$

A.4.3 Matrix Condition Number

The *condition number* of a matrix A, based on the *2-norm*, is defined as

$$\text{cond}(A) = \kappa = \|A\| \|A^{-1}\|.$$

Alternatively, it is equal to the ratio of the largest singular value of A to the smallest singular value, that is,

$$\kappa = \frac{\sigma_{max}}{\sigma_{min}}.$$

If A is normal, that is, $AA^* = A^*A$, then it follows that

$$\kappa = \frac{|\lambda(A)|_{max}}{|\lambda(A)|_{min}}.$$

The condition number is related to how numerically *well-posed* the problem is. Alternatively, the condition number, κ, can be taken to be a measure of the sensitivity of a matrix when subjected to to numerical operations, that is, it provides a sensitivity bound for the solution of a linear equation. For example, if $Ax = b$ and $A(x + \Delta x) = b + \varepsilon$, then $\|\Delta x\|/\|x\| \leq \kappa \|\varepsilon\|/\|b\|$.

A problem having a low condition number, $\kappa \simeq 1$, is said to be well-conditioned, whereas a problem having a high condition number, $\kappa \gtrsim 10^6$, for single precision or $\kappa \gtrsim 10^{12}$ for double precision, is said to be ill-conditioned [Pre-92]. Thus, a large condition number can result in large errors in x even though the errors in b are small. On the other hand, a small condition number means that the errors in x will be of similar relative magnitude to those in b.

A.5 DIFFERENTIATION AND INTEGRATION

A.5.1 General

Smooth functions: $f \in C^k$. A function f that has k successive derivatives is defined as being k times differentiable. If the k_{th} derivative is also continuous, then the function is said to belong to the *differentiability class* C^k. If f is infinitely differentiable, then $f \in C^\infty$.

Lipschitz continuous: A function $f(x)$ is said to be Lipschitz continuous if there exists a constant L, the *Lipschitz constant*, such that

$$|f(a) - f(b)| \leq L|a - b|.$$

See also Section A.7.6.

A.5.2 Operations on Vectors and Matrices

Differentiation and integration operations on vectors and matrices are carried out element by element; that is, for vector $x \in \mathbb{R}^m$ and matrix $A \in \mathbb{R}^{m \times m}$, we have as follows:

- $\dfrac{dv}{dt} = \left[\dfrac{dv_1}{dt}, \dfrac{dv_2}{dt}, \ldots, \dfrac{dv_m}{dt}\right]^T$
- $\int v \, dt = \left[\int v_1 dt, \int v_2 dt, \ldots, \int v_m dt\right]^T$
- $\dfrac{dA}{dt} = \dfrac{d}{dt}\{a_{ij}\}$
- $\int A \, dt = \int \{a_{ij}\} \, dt$

A.5.3 Taylor Series

A Taylor series is a series expansion of a function $f(x)$ about a point, say, x_0, given by

$$f(x) = \sum_{n=0}^{\infty} \frac{f^n(x_0)}{n!}(x - x_0)^n, \tag{A.4}$$

where $n!$ represents factorial n and f^n the nth derivative of f. If $x_0 = 0$, the series is known as the *Maclaurin series*.

A.5.4 Multi-index Operation

The multi-index differential operator is defined as

$$\partial^\alpha u = \frac{\partial^{|\alpha|}}{\partial x_1^{\alpha_1} \partial x_2^{\alpha_2} \cdots \partial x_n^{\alpha_n}} u,$$

where $|\alpha| = \sum_{i=1}^{n} \alpha_i$ and u is a smooth function such that the order in which derivatives are taken is arbitrary. Thus, if $u = u(x, y, z)$ and $\alpha = (1, 3, 2)$, we get

$$\partial^\alpha u = \frac{\partial^{|\alpha|}}{\partial x \partial y \partial z} u = \partial_x \partial_y^3 \partial_z^2 = \frac{\partial^6 u}{\partial x \partial y^3 \partial z^2}.$$

A.6 DIFFERENCE OPERATORS

Difference operators are used in finite difference schemes and are defined as follows:

- *Forward-difference operator*, Δ

$$\Delta f(x) = f(x+h) - f(x), \tag{A.5}$$

where h is the step size in x. Alternatively, if we define $x = x_k$ as the value of x at the kth step with $y_k = f(x_k)$, we obtain the following equivalent form:

$$\Delta y_k = y_{k+1} - y_k. \tag{A.6}$$

The following recurrence relationship follows:

$$\begin{aligned}\Delta^r y_k &= \Delta^{r-1} y_{k+1} - \Delta^{r-1} y_k, \quad r = 1, 2, \ldots, \\ &= \sum_{j=0}^{r} (-1)^j \binom{r}{j} y_{k+r-j},\end{aligned} \tag{A.7}$$

with $\Delta^0 y_k \triangleq y_k$,

where $\binom{r}{j}$ is a binomial coefficient.

- *Backward-difference operator*, ∇,

$$\nabla f(x) = f(x) - f(x-h), \tag{A.8}$$

where h is the step size in x. Alternatively, as for forward differences, if we define $x = x_k$ as the value of x at the kth step with $y_k = f(x_k)$, we obtain the following equivalent form:

$$\nabla y_k = y_k - y_{k-1}. \tag{A.9}$$

The following recurrence relationship follows:

$$\begin{aligned}\nabla^r y_k &= \nabla^{r-1} y_k - \nabla^{r-1} y_{k-1}, \quad r = 1, 2, \ldots, \\ &= \sum_{j=0}^{r} (-1)^j \binom{r}{j} y_{k-j},\end{aligned} \tag{A.10}$$

with $\nabla^0 y_k \triangleq y_k$.

A.7 GENERAL DEFINITIONS

A.7.1 Stability

Stability can be considered to be an intrinsic property of a dynamical system's behavior when its equilibrium is disturbed. A system's equilibrium state can be classified as *unstable*, *marginally stable*, or *stable*, as follows:

- An *unstable system* is one that when disturbed, becomes unbounded. Mathematically, this implies that one or more of the system eigenvalues has a positive real part.

- A *marginally stable system* is one that when disturbed continues to change and does not settle to a constant state; but it does remain bounded. Mathematically, this implies that one or more of the system eigenvalues has a real part equal to zero, with the other eigenvalues having negative real parts.
- A *stable system* is one that when disturbed either returns to the original equilibrium state or moves to a new equilibrium state. Mathematically, this implies that all of the system eigenvalues have negative real parts.

A.7.2 Well-Posedness

Well posed problems are those that are both solvable and uniquely solvable. Mathematical models of continuum physical processes should be required to exhibit the following characteristics [Had-02]:

- a solution exists
- the solution is unique
- the solution depends continuously on the initial data, that is, a small change in the data results in a small change in the solution

Alternatively, but less general, we can say that an initial-value problem is *well posed* if the system, together with any bounded initial conditions, is *marginally stable*.

If a problem is not well-posed, it needs to be reformulated, which typically includes adding additional assumptions, such as *smoothness of solution* or bounds on the *total variation* (e.g., total variation diminishing). This process is known as *regularization*. To solve a well posed *continuum problem* numerically requires the problem to be transformed into a well posed *discrete problem*. This may require additional constraints to be imposed, such as a maximum step size or maximum grid spacing.

Problems that are not well posed are called *ill-posed*.

A.7.3 Lax–Richtmyer Equivalence Theorem

A consistent finite difference scheme for a partial differential equation for which the initial value problem is well-posed is convergent if and only if it is stable.

A.7.4 Lax Equivalence

Given a properly posed initial value problem and a finite-difference approximation to it that satisfies the consistency condition, stability is the necessary and sufficient condition for convergence [Ric-67], that is,

$$\text{consistency} + \text{stability} \Leftrightarrow \text{convergence}.$$

A.7.5 Initial Value Problem (Cauchy Problem)

An equation composed of a differential equation with an initial condition, that is,

$$\frac{dx}{dt} = f(t, x(t)), \quad x \in \mathbb{R}^m, \ f : \mathbb{R} \times \mathbb{R}^m \to \mathbb{R}^m, \ x(t_0) = x_0, \ t \geq t_0,$$

is called an *initial value problem* (IVP), or, equivalently, a *Cauchy problem*, if $x(t)$ is differentiable and the differential equation has a unique solution [Lam-91]. The function f need not be *analytic* but must be *smooth*. The term was coined by J. Hadamard [Had-21].

A.7.6 Lipschitz Condition

The *initial value problem* (IVP)

$$\frac{dx}{dt} = f(t, x(t)), \quad x \in \mathbb{R}^m, \; f : \mathbb{R} \times \mathbb{R}^m \to \mathbb{R}^m, \; x(t_0) = x_0, \; t \geq t_0,$$

is said to satisfy the *Lipschitz condition* or to be *Lipschitz continuous* of order n with respect to $x \in \mathbb{R}^m$ if there exists a *finite constant* $L \geq 0$ such that for any two points (t, x) and (t, x^*) in the solution domain $D = \{t_0 \leq 0 \leq T, \|x\| < \infty\}$ the relationship

$$\|f(t, x) - f(t, x^*)\| \leq L \|x - x^*\|^n$$

holds for all $x, x^* \in \mathbb{R}^m$, $t \geq t_0$, $n > 0$, and T finite.

- If the function $f(t, x(t))$ of the ODE is continuous with respect to (t, x) and is continuously differentiable with respect to x, then f is *locally Lipschitz*.
- If the function $f(t, x(t))$ is defined and continuous on D and also satisfies the Lipschitz condition, then $f(t, x(t))$ has a *unique solution* in D.

In practice, the size of the *Lipschitz constant* L is significant for numerical approximations. Its size is an indicator as to whether a solution curve is likely to stay close to the true solution or will diverge.

REFERENCES

[Bur-93] Burden, R. L. and J. D. Faires (1993), *Numerical Analysis*, PWS.
[Elg-67] Elgerd, O. I. (1967), *Control Systems Theory*, McGraw-Hill.
[Gan-60] Gantmancher, F. R. (1960), *Matrix Theory*, Vol. 2, Chelsea.
[Gan-77] Gantmancher, F. R. (1977), *Matrix Theory*, Vol. 1, Chelsea.
[Gol-96] Golab, G. H. and C. F. van Loan (1996), *Matrix Computations*, The Johns Hopkins University Press.
[Gra-96] Gray, R. M. (1996), *Toeplitz and Circulant Matrices: A Review*, Now Publishers, available online at http://www-ee.stanford.edu/~gray/toeplitz.pdf.
[Had-02] Hadamard, J. (1902), Sur les problèmes aux dérivées partielles et leur signification physique. *Princeton University Bulletin* **13**, 49–52.
[Had-21] Hadamard, J. (1921), *Lectures on Cauchy's Problem in Linear Partial Differential Equations*, Yale University.
[Jen-92] Jennings, A. and J. J. McKeown (1992), *Matrix Computation*, John Wiley.
[Lam-91] Lambert, J. D. (1991), *Numerical Methods for Ordinary Differential Systems—The Initial Value Problem*, John Wiley.
[Mol-03] Mole, C. and C. van Loan (2003), Nineteen Dubious Ways to Compute the Exponential of a Matrix, Twenty-Five Years Later, *SIAM Review* **45**-1, available online at http://www.siam.org/journals/sirev/45-1/41801.html http://www.siam.org/journals/sirev/45-1/41801.html.
[Pre-92] Press, W. H., S. A. Teukolsky, W. T. Vetterling and B. P. Flannery (1992), *Numerical Recipes in C*, Cambridge University Press.
[Ric-67] Richtmyer, R. D. and K. W. Morton (1967), *Difference Methods for Initial Value Problems*, 2nd ed., John Wiley.

[Shi-75] Shinners, S. M. (1975), *Modern Control System Theory and Application*, Addison-Wesley.
[Tee-05] Tee, G. J. (2005), Eigenvectors of Block Circulant and Alternating Circulant Matrices, *Research Letters in the Information and Mathematical Sciences*, **8**, 123–142. http://mro.massey.ac.nz/handle/10179/4456.
[Tre-97] Trefethen, L. N. and D. Bau (1997), *Numerical Linear Algebra*, SIAM.

Index

<-, *see* assignment symbol
<<-, *see* super-assignment symbol
`interp1`, 556
`lsodes`, 540
`roots`, 549

absorptivity, 562, 566
acidity
   surface seawater, 552
Acknowledgments, xix
Adams Bashforth method, 97
Adams Moulton method, 97
advection equation, 228, 233, 234
Air Mass, 562, 564
air-ocean exchange, 544–546
albedo, 560
alkalinity (Alk), 543
Allen–Cahn equation, 415
`alphashape3d`, 449
ambient conditions, 480, 490, 491, 492
   data, 481
amplification factor, 265
amplification ratio, 244
angular momentum, 488
ansatz, 8
Antoine equation, 481
assignment symbol, xvi
atomic explosion, 508
   fire ball, 518
   photographic evidence, *see* time-lapsed photographs
   systems analysis, 508
   time-lapsed photographs, 508, 516, 531
      distance scale, 508, 516
   yield, 520
avoirdupois system, 480

$B(OH)_4^-$, tetrahydroxyborate, 548

$B(OH)_3$, boric acid, 548
Backau, the, 476
basis
   function, 380
BDF method, 89
BDFs, *see* Linear multistep methods, Backwards differentiation formulas
Beer's Law, 562
black body, 558
Black-Scholes model, 10n7
blending function, *see* basis function
Boltzmann's constant, 558
boron
   dissociation constant, K_B, 548, 576
   total, B_T, 548
boundary condition, 103–105, 409
   absorbed, 105
   Dirichlet, 104, 120, 121, 122, 124, 130, 136, 141, 148, 150, 153, 170, 178, 185, 220, 221, 406, 409, 415, 419, 430, 435, 441, 443, 446
   discontinuous with IC, 105
   Neumann, 104, 122, 130, 136, 148, 150, 170, 217, 220, 221, 409
   periodic, 104, 225, 233, 235, 236
   Robin, 104
Boussinesq equation, 468

$CaCO_3$, calcium carbonate, 552
   counter pump, 552
   supersaturated, 553
   under-saturated, 553
calculus, 1
Carbon Cycle, 539
carbonate chemistry, 546–552
carbonic acid
   dissociation constants, K_{a1} and K_{a2}, 547, 577
Cauchy problem, *see* initial value problem

Index

Cauchy-Schwarz inequality, 227, 228, 244, 254, 600
circulation, 476
climate change, 539
climate model, 539–571
   acidity of surface seawater, 552
   air-ocean exchange, 544
   atmosphere, 542
   carbonate chemistry, 546
   climate sensitivity, 566
   correlations—thermodynamic, 576
   Earth's radiation balance, 557
   emission profiles, 554–557
      calculations, 554
   fossil fuel emissions, *see* climate model, emission profiles
   ocean circulation, 553
   oceans, 543
   ODEs, 542–544
   parameter values, 539
   piston velocity, 542, 545
   radiation from the Earth, 560
   radiation from the Sun, 557
   sea-level rise, 570
   simulation results, 571–574
      atmosphere-seawater fluxes, 572
      carbon build-up in surface seawater, 572
      carbon build-up in the atmosphere, 571
      deep ocean dissolved inorganic carbon concentration and pH, 573
      Revelle factor, 574
      seawater buffering, 574
      seawater dissolved inorganic carbon distribution, 573
      surface seawater dissolved inorganic carbon and pH, 573
      surface temperature changes, 575
      temperature anomaly, 575
      uptake factor, 574
   single layer radiation model, 565
      example calculation, 566
$CO_2(aq)$, dissolved carbon dioxide, 547
CO_3^{2-}, carbonate, 547
collocation point, 384
compressibility factor
   air, 481
conservation law, 107, 126, 161, 210, 287, 288, 289, 290, 292, 298, 302, 303, 304, 312, 316, 324, 325, 337, 457
   Boussinesq equation, 468
   conserved quantity, 458
   constant of motion, 458
   decay conditions, 458, 462, 464, 466, 468
      at infinity, 457, 458
   density, the, 458
   energy, 457
   flux,the, 458
   infinity of, 461, 465
   invariant, 458
   KdV equation with variable coefficients, 464
   Korteweg–de Vries equation, 459
      equations of motion, 463
      first conservation law, u, 459
      invariant, 463
      second conservation law, u^2, 459
      third conservation law, $u^3 + \frac{1}{2}u_x^2$, 460
      undefined KdV conservation law, 461
   Korteweg–de Vries equation: 2D, 463
   local conservation laws, 457
   mass, 457, 466, 468
   momentum, 457, 467, 468
   Nonlinear Schrödinger equation, 466
conservation of energy, 483
conservation of momentum, 483
continuity equation, 511
convection–diffusion equation, 236
Coriolis force, 554
corneal curvature, 129
correlations—chemical equilibrium and dissociation constants, 576
cross product, 478, 503
   right hand rule, 479
CSRBF, *see* radial basis function, compact support

Dahlquist
   first barrier theorem, 75
   second barrier theorem, 75
   test problem, 75
DE, *see* differential equation
delaunayn, 449
derivative, 1
   analytical, 4
   lookup table, 4
   single-sided, 1
   two-sided, 2
derivative matrix coefficients, 211
   first derivative schemes, 211
   fourth derivative schemes, 216
   second derivative schemes, 213
   third derivative schemes, 215
derivative matrix library, 217–221
derivative options, 10
deSolve, 70, 124, 411, 431
determinant, 478
deterministic process, 10
difference operators, 602
differential equation, 1
   matrix form, 10
   terms used to classify, 7
      homogeneous, 7
      linear, 7
      nonhomogeneous, 7
      order, 7
differential operator
   linear, 406

differentiation and integration, 601
  multi-index operation, 601
  operations on vectors and matrices, 601
  Taylor series, 601
dimensional analysis, *see* similarity analysis
dispersion, 267
  errors, 269–277
dispersion relation, 264
dissipation, 266
  errors, 269–277
dissolved inorganic carbon (DIC), 543
drag
  coefficients, 479

Earth's radiation balance, 557–571
eigenvalues, 425
electromagnetic spectrum, 559
electrostatics problem, 446
  charge density, 446
elliptical domain, 446
emissivity
  effective, 560
  surface, 558
environment, xvi
  hierarchy, xvi
equation of state, 509, 511
Euler equations, 230, 529
  conservative form with source, 529
  eigenvalues, 351
Euler methods, 76
  backward, 12
  forward, 11

false transients, method of, 130, 148, 150, 185
`fields`, 441
finite difference approximation
  backward, 118
  central, 118
  forward, 118
finite difference matrices, 115–123
finite volume method, 207–210
  1D conservative system, 208
  general conservation law, 210
Fisher–Kolmogorov equation, 403, 418, 430
Fisher–Kolmogorov-Petrovsky-Piscunov equation,
    see Fisher–Kolmogorov equation
FK, *see* Fisher–Kolmogorov equation
FKPP, *see* Fisher–Kolmogorov-Petrovsky-Piscunov
    equation
Flettner rotor
  the Backau, 476
  the E-ship, 476
Flettner, Anton, 476
flow
  laminar, 472
flux limiter, 292–298
  admissible limiter region, 295

Euler equation problems
  Sod's shock-tube, 317
  Taylor-Sedov blast wave, 322
  Woodward-Colella interacting blast wave,
    319
  how they work, 293
  limiter functions
    HCUS, 294
    HQUICK, 294
    Koren, 294
    minmod, 294
    monotonized central (MC), 294
    Osher, 294
    ospre, 294
    smart, 294
    superbee, 294
    Sweby, 294
    UMIST, 294
    van Albada 1, 294
    van Albada 2, 295
    van Leer 1, 295
    van Leer 2, 295
  scalar problems
    advection, 305, 314
    Buckley-Leverett equation, 309, 315
    Burgers equation, 309, 315
  Sweby diagram, 295
force
  drag, 472
    coefficient, 472, 475
    friction drag, 472
    pressure drag, 472
    sheer stress, 472
  gravitational, 479
  inertial, 472
  Magnus, 476
  lift, 476
  Newton, 476
  viscous, 472
fossil fuel reserves, 555
Fourier series, 380, 441
Fourier transform, 227
  pairs, 262
  properties, 261
Franke's function, 389, 452
  defined, 452
frequency spectrum, 243
frontogenesis, 348, *see* partial differential
    equation
fugacity, 547
fully discrete methods, 194–207
  backward in time central in space,
    BTCS—implicit, 195
  Beam-Warming (B-W), 196
  central in time central in space, CTCS or leapfrog,
    124, 196
  Crank-Nicholson, 196

fully discrete methods (*cont.*)
   forward in time backward in space, FTBS, 194
   forward in time backwards in space, FTBS—implicit, 194
   forward in time central in space, FTCS, 195
   Lax-Friedrichs (LxF), 196
   Lax-Wendroff (LxW), 196

gain factor, 244
Gardner transformation, 462
gas laws, 510
   Rankine-Hugonoit conditions, 512
      continuity, 512
      energy, 512
      momentum, 512
      shock wave conditions, 514
   useful relations, 512
      calorically perfect gas, 512
      density, 512
      enthalpy, 512
      ideal gas law, 512
      internal energy, 512
      Mach number, 513
      ratio of specific heats, 512
      specific heat, 512
      specific volume, 512
      speed of sound, 512
gets, xvi
Godunov method, 288
Godunov's order barrier theorem, *see* Godunov's theorem
Godunov's theorem, 290
golf ball
   Balata, 480
   coefficient of restitution, 483
   compression, 483
   construction, 480
   drag effect at ground level, 497
   moment of inertia, 488
   size, 480
   Surlyn, 480
   weight, 480
   wind tunnel tests, 477
golf ball flight, 470–506
   aerodynamic torque, 484
   air speed, 474
   ambient conditions, 480, 490, 491, 492, 493
   angular momentum, 488, 503
   back spin, 483
   bounce and roll, 485
   carry distance, 486
   completing the mathematical description, 487
   computer simulation, 489
      driver shots, 490
      iron shots, 491
      wood shots, 491
   conservation of energy, 483
   conservation of momentum, 483
   drag/lift test, 497
   draw, 484
   draw/fade/hook/slice, 495
   fade, 484
   gyroscopic effect of rotation, 475
   hook, 484
   inclined spin axis, 495
   launch angle, 484
   launch speed, 484, 486
   lift, 476
   peripheral speed, 474
   push/pull, 495
   shot statistics, 486
      carry data, 489
      compared to simulation, 489
      PGA tour averages, 486
   slice, 484
   spin, 484
   spin rate, 483
   spin-rate decay, 484
   trajectory problem, the, 488
   velocity vector, 478
   wind
      atmospheric boundary layer, 498
      effect of, 488, 492
      mean velocity distribution, 498
golf club
   dynamic loft, 484
   effective loft, 484
   friction force, 484
   launch data, 486
   loft, 484
   shaft flex, 484
   strike force, 484
golf course
   St. Andrews Old Course, 470
golf player
   left-handed, 489
   right-handed, 489
   class, 486
   professional
      Justin Leonard, 486
      Phil Michelson, 486
      Tiger Woods, 486
      Vijay Singh, 486
gray body, 562
greenhouse effect, 539
group and phase velocities, 277–282

H^+, positive hydrogen ion, 547
H_2CO_3, carbonic acid, 547
`halton`, 452
Halton sequence, 385, 387, 393, 452
   low discrepancy, 452
   quasi-random, 452
HCO_3^-, bi-carbonate, 547
heat equation, 229, 424
   eigenvalues, 425

Henry's constant, K_H, 545, 547, 577
Henry's Law, 547
high resolution scheme, 285
hydroxyl dissociation constant, K_W, 549, 577

image.plot, 441
ImageMagick, 143
initial condition, 409
initial condition, IC, 103
initial value problem, 603
integration, 4
   A-stable, 74
   $A(\alpha)$-stable, 74
   analytical, 6
   anti-derivative, 4
   definite, 4
   global order of accuracy, 74
   indefinite, 4
   look-up table, 7
   numerical, 10
      choosing method, 69
      consistent scheme, 62
      order of accuracy, 62
      order of convergence, 62
      TVD, 287
      verification of order, 66
   quadrature, 4
   zero-stable, 75
interpolation, 384–398
   1D example, 385
   2D example, 387
   3D example, 393
   estimation points, 384
   interpolation points, 384
   polynomial precision, with, 397
   scattered data, 384
      multivariate, 397
interpretation of results
   truncation error, 211
   validation, 211
   verification, 210
intervals
   closed, 585
   left-closed,right-open, 585
   left-open, right-closed, 585
   open, 585
invariant, 126, 458, 460, 463
irregular shaped domain, 446
isentropic
   exponent, 509
   process, 509

Kadomtsev–Petviashvili equation, 464
KdV, *see* Korteweg–de Vries equation
KdV equation with variable coefficients, 464
Kirchoff's Law, 562
Korteweg–de Vries equation, 123, 410, 459, 463
   2-soliton solution, 410

   3-soliton solution, 123
   conservation law, 126
   invariant, 126
KP, *see* Kadomtsev–Petviashvili equation
Kurganov and Tadmor semi-discrete scheme, 304
Kutta-Joukowski theory, 476

l'Hôpital's rule, 4
Lagrange polynomial, 40
Lambert function, 52
Laplace equation, 434, 441
Lax equivalence, 226, 603
Lax-Richtmyer equivalence theorem, 603
leapfrog, *see* fully discrete methods
Leibnitz
   Gottfried Wilhelm von, 1
Lengyel-Epstein model, 226
lift
   coefficients, 479
limit
   l'Hôpital's rule, 4
Linear multistep methods, 37–61, 87–101
   Adams methods, 60, 97
      Adams Bashforth, 60
      Adams Basforth stability contours, 98
      Adams Moulton, 61
      Adams Moulton stability contours, 98
      coefficients, 60t, 61t
   Backwards differentiation formulas, 38, 89
      1st order, 38, 90
      2nd order, 39
      3rd order, 39
      4th order, 39, 91
      5th order, 39
      coefficients, 38, 89, 90
      coefficients derived from backward differences, 43
      coefficients derived from Lagrange polynomials, 40
      stability contours, 94
      standard form, 38
   consistency, 88
   convergence, 46, 88
      iteration starting values, 50
      Levenberg-Marquardt method, 46
      Newton's method, 46
   Numerical differentiation formulas, 44, 95
      1st order, 45
      2nd order, 45
      3rd order, 45
      4th order, 45
      5th order, 45
      coefficients, 45, 95
      stability contours, 94
   variable step size
      combustion example, 51
      nonlinear pendulum example, 56

Linear multistep methods (*cont.*)
   region of absolute stability, 88
   truncation error, 89
   zero stability, 88
Lipschitz condition, 604
LMM's, *see* linear multistep methods
local Lax-Friedrichs (LLF) flux, 304
logarithm tables
   antilogarithm, 518
   over bar, 518
   use of, 518
long ton, 520
`lsodes`, 411, 431
LTE, *see* truncation error, local

Magnus, Heinrich Gustav, 476
Manhattan Engineer District (MED), 518
Manhattan project, *see* Trinity project
Maple, xix, 116, 335, 337, 352, 469, 511, 514, 531, 534
mass action, 580
matrices, 587–599
   Cayley-Hamilton theorem, 598
   circulant, 590
   condition number, 600
   determinant, 591
   diagonal dominance, 594
   diagonalization, 595
   eigenvalues, 592
   eigenvectors, 592
   exponential, 596
   Gershgorin's theorem, 594
   Hankel, 589
   inverse, 590
   Jordan blocks, 597
   Jordan canonical form, 598
   persymmetric, 589
   powers, 595
   pseudoinverse, 593
   singular value decomposition, 593
   singular values, 592
   spectral abscissa, 593
   spectral radius, 593
   symmetric, 589
   toeplitz, 589
   trace, 588
   transpose, 589
   useful properties, 591
`Matrix`, 118, 419, 430
matrix
   condition number, 403
   Euclidean norm, 404
   factorize, 404
   sparse, 118, 419, 430, 431
matrix norms, 599
MAUD, *see* Military Application of Uranium Detonation
Maxima, xix, 146, 335, 337, 352, 469, 511, 514, 531, 534

memory
   conserving, 431
meshfree, *see* meshless method
meshless method, 380–455
   Kansa, 406
method of lines, 114–187, 408
   1D—Cartesian coordinates, 123
      convection-diffusion-reaction equation, 134
      corneal curvature, 129
      Korteweg–de Vries (KDV) equation, 123
   2D—Cartesian coordinates, 141
      diffusion equation, 141
      frontogenesis, 169
      Laplace equation, 148
      Poisson equation, 149
      rotation of a solid body, 153
      wave equation, 161
   2D—polar coordinates, 175
      diffusion equation, 178
      Poisson equation, 185
      wave equation, 186
   finite difference matrices, 115
      coefficients, 211
Military Application of Uranium Detonation committee, 508
Miura transformation, 461
modified KdV equation, 461
modified PDEs, 282
MOL, *see* method of lines
molarity, 580
momentum equation, 510
monotone upstream-centered schemes for conservation laws, 292, 298–324
   high resolution, TVD discretization scheme, 302
   Kurganov and Tadmor scheme, 303
      local propagation speed, 304
   linear reconstruction, 298
   parabolic reconstruction, 312
MUSCL, *see* monotone upstream-centered schemes for conservation laws

NDF method, 95
NDFs, *see* Linear multistep methods, Numerical differentiation formulas
New Mexico, 508, 518
Newton
   Principia, 1
   Sir Isaac, 1
Newton's law of restitution, 483
NLS, *see* Nonlinear Schrödinger (NLS) equation
Nonlinear Schrödinger equation, 466
number system
   integer, 585
   natural, 585
   rational, 585
   real, 585
numerical integration
   θ-method, 407

lsodes, 124, 150, 179, 185, 411, 431
  explicit Euler, 406
  Runge–Kutta, fourth order, 411, 419, 424, 426, 430
  SHK, fourth order, 415
  weighted average, 407

object.size, 431
ocean circulation, 553–554
ocean layers, 554
ODE, *see* ordinary differential equation
OH^-, hydroxide ion, 549
optical thickness, 563
ordinary differential equation, 8
  solution
    general, 8
    homogeneous part, 8
    nonhomogeneous part, 8
    particular, 8
organic carbon pump, 552

Painlevé integrable condition, 465
PAL, *see* preindustrial activity level
Parseval's theorem, 244, 253, 255, 261
partial differential equation, 8
  advection, 228, 233, 234
  advective-diffusion, 381
  Allen-Cahn, 415
  basics, some, 102
  boundary condition, 409
  classification, 107
  conservation law, 107, 126, 161, 210, 287, 288, 289, 290, 292, 298, 302, 303, 304, 312, 316, 324, 325, 337
  consistent scheme, 226
  convection–diffusion, 236
  convection-diffusion-reaction, 134
  diffusion, 102, 108, 109, 130, 134, 141, 148, 150, 178, 185
  diffusion equation, 8
  discretization, 109
    Courant-Friedrichs-Lewy number, Co, 112
    finite difference terminology, 109
    mesh, the, 111
    non-uniform grid spacing, 112
    stencil, the, 112
    upwinding, 113
  dispersion, 267
  dissipation, 266
  elliptic, 107, 108, 148, 149, 185
  Euler, 230
  Fisher–Kolmogorov, 403, 418, 430
  frontogenesis, 169
  general system, 106
  group and phase velocities, 277–282
  harmonic function, 148
  heat, 229
  heat equation, 424
  hyperbolic, 107, 108, 113, 194–207
  hyperbolic-parabolic, 108, 114, 220
  initial and boundary conditions, 103–105
  initial condition, 409
  invariant, 126
  Laplace, 108, 146t, 148, 434, 441
  Lengyel-Epstein, 226
  modified, 282
  parabolic, 107, 108, 130, 141, 148
  Poisson, 109, 146t, 149, 151, 185, 442, 446
  reaction-diffusion, 415
  stability
    fully-discrete systems, 243–253
    matrix method, 231–242
    semi-discrete systems, 231, 253–260
    unstructured grids, 260
    von Neumann method, 242–260
  subscript notation, 105
  time-dependent, 410
  time-independent, 434
  types of solution, 105
  wave equation, 186
  well posed problem, the, 226
partial pressure, 546
particulate organic carbon (POC), 543
PDE, *see* partial differential equation
Péclet number, 237
persp3d, xx, 441
pH, 552
Plank's constant, 558
plot3D
  3D slice, 393
  iso-surface contours, 395
point detonation, 531
Poisson equation, 442, 446
potential of hydrogen, *see* pH
pracma, 534
preindustrial activity level (PAL), 542
projectile
  form drag, 473
  frictional drag, 473
  induced drag, 473
  liftinduced drag, 473

quad, 534

radial basis function, 381
  centers, 384
  compact support, 380, 382, 431
    positive definite examples, 455
    Wendland, 383, 455
  cutoff function, 383
  differential operator
    linear, 406
  differentiation, 398
    example 1D, 399
    Laplacian, 398
    mixed partials, 398

radial basis function (*cont.*)
  differentiation matrix, 399
    algorithm, 406
  Euclidean distance, 384
  evaluation points, 384
  global support, 382
    positive definite examples, 454
  interpolation
    1D example, 385
    2D example, 387
    3D example, 393
    polynomial precision, with, 397
    scattered data, 384
  inverse multiquadratic, 435, 443, 449
  irregular shaped domain, 446
  local RBFs, 380, 401, 419
    allocating stencil nodes, 403
    stencil, 402
  method of lines, 408
  multiquadratic, 381, 382, 387, 393, 415, 424, 430, 435, 441, 443, 449
    first derivative, 400
    third derivative, 411
  nodes
    exterior, 407
    interior, 407
  partial differential equation, 406–451
    Allen-Cahn, 415
    Fisher–Kolmogorov, 403, 418, 430
    heat equation, 424
    Korteweg–de Vries, 410
    Laplace, 434, 441
    Poisson, 442, 446
    time dependent, 410
    time-independent, 434
  PDEs with nonlinear terms, 408
    linearization, 408
  positive definite RBFs, 382
  shape parameter, 381
    choosing value, 404
    trial and error, 386
  stability, 410, 425
  support, 382
  support radius, 383
  trial function, 381
  Wendland, 435
radial symmetry, 380
radially invariant, *see* radial symmetry
radiation
  emittance, 558
  energy spectrum density, 558
  exchange area, 559
  exitance, 558
  flux, 558
  frequency, 558
  inverse-square-law, 559
  Planck's equation, 558
  power, 558
  solar, 557
  spectrum data, 559
  terrestrial, 560
  view factor, 559
  wavelength, 558
radiative forcing, 566
  carbon dioxide, due to, 569
RAM, 431
Rankine–Hugoniot conditions, 317, 512
RBFs, *see* radial basis function
reflectivity, 562, 566
residence time, 579
restitution
  coefficient of, 483
  Newton's law of, 483
Revelle factor, 577
Reynold's number, 473
rgl, xx, 441, 449, 499
Riemann, 4
  Georg Friedrich Bernhard, 4
  Riemann sum, 4
    center, 5
    left, 5
    right, 5
Riemann problem, 285
RKF, *see* Runge–Kutta, Fehlberg
rm, 431
RSEIS, 499
RStudio, xvi
rSymPy, xv
rSymPy, 32, 52
  installation, 71
Runge–Kutta, 12–37, 76
  1st order, 15
  2nd order, 15
  3rd order, 16
  4th order, 13, 16
  Bogacki-Shampine, order 2/3, 26*t*
  Butcher tableau, 13
    modified for local error estimate, 26*t*
  Cash-Karp, order 5/4, 31
  consistency conditions, 14
  Dormand-Prince, order 5/4, 29
  Dormand-Prince, order 8/7, 33
  embedded solutions, 21
  Felberg, order 4/5, 26
  general form, 13
  order conditions, 14
  Sommeijer, van der Houwen, and Kok method, 36
  stability
    1st order, 76
    2nd order, 79
    4th order, 80
    contours, 77, 85, 426
    Fehlberg, RKF54, 83
    region of absolute, 76
    Sommeijer, van der Houwen, and Kok (SHK), 85

variable step size methods, 19
  example, 31
  Richardson extrapolation, 19
Rusanov flux, 304
`Ryacas`, xv, 84
`rSymPy`, 84
`Ryacas`, 32, 52
  installation, 70

S3 class, 402
S4 class, 402
Schimdt number, 545
sea level
  global mean sea-level (GMSL), 571
Sedov, Leonid Ivanovitch, 508
semi-discrete scheme, *see* method of lines
SHK, *see* Runge–Kutta—Sommeijer, van der Houwen, and Kok method, *see* Runge–Kutta
shock tube, 512
  moving piston, 512
similarity analysis, 530
singular value decomposition, 403, 404
slope limiter, 293
solar constant, 560n8
solar irradiance, 560n8
soliton, 123, 410
  2-soliton solution, 410
  3-soliton solution, 123
  breather, 123
  compacton, 123
  hump, 123
  kink, 123
  peakon, 123
sparse matrix, 118
`sparseMatrix`, 419, 430
spectral radius, 244, 250
`splinefun`, 503
stability, 602
  fully discrete method
    Beam-Warming (BW), 250
    Crank-Nicholson, 252
    CTCS, leapfrog, 249
    FTBS, 245
    FTBS, implicit, 248
    FTCS, 248
    FTCS, Implicit, 248
    Lax–Friedrichs (LxF), 251
    Lax–Wendroff (LxW), 251
    Lax–Wendroff (LxW), 2D, 251
  matrix method, 231–242
  radial basis function, 410
  region of, 244
  semi-discrete method
    first order upwind, 1D, 256
    first order upwind, 2D, 259
    second order upwind, 1D, 258
  unstructured grids, 260
  von Neumann method, 242–260

Stefan-Boltzmann
  constant, 558
  law, 558
STEM, 1
step-by-step, *see* numerical integration
stiffness, 69
stochastic process, 10
Stokes law, 472
super-assignment symbol, xvi
supernova, 508
SVD, *see* singular value decomposition
symmetry
  cylindrical, 509, 530
  planar, 509, 530
  spherical, 509, 530

Tait, F. G., 470
Tait, P. G., 470
Taylor series, 11
Taylor, Sir Geoffrey Ingram, 508
Taylor-Sedov blast wave, 508–537
  additional complexity, 537
  analytical solution, 531
  closed form solution, 533
    internal energy, 534
    kinetic energy, 534
    total energy, 534
  density, 509
  energy, 515
    heat, 515
    kinetic, 515
    total, 515
  pressure, 509
  radius, 509
  shock wave, 510, 514
    boundary, 514
    conditions, 514
    strong shock conditions, 514
  velocity, 509
TDV, *see* total variation diminishing
temperature
  effective global, 560
  globally averaged, 560
temperature anomaly, 569
thermohaline, 554
Thompson, J. J., 470
time constant, 579
TNT, 520
top of atmosphere, TOA, 559
total energy, 510
total variation diminishing, 286
  Godunov's theorem, 287
  monotonicity preserving, 287
  numerical integration, 287–288
    2nd order Runge–Kutta, 288
    3rd order Runge–Kutta, 288
    general Runge–Kutta method, 287
transmitivity, 562, 566

Trinity
  project, 508, 518
  site, 508, 518
    conditions, 518
truncation error, 61–65
  LMM, 62
    BDF, 63
    NDF, 65
  local, 61
  order, 62

uptake factor, 577
upwelling and downwelling, 554

variable separable method, 555
vcKdV, *see* KdV equation with variable coefficients
vector norms, 599
vector product, 478
vectors
  cross product, 586
  dot product, 586
  outer product, 586
von Neumann
  method, 242–260
von Neumann, John, 508

wave equation, 186
wavenumber, 242–243
  spectrum, 243
weighted essentially non-oscillatory method, 324–350
  2D problems
    frontogenesis, 348
    rotation of a solid-body, 347

  alternative sub-stencil reconstruction, 332
  Euler equation problems
    Sod's shock tube, 344
    Taylor-Sedov blast wave, 346
    Woodward-Colella interacting blast wave, 345
  finite volume
    implementation, 337
  flux splitting, 337, 339
  polynomial coefficients, 327
  polynomial reconstruction
    finite difference, 331
    finite volume, 325
  reconstructed edge values, 339
  scalar problems
    advection of composite waveform, 342
    advection of step and cosine pulses, 342
  smoothness indicators, 336
    coefficients, 336
  stencil weights, 338
  sub-polynomials, 338
  weights, 335
  WENO construction, 331
well-posed, 603
WENO, *see* weighted essentially non-oscillatory method
Wien's displacement law, 558
wind
  atmospheric boundary layer, 499
  mean velocity distribution, 499
wxMaxima, xix

xprod, 503